As Time Goes By

As Time Goes By

A Comprehensive Guide to Time and Frequency Measurements

Michael A. Lombardi

Jenny Stanford Publishing

Published by

Jenny Stanford Publishing Pte. Ltd.
101 Thomson Road
#06-01, United Square
Singapore 307591

Email: editorial@jennystanford.com
Web: www.jennystanford.com

British Library Cataloguing-in-Publication Data
A catalogue record for this book is available from the British Library.

As Time Goes By: A Comprehensive Guide to Time and Frequency Measurements

ISBN 978-981-5352-28-3 (Hardcover)
ISBN 978-1-042-01532-0 (eBook)
DOI 10.1201/9781042015320

To my lovely wife, Susan, for all the time we've shared.

Contents

Preface

I was blessed with the privilege of working in the Time and Frequency division of the National Institute of Standards and Technology (NIST)—originally called the National Bureau of Standards (NBS)—for more than 40 years. The people always make the place, and I'm especially grateful for the friendship of colleagues both inside and outside of NIST. They continually taught me new things, and they inspired me to learn new things on my own. Many of my colleagues are mentioned and cited in the pages that follow.

Three special people, all now deceased, made it possible for me to write this book. First and foremost was my mother, Rita Lombardi, who always encouraged me to read by giving me books and taking me to the library. My love for books came from her, and my first thoughts about writing a book took place many years ago, when I was a small boy wandering through the stacks of the public library in Lafayette, Colorado.

Another person who made it possible was Sister Mary Faith Schuster, who was an English professor at Benedictine College in Atchison, Kansas. Sister Faith was an outstanding teacher who cared deeply about her students, and her gentle combination of criticism and praise made us better writers. She didn't hesitate to correct my mistakes, and she always encouraged me to keep writing. She once told me, with a nod to my Italian ancestry, that I could be the next Dante. That didn't happen, of course, but the fact that she said it makes me feel good to this day.

My technical career was made possible by George Kamas, the electrical engineer who hired me at NBS. George hired me even though I then knew very little about science and almost nothing about measurements. He found that I was interested in radio, so he taught me the basics of both radio communications and time and frequency measurements, explaining to me how they were closely related. I learned slowly, but George was a patient teacher who only cared about effort. He knew that my effort was sincere, so he let me ask the same questions over and over, and he kept answering them until I understood.

My career was rewarding, and I wrote this book to try and give back some of what's been given to me. My goal was to present the "big picture" of time and frequency metrology, based on the things I've learned over the course of many years. Like most metrologists, I spent much of my career focusing on small details, which can make the big picture difficult to see. My biggest challenge while writing this book was remembering to step back from some of the small details, so the big picture could come into focus. I hope that the focus is sharp enough to capture the beauty of the subject, and that you find this book to be useful and enjoyable.

Michael A. Lombardi
Broomfield, Colorado
May 2026

List of Acronyms

2SOPS	2nd Space Operations Squadron of the United States Space Force
ABC	American Broadcasting Corporation
AC	alternating current
ACTS	Automated Computer Time Service
AD	*anno domini* (in the context of the Gregorian calendar)
ADC	analog-to-digital converter
ADEV	Allan deviation
AF	audio frequency
a.m.	ante meridiem
AM	amplitude modulation
AOG	auxiliary output generator
AOM	acousto-optic modulator
AOS	Academy of Sciences, Poland
APL	Applied Physics Laboratory of Johns Hopkins University
ARPA	Advanced Research Projects Agency
ASCII	American Standard Code for Information Interchange
ASK	amplitude shift keying
AT	Atomic Time
AT-cut	temperature compensation cut (implemented in quartz crystal resonators)
ATSC	Advanced Television Systems Committee
BACON	Boulder Atomic Clock Optical Network
BC	Before Christ (in the context of the Gregorian calendar)
BC	boundary clock (in the context of the Precision Time Protocol)
BCD	binary coded decimal
BDGIM	BeiDou Global Ionospheric Model
BDT	BeiDou Time
BFO	beat frequency oscillator
BIH	International Time Bureau
BIPM	International Bureau of Weights and Measures, France
BNC	Baby Neill Constant, a type of coaxial cable connector
BOC	binary offset carrier
BPF	band pass filter
bps	bit(s) per second

BPSK	binary phase shift keying
BST	British Summer Time
BTS	base transceiver station
C/A	coarse acquisition
CA	carrier aggregation
CCTF	Consultative Committee for Time and Frequency
CD	compact disc
CDMA	code-division multiple access
CEH	Electronic Watch Center, Switzerland
CENAM	National Metrology Center, Mexico
CEO	carrier-envelope offset
CERN	European Organization for Nuclear Research, Switzerland
CEST	Central European Summer Time
CGGTTS	Common GPS GLONASS Time Transfer Standard
CGPM	General Conference on Weights and Measures
CGS	centimeter-gram-second system of units
CIO	carrier insertion oscillator
CIPM	International Committee for Weights and Measures
CMC	calibration and measurement capability
CMOS	complementary metal oxide semiconductor
CNAV	civil navigation
CNES	National Center for Space Studies, France
CNSA	Chinese National Space Administration
COPL	Center for Optics, Photonics and Lasers, Canada
CP	cyclic prefix
CPFSK	continuous-phase frequency shift keying
CPT	coherent population trapping
CSAC	chip scale atomic clock
CUT	cable under test
CVDC	common-view disciplined clock
CVS	common view signal
CW	continuous wave
DARPA	Defense Advanced Research Projects Agency
DBM	double balanced mixer
DC	direct current
DCP	distributed cavity phase
DDC	digital down converter
DDS	direct digital synthesizer
DFB	distributed feedback

DI	designated institute
DLL	delay locked loop
DMTD	dual mixer time difference
DNS	domain name system
DoD	Department of Defense, United States
DS1	Digital Signal 1
DSP	digital signal processing
DST	daylight saving time
DTV	digital television
DUT	device under test
DUT1	time difference between UT1 and UTC
DVB	digital video broadcasting
DVL	Doppler Velocity Log
EAL	free atomic time scale
ECEF	Earth-centered, Earth-fixed
EDFA	erbium-doped fiber amplifier
EGNOS	European Geostationary Navigation Overlay Service
eLoran	enhanced long range navigation
EMI	electromagnetic interference
ENT	EGNOS Network Time
EOM	electro-optic modulator
EPRC	enhanced primary reference clock
ESA	European Space Agency
ESMA	European Securities and Market Authority
ET	Ephemeris Time
EUSPA	European Union Agency for the Space Program
FAA	Federal Aviation Administration
FC	frequency counter
FCC	Federal Communications Commission, United States
FDD	frequency division duplex
FDMA	frequency-division multiple access
FINRA	Financial Industry Regulatory Authority, United States
FFT	fast Fourier transform
FLL	frequency locked loop
FM	frequency modulation
FSK	frequency shift keying
Gbps	gigabits per second
GCT	Greenwich Civil Time
GEO	geostationary Earth orbit

GGTO	Galileo/GPS Time Offset
GHA	Greenwich Hour Angle
GMAT	Greenwich Mean Astronomical Time
GMC	grandmaster clock
GMT	Greenwich Mean Time
GNSS	global navigation satellite system
GOES	Geostationary Operational Environmental Satellites
GPS	Global Positioning System
GPSDC	Global Positioning System disciplined clock
GPSDO	Global Positioning System disciplined oscillator
GPS MC	GPS multi-channel time transfer
GPS SC	GPS single-channel time transfer
GPST	Global Positioning System Time
GRI	group repetition interval
GSM	Global System for Mobile Communications
GST	Galileo System Time
GUM	Guide to the Expression of Uncertainty in Measurement
HF	high frequency
HFT	high frequency trading
HTML	Hypertext Markup Language
HTTP	Hypertext Transfer Protocol
HTTPS	Hypertext Transfer Protocol Secure
IAU	International Astronomical Union
IC	integrated circuit
ICAO	International Civil Aviation Organization
ICMP	Internet Control Message Protocol
IEC	International Electrotechnical Commission
IEEE	Institute of Electrical and Electronics Engineers
IERS	International Earth Rotation Service
IETF	Internet Engineering Task Force
IF	intermediate frequency
IGS	International GNSS Service
IGSO	inclined geosynchronous orbit
ILAC	International Laboratory Accreditation Cooperation
INSAT	Indian National Satellite System
IoT	Internet of Things
IP	Internet Protocol
IRIG	Inter-Range Instrumentation Group
IRNSS	Indian Regional Navigation Satellite System

ISRO	Indian Space Research Organization
I/Q	in-phase/quadrature
ISO	International Organization for Standardization
ITS	Internet Time Service
ITU	International Telecommunications Union
JCGM	Joint Committee for Guides in Metrology
JILA	Joint Institute of the University of Colorado and NIST
JND	just noticeable difference
JPL	Jet Propulsion Laboratory, a division of NASA
JPO	Joint Program Office
JSON	JavaScript Object Notation
KRISS	Korean Research Institute of Standards and Science
LAN	local area network (in the context of computer networks)
LAN	local apparent noon (in the context of celestial navigation)
Laser	light amplification by stimulated emission of radiation
LEO	low Earth orbit
LHCP	left hand circularly polarized
LPF	low pass filter
LF	low frequency
LLS	low level signaling
LNA	low noise amplifier
LO	local oscillator
LOP	line of position
Loran	long range navigation
LSB	least significant bit (in the context of digital information)
LSB	lower side band (in the context of radio communications)
LTE	Long Term Evolution
MAC	miniature atomic clock
Maser	microwave amplification by stimulated emission of radiation
Mbps	megabits per second
MEMS	micro-electro-mechanical systems
MEO	medium Earth orbit
MF	medium frequency
MIT	Massachusetts Institute of Technology, United States
MJD	Modified Julian Date
MKS	meter-kilogram-second system of units
MKSA	meter-kilogram-second-ampere system of units
MLL	mode-locked laser
Mod ADEV	Modified Allan deviation

MOT	magneto-optical trap
MPQ	Max Planck Institute of Quantum Optics, Germany
MPU	microprocessor unit
MSB	most significant bit
MSCVDC	multi-source common-view disciplined clock
MSSD	mean square successive difference
MTIE	maximum time interval error
NASA	National Aeronautics and Space Administration, United States
NATO	North Atlantic Treaty Organization
NavIC	Navigation with Indian Constellation
NBS	National Bureau of Standards, later renamed NIST, United States
NCO	numerically controlled oscillator
NDB	non-directional beacon
NERC	North American Electric Reliability Corporation
NGA	National Geospatial-Intelligence Agency, United States
NICT	National Institute of Information and Communications Technology, Japan
NIST	National Institute of Standards and Technology, formerly NBS, United States
NISTDC	National Institute of Standards and Technology disciplined clock
NITZ	Network Identity and Time Zone
NMI	national metrology institute
NNSS	Navy Navigation Satellite System
NOAA	National Oceanic and Atmospheric Administration
NPL	National Physical Laboratory, United Kingdom
NPLI	National Physical Laboratory, India
NRC	National Research Council, Canada
NRL	Naval Research Laboratory, United States
NTP	Network Time Protocol
NTS	Navigation Technology Satellite
NTSC	National Television System Committee
NTSC	National Time Service Center, China
OATS	order audit trail system
OC	ordinary clock
OCXO	oven controlled crystal oscillator
OFC	optical frequency comb
OFDMA	orthogonal frequency-division multiple access
ONRJ	National Observatory Rio de Janeiro, Brazil
OS	operating system
OTA	over-the-air
OTDR	optical time domain reflectometer

OTM	on-time marker
OTN	optical transport network
OWD	one-way delay
PBS	Public Broadcasting System
PC	personal computer
PD	phase detector
PE	polyethylene
PI	proportional-integral
PID	proportional-integral-derivative
PIT	programmable interrupt timer
PLL	phase locked loop
p.m.	post meridiem
PM	phase modulation
PMD	polarization mode dispersion
PMU	phasor measurement unit
PNT	positioning, navigation, and timing
ppb	parts per billion
ppm	parts per million
PPP	precise point positioning
pps	pulse per second
PRC	primary reference clock
PRN	pseudorandom noise
PRPSK	pseudorandom phase shift keying
PSD	power spectral density
PSIP	program and system information protocol
PTB	Physikalisch-Technische Bundesanstalt, Germany
PTP	Precision Time Protocol
PVC	polyvinyl chloride
Q	quality factor
QZSS	Quasi-Zenith Satellite System
QZSST	Quasi-Zenith Satellite System Time
Radar	radio detection and ranging
RAIM	receiver autonomous integrity monitoring
RBDS	Radio Broadcast Data System
RCC	radio controlled clock
RDS	Radio Data System
RF	radio frequency
RFC	Request for Comments, a document that defines an Internet protocol
RG	radio guide, a classification system for coaxial cable

RGO	Royal Greenwich Observatory
RHCP	right hand circularly polarized
RINEX	Receiver Independent Exchange
RMO	regional metrology organization
RMS	root mean square
RNNS	regional navigation satellite systems
RSS	root sum of squares
RTC	real time clock
RTD	round trip delay
Rx	signal reception
SA	selective availability
SBAS	satellite-based augmentation system
SC-cut	stress compensation cut (implemented in quartz crystal resonators)
SC-FDMA	single-carrier frequency-division multiple access
SDARS	Satellite Digital Audio Radio Service
SDR	software defined radio
SEC	Securities and Exchange Commission, United States
SHA	Sidereal Hour Angle
SHG	second-harmonic generator
SI	International System of units
SIM	Inter-American Metrology System
SIMT	SIM Time Scale
SIMTN	SIM Time Network
SLR	satellite laser ranging
SMD	surface mount device
SMTPE	Society of Motion Picture and Television Engineers
SNTP	Simple Network Time Protocol
Sonar	sound navigation and ranging
SPBA	single path bidirectional amplifier
SSB	single side band
ST	summer time
STL	Satellite Time and Location
STT	system time table
SyncE	synchronous Ethernet
SYRTE	Time-Space Reference Systems Laboratory, France
T2L2	Time Transfer by Laser Link
TAE	time alignment error
TAI	International Atomic Time
TAR	test accuracy ratio

TC	transparent clock
TCG	time code generator
TCP	Transmission Control Protocol
TCXO	temperature compensated crystal oscillator
TD	time difference
TDD	time division duplex
TDEV	time deviation
TDM	time division multiplexing
TDMA	time-division multiple access
TDOA	time difference of arrival
TDR	time domain reflectometer
TEC	total electron content (in the context of the ionosphere)
TEC	time error correction (in the context of AC power generation)
THD	total harmonic distortion
TIC	time interval counter
TIMATION	Time Navigation satellites
TLS	Transport Layer Security
TOA	time of arrival
TOC	time of coincidence
TOF	time of flight
TOW	time of week
TPFC	time-programmable frequency comb
TRF	tuned radio frequency
TTL	transistor-transistor logic
TUR	test uncertainty ratio
TVE	total vector error
TVO	triggered Vernier oscillator
TWSTFT	two-way satellite time and frequency transfer
TWTT	two-way time transfer
Tx	signal transmission
UDP	User Datagram Protocol
UHF	ultra high frequency
USAF	United States Air Force
USB	upper side band
USCG	United States Coast Guard
USNO	United States Naval Observatory
USSF	United States Space Force
UT	Universal Time
UT1	Universal Time scale 1, based on the mean solar second

UT2	Universal Time scale 2, a smoothed version of UT1
UTC	Coordinated Universal Time
UTCr	Rapid Coordinated Universal Time
VBI	vertical blanking interval
VCO	voltage controlled oscillator
VCR	videocassette recorder
VCSEL	vertical-cavity surface-emitting laser
VCXO	voltage controlled crystal oscillator
VF	velocity factor
VFO	variable frequency oscillator
VHF	very high frequency
VIM	International Vocabulary of Metrology
VLF	very low frequency
VNA	vector network analyzer
VNIIFTRI	Russian Metrological Institute of Technical Physics and Radio Engineering
VOIP	Voice over Internet Protocol
WAAS	Wide Area Augmentation System
WAN	wide area network
WCDMA	wideband code-division multiple access
WDM	wavelength division multiplexing
WLAN	wireless local area network
WNT	WAAS Network Time
WR	White Rabbit
WWW	World Wide Web
XDS	Extended Data Service
XML	Extensible Markup Language
XO	quartz crystal oscillator
ZCD	zero crossing detector

Introduction

I would stop it from swinging like a pendulum
Just to hold time in my hand

— Mark Knopfler, "Golden Heart"

There are few subjects, if any, that are talked about more than time. We all seem to realize that being aware of time is an essential part of being alive. "What time is it?" is probably the world's most commonly asked question. We constantly ask that question, either by verbally asking someone else or by simply glancing at a clock. We organize our lives around time, with our calendars and clocks providing constant reminders of our schedules and appointments. We observe the days that repeat every year—the birthdays, holidays, and anniversaries. We reflect on the past, live in the present, and look toward the future. We treasure the good times and we lament the wasted or lost times. We sometimes hope that time will pass more quickly or more slowly, and we often wish that we had more time.

Despite our preoccupation with time, we seem to know very little about it. Time is elusive and intangible, and it's hard to define and understand. We can try to make it less elusive and more tangible, but it seems that our efforts always fall short, because time seems to reside outside of the reach of our traditional senses. We can't see it, hear it, smell it, or touch it, but we can still observe and feel its effects. For example, we can observe and feel the passage of time as we watch night turn to day, watch summer turn to autumn, or watch a child grow up. These experiences force us to accept—sometimes grudgingly as we get older—that time is something we can't change or control.

We can, however, do something about time, and we can do it exceptionally well. We can measure time! We can measure time more accurately, with more resolution and with less uncertainty, than we can measure anything else. It's gratifying to know that the measurement of time isn't just an academic pursuit, but instead is a practical activity that provides immense benefits to society. The technologies we rely upon every day, including our electric power grids, our smartphones, and our satellite navigation systems, wouldn't work at all without our ability to accurately measure time.

This book is a practical and comprehensive guide to time and frequency measurements. The chapters are divided into five parts. Part I discusses clocks, the instruments that keep time by measuring frequency, and it describes how oscillators provide the frequency that gives clocks their "heartbeat." Parts II and III explain how clocks and oscillators are measured, how time is kept internationally, and how radio and network signals are used to transfer time from one place to another. Part IV describes the current research areas and state-of-the-art techniques for measuring and transferring time. Part V explains how accurate time and frequency measurements apply to our everyday lives, and why they are an essential part of numerous industries and technologies. In each of the five parts, the narrative briefly introduces some of the people who have advanced measurement science and brought us to where we are today.

DOI: 10.1201/9781042015320-1

PART I

Clocks and Oscillators

Chapter 1

Shadow, Flow, and Mechanical Clocks

Hard times, baby well they come to us all
Sure as the ticking of the clock on the wall
Sure as the turning of the night into day

— Bruce Springsteen, "Waiting on a Sunny Day"

Time measurements are based on comparisons. That shouldn't be a surprise, because all measurements are based on comparisons. We measure unknown quantities that we want to learn more about by comparing them to known quantities that we've learned to trust. The trusted quantities serve as measurement references or standards. A large part of measurement science, the discipline called metrology, consists of agreeing upon standards.

Time, as we know it, is entirely based on agreed-upon standards. Before we could keep time, we first had to define the base quantities, or units, of time interval. Once the base units were defined, we could develop ways to measure them. We could also measure shorter time intervals by dividing base units, or measure longer time intervals by counting base units. The instruments that record the passage of time by measuring and counting time interval units are called clocks, and clocks have taken many forms since they first appeared thousands of years ago. A clock can show us what time it is now, and it can also measure the time that elapses between two events.

It's common to think of measurement instruments as meters, the devices that measure a specific quantity and display the result as an analog or digital reading. Therefore, it seems to follow that we can think of a clock as a meter that measures time. In fact, the term chronometer—which is still used to describe some types of mechanical clocks—literally means time meter, because *chrono* is the Greek word for time. We can choose to think of clocks as meters, but it's important to remember that time measurements differ from other measurements because the meter is always running.

To visualize this, consider that a water meter is designed to measure a specific flow of water. When the flow stops, the meter stops, because there is nothing for it to measure. In contrast, all clocks measure the same "flow" of time, and they always have something to measure because the flow of time never stops. Even though time doesn't stop, an individual clock can, which of course renders it temporarily useless.

As Time Goes By: A Comprehensive Guide to Time and Frequency Measurements
Michael A. Lombardi

ISBN 978-981-5352-28-3 (Hardcover), 978-1-042-01532-0 (eBook)
www.jennystanford.com

DOI: 10.1201/9781042015320-3

However, once a clock is restarted, it can be synchronized to agree with a reference clock that kept time during its absence, and it can start measuring time again.

The quest to design and build new clocks that measure time more accurately than their predecessors continues to challenge scientists, just as it has for thousands of years. We'll begin our discussion of this perpetual quest by describing some of the methods that humans first used to measure the passage of time.

1.1 Counting Days, Months, and Years

Long before they gained understanding of the basic principles of astronomy—including the Earth's rotation upon its axis, the Earth's orbit around the Sun, or the Moon's orbit around the Earth—humans were undoubtedly aware of the repetitive cycle where light turned to darkness and then back to light. This cycle marked a period of time that was part of their natural world, and that period became known as the day. It seems likely that the first attempts to measure time simply involved recording the number of elapsed days by counting either sunrises or sunsets. The counting of days eventually led to calendars, which were systems that organized days into longer time periods that came to be known as months and years.

The period of the day was eventually found to be the period of one rotation of the Earth on its axis, and it provided a natural foundation for an agreed-upon system of timekeeping. The Earth's rotational axis can be thought of as an imaginary line that passes through the North and South poles, and its orbital axis is a line perpendicular to an imaginary orbital plane that the Earth moves through as it revolves around the Sun. As shown in Figure 1.1, the rotational axis is tilted with respect to the orbital axis by about 23.4°. As we'll continue to discuss in later chapters, the rotational rate of the Earth was considered to be the most accurate time standard available for thousands of years, and it provided the reference for all clocks and systems of timekeeping from antiquity until the early part of the 20th century.

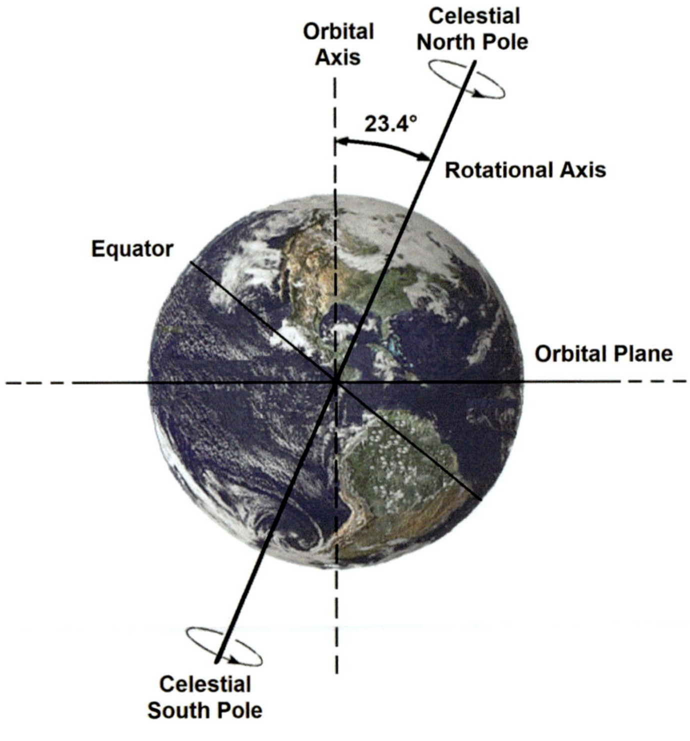

Figure 1.1 The rotational and orbital axes of the Earth.

Ancient civilizations were also undoubtedly aware of the cycle of the changing seasons. The seasons were caused by the Earth's axial tilt, which caused the Sun to shine on different latitudes of the Earth during different periods, leading to changes in the weather. The cycle of seasons is predictable and repetitive, much like the cycle of the day, but with a much longer period. It was soon learned that the change in seasons could be marked and anticipated by observing the changing phases of the Moon.

Telescopes weren't invented until the 17th century, but they weren't needed to observe the Moon, which is the largest and brightest object in the night sky. The Moon's large size is due to its proximity to Earth, and its brightness is due to the light it reflects from the Sun. When viewed from Earth, the Moon appears as a circle called the lunar disc. During the Moon's orbit, the same side of the Moon always faces Earth, and the visible portion of the lunar disc is the portion that's reflecting sunlight. When the Moon's orbit brings it closest to the Sun, it briefly has the same celestial longitude as the Sun. During this period, the Earth-facing side of the Moon can't reflect any sunlight, so the lunar disc is dark and invisible. This event is called a new moon. Conversely, when the Moon's orbit brings it furthest from the Sun, the entire Earth-facing side of the Moon can and does reflect sunlight, and the lunar disc is fully illuminated. This event is called a full moon (Figure 1.2).

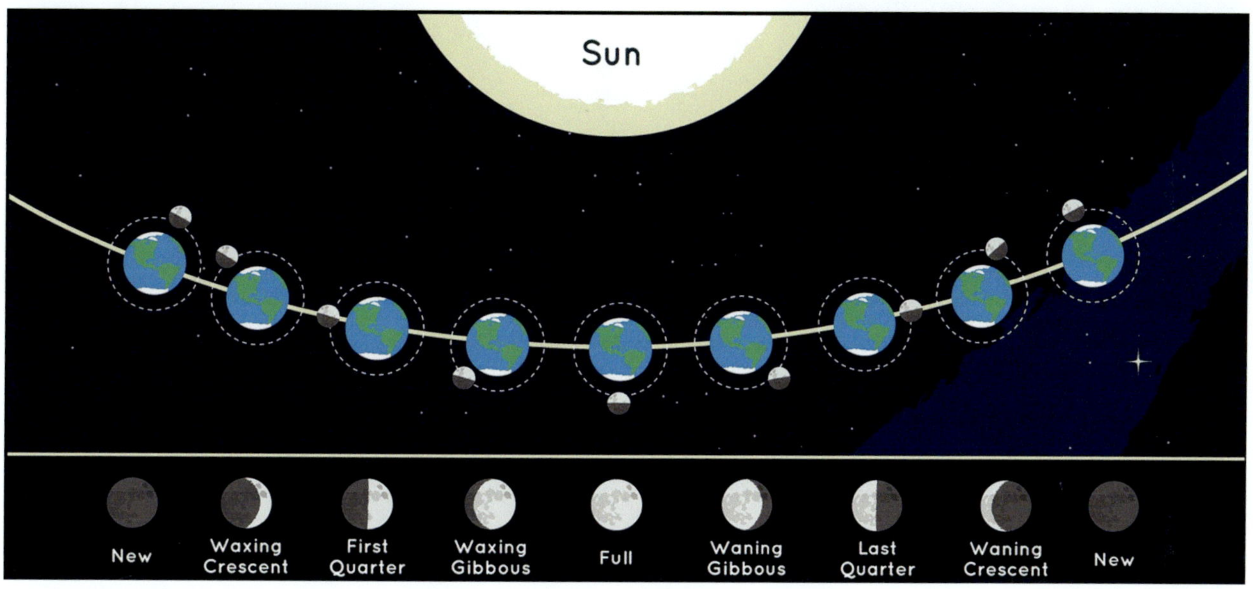

Figure 1.2 The Moon as seen from Earth (courtesy of NASA).

The lunar cycle, which is the period between new moons, full moons, or any other recurrence of the same Moon phase, has a duration of about 29.53 days. The first calendar, and the calendars still used by some religions and cultures, defined the month as the period of one lunar cycle. A lunar year consists of 12 lunar months, and therefore has a period of about 354.36 days. Depending on the cultural or religious practices being followed, the first day of a lunar month could be indicated by either a new moon or a full moon. In fact, the English word "calendar" is derived from the Latin word *kalendae*, the name that was given to the first day of the month in the Roman calendar. The word *kalendae* was derived from the ancient verb *calare*, which means to call out or announce, which referred to the announcement of the new moon when it was first observed [1].

The calendar that's universally known in today's world isn't a lunar calendar, but instead is a solar calendar. Solar calendars define the year as the period of one revolution of the Earth around the Sun, without any reference made to the Moon. As shown in Figure 1.3, the Earth's orbit is elliptical, or oval-shaped, with the aphelion and perihelion the respective points where the Earth is furthest and closest to the

Sun. The aphelion is reached in early July, and the perihelion in early January. The illustration exaggerates the difference in distance between the aphelion and perihelion, which in absolute terms is about 5,000,000 km, but in relative terms is only slightly more than 3%.

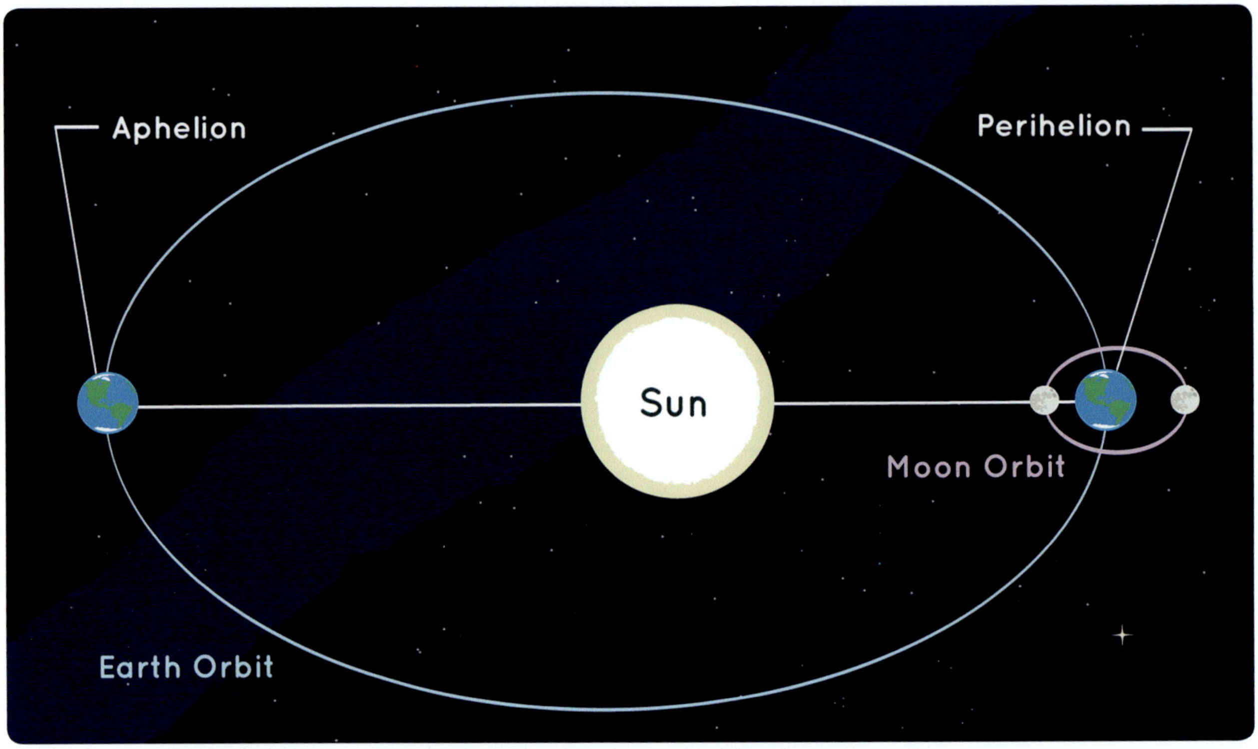

Figure 1.3 The Earth's revolution around the Sun that defines the period of a year (courtesy of NASA).

The period of the solar year is about 365.2422 days, or almost 11 days longer than the lunar year. The 12 months in the modern solar calendar are unequal divisions of the solar year, with periods ranging from 28 to 31 days. Interestingly enough, the period of a month still approximates the period of a lunar cycle to within ±1.5 days.

It took many centuries, however, before solar calendars evolved to their current form. The Egyptians are believed to have been the first to use solar calendars, perhaps beginning in about 2782 BC. The Egyptians marked the beginning of the solar year as the time when Sirius, the brightest star in the night sky, became visible on the eastern horizon just before sunrise. The annual reappearance of Sirius coincided with the annual flooding of the Nile River. The ancient Egyptian calendar consisted of 12 months of 30 days each, with five extra days added. The period of the equal length months was a fairly close approximation of the lunar cycle, short by about half a day per month; and the five extra days made the Egyptian calendar year almost equal to the period that we now know as the solar year, short by about a quarter of a day. The five extra days are thought to have been holidays celebrating gods who the Eqyptians considered to be immortal, and whose births were thought to have occurred outside of the 12 conventional months of the calendar [2].

It was of course not known to ancient civilizations that the Earth revolved around the Sun—instead it was believed that the Earth was the center of the universe, and that the Sun revolved around the Earth. From the perspective of an observer on Earth, this geocentric view of the universe seems logical because we can't feel the Earth move, and because its entire landscape seems to be stationary. In contrast, the Sun appears to be moving across the sky, going from one horizon to the other in the period between sunrise and sunset. The Greek astronomer Aristarchus (c. 310 to c. 230 BC) was the first to present a heliocentric model of what we now call the solar system, placing the Sun at the center in a fixed position, with the Earth revolving around it. Aristarchus apparently left no theories, tables, or predictions to support his heliocentric model, but his hypothesis was well known to Archimedes (c. 287 to c. 212 BC), who discussed it in his own writings [3].

Despite the work of Aristarchus and other astronomers, the geocentric view of the universe persisted until the 16th century, when a complete mathematical model of a heliocentric system was formulated and published by the Polish astronomer Nicolaus Copernicus (1473–1543). Copernicus likely began working on his heliocentric theories no later than 1514 [3], but the complete and final version of his theories, entitled *De Revolutionibus Orbium Cœlestium, Libri VI* (*Six Books on the Revolutions of the Heavenly Spheres*), was published just before his death in 1543. In this work, often considered as one of the greatest achievements in the history of science, Copernicus proposed that the Sun provided the fixed reference point for the motion of the Earth and the other planets. He described the Earth as a sphere that rotates on its axis, with the solar day being the period of one rotation. The rotating Earth was simultaneously revolving around the Sun along with the other planets, with one revolution around the Sun being the period of a solar year. Copernicus also explained why the periods of both the solar day and solar year had variations [4]. He supported his clear explanations with rigorous mathematical detail, which made the heliocentric model difficult to ignore or refute from that point forward. In fact, the shift in astronomy from a geocentric to a heliocentric understanding of the universe was so impactful that it became known to historians as the Copernican Revolution.

The solar calendar that we use today can be attributed to two reformations of the Roman calendar, one occurring almost 16 centuries before Copernicus published his landmark work, and the other occurring about four decades afterwards. The Julian reformation of the calendar was proposed in 46 BC, and was ordered into effect on the first day of the following year by the Roman general and statesman Julius Caeser (100–44 BC). The reformation was necessary because the Roman calendar, which went into use shortly after the Roman republic was founded in 753 BC, was causing considerable confusion in Roman society by the 1st century BC. The Roman calendar was based on the lunar calendar, but it had 355 days instead of the usual 354, and none of its months had an even total of 30 days. The Julian calendar fixed these problems by assigning an average period of 365.25 days to the year. Eleven of the 12 months had either 30 or 31 days, but February had only 28. However, to accommodate for the extra quarter of a day in the calendar year, a 29th day was added to February once every four years. This day was called a leap day, and it resulted in a leap year with a period of 366 days, instead of the usual 365.

The Julian calendar was in general use for more than 1,600 years, and is still observed today by some religions and cultures, but it had a significant problem. The 365.25 day period it assigned to the average year was about 11 minutes longer than the period of the Earth's average revolution around the Sun. This meant that the Julian calendar was gradually slipping behind the time scale referenced to the Earth's orbit, at a rate of about three days every 400 years. By the year 1582, this slippage had resulted in an accumulated error of well over 10 days since the Julian calendar reform, but about 10 full days since the First Council of Nicaea, which was an ecumenical council of Christian bishops that had convened in the year 325. Among other things, the council had mandated that Easter would be celebrated on the same day by all Christians worldwide. To correct the error that was impacting the calculation of the date of Easter, Pope Gregory XIII (1502–1585) instituted the Gregorian calendar that's used in most parts of the world today. The Gregorian calendar immediately moved the date forward by 10 days when it went into effect in October 1582. The date of October 4, 1582, on the Julian calendar was immediately followed by October 15, 1582, on the Gregorian calendar, and the days in between were completely skipped.

To guard against the future slippage of dates, it was decided that the leap years would not be observed during the first year of a new century unless the year number was evenly divisible by 400. This meant that 1600 was a leap year, as was 2000, but that 1700, 1800, and 1900 were not. The omission of three leap days every four centuries corrects nearly all of the error, but the average Gregorian year of 365.2425 days is still longer than the average solar year by almost 26 seconds, which is enough to cause a slippage of one calendar day every 3,333 years [1, 2].

The development of calendars was absolutely essential to all civilizations, but calendars have obvious limitations. They can only mark the passage of days, and not the passage of shorter intervals. Therefore, calendars can't schedule the activities that take place during a day. That task was left to clocks.

1.2 Sundials and the Concept of the Hour

The Egyptians are credited by most historians as being the first civilization to use clocks. Their first clocks were probably nothing more than sticks placed in the ground that indicated time by both the length and direction of their shadow, which changed based on the Sun's position in the sky [5]. The object casting the shadow is called a gnomon, derived from an ancient Greek word meaning "one who knows" [6]. Some historians believe that shadow clocks were in use before 3000 BC. Before sticks were placed in the ground, it seems likely that the shadows of structures, trees, and a person's own body were used as gnomons to provide some indication of the time of day [7].

By about 1500 BC, the Egyptians had developed a more advanced shadow clock. This clock had not only a gnomon, but markings to indicate the time of day, which perhaps made it the first true sundial (Figure 1.4). This T-shaped instrument was placed in a sunlit area on the ground. In the morning, the crosspiece (AA) was set to face east and then rotated in the afternoon to face west. The long arm (BB) was divided into six parts, marked by a scale that formed perhaps the earliest representation of what we now call the hour. The morning and afternoon hours combined to form a 12-hour day, reflecting Egypt's early use of the duodecimal (base 12) number system [5, 8].

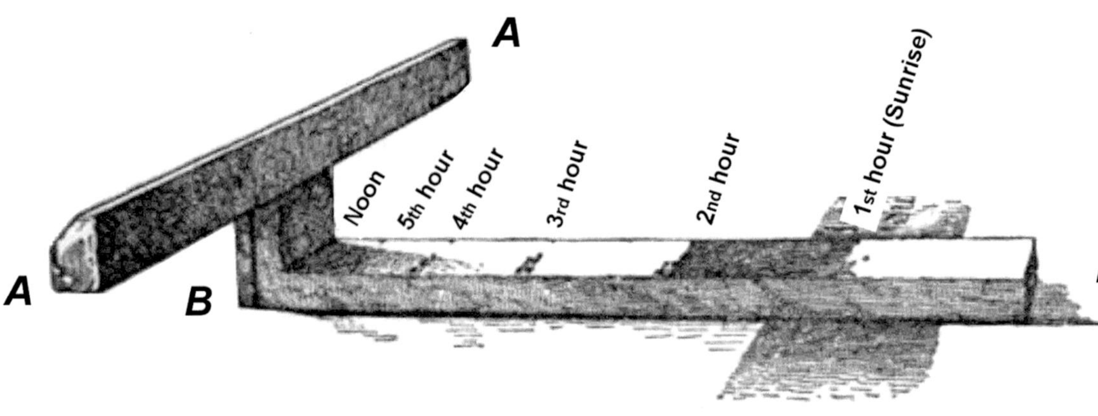

Figure 1.4 Early Egyptian Shadow Clock (c. 1500 BC).

There are at least several possible reasons why the number 12 and the duodecimal system were considered to be so important to the Egyptians and other early civilizations. There were 12 signs in the zodiac, and as previously discussed, 12 lunar cycles were counted to determine the period of the lunar year. Twelve also has the advantage of being evenly divisible by 2, 3, 4, and 6, making it easy to determine halves, thirds, fourths, and even sixths of a quantity. Finally, 12 is equal to the number of phalanges, or bone segments, in the four fingers of the hand (excluding the thumb, which has two more phalanges). By using the thumb to count the phalanges of the other four fingers, we can count to 12 with one hand [9].

A sundial is a shadow clock where the gnomon casts its shadow on a plate called the dial, which is usually circular and flat. The numbers marked on the dial indicate the time of day. As described in the previous section, before the work of Copernicus it was commonly thought that the Sun was moving, rather than the Earth, and that time was referenced to the Sun's motion. After it was finally proven that the Sun's position is fixed, it became known that the time kept by sundials was not referenced to the actual motion of the Sun, but to the apparent motion of the fixed Sun as observed from the moving Earth. It was also learned that the tilt in the Earth's rotational axis (Figure 1.1) results in a non-circular orbit, which prevents sundials from being accurate clocks.

To a sundial, noon is the instant when the Sun reaches its highest point in the sky. However, due to the Earth's non-circular orbit, the Sun's highest point in the sky is reached at different times during the course of a year. Therefore, the apparent solar time displayed by sundials is offset with respect to mean solar time, which is the approximate time displayed by all modern clocks. Mean solar time is referenced to the rotational rate of the Earth, and it can be accurately measured by comparing the Earth's rotational rate to celestial objects that are further away than the Sun. The time offset of apparent solar time varies throughout the year in a smooth, periodic fashion that can be determined from a series of calculations known as the equation of time [10]. Figure 1.5 is a graph of the daily offset of apparent solar time for a one-year period, obtained with the equation of time.

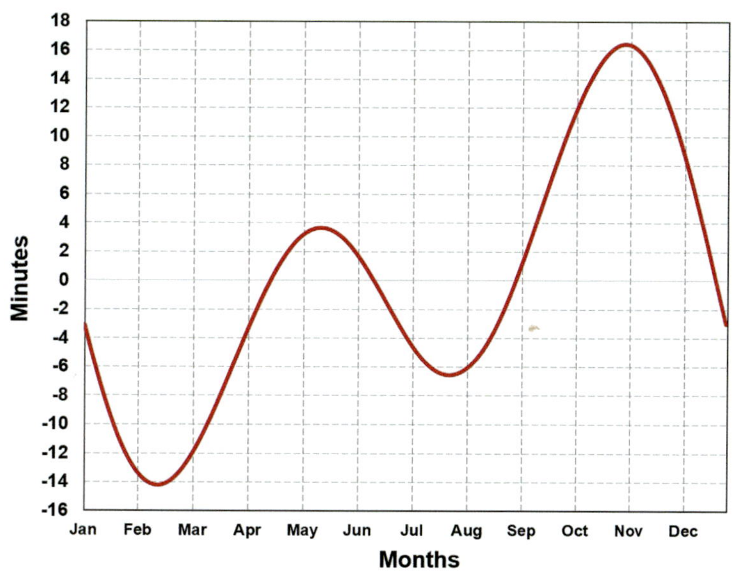

Figure 1.5 A graph of the equation of time.

The graph in Figure 1.5 is for the year 2024 and for a longitude of 0°, but it would look essentially the same for all years and all longitudes. As the graph shows, a sundial will be ahead of mean solar time, or fast, by more than 16 minutes in early November, but behind mean solar time, or slow, by more than 14 minutes in the first half of February, which is a swing of about 30 minutes in about three months. This is because the Sun appears to be moving the fastest when it is closest to the Earth, as it approaches and then moves away from the point of perihelion in January. The Sun appears to move slowest when it is furthest from the Earth, as it approaches and moves away from the point of aphelion in July (Figure 1.3). Therefore, during the period from about April until about mid-September, the apparent solar time offset goes up and down as always, but it stays within a range of about 10 minutes. The equation of time graph is often engraved on the base or face of sundials, and it serves as a reference for converting apparent solar time to mean solar time. In most cases, the graphs on sundials are inverted from the one shown in Figure 1.5. This makes it easier for observers to determine how fast or slow their watch should be when compared to the sundial.

An equatorial sundial, which was given its name because it's dial plate is parallel to the Earth's equatorial plane, can work at any latitude when properly positioned. The gnomon on an equatorial sundial is a rod perpendicular to the dial plate that passes through its center. The gnomon is positioned parallel to the Earth's rotational axis, and it points towards either the Earth's celestial north or south pole, depending on whether the sundial is located north or south of the equator. Hour marks are placed at intervals of 15° around the gnomon on both sides of the dial plate. The hour marks run clockwise on the top side of the dial plate, and counterclockwise on the bottom side, with the number 12 representing the hour of noon at the bottom of the dial on both sides.

The Sun casts its shadow on the top side of the dial plate during the spring and summer months, and on the bottom side of the dial plate during the fall and winter months. Therefore, assuming that sunlight is present, each side of the sundial will be usable for six months out of the year [6]. Figure 1.6 shows the front (a) and back (b) of a large equatorial sundial located on the University of Colorado campus in Boulder, Colorado. A six-month graph of the equation of time is located on each side of the sundial, indicating the months when the top and bottom sides of the dial plate should be read.

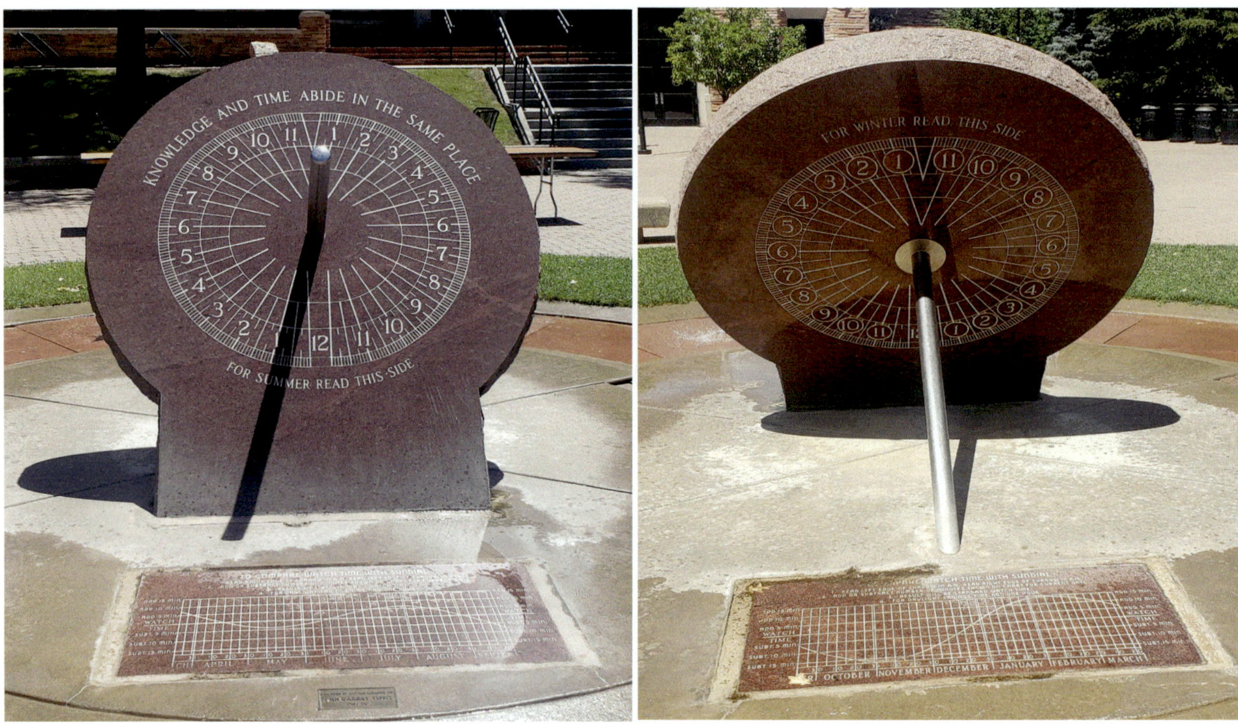

(a) **(b)**

Figure 1.6 The front (a) and back (b) of an equatorial sundial.

As timekeeping advanced, applying and understanding the equation of time was essential for ensuring a uniform timekeeping system where the time kept by sundials agreed with the time kept by more modern clocks. The basic concepts of mean solar time, along with the principles for determining the equation of time, extend back to the work of the Greek astronomer and mathematician Claudius Ptolemy (c. 100 to c. 170 AD). However, until the early 19th century, apparent solar time remained in general use for situations where the only way to obtain accurate time was by astronomical observations. Mean solar time, when needed, was obtained by applying the equation of time to the apparent solar time [11].

Sundials required sunlight to work, so they obviously couldn't work indoors. They were also unusable during cloudy days and during the nighttime hours between sunset and sunrise. The first attempts to measure the passage of time during the night are thought to have occurred around 2000 BC, when Egyptian astronomers developed a method that was based on observations of a set of 36 groups of stars. These groups divided the circle of the sky into approximately equal parts, and they appeared to rise consecutively on the horizon. The first group to be visible at night would change every 10 days, when a new first group appeared above the horizon, and the previous group moved higher in the sky. The star groups were called decans, a name derived from *dekanoi*, the Greek word for "tenths."

A full day would correspond to the rising of all 36 decans, but the stars that rose above the horizon at night could only be observed during periods of darkness. If the periods of light and darkness were equal, and if the sky was totally dark from the moment of sunset to the moment of sunrise, then the passage of night would be marked by the rising of 18 decans. This wasn't the case, however, because there were periods of twilight when the sky wasn't dark enough to observe the decans, and the duration of the period of total darkness varied throughout the year. The shortest period of darkness was in the summer, when the rising of only 12 decans could be observed. These summertime observations of decan stars, coupled with the Egyptian's use of the duodecimal system, resulted in the night being divided into 12 parts. This established the concept of nighttime hours. During the period from about 2000 to 1800 BC, the coffin lids of some Egyptian noblemen were inscribed with illustrations that depicted the decan stars and the 12 divisions of the night [12].

Water clocks, also known as clepsydras, were a more convenient and practical way of measuring time during both the day and night, and they were likely the most accurate clocks of the ancient world. A specimen found at the Egyptian temple of Amun-Re in Karnak dates back to about 1400 BC. This clock consisted of a vessel with slanted interior surfaces that were inscribed with markings that divided the night into 12 parts during various months (Figure 1.7). Water clocks are flow clocks, and in this example a flow of water slowly dripped out of a hole at the bottom of the vessel, causing the water level to drop from one hour line to the next. We now know that clocks of this type could measure time with an uncertainty of about 15 minutes per day [2, 7].

Figure 1.7 Egyptian Water Clock, c. 1400 BC (courtesy of the Science Museum, London).

Dividing the periods of both light and darkness into 12 parts established the modern concept of a 24-hour day. However, the period of an hour still varied depending on how it was measured. Hours measured with sundials or decan stars had periods that varied seasonally, and hours measured with flow clocks had periods that varied based on the type of device used. The concept of hours with equal durations didn't originate until Greek astronomers needed such a system for their theoretical calculations. The Greek astronomer Hipparchus (c. 190 to c. 120 BC), proposed dividing the day into 24 hours of equal duration, called equinoctial hours, on the days when the Sun appeared directly above the Earth's equator, an event called a solar equinox. The word equinox is derived from the Latin word *aequinoctium*, which means "equal night," and the periods of night and day are essentially equal during a solar equinox. Solar equinoxes occur twice a year, on dates around March 20th and September 23th.

Astronomers performed extensive calculations to estimate the period of an equinoctial hours, and flow clocks became increasingly sophisticated. Despite these efforts, time measurements based on hours of equal duration weren't common or practical until the age of mechanical clocks.

1.3 Minutes and Seconds

Many of the astronomical calculations used by the early Greek astronomers had been previously developed by the Babylonians. The Babylonians relied on the sexagesimal (base 60) number system that they had inherited from the Sumerians, who had developed it around 2000 BC. The sexagesimal system was convenient for expressing fractions, because 60 is the smallest number divisible by the first six counting numbers, and it is also divisible by 10, 12, 15, 20, and 30. Although it is no longer used for general computation, the sexagesimal system is still used to measure angles, geographic coordinates, and time.

The Greek astronomer Eratosthenes (c. 276 to c. 194 BC) used a sexagesimal system to divide a circle into 60 parts. He then devised an early geographic system of latitude, measuring distance north or south of the equator, with his irregularly spaced latitude lines running through well-known places on the Earth. Hipparchus improved on this system about a century later by rationalizing the lines of latitude, which made them parallel and obedient to the Earth's geometry instead of to its familiar places. Hipparchus also devised a system of longitude that measured distance east or west of a selected meridian, which was an imaginary line that connected the North and South poles. His latitude and longitude systems each encompassed 360°, employing the methods of the Babylonians, who had divided the arc of a circle into 360 parts.

Claudius Ptolemy made further refinements to this coordinate system. By doing so, he eventually gave names to the time units of minutes and seconds. Around 150 AD, Ptolemy completed a mathematical treatise now known as the *Almagest* [13]. The title, which means "the greatest" when translated to English, didn't originate until Arabic scholars translated the text in the 9th century, but it continued to be used in the Latin translations that came afterwards. Consisting of 13 books, the *Almagest* was a complete exposition of mathematical astronomy as understood in Greece at that time, and it became a standard astronomy textbook that remained in use for more than one thousand years. In Book I, Ptolemy describes subdividing each of the 360° of latitude and longitude into 60 parts, which were again subdivided into 60 smaller parts. Ironically, Ptolemy was explaining the chords of a circle and not discussing time, and he didn't use the terminology from which "minutes" or "seconds" could be derived, as those words weren't even known in his language.

Ptolemy wrote the *Almagest* in Koine Greek, which was the same common Greek dialect that was used to write the New Testament of the Bible. He described the two divisions by 60 as *prota heksēkosta* and *deutera heksēkosta,* which loosely translated means the "the first sixtieth" and "the second sixtieth." Latin translations of both the Arabic and Greek texts of the *Almagest* were completed in the 12th century, about 1,000 years after the original work was completed, and it was in the Latin translations where the familiar words we know as minute and second first appeared. The phrasing, word placement, and alphabet vary in different transcriptions of the Latin translations, but the terms used to describe the first subdivision included *partes minutœ primœ* and *pars minuta prima*, which loosely translated means "the first small part." The first small part is what we now call the minute, or 1/60th of an hour. The second subdivision was called *partes minutœ secundœ* or *pars minuta secunda*, meaning "the second small part." The second small part is what we now call the second, or 1/60th of a minute. We can therefore trace both the circular face of a clock and the sphere of a globe to the sexagesimal system, a Babylonian number system that is now about 4,000 years old [14].

After the Latin translations of the *Almagest*, at least several more centuries passed before minutes and seconds became part of everyday timekeeping. Before clocks divided the hour by 60, they often divided the hour into halves, thirds, quarters, or sometimes even 12 parts, and the hour wasn't understood to be the duration of 60 minutes. Measuring minutes wasn't practical until near the end of the 16th century, when mechanical clocks with minute hands began to appear [7]. All modern clocks, both analog and digital, display minutes in the form of 1/60th of an hour. However, some modern analog clocks still lack a second hand, and most modern digital clocks still lack the additional digits needed to display seconds.

1.4 Clocks in the Middle Ages

Numerous clocks were invented during the Middle Ages that kept time by measuring a continuous process, such as the flow rate of water. We've previously mentioned the ancient Egyptian water clock, and the more elaborate water clocks that followed were perhaps the world's most accurate timekeeping instruments until about the end of the 15th century. Water clocks were calibrated so that the time they kept agreed with the flow rate of water, and some utilized multiple vessels, siphons, and various types of hydraulic mechanisms. Others even marked the passage of time by moving dials or ringing bells. All water clocks had limitations, however, because water froze in cold climates and evaporated in warm climates, and sloshing water didn't line up well with hour marks, which often make it difficult to get an accurate time reading. A more serious limitation was the fact that few people had access to water clocks, they were mostly found in palaces and other places of wealth [2, 5, 7].

Figure 1.8 A sandglass that measured a one-hour period (c. 1659, courtesy of the Concord Museum in Massachusetts).

The sandglass is another type of flow clock. Its origins are unknown, but it seems that very few existed before the 14th century. A sandglass consists of two vertical glass bulbs that are connected by a small opening. The opening allows a trickle of sand to flow from the top bulb to the bottom bulb (Figure 1.8). A sandglass is usually less accurate than a water clock, simply because the flow rate of sand is less predictable than the flow rate of water. The friction of the sand could also gradually widen the opening, so a sandglass often became less accurate the more often it was used. Even so, sandglasses had many advantages over water clocks—they were more affordable, more portable, and less affected by temperature. They also worked on ships, where the motion of the sea made water clocks unusable, and mariners used sandglasses until the 18th century to help them measure a ship's speed and the distance that it traveled. Of course, sandglasses were never well suited for continuous timekeeping—a device that contained enough sand to run through the day or night would be too large and heavy for a person to turn over. Their main use was to measure short periods of time, with a good example being the three-minute egg timers that are still occasionally found in today's kitchens [2, 5, 7, 15].

Candle clocks measured time based on melting rate, rather than flow rate, but the principle was the same. The candle was calibrated so that its approximate melting rate was known, and either the candle itself or a nearby chart was marked with a scale that indicated the passage of time. The candle's height indicated the current time. By sticking a nail in the candle at a desired time interval, a candle clock could be used as an interval timer in much the same fashion as a sandglass. When the part of the candle surrounding the nail melted, the nail would fall and clatter on to a plate below.

Many historians credit the invention of the candle clock to King Alfred the Great of England (849–899) in the latter part of the 9th century, although they were almost certainly used before then. King Alfred's clocks consisted of six 12-inch candles, each divided into 1-inch sections. It would take about four hours for one candle to burn, with the burning of one inch of candle equal to about 20 minutes, and the burning of six candles equal to one day. The candles allowed the king to divide his day into three eight-hour segments, one for study and prayer, one for official business, and one for rest [7].

Because they were used for centuries to measure time, both sandglasses and candles remain enduring symbols of the passage of time. The image of a sandglass is often used as an icon for timekeeping, and a candle that melts down to nothing or goes out prematurely is often used as a metaphor for the end of life.

Despite the ingenuity displayed by the inventors of shadow and flow clocks, time measurement was almost unknown to most citizens during the Middle Ages. According to the French medieval historian Marc Bloch (1886–1944), the people of that era

> *"…. lived in a world in which the passage of time escaped their grasp all the more because they were so ill-equipped to measure it. Water-clocks, which were costly and cumbersome, were very rare. Hourglasses were little used. The inadequacy of sundials, especially under skies quickly clouded over, was notorious."* [16]

All of this changed with the invention of the mechanical clock.

1.5 Mechanical Clocks

Mechanical clocks measure time using a different method than the shadow and flow clocks that preceded them. They keep time by counting and measuring the repetitive motion of oscillators, which are devices that swing back and forth, or oscillate, at a nearly constant rate or frequency. This concept permanently changed time measurement. Since the invention of the mechanical clock, every new clock design has been based on some type of oscillator, and oscillators provide clocks with both their "heartbeat" and their accuracy.

There is no historical consensus on who invented the first mechanical clock, nor is it known where or when they first appeared. However, mechanical clocks may have first appeared in the late 13th century, perhaps in northern Italy or southern Germany, and most historians agree that they did exist in Italy by the early part of the 14th century. The Italian poet Dante Alighieri (1265–1321) made allusions to mechanical clocks in his *Divine Comedy*, which was completed between 1315 and 1321, and there is other evidence that public tower clocks existed in Italy around that same period [7].

The first mechanical clocks didn't have dials, but instead indicated the time by ringing one or more bells. In fact, the word clock is derived from the medieval Latin word *clocca*, which means bell. Other European languages inherited words for "bell" that are similar to "clock." There are many examples—including the Dutch word *klok*, the French word *cloche*, the German word *glocke*, and the Swedish word *klocka*.

The first clocks with dials and a hand that pointed to the hour likely appeared in the latter part of the 14th century. By the 15th century, dials were being added to some existing bell clocks that previously didn't have them [7].

The escapement of a mechanical clock is the machinery that links the back-and-forth motion of the clock's oscillator to the rotational motion of the dial that displays the time. Because this machinery is always in motion, it produces a clock's familiar "tick-tock" sound. The escapement provides a force or impulse to the clock's oscillating element, which is usually some type of wheel or pendulum. This force periodically allows a gear train to "escape" and move forward. This movement in turn moves the hands on the dial, and a mechanical clock keeps time by advancing the hands at a steady and continuous rate.

The verge escapement was the first type of escapement used in mechanical clocks. As shown in Figure 1.9, the verge is a vertical rod that's similar to an axle. Two metal plates, called pallets, are connected to the verge. A crown wheel, with sawtooth-shaped teeth, was typically rotated by energy from a slowly falling weight on a rope. As the crown wheel rotates, its teeth catch the first pallet, and this turns the verge in one direction until the second pallet is in the path of the wheel's teeth. The teeth then catch the second pallet and reverse the direction of the verge, and the cycle continuously repeats. In most early mechanical clocks, the verge would swing a horizontal bar back and forth. This bar, called the foliot, had weights hanging from each end, and the clock's rate could be adjusted by moving the weights either closer to the verge, or further away.

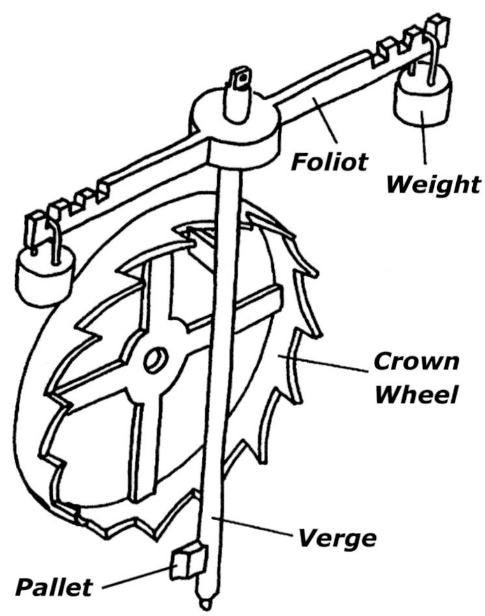

Figure 1.9 Block diagram of a verge escapement with a foliot.

The oldest surviving example of a verge and foliot clock is the Salisbury Cathedral clock in England, which is believed to have been built in 1386 (Figure 1.10). The early verge and foliot clocks were not accurate, even less accurate than the water clocks that preceded them, with some gaining or losing more than one hour per day. Later versions were more accurate, but they still probably gained or lost at least 15 minutes per day [5]. This large error wasn't due to the verge escapement, but was instead due to the foliot oscillator.

The invention of the pendulum oscillator immediately improved mechanical clock accuracy by more than a factor of 10. The first scientist to study pendulums was the Italian astronomer and physicist Galileo Galilei (1564–1642). The first experiments began around 1583 when Galileo, who was then studying medicine, used the human pulse rate to check the regularity of a pendulum. He later applied the reverse concept, and measured pulse rates with the help of a pendulum [7, 18]. Galileo discovered that the period of a pendulum's swings was approximately constant. This periodicity was mostly independent of the width of a pendulum's arc or the angle of its swing, although it was later found that there was less variation in the period when the swings were limited to very small angles. The duration of the period depended on the length of the pendulum and the local strength of gravity.

Figure 1.10 Salisbury Cathedral Clock.

Galileo's pendulum experiments were invaluable to science, and he was the first scientist to realize that the natural periodicity of a pendulum could be applied to clocks. Near the end of his life, when he was blind and in failing health, Galileo designed an escapement and made plans to build a mechanical clock with a pendulum oscillator. He described the device to his son Vincenzio, who drew a sketch of the design and later began building a prototype. However, both men unfortunately died before their work was completed.

Galileo's experiments likely influenced the Dutch scientist Christiaan Huygens (1629–1695), who is considered by most historians to be the inventor of the pendulum clock. Huygens was a key figure in the development of modern science, and his numerous other accomplishments include discovering the largest moon of Saturn, improving the lenses and designs of telescopes, and in his later years formulating the wave theory of light. After designing his first pendulum clock, Huygens contracted a Dutch clockmaker named Saloman Coster (1620–1659) to build it, and the clock was completed and patented in 1657. The clock still used a verge escapement, but the foliot was replaced with a pendulum. Figure 1.11 is a diagram of the internals of a pendulum clock that Huygens published in his *Horologium Oscillatorium* (in English, *The Pendulum Clock*) in 1673 [19]. In this design, the verge escapement was rotated so that the crown wheel faces up.

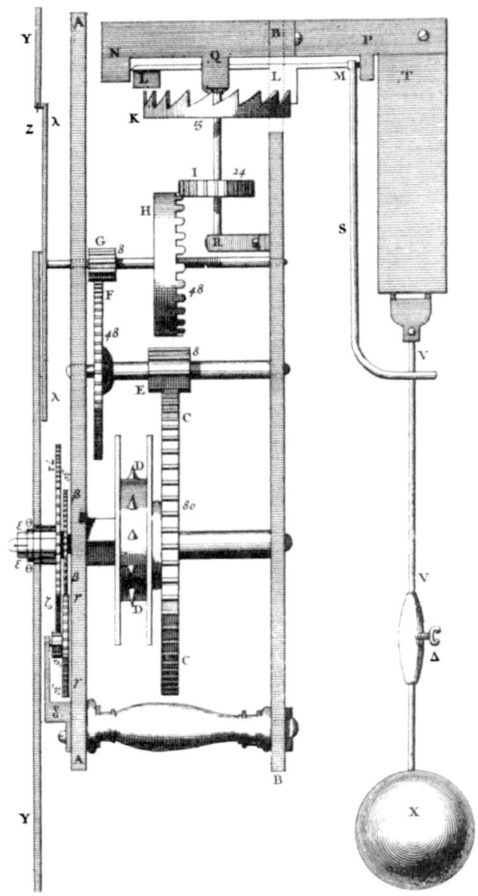

Figure 1.11 Huygen's pendulum clock mechanism, 1673.

Pendulums immediately improved the accuracy of mechanical clocks to about one minute per day, with later refinements resulting in more improvement. As a result, many of the verge and foliot clocks of Huygen's era were taken apart and rebuilt with pendulums, and new pendulum clocks soon appeared throughout Europe. Many had prominent clock dials that brought attention to the new levels of accuracy [20]. Some even had second hands. A few pre-pendulum clocks also had second hands, but they were known to be inaccurate, and their purpose was simply to show that the clock was running.

After the foliot was replaced by the pendulum, clockmakers soon focused on the verge escapement as the main source of inaccuracy. This led to a new mechanism, called the anchor escapement, that made it easier to build an accurate pendulum clock. The anchor escapement was likely invented around 1666 by the British scientist and philosopher Robert Hooke (1635–1703), but it first appeared in a clock built by William Clement (1638–1704) of London in 1671 [5, 17, 18]. The anchor escapement reduced the wide swings of the verge escapement, keeping them within a range of about 3° to 5°. This narrow range reduced the variations in the period of the pendulum's swings. It also allowed the use of longer, slower-moving pendulums that required less energy to sustain oscillation, and that caused less wear on the clock's other moving parts.

The tall and narrow shape of a longcase, or grandfather clock, results from the use of an anchor escapement with a long pendulum. The accuracy of the anchor escapement was superior to the verge escapement, but it was subsequently improved upon by other escapement types. The escapement types included the cylinder, deadbeat, detent, lever, and pin wheel, as well as other types that were based on a combination of features from various methods. Despite these advancements, most mechanical clocks continued to be built with verge escapements until the latter part of the 19th century [17, 18, 20].

1.6 John Harrison and the Problem of Determining Longitude at Sea

The British clockmaker John Harrison (1693–1776) is the main character in what is probably the most often told story in the annals of time measurement. Harrison was self-educated and a carpenter by trade, but he was naturally drawn to clockmaking. He began repairing clocks at an early age, and built his first clock—a longcase clock with a wooden mechanism—in 1713. By 1722, Harrison had invented a low-friction escapement for pendulum clocks that became known as the grasshopper escapement, a name that probably arose from the kicking action of its pallets. The invention allowed him, with some assistance from his brother James, to build a series of clocks in the mid-1720s that kept time to within fractions of a second per day. The clocks that Harrison built for use on land were the most accurate in the world, and their accuracy wasn't equaled until the late 19th century.

Harrison spent most of his life working on the greatest scientific problem that was then faced by society, which was determining the longitude of a ship at sea. It was a problem with huge economic implications, because trade with other countries depended upon sailing. More importantly, it was a problem of life and death. There were no landmarks in the ocean, and thousands of sailors had died due to ships being lost or colliding with shorelines. Latitude, the position in the north-south direction, could be accurately estimated by using an angle-measuring instrument, such as a quadrant or sextant, to measure the elevation angle of the North Star in the northern hemisphere or the Sun in the southern hemisphere (Section 15.1). Determining longitude, or position in the east-west direction, was more complicated. Because the Earth continually turns on its axis, there was nothing visible from one longitude that wasn't seen during the same day from every other longitude. This meant that mariners would see the same Sun, the same Moon, and the same stars regardless of their location. However, mariners would see these celestial objects at different times based on their location, so the key to determining longitude was an accurate clock.

Harrison knew that if mariners carried a shipboard clock of sufficient accuracy, they could navigate by comparing their local time to the ship's clock. The ship's clock would first need to be synchronized at a reference longitude line, for example, at the 0° longitude line that was later known as the prime meridian. Local time could be obtained from the sky. For example, local noon could be determined by waiting for the Sun to be directly overhead. Every four minutes of difference between the local time and the ship's clock equaled 1° of longitude, in a direction either east or west of the reference longitude.

In an effort to solve the longitude problem, the British Board of Longitude began offering prizes in 1714. The top prize was £20,000, which was an enormous sum for that era. It was to be awarded for the development of a method that could continuously determining longitude to within 30 nautical miles. One degree of longitude equaled about 60 nautical miles at the equator, so the prize requirement meant that longitude needed to be known to within one half of a degree. The terms of the prize also stated that this accuracy would have to be held during an entire outward voyage from England to the West Indies, a trip that would probably take at least 60 days. Half a degree translated into two minutes of time, so the clock could gain or lose no more than two seconds per day during the 60-day period. Harrison had already built clocks that could easily do this on land, but they were unable to function on ships tossed about by the sea.

In one of the greatest examples of perseverance in scientific history, Harrison pursued the longitude problem for more than 40 years. He built a series of five clocks, now known as H1 through H5. He invested nearly 30 years into building the first three clocks, and H3 alone took him 19 years to complete. Unable to meet the prize requirements, Harrison finally abandoned the idea of building a large sea clock and focused on building H4, a much smaller clock than its predecessors, which he completed in 1759. The dial of H4 was just 13 cm in diameter, so it looked like an oversized pocket watch (Figure 1.12). It used a heavily modified verge escapement with pallets made of diamonds, and the oscillator was a spring and balance wheel that was compensated for temperature changes. Harrison wrote of H4 that there was nothing "more beautiful or curious in texture than this my watch or Timekeeper for the Longitude .... I heartily thank Almighty God that I lived so long, as in some measure to complete it."

Figure 1.12 John Harrison's marine chronometer, H4 (courtesy of Royal Museums Greenwich).

The accuracy of H4 was so advanced for its era that it remains the most famous of all man-made clocks. Its first sea trial was a round trip from England to Jamaica that began in November 1761 and lasted for 147 days. John Harrison had by then grown too old for sea travel, so his son William accompanied H4 on the voyage. Adjusted for rate, H4 lost only one minute and 54.5 seconds during the voyage, which meant that it maintained longitude within half a degree for twice the duration and distance specified in the terms for the prize. However, there was dispute about the clock's rate correction of 2.66 seconds per day, which had not been declared before the voyage. As a result of this dispute, Harrison was denied the prize.

After much negotiating, a second sea trial finally began in March 1764. William Harrison again sailed with H4, this time on a voyage to Barbados. The rate correction was declared prior to the Barbados voyage, and to avoid any other disputes, four different mathematicians were selected to independently calculate the results. During the 46-day voyage, the average estimated time error was just 39.1 seconds, amounting to less than 10 nautical miles of longitude, which was more than three times better than the prize requirements. The Board of Longitude accepted these results, but they still felt that the results were a fluke, and that H4 and its accuracy might never be replicated. After Harrison agreed to let H4 be copied, he was given £10,000, half of the prize amount, in 1765.

The Harrisons were deeply frustrated by the Board of Longitude's actions, but they continued to persevere as always. While waiting for the copies of H4 to be made, they built another clock called H5, which improved on H4's accuracy but took five years to build and test. They then appealed to King George III (1738–1820) for help in collecting the rest of the prize, and the King intervened after he personally supervised measurements of H5 at his private observatory. As a result of the King's help, John Harrison received £8,750 in 1773 when he was 80 years old, which was most but not all of the remaining prize [21–24].

1.7 The Mass Production of Mechanical Clocks and Watches

Mechanical clocks are complex and elaborate machines. Their finely tuned mechanisms consist of a carefully assembled collection of gears, springs, weights, and balances. These parts all work together to move the hands of a clock at a regulated rate that approximates the rotational rate of the Earth. In a grand, imaginative sense, mechanical clocks can be thought of as geared models of the universe.

As we can glean from the story of John Harrison, clockmaking was a profession that required a considerable amount of skill. This skill included both a strong understanding of science and a highly advanced level of craftmanship. In fact, from the advent of mechanical clocks in the late 13th or early 14th century, to perhaps the end of the 18th century, the advances made by clockmakers remained near the cutting edge of technical achievement [15]. During this period, mechanical clocks were produced only in small numbers. Their construction was labor intensive and many of their parts were made out of costly materials. This made clocks too rare and expensive for the average person to own. For the few that did own them, clocks were seen as a symbol of their social standing and wealth. For the most part, mechanical clocks were communal devices, located in places such as government buildings, cathedrals, and town squares, where they could be seen by those who were passing by.

In the early 19th century, the mass production of mechanical clocks finally made it feasible for middle-class families to purchase clocks and bring them into their homes. The mass-produced clocks had several forms, including longcase clocks that stood on the floor, clocks that hung on walls, and mantel clocks that were placed above a fireplace, or on a shelf, dresser, or table. Due to their affordability, wall and mantel clocks were more common than longcase clocks. Ironically, the mass production effort wasn't dominated by England, Germany, or Switzerland, or any of the other European nations who already had centuries of clockmaking experience by the start of the 19th century. Instead, mass production was led by the United States of America, a then-new nation that had won its independence from England in 1776. Generally speaking, the clockmakers in Europe were focused on making clocks from the highest quality materials, and they built high-cost products that were produced in relatively low quantities. This differed from the U. S. clock industry, which from its inception was devoted to producing high quantities of low-cost products. The early American clocks were less ornate and likely considered to be less attractive than the clocks made in Europe, but despite their low cost, they were usually sturdy, accurate, and repairable [18].

The American inventor and clockmaker Eli Terry (1772–1852) was a pioneer in the mass production of house clocks. Terry devised methods where clocks were made out of mass-produced, uniform, and interchangeable parts. These methods were eventually applied to other types of products, and came to be known as the "American System" of manufacturing [15]. Terry began making clocks in Plymouth, Connecticut in 1793. At first his clocks were made by hand, then delivered on horseback to purchasers who had ordered them in advance. By 1803, Terry had found that his clocks could be sold without delivering them in person, and he then began efforts to manufacture clocks on a large scale, ignoring the comments of those who felt that it wasn't possible. He set up a factory in Plymouth where a water mill powered his clock-making machinery [26]. In 1807, Terry accepted a contract from merchants in Waterbury, Connecticut to build 4,000 wooden clock movements for longcase clocks over a three-year period. The contract was successfully fulfilled, and the clock movements that Terry delivered were the first mass-produced clocks with interchangeable parts in the United States [15, 20].

From that point forward, Terry was determined to make clocks that were as affordable as possible. In 1816, he patented and began producing a simple wooden box clock with a glass door that was designed to be set on a mantel or shelf (Figure 1.13). The clock dial consisted of numbers that were reverse-painted on a glass door. The gears in the clock movement were also made of wood to reduce cost. By 1822, Terry's simple box clock had been refined to make it more attractive, with slender pillars on the side and scrollwork on the top, supported by a set of graceful feet. This type of clock, shown in Figure 1.14, became widely known as the Connecticut pillar-and-scroll clock.

Pillar-and-scroll clocks took their name from their outward appearance, but those designed by Eli Terry also some distinctive characteristics on the inside. The movement was attached to the clock dial, and the clock dial was directly secured to the case. The pendulum and escapement were in front of the clock dial, outside of the clock movement, a style known as an outside escapement. The dial and pendulum were covered by a glass door, and a painting on the bottom portion of the glass left an opening where the pendulum could be viewed. The clocks needed to be wound at least once every 30 hours. By 1825, Terry was operating three factories and producing two to three thousand pillar-and-scroll clocks per year. They were sold to peddlers and store merchants at wholesale prices that started at $12, but that later dropped to well below $10. Traveling peddlers sold the clocks by going door-to-door, typically for a price of about $15 [15].

Figure 1.13 Eli Terry Mass-Produced Box Clock, 1816 (courtesy of National Museum of American History, Smithsonian Institution).

Figure 1.14 Eli Terry Pillar-and-Scroll Type Shelf Clock, c. 1822 (courtesy of Henry Ford Museum).

In addition to house clocks, 19th-century mass production also made it possible for the average person to own a watch, which was the name given to a small, portable clock. Watches date back to 16th century Europe, and were usually either worn on the wrist or carried in the pocket, although some were worn as pendants. The word "watch" was likely derived from the Anglo-Saxon word *wacian* which means to be watchful, awake, and vigilant. It likely refers to the practice of workers who kept the night watch, either in the streets of a town or onboard a ship. Carrying a watch allowed them to know what time it was during their shift, and to announce the time when they were asked to do so [25].

Pocket watches were mass produced long before wristwatches. The first wristwatches were almost exclusively worn by women, and they were not especially common. They were usually decorative bracelets containing small timepieces, elaborately designed and expensive to own, and were likely thought of more as jewelry than as clocks. The men who carried clocks wore pocket watches, usually kept in their vest pocket and connected to a chain. For the most part, men didn't wear wristwatches until the early years of the 20th century, probably because they were considered to be feminine jewelry. The perception changed when soldiers and military officers began wearing wristwatches, due to the obvious advantage of being able to check the time while keeping both hands free. By the end of World War I, almost all enlisted men wore a wristwatch, and the fashion caught on amongst civilians. In the years following the war's end, the men's wristwatch became common, and the pocket watch became rare [15].

The first mass-produced pocket watches were manufactured by the American Watch Company in Waltham, Massachusetts. After several years of experimentation and small production runs, the company began mass production of a low-cost pocket watch in 1861, shortly after the start of the American Civil War. The first industrialized model of the watch was known as the William Ellery grade M57, shown in Figure 1.15. The name was chosen to honor William Ellery (1727–1820), an abolitionist who had signed the Declaration of Independence in 1776, and the number 57 referred to 1857, the year when the watch was first designed [27]. The William Ellery model cost just $13 and became a favorite of Union soldiers during the Civil War [15]. Years later, after numerous versions of the watch were made—all designed so that "any movement could fit any case, and any part of one unit could be interchanged with any other"—the watches were collectively categorized as the Waltham Model 57. More than 830,000 Model 57s were made before production stopped in the 1880s [27].

Figure 1.15 William Ellery grade M57 pocket watch, made by the American Watch Company, 1864 (courtesy of National Museum of American History, Smithsonian Institution).

All versions of the Model 57 pocket watches used an English-style, side lever escapement [27]. The lever escapement was invented in 1755 by the British clockmaker Thomas Mudge (1715–1794), but it wasn't widely used in watches outside of England until the mid-19th century. However, it was found to be both accurate and reliable—watches with lever escapements seldom stopped when they were bumped or jolted—and by the start of the 20th century it was found in nearly all new watches being made.

The lever in the Model 57 was powered by the mainspring and gear train of the watch. The lever had two pallets that were made from jewels to reduce friction. As the lever rocked back and forth, its pallets would catch the teeth on the balance wheel, giving it a brief push before releasing it. The lever would then get out of the way, until the balance wheel returned for another energizing push. In between pushes, the balance wheel had the advantage of being able to freely swing. The role of the lever was to convert the gear train's linear motion into the angular, oscillatory motion of the balance wheel [20].

Continuous improvements made to the United States manufacturing methods eventually led to pocket watches that sold for just $1. The most popular dollar watches were sold by the Ingersoll Watch Company beginning in 1892. Ingersoll was then a mail order company in New York City that marketed watches made by the Waterbury Clock Company in Connecticut, but they later made their own watches, which were manufactured both in Waterbury and in Trenton, New Jersey [28]. The Ingersoll watches became famous in 1893, when 85,000 were reportedly sold at the Chicago World's Fair for $1.50 each, a price that was later reduced to an even dollar. The dollar watches were sturdy, but there were, of course, some compromises in their design. They used a pin-lever escapement, where the lever's pallets were metal pins instead of jewels. This reduced manufacturing costs, but it added friction to the movement and made a dollar watch less accurate than a more expensive jeweled watch. Dollar watches were also difficult to disassemble and repair, but they were guaranteed for a year, and if they failed after that they could be thrown away, and replaced by buying another one. Their low cost made them accessible to almost anyone—in fact they were so cheap that they could be given away as promotional items and even marketed as presents for children [15]. Ingersoll heavily advertised their best known model, called the Yankee, with some catchy slogans including "It Runs with the Sun!" and "The Watch that Made the Dollar Famous." The late-19th advertisement pictured in Figure 1.16 states that 3,000 watches were being produced per day, more than one million per year. Production continued to increased, and Ingersoll produced 16,000 dollar watches per day in 1916 [28].

Figure 1.16 A late-19th century advertisement for the Ingersoll dollar watch.

1.8 The Shortt Free Pendulum Clock

During the same years when some clockmakers were focused on making clocks of modest accuracy affordable for everyone, other clockmakers were engaged in efforts to build the most accurate clocks possible, regardless of cost. The pinnacle of mechanical timekeeping accuracy was reached in 1921, when the British railway engineer William Hamilton Shortt (1881–1971) invented a new type of clock that included two pendulums. Only about 100 of Shortt's clocks were made, but they were produced for more than 30 years, and served as time references at astronomical observatories from the 1920s until the 1950s.

The Shortt Free Pendulum clock was marketed by the Synchronome Company in London. Their 1928 advertisement, shown in Figure 1.17, called the Shortt Free Pendulum the "The Perfect Clock," and noted that "over a period of nearly 12 months where it was under the closest observation, no change of rate could be detected" (Figure 1.17). The ad copy could be considered as overenthusiastic, but there was a definite reason for the enthusiasm. The Shortt clock wasn't perfect, but it was accurate enough to reveal that the Earth's rotation rate also wasn't perfect. The Shortt clock ran at a rate that was very close to the rate of the Earth's rotation, so close that it wasn't easy to tell if the differences could be attributed to the Shortt clock, or to the Earth itself [29, 30].

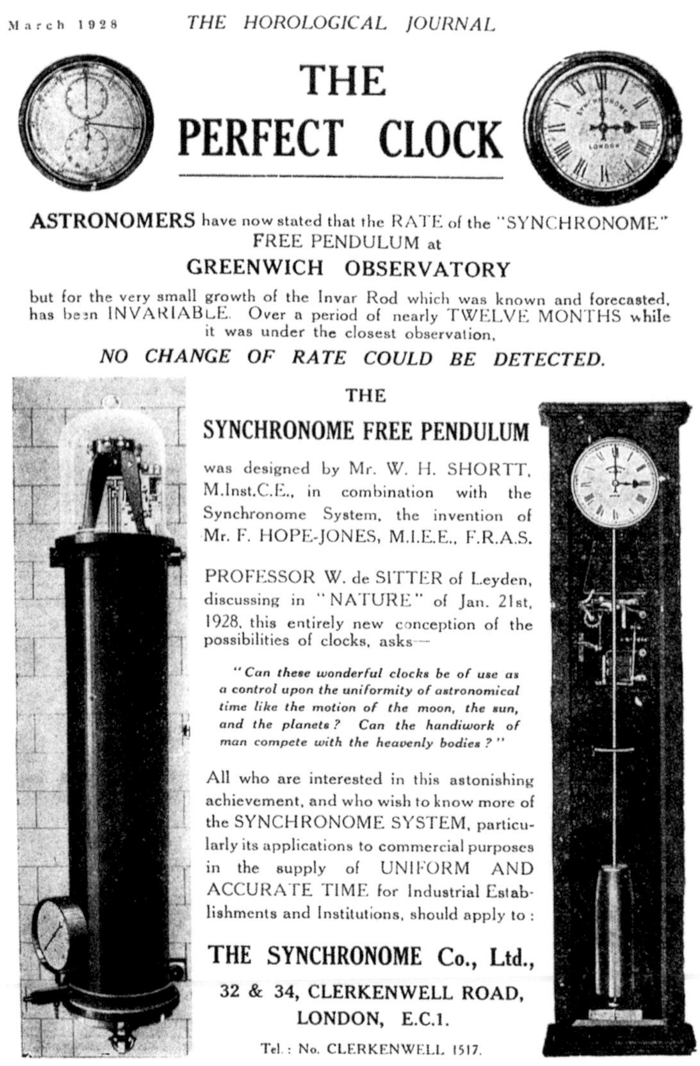

Figure 1.17 A 1928 advertisement for a dual-pendulum Shortt clock.

The Shortt clock was a complex electromechanical device that had both a primary and a secondary pendulum (respectively shown on the left and right sides of the photograph in Figure 1.18). Both pendulums were about one meter long, and both were seconds pendulums, which meant that their period was two seconds, one second for a swing in one direction, and another second for the return swing. The use of two pendulums distinguished the Shortt clock from traditional pendulum clocks, where a single pendulum is linked to the clock mechanism, which disturbs its natural swing.

Figure 1.18 The Shortt Free Pendulum Clock (courtesy of NIST museum).

In the Shortt clock, only the secondary pendulum was linked to the clock mechanism. It was housed in a cabinet that made it look similar to a traditional longcase pendulum clock. The primary pendulum was the free pendulum, and it was free in more than one sense. It was free from air resistance because it was housed inside a vacuum chamber. This allowed it to move without friction, and it reduced the amount of energy that was needed to sustain oscillation. The primary pendulum was also free because it was left alone most of the time. It was allowed to swing freely for 30 seconds—exactly 30 pendulum swings—before being briefly disturbed by a gravity lever escapement that gave it the energy to keep swinging. The motion of the gravity lever then sent an electrical pulse to a mechanism that kept the secondary pendulum synchronized. Two clock dials on the case of the secondary pendulum displayed the time kept by both pendulums, to verify that they were synchronized [31].

The clever design of the Shortt Free Pendulum made it more accurate than all previous clocks. During the decade following its introduction, many were on record as gaining or losing less than one second per year [29, 30, 32]. This accuracy was confirmed decades later, when a Shortt clock was compared to atomic clocks in 1984. The measurements showed that the Shortt clock was not only capable of accuracy much better than one second per year, but that its pendulum was sensitive enough to detect the slight gravitational changes on Earth that are caused by the Sun and Moon [33].

Much like Harrison's H4, the Shortt Free Pendulum was a truly elegant instrument, and a shining example of the remarkable skill and ingenuity of clockmakers. It wasn't perfect, contrary to its marketing claims, but it was the most accurate clock in the world for about a decade. Its accuracy was finally surpassed by clocks referenced to quartz crystal oscillators, the topic of Chapter 2.

References

1. L. Holford-Stevens, *The History of Time: A Very Short Introduction*, Oxford University Press, 2005.
2. C. Orzel, *A Brief History of Timekeeping*, BenBella Books, 2022.
3. O. Gingerich, "Did Copernicus Owe a Debt to Aristarchus?," *Journal for the History of Astronomy*, vol. 16, no. 1, pp. 37–41, February 1985.
4. N. Copernicus, *De Revolutionibus Orbium Cœlestium, Libri VI* (*Six Books on the Revolutions of the Heavenly Spheres*), translated by E. Rosen, Warsaw: Polish Scientific Publications, 1978.
5. J. E. Barnett, *Time's Pendulum*, Harcourt Brace & Company, 1998.
6. A. E. Waugh, *Sundials: Their Theory and Construction*, New York: Dover Publications, 1973.
7. G. Dohrn-Van Rossum, *History of the Hour: Clocks and Modern Temporal Orders*, University of Chicago Press, 1996.
8. J. H. Breasted, *Ancient Times: A History of the Early World*, Chicago: Ginn and Company, 1916.
9. G. Ifrah, *The Universal History of Numbers: From Prehistory to the Invention of the Computer*, John Wiley & Sons, 2000.
10. D. W. Hughes, B. D. Yallop, and C. Y. Hohenkerk, "The Equation of Time," *Monthly Notices of the Royal Astronomical Society*, vol. 238, pp. 1529–1535, June 1989.
11. P. K. Seidelmann, *Explanatory Supplement to the Astronomical Almanac*, Mill Valley, California: University Science Books, 1992.
12. O. Neugebauer, *The Exact Sciences in Antiquity*, 2nd edition, Brown University Press, 1957.
13. Ptomely, *Ptomely's Almagest*, translated by G. J. Toomer, Princeton University Press, 2nd edition, 1998.
14. F. Cajori, *A History of Mathematics*, MacMillan and Co., 1893.
15. C. E. Stephens, *On Time: How America Has Learned to Live by the Clock*, Smithsonian Institution, Bullfinch Press, 2002.
16. M. Bloch, *Feudal Society*, translated by L. A. Manyon, University of Chicago Press, 1961.
17. D. S. Landes, *Revolution in Time: Clocks and the Making of the Modern World*, Cambridge: Harvard University Press, 2000.
18. A. Bird, *Illustrated Guide to House Clocks*, New York: Arco Publishing Company, 1973.
19. J. G. Yoder, "Christiaan Huygens, Book on the Pendulum Clock (1673)," Chapter 3 in *Landmark Writings in Western Mathematics: 1640–1940*, pp. 33–44, Elsevier Science, 2005.
20. D. Christianson, *Timepieces: Masterpieces of Chronometry*, Toronto: Firefly Books, 2002.
21. H. Quill, *John Harrison: The Man who found Longitude*, New York: Humanities Press, 1966.
22. D. Sobel, *Longitude: The True Story of a Lone Genius Who Solved the Greatest Scientific Problem of His Time*, New York: Walker Publishing, 1995.
23. D. Howse, *Greenwich Time and the Longitude*, London: Philip Wilson Publishers, Ltd., 1997.
24. W. J. H. Andrewes, ed., "The Quest for Longitude," *Proceedings of the Longitude Symposium*, Harvard University, November 1993.
25. J. Jespersen and J. Fitz-Randolph, "From Sundials to Atomic Clocks: Understanding Time and Frequency," *National Institute of Standards and Technology Monograph 155*, 304 p., 1999 edition, March 1999.

26. F. Atwater, *History of the Town of Plymouth, Connecticut*, The Journal Publishing Company, 1895.
27. R. Price, "Origins of the Waltham Model 57: Evolution of the First Successful Industrialized Watch," *NAWCC Special Order Supplement No. 7*, 2005.
28. C. Shugart and T. Engle, *The Complete Guide to American Pocket Watches*, 2nd ed., Overstreet Publications, Inc., 1982.
29. J. Jackson and W. Bowyer, "The Shortt Clocks at the Royal Observatory, Greenwich," *Monthly Notices of the Royal Astronomical Society*, vol. 88, no. 5, pp. 465–481, March 1928.
30. J. Jackson, "Shortt Clocks and the Earth's Rotation," *Monthly Notices of the Royal Astronomical Society*, vol. 89, no. 3, pp. 239–250, January 1929.
31. F. Hope-Jones, *Electrical Timekeeping*, London: N. A. G. Press, 1940.
32. E. W. Brown and D. Brouwer, "Analysis of Records Made on the Loomis Chronograph by Three Shortt Clocks and a Crystal Oscillator," *Monthly Notices of the Royal Astronomical Society*, vol. 91, no. 5, pp. 575–591, March 1931.
33. P. H. Boucheron, "Effects of the Gravitational Attractions of the Sun and Moon on the Period of a Pendulum," *Antiquarian Horology*, vol. 16, no. 1, pp. 53–65, March 1986.

Chapter 2

Quartz Clocks

The apparatus was designed especially as a reference standard of frequency for the Bell System, but it was recognized that it might also serve as a reference standard of time .... Some types of time measurement can be made more readily by means of a time piece of the crystal clock type than by any other clock mechanism.

— Warren Marrison, "The Quartz Crystal Clock," 1930

Pendulum clocks, described in the previous chapter, are especially useful for illustrating the close relationship between time and frequency. For example, we noted earlier that oscillators provide the heartbeat of all modern clocks. Oscillators generate a frequency, f, that is the reciprocal of the period of oscillation, T, therefore $f = \frac{1}{T}$. Conversely, the period is the reciprocal of the frequency, so $T = \frac{1}{f}$.

The relationship between time and frequency applies to all clocks, but it's especially easy to see with a pendulum clock. If a pendulum swings back and forth once per second, the period, T, is one second, abbreviated in the International System of units (SI) as 1 s, and the frequency, f, is 1 cycle per second, or 1 hertz, abbreviated as 1 Hz. We initially said that a clock is a device that measures and counts time intervals, and this is again easy to see with a pendulum clock. Because the period of the pendulum swings is 1 s, a clock can keep time by simply counting the swings to measure seconds, minutes, and hours. If the frequency of the pendulum isn't exactly 1 Hz, the clock will either gain or lose time. The reason for this is simple—if the pendulum swings have a duration that's either shorter or longer than the duration of a standard second, then the clock will run too fast or too slow. We'll go into much more detail later on about the importance of frequency to accurate timekeeping, but the important thing to remember for now is that a clock can only be as good as its oscillator.

The words clock and oscillator are often used interchangeably by time and frequency metrologists, and that practice continues in parts of this book. More people are interested in time than frequency, so it follows that "clock" is a more familiar word than "oscillator," and oscillators are often called clocks. There is, however, a technical distinction between clocks and oscillators that we've already touched upon, and that we can expand upon here. Technically speaking, the device inside of a clock that produces a repetitive, periodic signal is called a resonator, and the frequency of the signal it produces is called the resonance

As Time Goes By: A Comprehensive Guide to Time and Frequency Measurements
Michael A. Lombardi

ISBN 978-981-5352-28-3 (Hardcover), 978-1-042-01532-0 (eBook)
www.jennystanford.com

DOI: 10.1201/9781042015320-4

frequency. In our simple example of the pendulum clock, the pendulum is the resonator and the resonance frequency is 1 Hz. Turning a resonator into an oscillator requires an energy source, such as a mainspring, a motor, or some form of electrical current, so that the resonance frequency can be sustained. Taken together, the energy source and the resonator form an oscillator. To make a clock, we need more than just an oscillator. We also need a mechanism that counts the cycles or events produced by the oscillator and converts them to time units. To complete the clock, we need a way to display or record the timekeeping results.

Quartz crystal oscillators–invented in the 1920s—didn't change these basic principles, but they did change nearly every aspect of time and frequency metrology. By running at much higher frequencies than previous oscillators, they enabled the development of the first frequency standards, and frequency metrology as we know it today couldn't exist without them.

Beginning in the 1930s, quartz oscillators were used to build quartz clocks that easily surpassed the best pendulum clocks in just about every category, providing better accuracy and reliability in a smaller package at a lower cost. Quartz devices have since become the most common timekeepers of all. Billions of quartz oscillators are produced each year to provide the frequency and time reference for just about every type of electronic device.

This chapter discusses the evolution of quartz oscillators and clocks, beginning with the discovery of piezoelectricity in Section 2.1. The essential role that quartz oscillators played in the early days of the radio industry is described in Section 2.2. The first quartz clocks are described in Section 2.3, followed by a discussion of synthetic quartz crystals (Section 2.4) and quartz wristwatches (Section 2.5). Section 2.6 covers the basics of quartz oscillator design, and Section 2.7 concludes the chapter by discussing the limitations of quartz oscillator performance.

2.1 Piezoelectricity

It has been long known, perhaps even to ancient civilizations, that crystals of certain minerals have electrical properties. One often cited example involves tourmaline crystals, which were once brought from Ceylon (Sri Lanka) to Europe in large quantities to be sold as gemstones. In 1703, a Dutch merchant noted that a heated tourmaline crystal would both attract and repel small particles of ash. Tourmaline then became known as the "electric stone" or the Ceylon magnet [1]. This resulted in more than a century of scientific experiments and theories. The Scottish scientist David Brewster (1781–1868), who had observed the effect with various kinds of crystals, called it pyroelectricity in 1824 [2].

While studying pyroelectricity, the French physicists Pierre Curie (1859–1906) and Paul-Jacques Curie (1856–1941) demonstrated in 1880 that certain crystals would also produce a charge when mechanical pressure was applied. These crystals would show positive and negative charges on portions of their surface when they were compressed in particular directions. The amount of the electrical charge was proportional to the pressure, and the charge disappeared when the pressure was withdrawn. This effect was called piezoelectricity, or electricity resulting from pressure. The term is derived from the Greek word *piezin*, which means to squeeze or press. Among the crystals tested by the Curie brothers, the most piezoelectricity was exhibited by quartz crystals (Figure 2.1) and Rochelle salt crystals, which are less durable than quartz and will dissolve in water [3].

The Curies' initial experiments only demonstrated the direct piezoelectric effect, where the crystal converted physical pressure into electrical energy. In 1881, based on a theoretical suggestion by the French physicist Gabriel Lippman (1845–1921), the Curies demonstrated that piezoelectricity also had an inverse effect. When an electrical charge was applied to the crystal, it would physically strain and change its shape. This meant that piezoelectric crystals were transducers, or devices that could convert energy from one form to another. A crystal could convert mechanical energy to electrical energy and vice versa, although some of the energy was lost during each conversion.

Figure 2.1 A cluster of natural quartz crystals.

Piezoelectricity was an important scientific discovery, but for several decades after the Curies' experiments, it was regarded as little more than a curiosity. During those years, it received even less attention than pyroelectricity, even though it was the chief cause of many pyroelectric effects [2]. Piezoelectricity finally found an application when the Allied forces of World War I desperately needed a way to detect the presence of enemy submarines. By 1917, the French physicist Paul Langevin (1872–1946) and the American scientist Alexander McLean Nicolson (1869–1950) had independently begun work on piezoelectric instruments that could transmit and receive sound waves underwater, determining the presence of submarines by a method that was then called echo-ranging. Langevin's devices employed thin crystals of quartz glued between steel plates. His designs didn't become mature until the very end of the war, but Langevin's work had a huge impact in future years. He became known as one of the key inventors of both sound navigation and ranging (sonar) and the field of ultrasonics, which is a branch of acoustic science that involves generating sound waves at frequencies higher than the human ear can hear.

Nicolson went on to develop other types of piezoelectric devices, including microphones, loudspeakers, and phonograph cartridges [4]. He also constructed the first piezoelectric oscillator, using Rochelle salt crystals while working at Bell Telephone Laboratories [5]. The work of Langevin and Nicolson indirectly led to the quartz oscillator, a device that revolutionized both radio communications and timekeeping.

2.2 Frequency Measurement and the Early Days of Radio

As we saw in Chapter 1, time measurement has always been a topic of human interest, but frequency measurements were seldom discussed until the invention of radio, when measuring frequency became essential. Successful radio communication requires transmitters and receivers to each tune to the same frequency, so that the information sent "over the air" can reach its audience. For example, if a radio station has an assigned frequency of 97.3 MHz, we expect to hear that station when we tune the dial to 97.3. If we do hear it, then we know that the station is broadcasting on its assigned frequency, and we also know that the reading on our radio dial is relatively accurate. Of course, it's much more important for the transmitter to be on frequency than the receiver. If the receiver dial is misaligned or inaccurate, as is often the case on an old radio, we can still find the station by adjusting the dial until the signal is heard. However, if the transmitter strays from its assigned frequency, it can interfere with other stations and potentially wreak havoc. In the early days of radio, the lack of available methods for generating and measuring radio frequency (RF) signals was one of the great measurement problems of the 20th century. Finding solutions to this problem launched the field of frequency metrology as we know it today.

By the end of World War I, military and amateur radio communication had been ongoing for more than a decade, and by 1920 the first commercial radio stations were on the air [6]. The new radio industry grew quickly, and to support its needs, the National Bureau of Standards (NBS) in the United States began searching for ways to accurately measure frequency. The NBS effort was led by the American engineer John H. Dellinger (1886–1962), who clearly explained the problem in a 1923 paper entitled "Reducing the Guesswork in Tuning." Dellinger wrote that

> *"The waves used by the broadcasting stations are spaced 10 kilocycles apart (3 meters at a wavelength of 300 meters). Thus one station is on 990 kilocycles, another on 1000, and another on 1,010 kilocycles.... A variation of the frequency of 1 per cent, for example, would be a variation of 10 kilocycles and could cause one station to be using exactly the wave that had been assigned to another. The whole success of American broadcasting is thus tied up with the placing of broadcasting stations on the correct frequencies to an accuracy approaching 99.9 per cent. Since receiving sets are now available by which an individual can hear the stations from all over the United States on the same night, the importance of this accuracy is apparent."* [7]

At the time of Dellinger's paper, 570 broadcast stations in the United States were already on the air, and there was no satisfactory way to measure or regulate the frequency of their transmitted signals [8]. A common frequency measurement method then in use involved viewing Lissajous patterns on cathode-ray oscilloscopes, which could show the ratio of an RF signal with respect to an audio frequency produced by a tuning fork (Section 5.2.1). However, that method couldn't meet the 99.9% accuracy requirement. More accurate frequency measurements involved wavemeters, which were instruments that employed several inductance coils of different values and a variable capacitor. Each inductance coil enabled the measurement of a range of wavelengths. Wavemeters could just barely meet the 99.9% accuracy requirement, but would be unable to meet future requirements, which were sure to become more stringent as more stations went on the air. More importantly, a wavemeter was only a measurement instrument and not a source of frequency, and therefore it couldn't be used to control the frequency of a radio transmission.

The invention of the quartz oscillator—the first reliable and accurate standard of radio frequency—was a game changer for both frequency metrology and the radio industry. The invention is generally credited to Walter Cady (1874–1974), an American physicist who received his doctorate degree from the University of Berlin. After returning to the United States, Cady became a professor at Wesleyan University in Connecticut, a position he held for about 45 years. Like Langevin, Cady worked on anti-submarine systems during the war, but afterwards focused his attention on building a frequency standard. Like Nicolson, Cady began by experimenting with Rochelle salt crystals, but turned his attention to quartz crystals in November 1919 [2].

When applying an AC voltage over a wide range of frequencies, Cady noticed that both Rochelle salt and quartz crystals would reach their maximum amplitude oscillation, or resonance, over a very narrow range of frequencies. Quartz, however, had a sharper resonance curve than Rochelle salt. The quartz resonance curve was similar to that of a tuning fork, which had long been used as a frequency standard for sound waves (Section 5.1.2). Cady quickly realized that quartz could be made to oscillate at higher frequencies than a tuning fork—perhaps high enough to be a frequency standard for radio waves [9].

Cady applied for a patent for his "Piezo-Electric Resonator" in January 1920 [10], which was followed by an application for a "Method of Maintaining Electrical Currents of Constant Frequency" in May 1921 [11]. Figure 2.2, taken from the second patent, is thought by some historians to be a schematic of the first quartz oscillator. The output of a three-stage vacuum-tube amplifier was used to drive a rod-shaped quartz crystal (component #12 in Figure 2.2) at its natural frequency through one pair of electrodes that made use of the inverse piezoelectric effect. The amplifier was powered by the direct piezoelectric effect from the other pair of electrodes. The feedback to sustain oscillations in the electrical circuit was obtained only through the vibration of the quartz rod. Cady described it as "virtually a mechanically-tuned feedback" circuit [12].

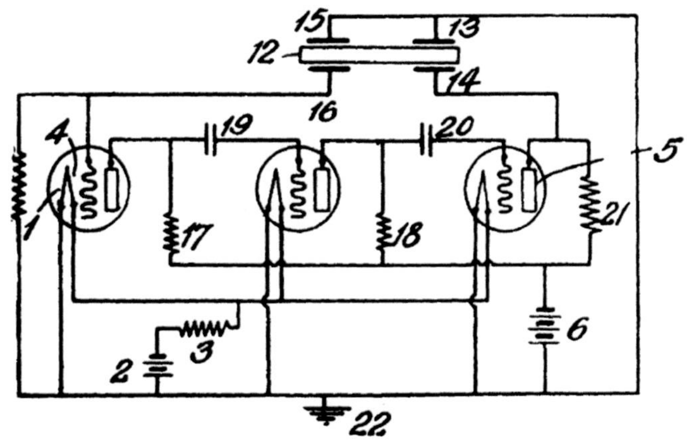

Figure 2.2 Cady's original quartz oscillator circuit (from U. S. Patent 1,472,583, [11]).

George Washington Pierce (1872–1956), who was a friend of Cady's and an electrical engineering professor at Harvard, improved upon this work. Pierce demonstrated that plates of quartz cut a certain way could be made to vibrate at specific frequencies that were proportional to the thickness of the plates. Cutting, grinding, and mounting quartz crystals to generate specific frequencies soon became a huge area of research, development, and production. Pierce also simplified and improved Cady's amplifier circuit (Figure 2.3). He used just one vacuum-tube instead of Cady's three, and connected just one pair of electrodes to the crystal (component #2 in Figure 2.3). Pierce's circuit could sustain oscillation with a constant frequency that was determined by the crystal alone, and it could oscillate at much higher frequencies than the previous circuit [13].

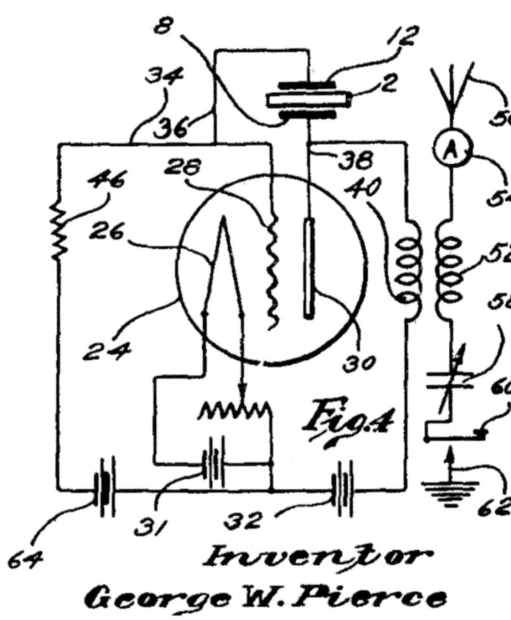

Figure 2.3 The Pierce oscillator circuit in a wireless telegraphy system (from U. S. Patent 1,789,496).

In his 1924 patent application, Pierce wrote that he had constructed apparatuses for "fundamental frequencies ranging from 35 kilocycles per second to 3,000 kilocycles per second" and that he had "utilized harmonics of the device at 20,000 kilocycles per second" [14]. Harmonics—which have been familiar to musicians since ancient times and perhaps first described by the Greek philosopher Pythagoras (570–495 BC)—are positive integer multiples of the fundamental frequency, f, which is also called the first harmonic. The second harmonic is $2f$, the third harmonic is $3f$, and so on. The generation of higher frequencies using harmonics was exactly what the radio industry needed. Within a few months, Pierce's method was being used to construct crystal-controlled radio transmitters.

The work of Cady and Pierce was synergistic—Cady proved the concept of the quartz oscillator, and Pierce made its use practical by designing a reliable and easy-to-use circuit [9]. Their work quickly led to commercial products, and by the end of 1924 radio engineers could purchase quartz oscillators from the General Radio Company in Cambridge, Massachusetts. The first commercial quartz oscillator, the General Radio Type 275, measured 254 × 279 × 203 mm and weighed about 9 kg (Figure 2.4). It sold for a price of $145, and its quartz plate was ground to a frequency that was specified by the buyer [15].

Figure 2.4 The first commercial quartz oscillator, the General Radio Type 275 (courtesy of Collection of Historical Scientific Instruments, Harvard University).

Radio engineers instantly recognized the value of the quartz oscillator, but it was several years before the standards of frequency measurement were changed. In April 1920, shortly after filing his patent application, Cady sent four crystal oscillators to John Dellinger's group at NBS for calibration. Later that same year, Dellinger wrote in an NBS annual report that "measurements have been made on a remarkable new type of wavemeter consisting of a quartz crystal" [8]. Even so, when NBS began broadcasting standard frequency signals from radio station WWV in March 1923, the radio signals weren't quartz controlled. This finally changed in 1927 when the WWV broadcasts were referenced to a 50 kHz quartz crystal oscillator located at the transmitter site. This greatly improved the accuracy of the transmitted signals, which went from parts per thousand (10^3) to parts per million (10^6).

The improved accuracy of WWV was fortuitous, because the accuracy of commercial radio stations was regulated beginning in 1927. The newly formed Federal Radio Commission, later called the Federal

Communications Commission (FCC), required all radio broadcasters to stay within 500 Hz of their assigned frequency [8]. The requirement was tightened to 50 Hz in 1932, but by then meeting the requirement was less of an issue, because quartz oscillator technology had improved. In 1928, General Radio began selling a new quartz frequency standard, the Model C-21-H, that was accurate to within 1×10^{-6} [16]. This meant that a 1 MHz radio signal could be accurate to within 1 Hz.

By 1929, the national frequency standard at NBS that controlled the frequency of WWV was accurate to 1×10^{-7}. The standard consisted of four quartz oscillators housed in temperature-controlled chambers (Figure 2.5). The oscillators were manufactured by Bell Telephone Laboratories and operated at a resonance frequency of 100 kHz [8, 17]. By this time, the rapid developments in quartz oscillator technology had made it easy for the radio industry to obtain the accuracy it needed, and a daunting measurement problem had been solved in less than a decade.

Figure 2.5 The United States national frequency standard in 1929, which consisted of a group of four quartz oscillators.

2.3 The First Quartz Clocks

From the day they were invented, quartz oscillators ran at frequencies that were much higher than the resonators in mechanical clocks. For example, some marine chronometers would beat twice a second (2 Hz) or five times every 2 s (2.5 Hz), and some clocks with long pendulums completed just one-half of an oscillation per second (0.5 Hz). In sharp contrast, quartz oscillators ran at frequencies of thousands of hertz. Therefore, building a clock based on a quartz oscillator required developing methods to measure and count these much higher frequencies.

The invention of the quartz clock can be credited to the engineer Warren Marrison (1896–1980). Born in Canada, Marrison worked on radio communications in the Royal Flying Corps during World War I. After the war, he attended Harvard University and then went to work in New York City, first for Western Electric and then at Bell Telephone Laboratories. At Bell, he worked on quartz frequency standards with Joseph Horton (1889–1967), his supervisor and colleague. Horton and Marrison presented a paper in October 1927 that described two new clocks—one referenced to the 100 Hz frequency of an electric tuning fork, and the other to the 50 kHz frequency of a quartz crystal oscillator [18]. This was likely the first public mention of a quartz clock, but Marrison's notes suggest that he had conceived of the idea several years earlier, in November 1924 [19].

Over a period of several years, Marrison refined his quartz clock to use a 100 kHz oscillator and a frequency divider, that as Marrison wrote, functioned as a "reducing gear." The frequency was divided twice, each time by a factor of 10, so that one cycle in the output of the divider corresponded to 100 vibrations of the crystal. The resulting 1 kHz frequency was used to control the speed of a small 1000-cycle synchronous motor, which locked the motor's frequency to the frequency of the crystal. Marrison noted that when the crystal produced "its nominal frequency exactly, the clock keeps accurate time," but that the frequency of the crystal could vary due to external conditions such as pressure, temperature, and vibration. Figure 2.6 shows the crystal oscillator (left) and the back of the clock face with the synchronous motor exposed (right).

Figure 2.6 The crystal oscillator (left) and motor and display assembly (right) of Marrison's quartz clock [20].

To determine the accuracy of his new quartz clock, Marrison compared it to United States Naval Observatory (USNO) time signals during December 1929 and January 1930. He found that the clock was initially accurate to within about 0.2 s per day. However, its time error was gradually changing. It changed by 0.14 s over a period of about 47 days, which was about three milliseconds per day on average. Marrison correctly attributed these changes to the "aging effect" of the crystal [20]. We can deduce from his measurements that the frequency of the crystal was changing by a few millihertz each day on average, because a 1.16 mHz frequency change at 100 kHz would result in a time change of one millisecond.

Horton left Bell Labs to become the chief engineer of General Radio in 1928, and it was there that the first commercial quartz clocks were sold. He designed a series of *Syncro-Clocks* that avoided infringing on the patents that Marrison had obtained by utilizing a different frequency divider circuit [9]. These clocks, which were introduced in the early 1930s, didn't come with their own oscillators. They were intended to be used with quartz frequency standards, which then often filled large equipment racks.

As the technology advanced, quartz clocks became both smaller and more accurate. The first "portable" quartz clock was probably the Rohde & Schwarz Model CFQ, which was manufactured in Germany beginning in 1938 (Figure 2.7). The Model CFQ weighed 46 kg and was accurate to within ±0.004 s per day (5×10^{-8}). This clock, and other similar products, established quartz clocks as the world's most accurate timekeepers. Their accuracy surpassed all pendulum clocks, including the Shortt Free Pendulum described in Section 1.8.

Figure 2.7 Rohde & Schwarz Model CFQ, the first portable quartz crystal clock.

2.4 Growing Synthetic Quartz Crystals

Quartz is the second most common mineral found on Earth, only feldspar is more common. Even so, quartz quickly became so important to radio communications and timekeeping that a shortage was feared, and the United States government declared quartz a critical national resource during World War II. To support the war effort, at least 30 million quartz crystal units were produced for military radios from 1942 to 1945. Nearly all of the quartz used during the war was mined in Brazil, but deliveries were slow, and there were always questions about the amount of quartz that was available. Therefore, when the war ended, the United States Army Signal Corps and private industry began investing in research aimed at finding a substitute for natural quartz.

The task of growing synthetic, or "cultured," quartz crystals had been pursued prior to World War II with unsatisfactory results. The key players in the postwar effort were Bell Telephone Laboratories and the Brush Development Company, a company led by the American physicist and businessman Charles Baldwin Sawyer (1894–1964). Bell Labs and Brush Development were both focused on the hydrothermal method of crystal growth. This method grew crystals under high pressure and high temperature in an enclosure called an autoclave, using natural quartz as a nutrient supply to "seed" the growth of the synthetic quartz. This method worked, but it had two problems—the crystal growth was either too slow, or it was unsustainable for more than a day. However, the postwar research refined the method, and by the end of 1948 Brush had grown a flawless crystal of 38 cm in diameter. Bell Telephone Laboratories made further advances by developing new, higher-pressure autoclaves that allowed even larger crystals to be grown at a predictable rate [9, 21].

In 1953, funded by the Army Signal Corps, Brush built the first manufacturing plant for synthetic quartz in Bedford, Ohio. After Brush merged with another company, Sawyer founded a new company named Sawyer Research Products in 1956. They, along with other manufacturers, soon began selling synthetic quartz, not only to the military but also to private industry. New markets for quartz oscillators emerged, including color television. Every color television required a quartz oscillator to generate a 3.579545 MHz color burst frequency that was used to superimpose color onto the standard black-and-white signal. This market was slow to develop, but when the color television industry first reached high volume sales during the 1960s, most television sets included oscillators made of synthetic, rather than natural quartz [9].

By 1971, synthetic quartz was more common than natural quartz in electronic circuits, and today nearly all quartz oscillators contain synthetic crystals. The hydrothermal synthesis method of growing crystals has been continuously refined, but the basic methods developed during the postwar effort are still in use [9, 21].

2.5 The Quartz Watch

Until the latter part of the 1950s, all wristwatches were powered by a mainspring that had to be hand-wound daily. This changed in January 1957 when the Hamilton Watch company of Lancaster, Pennsylvania introduced the world's first battery-powered watch [22]. The watch was called the Electric 500. It ran at the same 5 Hz frequency of a typical mechanical watch, but its frequency was derived from tiny electrical contacts that opened and closed five times per second.

When battery-powered watches first appeared, wristwatch designers were already aware that new technology—including the recent invention of the transistor—had made it possible to build a quartz watch. Seven different programs to build a quartz watch were launched between 1955 and 1969 [23]. Before any of these programs had succeeded, the Bulova Watch Company of New York introduced its Accutron watch in October 1960. Designed by the Swiss electrical engineer Max Hetzel (1921–2004), the Accutron wasn't a quartz watch, but instead substituted a tuning fork for the traditional balance wheel found in mechanical watches. The tuning fork was powered by a one-transistor oscillation circuit that ran at a frequency of 360 Hz, making the watch hum instead of tick. The Accutron was accurate to within 2 s per day, which was about 10× more accurate than the best mechanical watches [22, 23].

The Accutron captured the public's imagination and posed a serious threat to the dominant Swiss watch industry, which had produced well over half of the world's watches during the 1950s. To combat the threat, the Electronic Watch Center (*Centre Electronique Horloger*, CEH) was formed in Neuchâtel, Switzerland, in 1962 with the intent of developing a quartz watch that would surpass the accuracy of the Accutron. Two key members of the CEH team were Armin Frei (1931–2012) and Rolf Bruno Lochinger (b. 1932). Both had received doctoral degrees from the Swiss Federal Institute of Technology, and both had worked in the United States before returning to Switzerland. Beginning in May 1965, Frei focused on developing a miniature quartz oscillator, and Lochinger began developing frequency dividers and driver circuits for the stepping motors that would move the watch hands [23, 24].

Frei had a miniature quartz oscillator working by early 1966 (Figure 2.8). He also developed a driver circuit that consumed less than 4 μA of power (the black IC with the red dot), and a circuit for fine-tuning the quartz frequency (the black IC with no dot). The resonator was a 24 mm bar-shaped quartz crystal hidden inside the 27 mm oscillator case. To keep the electronics simple, the oscillator frequency had to equal a power of two in order to produce a pulse with a 1 s period at the end of the divider chain. Frei chose 2^{13}, which was a frequency of 8192 Hz. This meant that by using 13 binary flip-flop stages, the frequency could be reduced to 1 Hz, and a stepping motor could be used to drive the second hand.

The 13-stage divider scheme was utilized in the first quartz oscillator movement produced by CEH, which was called Beta 1. However, the divider stages placed a large load on the battery, limiting its lifetime to less than one year. For this reason, CEH produced a second movement, Beta 2, which used the same oscillator but had only five divider stages. Beta 2 only divided the quartz frequency to 256 Hz, and utilized a vibrating motor and ratchet wheel to move the second hand. Its only advantage over Beta 1 was that it used less power [24].

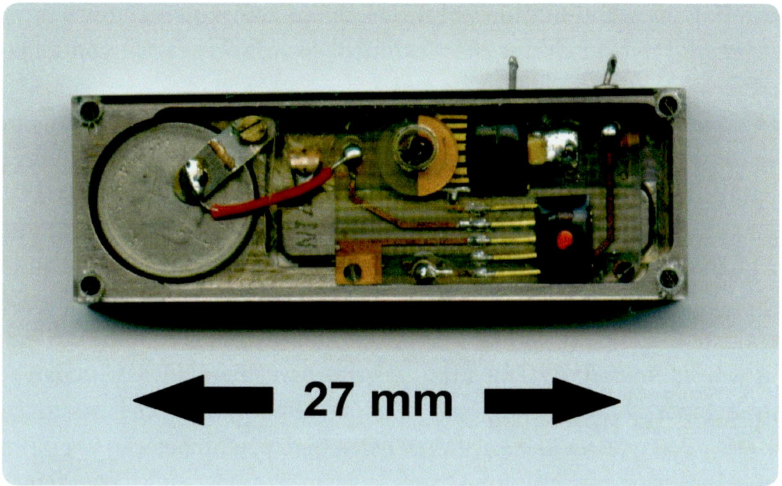

Figure 2.8 The prototype 8192 Hz oscillator used in the first quartz watch.

The first Beta 1 prototype was assembled in July 1967 and qualifies as the first quartz watch. The first Beta 2 prototype was assembled about one month later [24]. During August 1967, six Beta 2 and five Beta 1 movements were tested at the Observatory in Neuchâtel and were found to be accurate within a range of 0.1 to 0.2 s per day, which was a remarkable feat for a clock small enough to wear on your wrist [23]. Because its battery could last more than one year, CEH elected to manufacture the Beta 2 movement instead of Beta 1. They called the commercial version of the movement Beta-21, and about 20 Swiss firms introduced watches with Beta-21 movements on April 10, 1970 [24].

The Swiss watchmakers had been beaten to the marketplace about four months earlier by the Japanese company Seiko, which introduced a quartz watch called the Astron on Christmas day of 1969. The Astron, pictured in Figure 2.9, was the first commercially available quartz watch. It was priced at $1250 USD, which in 1969 was equivalent to the price of an economy car.

Figure 2.9 The first commercially available quartz watch, the Seiko Astron (courtesy of Seiko).

A team of 20 people worked on the design of the Seiko Astron, led by the engineer Tsuneya Nakamura (1923–2018). It was designed differently than Beta 1 and Beta 2, and it was slightly less accurate. Like Beta 1, the Astron ran at 8192 Hz with a 13-stage binary divider. However, it used a quartz tuning fork oscillator, rather than a bar. In addition, the Swiss had manufactured their own integrated circuits (ICs), whereas Seiko had designed a hybrid circuit that included 76 transistors, 29 condensers, and 84 resistors, each hand-soldered to the watch movement [22].

Due to their high cost and some manufacturing issues, sales of the first quartz watches were small, but the technology rapidly improved. The problem of excessive battery consumption was solved by the use of low-power complementary metal oxide semiconductor (CMOS) components. Frank Wanlass (1933–2010) of Fairchild Semiconductor in the United States had invented CMOS logic in 1963, and CMOS components found their way into quartz watches by about 1970. The approach used by the Beta 21 movement was quickly abandoned, and dividing to 1 Hz and using a stepper motor became standard practice.

By 1972, the Swiss watch company Girard-Perregaux had introduced a quartz watch movement called the GP350 that ran at a frequency of 32.768 kHz (2^{15} Hz) and included a 15-stage CMOS divider circuit. Over 50,000 GP350 movements were sold, and 32.768 kHz became the standard frequency for the quartz watch industry [23]. Today, even the least expensive quartz watches can keep time to within 0.5 s per day (5×10^{-6}) and their batteries can last for several years. The best quartz watches can now keep time to within about 10 s per year (3×10^{-8}).

2.6 Quartz Oscillator Design Basics

The basic elements of a quartz crystal oscillator, abbreviated as XO, are shown in the simplified circuit diagram in Figure 2.10. A quartz crystal serves as the resonator. The crystal is strained when a voltage is applied, which causes it to expand or contract. Due to the piezoelectric effect, reversing the polarity of the applied voltage reverses the strain and forces the crystal to mechanically oscillate. The energy needed to sustain oscillation is obtained by taking a voltage signal from the resonator, amplifying it, and then feeding it back to the resonator. A quartz clock could technically be classified as a mechanical clock, like those described in Chapter 1, because the crystal resonator is a mechanically vibrating device. However, the piezoelectric effect links the quartz clock to the electrical world [25, 26].

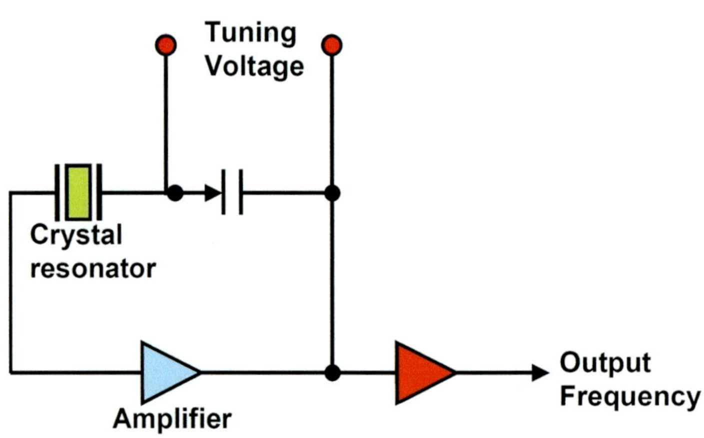

Figure 2.10 A simplified diagram showing the basic elements of a quartz oscillator.

Quartz crystals are formed from silicon and oxygen (SiO_2) and have a three-dimensional geometric body. They are anisotropic, which means that their physical properties have different values when measured in different directions [27]. Wood is a common example of an anisotropic material, because it is much easier to split along the grain than across the grain. Woodworkers know this, and rectangular tables are usually cut so that the wood grain runs in the direction of the length of the table, and not in the direction of its width. This allows the load-bearing part of the table to support more weight. In a similar fashion, the cut of a crystal determines how well it performs as a resonator, because the way it expands and contracts, and the way it responds to temperature changes, will vary with the crystallographic direction.

The natural quartz crystals chosen for piezoelectric applications were usually hexagonal, having six side faces and a six-sided pyramid at one or both ends. Crystals are mapped in an orthogonal *X*, *Y*, *Z* coordinate system, with the *Z*-axis being the single axis that passes longitudinally through the center of the crystal. Rays of light can pass through the *Z*-axis without being split, so the *Z*-axis is called the optical axis. Three *X*-axes and three *Y*-axes are separated by 120° increments around the *Z*-axis [27]. The *X*-axes join two points at opposite corners or angles of the hexagon and are called the electrical axes, because electric polarization occurs in this direction when mechanical pressure is applied. The *Y*-axes are perpendicular to the faces of the crystal, and join two faces at opposite sides of the hexagon. The *Y*-axes are called the mechanical axes because the deformation, or change in shape of the crystal caused by the electric field, occurs in this direction. Figure 2.11 is a simplified diagram showing the top and side view of a hexagonal crystal.

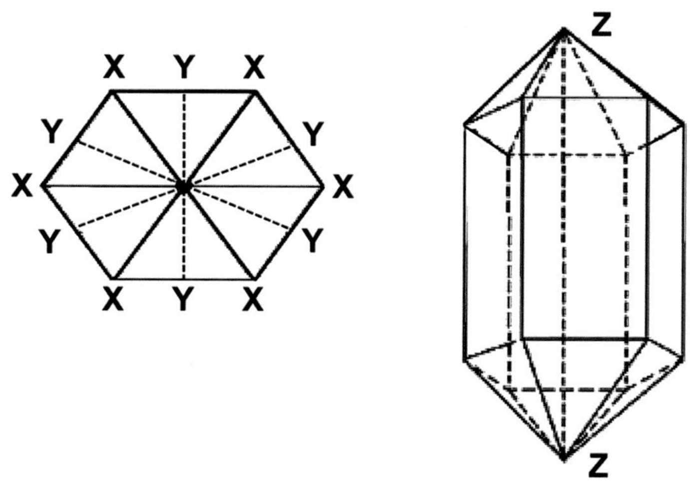

Figure 2.11 The top and side view of a hexagonal crystal.

Natural quartz crystals (Figure 2.1) vary in size and shape, and they contain chemical impurities. In contrast, synthetic crystals are grown to strict dimensional specifications to meet the needs of oscillator manufacturers, and the chemical impurities are eliminated. The seed crystal used to grow quartz (Section 2.4) is often *Y*-oriented, so that crystal grows in the direction of the *Y*-axis instead of the *Z*-axis. To illustrate this, Figure 2.12 includes a photograph of a typical synthetic crystal, along with a diagram showing its orientation with respect to a natural quartz crystal.

The crystal resonator in an oscillator circuit is a small piece of quartz cut from the crystal, which is mounted in a holder and connected to electrodes so that electricity can be applied. The piece of quartz can have various shapes—including the shape of a square, circle, or rectangle. It can also be known by various names—including plate, disc, blank, and wafer. When quartz crystals were first cut to make resonators, the cuts were named according to the axes to which their plane, or face, were perpendicular. For example, for an *X*-cut the face of the plate was perpendicular to the *X*-axis, and for a *Y*-cut the face of the plate was perpendicular to the *Y*-axis.

Variations of these simple cuts are still widely used for oscillators that produce low frequencies. For example, *XY*-cut resonators, with their face perpendicular to the *X*-axis in width and the *Y*-axis in length, are often cut in the shape of a tuning fork. They vibrate in flexure mode, which means that they move up and down as a whole, similar to the way a violin string moves when producing its lowest or fundamental tone. This makes *XY*-cut resonators suitable for designing low-frequency oscillators that only resonate at their fundamental frequency. They are most commonly found inside the small and inexpensive 32.768 kHz oscillators used in wristwatches, but are also found in other oscillator designs that produce frequencies ranging from about 3 kHz to 85 kHz [28].

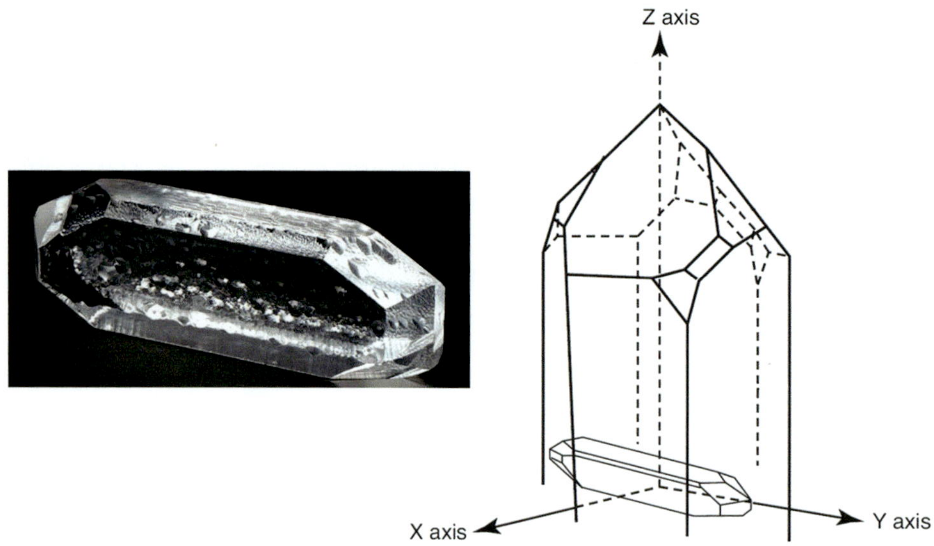

Figure 2.12 Photograph of synthetic quartz crystal (left) and diagram showing its orientation with respect to a natural quartz crystal (right).

In the tuning fork type oscillator, the flexure mode can be coupled with torsional mode vibrations to form a twin-mode resonator that lowers the temperature coefficient and improves performance [29]. Figure 2.13 shows a 32.768 kHz "tuning fork" quartz resonator that's housed in a ceramic surface mount device (SMD) package. The entire package is less than 3 mm in length.

Figure 2.13 A 32.768 kHz "tuning fork" quartz resonator in a ceramic SMD package.

Quartz oscillators designed to produce the higher frequencies needed for radio applications originally used the *X*-cut or *Y*-cut, and still do in some instances, but more complex crystal cuts were developed as designs evolved. These cuts were usually given a two-letter name that usually doesn't refer to *X*, *Y*, or *Z*, including the AT, BT, CT, DT, GT, IT, and SC cuts. The "T" in the various names indicates a cut that compensates for temperature variations, whereas the name "SC" is an acronym for stress compensation. Excluding the low-frequency oscillators just described, the AT and SC cuts are the most common cuts employed in quartz oscillators, so we'll briefly discuss both of them.

From the beginning of quartz oscillator development, it was known that temperature changes had a large effect on the frequency accuracy of a crystal resonator, and this had a negative impact on radio communications. Therefore, by the late 1920s, researchers in the United States, Germany, Japan, and elsewhere had begun working independently to develop new methods of cutting crystals that would reduce their temperature coefficients. Marrison had used an *X*-cut oscillator in his 50 kHz quartz clock, and did extensive work at Bell Labs to reduce the temperature coefficients of the *X* and *Y* cuts [30, 31]. The Japanese scientist and inventor Issac Koga (1899–1982), working at the Tokyo Institute of Technology, further advanced the crystal cutting art by inventing and patenting what he named the R-cut and R'-cut resonators in 1932 [32]. Koga's work was soon followed by the AT-cut, which became the first practical and widely deployed technique for reducing the temperature coefficient of a crystal resonator. The AT-cut is still the dominant form of crystal cut today, employed by perhaps 90% of the quartz oscillators that operate at frequencies above 500 kHz.

The AT-cut was introduced in 1934 by Frederick Lack (1895–1988), Gerald Willard (1901–1955), and Irvin Fair (1907–2005) of Bell Telephone Laboratories. Lack had a long and notable career in the radio industry. He served overseas in the Army Signal Corps in World War I, prior to spending 13 years at Bell. He later became vice president and head of the radio products division at Western Electric until his retirement in 1958, interrupted by a stint during World War II, when he directed the electronics production agency for the Army and Navy. The work he and colleagues performed at Bell Labs was motivated by the need to improve upon the *Y*-cut oscillators then used in radio communications, not only by reducing their temperature coefficient, but also by allowing them to operate at higher power levels and frequencies (originally up to 20 MHz) without fracturing the crystal, resulting in a "considerable simplification of future short wave radio equipment" [33].

Lack and his colleagues had measured the temperature coefficient of the *Y*-cut resonators as 85 parts per million per 1 °C (85×10^{-6}). In a series of experiments, they found that they could dramatically reduce the temperature coefficient by using crystal cuts rotated around the *X*-axis. They reported that both the AT-cut, rotated 35° 15' from the *Z*-axis, and the BT-cut, rotated 49° in the opposite direction (Figure 2.14), would result in a "zero temperature coefficient" [33]. Of course, what they called "zero" wasn't actually zero, but it was less than one part per million (1×10^{-6}), which was too small a frequency difference for them to measure at that time.

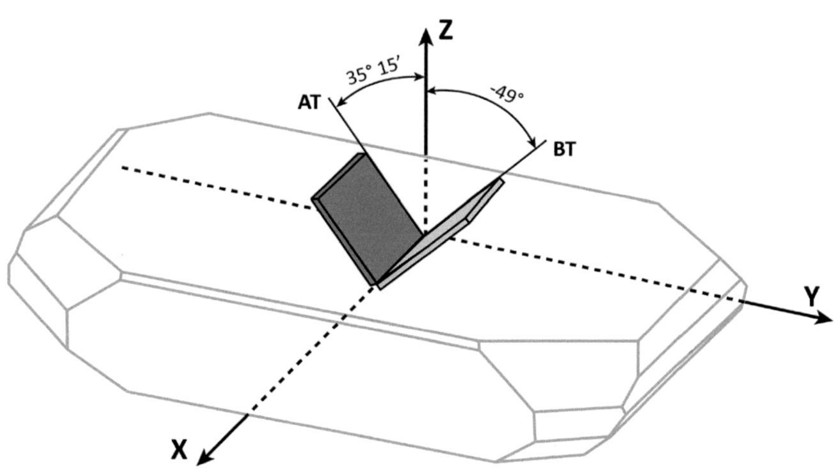

Figure 2.14 The AT and BT crystal cuts utilized in quartz resonators (courtesy of Epson).

An AT-cut resonator vibrates in thickness shear mode, a mode of vibration where an electric potential applied through the thickness of the crystal plate produces a shear stress that causes the top and bottom half of the plate to move in opposite directions. The plate is supported by the zone around its edge, so the vibrating area is usually near the center of the plate. The fundamental frequency of the resonator is inversely proportional to its thickness, which means that thinner crystals produce higher frequencies. The frequency constant is approximately 1.664 MHz per millimeter of thickness [34], so a crystal plate with a fundamental frequency of 5 MHz is about 0.33 mm thick. Thin crystal plates are difficult to manufacture and can easily fracture, so a plate's fundamental frequency is usually less than 50 MHz. A 50 MHz crystal plate is just 33 μm thick, or about two-thirds the thickness of a human hair.

We mentioned earlier that the *XY*-cut resonators found in wristwatches run at their fundamental frequency. Other quartz oscillators also run at their fundamental frequency, and there are even examples where the fundamental frequency is higher than 50 MHz. However, because of the problems associated with thin crystal plates, it's common for quartz oscillators that output frequencies higher than 20 MHz to operate at a harmonic of their fundamental frequency, which is called an overtone. Some of the overtone frequencies are spurious signals that have to be suppressed, and the third and fifth overtones are usually the most reliable and controllable frequencies to use as the oscillator's output. Therefore, the highest output frequency for a AT-cut resonator is usually about 250 MHz, which is the fifth overtone of 50 MHz.

The theory and name for the "stress compensated" SC-cut were introduced in 1975 [35] by the electrical engineer Errol Eernisse (b. 1940), who was then working at Sandia National Laboratories in Albuquerque, New Mexico. Eernisse's work followed the work of Richard Holland (b. 1940), an electrical engineer at Mission Research Corporation in Albuquerque, who formerly had worked at Sandia. Holland published a theoretical description of what he called the TS-cut in 1974, a crystal orientation very similar to the SC-cut [36]. Eernisse later collaborated with Arthur Ballato (b. 1936) and Theodore Lukaszek (b. 1935) of the U. S. Army Research and Development Laboratory to experimentally verify the SC-cut, and their measurements published in 1978 closely agreed with the theory [37].

The SC-cut is a double-rotated cut, rotated on both the *X* and the *Z*-axis, and somewhat similar to the compound miter cut known to woodworkers (Figure 2.15). The cut angles are 22.4° for the *X*-axis and 33.88° for the *Z*-axis, but these angles can vary by much as one or two degrees in some implementations [38]. Like the AT-cut, the SC-cut operates in thickness shear mode, but stress patterns in the resonator plate are decoupled from the thickness shear vibrations. This meant that in addition to the resonator being less sensitive to temperature, it also made the resonator less sensitive to stress along the edge of the crystal caused by the crystal mount and the electrodes [35, 39].

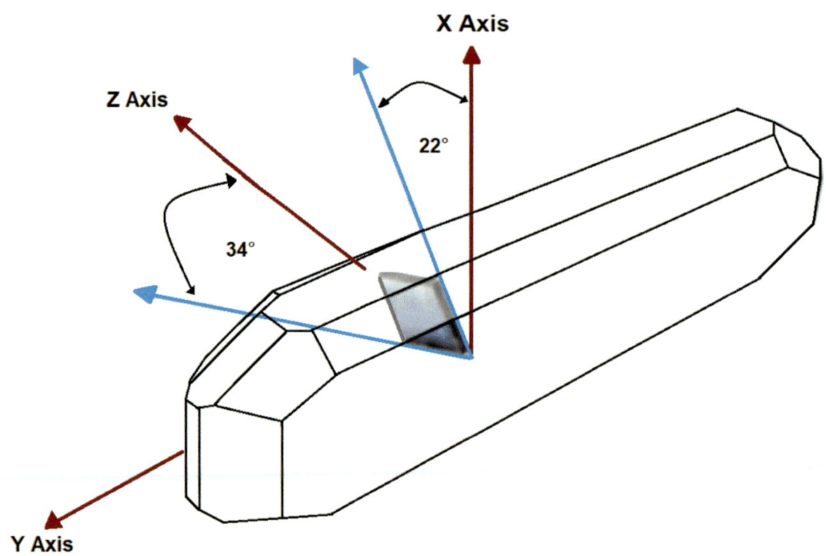

Figure 2.15 The SC crystal cuts utilized in quartz resonators.

Resonators based on the SC-cut operate at about the same frequencies as AT-cut resonators, but they are more difficult to manufacture than AT-cut resonators, which makes them more expensive and less common. However, they outperform AT-cut resonators in most cases. The crystals are slightly thicker, with a frequency constant of approximately 1.8 MHz per millimeter of thickness [34], meaning that a SC-cut resonator with a fundamental frequency of 5 MHz is about 0.36 mm thick. An SC-cut resonator warms up faster than an AT-cut resonator when first turned on, and because SC-cut resonators are less sensitive to mechanical stress, they are also less sensitive to vibrations.

The temperature coefficients of the AT and SC cuts are compared in Figure 2.16, with the curves showing four examples of each resonator type. The frequency-temperature curve for an AT-cut resonator is almost sinusoidal from −40 °C to 120 °C. The inflection point, where the direction of the curve changes and where its slope is near 0, is about 25 °C. In contrast, the frequency-temperature curve for the SC-cut slopes down rapidly at temperatures below 50 °C. However, unlike the AT-cut whose frequency is only stable near its inflection point, the SC-cut has relatively good frequency stability at a wide range of temperatures above 50 °C, with an inflection point of about 95 °C [25, 26, 34, 40].

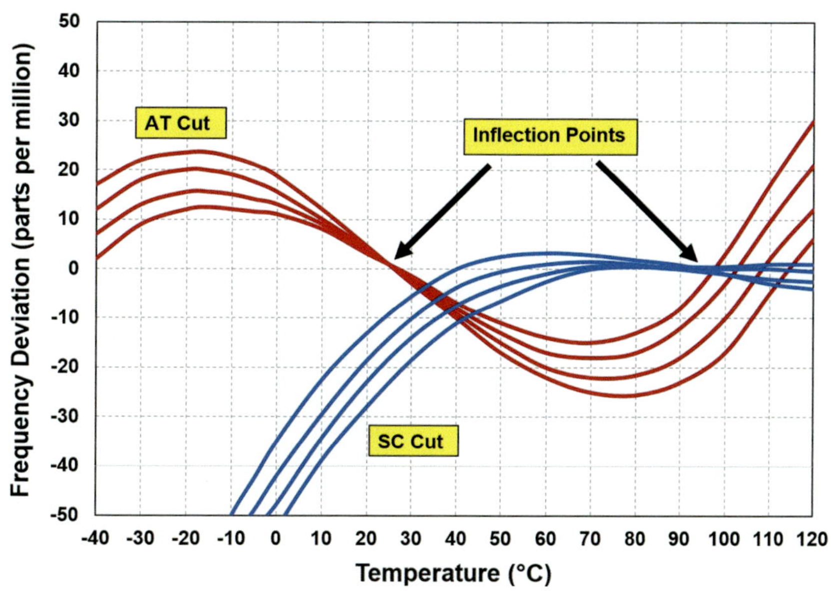

Figure 2.16 An example of a temperature coefficient graph for AT-cut and SC-cut crystal resonators.

The AT and SC crystal cuts are ingenious techniques, but as Figure 2.16 illustrates, neither completely solves the temperature problem. For applications that require more stable frequency, both types of resonators require their operating temperature to either be compensated or controlled. There are two common design methods that accomplish this. The first method, the temperature-compensated crystal oscillator (TCXO), involves first measuring the temperature, and then adjusting the resonator frequency to compensate for temperature changes. The second method, the oven-controlled crystal oscillator (OCXO), simply protects the resonator from being exposed to temperature changes.

The TCXO was made possible by the inventions of the thermistor, a type of resistor that responds in a predictable way to temperature changes, and the varactor diode, a voltage-controlled capacitor. Many of the early TCXO designs were proprietary and not published, so it's difficult to trace its origin to a single source. However, as is usually the case with quartz oscillator designs, the work was motivated by the need for better radio communications, and it's likely that much of the early work took place at the Collins Radio Company in Cedar Rapids, Iowa. By the early 1960s, Collins had introduced stable quartz oscillators covering a wide range of temperatures for use in their portable single side band (SSB) military radios [41, 42].

Long before the development of the TCXO, it was known that the frequency of a crystal resonator could be shifted or "pulled" by changing the load capacitance. The varactor diode made it practical to do this, because its capacitance is varied by varying its reverse voltage. If a varactor diode was placed in series with the resonator, and if the input voltage to the diode was changed, the frequency of the resonator would also change. Implementing this technique results in a voltage-controlled crystal oscillator (VCXO). A TCXO uses a thermistor-based temperature sensor that automatically supplies a correction voltage to the varactor diode [43].

Figure 2.17 shows a simplified block diagram of a TCXO. All of the applied voltages are regulated to protect against outside noise, such as noise from the power supply. The temperature sensor supplies measurements to a compensation network that determines and supplies the necessary correction voltage to the pulling network. In pure analog designs, multiple thermistors and other electronics are needed to create a compensation network that can generate the required voltage-temperature correction curve, but digital techniques can utilize stored memory devices with look-up tables and microprocessors to perform the same function. The pulling network then corrects the frequency of the resonator by varying its load capacitance, and this produces a frequency change equal and opposite to the frequency change produced by temperature (Figure 2.18). The resulting signal then passes through the final output state [25, 26, 43]. The resonator is usually an AT-cut crystal, as SC-cut resonators are generally not used in TCXO or VCXO designs. This is partially due to cost, but it's mostly because their motional capacitance is several times lower than the AT-cut, which makes them harder to tune with a variable-capacitance pulling network.

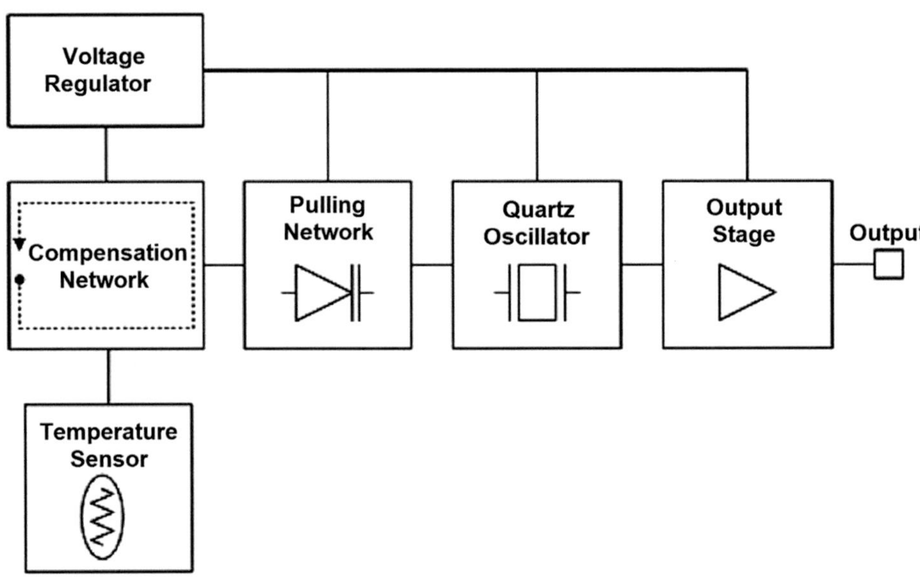

Figure 2.17 Simplified block diagram of a TCXO.

The OCXO is the most stable type of quartz oscillator. It encases the crystal resonator and its associated electronics in a temperature-controlled oven, so that the resonator temperature is held constant even when the outside temperature varies. This basic principle is old, practical, and effective, and housing crystal resonators in ovens was prevalent even before the development of the AT-cut [33, 44]. A closed-loop feedback network consisting of a temperature sensor (typically a thermistor), a heating element, and an oven control circuit controls the temperature of the oven (Figure 2.19). Unlike a TCXO, which reaches its specified frequency stability almost instantly after being turned on, an OCXO goes through a warm-upperiod of a few minutes before reaching its normal operating temperature and allowing the crystal to stabilize.

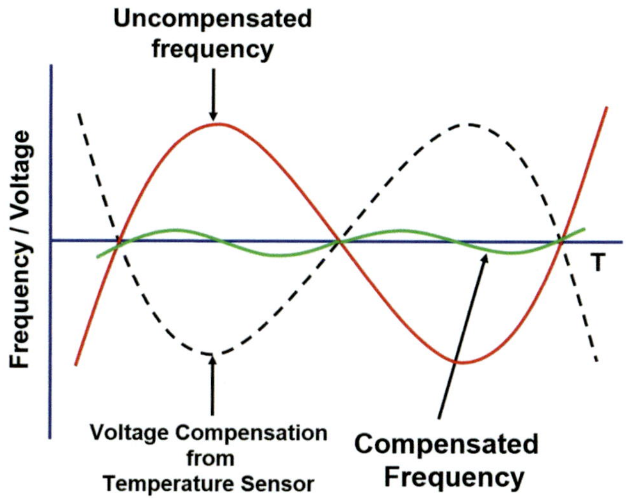

Figure 2.18 Frequency compensation of a TCXO.

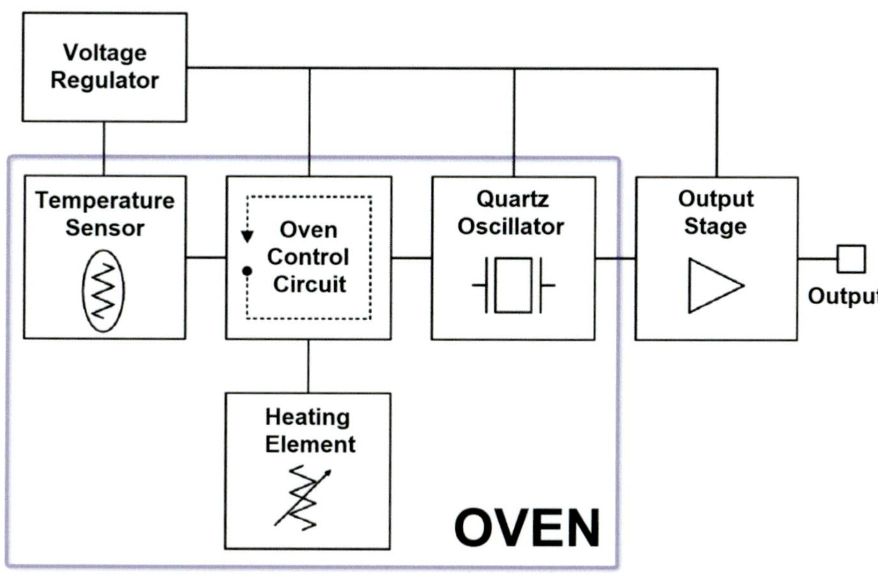

Figure 2.19 Simplified block diagram of an OCXO.

Low-cost OCXOs with AT-cut resonators are common, but the most stable OCXOs use SC-cut resonators. As shown in Figure 2.16, an SC-cut resonator doesn't operate especially well at ambient room temperatures, and its performance rapidly degrades as it gets colder. However, it's very stable at typical oven temperatures, which are typically set in a range between 75 °C to 100 °C, depending on the application. The oven temperature is usually set about 10 °C higher than the highest ambient temperature where the OCXO can be used. For example, if the OCXO is specified to work in ambient temperatures of up to 70 °C, the oven temperature is usually set to 80 °C. If the ambient temperature were to exceed the oven temperature, the temperature regulation of the oven would stop working, and the resonator frequency would become less stable [25, 26]. For SC-cut resonators, the oven temperature is often near the inflection point, which is ideal.

Due to their higher cost, larger size, and greater power consumption, OCXOs were originally only found inside the highest quality test instruments and radio equipment, and lower-cost instruments mostly relied on XOs and TCXOs. This is still true, but as the cost and size of OCXOs have gone down, their use has become more common. The OCXOs found in instruments requiring the best frequency stability can reach dimensions of perhaps 100 × 75 × 50 mm, but OCXOs with dimensions as small as 5 × 3.2 × 2.5 mm have been demonstrated for telecommunication applications [45].

Figure 2.20 shows a small OCXO with dimensions of 25 × 25 × 13 mm that was designed to operate onboard a satellite. Operating in the demanding space environment requires this device to have a low sensitivity to acceleration, and the ability to withstand a temperature range as wide as –55 to +85 °C, while being subjected to high levels of vibration and solar radiation. An SC-cut resonator is housed inside its steel case, which was designed to be mounted on a printed circuit board, where it is insulated from the other components by glass standoffs. Four pins on the bottom of the case provide a ground connection, inputs for DC power and oscillator tuning, and a fixed RF output that can be customized to produce frequencies ranging from 25 MHz to 320 MHz. After operating for 30 days, the daily aging rate of this OCXO is typically less than 5×10^{-9}.

Figure 2.20 An OCXO designed for space applications (courtesy of Quantic Wenzel).

2.7 Quartz Oscillator Performance and Limitations

As we have seen, quartz oscillators revolutionized high accuracy timekeeping shortly after their invention. A key reason why was their quality factor, or Q, which was much higher than that of the mechanical resonators that preceded them. The Q of a resonator is its resonance frequency divided by its resonance width. The resonance frequency is the resonator's natural frequency of oscillation. The resonance width is the range of possible frequencies where the resonator will oscillate. Raising the Q factor requires either increasing the resonance frequency, narrowing the resonance width, or doing both.

We'll take a detailed look at how the accuracy and stability of clocks and oscillators are measured and estimated in Chapters 6 and 7, but for now the important thing to remember is that raising the Q of an oscillator usually leads to better accuracy and stability, because a high Q means that an oscillator can only oscillate at frequencies that are close to its natural resonance frequency.

The Q of a quartz oscillator typically ranges from about 10^4 for low-cost oscillators like those found in wristwatches, to more than 10^6 for the highest quality devices. For example, an oscillator that runs at a frequency of 32.768 kHz has an adjustment range of a few hertz or less. This means that its Q should be at least 10,000 (10^4). From empirical data, the maximum Q for a quartz oscillator has been estimated as $\frac{1.6\times10^7}{f}$, where f is the resonance frequency in megahertz [25, 26]. Therefore, the Q of a 5 MHz quartz frequency standard can exceed 3×10^6 [46].

After easily meeting the accuracy requirements for the radio applications discussed in Section 2.2, the accuracy of quartz oscillators continued to improve for a number of years. By the early 1950s, the regularly adjusted quartz oscillators at NBS, with GT-cut resonators and a Q factor of about 2×10^6, were producing frequency accurate to about 1×10^{-10} [47]. Today's quartz oscillators are smaller, more reliable, and less expensive than they were in the 1950s, but they aren't significantly more accurate, and the best devices available today still require constant adjustments to maintain frequency accurate to within parts in 10^{10}. The frequency adjustments can be made manually, but frequency control systems called disciplined oscillators usually make them automatically (Section 5.3.2). Continuous frequency adjustments are always needed because a quartz oscillator's frequency is always changing.

The changes in a quartz oscillator's frequency are caused by environmental effects that are both external and internal to the oscillator. As we've discussed, temperature changes are the largest external environmental problem, and decades of research and development have gone into designing optimal crystal resonator cuts and electronic circuits to try and compensate for the effects of temperature. However, the effects that occur inside of a quartz oscillator can be even more difficult to predict or prevent. As Marrison observed in his early timekeeping experiments, the effect of aging can change the frequency of a quartz resonator. The term aging is sometimes used interchangeably with frequency drift, but aging is just one component of drift. Frequency aging occurs when the effects that cause the frequency to change are internal to the oscillator. Frequency drift is aging plus the frequency changes caused by effects external to the oscillator [48]. Therefore, aging is just one part of frequency drift, but it eventually becomes the dominant part [49].

The factors that cause frequency aging are not completely understood. For example, aging can be either positive or negative, and in some cases can reverse its direction. The aging rate is highest when the oscillator is first turned on, and it slows down and becomes more linear after a period of days or weeks. The primary causes of aging are thought to be contamination of the resonator's surfaces, stress relief in the mounting structure of the crystal, leaks in the resonator packaging, and changes in the quartz material itself. This tells us that aging is related to the way that the resonator is packaged and mounted. The quartz oscillators with the lowest aging rates are usually OCXOs with SC-cut crystals, but no matter how well a quartz resonator is designed and constructed, some frequency drift due to aging is unavoidable [25, 26, 40, 49].

Despite their limitations, the importance of quartz devices to time and frequency metrology is impossible to overstate. The quartz clocks on our wrists, in our pockets, and on our walls, are more than accurate enough for everyday timekeeping. Quartz oscillators are found inside measurement and test instruments, inside atomic clocks, and inside clocks controlled by radio, network, and satellite navigation systems, as we'll see in the following chapters. The now multi-billion-dollar quartz oscillator industry is still expanding, largely because of the proliferation of portable electronic devices. A smartphone, for example, has been known to contain as many as six quartz oscillators, with some serving as timing references and others regulating the transmit and receive frequencies. Smartphones continue to add new features that require frequency sources, but the total number of oscillators needed may decrease due to designs where single oscillators support multiple functions [50, 51]. Even so, the ubiquitous smartphone will rely heavily on quartz oscillator manufacturing and technology for the foreseeable future.

Quartz clocks will likely always have an important role in world timekeeping, but many of today's technologies require far more accuracy than quartz clocks can provide. The perpetual quest for more accurate time led to the development of atomic clocks, the topic of Chapter 3.

References

1. "Testing Explosive Force with the Piezometer," *Scientific American Monthly*, vol. 3, no. 4, pp. 337–338, April 1921.
2. W. G. Cady, *Piezoelectricity: An Introduction to the Theory and Applications of Electromechanical Phenomena in Crystals*, New York: McGraw-Hill, 1946.

3. J. Curie and P. Curie, "Développement, par pression, de l'électricité polaire dans les cristaux hémièdres à faces inclinées (Development, by pressure, of electrostatic polarization in hemihedral crystals with inclined faces)," *Comptes Rendus de l'Académie des Sciences*, vol. 91, pp. 294–295, August 1880.
4. R. A. Heising, *Quartz Crystals for Electrical Circuits: Their Design and Manufacture*, New York: Van Nostrand, 1946.
5. A. M. Nicolson, "Generating and Transmitting Electric Currents," *United States Patent 2,212,845*, filed April 10, 1918, granted August 27, 1940.
6. T. Lewis, *Empire of the Air: The Men Who Made Radio*, New York: HarperCollins, 1991.
7. J. H. Dellinger, "Reducing the Guesswork in Tuning," *Radio Broadcast,* vol. 3, pp. 241–245, December 1923.
8. W. F. Snyder and C. L. Bragaw, "Achievement in Radio: Seventy Years of Radio Science, Technology, Standards, and Measurement at the National Bureau of Standards," *National Bureau of Standards Special Publication* 555, October 1986.
9. C. S. McGahey, *Harnessing Nature's Timekeeper: A History of the Piezoelectric Quartz Crystal Technology Community (1880–1959)*, Doctoral Dissertation, Georgia Institute of Technology, May 2009.
10. W. G. Cady, "Piezo-Electric Resonator," *United States Patent 1,450,246*, filed January 28, 1920, granted April 3, 1923.
11. W. G. Cady, "Method of Maintaining Electric Currents of Constant Frequency," *United States Patent 1,472,583*, filed May 28, 1921, granted October 30, 1923.
12. W. G. Cady, "The Piezo-Electric Resonator," *Proceedings of the Institute of Radio Engineers (IRE)*, vol. 10, pp. 83–114, April 1922.
13. G. W. Pierce, "Piezoelectric Crystal Resonators and Crystal Oscillators Applied to the Precision Calibration of Wavemeters," *Proceedings of the American Academy of Arts and Sciences,* vol. 59, no. 4, pp. 81–106, October 1923.
14. G. W. Pierce, "Electrical System," *United States Patent 1,789,496*, filed February 15, 1924, granted January 20, 1931. Also see *United States Patent 2,133,642*.
15. *General Radio Company Bulletin 718*, Cambridge, Massachusetts, pp. 767–770, March 1926.
16. "A Review of Twenty Years of Progress in Communication-Frequency Measurements," *The General Radio Experimenter*, vol. X, No. 1, June 1935.
17. "New Fundamental Frequency Standard," *Technical News Bulletin of the Bureau of Standards*, no. 157, p. 43, May 1930.
18. J. W. Horton and W. A. Marrison, "Precise Determination of Frequency," *Proceedings of the Institute of Radio Engineers (IRE)*, vol. 16, no. 2, pp. 137–154, February 1928.
19. W. A. Marrison, "The Evolution of the Quartz Crystal Clock," *Bell System Technical Journal*, vol. 27, no. 3, pp. 510–588, 1948.
20. W. A. Marrison, "The Crystal Clock," *Proceedings of the National Academy of Sciences of the United States of America*, vol. 16, no. 7, pp. 496–507, July 1930.
21. F. Iwasaki and H. Iwasaki, "Historical Review of Quartz Crystal Growth," *Journal of Crystal Growth*, vol. 237–239, pp. 820–827, April 2002.
22. C. Stephens and M. Dennis, "Engineering Time: Inventing the Electronic Wristwatch," *British Journal for the History of Science*, vol. 33, pp. 477–497, 2000.
23. L. F. Trueb, *The World of Watches*, New York: Ebner Publishing, 2005.
24. A. H. Frei, "First-Hand: The First Quartz Wristwatch," *Engineering and Technology History Wiki* (etwh.org), 2009.
25. J. R. Vig, "Introduction to Quartz Frequency Standards," Army Research Laboratory, *Research and Development Technical Report SLCET-TR-92-1 (Rev. 1)*, 56 p., October 1992.
26. J. R. Vig and A. Ballato, "Frequency Control Devices," Chapter 7 in *Ultrasonic Instruments and Devices*, pp. 637–701, Academic Press, 1999.
27. Hewlett-Packard Company, "Fundamentals of Quartz Oscillators," *HP Application Note 200-2*, 1997.
28. V. Matko and M. Milanovič, "Detection Principles of Temperature Compensated Oscillators with Reactance Influence on Piezoelectric Resonator," *Sensors*, vol. 20, 802, 18 p., February 2020.

29. E. Momosaki, "A Brief Review of Progress in Quartz Tuning Fork Resonators," *Proceedings of the 1997 IEEE International Frequency Control Symposium*, pp. 552–565, May 1997.

30. W. A. Marrison, "A High Precision Standard of Frequency," *Bell System Technical Journal*, vol. 8, no. 3, pp. 493–514, July 1929.

31. F. R. Lack, "Observations on Modes of Vibration and Temperature Coefficients of Quartz Crystal Plates," *Proceedings of the Institute of Radio Engineers (IRE)*, vol. 17, no. 7, pp. 1123–1141, July 1929.

32. I. Koga, "Notes on Piezoelectric Quartz Crystals," *Proceedings of the Institute of Radio Engineers (IRE)*, vol. 24, no. 3, pp. 510–531, March 1936.

33. F. R. Lack, G. W. Willard, and I. E. Flair, "Some Improvements in Quartz Crystal Circuit Elements," *The Bell System Technical Journal*, vol. 13, no. 3, pp. 453–463, July 1934.

34. V. E. Bottom, *Introduction to Quartz Crystal Unit Design*, New York: Van Nostrand Reinhold Company, 1982.

35. E. P. EerNisse, "Quartz Resonator Frequency Shifts Arising from Electrode Stress," *Proceedings of the 29th Annual Symposium on Frequency Control*, pp. 1–4, May 1975.

36. R. Holland, "Nonuniformly Heated Anisotropic Plates: II. Frequency Transients in AT and BT Quartz Plates," *Proceedings of the 1978 Ultrasonics Symposium*, pp. 592–598, November 1974.

37. A. Ballato, E. P. EerNisse, and T. J. Lukaszek, "Experimental Verification of Stress Compensation in the SC Cut," *Proceedings of the 1978 Ultrasonics Symposium*, pp. 144–147, September 1978.

38. IEEE, "IEEE Standard on Piezoelectricity," *ANSI/IEEE Standard 176–1987*, 1987.

39. J. A. Kusters, "The SC Cut Crystal – An Overview," *Proceedings of the 1981 Ultrasonics Symposium*, pp. 402–409, October 1981.

40. F. L. Walls and J. R. Vig, "Fundamental Limits on the Frequency Stabilities of Crystal Oscillators," *IEEE Transactions on Ultrasonics, Ferroelectrics, and Frequency Control*, vol. 42, no. 4, pp. 576–589, July 1995.

41. M. E. Frerking, "Fifty Years of Progress in Quartz Crystal Frequency Standards," *Proceedings of the 1996 IEEE International Frequency Control Symposium*, pp. 33–46, June 1996.

42. G. R. Hykes and V. E. Newell, "A Temperature-Compensated Frequency Standard," *Proceedings of the 15th Annual Symposium on Frequency Control*, pp. 298–317, May 1961.

43. M. E. Frerking, *Crystal Oscillator Design and Temperature Compensation*, New York: Van Nostrand Reinhold Company, 1978.

44. C. L. Watson, "Constant Temperature Oven," *United States Patent 2,000,266*, filed April 18, 1932, granted May 7, 1935.

45. W.-L. Hsieh, W.-C. Wang, E.-S. Hsu, S.-H. Kao, and M.-H. Wang, "The World's Smallest OCXO for 5G Synchronization Applications," *Proceedings of the 2021 Joint Conference of the European Frequency and Time Forum and IEEE International Frequency Control Symposium (EFTF/IFCS)*, 4 p., July 2021.

46. J. R. Norton and J. M Cloeren, "Precision Quartz Oscillators and Their Use Aboard Satellites," *Johns Hopkins APL Technical Digest*, vol. 15, no. 1, pp. 30–37, 1994.

47. J. M. Shaull and J. H. Shoaf, "Precision Quartz Resonator Frequency Standards," *Proceedings of the Institute of Radio Engineers (IRE)*, vol. 42, pp. 1300–1306, August 1954.

48. IEEE Ultrasonics, Ferroelectrics, and Frequency Control Committee, "IEEE Guide for Measurement of Environmental Sensitivities of Standard Frequency Generators," *IEEE Standard 1193–2022*, 2022.

49. J. R. Vig and T. R. Meeker, "The Aging of Bulk Acoustic Wave Resonators, Filters and Oscillators," *Proceedings of the 45th Annual Symposium on Frequency Control*, pp. 77–101, May 1991.

50. C. S. Lam, "A Review of the Recent Development of MEMS and Crystal Oscillators and Their Impacts on the Frequency Control Products Industry," *IEEE Ultrasonics Symposium (IUS)*, pp. 694–704, November 2008.

51. C. S. Lam, "A Review of the Timing and Filtering Technologies in Smartphones," *Proceedings of the 2016 IEEE International Frequency Control Symposium*, 6 p., May 2016.

Chapter 3

Atomic Clocks

> *When it became evident that such spectrum lines might eventually provide new primary frequency standards, we scientists at the NBS began seeking a means of utilizing one of these lines to control an oscillator which in turn could be used to drive a clock.*
>
> — Harold Lyons, "The Atomic Clock," 1949

Quartz clocks were the world's most accurate timekeepers from the mid-1930s until the early 1950s, but they were never designated as an official time standard. That distinction was directly passed from clocks that kept time based on the observations of celestial bodies, to clocks that kept time based on the observations of atoms—the atomic clocks that we'll discuss in this chapter.

Before starting our discussion, let's briefly review what it means to be a standard. As noted at the beginning of this book, metrology largely consists of comparing an unknown quantity to a known quantity, and agreeing on the quantities that can be trusted to serve as standards. The standards are used to define the quantity. Metrologists—the scientists who study and practice measurements—need standard unit definitions for every quantity that they measure. These definitions should be accepted internationally, so that measurements made in different parts of the world can be compared. Once a unit definition is accepted internationally, it represents an ideal measurement reference that has an uncertainty of zero. In other words, the unit definitions are accepted as truth.

The International System (SI) of units provides these definitions, and the second—the base unit of time—is one of the seven base units of the SI [1]. Before atomic clocks were invented, the second was always defined as a fraction of an astronomical period. For example, the second was once defined as a fraction of the day, the period of one rotation of the Earth on its axis. As discussed in Section 1.2, the length of a day varied seasonally, so it was eventually found that accurate timekeeping required defining the second as a fraction of the period of the mean solar day. Therefore, the mean solar second was defined as the mean solar day divided by 86,400, which was the number of seconds in a day. The mean solar second was accepted as the base unit of time for centuries.

The second was redefined in 1956 by dividing the tropical year 1900 by 31,556,925.9747. This cumbersome definition, known as the ephemeris second, was the base unit of time when the SI was formally established in 1960. Ephemeris seconds were more stable than mean solar seconds, but they

As Time Goes By: A Comprehensive Guide to Time and Frequency Measurements
Michael A. Lombardi

ISBN 978-981-5352-28-3 (Hardcover), 978-1-042-01532-0 (eBook)
www.jennystanford.com

DOI: 10.1201/9781042015320-5

were almost impossible for metrologists or timekeepers to use as a standard. The era of the ephemeris second was mercifully short, and a more accurate and usable definition of the second was soon made possible by atomic clocks.

3.1 The First Suggestion of Atomic Clocks and Early Research Efforts

The first suggestion of an atomic clock was remarkably made by the Scottish physicist James Clerk Maxwell (1831–1879), who lived during an era when quartz clocks were still decades away from being invented. A true visionary, Maxwell not only foresaw the future development of quartz clocks, but he also prophetically suggested that the vibrations of atoms would be an even better time standard. Maxwell directed his suggestions to Peter Guthrie Tait (1831–1901), his lifelong friend, and to William Thomson (1824–907), who was later known as Lord Kelvin. In 1879, which was the year that Maxwell passed away, Tait and Thomson published the second edition of their *Elements of Natural Philosophy*, and included comments attributed to Maxwell that weren't included in the first edition [2]. On the subject of quartz crystals, Maxwell had suggested that

> *"The period of vibration of a piece of quartz crystal of specified shape and size and at a stated temperature (a tuning-fork, or bar, as one of the bars of glass used in the 'musical glasses') gives us a unit of time which is constant through all space and all time, and independent of the earth. A unit of force founded on such a unit of time would be better entitled to the designation absolute than is the 'absolute unit' now generally adopted, which is founded on the mean solar second."* [3]

Even so, a quartz standard would still essentially depend

> *"... on one particular piece of matter, and is therefore liable to all the accidents, etc. which affect so-called National Standards however carefully they may be preserved, as well as the almost insuperable practical difficulties which are experienced when we attempt to make exact copies of them."* [3]

Therefore, Maxwell felt that the best timekeeper of all would be

> *"... natural standard pieces of matter such as atoms of hydrogen or sodium, ready made in infinite numbers, all absolutely alike in every physical property. The time of vibration of a sodium particle corresponding to any one of its modes of vibration is known to be absolutely independent of its position in the universe, and will probably remain the same so long as the particle itself exists."* [3]

The world, of course, wasn't yet ready for Maxwell's ideas, and the first atomic clock experiments didn't begin until about six decades afterwards. The experiments were finally made possible by the rapid advances in quantum mechanics and microwave electronics that took place before, during, and after World War II.

Most of the fundamental concepts that led to atomic clocks were developed by the physicist Isidor Isaac Rabi (1898–1988) and his colleagues at Columbia University in New York City [4, 5]. Rabi was born in Galicia, then part of Austria-Hungary (now Poland), and emigrated to the United States as an infant. He grew up in New York City, and initially studied electrical engineering and chemistry at Columbia before receiving a doctorate in physics. After joining Columbia's faculty in 1929, Rabi built a molecular beam apparatus and started a laboratory that soon attracted other young physicists, including Sidney Millman (1904–2006), Jerrold Zacharias (1905–1986), Polykarp Kusch (1911–1993), and Norman Ramsey (1915–2011). Each of them would play key roles in atomic clock history.

Rabi invented the atomic and molecular beam resonance method of observing atomic spectra in 1937, and this scientific breakthrough earned him a Nobel Prize in physics in 1944. He had some informal discussions with scientists at the National Bureau of Standards (NBS) about applying his methods to a time standard, but this research was halted when World War II intervened. During the war, Rabi worked on radar systems at the Massachusetts Institute of Technology (MIT) and was also involved in the Manhattan Project. When the war neared its end, Rabi publicly discussed the possibility of atomic clocks for the first time, in a lecture given to the American Physical Society on January 20, 1945 [6]. The lecture was attended by a *New York Times* reporter, and a story about "the most accurate clock in the universe" was published the following day (Figure 3.1).

'COSMIC PENDULUM' FOR CLOCK PLANNED

Radio Frequencies in Hearts of Atoms Would Be Used in Most Accurate of Timepieces

DESIGN TERMED FEASIBLE

Prof. I. I. Rabi, 1944 Nobel Prize Winner, Tells of Newest Developments

By WILLIAM L. LAURENCE

Blueprints for the most accurate clock in the universe, tuning in on radio frequencies in the hearts of atoms and thus beating in harmony with the "cosmic pendulum," were outlined yesterday at the annual New York meeting of the American Physical Society, at Columbia University, by Prof. I. I. Rabi, who delivered the Richtmyer Memorial Lecture under the auspices of the American Association of Physics Teachers.

Figure 3.1 The first public suggestion of an atomic clock, *New York Times*, January 21, 1945, p. 34 (first paragraph of article).

The reporter eloquently stated the atomic clock would be "beating in harmony with the 'cosmic pendulum'" [7]. This metaphorical reference correctly acknowledged that the basic principles of atomic clocks would be similar to the pendulum clocks based on mechanical resonators. The energy transitions of an atom could serve as a nearly perfect pendulum, because—ignoring isotopic differences—all atoms of a specific element are identical. Due to their identical nature, the atoms of a specific element should all undergo the same transitions in their energy levels when stimulated with an electromagnetic signal at a specific frequency. This frequency could be used to define the second and to keep time.

The concept seemed simple, but building atomic clocks has always been difficult, and it was exceptionally difficult with the technology that existed in the 1940s. The radio frequencies that were then being generated and measured were much lower than the microwave frequencies that were needed to cause atomic energy transitions. This meant that new instruments had to be developed that could generate and measure microwave frequencies. Even if this was done, the atoms wouldn't be able to produce usable signals by themselves. To produce usable signals, the frequency of a quartz oscillator would need to be locked to the atomic resonance, and the frequency of the locked quartz would then need to be divided to a frequency that was low enough to be used for timekeeping. This required developing methods to control and divide frequency.

The first obstacle for early atomic clock researchers was finding ways to measure the resonance frequency of an atom. The theory was already in place, because the quantum theory of physics, also known as quantum mechanics, had been originated by the German physicist Max Planck (1858–1947) in 1900, resulting in a Nobel Prize in 1918. Planck postulated that electromagnetic energy, E, could only be emitted in a quantized form, as a multiple of the frequency of the radiation. Therefore,

$$E = hf, \tag{3.1}$$

where f is the frequency, and h is the Planck constant, a fundamental physical constant defined in the SI as $6.62607015 \times 10^{-34}$ J·s [1].

By generating an electromagnetic field at a specific frequency, an atom can be forced to absorb energy, which raises its low energy level to a higher one. Or, it can be forced to release energy, which drops its high energy level to a lower one. The resonance frequency of an atomic clock transition, f_r, is the difference between the two energy levels, E_1 and E_2, divided by the Planck constant,

$$f_r = \frac{E_2 - E_1}{h}. \tag{3.2}$$

It's true, of course, that atoms don't absorb or release their energy at a single frequency, but rather across a spread or bandwidth of frequencies that surrounds f_r. For mechanical clocks, the bandwidth surrounding the resonance frequency, Δf_r is usually called the resonance width, but for atomic clocks it's usually called the linewidth. The quality factor, Q, for a clock's oscillator is the ratio of its resonance frequency to its linewidth, or

$$Q = \frac{f_r}{\Delta f_r}. \tag{3.3}$$

Because their resonance frequencies were so high—hundreds of thousands of times higher than quartz clocks—it was obvious from the beginning that atomic clocks could have extraordinarily high Qs if their atomic linewidths could be controlled. This gave them a potentially huge timekeeping advantage over all previous clocks.

Before building an atomic clock, however, researchers needed to choose a specific chemical element and then measure its atomic transition frequency. It seems likely that Rabi and his colleagues expected the energy transition in cesium (^{133}Cs) atoms to be the resonator for the first atomic clock [8]. In a series of experiments conducted at Columbia in 1940, Millman and Kusch measured the resonance frequencies of sodium, two rubidium isotopes, and cesium. These atoms were all considered attractive candidates for building an atomic clock, because each had just a single electron in their outer shell. However, cesium had the advantage of having the highest resonance frequency of the four. Their measurements of the ^{133}Cs transition frequency yielded a result of 9191.4 MHz (±0.9 MHz), which was close to the value that later defined the second [9]. Even so, the first atomic clock wasn't based on cesium atoms, nor was it based on individual atoms. Instead, it was an ammonia molecule absorption clock built at NBS.

3.2 The First Atomic Clock, 1948

The first atomic clock was built in Washington, DC by the American physicist Harold Lyons (1913–1998) and his colleagues at NBS. Lyons was born in Buffalo, New York. He obtained a doctorate in nuclear physics from the University of Michigan in 1939 and was still in graduate school when Rabi and his colleagues at Columbia first discussed employing their molecular beam resonance technique as a time standard.

Lyons joined NBS in 1941, and he became the head of the microwave standards group in 1948. His group then moved rapidly past the pure experimental stage and began building a working atomic clock. An intense period of research and development began in the spring of 1948, and the new clock was first operated on August 12th. The clock was demonstrated to the public in January 1949 [10].

The photograph in Figure 3.2 shows Lyons and NBS director Edward Uhler Condon (1902–1974) standing in front of the clock, with Condon examining a model of the ammonia molecule while Lyons observes. The clock's electronics were enclosed in the two large equipment racks shown in the photo. The display on top of the equipment racks was mainly there for cosmetic reasons, so that the world would know that the new invention was a clock. However, the clock display did work—it was driven by a 50 Hz signal obtained by dividing the signal from a quartz oscillator that was locked to the ammonia absorption frequency.

Figure 3.2 The first atomic clock, along with its inventor Harold Lyons (right) and NBS director Edward Uhler Condon (left). Condon is examining a model of the ammonia molecule.

Lyons and his group overcame many obstacles while building the clock. Because they had chosen the absorption line of ammonia gas as the source of frequency, their first obstacle was measuring the inversion transition frequency, or what Lyons called "the constant natural frequency associated with vibrations of the atoms in the ammonia molecule" [11]. The ammonia gas was stored in an absorption cell, which was designed as a 9 m long copper tube that was wrapped into a spiral. The measurement involved multiplying a 100 kHz signal from quartz oscillator in several stages. The first stage consisted of a frequency multiplication chain built from vacuum tubes that produced a frequency of about 270 MHz. The second stage consisted of a frequency multiplying klystron, a linear beam vacuum tube, where the frequency was modulated by a 13.8 MHz oscillator that multiplied the signal to about 2.983 GHz. In the final stage, the signal was multiplied to approximately 23.87 GHz by use of a silicon crystal rectifier. This frequency was fed to the ammonia absorption cell and tuned until the signal reaching a silicon crystal detector at the end of the cell showed a spectral line (in the form of a dip, or inverted peak) when displayed on an oscilloscope (Figure 3.3). The spectral line indicated that the incoming frequency from the multiplication chain now matched the ammonia absorption frequency, and when the spectral line occurred, an electrical pulse was generated [11–14].

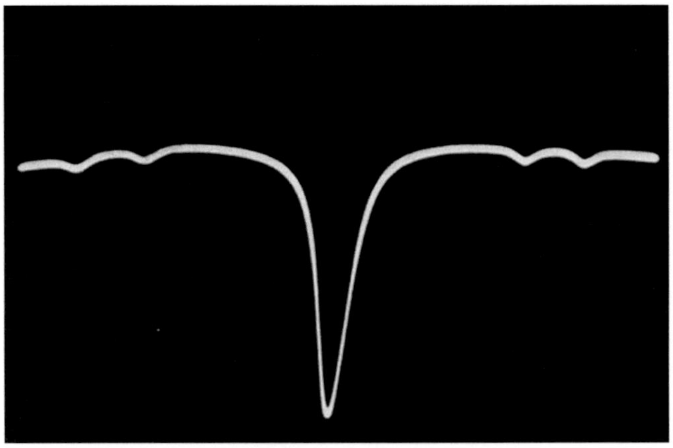

Figure 3.3 The spectral line of the first atomic clock (shown on an oscilloscope). The line occurred when the incoming microwave signal matched the absorption frequency of the ammonia molecule.

The next obstacle was developing a method to automatically adjust the quartz oscillator frequency until it matched the frequency of the ammonia absorption line. This was done by taking a 12.5 MHz frequency from the quartz oscillator frequency multiplication chain and mixing it with the signal from the 13.8 MHz frequency-modulated oscillator. When the frequency from the mixer matched its expected value, another electrical pulse was generated. The time interval between the two generated pulses—one from the absorption cell and the other from the mixer—was continuously measured. If the time interval was increasing, it indicated that the quartz oscillator frequency was increasing, and vice versa. The time interval measurements generated a control signal that adjusted the quartz oscillator whenever necessary, which kept it locked to the ammonia absorption frequency [11, 13].

Figure 3.4 shows a strip chart recording where the quartz oscillator was locked to the ammonia absorption frequency from about 5 p.m. to about 8 a.m., generating what was nearly a straight line. Before and after this period, the quartz oscillator was unlocked and the frequency was unstable. Each small division on the strip chart represents a frequency variation of less than 1×10^{-7}.

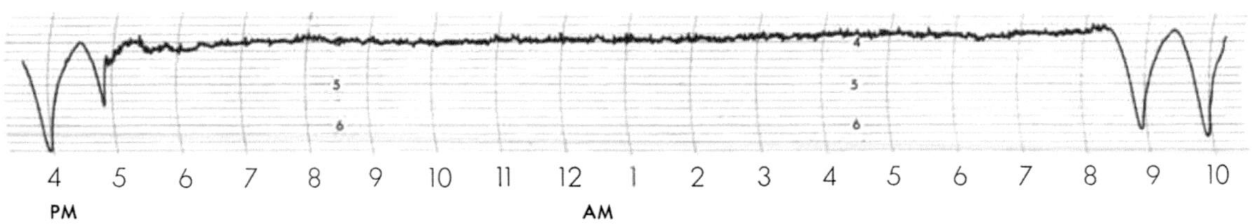

Figure 3.4 A strip chart recording showing the period (from 5 p.m. to 8 a.m.) when the quartz oscillator was locked to the ammonia absorption frequency.

A third obstacle involved dividing the locked quartz oscillator frequency of 100 kHz into lower frequencies that could be used for timekeeping [15]. The clock included dividers that produced two output signals—the 50 Hz signal that drove the synchronous motor clock on top of the equipment racks, and a 1 kHz signal that was used to drive a special synchronous motor clock that was designed for comparisons with astronomical time. Figure 3.5 is a simplified block diagram showing how the clock worked.

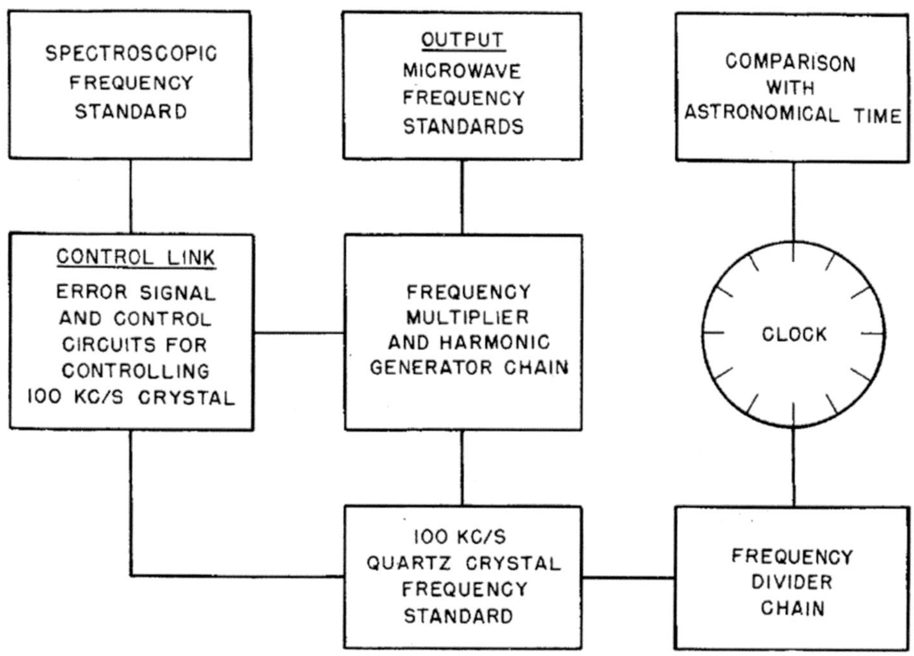

Figure 3.5 A simplified block diagram of the first atomic clock.

Once the clock was running, Lyons and his group focused their efforts on making it more accurate by increasing the Q factor. The ammonia clock had a resonance frequency of 23.8 GHz, about 2.6× higher than cesium, but the linewidth wasn't particularly narrow. Lyons and his colleagues tried to reduce the linewidth to increase the Q, but they had little success, and the Q factor of the ammonia clock ranged from about 50,000 to 500,000 in various experiments [12]. This was in the same general range as the best quartz oscillators of that era, but the new atomic clock could maintain a constant frequency for a much longer period than a quartz clock.

Lyons and his colleagues had advantages over university researchers when they built their atomic clock, because NBS wasn't just a research facility, but was also a metrology laboratory that had several decades of experience in generating, distributing, and measuring standard frequency signals. This allowed them to compare their new clock to the national frequency standard—the quartz oscillators at NBS that were periodically calibrated against the astronomical time standards maintained by the USNO. The group quickly learned from these comparisons that the potential accuracy of the ammonia clock was limited.

Two versions of the ammonia clock were built, and the best reported timekeeping accuracy was about 2 ms per day (2×10^{-8}). Work on a third version was halted when it became apparent that cesium beam techniques were likely to provide much better accuracy. Even though the cesium resonance frequency was much lower, it provided better options for narrowing the linewidth and increasing the Q factor. Therefore, Lyons and his NBS colleagues turned their attention away from ammonia molecules. With technical guidance from Kusch at Columbia, they began building a clock based on cesium atoms in the latter part of 1949 [16].

3.3 Cesium Atomic Clocks

Cesium (Cs) was discovered in 1860 by the German scientists Robert Bunsen (1811–1899) and Gustav Kirchhoff (1824–1887). Bunsen was a chemist, perhaps best known for the Bunsen burner that he helped develop, and Kirchhoff was a physicist. Their discovery of cesium occurred at the University of Heidelberg, where they were performing experiments with a prototype spectroscope that they had invented. After successfully using their spectroscope to identify the characteristic spectra of sodium, lithium, and potassium, they observed previously unknown blue spectral emission lines in samples of mineral water. The name they chose for this newly discovered element was derived from the Latin word *caesius*. It referred to the color of the spectral lines, and means "bluish gray" or "sky blue," depending upon the translation.

Cesium has several characteristics that made it attractive to early atomic clock researchers. A soft, pale gold metal, cesium is one of the six alkali metals, along with lithium (Li), sodium (Na), potassium (K), rubidium (Rb), and francium (Fr). All of the alkali metals are soft and reactive at standard temperatures, but cesium is the softest element listed on the Mohs hardness scale (0.2 MPa), and its low melting point of 28.5 °C means that it becomes a liquid at near room temperature. Excluding francium (which has a half-life of only 22 minutes), only mercury (Hg) has a lower melting point (−38.8 °C) than cesium. The low melting point is an advantage, because cesium can be vaporized without excessively high oven temperatures. Another advantage is the fact that cesium atoms are relatively heavy, with an atomic mass of about 133, and this causes them to typically move at a speed of less than 150 m/s [17, 18]. This is roughly a third of the speed of air molecules at room temperature (~500 m/s), and it allows cesium atoms to be observed for longer periods than lighter atoms, which is a major benefit for an atomic clock.

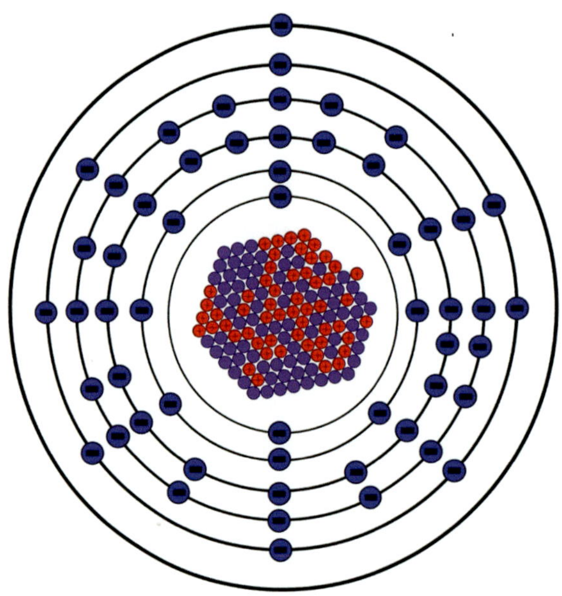

Figure 3.6 A cesium atom, showing the nucleus and the rings of outer electrons.

All naturally occurring cesium atoms are identical. With the exception of the previously mentioned francium, which doesn't exist in a stable form, cesium is the largest atom. It's radius—which is the distance from the nucleus of the atom to the outer edge of its electron cloud—is 0.267 nm. The nucleus consists of 55 protons and 78 neutrons; therefore, cesium has an atomic number of 133. Its electron cloud consists of 55 electrons in six orbital shells that bind to the nucleus. The five inner shells contain an even number of electrons, but as is the case with the other alkali metals, the outermost shell has just a single, unpaired electron (Figure 3.6). The outermost electron, known to chemists and physicists as the valence electron, spins on its axis. The nucleus also spins, and both the electron and the nucleus produce a magnetic field. This results in two energy states, one occurring when the two "magnets" point in the same direction, and the other occurring when they point in opposite directions. The resonance frequency of cesium is obtained from these energy state transitions. The cesium resonance frequency of about 9.2 GHz isn't as high as the 23.8 GHz resonance frequency of the ammonia molecule that Lyons had focused on earlier, but it's considerably higher than the two other atoms that were later commonly utilized in atomic clocks, rubidium (~6.8 GHz) and hydrogen (~1.4 GHz).

By 1952, a team at NBS led by Lyons and Jesse Sherwood (1922–1995) had built the first prototype of a cesium clock. The clock was based on Rabi's magnetic resonance technique, which interrogated the atoms with one microwave pulse, or oscillatory field, that was uniformly distributed throughout the atomic transition region. The team first observed the hyperfine transition frequency of cesium on August 15, 1951 after almost two years of work [10]. Sherwood reported initial results at a 1952 meeting of the American Physical Society [19], but the linewidth was too wide (~30 kHz) to serve as an accurate time standard, resulting in a Q of only 300,000. The research was again focused on narrowing the linewidth, which would require a better way of interrogating the cesium atoms.

A major breakthrough in narrowing linewidth had occurred in 1949 when Norman Ramsey, a former student of Rabi who was then a professor at Harvard University, invented the separated oscillatory field method [5, 20]. Ramsey eventually received a Nobel Prize in physics for his invention in 1989. His method consisted of interrogating the atoms with two short microwave pulses or oscillatory fields, coherent and in phase, that were separated by some distance in an atomic beam. Applying the oscillating field in two steps accomplished the goal of narrowing the linewidth, and it reduced an atomic clock's sensitivity to microwave power fluctuations and magnetic fields by factors of 10 to 100 or more.

Ramsey's method also eliminated most of the Doppler frequency shift, the change in the measured frequency of a signal due caused by its relative motion with respect to the observer. This effect is named after the Austrian physicist Christian Doppler (1803–1853), who first described it in 1842. A classic example of a Doppler frequency shift is the way a train whistle sounds to a listener standing alongside the railroad tracks. Even though the whistle is producing an audio tone at a constant frequency, the listener hears a higher-pitched tone as the train approaches, and a lower-pitched tone after the train has gone by. The listener only hears the actual tone produced by the whistle at the instant when the train is passing them. The separated oscillatory field method nearly eliminated the Doppler frequency shift because the two phase-coherent microwave fields were transmitted at right angles to the beam of atoms. This meant that the atoms were never moving away from or towards the microwave fields, but instead were moving across them.

After learning about Ramsey's new method, the NBS team quickly redesigned their cesium clock. They applied microwave pulses in two regions along the atomic beam, separated by 50 cm [10]. As they had anticipated, this narrowed the linewidth by 100×, from 30 kHz to just 300 Hz, increasing the Q to about 30 million. Lyons was encouraged and wrote that "these results indicate an accuracy of one part in 10 billion may be achieved" [21].

His prediction of 1×10^{-10} accuracy would eventually come true, but it happened much later than expected. NBS interrupted their atomic clock program in 1953, a decision that was partially made for budgetary and political reasons, but also because the agency had elected to focus on other areas. Sherwood took a position at Oak Ridge National Laboratories that same year, and Lyons moved to the new NBS laboratories in Boulder, Colorado in 1954. A year later, Lyons also left NBS, accepting a position at Hughes Aircraft Company [8, 10].

The cesium clock was disassembled and moved from Washington to Boulder, where it was reassembled by a new team of researchers led by Richard Mockler (1925–2011) and Roger Beehler (b. 1934), and eventually became known as NBS-1. The clock had a microwave cavity—now commonly known as a Ramsey cavity—that was 55 cm long [22–24]. The process took several years, but by the spring of 1959 the clock's accuracy was reported as 8.5×10^{-11}, and NBS-1 was designated as the national frequency standard for the United States [25]. Beehler is shown making an adjustment to NBS-1 in Figure 3.7.

Figure 3.7 Roger Beehler of NBS with NBS-1.

On January 1, 1960, NBS-1 was replaced as the national frequency standard by NBS-2, but it continued to be operated as a secondary standard until 1962, when its reported accuracy was 1×10^{-11}, or about one microsecond per day [26]. It's interesting to note that the popular syndicated cartoon, Ripley's *Believe it or Not!*, credited Lyons with the invention of the atomic clock in 1953 (Figure 3.8). Ironically, and perhaps appropriately, the clock pictured in the cartoon wasn't the ammonia absorption clock that had been invented five years earlier, but instead was NBS-1, the cesium beam clock that was started in 1949 but not finished until about a decade later.

Figure 3.8 A 1953 cartoon featuring the first cesium clock developed at NBS.

The disruption of the atomic clock program in 1953 not only delayed NBS-1 by at least several years, but it also caused NBS to lose their large early lead in the cesium clock race. This allowed the National Physical Laboratory (NPL) in the United Kingdom to be the first laboratory to build an atomic clock that was reliable enough to be used as a time standard. The NPL effort was led by the British physicist Louis Essen (1908–1997). Essen was an accomplished scientist who had long been interested in frequency and time metrology—he had previously developed a quartz ring clock as a standard for timekeeping, as well as a cavity resonance wavemeter that he used to measure the speed of light.

Beginning in 1949, Essen made several trips to the United States to meet with atomic clock researchers, visiting with Rabi and his colleagues at Columbia, with Jerrold Zacharias of MIT, and with Lyon's group at NBS. Essen was anxious to build a cesium clock, so he encouraged the NPL director—the British geophysicist Edward Bullard (1907–1980)—to start an atomic clock program. There were other priorities and several delays, but the program began in the summer of 1953, which was the same year when cesium clock development was halted at NBS [8, 27].

The work at NPL progressed rapidly over the next two years, and by mid-1955, Essen and his colleague, the Welsh physicist Jack Parry (1923–1995), had built and demonstrated a working cesium clock. The first published measurements showed that the clock had a linewidth of about 340 Hz and an accuracy of about 1×10^{-9} [28]. The NPL cesium clock, pictured along with Essen and Parry in Figure 3.9, allowed atomic time to be compared to astronomical time, as described in the next section.

Figure 3.9 Louis Essen (right) and Jack Parry (left) with the original NPL cesium clock.

3.4 Defining the Atomic Second by Comparison to Ephemeris Time

When atomic clocks first worked well enough to be compared to astronomical time, the base unit of time was the ephemeris second, which had recently replaced the mean solar second. As noted at the beginning of this chapter, the ephemeris second was defined in 1956 as 1/31,556,925.9747 of the tropical year 1900. In retrospect, it was unfortunate that a new astronomical definition of the second was accepted during a period when working atomic clocks were already being built [29, 30]. A clean transition from the mean solar second to the atomic second would have certainly made more sense. The era of the ephemeris second was fortunately short, and had it not served as the comparison reference for the atomic second, it would likely be easy for historians and time metrologists to completely dismiss.

Ephemeris time (ET) was difficult to measure. It was determined by measuring the position of the Moon with respect to several surrounding stars. The most accurate Moon observations had been recorded at the USNO by the astronomer William Markowitz (1907–1998). Markowitz was born in Austria, but his family emigrated to the United States when he was three years old. He obtained a doctorate in astronomy from the University of Chicago in 1931, and worked at the USNO from 1936 until 1966. By 1952, Markowitz was performing Moon observations with a sophisticated dual-rate Moon camera of his own design [30, 31].

Collecting Moon observations was a tedious task and results were obtained very slowly. Bullard wrote in 1955 that it would take four years of Moon observations to determine time with the same accuracy as their new cesium clock. He also noted that "atomic clocks will be improved, probably by a greater factor than the astronomical determinations," which in retrospect was a vast understatement [32]. To Bullard and to many others, it was clear by the mid-1950s that atomic clocks represented the future of timekeeping.

In June 1955, NPL and the USNO began collaborating on a program where the goal was to determine the frequency of cesium with respect to the ephemeris second [33, 34]. It was a worthy goal, because researchers knew that redefining the second based on an atomic transition would allow large numbers of laboratories to simultaneously produce physical realizations of the second with their own clocks, and that these clocks could be the reference for other time and frequency measurements. It was also an ambitious goal, and to achieve it NPL and the USNO would need to compare their clocks and measure their time differences over a lengthy period. That period turned out to be about 2.75 years, from mid-1955 until the end of the first quarter of 1958.

The standard for ephemeris time was a USNO clock based on a 2.5 MHz quartz oscillator manufactured by Western Electric. The clock was steered to ET by applying corrections obtained with the dual-rate Moon camera. The standard for atomic time was the NPL cesium clock, which by then was considered to be accurate to within 5×10^{-10}. Comparing the ephemeris clock to the atomic clock was difficult, because the two clocks were located on opposite sides of the Atlantic Ocean. The measurement was performed by simultaneously comparing both clocks to radio signals that could be received at both laboratories, a technique that's now called common-view time transfer (Section 4.3.3). Several radio stations that were dedicated to broadcasting time signals were involved in the common-view measurements, including WWV, the NBS station that was then located in Beltsville, Maryland, and the stations MSF and GBR, which were both located in the United Kingdom [33].

To make the measurement uncertainty as small as possible, four different solutions were calculated to determine the effects of using different sets of data. The final measurement result, published in August 1958, was the average of the four solutions. The frequency of the atomic second was determined to be 9,192,631,770 cycles/s with an estimated measurement uncertainty of ±20 cycles/s [34]. The measurement uncertainty was limited not by the cesium clock, but rather due to the many complexities involved in measuring ET. There were multiple sources of uncertainty, including uncertainty in the radio signals used in the common-view time transfer, and uncertainties in the Moon camera observations. This meant that the necessary resolution could only be achieved by averaging data for periods of at least 100 days. If we stop to consider the limitations of the technology that existed in the 1950s, and the large number of complex steps that were required, the measurement results were incredibly accurate. This was evident to the Italian scientist and time metrologist Sigfrido Leschiutta (1933–2011), who wrote the following thoughtful words in 2005 –

> *"As a personal remark, taking into account the capabilities of the timing emissions at the moment, of the frequency standards available, of the inevitable scatter of the moon camera, and some other factors, not least the widespread use and abuse in 'touching' the piezo-oscillator, it is almost impossible to explain the accuracy of the Markowitz determination. Similar events, i.e. results surpassing the capabilities of the moment, are not uncommon in the history of science that sometimes is prone to accepting the intervention of a serendipity principle. The other possible explanation calls for a first class understanding of physics, coupled with scientific integrity."* [33]

After the Markowitz/Essen measurement results were published, it was inevitable that the atomic second would eventually become the SI second. Nearly a decade passed, however, before the SI definition was changed. At least part of the delay was because atomic clocks other than cesium—specifically clocks based on hydrogen and thallium—were still being considered as the basis for the atomic second. As it still is today, the SI was then coordinated by the International Bureau of Weights and Measures (*Bureau International des Poids et Mesures*, BIPM), located in the suburbs of Paris, France. The BIPM director Jean Terrien (1907–1992) wrote in August 1967 that the definition of the second had nearly been changed in 1964, but

> *"... it was considered premature in 1964 to redefine the second in terms of the caesium frequency because the hydrogen maser was expected to offer, within a short while, a better base for such a redefinition.*
>
> *Contrary to expectation, it happened that the hydrogen maser, though potentially slightly superior to the caesium beam frequency standard, progressed only slowly (one of its uncertainties comes from the wall shift), and that caesium standards were so much improved that an almost perfect confidence was acquired in their reliability, within the most severe present needs of precision. Though it was agreed that the improvement of other atomic frequency standards, using for example hydrogen or thallium atoms, ought to be pursued actively, the Comité Consultatif pour la Définition de la Seconde, at its meeting of 12–13 July 1967, recommended unanimously to the Comité International des Poids et Mesures not to wait any longer ..."* [35].

Therefore, the waiting finally ended in October 1967 when the SI second was formally redefined as

> *"the duration of 9 192 631 770 periods of the radiation corresponding to the transition between the two hyperfine levels of the ground state of the caesium 133 atom"* [36].

The atomic definition of the second fundamentally changed how time was measured and kept. Previously, astronomical periods—such as the mean solar day or the tropical year—were divided into shorter intervals to obtain hours, minutes, and seconds. The new definition meant that everything was based on the second. The duration of a second was now determined by counting oscillations of RF signals that caused atoms to change their energy state. Longer time intervals, such as minutes, hours, and days, were now obtained by counting seconds. Furthermore, because time could now be measured with more resolution and less uncertainty than any other physical quantity, the second attained a prominent position in the metrology world that it continues to hold.

The definition of the SI second remains essentially the same as of 2025, modified only by a slight amendment made in 1997. Calculations made in the early 1980s by the NBS physicist Wayne Itano (b. 1951) revealed that blackbody radiation can cause noticeable frequency shifts in cesium standards [37], and this work eventually resulted in an addendum to the definition. The CIPM affirmed in 1997 that the definition refers to "a cesium atom at rest at a thermodynamic temperature of 0 K." This means that an optimal realization of the second requires the cesium atoms to be in a field-free environment where the temperature is absolute zero, and where the atoms have no residual velocity.

3.5 Leap Seconds

When the atomic second was accepted as the SI second in October 1967, it was of course known that atomic time would continuously diverge from astronomical time from that point forward. The new atomic time scale, called Coordinated Universal Time (UTC), would need to be periodically corrected to keep it in agreement with UT1, an astronomical time scale that's based on the mean solar second. We'll discuss how UTC is determined in Chapter 9, but for now we can note that the frequency adjustments to atomic time began in 1960, about seven years before the second was redefined. To keep track of these adjustments, the International Time Bureau (*Bureau International de l'Heure*, BIH) at the Paris Observatory began overseeing the international coordination of atomic time in 1961, a responsibility that was passed to the BIPM in 1988 [30, 38].

The frequency adjustments continued for almost five years after the redefinition, but it was eventually decided that periodic time steps would be a better method of keeping UTC in agreement with UT1. The time steps were integer seconds called leap seconds, and the first leap second was inserted on June 30, 1972. A total of 27 leap seconds have been added to UTC as of 2025, an average rate of about one leap second every two years. So far, every leap second has been positive, which means that extra seconds have been added to UTC.

Due to leap seconds, UTC is always within ±0.9 s of UT1. When a leap second is added, it pauses UTC for 1 s, which allows UT1 to catch up. Leap seconds are only inserted on the last day of either June or December, and the extra second is added to the last minute of the day. The UTC sequence for a positive leap second is:

23:59:59
23:59:60 (leap second)
00:00:00

It's not easy to model or predict the divergence between astronomical and atomic time, but there are two general reasons why leap seconds are periodically needed. Reason one can be traced to the original Markowitz/Essen measurement—the atomic second was defined by comparison to the ephemeris second instead of the mean solar second, and the mean solar second is the basis for UT1. Ephemeris seconds were already slightly shorter than mean solar seconds, and this characteristic was inherited by the atomic second. The reason that ephemeris seconds were slightly shorter was that their duration had been made approximately equal to the mean solar second at the beginning of the year 1900, but by 1956 the mean solar day had lengthened and UT1 was running about 31 s behind ET. Therefore, when Markowitz and Essen began their measurements, the existing SI second was simply too short to match the Earth's rotation rate [39].

If the atomic second had been defined by comparison to the mean solar second instead of the ephemeris second, leap seconds would have occurred much less frequently [40]. However, some leap seconds would still have been needed due to reason number two, which is the fact that the Earth's average rotational rate is gradually slowing down. The gradual slowing of the Earth's rotation is commonly cited as the main reason for leap seconds, but the way the atomic second was defined has so far had a larger effect on the number of leap seconds than the slowing of the Earth. Ironically, the Earth has begun to spin faster in recent years, probably due to climate change. If this trend continues, it could eventually result in the first negative leap second, which would involve subtracting a second from UTC.

Leap seconds temporarily disrupt the normal conventions of timekeeping, and this disruption has caused some timing systems to fail. Therefore, based on a vote conducted in November 2022, the BIPM has resolved to work with the International Telecommunications Union (ITU) and other organizations to stop the practice of inserting leap seconds by, or before, the year 2035. After the practice of leap seconds is stopped, the time difference between UTC and UT1 will continuously diverge and be allowed to exceed 1 s. However, at some point in the future, UTC will still need to be corrected, perhaps by inserting a leap minute or a leap hour. We'll provide more information about leap seconds when we discuss traceability in Chapter 9.

3.6 Commercial Cesium Clocks

Commercial cesium clocks were available before the publication of the Markowitz/Essen measurement in 1958, and several different models were in use in laboratories before the SI definition of the second changed in 1967. The availability of commercial cesium clocks, as well as the data and experience gained from their use, undoubtedly sped up the acceptance of the atomic second.

The first commercial cesium clock was introduced in October 1956, just slightly more than a year after Essen's cesium clock became operational at NPL. The commercial clock was named the Atomichron, and it was manufactured by the National Company of Malden, Massachusetts. It was developed by a team led by Jerrold Zacharias of MIT [8, 41], an experienced researcher who had previously collaborated on early molecular beam experiments with Rabi at Columbia [4]. The National Company sold about 50 Atomichrons between 1956 and 1960. Most were sold to the military, with the first unit delivered to the Naval Research Laboratory (NRL) in Washington, D.C., in September 1956.

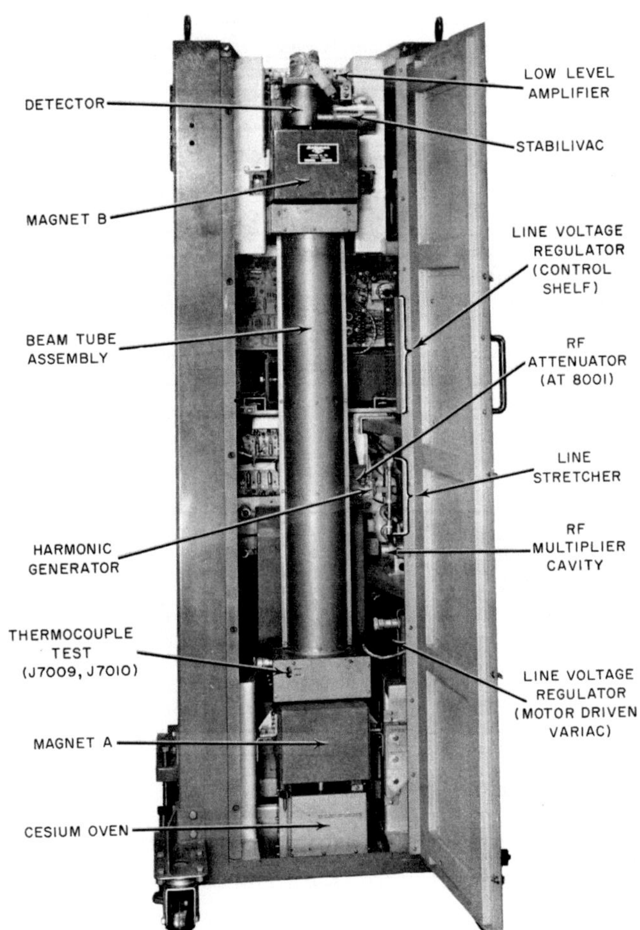

Figure 3.10 The interior of the Atomichron NC2001, showing the vertical cesium beam tube (courtesy of Tom Van Baak, leapsecond.com).

The first two models of the Atomichron, numbered as NC1001 and NC2001, were housed in equipment racks about 2.1 m tall and weighed more than 350 kg. As shown in Figure 3.10, the cesium beam tube was vertically mounted and nearly as tall as the equipment rack [8, 40]. The resonance frequency of the first model was 9,192,631,830 cycles/s (±10 cycle/s), which was 60 cycles higher than the result later published by Markowitz and Essen [41].

At least two Atomichrons were in operation at NBS in 1958 and 1959, where they were successfully compared to NBS-1 to within parts in 10^{10} [25]. Two Atomichrons were also delivered to England, where comparisons with Essen's clock at NPL yielded similar results [43]. Subsequent versions of the Atomichron were mainly produced for the United States Air Force and were much smaller than the first models. The National Company ultimately failed due to financial difficulties, and its assets, intellectual property, and trade names were sold to Frequency Electronics in 1969 [8, 42].

A far more successful line of cesium clocks was developed through the efforts of Leonard Cutler (1928–2006) and his colleagues at the Hewlett-Packard Company in Santa Clara, California. After earning a doctorate from Stanford University, Cutler spent over four decades working for Hewlett-Packard and its successor, Agilent Technologies, where he designed or co-designed numerous frequency standards and clocks. During Cutler's tenure, Hewlett-Packard produced a series of reliable cesium clocks that could run continuously for years and that were small enough to fit into standard equipment racks. The first of these clocks was the Hewlett-Packard 5060, pictured in Figure 3.11, which was introduced in 1964 and had a specified accuracy of 1×10^{-11} by 1966 [23]. The 5060 was the first cesium clock to feature all solid-state electronics, which helped to reduce its size. Its Ramsey cavity was just 12.4 cm long and it weighed less than 30 kg [42, 44].

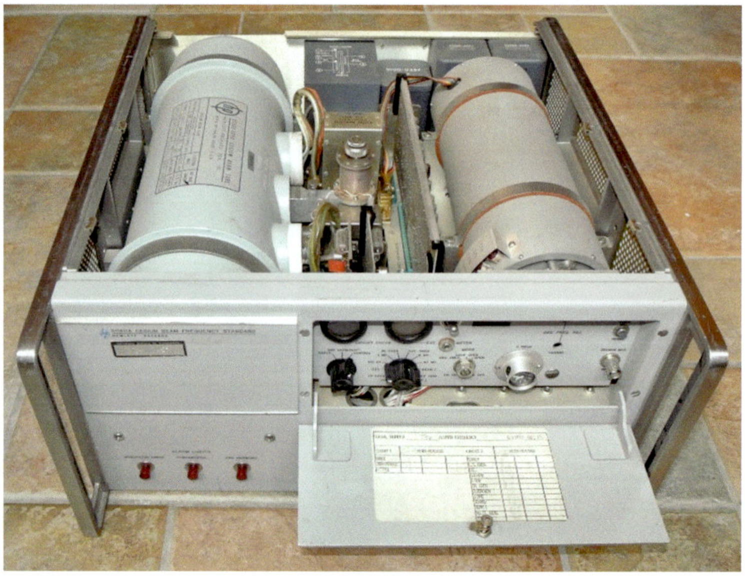

Figure 3.11 Hewlett-Packard 5060A cesium beam atomic clock with the cover removed. The beam tube is on the left side (courtesy of Tom Van Baak, leapsecond.com).

The 5060 was followed by the 5061, which was manufactured from 1967 until the early 1990s. The 5071A, which first appeared in the Hewlett-Packard catalog in 1993, added an internal microprocessor and an improved cesium beam tube, and was more stable and accurate than its predecessors [42]. By 1999, the specified frequency accuracy of the 5071A was 5×10^{-13}, which translated to a timekeeping accuracy of about 50 ns/day. The 5071A was later manufactured by Agilent, Symmetricom, and Microsemi, and is still being manufactured by Microchip as of 2025 (Figure 3.12). It continues to serve as a frequency and time standard at numerous government and commercial laboratories.

Figure 3.12 Microchip 5071A cesium beam atomic clock (courtesy of Microchip).

Commercial cesium clocks are very reliable, but they are expensive, with prices approaching \$100,000 USD as of 2025, and they also require periodic replacement of the beam tube. The initial cost and the cost of replacement tubes prevent them from deployed in large quantities. For example, a beam tube typically lasts for about 10 years before it runs out of cesium and needs replacement, and replacing the beam tube can sometimes cost nearly as much as replacing the entire clock. Only a few hundred cesium beam clocks are sold each year on average, and the typical customers are national timing laboratories, calibration laboratories, and military installations.

3.7 Principles of a Cesium Clock

The SI definition of the second mentions that the cesium clock transition occurs between "the two hyperfine levels of the ground state" of the ^{133}Cs atom. To understand what that means, let's first consider the fact that any object moving with mass possesses momentum. Angular momentum is the momentum associated with a rotating or spinning object. It exists in two forms, one where the object is accelerating around a fixed point, such as the Earth revolving around the Sun. This is known as orbital angular momentum. The second form is where the object is rotating about a fixed point, such as the Earth rotating on its axis. This is known as spin angular momentum. The motion of the outer electron around the nucleus of an atom has an orbital angular momentum, associated with the quantum number L. However, the ground state is the state where the energy of the outer electron is at its minimum, with no orbital angular momentum. An excited state is any state with energy greater than the ground state. If we only consider two values of L, the ground state is where $L = 0$ and the excited state is where $L = 1$ [29, 30].

The ground state of cesium is denoted by the spectroscopic notation $^2S_{1/2}$ (Figure 3.13). In this notation, the letter S is used when $L = 0$ (ground state) and the letter P is used when $L = 1$ (excited state). The subscript following the letter S is the sum of the orbital and spin angular momentum of the outer electron, or s. As was just noted, there is no orbital angular momentum in the ground state, but the electron does have a spin angular momentum of 1/2, which is similar to the angular momentum of a spinning top, therefore $s = 1/2$. The superscript to the left of the letter S specifies the quantity $2s + 1$ [30, 45].

We noted in Section 3.3 that both the nucleus and the outer electron spin, that their spins produce a magnetic field, and that there are two energy states—one occurring when the two "magnets" point in the same direction, and the other occurring when they point in opposite directions. The two energy states are the two hyperfine levels of the ground state of cesium that are referred to in the SI definition.

The nucleus spin of ^{133}Cs is equal to 7/2 and the electron spin is equal to 1/2. If the nucleus and electron spins point in the same direction (parallel), then the hyperfine level specified by the quantum number F is equal to the nucleus spin plus the electron spin, $\frac{7}{2}+\frac{1}{2}$, or $F = 4$. If the nucleus and electron spins point in opposite directions (anti-parallel), then the hyperfine level equals the nucleus spin minus the electron spin, $\frac{7}{2}-\frac{1}{2}$, or $F = 3$.

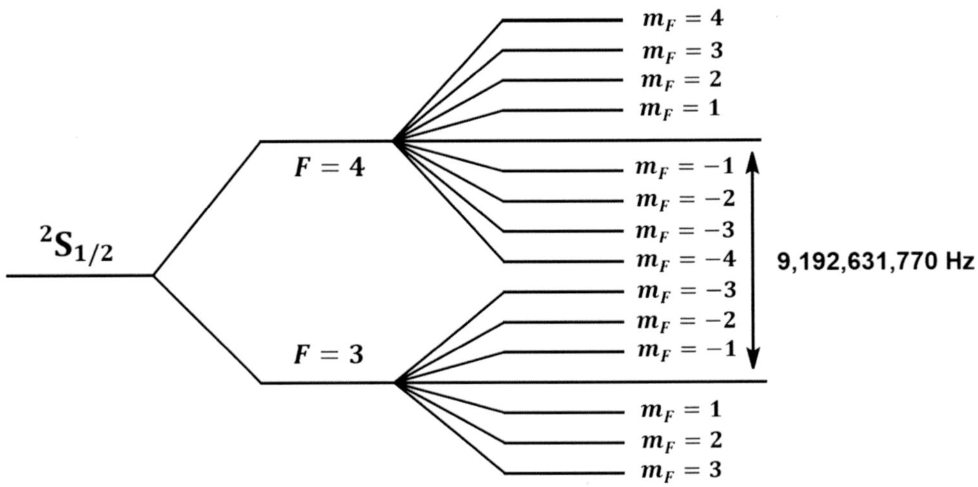

Figure 3.13 The cesium clock transition.

The two hyperfine levels of the cesium ground state are characterized not only by the quantum number F, but also by a magnetic quantum number, m_F, which has integer values between $-F$ and $+F$. The splitting of the $F = 3$ and $F = 4$ states into the various m_F sublevels is known as hyperfine splitting, and it occurs when a static magnetic field is applied.

The effect of splitting a spectral line with a magnetic field is called the Zeeman effect, named in honor of the Dutch physicist Pieter Zeeman (1865–1943) whose discovery resulted in a Nobel Prize in physics in 1902. The Zeeman effect results in 16 possible sublevels in the ground state of cesium, seven sublevels of $F = 3$ ranging from −3 to +3, and nine sublevels of $F = 4$ ranging from −4 to +4. The cesium clock transitions occur at sublevel $m_F = 0$, therefore the ground state hyperfine transition that defines the SI second is denoted as $|4,0\rangle \leftrightarrow |3,0\rangle$. This transition was relatively easy to detect with the electronic systems that were available when Rabi and his colleagues began their experiments, but it also has other advantages that make it well suited for an atomic clock. For example, the probability of a spontaneous transition occurring during the observation time is very small. Perhaps more importantly, the energy level at the $m_F = 0$ sublevel changes very slowly when a small magnetic field is applied, therefore the presence of small magnetic fields near a cesium clock will have little effect on its frequency [29].

Figure 3.14 provides a simplified diagram of a cesium beam clock. As shown on the left side of the figure, ^{133}Cs atoms are heated to a gaseous state in an oven. A beam of atoms emerges from the oven at a temperature near 100 °C and travels through a spatially varying magnetic field, where the beam is split into two atomic beams with different magnetic states. The two beams now have different trajectories. One beam is absorbed by the getter and is of no further interest. The other beam is deflected into the microwave interrogation cavity (commonly known as the Ramsey cavity), where the atoms are exposed to microwaves with two spatially separated pulses.

The microwave signals are generated by a frequency synthesizer driven by a quartz oscillator. If the synthesized frequency exactly equals the cesium resonance frequency, the atoms will change their magnetic state while still inside of the Ramsey cavity, undergoing the desired $|4,0\rangle \leftrightarrow |3,0\rangle$ energy transition. After leaving the cavity, the atoms pass through a second spatially varying magnetic field. These magnets direct only the atoms that changed state to the detector; the other atoms are directed to a getter and absorbed. The detector sends a feedback signal to a servo circuit that continually tunes the synthesizer so that a maximum number of atoms reaches the detector. This keeps the microwave source locked to the cesium resonance frequency. The process of keeping the frequency synthesizer locked to cesium resonance is analogous to carefully and continuously tuning a radio dial to ensure that the loudest and clearest signal is always heard [45–47]. The microwave frequency locked to cesium resonance is divided to produce standard outputs for frequency and time, typically in the form of 10 MHz and 1 pulse per second (pps) signals.

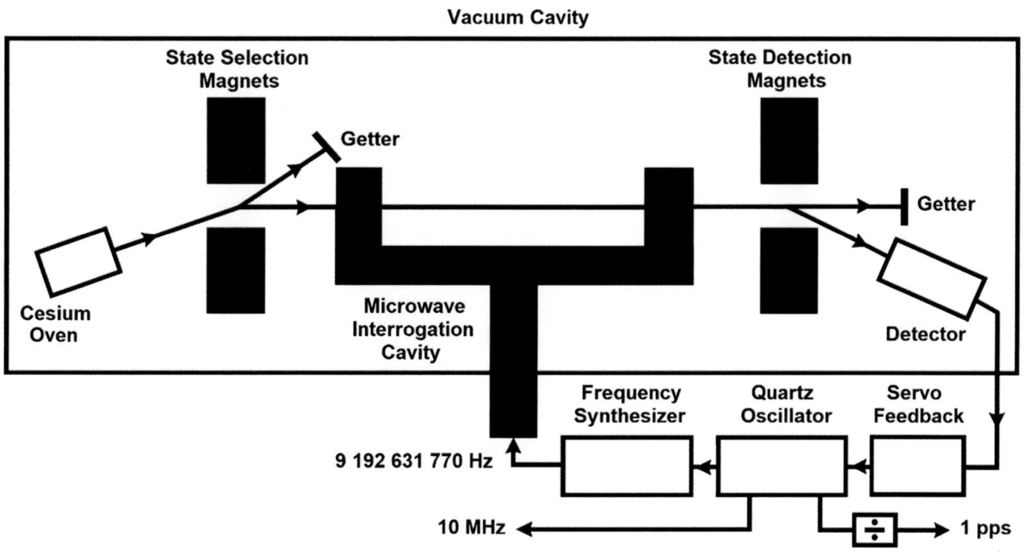

Figure 3.14 A simplified diagram of a cesium beam clock.

3.8 Cesium Fountain Clocks

The most accurate cesium clocks are not sold commercially, and only a small number of them exist. These clocks, called cesium fountains, were specifically designed so that the atoms could be observed for a longer period than a cesium beam clock allows. Cesium fountain clocks are typically designed and built by researchers at national timing laboratories to serve as primary frequency standards. Producing a fountain clock that runs continuously, or nearly continuously, is a significant scientific accomplishment.

The idea of a fountain clock was introduced by Zacharias in the 1950's, shortly before he developed the Atomichron [8, 48]. His original idea was simply to build a vertical cesium beam clock with one microwave cavity. This would allow atoms from the cesium oven to first pass through the cavity while traveling upward. Then, influenced by gravity like the water in a fountain, the atoms would stop, reverse their direction, and pass through the cavity again on their way down. The two passes through the cavity reproduced Ramsey's two-pulse interrogation method, and it also increased the observation period. If the flight of the atoms reached a height of about 1 m, the atoms could be observed for nearly a full second, which was 100× longer than the observation period of a cesium beam clock. Unfortunately, Zacharias was never able to make his fountain clock work properly. The atoms collided in the beam, and the slow atoms were scattered out of the beam by the fast atoms that overtook them [48, 49].

The fountain idea was revived several decades later by the American physicist Steven Chu (b. 1948) and his colleagues at Stanford University in California, who succeeded in building a working fountain in the late 1980s. This work contributed to a Nobel Prize in physics that Chu shared in 1997 for the "development of methods to cool and trap atoms with laser light." The first fountain built by Chu's group was based on sodium atoms [50], but measurements of a cesium fountain—proposed for use as a frequency standard—were published in 1993 [51]. The following year, a cesium fountain that did operate as a primary frequency standard was built by scientists at the Primary Time and Frequency Laboratory of the National Bureau of Metrology (*Primaire du Temps et des Fréquences—Bureau National de Métrologie*, LPTF-BNM) in Paris [52].

The key to making a fountain clock work is laser cooling, a technique that cools atoms down to a temperature near absolute zero. Laser cooling was first proposed in 1975 [53], and was successfully demonstrated at NBS by the American physicists David Wineland (b. 1944), Robert Drullinger (b. 1944), and Fred Walls (b. 1940) in 1978 [54]. There are many types of laser cooling, but fountain clocks generally implement a method called optical molasses. This method exerts a damping force on the atoms by using three pairs of oppositely directed lasers, as shown in Figure 3.15. The lasers are tuned to a frequency slightly below the optical resonance of the atoms, and atoms at the intersection of the six laser beams are cooled to a temperature of less than 1 μK in a few tenths of a second. The cold cesium atoms are slowed down to a speed of about 1 cm/s, which is a tiny fraction of their approximate 100 m/s speed in a cesium beam clock. The reduced speed allows a large "ball" of atoms to be gathered together and confined in one place.

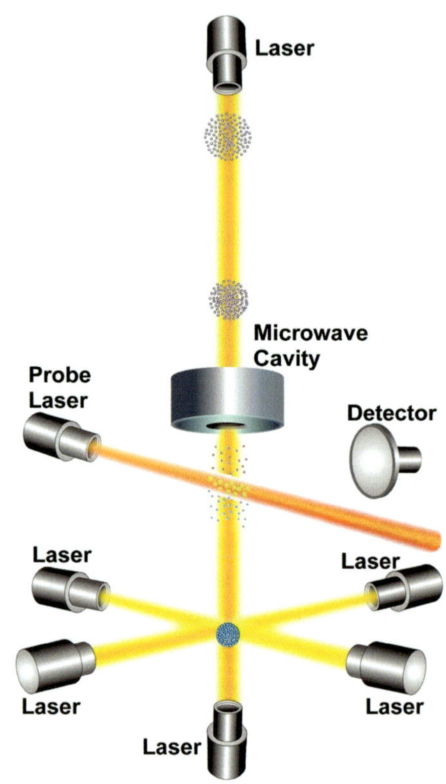

Figure 3.15 A simplified diagram of a laser-cooled cesium fountain clock.

The lasers provide an opposing force to the atom's natural motion. Atoms absorb photons from the laser beam they are moving towards, and each absorbed photon carries momentum in the direction opposite to the atom's motion. Atoms reemit photons in a random direction, and they reemit slightly more energy than they absorb, because the laser is tuned below the resonance frequency. The process can be thought of as a bowling ball that's being bombarded by a stream of ping pong balls. The opposing force that the atom receives from each scattered photon is so small that it appears to have no effect. However, if this tiny force is repeated often enough, the velocity of the atoms is gradually reduced to almost zero.

Slowing down the atoms increases their observation period. This in turn increases the oscillator Q because it effectively narrows the linewidth (Eq. 3.3). As we've continued to emphasize, oscillators with higher Qs are potentially more accurate and stable, so it's always advantageous to increase Q by either using an atomic transition where f_r is as high as possible, or by making Δf_r as narrow as possible [45]. Of course, f_r is the same for both cesium beam and fountain clocks, but fountain clocks have a narrower linewidth due to their longer atomic observation period. We can look at the advantage of a long observation period from another viewpoint, by remembering that atomic clocks work by locking a synthesized frequency to the atomic resonance frequency. The locking process involves continuously tuning the synthesized frequency to match f_r, and when the atoms are observed for a longer period, the tuning can be done more slowly and with less uncertainty.

Since 1998, the United States national primary frequency standard has been a cesium fountain clock. A series of cesium fountains, numbered in sequence as NIST-F1 through NIST-F4, have been built in Boulder, Colorado at the National Institute of Standards and Technology (NIST), which was the new name given to NBS in 1988. The first of the NIST cesium fountain clocks, called NIST-F1 (Figure 3.16), was designed and operated by a group of American physicists led by Steven Jefferts (b. 1959). The cesium fountain group originally included Dawn Meekhof (b. 1965), and later included Thomas Heavner (b. 1966) and Elizabeth Donley (b. 1970).

Figure 3.16 The NIST-F1 cesium fountain clock as it appeared in 2005 (courtesy of NIST).

NIST-F1 worked by laser cooling a sample of about 10^8 cesium atoms at the intersection of six laser beams (Figure 3.15). The atoms were launched upwards at a velocity of about 4 m/s to make a moving optical molasses, a ball of atoms that was about 1 cm in diameter. The atoms were typically in the $F = 4$ ground state, but all of the m_F levels were populated. A short microwave pulse drove the $|4, 0\rangle$ atoms into the |3, 0⟩ state, and left the other $F = 4$ atoms unperturbed. The remaining $F = 4$ atoms were then removed from the cloud with a short optical blast. At this point the remaining cesium atoms, all in the |3, 0⟩ state, entered the microwave cavity while moving at a velocity of about 3 m/s. After reaching a height of about 1 m above the cavity, the atoms fell back down due to the effects of gravity. The falling atoms then passed through the cavity again about 1 s after their first passage.

Just as Zacharias originally expected, this long interaction period allows cesium fountain clocks to be far more accurate than cesium beam clocks. Cesium fountain clocks have a linewidth of about 1 Hz and a Q factor of about 10^{10}. These numbers are about 100× better than commercial cesium beam clocks, and fountain clocks are at least 100× more accurate, with their best accuracies approaching 1×10^{-16}, or about 0.01 ns per day [55–58]. However, microwave atomic clock accuracy has essentially reached its limits, and any future accuracy gains for cesium fountain clocks will likely be small. All significant future advances in timekeeping accuracy will come from optical atomic clocks, covered in Chapter 13, which operate at frequencies about 100,000× higher than a cesium clock.

3.9 Rubidium Clocks

Rubidium clocks are by far the most common type of atomic clock, with many thousands of devices manufactured and sold every year. They are less stable and accurate than cesium clocks, but have the advantage of being smaller, less expensive, and more reliable, often operating for decades without any maintenance. These attributes allow rubidium clocks to be deployed in large numbers. They fly onboard global navigation satellite systems (GNSS), they're controlled by GNSS signals in disciplined oscillator and clock instruments (Chapter 12), and they're even embedded into test and measurement equipment as time base oscillators, when more accuracy than a quartz time base is required. Figure 3.17 shows a typical rubidium clock module with dimensions of 10.1 × 7.6 × 5.1 cm.

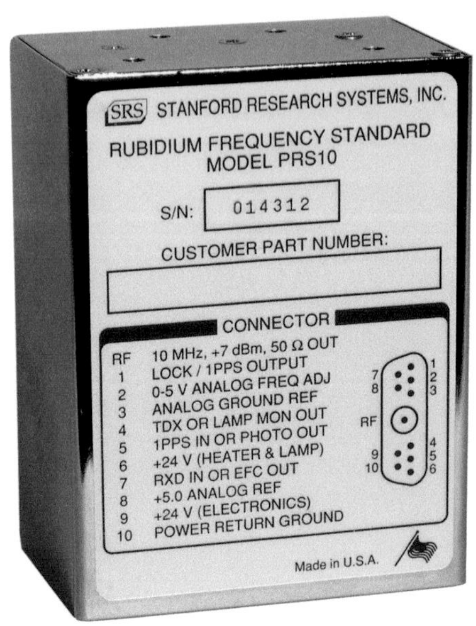

Figure 3.17 A rubidium oscillator module (courtesy of Stanford Research Systems).

Rubidium is an alkali metal that was also discovered by Bunsen and Kirchhoff in 1861, shortly after they discovered cesium. Its pair of spectral lines are red, and it was named based on the Latin word *rubidus*, which means bright or ruby red. Rubidium has five electron shells binding to the nucleus, and like cesium is well suited for use in an atomic clock, because its outer shell contains just a single, unpaired electron

(Figure 3.18). However, unlike cesium, where all naturally occurring atoms are identical, rubidium naturally occurs in a mixture of two isotopes. The two forms of rubidium contain the same number of protons (37) but have different numbers of neutrons in their nucleus, and they therefore differ in atomic mass. About 72% of rubidium naturally occurs as ^{85}Rb with 48 neutrons and a resonance frequency of about 3.035 GHz. The other 28% naturally occurs as ^{87}Rb, with two additional neutrons and a resonance frequency of about 6.835 GHz [40]. The resonance frequency for rubidium atomic clocks is obtained from the hyperfine transitions of ^{87}Rb, the less common of the two isotopes.

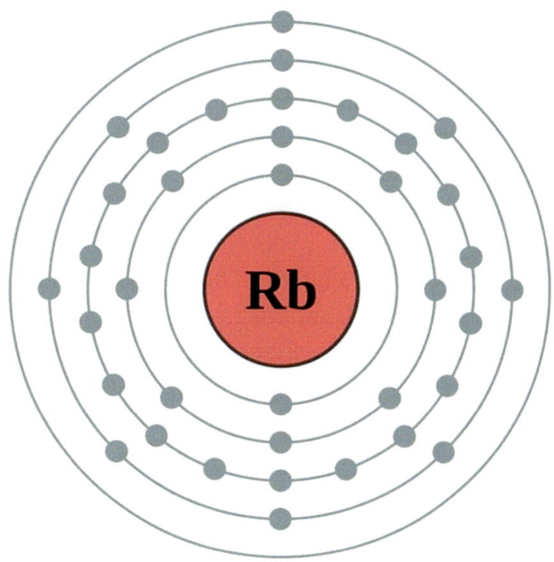

Figure 3.18 A rubidium atom, showing the nucleus and the rings of outer electrons.

Although rubidium and cesium are both alkali metals, the operation of a rubidium clock is quite different from that of a cesium beam clock. As we saw in Section 3.7, the cesium resonance frequency is obtained by applying a microwave frequency to a beam of atoms as they pass through a cavity, and then detecting which atoms have changed their trajectories and magnetic states. In contrast, rubidium clocks are vapor cell clocks—often called gas cell clocks—that employ a double-resonance technique. This means that they use both optical and microwave signals to detect the hyperfine energy transitions of atoms within a sealed gas cell.

Optical pumping, or the use of light to raise or "pump" the energy state of an atom from a lower level to a higher one, is fundamental to the operation of gas cell clocks. The concepts of optical pumping and double-resonance can be traced to experiments performed independently by the American physicist Francis Bitter (1902–1967) and the French physicist Alfred Kastler (1902–1984) in the late 1940s and early 1950s, which resulted in a Nobel Prize for Kastler in 1966 [23, 59]. The method of monitoring the intensity of light transmitted through a gas cell as a means of detecting the hyperfine energy state transitions, called optical detection, is also fundamental to gas cell clocks. The optical detection method can be traced to the mid-1950s work of another Nobel laureate, the German-American physicist Hans Dehmelt (1922–2017), who was then working with sodium atoms [60].

Much of the early work on gas cell atomic frequency standards was focused on devices based on cesium and sodium atoms. The first experiments directed towards developing a gas cell frequency standard based on ^{87}Rb atoms can be attributed to Thomas Carver (1929–1981), an American physicist at Princeton University. Carver utilized optical pumping beginning in about 1956, but he used microwave absorption to detect the hyperfine energy state transitions of the atoms, noting that an optical detection system was still being designed [61]. Shortly afterwards, an optical detection system that became the basis for future rubidium clock development was developed by the physicists Peter Bender (1930–2024) and Earl Beaty (1930–2018) of NBS, and by Andrew Chi (1920–1982) of NRL [62].

It's difficult to credit any individual as being the developer of the first complete rubidium frequency standard, but the strongest candidate for this distinction is probably Maurice Arditi (1913–2011), who was a physicist at the International Telephone and Telegraph Corporation (ITT) in New Jersey. Arditi, who was a frequent collaborator with Carver at nearby Princeton, led the development efforts for numerous gas cell frequency standards. His design efforts started with sodium, moved to cesium, and then finally moved to rubidium [59, 63].

Beginning in the 1960s, rubidium frequency standards were manufactured by companies including Varian Associates, General Technology Corporation [59, 64], and Hewlett-Packard [65]. They were originally manufactured in the form of large rack-mount instruments, but advances in electronics soon made rubidium clocks more compact and less expensive. These advances were motivated by the emergence of new technologies that required atomic clock accuracy—especially telecommunication networks. The demand for rubidium clocks has steadily increased for several decades and is continuing to grow.

Figure 3.19 provides a simplified diagram of a rubidium clock that's based on the double-resonance method, utilizing both optical and microwave signals. The main components of a rubidium clock are a discharge lamp, a filter cell, an absorption cell (sometimes called the resonance cell), a photo cell detector, and a quartz oscillator that drives a microwave frequency synthesizer. The lamp typically contains about 1 mg of ^{87}Rb, along with a noble gas such as krypton (Kr) or argon (Ar). It emits light containing twin red spectral lines, known as D1 and D2, with respective wavelengths of 795 nm and 780 nm. The light passes through a filter cell containing ^{85}Rb gas at lower temperatures, along with a noble gas. Due to differences in the hyperfine structures of the ground states of the two rubidium isotopes, the light that emerges from the filter will be absorbed by ^{87}Rb atoms in the lower hyperfine state (F = 1), but not by atoms in the upper hyperfine state (F = 2).

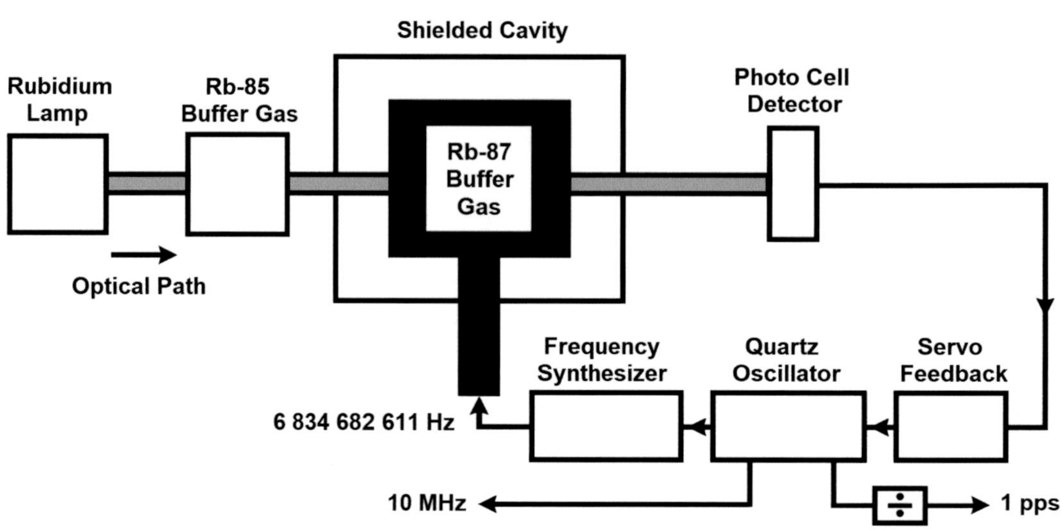

Figure 3.19 A simplified diagram of a rubidium clock.

After the light has gone through this hyperfine filtration process, it passes through a magnetically shielded absorption cell. The absorption cell contains ^{87}Rb along with a low-pressure buffer gas that reduces the number of collisions between the rubidium atoms and the cell walls. The continuous optical pumping raises the energy state of the atoms and would eventually bring all of them to their F = 2 state, making them unable to absorb light and causing the absorption cell to be lit and transparent. However, injecting a microwave signal into the absorption cell causes some of the atoms to return to their F = 1 state, allowing them to again absorb light, and causing the cell to go dark. The microwave signal is provided by a quartz-based frequency synthesizer. The synthesizer receives continuous feedback from a photo cell detector that measures how much light has passed through the absorption cell. Its frequency is then adjusted so that it stays locked to the rubidium resonance, which is the frequency where the maximum amount of light is absorbed, and where the minimum amount of light reaches the photo cell. Standard frequency and time signals are derived from the locked synthesizer and provided as outputs [23, 29, 40, 46, 47, 64].

The BIPM recognizes the ground-state, hyperfine transition frequency of the ^{87}Rb atom as a secondary representation of the second, with "secondary" meaning that its frequency must be calibrated with respect to a primary standard such as a cesium fountain before it can be used to realize the second. The recommended value, or the best approximation of the rubidium resonance frequency, is 6,834,682,610.9043126 Hz [66]. This value was obtained by averaging measurements that compared rubidium fountain clocks to cesium fountain clocks [67]. It differs by only a few hertz from the measurements that Bender and his colleagues made in the late 1950s [62].

Obtaining a Q factor that exceeds 10^7 for a rubidium clock requires a linewidth of less than 683 Hz, which is typically accomplished. Some devices narrow the linewidth to less than 200 Hz, and achieve Q factors greater than 3×10^7, which is an order of magnitude higher than the best quartz clocks but still much lower than cesium beam clocks.

Rubidium clocks can have excellent short-term stability, which will be further discussed in Chapter 7, but they typically have no guaranteed specification for accuracy. A typical rubidium clock is accurate to within parts in 10^{10} after a warm-up period of a few minutes, and some devices can achieve accuracies of parts in 10^{12}. However, rubidium clocks aren't suited for applications with strict accuracy requirements unless their frequency is periodically calibrated. Due to aging, their accuracy will change by about 5×10^{-11} per month until it eventually reaches a limit of about 5×10^{-9}, which means that timekeeping accuracy better than about 500 μs per day can't be assumed. This is at least 5,000× worse than the accuracy that can be assumed from a cesium beam clock, but at least 100× better than the accuracy that can be assumed for an unadjusted quartz clock.

3.10 Hydrogen Masers

Hydrogen masers are designed for applications that require the best possible frequency stability. They are much larger and far less common than rubidium or cesium beam clocks, and only a small number are sold each year. The cost of the most stable masers is typically three to four times higher than a cesium beam clock (about $300,000 USD as of 2025). Despite their high cost, hydrogen standards are not considered to be a primary standard, because the SI second is defined based on cesium. However, they still play a key role in the world's timekeeping infrastructure because their exceptional frequency stability greatly improves the performance of UTC, as discussed in Chapter 9. Figure 3.20 shows a commercially available hydrogen maser, with dimensions of 107 × 46 × 76 cm and a weight of 216 kg.

The word maser is now recognized as a common noun, but it originated as an acronym. It stands for "microwave amplification by stimulated emission of radiation," which is nearly identical to the words in the acronym for laser, except that "microwave" is substituted for "light." Although far less known, the maser preceded the laser by several years. The invention of the maser, and much of the theory that led to the laser, can be credited to the American physicist Charles Townes (1915–2015).

Townes conceived the idea of the maser while working at Columbia University in 1951. His goal wasn't to build a clock, but instead to build an amplifier that could generate and amplify electromagnetic waves in the microwave region of the radio spectrum. The prototype maser was built by Townes, James Gordon (1928–2013), and Herbert Zeiger (1925–2011). It was based on a beam of ammonia

Figure 3.20 A Microchip MHM 2020 active hydrogen maser (courtesy of Microchip).

molecules and first demonstrated in late 1953 after more than two years of work. Although the maser was simply an amplifier with enough positive feedback to make it oscillate, it was soon found that masers had characteristics that would make good atomic clocks. Masers produced a constant frequency and only oscillated in a narrow range that was within the linewidth of the particular atom or molecule chosen [68]. In their first publication about the new device, Townes, Gordon, and Zieger wrote that,

"As an oscillator, the apparatus promises to be a rather simple source of a very stable frequency." [69]

The maser concept was applied to atomic clocks using hydrogen atoms rather than ammonia molecules. In many ways, hydrogen (H) is ideally suited for use in an atomic clock. Hydrogen is the simplest and lightest of all atoms, having just one electron and a nucleus consisting of a single proton and no neutrons (Figure 3.21). Therefore, hydrogen has an atomic number of 1, and because there are no neutrons in its nucleus, its atomic mass is very close to 1, about 1.008 [70]. A hydrogen atom has a radius of 0.053 nm, much smaller than cesium, and as you might expect, the hyperfine transition of the hydrogen ground state is less complex than the cesium hyperfine transition that was previously shown in Figure 3.13.

For a hydrogen atom in its L = 0 ground state, there are just two possible values of total angular momentum, because the single proton nucleus and the electron each have a spin of $\frac{1}{2}$. In the higher energy state, the hyperfine level equals the proton spin plus the electron spin, $\frac{1}{2}+\frac{1}{2}$, or F = 1. In the lower energy state, the hyperfine level equals the proton spin minus the electron spin, $\frac{1}{2}-\frac{1}{2}$, or F = 0. Because the proton and electron have opposite charges, energy states with parallel spins have anti-parallel magnetic moments, and vice versa. Therefore, the higher energy F = 1 state occurs when the magnetic orientations are opposed and the spin is parallel. The lower energy F = 0 state occurs when the magnetic orientations are aligned and the spin is anti-parallel. As shown in Figure 3.22, the F = 1 level splits into three sublevels whose energy varies as a function of the applied magnetic field, but the hydrogen maser clock transition, denoted as $|1,0\rangle \leftrightarrow |0,0\rangle$, occurs where the applied magnetic field is near zero [29, 45].

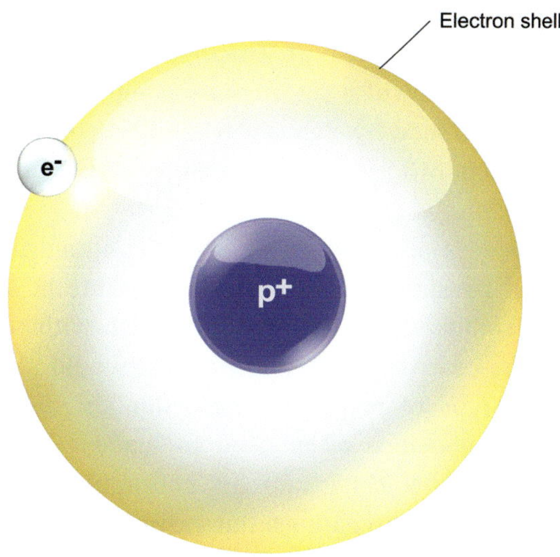

Figure 3.21 A hydrogen atom.

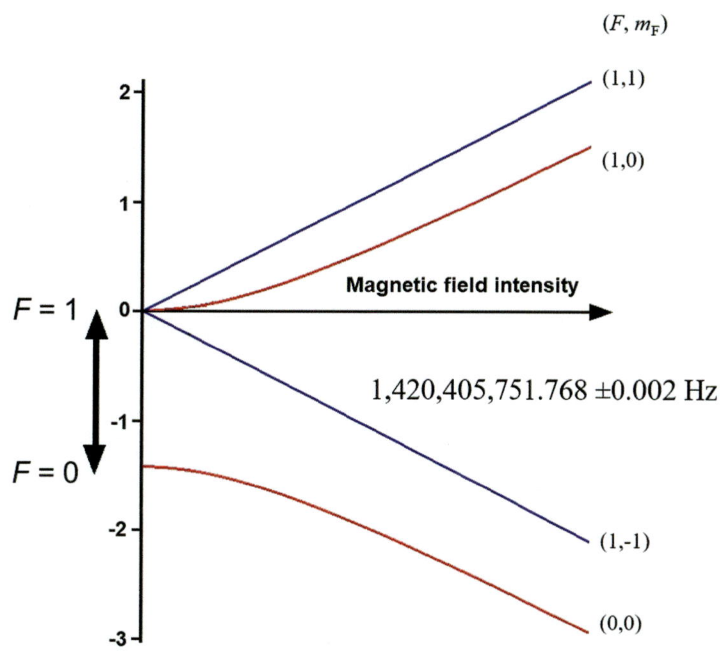

Figure 3.22 The hydrogen maser clock transition.

Interestingly, the hyperfine transition frequency of ground-state hydrogen atoms was known to astronomers long before hydrogen clock experiments began. It was known as early as the 1930s that signals thought to originate from galactic sources could produce a hiss or static that could be heard with radio receivers. In 1944, the Dutch astronomer Hendrik van de Hulst (1918–2000) predicted that neutral hydrogen could be the source of this radio noise, as it could produce radiation near 1420.4 MHz due to the hydrogen ground state transition. Referred to by its wavelength, the 21 cm hydrogen line was first detected at Harvard University in 1951 by the American physicist Edward Mills Purcell (1912–1997) and his student Harold Irving Ewen (1922–2015) [71]. Using a receiver and horn antenna that Ewen designed, the pair successfully detected 21 cm emissions from neutral hydrogen in the Milky Way, and their discovery helped launch the field of radio astronomy.

The hydrogen maser was invented at Harvard in 1960 by Norman Ramsey, Harold Mark Goldenberg (1935–1978), and Daniel Kleppner (1932–2025) [72]. The invention at least partially resulted from a search for methods to reduce the resonance linewidth by increasing the observation time of the atoms [6, 73]. This was the same goal previously pursued by Ramsey when he invented the separated oscillatory fields method in cesium beam clocks (Section 3.3), and that was later pursued by vertically launching laser-cooled atoms in cesium fountain clocks.

The maser design, however, used a different technique. It confined the atoms in what was originally called a "storage box" [72], which consisted of a quartz bulb inside of a tuned microwave cavity. Once inside the bulb, the atoms would collide with each other, and also with the walls of the bulb. This meant that the walls of the bulb had to be coated with a material that would allow the atoms to survive a large number of collisions without any type of magnetic interaction. In the original experiments, the walls were coated with paraffin, a wax substance often used to make candles or crayons, and observation periods of up to 0.3 s were recorded. Ramsey and his colleagues continued their experiments using cesium atoms and a variety of different bulb coatings, but cesium was found to not work well, because its atoms were easily polarized by collisions with other atoms and the bulb walls, and that caused unwanted magnetic state changes that made it difficult to measure the resonance frequency [45, 74]. They found that hydrogen atoms could survive far more collisions without polarization, allowing much longer storage periods, so they focused their experiments on hydrogen.

From there, hydrogen maser theory and development rapidly advanced at Harvard and was mostly in place by the end of 1961 [75]. By then, Ramsey and his colleagues had constructed a device with a cell containing molecular hydrogen gas that was heated until the molecules split into atoms. The atoms then passed through a state selecting magnetic field generated by a six-pole permanent magnet, called a hexapole magnet, with alternating north and south poles that were arranged around an axis. The magnetic field forced atoms in the upper $F = 1$ states (where m_F was either 1 or 0) into a quartz storage bulb housed inside a microwave cavity.

The insides of the bulb were now coated with Teflon, a material best known for its use in coating non-stick frying pans. Teflon, the brand name for polytetrafluoroethylene (PTFE), had been discovered by the chemist Roy Plunkett (1910–1994), while working at Dupont in 1938. After entering the bulb through a small aperture, the atoms would randomly collide with the inner walls of the bulb, but the Teflon coating prevented the atoms from being seriously perturbed by the collisions. The atoms were able to stay inside the bulb for about 1 s, a remarkably long period in an atomic context, undergoing about 100,000 collisions with the Teflon-coated walls before losing coherence and exiting the storage bulb through the same aperture where they entered.

Because the microwave cavity was tuned to be resonant at the hydrogen hyperfine transition frequency (~1.4204 GHz), some atoms in the storage bulb would drop from the $F = 1$ to the $F = 0$ energy state. When the atoms changed their energy state, they released radiation in the form of a photon, which was at the same frequency as the microwave radiation. The atoms that released photons stimulated other atoms to also release photons, amplifying the signal and causing the cavity to oscillate. The atomic radiation that was generated was then coupled to a loop or antenna inside the cavity, producing an output signal that had enough power to control the frequency of a quartz oscillator [6, 45, 75–79]. From this, we can see that the words contained in the maser acronym, "microwave amplification by stimulated emission of radiation," are self-descriptive.

Varian Associates in Beverly, Massachusetts, working closely with Ramsey and Kleppner at nearby Harvard [79], began manufacturing commercial hydrogen masers in the mid-1960s [64, 80]. Shortly afterwards, an extensive hydrogen maser research and development program began at Goddard Space Flight Center of the National Aeronautics and Space Administration (NASA) in 1966 [81]. A central figure in hydrogen maser design and manufacturing was the American engineer Harry Peters (1923–2010), who played a key role in hydrogen maser research and development at both Varian and NASA. Peters retired from NASA in 1975 and founded a company dedicated to hydrogen maser development, the Sigma Tau Standards Corporation in Tuscaloosa, Alabama. Through a series of acquisitions and mergers, Sigma Tau is now part of Microchip, and its masers remain in production as of 2025 (Figure 3.20).

The original masers built at Harvard, the prototype masers built at NASA, and the commercial masers manufactured by Varian, Sigma Tau, and Microchip, can all be classified as active hydrogen masers. As discussed, masers work because atoms in a microwave cavity that is resonant at the hydrogen hyperfine transition frequency will change their energy state and emit atomic radiation, amplifying the signal. The amount of amplification depends on the number of atoms that emit radiation. If enough atoms emit radiation to exceed a level called the atomic flux threshold, the amplification will be large enough to make the cavity self-oscillate. The cavity will continue to oscillate as long as the atomic flux threshold is exceeded, or, in other words, as long as the atomic beam carries enough energy to compensate for the cavity's energy losses. A quartz-based synthesizer is then locked to the frequency of the cavity's self-oscillations.

A second type of hydrogen maser doesn't rely on self-oscillation of the cavity. These devices, called passive hydrogen masers, allow the use of lower atom densities. Because the atomic beam has less energy, the atomic flux threshold isn't exceeded, and self-oscillation doesn't occur. However, the atomic signal is still amplified when an external microwave signal at the hydrogen hyperfine transition frequency is fed to the cavity. Then, in a fashion similar to the way that a rubidium or cesium beam clock works, the synthesizer is tuned to the microwave frequency where the largest number of atoms are changing state and releasing radiation, which is simply the frequency where the power level inside the cavity is at its peak. Passive hydrogen maser prototypes were constructed at NBS in the 1970s, originally built with a storage beam tube that had both an entrance and an exit aperture [82], and then later with a single-aperture storage bulb [83]. The production of passive hydrogen masers for industrial use began in Russia in the late 1980s [84].

Active hydrogen masers have the advantage of deriving their frequency directly from the atomic radiation, which makes them more stable than passive masers in both the short and long term. However, passive masers have the advantage of being smaller and less expensive to build, and they are now flown in space onboard the Galileo (Section 12.8.2) and BeiDou (Section 12.8.3) navigation satellites.

Figure 3.23 provides a simplified diagram of an active hydrogen maser. The box labeled as the "hydrogen pump" allows a controlled amount of gas to leak from a storage bottle of molecular hydrogen (H_2) to a source discharge bulb. The bulb is inside of an electric field tuned to a frequency of about 100 to 200 MHz. Through a chemical process known as disassociation, the discharge bulb splits or separates the molecules into individual hydrogen atoms. The atoms emerge through a small elongated hole called the source collimator and then pass through a magnetic field, formed by a permanent multipole magnet, such as the hexapole magnet described earlier. The magnetic field forms a state selection gate that separates the atoms in the $F = 1$ state from atoms in the $F = 0$ state, allowing the atoms in the upper energy state to enter the storage bulb. The storage bulb is typically made of fused quartz and its insides are coated with Teflon or a similar material. The diameter of the bulb is typically 10 to 15 cm. The bulb resides within a microwave cavity, typically made of copper or a silver-plated ceramic, that is resonant at the hydrogen transition frequency. The axis of the cavity lies along the axis of the atomic beam. The cavity is protected from electromagnetic interference by several shielding layers, and from temperature variations by a thermal control system [29, 45, 77].

The atoms that enter the storage bulb stay for about 1 s. During this period, some atoms drop to a lower energy level, releasing photons of microwave frequency. These photons stimulate other atoms to drop their energy level, and they in turn release additional photons. In this manner, a self-sustaining microwave field builds up in the bulb and causes the cavity to oscillate. A small loop couples the microwave signal generated by the cavity to a heterodyne receiver system. The output signal received from the cavity is used to lock the output of frequency synthesizer, driven by a quartz oscillator, to the hydrogen resonance frequency. Through the use of frequency multipliers and dividers, standard frequency and time signals such as 10 MHz and 1 pps are derived from the locked quartz oscillator and provided as outputs.

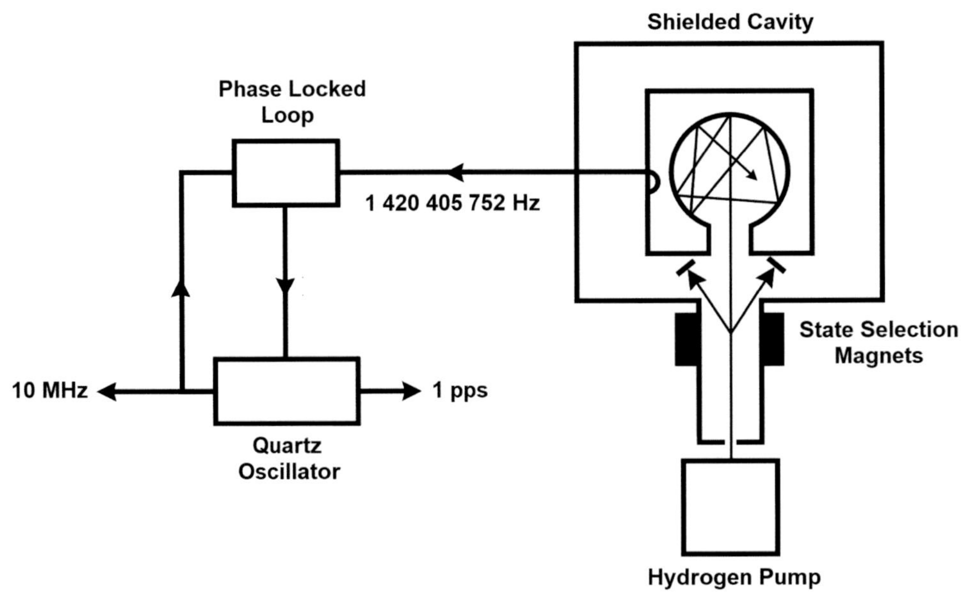

Figure 3.23 A simplified diagram of an active hydrogen maser.

The use of active and passive hydrogen masers isn't limited to serving as frequency standards in metrology laboratories. They also serve a variety of space science applications, including spacecraft tracking and detection, global navigation satellite systems, and radio astronomy [85]. Their stability exceeds all other commercially available frequency standards at short and medium intervals. For example, at an averaging period of 10,000 s, the frequency stability of the active hydrogen maser pictured in Figure 3.20 is specified as 2×10^{-15}, which is about 40× more stable than the cesium beam clock pictured in Figure 3.12 during the same averaging period. Over the course of a year, however, the frequency stability of hydrogen masers has been shown to be limited to about 1×10^{-14}, whereas some cesium beam clocks can reach stabilities of parts in 10^{15} [86]. This indicates that the frequency of an active maser's tuned microwave cavity changes slowly over time due to a complex set of factors. It also might relate to the fact that hydrogen masers lack inherent accuracy, because the definition of the second is based on cesium resonance.

Unlike cesium and rubidium, the BIPM doesn't publish a definition for the hydrogen maser resonance frequency, as it is not considered to be a secondary representation of the second. However, the hydrogen resonance frequency was measured at NBS as 1,420,405,751.768 ±0.002 Hz, the value shown in Figure 3.22, during the period from November 1969 to February 1970 [87]. The hydrogen resonance frequency was again measured at the USNO as 1,420,405,751.7755 ±0.0031 Hz from February to April of 1972 [88]. When the uncertainties are applied, these measurements agree to within a few millihertz, and they remain in agreement with the resonance frequency specified by the manufacturers of today's commercial standards.

The resonance frequency of hydrogen is much lower than that of rubidium or cesium, but the linewidth of a hydrogen of a hydrogen maser is usually near 1 Hz, which is similar to a cesium fountain. Therefore, the Q factor for a hydrogen maser is about 10^9, about one order of magnitude better than a cesium beam clock, but about one order of magnitude worse than a cesium fountain.

3.11 Chip-Scale Atomic Clocks

Atomic clocks that were smaller, consumed less power, and dissipated less heat than any previous designs became commercially available in 2011, after about two decades of research and development [89]. These devices, known as either chip-scale atomic clocks (CSAC) or miniature atomic clocks (MAC), use either cesium or rubidium atoms as their source of atomic resonance. Due to their small size and low power consumption, CSACs have introduced atomic clock stability to instruments and applications that were previously limited to using quartz clocks. The cesium-based CSAC pictured in Figure 3.24 has dimensions of about 4.1 × 3.6 × 1.1 cm and weighs just 35 g.

Figure 3.24 A chip-scale atomic clock (courtesy of Microchip).

The development of the CSAC was funded by the Defense Advanced Research Projects Agency (DARPA), a division of the United States Department of Defense (DoD). Their goal was to develop a small, energy-efficient atomic clock for use in portable equipment, envisioning its use in both consumer and military applications. For example, in situations where Global Positioning System (GPS) disciplined clocks (Chapter 12) were unable to receive satellite signals, a CSAC could continue to provide time signals accurate enough to meet the requirements of the cellular telephone industry. In addition, a CSAC could meet the accuracy requirements of soldiers carrying battery-powered communication and navigation devices [90].

DARPA began their Miniature Time Standard program in the early 1990s, and the first attempts at building CSAC prototypes took place at Westinghouse in Pittsburgh, Pennsylvania in 1991. Like the smallest atomic clocks that were then in existence, the rubidium clocks described in Section 3.9, the CSAC was also designed as an optically pumped, vapor cell clock. The Westinghouse researchers recognized that a rubidium vapor cell clock could have potential advantages in stability, but still chose cesium due to its smaller cavity size [91]. To further reduce the size and power consumption, the researchers first replaced the bulky discharge lamp with a diode laser as the optical pumping source. Later, they settled on using a vertical-cavity surface-emitting laser (VCSEL), due to its lower power consumption [92].

The VCSEL became a standard component in future CSAC designs, and its power consumption was eventually reduced to just a few milliwatts. VCSELs are semiconductor diode lasers that emit a laser beam perpendicular to their top surface, and they are commonly found in computer products such as optical mice and laser printers. In a CSAC, the laser is emitted upward through the vapor cell, and the laser wavelength matches either the D1 or D2 optical transition frequency. As was described in Section 3.9, the rubidium D1 and D2 lines have respective wavelengths of 795 nm and 780 nm. For cesium, the respective D1 and D2 wavelengths are 895 nm and 852 nm. Both cesium and rubidium CSACs are more likely to use the D1 line due to a higher signal-to-noise ratio and a narrower linewidth [93, 94].

The next advance in CSAC technology was eliminating the microwave cavity. The cavity limited how small the clock could be, because the dimensions of the cavity needed to be half of the wavelength of the microwave frequency in order to be resonant, which was about 3.3 cm for cesium or 4.4 cm for rubidium [90]. In 1992, researchers at the Centre for Optics, Photonics and Lasers (COPL) at Laval University in Canada proposed a new method where the ground state resonance of rubidium atoms was observed entirely by optical means. Their method, accomplished by frequency modulating a diode laser, suggested that the microwave cavity could be eliminated [95].

Building CSACs without a microwave cavity was made possible by the use of coherent population trapping (CPT), a phenomenon first observed in 1976 by researchers at the University of Pisa in Italy [96]. Experimental research conducted in the 1980s had explored the use of CPT in atomic clocks, primarily involving sodium beams [97], but the work done at COPL was more applicable to future CSAC development because it involved rubidium vapor cell clocks. Later in the 1990s, the Canadian physicist Jacques Vanier (b. 1934) led research efforts that applied CPT to cesium vapor cell clocks [98, 99].

A CPT atomic clock works by interrogating the atoms in the vapor cell with two phase-coherent laser beams. In the case of the CSAC, the VCSEL allows both laser beams to be generated from the same source, because its modulation bandwidth is greater than 10 GHz. The two laser beams can be thought of as sidebands on the optical carrier, generated by frequency modulating the injection or drive current of the VCSEL with a microwave synthesizer. The synthesizer produces signals at a frequency equal to one half of the ground state transition frequency of the atom, and these signals are capacitively coupled to the VCSEL current leads to modulate the laser. The resulting optical signal is frequency modulated, by about ±4.6 GHz for cesium or ±3.4 GHz for rubidium.

When the two sidebands on the optical carrier are exactly separated by the ground state transition frequency of the atom, about 9.2 GHz for cesium or 6.8 GHz for rubidium, the atoms are placed in a quantum superposition of two hyperfine states, known as the dark state. The population of atoms "trapped" in the dark state are decoupled from both optical fields, which means that they don't emit or absorb any light. Because light is not being absorbed, the power transmitted through the vapor cell increases. The increase in power leads to a detectable resonance frequency that is used to lock a quartz oscillator [90, 98, 100].

Encouraged by the technical advancements made in the 1990s, DARPA funded a CSAC program in four phases from 2000 to 2010. Competitive contracts were awarded to multiple organizations, a group that included NIST, Symmetricom (now Microchip), Draper Laboratory, Sandia National Laboratories, Agilent, Teledyne Scientific, and Honeywell. The DARPA program goals were to develop and mass-produce atomic clocks with a volume of less than one cubic centimeter (1 cm^3), a total power consumption of less than 30 mW, and a frequency stability of less than 1×10^{-11} at an averaging period of one hour [93]. These exceptionally ambitious goals have not been reached as of 2025, but their pursuit led to the commercialization and mass production of CSACs.

The mass production of CSACs required a physics package that could be manufactured at reasonable costs using standard semiconductor fabrication techniques. The physics package is the heart of the CSAC, and it includes the VCSEL and the other optical components used to excite the atoms, the vapor cell where the atoms are stored, and the photodiode that detects the atomic resonance [90]. Ideally, the design would be simple enough to either fabricate the entire physics package on a single wafer, or to fabricate the individual components on wafers that are then integrated in a multi-layer stack using micro-electro-mechanical systems (MEMS) technology. In general terms, MEMS devices, sometimes called micro-machined devices, are miniaturized electro-mechanical devices that are made using microfabrication techniques.

One main criterion of MEMS is that at least some elements of the device will have mechanical functionality, regardless of whether or not those elements can actually move.

Work leading towards a MEMS-based physics package began at NIST in 1999, performed by a team of physicists that included Leo Hollberg (b. 1952), John Kitching (b. 1968) and Svenja Knappe (b. 1973), who was then with the Institute of Applied Physics at the University of Bonn in Germany. The NIST team demonstrated a VCSEL-driven cesium vapor device in 2000, noting that a physics package based on their design "could be compact, low power, and simple to implement" [101]. Shortly afterwards, Hollberg and Kitching patented a conceptual design for a miniature atomic clock with a micro-machined vapor cell [102].

A MEMS-based physics package (Figure 3.25) was fabricated and demonstrated at NIST in 2004. The device's components were integrated as vertically stacked layers. A VCSEL was mounted at the bottom of the device, on a substrate of gold traces. A series of optics layers were positioned over the VCSEL. The optics layers attenuated the power of the laser beam to about 10 µW, collimated the beam to a diameter of about 250 µm, and converted the light polarization from linear to circular. The cesium vapor cell and its associated cell heaters were mounted on top of the optics assembly. The external dimensions of the cell were 1.5 × 1.5 × 1.4 mm. The top layer of the stack was a photodiode assembly, that monitored the light transmitted by the cell. The complete physics page was 4.2 mm tall with a square cross-section of 1.5 mm, resulting in a total volume of less than 10 mm^3. It operated on 75 mW of electrical power, dominated by the power required to heat the vapor cell [90, 103].

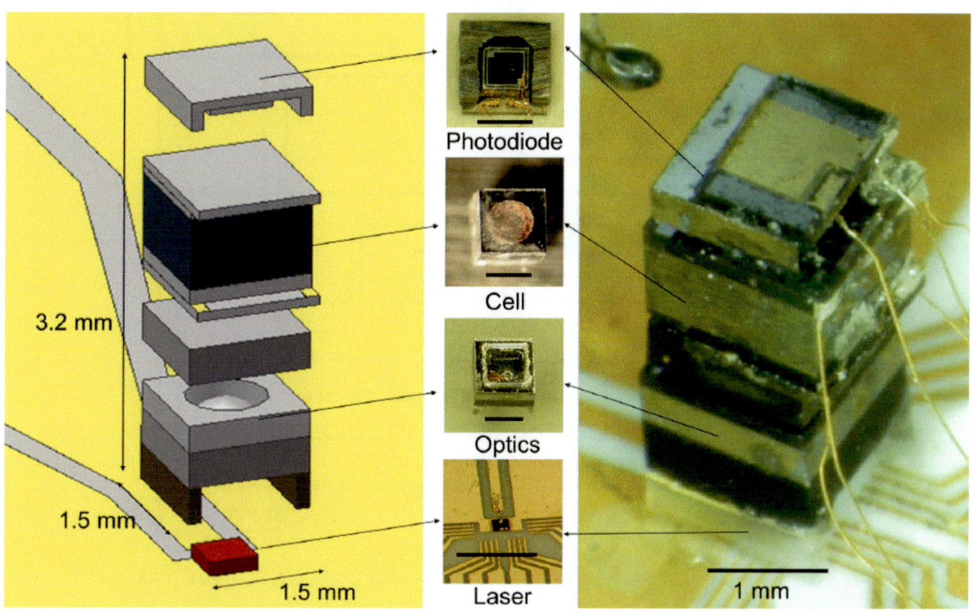

Figure 3.25 A MEMS-based CSAC physics package developed at NIST in 2004.

Later that same year, the combined development efforts at Symmetricom, Draper Laboratory, and Sandia National Laboratories, yielded a physics package that consumed less than 10 mW of power. After the laser beam from the VCSEL passed through the cesium vapor cell, it was reflected from a mirrored surface, and then passed through the cell a second time before reaching the photodetector. Because the dimensions of the vapor cell were matched to the natural divergence of the VCSEL output beam, the optical layer needed to collimate the laser beam was eliminated, reducing the size of the physics package [104].

CSACs became commercial products in 2011 when they were first sold by Symmetricom, and the American physicist Robert Lutwak (b. 1965) played a central role in their commercialization. Lutwak served as the principal investigator of the CSAC development program at Symmetricom, where he worked from 1998 to 2013. He later became a DARPA program manager, returning to Microchip (who had acquired Symmetricom) in 2019.

A simplified diagram of a complete CSAC is shown in Figure 3.26. In addition to the physics package, the CSAC also includes an RF package and an electronics package, both optimized for small size and low power consumption. The RF package includes an oscillator, or an oscillator combined with a frequency synthesizer, that supplies a microwave signal to the VCSEL. We can think of this device as the laser oscillator. It's tuned to a subharmonic of the hyperfine frequency, typically the first subharmonic (4.6 GHz for cesium or 3.4 GHz for rubidium), and its tuning range is wide enough to compensate for frequency variations due to aging or drift. It also outputs enough power, typically between 0.1 mW and 1 mW, to drive the VCSEL.

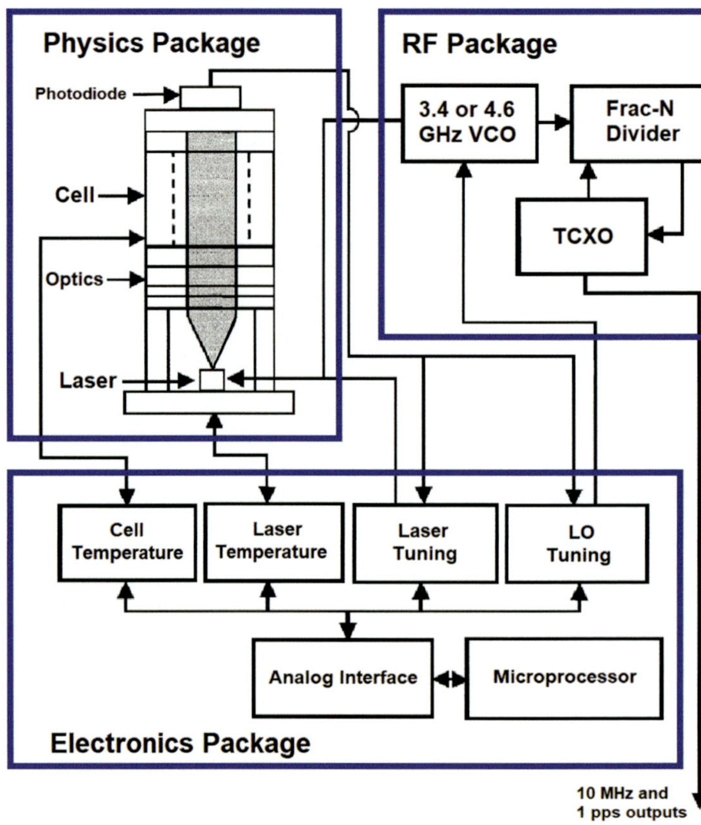

Figure 3.26 A simplified diagram of a rubidium chip-scale atomic clock (CSAC).

The heated cell vaporizes the cesium or rubidium alkali metal that it contains, and the VCSEL optically pumps the atoms by shining light through the cell. The input current to the VCSEL is modulated, which splits its output into two phase-coherent beams, or optical subcarriers. When the difference in frequency between the two optical subcarriers equals the hyperfine transition frequency of the atoms in the cell, the CPT phenomenon occurs and traps some of the atoms in their dark state. A photodiode on the other side of the vapor cell serves as a detector, by monitoring the power of the light that passes through the cell, and sending that information to a control loop in the CSAC's electronics package. When the power level increases, it means that more atoms are in their dark state; therefore, less light is being absorbed. The frequency of the laser is tuned, by either varying the laser's current or temperature, to the peak power level reported by the photodiode. A second oscillator in the RF package, typically a 10 MHz TCXO, is then phase locked to the laser oscillator with a control loop that includes frequency dividers.

As is the case with all of the atomic clocks described in this chapter, the locked quartz oscillator provides the CSAC with its output signals, which are typically 10 MHz and 1 pps [90].

Commercial CSACs, such as the device pictured in 3.24, have been made available with either cesium or rubidium vapor cells. The cesium devices are slightly smaller, weighing from about 35 to 75 g, with volumes ranging from about 17 to 45 cm^3. The smallest cesium-based CSACs are just slightly larger than a matchbox, which for comparison has a volume of about 15 cm^3. The rubidium devices weigh from about 75 to 100 g with volumes ranging from about 32 to 50 cm^3. The cesium devices also use much less power, ranging from about 120 to 300 mW, as opposed to 1.2 to 5 W for rubidium devices.

The small size and low power consumption of a CSAC provides obvious advantages, but it can compromise performance. When compared to a larger rubidium vapor cell clock, such as the one pictured in Figure 3.17, the use of a smaller cell increases the collisional rate of the atoms with the cell walls and with the buffer gas. These collisions broaden the linewidth and make the frequency less stable. The conventional rubidium clocks described in Section 3.9 are about twice as stable as rubidium-based CSACs, in both the short and long term, and about 10× more stable than cesium-based CSACs. However, due to strong demand and continued research efforts, future CSAC devices will likely be more stable and consume even less power [105].

Now that we've provided an overview of mechanical, quartz, and atomic clocks, we'll turn our attention to time and frequency measurements in Chapters 4 and 5. The use and performance of quartz and atomic clocks will continue to be discussed in those chapters, and in all of the remaining chapters of this book.

References

1. Bureau International des Poids et Mesures, "*The International System of Units (SI)*," 9th edition, ver. 2.01, December 2022.
2. W. F. Snyder, "Lord Kelvin on Atoms as Fundamental Natural Standards (for Base Units)," *IEEE Transactions on Instrumentation and Measurement*, vol. IM-22, p. 99, March 1973.
3. W. Thomson and P. G. Tait, *Elements of Natural Philosophy*, 2nd edition, pp. 61–62, Cambridge Press, 1879.
4. I. I. Rabi, J. R. Zacharias, S. Millman, and P. Kusch, "A New Method of Measuring Nuclear Magnetic Moment," *Physical Review*, vol. 53, p. 318, February 1938.
5. N. F. Ramsey, *Molecular Beams*, Clarendon Press, Oxford, 1956.
6. N. F. Ramsey, "History of Atomic Clocks," *Journal of Research of the National Bureau of Standards*, vol. 88, pp. 301–320, September-October 1983.
7. W. L. Laurence, "Cosmic Pendulum for Clock Planned," *New York Times*, p. 34, January 21, 1945.
8. P. Forman, "Atomichron™: The Atomic Clock from Concept to Commercial Product," *Proceedings of the IEEE*, vol. 73, pp. 1181–1204, July 1985.
9. S. Millman and P. Kusch, "On the Radiofrequency Spectra of Sodium, Rubidium, and Caesium," *Physical Review*, vol. 58, p. 438–445, September 1940.
10. P. Forman, "The first atomic clock program: NBS, 1947–1954," *Proceedings of the 17th Annual Precise Time and Time Interval Meeting (PTTI)*, pp. 1–17, December 1985.
11. National Bureau of Standards, "The Atomic Clock: An Atomic Standard of Frequency and Time," *NBS Technical Report 1320*, 27 p., January 1949.
12. H. Lyons, "The Atomic Clock," *Instruments*, vol. 22, pp. 133–135, December 1949.
13. H. Lyons, "Spectral Lines as Frequency Standards," *Annals of the New York Academy of Sciences*, vol. 55, pp. 831–871, November 1952.
14. H. Lyons and B. F. Huston, "Atomic Clock," *United States Patent 2,699,503* (Applied for April 30, 1949, granted January 11, 1955).

15. H. Lyons, "Microwave Frequency Dividers," *Journal of Applied Physics*, vol. 21, pp. 59–60, January 1950.
16. H. Lyons, "The Atomic Clock: A Universal Standard of Frequency and Time," *American Scholar*, vol. 19, pp. 159–168, April 1950.
17. D. A. Howe, "Velocity Distribution Methods of Cesium Beam Tubes," *Proceedings of the 30th Annual Symposium on Frequency Control*, pp. 451–456, June 1976.
18. S. K. Karuza, W. A. Johnson, M. F. Bottjer, and F. J. Voit, "An Automated System for Measuring Velocity Distribution in Cesium Beam Tubes," *Aerospace Corporation Report TR-0090(5925-08)-1*, 23 p., September 1992.
19. J. E. Sherwood, H. Lyons, R. H. McCracken, and P. Kusch, "High Frequency Lines in the hfs Spectrum of Cesium," *Bulletin of the American Physical Society*, vol. 27, p. 43, 1952.
20. N. F. Ramsey, "A Molecular Beam Resonance Method with Separated Oscillating Fields," *Physical Review*, vol. 78, pp. 695–699, June 1950.
21. H. Lyons, "Spectral Lines as Frequency Standards," *National Bureau of Standards Report 1848*, 71 p., August 8, 1952.
22. W. F. Snyder and C. L. Bragaw, "Achievement in Radio: Seventy Years of Radio Science, Technology, Standards, and Measurement at the National Bureau of Standards," *National Bureau of Standards Special Publication 555*, October 1986.
23. R. E. Beehler, "A Historical Review of Atomic Frequency Standards," *Proceedings of the IEEE*, vol. 55, pp. 792–805, June 1967.
24. R. C. Mockler, R. E. Beehler, and C. S. Snider, "Atomic Beam Frequency Standards," *IRE Transactions on Instrumentation*, vol. I-9, pp. 120–132, September 1960.
25. R. C. Mockler, R. E. Beehler, and J. A. Barnes, "An evaluation of a cesium beam frequency standard," *National Bureau of Standards Report 6075*, October 1959.
26. R. E. Beehler and D. J. Glaze, "The Performance and Capability of Cesium Beam Frequency Standards at the National Bureau of Standards," *IEEE Transactions on Instrumentation and Measurement*, vol. 15, no. 1/2, pp. 48–55, March-June 1966.
27. D. Henderson, "Essen and the National Physical Laboratory's Atomic Clock," *Metrologia*, vol. 42, no. 3, pp. S4–S9, June 2005.
28. L. Essen and J. V. L. Parry, "An Atomic Standard of Frequency and Time Interval: A Caesium Resonator," *Nature*, vol. 176, pp. 280–282, August 1955.
29. C. Audoin and B. Guinot, *The Measurement of Time: Time, Frequency and the Atomic Clock*, Cambridge University Press, 2001.
30. D. D. McCarthy and P. K. Seidelmann, *Time: From Earth Rotation to Atomic Physics*, Cambridge University Press, 2nd edition, 2018.
31. W. Markowitz, "Photographic Determination of the Moon's Position, and Applications to the Measure of Time, Rotation of the Earth, and Geodesy," *The Astronomical Journal*, vol. 59, pp. 69–73, March 1954.
32. E. C. Bullard, "An Atomic Standard of Frequency and Time Interval: Definition of the Second of Time," *Nature*, vol. 176, p. 282, August 1955.
33. S. Leschiutta, "The Definition of the 'Atomic' Second," *Metrologia*, vol. 42, no. 3, pp. S10–S19, June 2005.
34. W. Markowitz, R. Glenn Hall, L. Essen, and J. V. L. Parry, "Frequency of Cesium in Terms of Ephemeris Time," *Physical Review Letters*, vol. 1, no. 3, pp. 105–107, August 1958.
35. J. Terrien, "A Major Step towards the Redefinition of the Second, the Unit of Time in the International System of Units," *Metrologia*, vol. 3, no. 4, p. 130, October 1967.
36. Resolution 1 of the 13th Conference Generale des Poids et Mesures (CGPM), 1967.
37. W. M. Itano, L. L. Lewis, and D. J. Wineland, "Shift of $^2S_{1/2}$ hyperfine splittings due to blackbody radiation," *Physical Review A*, vol. 25, 1233–1235, February 1982.
38. B. Guinot, "History of the Bureau International de l'Heure," Polar Motion: Historical and Scientific Problems, IAU Colloquium 178, vol. 208, pp. 175–184, 2000.

39. D. D. McCarthy, C. Hackman, and R. A. Nelson, "The Physical Basis of the Leap Second," *The Astronomical Journal*, vol. 136, no. 5, pp. 1906–1908, November 2008.

40. T. Jones, *Splitting the Second: The Story of Atomic Time*, Bristol and Philadelphia: Institute of Physics Publishing, 2000.

41. "Atomichron – World's Most Accurate Clock," *Radio and Television News*, p. 63, January 1957.

42. L. S. Cutler, "Fifty Years of Commercial Caesium Clocks," *Metrologia*, vol. 42, no. 3, pp. S90–S99, June 2005.

43. A. O. McCoubrey, "Results of the Comparison: Atomichron – British Cesium Beam Standard," IRE Transactions on Instrumentation, vol. 1-7, no. 3/4, pp. 203–206, December 1958.

44. A. S. Bagley and L. S. Cutler, "A Modern Solid-State Portable Cesium Beam Frequency Standard," *Proceedings of the 18th Annual Symposium on Frequency Control*, pp. 344–365, May 1964.

45. F. G. Major, The Quantum Beat: The Physical Principles of Atomic Clocks, Springer-Verlag, New York, 1998.

46. W. M. Itano and N. F. Ramsey, "Accurate Measurement of Time," Scientific American, vol. 269, pp. 56–65, July 1993.

47. J. Jespersen and J. Fitz-Randolph, "From Sundials to Atomic Clocks: Understanding Time and Frequency," National Institute of Standards and Technology Monograph 155, 304 p., 1999 edition, March 1999.

48. N. F. Ramsey, "History of Early Atomic Clocks," *Metrologia*, vol. 42, no. 3, pp. S1–S3, June 2005.

49. S. R. Jefferts, T. P. Heavner, and E. A. Donley, "Cesium Primary Frequency References," *Japanese Journal of Applied Physics*, vol. 43, pp. 2803–2807, May 2004.

50. M. A. Kasevich, E. Riis, S. Chu, and R. G. DeVoe, "rf Spectroscopy in an Atomic Fountain," *Physical Review Letters*, vol. 63, pp. 612–615, August 1989.

51. K. Gibble and S. Chu, "Laser-Cooled Cs Frequency Standard and a Measurement of the Frequency Shift due to Ultracold Collisions," *Physical Review Letters*, vol. 70, no. 12, pp. 1771–1774, March 1993.

52. A. Clairon, P. Laurent, G. Santarelli, S. Ghezali, S. N. Lea, and M. Bahoura, "A Cesium Fountain Frequency Standard: Preliminary Results," *IEEE Transactions on Instrumentation and Measurement*, vol. 44, no. 2, pp. 128–131, April 1995.

53. Laser cooling was separately proposed in 1975 for both a gas of neutral atoms and for ions confined in an electromagnetic trap. See T. W. Hänsch and A. L. Schawlow, "Cooling of Gases by Laser Radiation," *Optics Communications*, vol. 13, no. 1, pp. 68–69, January 1975. Also see D. J. Wineland and H. G. Dehmelt, "Proposed 10^{14} $\Delta\nu < \nu$ Laser Fluorescence Spectroscopy on $T1^{+}$ Mono-Ion Oscillator III," *Bulletin of the American Physical Society*, vol. 20, p. 637, 1975.

54. D. J. Wineland, R. E. Drullinger, and F. L. Walls, "Radiation Pressure Cooling of Bound Resonant Absorbers," *Physical Review Letters*, vol. 40, pp. 1639–1642, June 1978.

55. S. R. Jefferts, D. M. Meekhof, J. H. Shirley, and T. E. Parker, "Preliminary Accuracy Evaluation of a Cesium Fountain Primary Frequency Standard at NIST," *Proceedings of the 1999 IEEE Frequency Control Symposium*, pp. 12–15, April 1999.

56. T. P. Heavner, T. E. Parker, J. H. Shirley, and S. R. Jefferts, "NIST F1 and F2," *Proceedings of the 7th Symposium on Frequency Standards and Metrology*, pp. 299–307, October 2008.

57. S. R. Jefferts, J. E. Shirley, T. E. Parker, T. P. Heavner, D. M. Meekhof, C. W. Nelson, F. Levi, G. Costanzo, A. DeMarchi, R. E. Drullinger, L. Hollberg, W. D. Lee, and F. L. Walls, "Accuracy evaluation of NIST-F1," Metrologia, vol. 39, no. 4, pp. 321–336, August 2002.

58. T. P. Heavner, S. R. Jefferts, E. A. Donley, J. H. Shirley, and T. E. Parker, "NIST-F1: Recent improvements and accuracy evaluations," *Metrologia*, vol. 42, no. 5, pp. 411–422, September 2005.

59. W. J. Riley, "A History of the Rubidium Frequency Standard," IEEE UFFC-S History (on-line PDF document available from https://ieee-uffc.org/media/history-rubidium-frequency-standard), 35 p., 2019.

60. H. G. Dehmelt, "Slow Spin Relaxation of Optically Polarized Sodium Atoms," *Physical Review*, vol. 105, no. 5, pp. 1487–1489, March 1957.

61. T. R. Carver, "Rubidium Oscillator Experiments," *Proceedings of the 11th Annual Symposium on Frequency Control*, pp. 307–317, May 1957.

62. P. L. Bender, E. C. Beaty, and A. R. Chi, "Optical Detection of Narrow Rb87 Hyperfine Absorption Lines," *Physical Review Letters*, vol. 1, no. 9, pp. 311–313, November 1958.

63. M. Arditi and T. R. Carver, "The Principles of the Double Resonance Method Applied to Gas Cell Frequency Standards," *Proceedings of the IEEE*, vol. 51, no. 1, pp. 190–202, January 1963.

64. A. O. McCoubrey, "A Survey of Atomic Frequency Standards," *Proceedings of the IEEE*, vol. 14, no. 2, pp. 116–135, February 1966.

65. D. H. Throne, "A Rubidium-Vapor Frequency Standard for Systems Requiring Superior Frequency Stability," *Hewlett-Packard Journal*, vol. 19, no. 11, pp. 8–16, July 1968.

66. BIPM Consultative Committee for Time and Frequency, "Rubidium 87 Atom ($f \sim 6.835$ GHz)," *Recommended Values of Standard Frequencies for Applications Including the Practical Realization of the Metre and Second Representations of the Definition of the Second*, 2 p., May 2022.

67. J. Guéna, S. Weyers, M. Abgrall, C. Grebing, V. Gerginov, P. Rosenbusch, S. Bize, B. Lipphardt, H. Denker, N. Quintin, S. M. F. Raupach, D. Nicolodi, F. Stefani, N. Chiodo, S. Koke, A. Kuhl, F. Wiotte, F. Meynadier, E. Camisard, C. Chardonnet, Y. Le Coq, M. Lours, G. Santarelli, A. Amy-Klein, R. Le Targat, O. Lopez, P. E. Pottie, and G. Grosche, "First International Comparison of Fountain Primary Frequency Standards via a Long Distance Optical Fiber Link," *Metrologia*, vol. 54, no. 3, pp. 348–354, May 2017.

68. C. H. Townes, *Making Waves*, American Institute of Physics Press, 1995.

69. J. P. Gordon, H. J. Zeiger, and C. H. Townes, "Molecular Microwave Oscillator and New Hyperfine Structure in the Microwave Spectrum of NH_3," *Physical Review*, vol. 95, no. 1, pp. 282–284, July 1954.

70. International Union of Pure and Applied Chemistry (IUPAC), *IUPAC Periodic Table of the Elements*, www.iupac.org, May 4, 2022.

71. H. I. Ewen and E. M. Purcell, "Observation of a Line in the Galactic Radio Spectrum: Radiation from Galactic Hydrogen at 1,420 M/s," *Nature*, vol. 168, p. 356, September 1951.

72. H. M. Goldenberg, D. Kleppner, and N. F. Ramsey, "Atomic Hydrogen Maser," *Physical Review Letters*, vol. 5, no. 8, pp. 361–362, October 1960.

73. R. F. C. Vessot, "The Atomic Hydrogen Maser Oscillator," *Metrologia*, vol. 42, no. 3, pp. S80–S89, June 2005 (erratum published in vol. 42, no. 5, pp. 468–470, October 2005).

74. H. M. Goldenberg, D. Kleppner, and N. F. Ramsey, "Atomic Beam Resonance Experiments with Stored Beams," *Physical Review*, vol. 123, no. 2, pp. 530–537, July 1961.

75. D. Kleppner, H. M. Goldenberg, and N. F. Ramsey, "Theory of the Hydrogen Maser," *Physical Review*, vol. 126, pp. 603–615, April 1962.

76. N. F. Ramsey, "The Atomic Hydrogen Maser," *Metrologia*, vol. 1, no. 1, pp. 7–15, January 1965.

77. J. Vanier and C. Audoin, *The Quantum Physics of Atomic Frequency Standards, Volume 2*, Adam Hilger, Bristol and Philadelphia, 1989.

78. M. Desaintfuscien, *Data Processing in Precise Time and Frequency Applications*, Springer, 2007.

79. D. Kleppner, H. C. Berg, S. B. Crampton, N. F. Ramsey, R. F. C. Vessot, H. E. Peters, and J. Vanier, "Hydrogen-Maser Principles and Techniques," *Physical Review*, vol. 138, no. 4A, pp. A972–A983, May 1965.

80. R. F. C. Vessot, H. E. Peters, and J. Vanier, "Recent Developments in Hydrogen Masers," *Proceedings of the 18th Annual Symposium on Frequency Control*, pp. 299–307, May 1964.

81. H. E. Peters, "Characteristics of Advanced Hydrogen Maser Frequency Standards," *NASA-TM-X-70579*, 34 p., December 1973.

82. H. Hellwig, "The Hydrogen Storage Beam Tube, a Proposal for a New Frequency Standard," *Metrologia*, vol. 6, no. 2, pp. 56–60, April 1970.

83. F. L. Walls and H. Hellwig, "A New Kind of Passively Operating Hydrogen Frequency Standard," *Proceedings of the 30th Annual Symposium on Frequency Control*, pp. 473–480, June 1976.

84. N. A. Demidov and A. A. Uljanov, "Design and Industrial Production of Frequency Standards in the USSR," *Proceedings of the 22*nd *Annual Precise Time and Time Interval Meeting (PTTI)*, pp. 187–207, December 1990.

85. J. Dai, T. Liu, Y. Cai, Z. Chen, and Q. Li, "Review of the Development of the Hydrogen Maser Technique and a Brief Introduction to Its Space Applications," *Frontiers in Physics*, vol. 10: 995361, pp. 1–15, September 2022.

86. T. E. Parker, "Performance of a Hydrogen Maser Ensemble and Characterization of Frequency Standards," *Proceedings of the 1999 Joint Meeting of the European Frequency and Time Forum and the IEEE International Frequency Control Symposium (EFTF/IFCS)*, pp. 173–176, April 1999.

87. H. Hellwig, R. F. C. Vessot, M. W. Levine, P. W. Zitzewitz, D. W. Allan, and D. J. Glaze, "Measurement of the Unperturbed Hydrogen Hyperfine Transition Frequency," *IEEE Transactions on Instrumentation and Measurement*, vol. IM-19, no. 4, pp. 200–209, November 1970.

88. H. E. Peters, R. G. Hall, and D. B. Percival, "Absolute Frequency of an Atomic Hydrogen Maser Clock," *Proceedings of the* 26th Annual Symposium on Frequency Control, pp. 319–323, June 1972.

89. R. Lutwak, "The SA.45S Chip-Scale Atomic Clock – Early Production Statistics," *Proceedings of the 43rd Annual Precise Time and Time Interval Meeting (PTTI)*, pp. 207–220, November 2011.

90. J. Kitching, "Chip-scale atomic devices," *Applied Physics Reviews*, vol. 5, no. 031302, pp. 1–38, August 2018.

91. P. J. Chantry, B. R. McAvoy, J. M. Zomp, and I. Liberman, "Towards a Miniature Laser-Pumped Cesium Cell Frequency Standard," *Proceedings of the 1992 IEEE International Frequency Control Symposium*, pp. 114–122, May 1992.

92. P. J. Chantry, I. Liberman, W. R. Verbanets, C. F. Petronio, R. L. Cather, and W. D. Partlow, "Miniature Laser-Pumped Cesium Cell Atomic Clock Oscillator," *Proceedings of the 1996 IEEE International Frequency Control Symposium*, pp. 1002–1010, June 1996.

93. R. Lutwak, D. Emmons, T. English, W. Riley, A. Duwel, M. Varghese, D. K. Serkland, and G. M. Peake, "The Chip-Scale Atomic Clock – Recent Development Progress," *Proceedings of the 35th Annual Precise Time and Time Interval Meeting (PTTI)*, pp. 467–478, December 2003.

94. M. Stähler, R. Wynands, S. Knappe, J. Kitching, L. Hollberg, A. Taichenachev, and V. Yudin, "Coherent population trapping resonances in thermal ^{85}Rb vapor: D_1 versus D_2 line excitation," *Optics Letters*, vol. 27, no. 16, pp. 1472–1474, August 2002.

95. N. Cyr, M. Tetu, and M. Breton, "All-Optical Microwave Frequency Standard: A Proposal," *IEEE Transactions on Instrumentation and Measurement*, vol. 42, no. 2, pp. 640–649, April 1993.

96. G. Alzetta, A. Gozzini, L. Moi, and Orriols, "An Experimental Method for the Observation of R. F. Transitions and Laser Beat Resonances in Oriented Na Vapour," *Il Nuovo Cimento B*, vol. 36, no. 1, pp. 5–20, November 1976.

97. J. E. Thomas, P. R. Hemmer, S. Ezekiel, C. C. Leiby, Jr., R. H. Piccard, and C. R. Willis, "Observation of Ramsey Fringes Using a Stimulated, Resonance Raman Transition in a Sodium Atomic Beam," *Physical Review Letters*, vol. 48, no. 13, pp. 867–870, March 1982.

98. J. Vanier, A. Godone, and F. Levi, "Coherent Population Trapping in Cesium: Dark Lines and Coherent Microwave Emission," *Physical Review A*, vol. 58, no. 3, pp. 2345–2358, September 1998.

99. J. Vanier, "Atomic Frequency Standard," *United States Patent 6,320,472 B1*, filed January 26, 1999, granted November 20, 2001.

100. R. Lutwak, D. Emmons, W. Riley, and R. M. Garvey, "The Chip-Scale Atomic Clock – Coherent Population Trapping vs. Conventional Interrogation," *Proceedings of the 34th Annual Precise Time and Time Interval Meeting (PTTI)*, pp. 539–550, December 2002.

101. J. Kitching, S. Knappe, N. Vukicevic, L. Hollberg, R. Wynands, and W. Weidmann, "A Microwave Frequency Reference Based on VCSEL-Driven Dark Line Resonances in Cs Vapor," *IEEE Transactions on Instrumentation and Measurement*, vol. 49, no. 6, pp. 1313–1317, December 2000.

102. L. Hollberg and J. Kitching, "Miniature Frequency Standard Based on All-Optical Excitation and a Micro-Machined Containment Vessel," *United States Patent 6,806,784 B2*, filed June 20, 2002, granted October 19, 2004.

103. S. Knappe, V. Shah, P. D. D. Schwindt, L. Hollberg, J. Kitching, L-A. Liew, and J. Moreland, "A microfabricated atomic clock," *Applied Physics Letters*, vol. 85, no. 9, pp. 1460–1462, August 2004.

104. R. Lutwak, J. Deng, W. Riley, M. Varghese, J. LeBlanc, G. Tepolt, M. Mescher, D. K. Serkland, K. M. Geib, and G. M. Peake, "The Chip-Scale Atomic Clock – Low-Power Physics Package," *Proceedings of the 36th Annual Precise Time and Time Interval Meeting (PTTI)*, pp. 339–352, December 2004.

105. P. Cash, W. Krzewick, P. Machado, K. R. Overstreet, M. Silveira, M. Stanczyk, D. Taylor, and X. Zhang, "Microsemi Chip Scale Atomic Clock (CSAC) Technical Status, Applications, and Future Plans," *Proceedings of the 2018 European Time and Frequency Forum (EFTF)*, pp. 65–71, 2018.

PART II

Measurements

Chapter 4

Time Measurements

I've been first and last
Look at how the time goes past

— Neil Young, "Old Man"

As we have learned in the first three chapters, all clocks measure time, but they do so with varying levels of accuracy. Metrologists are in the business of comparing clocks by measuring the time differences between them. Once the time difference of a clock is known with respect to a reference clock, a clock can be synchronized by adjusting its time until the time difference is near zero. A clock that's been synchronized to agree with a reference becomes a trusted source of time, and it can be used to record when events occur.

This chapter provides an overview of the methods that metrologists use to measure time. Section 4.1 describes the methods and instrumentation used to measure the time interval between two signals. Section 4.2 discusses time synchronization. The time transfer methods used to compare and synchronize remote clocks to a reference clock are explained in Section 4.3. Finally, Section 4.4 introduces the topic of ensemble time scales, and explains why a group of clocks can be a better time reference than a single clock.

Before delving into these topics, it's worthwhile to review some of the basic clock concepts introduced in previous chapters. It's important to remember that clocks measure seconds by counting a specific number of cycles produced by their oscillator. For example, to a pendulum clock, a second is one swing, or cycle, of its pendulum. To a quartz wristwatch, a second is 32,768 vibrations of a quartz crystal. To a cesium clock, a second is 9,192,631,770 periods of the transition between two hyperfine energy levels of the cesium atom. All modern clocks keep time by first measuring seconds, and then counting seconds to establish longer intervals like minutes and hours.

The official, internationally agreed-upon reference for world timekeeping is Coordinated Universal Time, which is abbreviated as UTC. Even though UTC is properly classified as a time scale, it works in much the same way as a clock. Like all clocks, UTC follows an agreed-upon system of keeping time. The first step is always to define the second, the base unit of time interval, and UTC is based on the atomic definition of the SI second. The next step is to measure the second, which UTC does by averaging data collected from hundreds of atomic clocks located around the world. Time is then kept by simply counting the seconds

As Time Goes By: A Comprehensive Guide to Time and Frequency Measurements
Michael A. Lombardi

ISBN 978-981-5352-28-3 (Hardcover), 978-1-042-01532-0 (eBook)
www.jennystanford.com

DOI: 10.1201/9781042015320-7

that have elapsed from an agreed-upon starting point, which is called an epoch. With these basic concepts in mind, let's begin our discussion of time interval measurements.

4.1 Time Interval Measurements

Time interval measurements determine the elapsed time between two events. The resolution of a single time interval measurement is limited by the smallest increment that the measurement instrument can detect and display. For example, if an instrument has 1 s resolution, it can produce readings of 10 s or 11 s, but not a reading of 10.5 s, because all of its readings are rounded to the nearest integer second.

It seems that much of human thought and perception has a time interval resolution of about one tenth of a second (0.1 s). There are many examples of this—when movies were invented, a new image had to be displayed at least every tenth of a second to form the illusion of continuous movement. A driver usually requires about a tenth of a second to press the brakes when an obstacle suddenly appears in front of them. Most humans cannot pronounce syllables in less than a tenth of a second, and only a small percentage of us can read more than 600 words per minute, which puts an upper limit on reading speed of about one word every tenth of a second. Due to the limits of "human resolution," experimental results published before the early 20th century sometimes included the scientist's personal equation, which was a correction factor that had to be applied before their data could be compared to data collected by other scientists. Because there were then no methods in place to make accurate measurements of short time intervals, it was debated for decades whether personal equation corrections were related to reaction time, to the speed of thought, or to the personal bias of the scientist [1].

When science was still being limited by the human resolution of time interval, it became apparent that more resolution would advance science, because it would allow us to see things that were previously unseen. Perhaps the most classic example of this involved the way humans perceived the motion of a galloping horse. The human eye isn't fast enough to process the action of a horse's motion, so artists had usually depicted a horse at full gallop as flying through the air, with its front legs fully extended forward, with its hind legs fully extended to the rear, and with all four of its hooves off the ground.

These notions of how a horse galloped were dispelled in 1878 by images captured by the English photographer Eadweard Muybridge (1830–1904). Using a group of 12 cameras installed at a race track in Palo Alto, California, Muybridge took sequential photographs of a horse galloping at a speed of about 62 km per hour. When the horse passed the group of cameras, it tripped wires connected to an electromagnetic circuit that released the camera's shutters. The photographs were taken at intervals of "about the twenty-fifth part of a second of time" with each having an exposure time of "less than the two-thousandth part of a second." A sequence of 12 of Muybridge's photographs entitled *The Horse in Motion* was enthusiastically and widely published (Fig. 4.1). The images revealed that there were indeed moments during a galloping horse's stride when all four of its hooves were off the ground. However, these moments didn't occur when a horse's legs were fully extended to the front and back, as artists had depicted, but instead occurred when a horse collected its legs beneath its body, as it switched from pulling with its front legs to pushing with its back legs. After Muybridge's photos were published, artistic depictions of horses seldom included the more graceful image of a flying gallop.

Modern scientific applications always measure time interval with sub-second resolution, ideally with as much resolution as possible, because insignificant digits can always be rounded or truncated if they are not needed. Table 4.1 provides the abbreviations and durations of the sub-second time interval units most commonly used by metrologists. A quick glance at the table indicates that as the range of time interval units goes from seconds to attoseconds, it covers a staggering 18 orders of magnitude. This unit range is needed because time metrology is a continuously advancing field that involves clocks, oscillators, and instrumentation at all levels of performance. As we'll learn in subsequent chapters, measurements of milliseconds or microseconds are required to support the requirements of the telecommunications and electric power industries, as well as world financial markets (Chapters 16 through 18). Milliseconds and microseconds are also the most common time units associated with measurements of clocks controlled by radio and network signals (Chapters 10 and 11). Units of nanoseconds and picoseconds are commonly associated with measurements of microwave atomic clocks (Chapter 3) and clocks controlled by global navigation satellite signals (Chapter 12). Units of femtoseconds and attoseconds are associated with measurements of state-of-the-art optical clocks and frequency combs (Chapter 13). The smallest time interval unit in the table, the zeptosecond, is typically only mentioned in cutting edge atomic physics research. The shortest time intervals measured so far are hundreds of zeptoseconds in duration, or about 10^{-19} s.

Figure 4.1 Muybridge's photographs of a galloping horse, 1878 (courtesy of the Library of Congress).

Table 4.1 Abbreviations and durations for sub-second time interval units

Unit name	Unit abbreviation	Duration (in seconds)
millisecond	ms	10^{-3}
microsecond	µs	10^{-6}
nanosecond	ns	10^{-9}
picosecond	ps	10^{-12}
femtosecond	fs	10^{-15}
attosecond	as	10^{-18}
zeptosecond	zs	10^{-21}

In calibration and metrology laboratories, the measured interval is usually the time that elapses between the arrival of two electrical pulses. One pulse originates from a reference clock. The other pulse originates from the clock being measured, known as the device under test (DUT). The pulses are usually generated at rate of one pulse per second, commonly abbreviated as 1 pps, so their frequency is 1 Hz.

Because both pulses have a period of 1 s, the time interval that elapses between their arrival is always within a range of ±0.5 s.

The pulses are usually transmitted at TTL (transistor-transistor logic) signal levels in the form of a square wave. The rise time of a pulse is the time required for the pulse to rise from its lower voltage threshold, or 10% of its peak amplitude, to its upper voltage threshold, or 90% of its peak amplitude. The duty cycle is the fraction of the 1 s period when the pulse is at or above its upper voltage threshold. For example, if a pulse has an amplitude that equals or exceeds 90% of its peak amplitude for 0.2 s during its 1 s period, it has a 20% duty cycle [2]. A larger duty cycle means a wider pulse, and vice versa. Figure 4.2 shows a series of pulses with a 2 V amplitude and a 50% duty cycle.

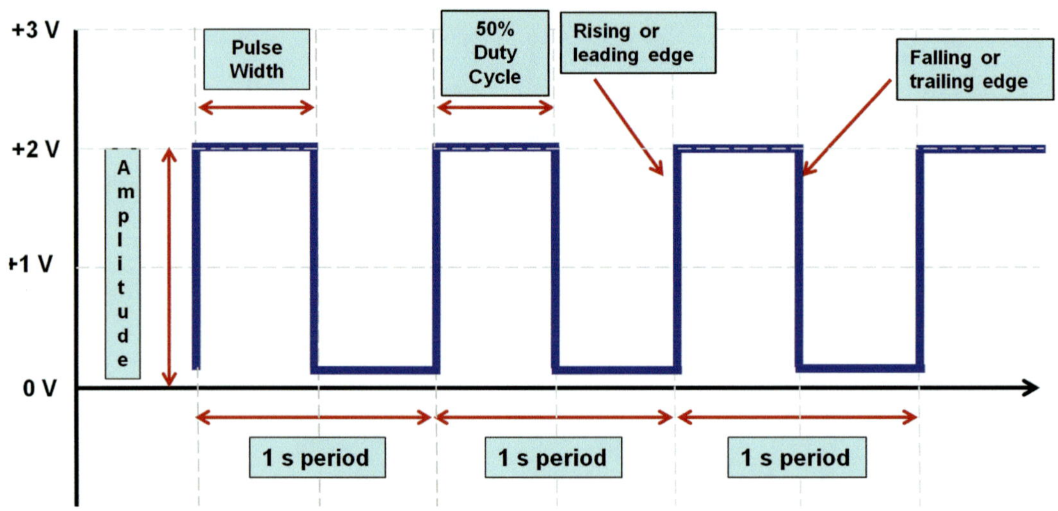

Figure 4.2 A timing signal transmitted at one pulse per second.

4.1.1 Measuring Time Interval with Oscilloscopes

Oscilloscopes are perhaps the most important of all electronic test instruments and are found in nearly every metrology laboratory. They allow viewing the waveform of a signal, and graphing the amplitude (voltage) of the signal as a function of time. They are also commonly used to measure time interval.

The German physicist and Nobel laureate Karl Ferdinand Braun (1850–1918) is generally credited with building both the first cathode-ray tube (CRT) and the first oscilloscope in 1897, and oscilloscopes intended for general purpose use have been available since the early 1930s. Since then, they have been an essential part of time and frequency metrology, with form factors that now range from handheld devices to large rack mount instruments, with an even wider range of features and capabilities.

An oscilloscope's ability to display pulse waveforms makes it easy to select the start and end points that define a time interval, which is helpful when dealing with odd-shaped or noisy waveforms. In some cases, just one pulse is displayed, while a different pulse is used to trigger the oscilloscope. A dual-trace oscilloscope is more useful, because it can simultaneously display both pulses. In this case, triggering can be achieved from either of the two displayed pulses, or from a third external pulse.

The divisions along the horizontal axis of an oscilloscope are in units of time, and many oscilloscopes have a delta-time function (Δt). This function measures the time interval between two cursors that can be moved along the horizontal axis, and provides a numeric readout on the display [3]. The oscilloscope display in Fig. 4.3 has time divisions of 50 ns, and is measuring a 358.0 ns time interval between two cursors that are aligned with the rising edge of two 1 pps signals.

Time interval measurements made with oscilloscopes are limited by their lack of resolution, and sometimes by their range. The measurement resolution is proportional to the time interval. For example, if the time interval is 10 ms, the best oscilloscopes can only resolve the interval to about 1 μs. However, if the pulses are only 100 μs apart, the resolution is 10 ns. In both cases, the resolution is about 1 part in 10,000 with respect to the interval, or 1×10^{-4}. The range of an oscilloscope specifies the minimum and maximum time intervals it can measure. Oscilloscope range varies from model to model, but some models have a very wide range and can scale, for example, from 50 ps to 10 s per time division.

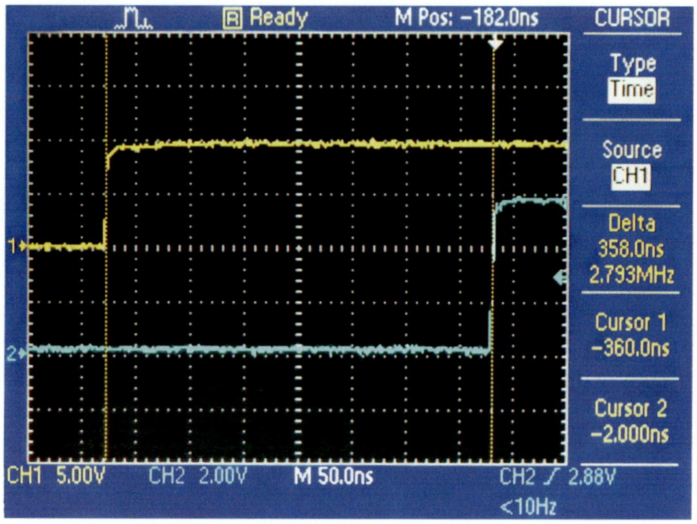

Figure 4.3 Measuring time interval with an oscilloscope.

4.1.2 Measuring Time Interval with Time Interval Counters

A time interval counter (TIC) is usually the instrument of choice for measuring all but the shortest time intervals. Nearly all timing systems include some type of TIC. In most cases, a universal counter, configured to work in time interval mode, serves as the TIC. However, single-function counters that only measure time interval are sometimes used.

Universal counters are most often used to measure frequency, but can also measure the period of a signal, count the total number of cycles of a signal, or the time interval between two input signals. As we'll further discuss in Section 5.2.2, the first frequency counter was introduced by Hewlett-Packard in 1950 and a time interval mode was added shortly afterwards [4, 5]. A modern universal counter that can be configured as either a frequency counter or a TIC is shown in Fig. 4.4.

Figure 4.4 A universal counter (courtesy of Keysight Technologies).

A TIC measures the time interval between a start pulse and a stop pulse. The arrival of the start pulse gates the TIC, meaning that it figuratively opens a gate so that time base cycles can pass through. When the stop pulse arrives, it closes the gate. While the gate is open, the TIC counts the cycles of the frequency produced by its time base oscillator. The time interval is measured by multiplying the number of cycles that pass through the gate by the period of the time base frequency (Fig. 4.5).

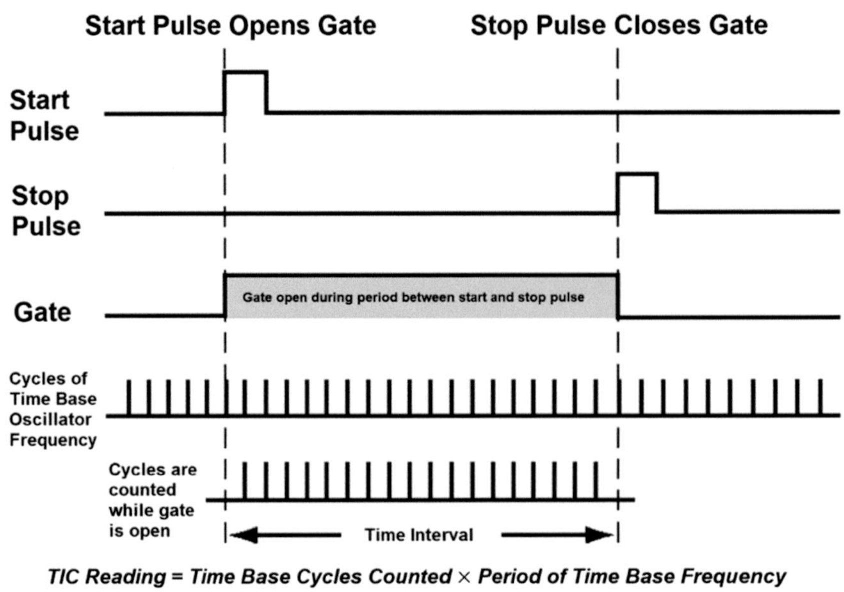

Figure 4.5 The basic functions of a time interval counter.

The simplest TICs can only count whole cycles. Therefore, their resolution is equivalent to the period of their time base frequency. If the time base frequency is 10 MHz, the resolution of a TIC than only counts whole cycles is limited to 100 ns. Some TICs use a higher frequency time base to get more resolution, in many cases by simply multiplying the frequency of the 10 MHz oscillator. For example, multiplying by 10 results in a 100 MHz time base and 10 ns resolution, and multiplying by 100 results in a 1 GHz time base and 1 ns resolution.

Making the time base frequency high enough to achieve nanosecond or sub-nanosecond resolution is not always practical, however, because it adds to the cost of a TIC and can also cause reliability issues. Therefore, most TICs obtain more resolution by using linear interpolation or digital processing methods. These methods work by using two measurement scales, a coarse scale that marks the occurrence of time base cycles, and a fine scale that divides the period of the time base frequency into smaller parts. The idea of using a fine or secondary scale is well known in metrology. It was conceived by the French mathematician Pierre Vernier (1580–1637), and the calipers used in dimensional metrology still bear his name. A Vernier scale is simply a way to take an accurate measurement between two markers on a linear scale by interpolating between the two markers, rather than by estimating. On a analog device such as a ruler or a tape measure, a Vernier scale adds more lines between the markers. On a digital device, a Vernier scale adds more digits to the display.

In 1957, not long after the first appearance of TICs, the American electrical engineer Robert Baron (1926–1981) described a method of measuring time interval "which is the electronic analog of the mechanical Vernier scale." Baron's method involved dividing the measurement into three parts. One part was the coarse scale, obtained by counting time base cycles that passed through the gate between the start and stop pulse. The other two parts were fine scales—one being the interval between the start pulse and the arrival of the first time base cycle, known as the start interval, and the other being the interval after the last time base cycle and the stop pulse, known as the stop interval. Each measurement required its own counter, what Baron called a coarse counter, a start Vernier counter, and a stop Vernier counter (Fig. 4.6). The three measurements were simply added together to get the total time interval measurement [6].

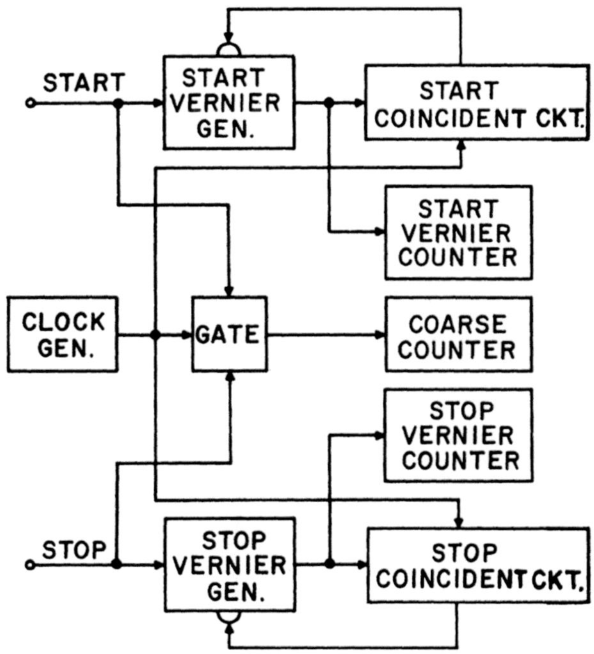

Figure 4.6 Simplified diagram of a Vernier time measurement system [5].

Various other methods have been implemented to measure the start interval and stop interval. One method, practiced since the early 1960s and still implemented today, involves the use of delay lines [7]. The start pulse is routed through multiple delay lines and compared to the phase of the time base oscillator. The first delay time that just exceeds the oscillator phase is the correct time difference for the start interval, and the same procedure is followed with the start pulse to obtain the stop interval. The resolution is simply T/N, where T is the period of the time base oscillator, and N is the number of delay lines, or the number of "cells" in a segmented delay line. Early models of TICs supported only a small number of delay lines, but modern instruments can support enough delay lines to obtain picosecond-level resolution [8].

Another widely used TIC method was invented in 1968 by Ronald Nutt (b. 1938), an electrical engineer at Oak Ridge National Laboratory. The Nutt method measures the start and stop intervals by charging and discharging capacitors. A charging current is applied to a capacitor at the arrival of the start pulse and the capacitor is discharged when the first cycle from the time base passes through the gate. A second capacitor is charged when the stop pulse arrives and discharged when the next time base cycle arrives. The charging rates of the capacitors are converted to units of time, and the start and stop intervals are determined. The total time interval is measured by adding the start interval to the main count and then subtracting the stop interval. Using a 20 MHz time base, the TIC resolution that Nutt originally demonstrated was 256× finer than the resolution of the coarse counter, dropping from 50 ns to less than 200 ps [9–11].

These methods, and many others, have been used to design TICs with more resolution than could be obtained by simply counting whole time base cycles. For the purposes of example, let's look in some detail at a single-function TIC designed at NIST by the electrical engineer Dick Davis (1932–2018). This instrument has a resolution near 25 ps and has served many applications, with several variations of its design integrated into NIST measurement systems [12, 13]. Without interpolation, obtaining 25 ps resolution would require a time base frequency of 40 GHz, which isn't practical. Instead, the NIST counter utilized a 10 MHz time base and three counting circuits; a main counter, a start counter, and a stop counter. The main counter provides the coarse resolution (100 ns), and the start and stop counters provide the fine resolution. The main counter counts the zero crossings of the time base cycles. The start counter measures the time interval between the start pulse and the first cycle zero crossing from the time base, δt_1. The stop counter measures the interval between the last cycle zero crossing from the time base and the stop pulse, δt_2.

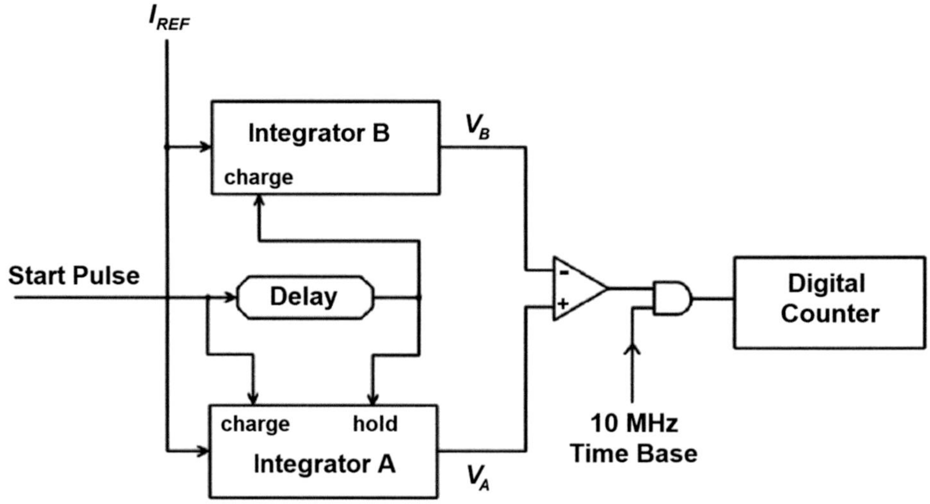

Figure 4.7 Start interpolator counter.

In addition to a counter, the start and stop interpolators each consist of two integrators (A and B), a delay cell, and a voltage comparator (Fig. 4.7). The integrators are charged with a constant current I_{REF}. Using the start interpolator as an example, the delay cell introduces a delay equal to $\delta t_1 + T$, where $T = 100$ ns and $\delta t_1 < 100$ ns. When the start pulse arrives, the charging current I_{REF}. is sent to integrator A, and the counter begins counting periods of the 10 MHz time base. The charging current is first sent to integrator A for the period of $\delta t_1 + T$, and then switched to integrator B. The voltage on integrator A, or U_R, is stored as a reference voltage for the comparator while integrator B is being charged to a voltage of V_B. When V_B is greater than U_R, the comparator blocks the connection between the 10 MHz time base and the digital counter (Fig. 4.8). The scaling factor for the interpolator, K, is then determined as

$$K = \frac{C_B}{C_A} + 1, \tag{4.1}$$

where C_A and C_B represent the current stored by the capacitors in integrators A and B.

The reading obtained from the start counter, N_{start}, equals the number of 10 MHz pulses counted, and is linearly proportional to the time interval of δt_1. The relationship between δt_1 and N_{start} is

$$\delta t_1 = A \times N_{start} + B, \tag{4.2}$$

where the slope A and the intercept B are determined during the interpolator calibration. The calibration software collects a series of readings is taken from the start counter and obtains the maximum and minimum values, ST_{max} and ST_{min}, by slewing the phase of a test signal with respect to the time base to simulate the range of possible values for δt_1.

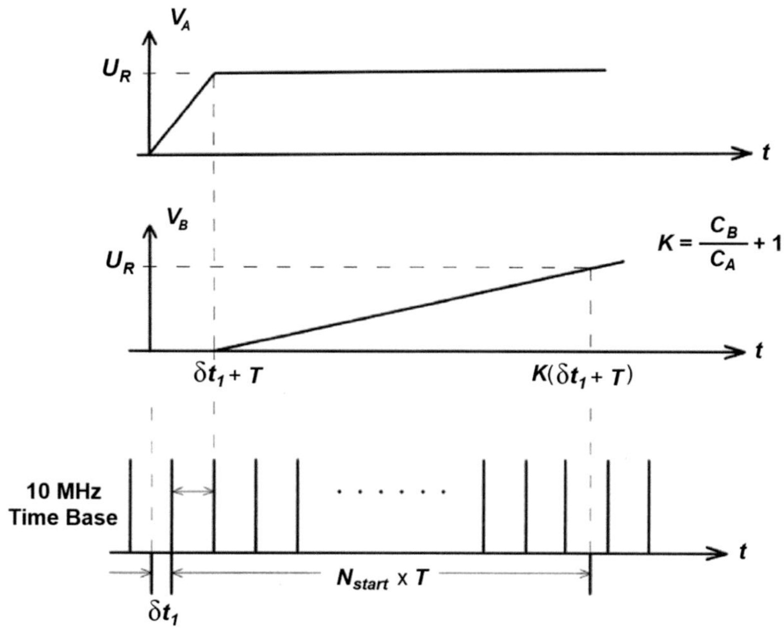

Figure 4.8 Charging the start interpolator counter.

Because the lowest possible value of δt_1 is 0, and because the highest possible value is *T*, the slope and intercept are calculated as

$$A = \frac{T}{ST_{\max} - ST_{\min}}, \tag{4.3}$$

and

$$B = \frac{ST_{\min} \times T}{ST_{\max} - ST_{\min}}. \tag{4.4}$$

Therefore,

$$\delta t_1 = A \times N_{\text{start}} + B = \frac{N_{\text{start}} - ST_{\min}}{ST_{\max} - ST_{\min}}, \tag{4.5}$$

and because

$$K = \frac{C_B}{C_A} + 1 = ST_{\max} - ST_{\min} \tag{4.6}$$

the resolution of the start counter is equal to T/K.

The NIST design utilizes a 10 MHz time base and 100 pF and 0.47 μF capacitors are used in the interpolator circuits. Thus, T = 100 ns, C_A = 100 pF, and C_B = 0.47 μF. The resolution is calculated as

$$\frac{T}{K} = \frac{100 \times 10^{-9}\text{s}}{4701} = 21 \times 10^{-12}\text{s} = 21\text{ps}. \tag{4.7}$$

In practice, sampling N_{start} across the entire range of 4701 potential counts isn't possible, and the calibration range is typically limited to about 4000 counts. This limits the actual resolution to about 25 ps, rather than 21 ps.

The stop counter is calibrated in the same fashion as the start counter and performs a nearly identical function, except that it measures the interval of $T - \delta t_2$. The start and stop counters have equivalent resolution.

The final part of the TIC calibration measures the signal path delay in the counter circuitry. This is done by dividing the 10 MHz time base frequency by 256 to obtain a frequency of 39062.5 Hz. This test frequency is then supplied to both the start and stop inputs of the counter, and a series of test measurements is recorded. The period of the test frequency is 25,600 ns and thus the average of the test measurements should equal this value. The difference between the measured value and the known value is the counter delay (D), which typically is less than 1 ns.

During the calibration, five values are stored as constants to be applied to future measurements; the maximum values of the start and stop interpolator counters (ST_{max} and SP_{max}), the minimum values of the start and stop interpolator counters (ST_{min} and SP_{min}), and the counter delay (D). During each time interval measurement, three values are obtained: the main counter reading (N_{main}), the start counter reading (N_{start}), and the stop counter reading (N_{stop}). These values are combined with the calibration constants to produce a single time interval reading, TIC_R, in units of nanoseconds,

$$TIC_R = N_{main}T + \frac{N_{start}T}{ST_{max} - ST_{min}} - \frac{N_{stop}T}{SP_{max} - SP_{min}} - D, \tag{4.8}$$

where T, once again, is the period of the time base frequency [12, 13]. The examples just given represent just one of many methods that can be used to obtain sub-nanosecond time interval resolution, and some methods can provide resolution of about 1 ps [14].

The uncertainty of TIC measurements is typically limited by factors including count ambiguity, trigger levels that aren't properly set, or uncertainty in the time base frequency [15, 16]. Like most test instruments, a TIC has a ±1 count ambiguity in the least significant digit of its readout. This is usually only a problem with single readings and it can be solved by averaging multiple readings. Because the count ambiguity is a nearly random process, the ±1 count error can be reduced to an insignificant level by averaging.

Trigger errors, or errors in selecting the signal level that causes the counter to open or close its gate, are perhaps the most significant problem with TIC measurements. The trigger levels can be set and adjusted on both the start and stop channel of a TIC, and trigger errors occur when the counter triggers at the wrong voltage level on the input signal. Trigger errors can be caused by levels that are set incorrectly, or by input signals that are noisier, larger, or smaller than expected. To ensure that the trigger levels are properly set, it's usually standard practice to first view the waveforms on an oscilloscope. Figure 4.9 illustrates a situation where triggering is desired on point A, but where noisy signals cause triggering on points B or C [15]. Depending upon the rise time and the pulse shape, trigger errors can result in time interval readings that are completely false. Some TICs eliminate most of the triggering problem with pulse shaping circuitry that forces both the start and stop signals to have a nearly identical shape and amplitude. Even then, errors can still be caused by input pulses that are too large or too small, and attenuators or amplifiers are sometimes needed to reduce or increase the pulse amplitude so that it falls within the TIC's usable range.

The instability and inaccuracy of the time base frequency can also add uncertainty to a TIC measurement. A TIC has an internal time base oscillator, which is usually a TCXO or an OCXO, as well as a provision for accepting an external time base. The internal time base will likely be accurate and stable enough to correctly measure short time intervals. However, if a better oscillator is available, then connecting an external time base is always a good idea. When an external time base signal is connected, a phase-locked loop (PLL) automatically locks the internal oscillator to the external reference, as described in Section 5.3.1.

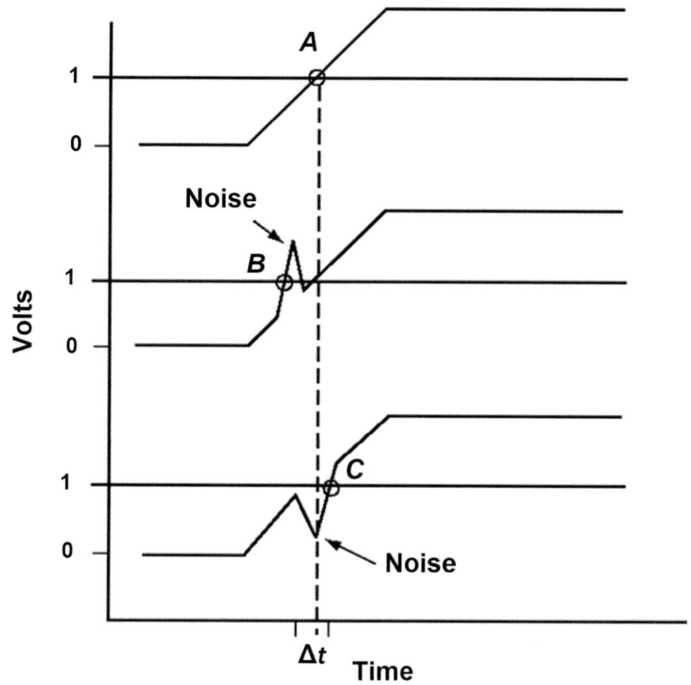

Figure 4.9 Trigger level errors.

Frequency stability will be discussed in more detail in Chapter 7, but for now, consider a situation where a TIC has an internal time base oscillator that is stable to 1×10^{-10} at 1 s. This simply means that the frequency of the time base won't change by more than $\pm 1 \times 10^{-10}$ during a 1 s period. When 1 pps signals are used to start and stop the counter, the average time interval between the pulses is 0.5 s, therefore the stability of the average TIC reading would be $0.5 \text{ s}/10^{10} = 5 \times 10^{-11} \text{ s} = 50$ ps. If the resolution of the TIC is coarser than 50 ps, which is often the case, then the instability of the time base will be insignificant.

Time base inaccuracy is more of a problem than time base instability. If the frequency of the time base oscillator is inaccurate, it introduces a corresponding inaccuracy in each time interval reading. Consider, for example, a typical OXCO time base with a frequency accuracy of 1×10^{-7}, meaning that it is only within 1 Hz of its 10 MHz nominal frequency. When measuring an average time interval of 0.5, this limits the accuracy of the readings to $0.5 \text{ s}/10^{7} = 5 \times 10^{-8} \text{ s} = 50$ ns. This error will be too large for most applications, but can be reduced by using a more accurate oscillator as an external time base.

4.1.3 Differential Delays in Time Interval Measurements

Accounting for delays in the signal path, especially the delays introduced by cables, is an essential part of making accurate time interval measurements. To understand why it's essential, let's consider an example where our goal is to measure the time difference between two rubidium clocks. Each clock has a 1 pps output, so we'll need two cables to connect the two clock outputs to the start and stop inputs of the TIC. If the two cables are of the same type and are the same length, the two signals will be delayed by approximately the same amount as they travel from the clocks to the TIC. In this case, the TIC reading will properly show the time difference between the two 1 pps outputs, even though we measured the signals at the cable ends, and not at the output connectors of the clocks.

If we use cables of different types or lengths to connect the rubidium clocks to the TIC, then the stop and signals will be delayed by different amounts. In this case, the TIC reading will still correctly show the time difference between the signals at the cable ends. However, the TIC reading won't show the actual time difference between the two clocks due to the differential delay between the two cables. If the stop cable has a larger delay the start cable, the TIC reading will be too small. If the start cable has a larger delay than the stop cable, the TIC reading will be too large. However, if we know the delay of both cables, we can still calculate the correct TIC reading, TIC_C, as

$$TIC_C = (TIC_D - D_{\text{STOP}}) + D_{\text{START}}, \tag{4.9}$$

where TIC_D is the reading displayed or recorded by the TIC, D_{STOP} is the delay introduced by the stop connector cable, and D_{START} is the delay introduced by the start connector cable.

Understanding cable delays is essential to making accurate time measurements, so it's important to know about the various types of cables. The cables used to connect the 1 pps signals to a TIC will likely be either coaxial cables or fiber optic cables. Coaxial cables, commonly called coax, are round cables used to transmit RF signals between two points. A coaxial cable has an inner or core conductor, which is either a copper wire or a copper-coated steel wire. This wire carries the RF signal, and the wire is surrounded by three layers. The first layer is a dielectric plastic insulator, usually made from either solid or foam polyethylene (PE), that keeps a constant distance between the inner conductor and the outer layers. The second layer is a metal shield that wraps around the insulator. It's typically made of braided copper or aluminum, and it shields the center internal from electromagnetic interference (EMI). The outer layer is a plastic jacket, typically made from polyvinyl chloride (PVC), that protects the internal parts of the cable from damage (Fig. 4.10).

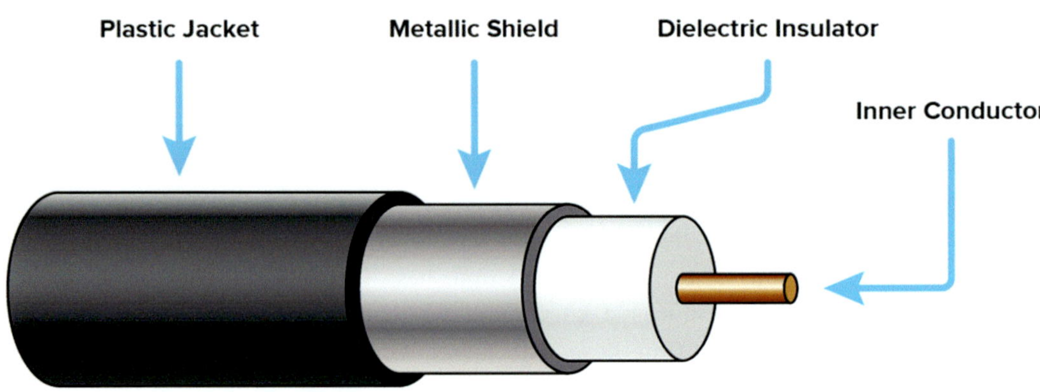

Figure 4.10 A coaxial cable.

The coaxial cable gets its name because the inner conductor that carries the signal runs along the same axis as the outer shield. The most common types of coaxial cable are labeled with a Radio Guide (RG) number, with RG-58 being the most common example. This numbering system dates back to U. S. military standards published in the 1940s. The current military standard uses a different scheme based on "M17" numbers, but still cross references the well-known RG numbers [17]. Coaxial cables with the LMR designation, a registered trademark of Times Microwave, are also commonly used to distribute 1 pps signals.

When the two clocks being compared are both located near the TIC, the use of coaxial cables is generally preferred. However, if one or both clocks are located far from the TIC, perhaps in another room or in a building that's hundreds of meters away, the use of fiber optic cables can be advantageous. Optical fiber has far less attenuation (signal loss) than coax, allowing cable runs to be much longer. It is also less sensitive to temperature variations, and is lighter and more flexible than coax, which makes it easier to route. Finally, because fiber is made out of glass, it doesn't conduct electricity and therefore isn't susceptible to EMI. Small unidirectional and bidirectional devices that convert signals at TTL levels to optical fiber signals and vice versa, are readily available (Fig. 4.11). These devices allow a clock that produces 1 pps signals at TTL levels to transmit their pulses over an optical fiber. A receiver at the other end of the fiber then converts the optical pulses back to TTL levels, so they can be measured with a TIC.

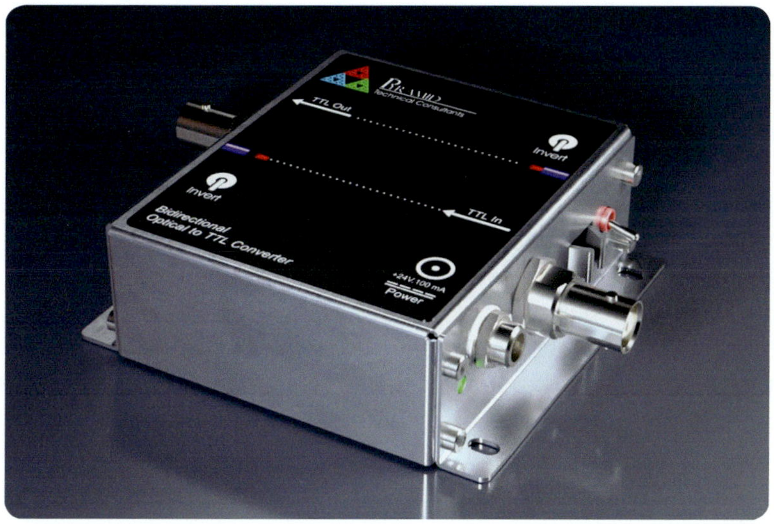

Figure 4.11 Bidirectional optical to TTL converter (courtesy of Pyramid).

A fiber optic cable (Fig. 4.12) has an assembly similar to a coaxial cable, but it carries light signals instead of RF signals. There are two glass layers inside the cable—the core that carries the light signal, and the cladding, which is the layer that surrounds the core. Because the cladding has a lower refractive index than the core it acts as a mirror, reflecting the light that reaches the boundary of the cladding back into the core. The cladding is surrounded by a plastic coating that doesn't affect the optical waveguide properties. Its purpose is to reinforce the fiber core and protect against excessive bending that could break the glass. Finally, the outer jacket protects the cable from environmental hazards.

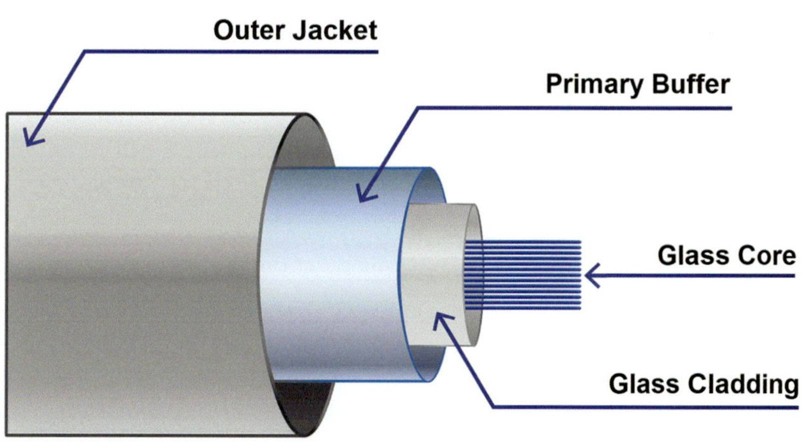

Figure 4.12 A fiber optic cable.

Fiber optic cables can be either single-mode or multi-mode. The mode refers to the possible paths where light can propagate through the fiber. A single-mode fiber supports only one path, but a multi-mode fiber supports multiple light paths. Single-mode fibers have much smaller cores, with typical diameters of 8 μm to 10 μm, as opposed to 50 μm or 62 μm for multi-mode fibers. Both single-mode and multi-mode optical fibers can be used with 1 pps signals, but it's important to note that they can't be mixed, as they use different connectors and electronics. For example, if a TTL/optical converter is used, it has to be compatible with the selected type of fiber.

The labeling on a fiber optic cable indicates the mode type (S for single-mode, M for multi-mode), the attenuation grade (1 or 2, with 2 having lower attenuation), the diameter of the core, and the diameter of the core plus the cladding. Therefore, cables labeled as OS2 9/125 are single-mode fibers with an attenuation grade of 2, a core diameter of 9 µm, and a core plus cladding diameter of 125 µm.

The speed of light in a vacuum, *c*, is a universal physical constant of 299,792,458 meters per second. An RF signal in a vacuum would travel at this same speed, but an RF signal in a cable will always travel at a slightly slower speed. The "slowest" cables use a solid dielectric material. The "fastest" cables use either a foam dielectric material that contains air, or use spiral spacers that maintain air space between the insulating material and the inner conductor, which is a type of cable called an air dielectric. However, air dielectric cables have the disadvantage of being larger in diameter and less flexible, which can make them difficult to use for TIC measurements.

The velocity factor, *VF*, indicates how fast signals travel in a cable. For signals traveling at the speed of light, *VF* would be equal to 1 (unity), but *VF* is always less than 1. If $VF = 0.75$, it means that signals travel through the cable at 75% of the speed of light. For coaxial cables, *VF* is always specified by the manufacturer. Fiber optic manufacturers usually don't specify *VF*, but they often specify the effective group index of refraction, n_{eff}, for a given wavelength. Because the speed of light constant is unity, the velocity factor of a fiber optic cable can be calculated as

$$VF = \frac{1}{n_{\text{eff}}}. \tag{4.10}$$

If the velocity factor is known, the cable delay in nanoseconds, D_{CAB}, can be calculated as

$$D_{\text{CAB}} = \frac{L \times 3.336}{VF}, \tag{4.11}$$

where L is the length of the cable in meters, 3.336 is the constant number of nanoseconds required for light in a vacuum to travel one meter (rounded to the nearest picosecond), and *VF* is the velocity factor.

There are many types of coaxial and fiber optic cables, and their specifications vary depending on the manufacturer. For the purposes of example, Table 4.2 provides a short list of some of the most common cable types. The table includes the outer diameter and velocity factor for each cable, as well as the time required for a signal to travel through a cable with a length of 10 m. As we can see from this information, it's important to know both the type and length of a cable when accounting for differential delays.

Table 4.2 The velocity of propagation through various cable types

Cable number	Type	Dielectric material	Outer diameter (mm)	Velocity factor	10 m cable delay (ns)
RG-8	Coaxial	Solid PE	10.3	0.66	50.5
RG-8	Coaxial	Foam PE	10.3	0.78	42.8
RG-58	Coaxial	Solid PE	5.0	0.66	50.5
RG-58	Coaxial	Foam PE	5.0	0.78	42.8
RG-174	Coaxial	Solid PE	2.8	0.66	50.5
RG-174	Coaxial	Foam PE	2.8	0.73	45.7
LMR-200	Coaxial	Foam PE	5.0	0.83	40.2
LMR-400	Coaxial	Foam PE	10.3	0.85	39.2
FSJ2-50	Coaxial	Foam PE	10.5	0.83	40.2
OS2 9/125	Optical fiber	Glass	3.0	0.68	49.1
OM1 62.5/125	Optical fiber	Glass	4.4	0.67	49.8

There are several methods used to measure the delay of a cable with just a TIC and a 1 pps signal source [18]. The signal source can be either a clock with a 1 pps output or a pulse generator. The trigger levels on the start and stop channels of the TIC should be set to the same level, and their impedance should be set to a 50 Ω load. It's a good idea to look at the pulse on an oscilloscope to determine its shape and amplitude so that a proper trigger level can be determined (Fig. 4.9), but triggering at either 0.5 V or 1 V is usually a safe choice.

After the TIC is configured, a simple and common method for measuring cable delay involves first placing a T-connector on the start channel. The 1 pps signal is connected to one side of the T-connector, and the cable under test (CUT) is then run from the other side of the T-connector to the TIC stop channel [15]. With this method, no calculations are necessary, because the same 1 pps signal starts and stops the TIC and the TIC reading equals the CUT delay. The problem with this method is that the T-connector changes the load impedance of the start channel, which introduces a small uncertainty to the measurement.

A better method involves configuring the TIC in the same way, but eliminating the T-connector. Two cables from the 1 pps source are connected to the start and stop channels of the TIC. The cables can be of arbitrary length, but to make it easy, the stop cable should be longer than the start cable so that the stop pulse arrives after the stop pulse. For example, if the stop cable has a delay that is 50 ns larger than the start cable, the TIC reading is 50 ns, but if the cable positions are reversed, the reading is 999,999,950 ns (1 s minus 50 ns). The TIC measurement of the delay between the two cables of arbitrary length is recorded as TIC_1. The CUT is then added to the stop cable to increase its delay, and a second measurement is recorded as TIC_2. For closure, the CUT cable is removed, and a third measurement, recorded as TIC_3, is a repeat of the first measurement. The cable delay, D_{CAB}, is then calculated as

$$D_{CAB} = TIC_2 - \frac{TIC_1 + TIC_3}{2}. \tag{4.12}$$

During each of the three measurement steps, the TIC readings should be averaged for a short interval, perhaps 30 s, before recording TIC_1, TIC_2, and TIC_3. Many TICs have a built-in averaging function that makes this easy to do. A variation of this method adds two more measurements, one where the CUT is added to the start cable, followed by another repeat of the original measurement for closure. This method has the advantage of testing both TIC channels, but has the disadvantage of taking longer to perform.

The TIC-based methods of measuring cable delays work well and are convenient to use. However, they require both ends of the cable to be accessible to the TIC, which won't be the case for cables that are already in place, such as long coaxial or fiber optic cables that run between rooms or buildings. In those cases, the time domain reflectometer (TDR) method, which is commonly used to measure impedance changes or locate faults in a transmission line, can also be used to measure cable delays. TDR measurements can be performed with a pulse generator and an oscilloscope, with a stand-alone TDR, or with a vector network analyzer (VNA) that has TDR capability.

The details vary based on the type of instrumentation used, but the basic principle of a TDR measurement is that a waveform propagating along an unmatched transmission line will reflect from the load back towards the source. An unmatched transmission line means that the impedance load at the end of the cable differs from the impedance of the cable itself. This causes the waveform to be reflected at the point where the impedance change occurs. Measurements are made by inputting a low voltage pulse to the CUT that serves as an incident waveform. The incident waveform is combined with the reflected waveform to form a pattern received at the input end of the cable. The pattern is either manually analyzed with an oscilloscope, or automatically analyzed by a dedicated instrument [2]. The pattern reveals the cable's round-trip delay (RTD), or the elapsed time between the release and return of the low voltage pulse.

The RTD is divided by two to obtain the cable delay. A comparison study, where a variety of TIC and TDR methods were used to measure the delays of several coaxial cables, showed that the results obtained with all methods agreed to within 1 ns. The smallest uncertainties, about 0.1 ns, were achieved with a VNA performing TDR measurements [19].

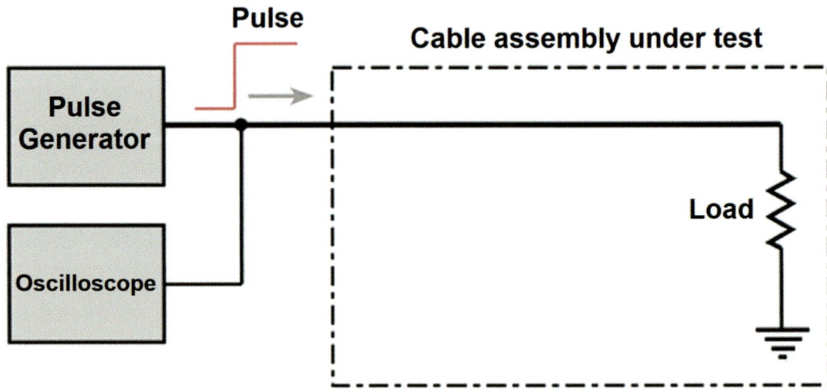

Figure 4.13 Simplified block diagram of a TDR measurement setup.

There are many types of cable connectors and adaptors, but female BNC connectors are typically found on the start and stop inputs of a TIC, including the one shown in Fig. 4.4. These connectors are compatible with the male BNC connectors that are typically found on the ends of coaxial cables. The unusual acronym BNC has an interesting history. It stands for "Baby Neill Constant," with the name honoring the American electrical engineer Paul Neill (1882–1968), who invented the threaded N-type connector at Bell Laboratories in the early 1940s. When the United States Navy began using smaller versions of the N-type connector, they referred to them as the "Baby N" or BN. The small threaded connectors couldn't be quickly connected or disconnected, so the Navy began using a bayonet locking mechanism that Neill had designed which made it faster and easier to connect cables. By 1948, the new connectors were called BNCs, with the "C" referring to their constant 50 Ω impedance that extends to frequencies higher than 1 GHz.

Cable connectors introduce small, sub-nanosecond delays to a signal, but these delays are already included in cable delay measurements, because the connectors are part of the cable. However, if the estimated delay of a cable of known length (Eq. 4.11) is compared to its measured delay (Eq. 4.12), the measured delay will likely be a fraction of a nanosecond larger, due to the delay added by the connectors.

Other components in the signal path can introduce much larger delays than connectors, including adaptors, attenuators, amplifiers, multiplexers, and fiber optic to TTL converters. To accurately determine the differential delay (Eq. 4.9), the delay of every component in a signal path should either be measured independently and added to the cable delay, or if possible, the cable delay can be measured with the components already in place. Another factor to consider are components that split a single 1 pps signal into multiple outputs, such as pulse distribution amplifiers. It can't always be assumed that every output of a distribution amplifier will have the same delay, so the delay of each output might need to be measured separately.

4.1.4 Stopwatch and Timer Measurements

We've so far only discussed measuring the elapsed time interval between two 1 pps signals. However, the most common time interval measurements are made with stopwatches and similar low-cost "timers." Like a TIC, a stopwatch measures the time interval between two signals, but stopwatch measurements differ in several ways. The signals that trigger a stopwatch aren't 1 pps signals, but instead are signals generated by the press of a mechanical button by a human hand, or by an automated sensor. The resolution of a stopwatch is much coarser than a TIC, typically just 0.01 s (10 ms). Stopwatches are also less accurate than a TIC, because they have lower quality time base oscillators and no provision for accepting an external time base. Finally, TICs usually measure sub-second intervals, and stopwatches usually measure intervals of multiple seconds, minutes, or hours.

Stopwatches have many industrial applications, but their use is perhaps most often associated with the timing of races—including athletic events, automobile races, and horse races. In fact, what was once believed to be the first stopwatch was invented in France for the sole purpose of timing horse races. The inventor, Nicolas Mathieu Rieussec (1781–1866), was a watchmaker for King Louis XVIII (1755–1824), and the king was a horse racing enthusiast. Rieussec called his invention a *chronograph*, deriving the word from the Greek words *chronos* (time) and *graphein* (to write), and his device was literally a "time writer." It worked by dropping ink on two rotating dials, one calibrated for seconds and the other for minutes, to mark the duration of a time interval to within 0.2 s. The device was successfully used to time horse races at Champs de Mars in 1821 and a patent was granted to Rieussec the following year [20].

An auction held in Geneva, Switzerland in 2012 unearthed an earlier chronograph invented in 1816 by another French watchmaker, Louis Moinet (1768–1853), who is now believed to be the inventor of the stopwatch. Moinet built his device (Fig. 4.14) for astronomical research, with the goal of accurately measuring the time interval of stars crossing the lens of a telescope. It was known as the *compteur de tierces* or "counter of thirds," because just as the hour and minute were subdivided into 60 parts; it also divided the second into 60 parts, resulting in 1/60 of a second resolution. Its mechanical oscillator ran at a frequency of 30 Hz, which was remarkable for that era, no other chronograph ran at a higher frequency for about a century [21].

Figure 4.14 Moniet's *compteur de tierces*, 1816, now believed to be the first stopwatch.

Most modern stopwatches are electronic devices with digital displays (Fig. 4.15) and a small internal quartz time base oscillator. Quartz based digital stopwatches first appeared in the mid-1970s, with perhaps the earliest examples produced in Switzerland by Heuer (now TAG Heuer). Their oscillators are similar to those found in quartz wristwatches (Section 2.5), typically operating at the same frequency of 32.768 kHz, although some run at multiples of that frequency, for example at 3.2768 MHz. The oscillator frequency is typically accurate to within ±5 × 10^{-6}, resulting in a time interval uncertainty of ±5 μs per second. This uncertainty isn't important when timing a short interval because the resolution of the stopwatch display is typically limited to 10 ms. However, when timing a long-distance race, such as a marathon, the oscillator inaccuracy might begin to affect the accuracy of a stopwatch's least significant digit. Consider, for example, a marathon that takes 10,000 s to complete. During the race, an oscillator accurate to 5 × 10^{-6} will accumulate a time error of 50 ms, which is 5× larger than the display resolution.

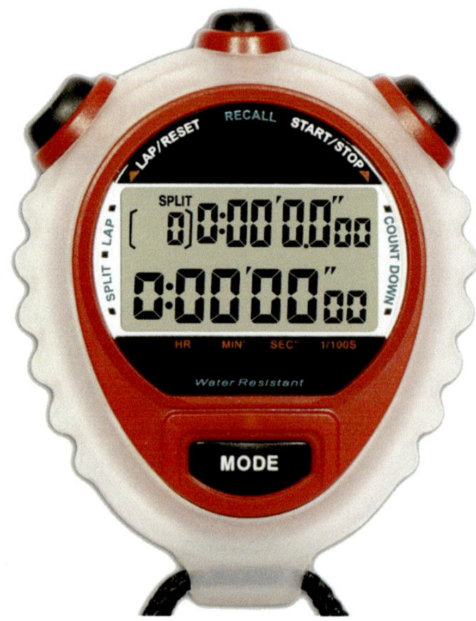

Figure 4.15 A digital stopwatch with a quartz time base oscillator.

Stopwatches need to be periodically calibrated to ensure they are working properly. The direct comparison method is the most common way to calibrate stopwatches because it's easy to understand and doesn't require any specialized equipment. The operator simply directly compares the stopwatch display to a to a reference time display or to a reference audio time signal, and starts and stops the stopwatch based on the signals that they see or hear. The uncertainty of these calibrations is limited by human reaction time, which is often no faster than 0.1 s (100 ms) when responding to a visual or audio signal. Even so, if the delays introduced by human reaction time were always the same, they wouldn't affect the time interval measurements at all. The differential delay, or the difference in human reaction time when starting and stopping the stopwatch, is all that matters. The variations in human reaction time have the same effect as using cables of different lengths when comparing two clocks with a TIC (Section 4.1.3).

Differential delays can contribute a large uncertainty to measurements made with the direct comparison method. For example, if a stopwatch operator takes 70 ms to respond to the start signal and 75 ms to respond to the stop signal, the uncertainty caused by the differential delay is ±5 ms. When measuring a time interval of 1 s, this ±5 ms uncertainty is 1,000× larger than the ±5 μs uncertainty introduced by a time base oscillator that's accurate to ±5 × 10^{-6}. For this reason, direct comparison stopwatch calibrations require a measurement period that's long enough to bring the differential delay uncertainty below the level of the time base oscillator uncertainty. Regulatory documents for stopwatch calibrations often specify a minimum measurement period of three hours (10,800 s) [22].

Fortunately, other stopwatch calibration methods can reduce or completely eliminate the uncertainty of human reaction time. The time base method uses an instrument, which is sometimes called a watch analyzer, that has acoustic or inductive sensors that can measure the time base frequency without opening the stopwatch case, much like a doctor uses a stethoscope to measure a patient's heartbeat. Once the time base frequency is known, it can be used to easily determine the accumulated time error for a given interval. For example, if the 32,768 Hz time base is offset in frequency by 1 Hz, the stopwatch would accumulate a time error of 0.305 s (305 ms) when measuring a 10,000 s interval, 10,000/32,768 = 0.305. The time base method has the advantage of producing a fast measurement result, without requiring the stopwatch to run for several hours. However, it has two disadvantages—it requires specialized equipment and it doesn't test the stopwatch display.

The totalize method does test the stopwatch display. However, it requires two measurement instruments, a universal counter and a signal generator, and both instruments should share a common external time base derived from a frequency reference (Fig. 4.16). The counter is placed in totalize mode, which is a mode that allows it to count the number of cycles of a gate frequency, which in this case will be the frequency produced by the signal generator. The period of the gate frequency should be at least 10× smaller than the resolution of the stopwatch, which means that a gate frequency of 1 kHz (1 ms period) will work when calibrating a typical stopwatch with 10 ms resolution. Once the gate frequency is connected, the totalize counter is started and stopped by rapidly pressing the start-stop button of the stopwatch against the start-stop button on the counter. The displays of both the stopwatch and counter are read, and the stopwatch uncertainty can be expressed as $\Delta t/T$, where Δt is the difference between the counter and stopwatch displays, and T is the duration of the calibration, which is obtained from the reading on the counter display. For example, if Δt = 20 ms and T = 1 hour (3,600,000 ms), the stopwatch uncertainty is 0.02 s per hour, or 5.6×10^{-6} [22]. The time base method can obtain the same information much faster than the totalize method, perhaps in 1 s rather than an hour, but the totalize method has the advantage of being able to test the stopwatch display.

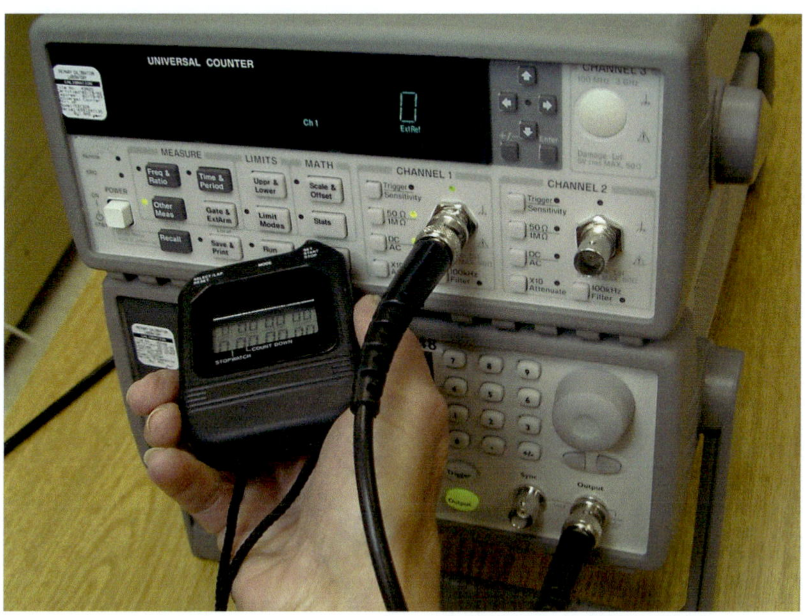

Figure 4.16 Calibrating a stopwatch with the totalize method.

Variations of the totalize method that involve photography have the potential for lower uncertainties. One such method, called the video totalize method, utilizes a camera that captures video at 1,000 frames per second, which is fast enough to record a display that updates every millisecond. In this method, two videos are recorded, one at the beginning and one at the end of the calibration. Both are recorded without stopping either the stopwatch or the counter, which eliminates any differences in latency between the buttons of the two devices. The start video is viewed frame-by-frame to search for an instant when the reading of the stopwatch changes, and a time difference measurement, TD_1, is recorded. The process is repeated with the end video to obtain TD_2. The stopwatch uncertainty can again be expressed as $\Delta t/T$, where $\Delta t = TD_2 - TD_1$, and T is the interval between the two measurements (obtained by subtracting the counter readings displayed when TD_1 and TD_2 were recorded). To get the lowest possible uncertainties out of this method, the counter has to update its display quickly enough to be synchronous with the gate frequency, and it has to simultaneously update each digit on the display. Therefore, a single-function totalize counter might produce better results than a universal counter operating in totalize mode [23].

4.2 Time Synchronization

Time synchronization is the process of setting or adjusting the time of a clock so that it agrees with a reference clock. In the most basic and most familiar form of synchronization, the time of the clock being synchronized is just forced or "set" to agree with the reference clock, without measuring the time difference between the clocks. This form of synchronization is practiced, for example, when we set a wristwatch. It's also practiced by the radio-controlled clocks we'll describe in Chapter 10.

In cases involving higher accuracy clocks, the first step in the synchronization process involves measuring the time difference between the clock being synchronized and the reference clock. If the clocks are located near each other, the time difference is measured using one of the methods just described in Section 4.1. If the clocks are far apart from each other, the time difference is measured with one of the time transfer methods that we'll discuss in Section 4.3. Once the time difference is known, we might decide that it's within an acceptable tolerance that meets our requirements. If so, the synchronization process is finished. However, if the time difference is outside our acceptable tolerance, the clock being synchronized needs to be adjusted. The adjustment might first involve a time step that immediately brings the clock closer to the reference time. Once the time difference is sufficiently small, it can be gradually reduced to near zero by adjusting the frequency of the clock's oscillator. This form of synchronization is practiced by disciplined clocks, which we'll describe in Section 5.3.2.

As we'll discuss in more detail in Chapters 8 and 9, the reference clock should have a known measurement uncertainty with respect to UTC, so that measurement traceability can be established. The reference clock also needs to output an on-time marker (OTM), which ideally is generated at a time coincident with the UTC second. If the reference clock generates 1 pps signals, the OTM is sent on either the rising or falling edge of the pulse (Fig. 4.2). The OTM can also be sent as part of a time code, typically at the beginning or end of a time code transmission. In this case, the time code provides a time stamp that indicates the time-of-day when the OTM was generated. Time codes are the topic of the next section.

4.2.1 Time Codes

A time code is a message containing time-of-day information. A time code can be used to label or time stamp when an event occurred. It can also be used to display the current time, or to synchronize a clock. Time codes contain information that includes the current UTC time-of-day and date. This information is sometimes provided in a format that doesn't have to be decoded—for example, the values for month, day, year, hours, minutes, and seconds might all be sent in separate fields. However, some time codes only send a single number, such as the number of elapsed seconds since a given epoch, and the date and time have to be calculated using this information. Time codes can also contain information about leap seconds (Section 3.5), about transitions to and from daylight saving time, about the current time difference between the UTC and UT1 time scales, and other information that's related to timekeeping. Time codes are often distributed via wired and wireless mediums as a free public service, for use by anyone within their coverage area. We'll describe many of those time codes in detail when we discuss radio and network controlled clocks (Chapters 10 and 11), and clocks controlled by satellite navigation systems (Chapter 12).

Earlier in this chapter, we discussed the resolution of the instruments that measure time interval. The resolution of a time code, sometimes called its granularity, is based on how closely it can label or time stamp the time-of-day. If a time code only transmits integer seconds, which is often the case with radio time codes, its resolution as received is limited to 1 s. However, some time codes achieve sub-second resolution, by including more digits to the right of the second's decimal point. Two examples of high-resolution time codes are those used by computers to time stamp synchrophasor measurements (Chapter 17) and financial transactions (Chapter 18). Achieving a time stamp resolution of 1 μs requires six digits to the right of the decimal point, and 1 ns resolution requires nine digits to the right of the decimal point (Fig. 4.17). To support a high-resolution time code, the time increments of the reference clock must be sufficiently small, and the transmission medium must be sufficiently stable. Otherwise, the least significant digits won't contain any meaningful information.

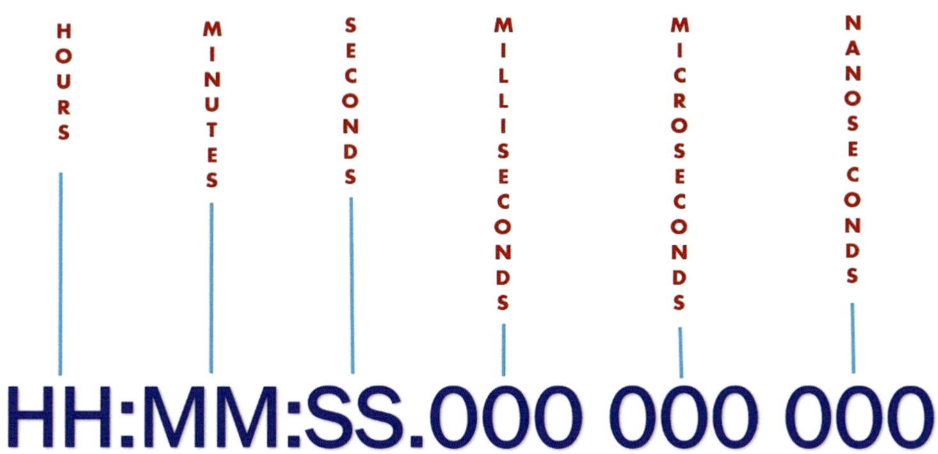

Figure 4.17 Time stamp resolution.

Standards for redistributing time codes within a facility were first developed in 1956 by the Inter-Range Instrumentation Group (IRIG), a group established by commanders of United States guided missile test ranges. The IRIG time code standards made it possible for different manufacturers to build compatible equipment. For example, the clocks controlled by satellite signals will often have an IRIG output that be connected to other devices using a coaxial cable. An IRIG output can control a time-of-day display, including the large displays sometimes seen by banks or schools, or it can provide a time reference to a computer that has compatible hardware. A diagram of the most common IRIG time code format, called IRIG-B, is shown in Fig. 4.18.

IRIG-B is usually distributed as either a DC level shift, pulse-width coded signal (unmodulated format), or as an amplitude modulated (AM) signal based on a sine wave carrier with a frequency of 1 kHz (modulated format). The time code frame has a period of 1 s, and 100 bits with a period of 10 ms each are transmitted during the frame. Every 10 ms a signal is sent by either raising the DC signal level or by sending a carrier. A 0 bit is sent by reducing the signal after 2 ms, and a 1 bit is sent by reducing the signal after 5 ms. If the signal is reduced after 8 ms, the bit serves as a frame synchronization marker. The bits convey time-of-day information (days, hours, minutes, seconds) in binary coded decimal (BCD) format or in straight binary format (seconds). Decoders that count carrier cycles can provide 1 ms resolution, equal to the period of the 1 kHz carrier [24, 25].

The SMPTE time code is less known to metrologists, but its use is essential to the entertainment, news, and recording industries. It was established by the Society of Motion Picture and Television Engineers in 1967 and it is revised when necessary to accommodate new media sources [26]. The SMPTE time code is used to label individual frames of video or film. The time code contains the hour, minute, second, and the frame number within the second, each represented by two digits that are sent in BCD format. The time codes added to every frame provide a time reference that makes it easy to locate and edit frames, and to synchronize video from multiple sources.

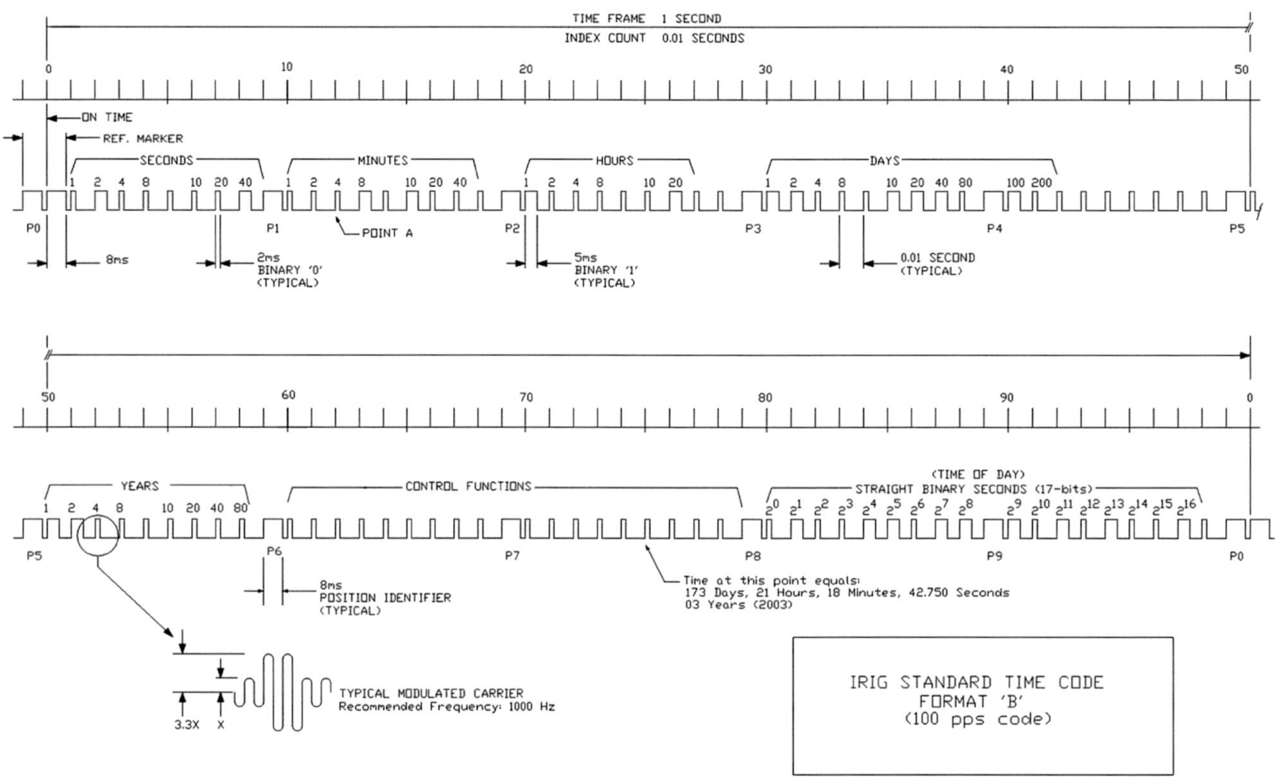

Figure 4.18 The IRIG-B Standard Time Code [24].

4.3 Time Transfer

Time transfer is the process of transferring time signals from one location to another, for the purpose of synchronizing or comparing clocks. In a typical time transfer system, the time kept by a reference clock at a given location is used to synchronize one or more clocks at remote locations. However, in some systems the remote clocks aren't synchronized, but instead are just compared to the reference clock to measure their time difference.

Time transfer measurements always have some uncertainty, and this uncertainty causes some of the accuracy of the reference clock to be lost during the comparison and synchronization processes. The amount of acceptable uncertainty depends on the application. For example, if our goal is to set the dashboard clock in our car to the time on our smartphone, we'll probably be happy if the clocks agree to within a few seconds, and we probably won't think about the uncertainty at all. However, if our goal is to use satellite signals to synchronize two clocks located in different countries to within 5 ns of each other, we'll have to think about uncertainty a lot. In that case, reducing the uncertainty to the acceptable number of nanoseconds will require compensating for every delay that's involved in the time transfer process.

The renowned physicist Albert Einstein (1879–1955) provided the foundation for time transfer, long before it became a well-researched area of metrology. In his classic 1905 paper on special relativity, "*Elektrodynamik bewegter Korper* (On the Electrodynamics of Moving Bodies)" [27], Einstein attempted to define the principle of simultaneity by describing a system where a central clock transmitted a time signal that was used to synchronize multiple secondary clocks. The time signals traveled at the constant speed of light, which meant that the secondary clocks closest to the central clock would be synchronized before the more distant clocks. Einstein objected to this, because it meant that true simultaneity was not achieved. The solution was to not set the secondary clocks to the central clock time, but to set them to the central clock time plus the travel time. Einstein proposed a system where the signal sent from the central clock to the distant clock was sent back to the central clock, which allowed the round-trip travel time to be measured. The distant clock could then be synchronized by adding one half of the round-trip travel time to the central clock time. This is the same synchronization method that's now commonplace in computer networks, as we'll discuss in detail in Chapter 11. Einstein realized that if we properly account for the travel time, the location of the clocks no longer matters when we attempt to achieve synchronization and true simultaneity [27, 28].

Of course, time transfer systems that didn't account for travel time existed before Einstein wrote about relativity, because sending time signals has historically been one of the first applications of a new communication medium. This was the case when Royal Observatory in Greenwich, England began sending time signals via telegraph lines in 1852, which was less than a decade after the first public telegraph service began [29]. And as we'll further describe in Chapter 10, the first radio time signals were sent by the United States Navy in 1904, which was just a few years after the Italian inventor Guglielmo Marconi (1874–1937) first sent wireless signals across the Atlantic Ocean [30].

Today's time transfer systems send signals through a variety of mediums using a variety of methods. The mediums include radio signals, coaxial cables, optical fibers, telephone lines, and computer networks. The methods can be organized into four general categories—one-way, loop-back, common-view, and two-way. The first two methods, one-way and loop-back, are the most common methods for the time-of-day synchronization of clocks. Therefore, they generally deliver both a time code and an OTM. The common-view and two-way methods are typically used for comparisons of high-accuracy clocks, and they generally only deliver an OTM in the form of a 1 pps signal.

As was noted earlier, the time difference between two unlabeled 1 pps signals can never exceed ±0.5 s, which means that the 1 pps signal from a clock under test can never be fast or slow by more than half a second with respect to a reference clock. Therefore, common-view and two-way time transfer (TWTT) systems usually operate with the assumption that the remote clock already has the correct time-of-day, and that its seconds are labeled correctly. To make this assumption hold true, common-view and TWTT systems sometimes obtain the time-of-day from a different time transfer system, such as an Internet time code service. The four time transfer methods are described in the next four sections.

4.3.1 One-Way Time Transfer

All time transfer systems have a reference clock at their source (point *A*). Information from the reference clock is encoded on a signal that is transmitted through a wired or wireless medium to the place (point *B*) where the remote clock is located. In the simplest form of time transfer, called the one-way method (Fig. 4.19), the remote clock is synchronized with the time from the reference clock. The one-way method is typically employed by systems that broadcast time to multiple receivers, and make time available to any receiver within the coverage area of the signal. The number of receivers is unknown to the transmitter and is virtually unlimited.

As previously noted, Einstein postulated about the "travel" time between clocks, which time metrologists usually call the path delay. In Fig. 4.19, the path delay through the medium between points *A* and *B* is indicated by the variable, d_{ab}. Even if the reference clock is a perfect time source, the accuracy of the time transferred to the remote clock can be no better than the uncertainty of the path delay measurement, which depends on how well we can measure d_{ab} [31]. This simple fact can be thought of as the first principle for all time transfer systems.

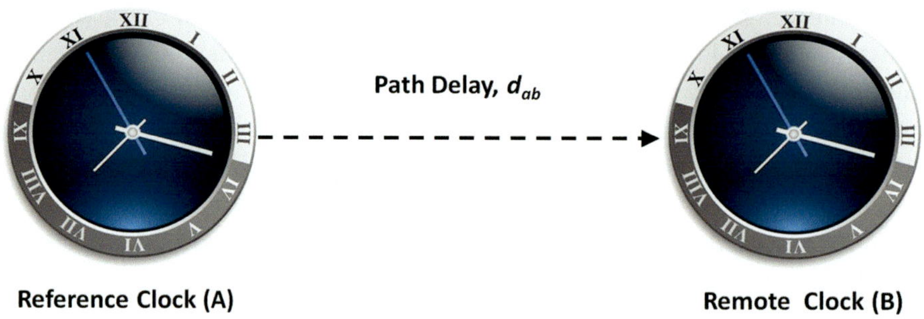

Figure 4.19 A one-way time transfer system.

To illustrate the concept of path delay, consider two clocks separated by a distance of 1,000 km. A radio signal from the reference clock travels across the 1,000 km path, where it is then used to synchronize a remote clock. All electromagnetic signals travel at the speed of light, *c*, which is a fundamental constant of 299,792,458 m/s for light in a vacuum (air reduces the speed by about 0.03%). The path delay, d_{ab}, is the path length divided by *c*, or about 3.3 ms in this example. This means that the reference clock time will be 3.3 ms late, or slow, when it's received by the remote clock.

For some types of one-way time transfer systems, the path delay isn't calculated at all, because it's considered to be small enough to ignore. For example, the purpose of a radio-controlled wall clock or wristwatch (Chapter 10) is often just to display the time-of-day, and it is unlikely that anyone viewing the clock will need time accurate to better than 1 s. If someone visually compares the clock display to that of a more accurate clock, the time difference caused by the path delay will likely be too small for their eyes to detect.

More sophisticated one-way time transfer systems compensate for the path delay to make the received time more accurate. One simple method is to send the OTM out early to compensate for at least some of the path delay. This method works best in systems where an average or minimum path delay can be estimated for all remote clocks. The OTM advance method was once implemented by NIST in a time code service operated from a geostationary satellite (Section 10.6). The reference clock was located on the ground, so the time signal had to travel from the Earth to the satellite and then back to Earth. The path delay depended on the remote clock's location, but was always within a range of 244 ms to 276 ms. Therefore, the OTM was sent 260 ms early to match the midpoint of the range, and this guaranteed that received time would be accurate to within ±16 ms throughout the signal's coverage area [32, 33]. Fixed OTM advances have also been implemented in telephone time transfer systems, where delays of tens of milliseconds are usually unavoidable [34]. Custom systems where a fixed path delay is introduced by a coaxial cable or fiber optic line can also benefit from this method. For example, if a time signal is sent via a coaxial cable between two buildings on the same campus, the delay of the cable can be measured or estimated, and the OTM can then be advanced by that amount.

The best one-way time transfer systems measure d_{ab}, instead of making an estimate, and use the measurement result to correct the time of the remote clock. Of course, this requires knowing the locations of the reference clock and of all of the remote clocks, so that we can calculate the delays between them. If the reference clock is stationary, a constant can be used for the clock position. If the reference clock is moving, such as a clock on-board a satellite, it must continuously broadcast its coordinates in addition to broadcasting a time signal. The same thing is true for the remote clocks. Fixed coordinates can be used if the clock is stationary, but the coordinates have to be continuously updated if the clock is moving.

We stated earlier that the accuracy of the time transferred to the remote clock can be no better than the uncertainty of the path delay measurement. Then, we provided a simple example where the distance between the clocks was 1,000 km and where the signal traveled at the speed of light, implying the remote clock should be adjusted by 3.3 ms to compensate for the path delay. However, that example involved many assumptions, including the assumption that 1,000 km represented the true distance between the clocks, which would need to be continuously verified, and also that the signal traveled the shortest possible path between the two clocks, which is a situation that will almost never be true. Some mediums are far more stable than others, but all path delays have some variability. The most advanced one-way time transfer systems, such as GPS, are capable of nanosecond-level accuracy; but achieving this requires measuring and compensating for each of the many factors that can add variability to the path delay. We'll look at those factors in detail in Chapter 12.

4.3.2 Loop-Back Time Transfer

In electronic systems—especially those used for telecommunications—such as serial interfaces, telephone circuits, and computer networks, loop-back tests are routinely used. A loop-back test routes an electronic signal or a digital data stream back to its source, usually without modifying the information that was originally sent. The purpose of a loop-back is generally just to test or debug a physical connection. For example, the transmit pin of a serial computer interface, such as RS-232, can be directly connected to the receive pin via a cable. This tests whether the device receives the same information that it transmitted, and by doing so it tests both the serial interface and the cable.

Loop-back tests are also routinely used in time transfer systems to measure round-trip path delay (Fig. 4.20). This is done by sending an OTM from a reference clock (*A*) to a remote clock (*B*) over the path d_{ab}. The remote clock (*B*) immediately sends the OTM back to (*A*) over the path d_{ba}. The RTD is equal to $d_{ab} + d_{ba}$, and the one-way path delay is assumed to be one half of the RTD. This method is common and easy to implement in computer networks.

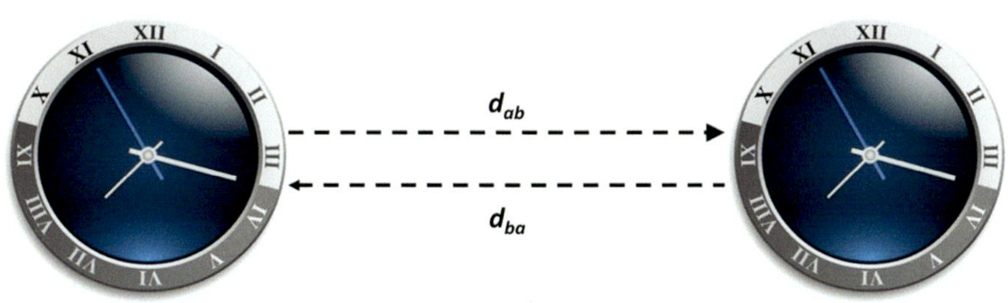

Figure 4.20 A loop-back time transfer system.

The method used to compensate for path delay varies slightly depending upon whether the RTD is known by the server clock or by the client clock. For example, the RTD was only known to the server clock (the reference clock) in a telephone time service operated by NIST. This service required the client computer to request the time by making a telephone call to the server. When the server answered the call, it sent a time code and an OTM to the client at a time T_1, and then waited for the client to return the OTM, marking its arrival at time T_2. The server could then estimate that RTD = $T_2 - T_1$, so it advanced the next OTM sent to the client by RTD/2, which was an estimate of the one-way path delay [34].

Time transfer systems that operate via the Internet, such as the Network Time Protocol (NTP), also work by having the client request the time [35]. However, the client records when the time was requested, T_1, and also records the time code and OTM were received, or T_2. Therefore, only the client (the remote clock) can estimate the RTD, and the client adds $(T_2 - T_1)/2$ to the received time to compensate for the one-way path delay. This method has an advantage, because the server doesn't have to advance the OTM for each client, allowing all of its time codes to stay synchronized to the reference clock. Time transfer via NTP and similar protocols will be discussed in detail in Chapter 11.

The accuracy of time transferred via the loop-back method is always limited by network asymmetry. If the network was symmetric, then the path delay would be the same in both directions, meaning that $d_{ab} = d_{ba}$, and dividing by two would provide a true estimate of the one-way path delay. However, in practice networks are asymmetric, meaning that the path delays are not the same in both directions. Due to asymmetry, estimating the one-way delay as half of the RTD will always introduce some uncertainty. In situations where time signals are sent over a wide area network (WAN) such as the public Internet, the outgoing and incoming signals might be routed over completely different paths, and an increase in network traffic on either path can slow down the signal. The upper limit of the time transfer uncertainty increases over a longer path, so the loop-back method usually works best when implemented over a local area network (LAN) where the RTD is small. The maximum amount of uncertainty that could occur in a loop-back system is 50% of the RTD. Of course, that could only occur in a hypothetical situation where 100% of the path delay was in one direction. In practice, the time transfer uncertainty is usually not more than a few percent of the RTD.

The loop-back method is sometimes confused with the TWTT method. It's true that the loop-back method is bidirectional, but the reference and remote clocks are sending messages to each other at different times over what can be very different paths. Therefore, the loop-back method is highly susceptible to network asymmetry. A true TWTT system requires the clocks at points A and B to simultaneously send and receive signals across the same path, which makes it much less susceptible to asymmetry. The TWTT method is discussed in Section 4.3.4.

4.3.3 Common-View Time Transfer

The one-way and loop-back methods generally send both an OTM and a time code to the remote clock so that it can be synchronized to agree with the reference clock. The common-view method can be used to synchronize the 1 pps output of a remote clock to a reference clock, but it doesn't deliver a time code. Its primary purpose is to compare clocks at two or more locations, and it does so by simultaneously comparing every clock involved in the comparison to a common-view signal (CVS) that can be received at all of the clock sites.

To better understand how the common-view method works, imagine two people, Juliana and Mary, who live in houses that are located in the same town. They each have a clock in their living room, and they decide one day to compare the times displayed by their two clocks. That would be easy enough to do if Juliana could take her clock to Mary's house or vice versa, because then they could just set the clocks next to each other and do a visual comparison. However, that's simply not practical, because the clocks are too large to move. After thinking about the problem, they have a clever idea. They know that every day around noon, plus or minus a minute or two, that a siren sounds in their town and that the siren is loud enough to be easily heard at both of their houses. They agree to write down the time displayed by their clock at the instant when they hear the siren, and they also agree to send each other a message that contains the time that they wrote down. By subtracting the time reading they sent from the time reading they received, they can each determine the time difference between their clocks. For example, if they know that Juliana's clock displayed 08:01:07 and Mary's clock displayed 08:01:17 at the instant the siren was heard, then they both know that time difference between the two clocks is 10 s. The sound of the siren is a CVS, and the CVS doesn't have to originate from a reference clock, nor does it have to be transmitted at any particular time. Its only requirement is to be receivable at all of the clock sites involved in the comparison.

The best ideas are often simple ideas, and common-view is a simple and effective time transfer method that's been practiced for decades. As discussed in Section 3.4, several different radio stations provided the CVS when the common-view method was used to compare ephemeris time to a cesium clock from 1955 to 1958, the series of measurements that led to the atomic definition of the second [36, 37]. Atomic clocks located in England, Switzerland, and the United States were routinely being compared via common-view by the early 1960s, with very low frequency (VLF) radio signals serving as the CVS [38]. Numerous other common-view systems followed, with a wide variety of signals serving as the CVS, including television signals, the AC power line, and even pulses from optical pulsars [39]. However, the common-view method didn't reach its full potential until after the launch of the first GPS satellites in 1978. It was soon found that the GPS satellites are an excellent CVS source because the path delays from the satellites can be accurately determined.

Common-view GPS time transfer was first demonstrated at NBS in 1980 [40] and it rapidly became the most widely used method for long-distance comparisons of atomic clocks [41]. There are many variations of the GPS common-view method. The code-based methods extract the CVS from the pseudo-random noise (PRN) codes broadcast by GPS, and the carrier-based methods extract the CVS from the GPS carrier frequency. Some methods use a single satellite that is receivable at the clock sites, other methods use all of the available satellites receivable at the clock sites and average the results. Other variations of the method—known as all-in-view methods—compare every clock to the average of all of the satellites that can be received at the site, even if some of those satellites can't be received at the other clock sites. In this case, some of the satellites involved in the comparison aren't actually in "common view," but the all-in-view method still works because the time signals from all of the GPS satellites closely agree with each other. The all-in-view method has the major advantage of allowing clocks located anywhere on Earth to be compared to each other, with no restrictions on the size of the coverage area.

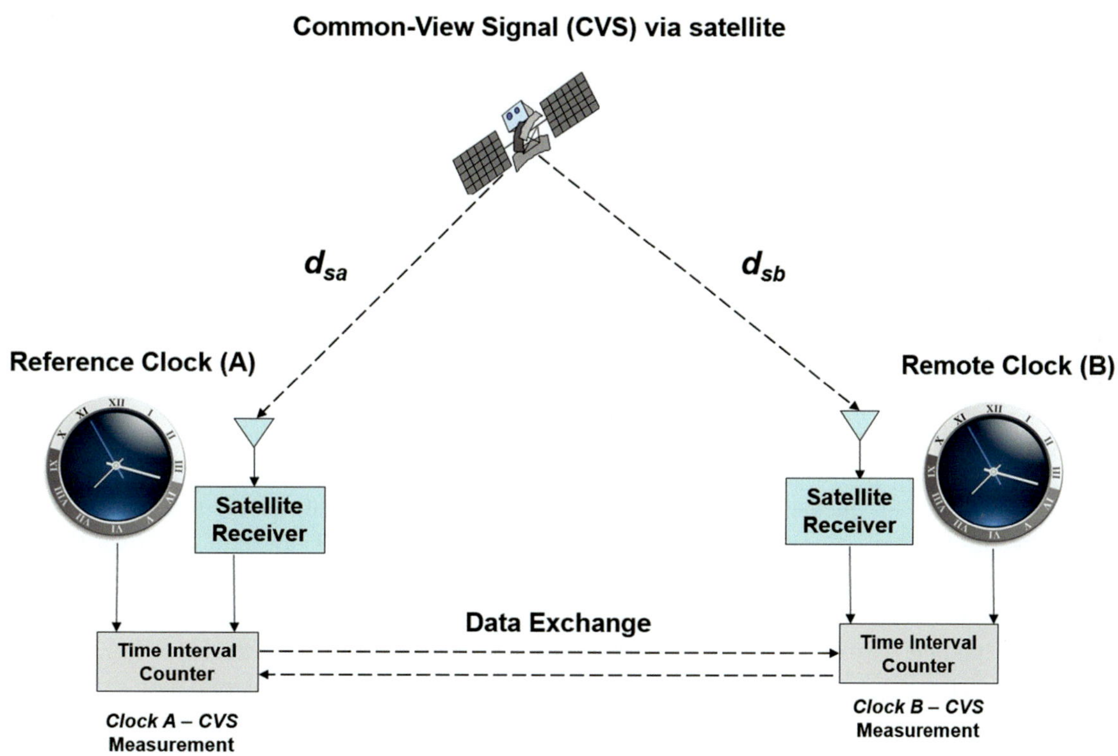

Figure 4.21 A common-view time transfer system.

Figure 4.21 shows a common-view time transfer system that compares a reference clock to a remote clock. The signals from a single satellite serve as the CVS and are simultaneously received at sites *A* and *B*. The clocks and receivers at both sites produce 1 pps signals, and the time interval between them is measured with a TIC. The site *A* measurement compares the CVS signal received over the path d_{SA} to the reference clock, producing the time difference *Clock A – CVS*. The site *B* measurement compares the

CVS signal received over the path d_{SB} to the local clock and produces the time difference *Clock B – CVS*. The two measurement results are then brought together, typically via the Internet, so they can be subtracted from each other. The difference between the two measurements is the time difference between the two clocks. Any delays that are common to both paths d_{SA} and d_{SB} will cancel even if they are unknown, but delays that aren't common to both paths contribute measurement uncertainty, resulting in an error term of d_{SA} – d_{SB}, which represents the differential delay between the two common-view systems. The time difference between clocks *A* and *B* is calculated as

$$Clock_A - Clock_B = (Clock_A - CVS) - (Clock_B - CVS) + (d_{SA} - d_{SB}). \tag{4.13}$$

The delays included in the d_{SA} – d_{SB} error term are either measured or estimated, and then applied as a correction to the measurement. The delays include not only delays between the CVS and the receiving antennas, but also delays that take place after the signal is received. For example, a common-view GPS system needs to account for delays added as the signal passes through the ionosphere and troposphere, for delays caused by multipath signal reflections, and for delays introduced by antenna coordinate errors. After the signal reaches the antenna, delays are introduced by the receiver, antenna, and antenna cable, and these delays also have to compensated to get the best results. The goal is to make d_{SA} – d_{SB} as small as possible. Some uncertainty will always be present, but the best common-view systems can routinely transfer time with uncertainties of less than 5 ns.

Common-view is a passive, receive-only method where the time signals travel in just one direction. The receiver in a common-view system, unlike the transceiver in a loop-back or TWTT system, doesn't send any signals back to the transmitter. However, at least one of the clock sites in a common-view system has to make its measurement data available. For example, if the remote clock sends its measurement data to the reference clock, the reference clock can compute the remote clock's time difference. Conversely, if the reference clock makes its own measurement data freely available, any number of remote clocks can compute their own time differences with respect to the reference clock. The latency of the data exchange determines how quickly common-view measurement results can be reported. If the common-view measurement data is made available in near real-time, it can serve the same purpose as the time signal in a one-way time transfer system, and be used to synchronize and control clocks [42]. More information about common-view time transfer is provided in Chapters 9 and 18.

4.3.4 Two-Way Time Transfer

The two-way method can measure and compensate for path delay with very little uncertainty, and can potentially outperform all other time transfer methods. In a TWTT system, the two clocks being compared each transmit their own time signal, and each receive a time signal sent by the other clock. The two time signals simultaneously travel across the same path through the same medium, although different communication channels may need to be used to prevent them from interfering with each other.

We noted earlier that sending time signals is often one of the first applications of a new communications medium. This was true in the case of satellites, because the TWTT method originated about five years after the first satellite, *Sputnik 1*, was launched in October 1957. In August 1962, an active-relay satellite named *Telstar 1* was used for the first two-way timing experiments, which allowed the United States Naval Observatory (USNO) and the Royal Greenwich Observatory (RGO) in the United Kingdom to perform a transatlantic comparison of their atomic clocks. These experiments demonstrated that clocks on either side of the Atlantic could be synchronized with an uncertainty of ±1 μs [43]. This was about 1,000× better than previous time transfer experiments, where the uncertainties were typically limited to milliseconds.

Modern TWTT systems that employ geostationary communication satellites can compare atomic clocks located on different continents with uncertainties near 1 ns [44]. Each clock is located near a satellite

Earth station that typically contains a spread spectrum satellite modem, a dish antenna, a TIC, and radio transmitting and receiving equipment (Fig. 4.22). Both Earth stations (*A* and *B*) simultaneously transmit time signals through a transponder on the same satellite. The transponder serves as a repeater, receiving the time signal from *A* and then retransmitting the signal so it can be received by *B*, and vice versa. By using their TIC, each station then measures the time difference between two 1 pps signals, one generated by the local clock and the other received via satellite from the remote clock. Station *A* records $TIC_A = CLK_A - (CLK_B + d_{ba})$, where CLK_A is the time from the local clock, CLK_B is the time from the remote clock, and d_{ba} is the path delay from *B* to *A*, which includes the delays introduced by the transmitting modem, the satellite uplink, the satellite transponder, the satellite downlink, and the receiving modem. Station *B* records $TIC_B = CLK_B - (CLK_A + d_{ab})$, where d_{ab} is the path delay from *A* to *B* [45].

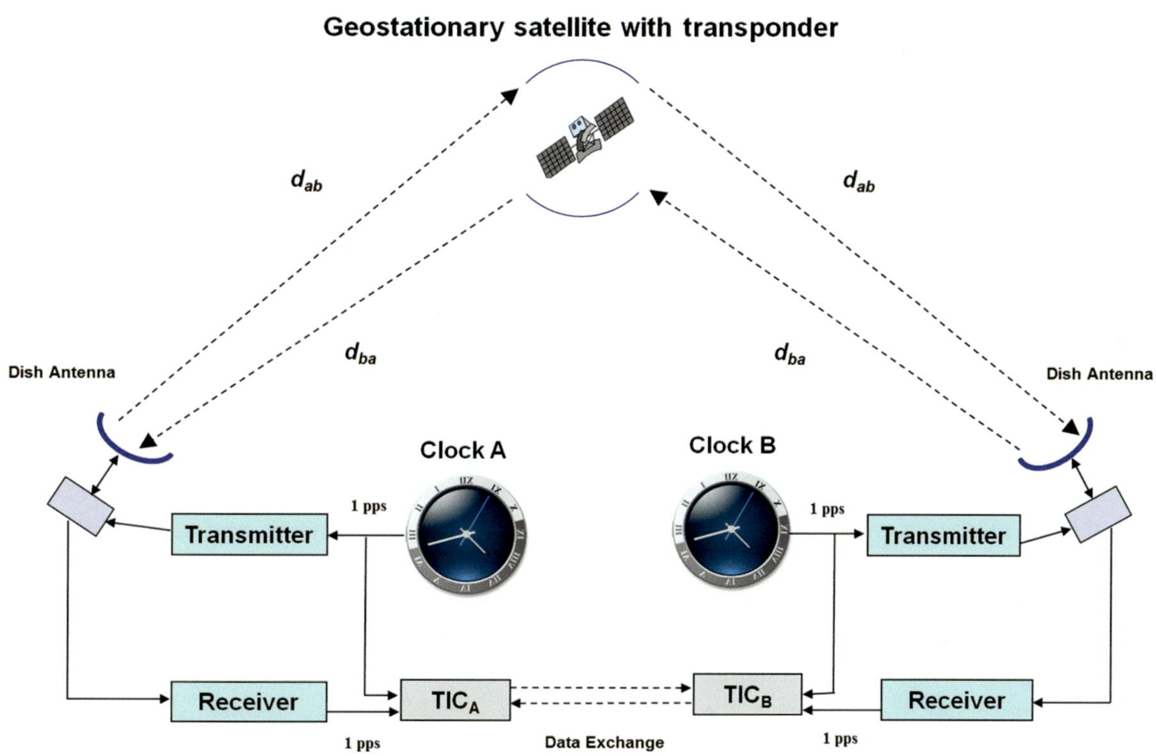

Figure 4.22 A two-way time transfer system that utilizes a geostationary satellite.

The two stations then exchange their measurements. The time difference between clocks *A* and *B* is calculated as

$$Clock_A - Clock_B = \frac{TIC_A - TIC_B}{2} - \frac{d_{ab} - d_{ba}}{2}. \tag{4.14}$$

If both paths had the same delay, only the left side of the equation, $\frac{TIC_A - TIC_B}{2}$, would be necessary. However, the right side of the equation, $\frac{d_{ab} - d_{ba}}{2}$, is needed to account for the differential delay between the two paths.

We noted in Section 4.3.2 that the loop-back method is sometimes confused with the two-way method, but the methods are quite different. The remote and reference clocks in a loop-back system exchange time signals at different times, a half-duplex method of communication. As illustrated in Fig. 4.22, TWTT systems are full-duplex, and require the simultaneous transmission of time signals from both clocks. As a result, the reciprocal path delays in TWTT systems are more symmetric than the path delays in loop-back systems. Nanosecond-level differences in the two path delays are still introduced by delays in the transmit and receive hardware, which are typically different at both sites, and by differences in RF propagation when different uplink and downlink frequencies are used. These delay differences have to be measured and compensated to obtain the lowest possible uncertainties [44–46].

The transmission of time signals also distinguishes the active two-way method from the passive common-view method (Section 4.3.3), where time signals are only received and not transmitted. When satellites are involved, the two-way method is far more expensive and complex to implement than common-view. More equipment is involved, including amplifiers and spread spectrum modems. Costs are involved, because transponder time on communication satellites has to paid for, and government licensing is usually required before signals can be transmitted. For these and other reasons, TWTT via satellite isn't especially common. It is used, however, by many of the national timing laboratories that contribute data to UTC (Section 9.4.4).

Implementing a TWTT system via a wired network connection is potentially much easier and less expensive than implementing a satellite system. For example, the two-way method can be utilized over a dedicated network to compare clocks located at neighboring laboratories or universities, or to transfer high-accuracy time throughout a building or campus. These types of networks often have smaller uncertainties than any satellite-based time transfer system. Time transfer uncertainties near or less than 100 ps (0.1 ns) have been demonstrated over short distances of hundreds of meters with coaxial cables [47], and over long distances of tens or hundreds of kilometers with optical fibers [48, 49]. Optical time transfer systems that utilize the two-way method are described in detail in Chapter 14.

4.4 Ensemble Time Scales

Throughout this chapter, we've described time measurements made with respect to a reference clock. For many time measurements, especially measurements made with respect to national or international time standards, the reference clock won't be a single clock. Instead, the reference clock will consist of a group of clocks called an ensemble time scale. When used as a reference clock, an ensemble time scale has several advantages over a single clock—it gives us more confidence that the time is correct, it's less likely to fail, and it's potentially more stable.

Our confidence that the time is correct increases with multiple clocks, because when we have just one clock, we no choice other than to accept and trust its time. The situation improves only slightly with two clocks, because if the two clocks disagree with each other, we have no way of knowing which clock is correct. However, if we have three clocks, we can form an ensemble time scale by continuously measuring the time differences between them. If one clock doesn't agree with the other two, we'll know that the disagreeing clock is wrong, and we can ignore its time. Having four or more clocks gives us even more confidence, because then we can continue our clock comparisons even if one or more clocks fail.

Ensemble time scales are more reliable than single-clock time scales for an obvious reason—a single clock can fail at any time, but barring an uncontrollable disaster, it's highly unlikely that every clock in an ensemble would fail at the same time. They are potentially more stable than a single clock because the individual clock signals can be statistically combined to produce a composite signal that outperforms the best clock in the ensemble [50]. The combining is done with a weighted-average algorithm. The algorithm first determines a performance model for each clock based on the time difference measurements. Each clock is then assigned a weight for the algorithm to use, based on its stability, with the sum of the clock weights equal to 1. Because future performance is easier to predict for a stable clock, the most stable clocks receive the most weight. The weighted-average algorithm predicts the future time error of each clock based on previous measurements. When the clocks are measured again, the predicted values are compared to the measured values, and the clock parameters are updated to keep the values stable [51].

Using the collected data, the time scale algorithm adjusts an output generator so that it produces signals that agree with the time scale's computation. The output signals serve as a time standard. The algorithm only adjusts the output generator, it doesn't adjust the individual clocks. The algorithm detects when an individual clock fails or has a bad reading. When this happens, the malfunctioning clock is removed from the weighted average, and its weight is redistributed amongst the other clocks. This prevents a malfunctioning clock from disturbing the ensemble's output signals [50, 51].

The concept of an ensemble time scale can be attributed to William Markowitz of the USNO, who was previously mentioned in Section 3.4 for his contributions to the atomic definition of the second. In 1959, Markowitz began calculating international atomic time from a weighted average of nine cesium clocks located at laboratories in North America and Europe, including the USNO and NBS. This time scale was called A.1, and can be thought of as an early version of UTC. Like UTC, A.1 was a virtual time scale that didn't produce physical signals. Instead, it calculated a time estimate based on measurements of a group of geographically dispersed clocks. Because A.1 was formed in the pre-satellite era, all of its clock data was collected with time transfer systems based on terrestrial radio signals [52].

The NBS statistician Edwin Crow (1916–2005), citing the work of Markowitz, published a general metrology study in 1964 on the use of composite standards, which were formed by combining data from several laboratory standards. Crow noted that it was well known, based on probability theory, that if independent observations of an instrument or object had a standard deviation, σ, it followed that the mean of n such observations would have a standard deviation of $\sigma/\sqrt{n}$. Therefore, it also followed that "if the goodness of a standard is judged by the standard deviation of realizations of the standard, the combination standard using the mean of n independent instrumentations is better than any single instrumentation." Crow went on to write that a properly weighted mean would have the advantage of being less affected by outlying measurements than a simple mean, where every contributing standard was given equal weight [53].

At the time Crow published his study, a local ensemble time scale was already operational at NBS in Boulder, Colorado, developed by a team led by the American physicist James Barnes (1933–2002), and first constructed in July 1962. Known as NBS-A, the ensemble consisted of five clocks, four quartz clocks and one rubidium, that were allowed to free run without adjustment. Each clock was compared for about 15 minutes each day to the U. S. frequency standard, a cesium beam clock built at NBS. Instead of a weighted-average software algorithm, NBS-A produced its output signals by coupling mechanical ball-and-disk integrators to a phase shifter that was referenced to a quartz oscillator. This resulted in a uniform time scale that was always stable to within a few microseconds in time and to within a few parts in 10^{12} in frequency, a significant accomplishment for that era [54].

By 1969, time scales based on cesium clocks and weighted-average algorithms were operational at both the USNO and NBS. The Austrian-American physicist Gernot Winkler (1922–2016) and his colleagues at the USNO built an atomic scale that produced a composite output based on the weighted average of 16 cesium clocks [55]. The statistic used to estimate clock stability and to determine clock weight factors was not the classical standard deviation. Instead, it was a similar statistic, better suited for atomic clock analysis, that was described by the American physicist David Allan (b. 1936) in 1966 while working at NBS [56]. This statistic, later known as the Allan deviation, is discussed in detail in Chapter 7.

The weighted-average time scale built by Barnes and Allan and their colleagues was named AT(NBS). It used the Allan deviation to estimate clock stability, and it included inputs for eight cesium clocks [57]. The source code for the AT(NBS) time scale originally consisted of less than 100 lines of FORTRAN code [58]. From these beginnings, the ensemble time scales at the USNO and NBS/NIST continued to evolve, becoming more sophisticated as clock technology, measurement system technology, and time scale algorithms constantly improved. As this was happening, ensemble time scales were being constructed at numerous other laboratories.

Ensemble time scales now serve as the national standards of frequency and time in numerous countries, including Belarus, Belgium, Brazil, Canada, China, the Czech Republic, Finland, France, Germany,

India, Israel, Italy, Japan, Mexico, the Netherlands, Norway, Panama, Poland, Russia, Singapore, South Korea, Spain, Sweden, Switzerland, Taiwan, Thailand, the United Arab Emirates, the United Kingdom, and the United States [59]. All include at least three atomic clocks, often all cesium clocks (Section 3.6), but they sometimes include a combination of cesium clocks and hydrogen masers (Section 3.10).

In addition to atomic clocks, ensemble time scales require other equipment—a measurement system, a computer and time scale algorithm software, a generator that outputs the frequency and time signals defined by the time scale's calculations, and a reference plane where the signals are distributed. An ensemble time scale must also be able to accept external corrections, so that its time can be initially set and then continuously coordinated to agree with UTC. Ensemble time scale designs have been patented [60] and marketed by several companies, but most national standards were designed in-house and customized as necessary to work with the specific clocks and measurement hardware already owned by the laboratory. Therefore, none of them are exactly alike. Even so, we can take a generalized look at each of the time scale components, going into a little more detail than we have so far.

The main function of a time scale measurement system is to continuously record the time differences between the clocks in the ensemble. It's not possible to measure the absolute time difference of any of the clocks, the only possible observation is to measure the time differences between them [61]. One clock is designated as a relay or pivot clock, and each of the other clocks are compared to the pivot clock at a fixed measurement interval. The pivot clock is often selected based on its dependability or longevity, and is not necessarily the best clock in the time scale [51, 62].

Several types of instruments can be used to measure the time differences between the clock, and the instrument that's chosen often depends on the resources available to the laboratory. If time outputs (1 pps) are measured, the measurement instrument is typically a TIC (Section 4.1.2). If a separate TIC is used to measure each clock, the clocks are measured simultaneously, with signals from the reference distributed to every TIC, typically to the stop input. If just one TIC is used, the clocks are measured in sequence. The reference is always connected to the TIC, typically to the stop input. The other clocks are connected to a multiplexer that switches between them, periodically sending each clock's signal to the TIC for comparison to the reference.

If frequency outputs (5 MHz or 10 MHz) are measured, the measurement instrument will likely be a dual-mixer time difference (DMTD) system (Section 5.2.4) or a digital sampling system (Section 5.2.5) that was designed to provide separate measurement channels for every clock, allowing simultaneous measurements. Figure 4.23 is an example of a time scale measurement system used at NIST that includes four DMTD systems, each with four channels, providing a total of 16 clock measurement channels. The chassis also contains a computer and signal distribution amplifiers.

Figure 4.23 A time scale measurement system with input channels for 16 clocks.

A time scale algorithm is a mathematical method of estimating the time error of each clock in the ensemble, correcting each clock's time by its estimated error, and then combining the clock signals with a weighted-average calculation to form a single composite signal that's made as stable as possible [63]. Ideally, the composite signal will be stable enough to serve as the oscillator for a local clock whose time can be coordinated to agree with UTC [64]. Perhaps not surprisingly, the time scale algorithms used for the local clock ensembles that are coordinated to agree with UTC are similar to those used to compute UTC [65].

Because time scales have to run continuously, any type of failure or interruption is considered unacceptable. Therefore, time scale algorithms are designed to handle every conceivable type of clock behavior. This makes time scale algorithms exceptionally robust, but it also makes them inherently complex, so we'll limit our discussion to some basic principles. For an ensemble time scale consisting of N atomic clocks, we can refer to the individual clocks as C_j, where $1 \leq j \leq N$. One clock, C_p, is selected as the pivot clock to provide access to the time scale. At the measurement time t_k, the time difference between C_j and C_p, is measured, or $x_{jp}(t_k)$. The time difference measurements allow us to construct a performance model for every clock that consists of three parameters, its time difference, x, its frequency difference, y, and its frequency aging rate, d. Because d is often insignificant for the short measurement intervals used by most time scales, it is typically either applied as a constant or ignored. Therefore, if we know x and y for a measurement of C_j made at the time t_k, we can estimate the time of C_j at the next measurement time, $t_k + 1$, as

$$C_j(t_{k+1}) = x_j(t_k) + y_j(t_k)\tau, \tag{4.15}$$

where $\tau = t_{k+1} - t_k$, which is simply the fixed interval between time scale measurements. Once all of the time differences $x_{jp}(t_k)$ are known, a clock model is also constructed for C_p. When the next measurement time, $t_k + 1$, actually occurs, the previously predicted values are compared to the new measurements, the clock times are corrected, and the clock models are updated [46, 51, 62, 65].

As previously noted, future performance is easier to predict for a stable clock, so it follows that the most stable clocks will have the smallest differences between their predicted time and their measured time. Therefore, the most stable clocks receive the most weight in the time scale average. If all of the clocks were equally weighted, they would each be given a weight of the inverse of N, or $1/N$, with the sum of their weights equaling 1. Instead, the weight of each clock, w_j, is determined by the inverse of either its time stability, $\sigma_x(\tau)$, or its frequency stability, $\sigma_y(\tau)$, where τ is the period during which the stability was estimated. This period is chosen to be long enough to reduce the measurement system noise to an acceptable level.

Two other factors can affect the clock weighting process. First, if the difference between a clock's predicted time and its measured time exceeds a pre-defined threshold, we can assume that something bad has happened to the clock that caused it to be unstable. The weight of the malfunctioning clock is then immediately set to 0 so that it can't harm the ensemble's output, and the weight that it previously had is redistributed to the other clocks. The algorithm will reassign weight to the clock when it determines that it's working properly. Second, most time scale algorithms specify a maximum clock weight, for example, limiting their weight to 0.2 or 20%, so the calculated weights of some clocks may need to be lowered. A maximum weight prevents a single clock from dominating the time scale, and in turn protects the time scale from excessive reliance on one clock. Once everything regarding the clock weights has been taken care of, the sum of the clock weights is normalized to equal 1 [51, 62]. The time output by the ensemble time scale, $TS(t)$, is then calculated as the weighted average of the corrected time $t_j(t)$ given by every clock C_j in the ensemble,

$$TS(t)=\sum_{j=1}^{N} w_j(t) \times t_j(t). \tag{4.16}$$

As this discussion has shown, estimating clock errors is essential to time scale algorithms, and some algorithms employ Kalman filters to help with the estimation [60, 63]. Kalman filters are named after the Hungarian-American electrical engineer Rudolf Kalman (1930–2016) who published his algorithms in 1960 [66]. They can accurately estimate the hidden states of a system in the presence of measurement uncertainty. They do so by providing a recursive solution that estimates the future system state based on past estimations, and they're widely used by tracking, guidance, and navigation systems.

Using Kalman filters to estimate the time of atomic clocks was perhaps first suggested by Allan and Barnes at NBS in 1980 [61], and their properties, especially their recursive solutions, have since been considered ideal for some time scale applications [67]. However, they are not always included in time scale algorithms, because situations such as non-stationary clock behavior caused by phase and frequency jumps, changes in clock weights, and the addition and removal of clocks to and from the ensemble, can be challenging to address with Kalman filtering. These challenges sometimes negate any potential performance gains [51, 67].

Now that we've provided an overview of the clocks, measurement systems, and algorithms included in an ensemble time scale, let's look at the entire system using the ensemble time scale at NIST in Boulder, Colorado as an example. Figure 4.24 provides a simplified diagram of this time scale, which is a local approximation of UTC known as UTC(NIST). The clock ensemble is shown on the left side of the diagram. It consists of a combination of cesium clocks and hydrogen masers that each output 5 MHz sine wave signals. These signals are measured with the multi-channel measurement system (MCMS) shown in Fig. 4.23. The measurements are recorded and processed by a time scale computer (TSPC) that runs the software that computes the time scale solution (Eq. 4.16). The TSPC then programs the output generator to produce clock signals that physically realize the time scale computations.

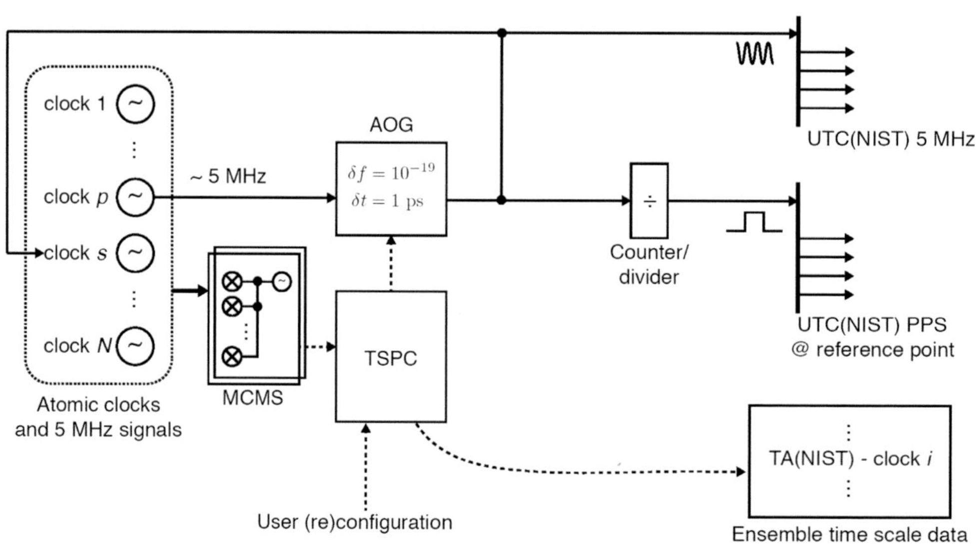

Figure 4.24 Simplified diagram of the UTC(NIST) time scale [64].

The output generator of an ensemble time scale can be a signal generator that uses an external atomic clock as its time base oscillator, or it can be an atomic clock that isn't part of the ensemble. In either case, it has to be able to accept digital time and frequency corrections, preferably with very high resolution, because the corrections are likely to be very small. The UTC(NIST) time scale uses a device called an auxiliary output generator (AOG) that accepts frequency corrections as small as 1×10^{-19}, and time corrections as small as 1 ps. The AOG outputs 5 MHz sine waves characterized by the time scale computations, that are split into two paths. One path goes directly to an amplifier that distributes 5 MHz signals as a frequency standard. The other path goes to a counter/divider unit that divides the 5 MHz to 1 pps. When the time scale was started, the 1 pps signals were synchronized to UTC. From that point forward, the UTC synchronization is maintained by small frequency or time corrections that are sent to the AOG as often as once per week. The synchronized time pulses, calibrated to compensate for all delays, are routed to distribution amplifiers (Fig. 4.25) that form the reference plane for UTC(NIST) [64].

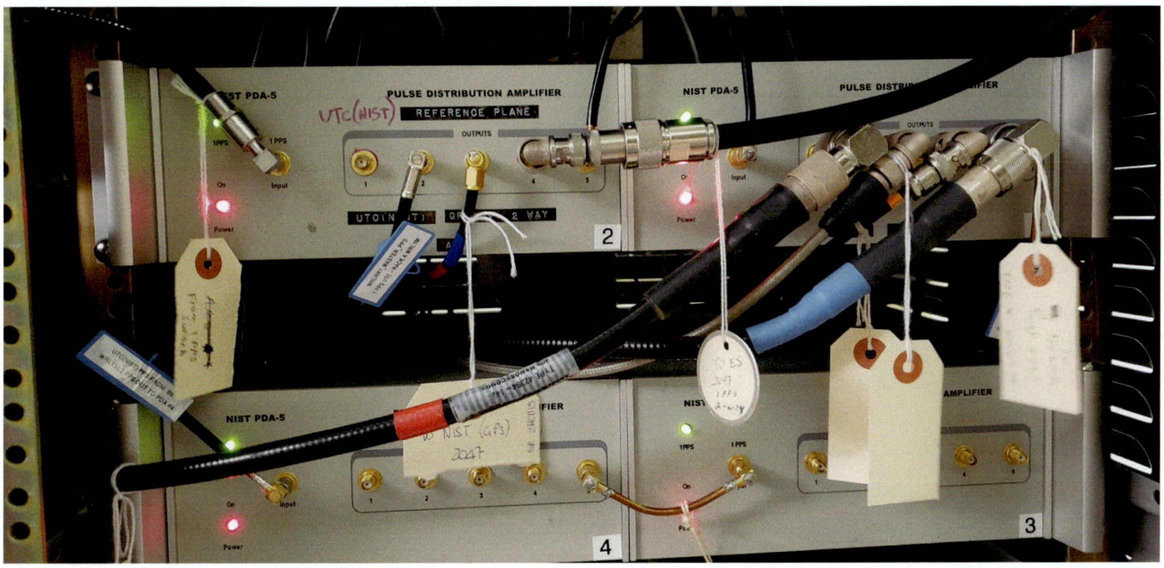

Figure 4.25 The pulse distribution amplifiers that form the UTC(NIST) reference plane.

The adjustments made to UTC(NIST) not only keeps its time synchronized to within a few nanoseconds of UTC, but also keeps its frequency in close agreement with the UTC frequency, a process called syntonization [68]. This allows UTC(NIST) to serve as a United States national standard for both time and frequency. Frequency measurements are the topic of Chapter 5.

References

1. J. Canales, *A Tenth of a Second*, The University of Chicago Press, 2009.
2. J. J. Carr, *Practical Radio Frequency Test & Measurement: A Technician's Handbook*, Newnes, 1999.
3. J. J. Carr, *Elements of Electronic Instrumentation and Measurement*, Prentice-Hall, 3rd edition, 1996.
4. A. S. Bagley, "The High-Speed Frequency Counter – A New Solution to Old Problems," *Hewlett-Packard Journal*, vol. 2, no. 5, pp. 1–4, January 1951.
5. A. S. Bagley, D. Hartke, and W. D. Myers, "Frequency and Time Measurements with the New-hp-High Speed Counter," *Hewlett-Packard Journal*, vol. 5., no. 7–8, pp. 1–8, March-April 1954.
6. R. G. Baron, "The Vernier Time-Measuring Technique," *Proceedings of the Institute of Radio Engineers (IRE)*, vol. 45, no. 1, pp. 21–30, January 1957.
7. E. D. Peterson and R. D. Valenti, "Time Interval Measuring System Employing Vernier Digital Means and Coarse Count Ambiguity Resolver," *United States Patent 3,218,553* (Applied for May 29, 1961, granted November 16, 1965).
8. K. Klepacki, R. Szplet, and R. Pelka, "A 7.5 ps Single-Shot Integrated Time Interval Counter with Segmented Delay Line," *Review of Scientific Instruments*, vol. 85, pp. 034703/1–10, March 2014.
9. R. Nutt, "Digital Time Intervalometer," *Review of Scientific Instruments*, vol. 39, no. 9, pp. 1342–1345, September 1968.

10. R. Nutt, "Digital Time Intervalometer with Analogue Vernier Timing," *United States Patent 3,541,448* (Applied for May 7, 1968, granted November 17, 1970).
11. R. Nutt, K. Milam, and C. W. Williams, "Digital Intervalometer," *United States Patent 3,983,481* (Applied for August 4, 1975, granted September 28, 1976).
12. V. S. Zhang, D. D. Davis, and M. A. Lombardi, "High Resolution Time Interval Counter," *Proceedings of the 1994 Precise Time and Time Interval Meeting*, pp. 191–200, December 1994.
13. M. A. Lombardi, "Remote Frequency Calibrations: The NIST Frequency Measurement and Analysis Service," *National Institute of Standards and Technology Special Publication 250–229*, 90 p., June 2004.
14. J. Kalisz, "Review of Methods for Time Interval Measurements with Picosecond Resolution," *Metrologia*, vol. 41, no. 1, pp. 17–32, December 2003.
15. Hewlett-Packard Company, "Fundamentals of Time Interval Measurement," *Hewlett-Packard Application Note 200–203*, 5965–7663E, 68 pp., June 1997.
16. Hewlett-Packard Company, "Fundamentals of the Electronic Counters," *Hewlett-Packard Application Note 200*, 5965–7660E, 44 pp., March 1997.
17. United States Department of Defense, "General Specification for Cables, Radio Frequency, Flexible and Semirigid," *MIL-DTL-17H*, including Supplement 1, August 2006.
18. G. D. Rovera, M. Siccardi, S. Romisch, and M. Abgrall, "Time Delay Measurements: Estimation of the Error Budget," *Metrologia*, vol. 56, no. 035004, pp. 1–9, May 2019.
19. D. Rovera, M. Abgrall, P. Ulrich, and M. Siccardi, "Techniques of Antenna Cable Delay Measurement for GPS Time Transfer," *Proceedings of the 2015 Joint Conference of the IEEE International Frequency Control Symposium and the European Frequency and Time Forum (IFCS/EFTF)*, pp. 239–244, April 2015.
20. "The Birth of the Chronograph: Rieussec," Montblanc Special Edition of *WatchTime*, pp. 14–21, 2011.
21. E. Doerr, "History Rebooted: The Chronograph's Inventor is...Louis Moinet!," *Forbes*, March 24, 2013.
22. J. C. Gust, R. M. Graham, and M. A. Lombardi, "Stopwatch and Timer Calibrations," *National Institute of Standards and Technology Special Publication 960–912*, 66 p., January 2009.
23. C. M. Tsui, Y. K. Yan, and H. M. Chan, "Calibration of Stopwatches by Utilizing Video Recordings and a Synchronous Counter," *NCSLI Measure: The Journal of Measurement Science*, vol. 6, no. 3, pp. 64–71, September 2011.
24. Range Commanders Council, Telecommunications and Timing Group, "IRIG Serial Time Code Formats," *IRIG Standard 200–216*, August 2016.
25. K. Behrendt and K. Fodero, "The Perfect Time: An Examination of Time-Synchronization Techniques," *Proceedings of the 33rd Annual Western Protective Relay Conference*, Spokane, WA, 19 p., October 2006.
26. SMPTE, "ST 12-1:2014 - SMPTE Standard – Time and Control Code," *SMPTE ST 12-1:2014,* February 2014.
27. A. Einstein, "Elektrodynamik bewegter Korper (On the Electrodynamics of Moving Bodies)," *Annalen der Physik*, vol. 17, pp. 891–921, 1905.
28. P. Gallison, *Einstein's Clocks, Poincare's Maps*, New York: W. W. Norton & Company, 2003.
29. D. Howse, *Greenwich Time and the Longitude*, London: Philip Wilson Publishers, Ltd., 1997.
30. L. S. Howeth, *History of Communications-Electronics in the United States Navy*, Bureau of Ships and Office of Naval History, U. S. Government Printing Office, 1963.
31. J. L. Jespersen, "A Survey of Time and Frequency Dissemination Techniques," *Proceedings of the 24th Annual Symposium on Frequency Control*, pp. 322–324, 1970.
32. J. V. Cateora, D. D. Davis, and D. W. Hanson, "Automatic Path Delay Corrections to GOES Satellite Time Broadcasts," *NBS Technical Note 1003*, 52 p., February 1978.
33. M. A. Lombardi and D. W. Hanson, "The GOES Time Code Service, 1974–2004: A Retrospective," *Journal of Research of the National Institute of Standards and Technology*, vol. 110, no. 2, pp. 79–96, March-April 2005.
34. J. Levine, M. Weiss, D. D. Davis, D. W. Allan, and D. B. Sullivan, "The NIST Automated Computer Time Service," *Journal of Research of the National Institute of Standards and Technology*, vol. 94, no. 5, pp. 311–321, September-October 1989.
35. D. L. Mills, *Computer Network Time Synchronization: The Network Time Protocol*, CRC Press, 2006.

36. S. Leschiutta, "The Definition of the 'Atomic' Second," *Metrologia*, vol. 42, no. 3, pp. S10–S19, June 2005.

37. T. Jones, *Splitting the Second: The Story of Atomic Time*, Bristol and Philadelphia: Institute of Physics Publishing, 2000.

38. A. M. J. Mitchell, "Frequency Comparison of Atomic Standards by Radio Links," *Nature*, vol. 198, pp. 1155–1158, June 22, 1963.

39. D. W. Allan, H. E. Machlan, and J. Marshall, "Time Transfer Using Nearly Simultaneous Reception Times of a Common Transmission," *Proceedings of the 26th Annual Symposium on Frequency Control*, pp. 309–316, June 1972.

40. D. W. Allan and M. A. Weiss, "Accurate Time and Frequency Transfer During Common-View of a GPS Satellite," *Proceedings of the 34th Annual Symposium on Frequency Control*, pp. 334–346, May 1980.

41. W. Lewandowski, J. Azoubib, and W. Klepczynski, "GPS: Primary Tool for Time Transfer," *Proceedings of the IEEE*, vol. 87, no. 1, pp. 163–172, January 1999.

42. M. A. Lombardi and A. P. Dahlen, "A Common-View Disciplined Oscillator," *Review of Scientific Instruments*, vol. 81, no. 5, pp. 055110/1-6, 2010.

43. J. M. A. Steele, W. Markowitz, and C. A. Lidback, "Telstar Time Synchronization," *IEEE Transactions on Instrumentation and Measurement*, vol. IM-13, pp. 164–170, 1964.

44. A. Bauch, "Time and frequency comparisons using radiofrequency signals from satellites," *Comptes Rendus Physique*, vol. 16, no. 5, pp. 471–479, June 2015.

45. D. W. Hanson, "Fundamentals of Two-Way Time Transfers by Satellite," *Proceedings of the 43rd Annual Symposium on Frequency Control*, pp. 174–178, May 1989.

46. M. Desaintfuscien, *Data Processing in Precise Time and Frequency Applications*, Springer, 2007.

47. P. Panek, J. Kodet, and I. Prochazka, "Accuracy of Two-Way Time Transfer via a Single Coaxial Cable," *Metrologia*, vol. 50, no. 1, pp. 60–65, 2013.

48. M. Rost, D. Piester, W. Yang, T. Feldmann, T. Wübbena, and A. Bauch, "Time Transfer through Optical Fibres Over a Distance of 73 km with an Uncertainty below 100 ps," *Metrologia*, vol. 49, no. 6, pp. 772–778, October 2012.

49. O. Lopez, A. Kanj, P. E. Pottie, D. Rovera., J. Achkar, C. Chardonnet, A. Amy-Klein, and G. Santarelli, "Simultaneous Remote Transfer of Accurate Timing and Optical Frequency Over a Public Fiber Network," *Applied Physics B*, vol. 110, pp. 3–6, October 2012.

50. D. W. Allan, N. Ashby, and C. C. Hodge, "The Science of Timekeeping," *Agilent Technical Note 1289*, 88 p., 2000.

51. J. Levine, "The Statistical Modeling of Atomic Clocks and the Design of Time Scales," *Review of Scientific Instruments*, vol. 83, pp. 021101/1–28, February 2012.

53. W. Markowitz, "The Atomic Time Scale," *IRE Transactions on Instrumentation*, vol. I–11, nos. 3–4, pp. 239–242, December 1962.

53. E. L. Crow, "The Statistical Construction of a Single Standard from Several Available Standards," *IEEE Transactions on Instrumentation and Measurement*, vol. IM-13, no. 4, pp. 180–185, December 1964.

54. J. A. Barnes, D. H. Andrews, and D. W. Allan, "The NBS-A Time Scale – Its Generation and Dissemination," *IEEE Transactions on Instrumentation and Measurement*, vol. IM-14, no. 4, pp. 228–232, December 1965.

55. G. M. R. Winkler, R. G. Hall, and D. B. Percival, "The U. S. Naval Observatory Clock Time Reference and the Performance of a Sample of Atomic Clocks," *Metrologia*, vol. 6, no. 4, pp. 126–134, 1970.

56. D. W. Allan, "Statistics of Atomic Frequency Standards," *Proceedings of the IEEE*, vol. 54, no. 2, pp. 221–230, February 1966.

57. D. W. Allan, J. E. Gray, and H. E. Machlan, "The National Bureau of Standards Time Scale: Generation, Dissemination, Stability, and Accuracy," *IEEE Transactions on Instrumentation and Measurement*, vol. IM-21, no. 4, pp. 388–391, November 1972.

58. D. W. Allan, J. E. Gray, and H. E. Machlan, "The National Bureau of Standards Time Scale: Generation, Stability, Accuracy and Accessibility," Chapter 9 in "Time and Frequency: Theory and Fundamentals," *National Bureau of Standards Monograph 140*, edited by B. E. Blair, pp. 205–231, May 1974.

59. Bureau International des Poids et Mesures, "BIPM Annual Report on Time Activities," vol. 15, 2020.

60. S. R. Stein, "Time Scale Computation System Including Complete and Weighted Ensemble Definition," *United States Patent 5,155,695*, filed June 15, 1990, granted October 13, 1992, and "Time Scale Computation System," *United States Patent 5,315,566*, filed June 22, 1993, granted May 24, 1994.
61. R. H. Jones and P. V. Tryon, "Estimating Time from Atomic Clocks," *Journal of Research of the National Bureau of Standards*, vol. 88, no. 1, pp. 17–24, January-February 1983.
62. J. Levine, "Introduction to Time and Frequency Metrology," *Review of Scientific Instruments*, vol. 70, no. 6, pp. 2567–2596, June 1999.
63. S. R. Stein, "Time Scales Demystified," *Proceedings of the 2003 Joint Conference of the IEEE International Frequency Control Symposium and the European Frequency and Time Forum (IFCS/EFTF)*, pp. 223–227, May 2003.
64. J. A. Sherman, L. Arissian, R. C. Brown, M. J. Deutch, E. A. Donley, V. Gerginov, J. Levine, G. K. Nelson, A. N. Novick, B. R. Patla, T. E. Parker, B. K. Stuhl, D. D. Sutton, J. Yao, W. C. Yates, V. Zhang, and M. A. Lombardi, "A Resilient Architecture for the Realization and Distribution of Coordinated Universal Time to Critical Infrastructure Systems in the United States: Methodologies and Recommendations from the National Institute of Standards and Technology (NIST)," *NIST Technical Note 2187*, 177 p., November 2021.
65. P. Tavella and C. Thomas, "Comparative Study of Time Scale Algorithms," *Metrologia*, vol. 28, no. 2, pp. 57–63, 1991.
66. R. E. Kalman, "A New Approach to Linear Filtering and Prediction Problems," *Transactions of the American Society of Mechanical Engineers – Journal of Basic Engineering,* vol. 82, no. 1, pp. 35–45, March 1960.
67. L. Galleani and P. Tavella, "Time and the Kalman Filter: Applications of Optimal Estimation to Atomic Timing," *IEEE Control Systems Magazine*, vol. 30, no. 2, pp. 44–65, April 2010.
68. D. D. McCarthy and P. K. Seidelmann, *Time: From Earth Rotation to Atomic Physics*, Cambridge University Press, 2nd edition, 2018.

Chapter 5

Frequency Measurements

> *If you will observe closely, you will notice that many events tend to recur in cycles at reasonably spaced time intervals.*
>
> — W. Clement Stone, "The Success System That Never Fails"

We organize our lives around clocks, calendars, and schedules, so we understandably pay far more attention to time measurements than to frequency measurements. However, as we stated at the start of the previous chapter and will emphasize here once more, all clocks are referenced to a frequency produced by an oscillator. Frequency is the rate of occurrence of a repetitive event, and repetitive events provide the basis for all timekeeping. It's important to note that the Greek letter Nu (ν) is typically used by physicists to represent frequency, and that f is more commonly used by engineers. The two notations are equivalent and can be used interchangeably. However, to keep things clear and consistent, we'll use f to represent frequency in nearly all of the examples in this book.

As we first described at the beginning of Chapter 2, frequency is the reciprocal of time interval or period; therefore

$$f = \frac{1}{T}, \tag{5.1}$$

and

$$T = \frac{1}{f}, \tag{5.2}$$

where T is the period of a repetitive event. The International System of Units (SI) states that the period should be expressed in units of seconds (s), and that the frequency should be expressed in units of hertz (Hz). As first noted in Chapter 3, the second is one of the seven base units of the SI, and the hertz is a derived unit, defined with respect to the base unit as $\text{Hz} = \text{s}^{-1}$. The frequency signals measured by metrologists are usually sine waves or square waves, and the continuous cycles of the waveforms are the repetitive events.

As Time Goes By: A Comprehensive Guide to Time and Frequency Measurements
Michael A. Lombardi

ISBN 978-981-5352-28-3 (Hardcover), 978-1-042-01532-0 (eBook)
www.jennystanford.com

DOI: 10.1201/9781042015320-8

Understanding cycles is essential to understanding frequency measurements. A glance at the dial of an antique radio will remind us that the unit of frequency was once called the cycle, and the units that we now know as kilohertz and megahertz were once called kilocycles and megacycles. This officially changed in 1930 when the International Electrical Congress (IEC) renamed the unit of frequency to honor Heinrich Hertz (1857–1894), the German physicist who first proved the existence of electromagnetic waves.

As we discussed in Chapter 4, time metrologists typically measure clock signals that repeat at a rate of one pulse per second (pps), a frequency of 1 Hz. Frequency metrologists sometimes divide frequencies down to 1 Hz prior to measurement, but they usually perform direct measurements of much higher frequencies. The names and abbreviations of the frequency units are listed in Table 5.1.

Table 5.1 The units of frequency

Unit name	Unit abbreviation	Cycles per second
hertz	Hz	1
kilohertz	kHz	10^3
megahertz	MHz	10^6
gigahertz	GHz	10^9
terahertz	THz	10^{12}
petahertz	PHz	10^{15}
exahertz	EHz	10^{18}

Frequency measurements can be loosely grouped into three broad categories. The first category, audio frequency (AF) measurements, involves measuring sound waves that are audible to the human ear. The AF spectrum extends from 20 Hz to 20 kHz. The second category involves measuring electromagnetic waves in the radio frequency (RF) spectrum. The International Telecommunications Union (ITU), somewhat loosely defines radio waves as "electromagnetic waves of frequencies arbitrarily lower than 3,000 GHz, propagated in space without artificial guide" [1]. Technically, this means that the RF spectrum includes electromagnetic waves at any frequency below 3,000 GHz, even though only the frequencies ranging from 9 kHz to 275 GHz have been allocated for radio use in the United States [2]. The frequencies below 20 kHz that overlap the audio spectrum are rarely used for radio transmissions, nor are the frequencies above the microwave region (300 MHz to 300 GHz). The frequencies from 300 GHz to 3,000 GHz are sometimes called microwaves, but are known to spectroscopic scientists as being part of the infrared region of the spectrum, which falls in between microwaves and the optical frequencies, which include visible and non-visible light. Optical frequency measurements, the third category of frequency measurements, aren't covered in this chapter but are discussed in Chapter 13.

Most of this chapter is about measuring RF signals, the area of greatest interest to both metrologists and timekeepers. However, sound waves introduce some basic concepts that can help us to better understand frequency, so we'll start this chapter with an introduction to AF measurements in Section 5.1. This is followed by a more detailed discussion of RF measurements in Section 5.2, which explains how frequency is measured with oscilloscopes, universal counters, heterodyne systems, and digital sampling systems. Section 5.3 concludes the chapter with a discussion of frequency control systems and disciplined oscillators.

5.1 Audio Frequency Measurements

Audio frequency (AF) measurements determine the frequency of sound waves that are audible to the human ear. Accurate AF measurements are required when designing or analyzing systems that produce or record sound. The AF spectrum is nearly always defined as 20 Hz to 20 kHz, which is the range of frequencies commonly specified by manufacturers of high-fidelity audio equipment [3]. The human ear is an imperfect instrument, however, and the specified AF range is often wider than the range of frequencies we can actually hear. Some infants can hear frequencies even slightly higher than 20 kHz, but we typically lose some of our ability to hear higher frequencies as we age. For most adults, the upper limit of the audible range will likely be closer to 15 kHz than to 20 kHz.

Sound waves that propagate outside of the AF range are classified as infrasonic (below 20 Hz) and ultrasonic (above 20 kHz). Measuring infrasonic and ultrasonic frequencies is not usually important to audio engineers producing sounds intended for the human ear, but it is important for other applications. Seismologists measure infrasonic frequencies while studying earthquakes [4], for example, and in Section 4.1.4 we described a stopwatch calibration system that uses an acoustic sensor to measure the ultrasonic sounds produced by a 32.768 kHz quartz oscillator. It's also interesting to note that some animals, including elephants and whales, can hear infrasonic frequencies, and numerous animals can hear ultrasonic frequencies. This is evidenced by the whistles commonly found in pet stores that emit ultrasonic sounds inaudible to humans, but audible to dogs, who can sometimes hear frequencies as high as 60 kHz. Cats can often hear frequencies as high as 80 kHz, and animals including bats, whales, and dolphins can hear frequencies higher than 100 kHz.

Unlike electromagnetic waves, which travel at a speed that's based on a fundamental physical constant, the velocity of sound waves isn't always the same. Sound waves are vibrations of kinetic energy that pass through a transmission medium, which can be a gas, a liquid, or a solid material. Their velocity mostly depends on the type of transmission medium, with a smaller dependence on the temperature of the medium. Sound waves travel through a medium by compression and rarefaction—compression presses molecules closer together, and rarefaction spreads molecules further apart. The speed of sound depends both on the density of the medium, a measure of how close the molecules are spaced; and the elasticity of the medium, a measure of how quickly compressed molecules can restore their position. The velocity of sound through air at room temperature (20 °C) is about 343 m/s, slowing to about 331 m/s at 0 °C. Sound travels more than 4× faster through water than air, with an average velocity of about 1,500 m/s, because water is both denser and more elastic than air. Sound travels several times faster through iron and steel that it does through water, with velocities exceeding 5,000 m/s.

Regardless of their medium, sound waves travel very slowly when compared to electromagnetic waves. As we noted in the discussion on time transfer in Section 4.3.1, electromagnetic waves travel 1 km in about 3.3 μs. A sound wave traveling through air at 20 °C takes about 2.9 s to travel that same distance. Lightning and thunder provide us with a familiar example of the speed disparity between light and sound. If we see lightning flash about 3 s before we hear thunder, we'll know that the lightning is about 1 km away.

Another distinction between electromagnetic waves and sound waves involves their wavelength, which is calculated in both the AF and RF spectrums as

$$\lambda = \frac{V}{f}, \tag{5.3}$$

where λ is the wavelength in meters, V is the velocity in meters per second, and f is the frequency in hertz. Because the velocity is a constant for electromagnetic waves, and nearly a constant for sound waves at room temperature, the wavelength always goes down when the frequency goes up. However, because the velocity is so high for electromagnetic waves, they have much longer wavelengths than sound waves at the same frequency. For example, a 15 kHz AF signal traveling through air has a wavelength of about 2 cm, compared to a wavelength of about 20,000 m for an RF signal at the same frequency.

The human ear is the preferred destination for the sound waves we produce and measure, and the ear has multiple components that allow it to receive sound. The external part of the ear is much larger than the ear's internal parts, and it reflects sound waves from the air into the ear canal, functioning much like a dish-shaped reflector that reflects radio waves from satellites to a smaller antenna. Once inside the ear, the sound waves are directed to the eardrum, which is a membrane made of thin connective tissue. The eardrum converts the sound waves from vibrations in the air to material vibrations in the ear.

The eardrum's vibrations are transmitted into the inner ear by a lever-based system consisting of the three smallest bones in the human body. These bones are formally called the malleus, incus, and stapes. However, due to their shapes, they are more commonly known as the hammer, anvil, and stirrup. One end of the hammer is attached to the eardrum, and the other end forms a lever-like hinge with the anvil and stirrup, which are fused together and act as one bone. The stirrup makes contact with the membrane covered opening to the inner ear, called the oval window. The stirrup moves back and forth, much like a piston in a car engine. As the oval window is pushed inward, another opening to the inner ear, called the round window, moves outward. This motion transfers the eardrum's vibrations to the cochlea canal, which is filled with fluid and small hair cells. The vibrations in the fluid cause the hair cells to vibrate, which activates thousands of nerve fibers that send messages from the cochlea directly to the brain. The brain integrates the information received from these messages to form an overall auditory pattern of sound [5, 6].

The cochlea, a spiral-shaped structure about 30 mm in length, is a frequency-selective mechanism. Waves propagate from the base of the cochlea, which is near the oval window, to its apex at the top or center of the spiral. The basilar membrane of the cochlea is thinnest and able to move the most near the base, allowing it to signal to the brain that it can sense high frequencies, potentially as high as 20 kHz. As the basilar membrane increases in thickness and decreases in motion along the path from the base to the apex, it senses lower and lower frequencies and sends corresponding messages to the brain, culminating with the ability to sense frequencies as low as 20 Hz [5, 6]. To better illustrate the frequency selection process from the base to the apex of the cochlea, Figure 5.1 expands and straightens the spiral-shaped cochlea structure.

The way that humans perceive sound is related to the frequency and wavelength of sound waves. As the frequency of the wave decreases and its wavelength increases, the sound heard by humans becomes deeper. Conversely, the sound heard by humans becomes shriller as the frequency increases and the wavelength decreases. The size of the object that produces the sound contributes to its deepness or shrillness. There are many familiar examples of this, such as the deepness of the sound of a church bell when compared to the sound of a sleigh bell, or the shrillness of a child's voice compared to the voice of an adult [8]. Another common example is a multi-driver loudspeaker, where a crossover circuit separates the audio frequencies into two or more ranges. The woofer, the driver that produces the lowest frequencies, is at least several times larger in diameter than the tweeter that produces the highest frequencies.

Music is perhaps the highest form of human expression, and it is sometimes defined as the art of organizing sounds and silence in a way that is pleasing to the ear. The sounds of music are produced by either the human voice or by musical instruments, and are organized by their frequency, as discussed in the following section.

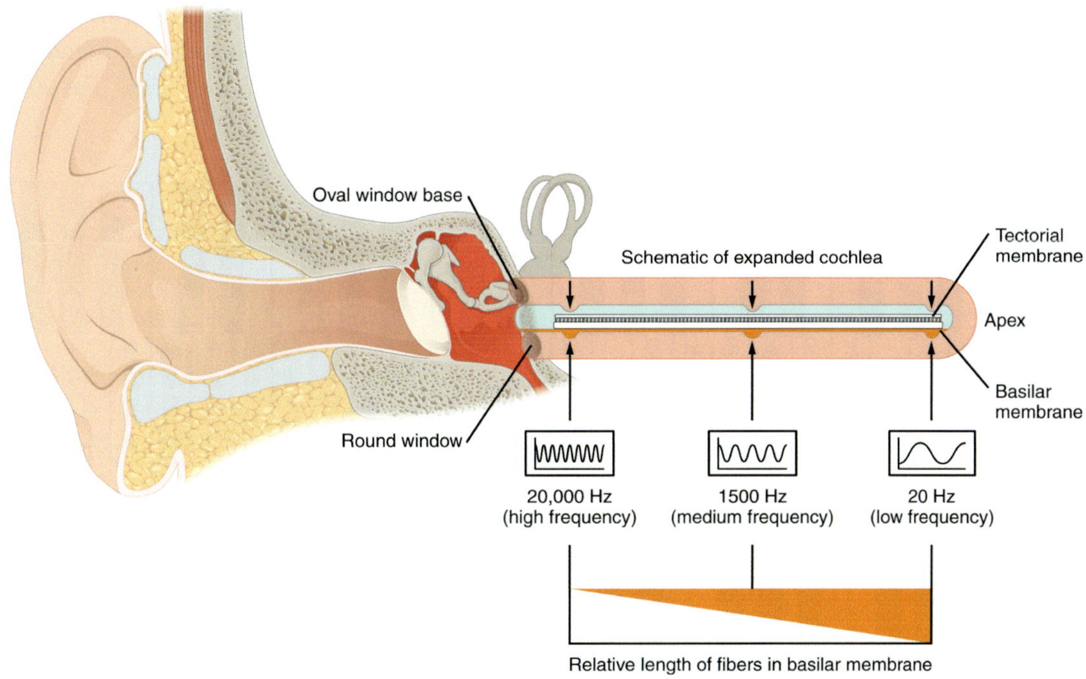

Figure 5.1 Frequency selection of the cochlea in the human ear.

5.1.1 Notes, Octaves, Scales, and a Standard for Musical Pitch

A pitch is a particular frequency of sound, for example, 220 Hz. A note is the name assigned to a musical pitch, which is A3 in the case of 220 Hz. An octave is the interval between one musical pitch and another pitch of either half or double its frequency. Therefore, the notes A2 and A4 are both separated from A3 by one octave, with A2 (110 Hz) half the frequency of A3, and A4 (440 Hz) double the frequency of A3. A scale is a series of sounds arranged from low to high frequencies by definite frequency intervals [5].

The piano is the musical instrument that best demonstrates pitch, notes, octaves, and scales. When a piano key is pressed, it activates a felt-covered hammer that strikes a wire string to produce sound. Every key on a piano is associated with a string of varying length and thickness. When moving from left to right on the keyboard, the length and the thickness of the string's decreases, and the frequency of the sound increases. A piano has 88 keys, 52 white keys for the notes A0 (27.5 Hz) through C8 (4186 Hz), and 36 black keys raised above the white keys and set further back, for sharps (#) and flats (♭). The range of frequencies covers slightly more than seven octaves. Each octave is arranged as a 12-tone, or chromatic scale, with seven white and five black keys. The seven complete octaves of a piano's keyboard, beginning with the notes C1 through C7, are shown in Figure 5.2.

The chromatic scales of a piano follow an equal temperament system, where two adjacent notes are separated by a frequency interval, called a semitone. The ratio of the interval between notes is always the same, equal to the 12th root of 2, $\sqrt[12]{2} = 1.059463$. The semitones are divided into 100 parts, called cents, a logarithmic unit of measure equal to the 1200th root of 2, or $\sqrt[1200]{2}$. Table 5.2 lists the musical notes, including their intervals and frequencies, for the octave starting at C4. Because this octave is in the middle of the seven complete octaves on a piano keyboard (Figure 5.1), C4 is commonly called middle C.

By international agreement, pianos and other musical instruments are tuned so that A4, or A above middle C, is equal to 440 Hz (highlighted in Table 5.2). To match its frequency, A4 is now better known as A440. The use of 440 Hz as an audio frequency standard has been practiced in both the United States and internationally since the 19th century. By the 1930s, several nations and organizations had formally recognized A440 as a standard for musical pitch, and the International Organization for Standardization (ISO) recognized A440 as an international standard in 1975 [9]. As a service to musicians, the NIST radio station WWV has provided scheduled broadcasts of 440 Hz tones since 1936 [10].

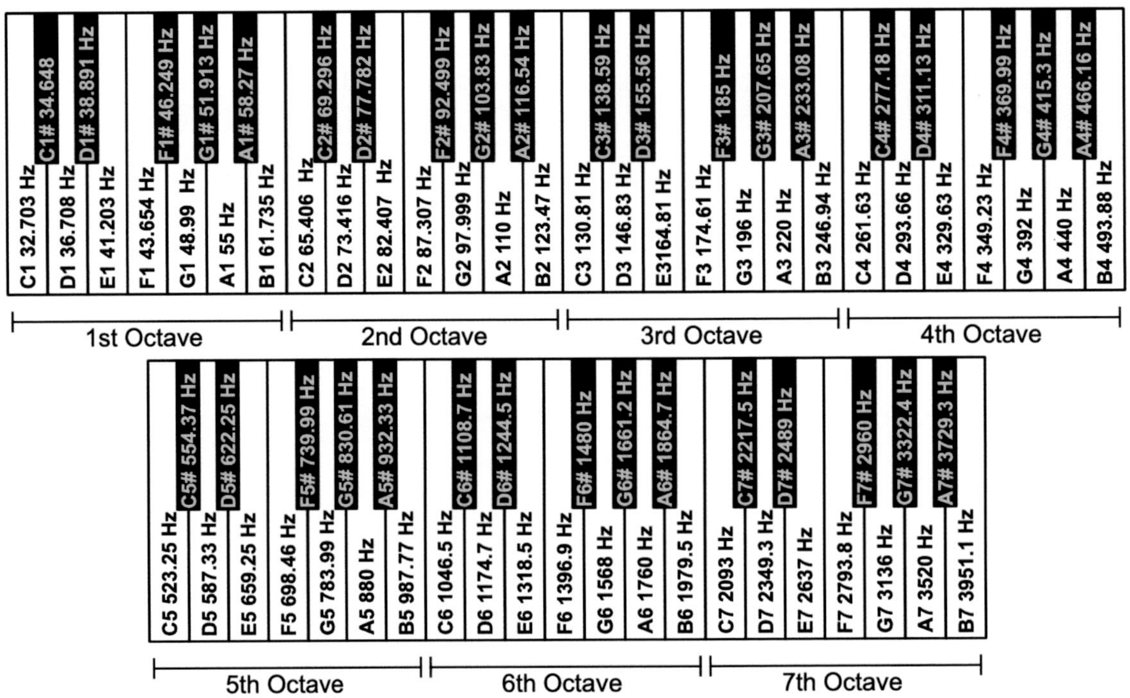

Figure 5.2 The seven complete octaves of a piano keyboard.

Table 5.2 Frequencies and intervals for the octave starting at C4 (middle C) on a piano

Musical note	Frequency interval with respect to C4				Frequency (Hz)
	Semitone name	Ratio	Multiplier	Cents	
C4	Unison	$2^{0/12}$	1.000000	0	261.63
C4#/D4♭	Minor second	$2^{1/12}$	1.059463	100	277.19
D4	Major second	$2^{2/12}$	1.122462	200	293.67
D4#/E4♭	Minor third	$2^{3/12}$	1.189207	300	311.13
E4	Major third	$2^{4/12}$	1.259921	400	329.63
F4	Perfect fourth	$2^{5/12}$	1.334840	500	349.23
F4#/G4♭	Tritone	$2^{6/12}$	1.414214	600	370.00
G4	Perfect fifth	$2^{7/12}$	1.498307	700	392.00
G4#/A4♭	Minor sixth	$2^{8/12}$	1.587401	800	415.31
A4	**Major sixth**	$\mathbf{2^{9/12}}$	**1.681793**	**900**	**440.00**
A4#/B4♭	Minor seventh	$2^{10/12}$	1.781797	1000	466.17
B4	Major seventh	$2^{11/12}$	1.887749	1100	493.89
C5	Octave	$2^{12/12}$	2.000000	1200	523.26

Tuning a piano is an excellent example of a frequency measurement and calibration that's performed with the human ear. The piano tuner listens to a standard audio tone of a known frequency and compares it to the note on the piano that is supposed to produce that same frequency. When the two tones are close in pitch but not identical, the difference in their frequency generates a beat frequency. The beat frequency is usually below the range of human hearing, but is still perceived as a periodic variation in the volume of the sound. The piano's frequency is then tuned by tightening or loosening the tension of the string to reduce the frequency difference. When the pitch of the piano has been properly adjusted to agree with the audio standard, the beat frequency and the variations in volume have been slowed down to the point where they are almost imperceptible.

To understand why changing the tension of a string changes the frequency, we should first consider that in all musical instruments that use vibrating strings to produce sounds, the physical length of the string determines its wavelength (λ). However, the fundamental frequency of the sound depends not only on the length of the string, but also on its velocity,

$$f = \frac{V}{\lambda}, \tag{5.4}$$

where V is calculated as

$$V = \sqrt{\frac{S_T}{S_{\text{LD}}}}, \tag{5.5}$$

where S_T is the string's tension and S_{LD} is the string's linear density, obtained by dividing the mass of the string by its length. Both the length and the mass of the string are hard to change (without replacing the string), so adjusting the tension is usually the only option available, but it's also the only option that's needed to adjust the frequency.

The smallest frequency difference that a listener can detect, known as the just noticeable difference (JND), is much smaller than the frequency interval between notes. To understand why, consider that if we add just three more octaves to the seven shown on a piano keyboard (Figure 5.1), one below the 1st octave and two after the 7th octave, we'll have a 10-octave equal temperament scale that essentially spans the entire range of human hearing. The 10 octaves cover the range of frequencies from 16.35 Hz (C0) to 16,744 Hz (C10) in just 120 notes or semitones (12,000 cents). A typical human ear can distinguish about 1,400 frequencies, meaning that the full pitch capabilities of the ear are not fully realized in music [5]. If we divide the 10-octave range in cents by the number of distinguishable frequencies, we can estimate that the average JND is about 8.6 cents.

In practice, the JND varies substantially from listener to listener. It's highly dependent upon several factors, including the frequency of the musical pitch, the ambient background noise, and the listener's hearing ability and musical training. For a 440 Hz tone the JND is typically estimated as 0.5%, which is very close to the estimated average of 8.6 cents, meaning that a listener can detect a frequency difference of 2.2 Hz. However, for skilled listeners the JND is sometimes estimated as 5 cents, obtained by multiplying A4 by $2^{5/1200}$, which reduces the detectable frequency difference to about 1.3 Hz. In either case, the human ear can determine frequency with an uncertainty of less than 1%.

The most common devices used as references for AF measurements are tuning forks and electronic AF generators. These devices are discussed in the next two sections.

5.1.2 Tuning Forks

Tuning forks are simple mechanical devices that produce a standard audio frequency. They are made out of metal and have a stem, two prongs of equal length, and a base where the two prongs meet. When struck against a hard surface, or when struck by another object, tuning forks vibrate. The vibrations produce an audio tone at the tuning fork's fundamental frequency, and the tone is sustained for at least several seconds. The fundamental frequency is determined by the length and mass of the two prongs, and is usually stamped or engraved on the stem or base of the tuning fork (Figure 5.3).

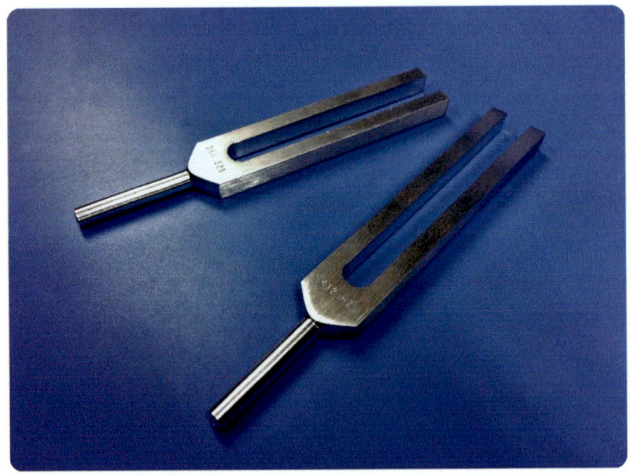

Figure 5.3 A 432 Hz and 528 Hz tuning fork.

The use of two prongs, both fixed at their base but free at the other end, provides two main advantages. The first advantage is that most of the vibrational energy is at the fundamental frequency. This allows tuning forks to produce a mostly pure tone. There are some harmonics, but they fade out quickly, leaving a pure tone at the fundamental frequency which is sustained longer. This differs from a vibrating string in a musical instrument, which produces tones that are less pure, because they always include some mixture of the fundamental frequency and its harmonics. The second advantage is that the two prongs vibrate in a symmetric mode, always moving in opposite directions, like mirror images of each other. When two waves traveling in opposite directions meet, they form a node where they cancel each other's energy, which is what happens at the base of the tuning fork. This energy cancellation is advantageous, because it means there is very little vibration in the base or stem. This allows a tuning fork to be held by its stem without damping its vibrations [5].

The tuning fork was invented in 1711 by the English musician John Shore (1662–1752). Shore was a respected court musician, well known during his lifetime for his ability to play both the trumpet and the lute. He developed the tuning fork, then called a pitch fork, so that he could tune the pitch of his lute. According to legend, Shore would humorously tell audiences, "I never go anywhere without my pitch fork," and then tune his instrument with a pitch fork before beginning a performance [11]. In 1712, the famous German composer George Frideric Handel (1685–1759) moved to London, where he wrote numerous trumpet parts for Shore, who became one of his favorite musicians. During this period, Shore gave Handel one of his tuning forks. Handel's tuning fork, which he used throughout the remainder of his career, is still in existence. It's been measured as producing the note C5 at 512 Hz, which is equivalent to A4 at 422.5 Hz. From this, we can estimate that modern orchestras play Handel's compositions at a pitch about 70 cents sharper than his original intention [7, 11].

An interesting application of tuning forks is their use in the calibration of the Doppler radar devices that law enforcement officers use to measure vehicle speed. These devices, commonly called radar guns, are microwave transceivers. They transmit a signal that's reflected from a moving vehicle, and they receive the reflected signal. The received signal is mixed with the transmitted signal to determine their frequency difference, or the Doppler frequency shift. This measurement is then converted to vehicle speed in either kilometers or miles per hour (kph or mph). The calibration of a radar gun is straightforward—

when a vibrating tuning fork is placed in front of the transceiver, its vibrations modulate the phase of the reflected signal. The vibrating tuning fork doesn't actually simulate a moving vehicle, but it does simulate the Doppler frequency shift that a moving vehicle would produce [12]. If a radar gun is properly calibrated, the speed that it indicates should match the speed labeled on the tuning fork. The tuning fork is usually also labeled with the radar band it was designed for—either X, K, or Ka-band, because the frequency of the tuning fork depends on the type of radar used. Tuning forks don't have to be accurate to perform accurate speed calibrations. For example, a K-band radar tuning fork can simulate 60 mph to within ±0.5 mph with frequencies ranging from 4285 Hz to 4357 Hz [13].

In this case of Doppler radar calibrations, only the frequency of a tuning fork's vibration is needed, and not its audio tone. The audio tones produced by tuning forks still remain useful for many applications in music, medicine, and science education. However, for sound applications that require more than just rudimentary accuracy, tuning forks have been mostly replaced by the electronic AF generators discussed in the next section.

5.1.3 Electronic Audio Frequency Generators and Analyzers

There are two basic types of electronic AF generators. The first type, called a chromatic tuner, is used to tune musical instruments or voices. Normally packaged as a small, battery-powered instrument, a chromatic tuner can essentially replicate the function of a large set of tuning forks. It allows the selection of a range of musical notes, which usually includes the piano's entire range from A0 to C8, and then plays the selected note through a built-in loudspeaker. Because the frequencies of its tones are referenced to an internal quartz oscillator, chromatic tuners are far more accurate than tuning forks. Computers and smartphones also have internal quartz oscillators, so numerous software applications to accurately generate audio tones have been developed, making stand-alone chromatic tuners less common. All devices should work equally as well, but the small loudspeakers found in portable devices might limit a chromatic tuner's ability to produce tones below about 100 Hz.

The second type of AF generator is used to test and measure electronic systems that produce audio, most notably the components in high-fidelity sound systems, but also telephones, radio transceivers, and so forth. The device under test (DUT) can be an audio amplifier, a loudspeaker, a microphone, a signal source such as a compact disc (CD) player or record player, or any type of related audio equipment. The AF generator uses a low-noise oscillator to produce sine wave or square wave signals. The signals are produced at either a continuous frequency, or swept across a range of frequencies, or bandwidth, using a sweep generator [3, 14]. When the frequency response of the DUT is tested across its entire usable audio spectrum, the signal amplitude is usually held constant. The typical usable bandwidth for an audio device is often less than the full AF range, as shown in Table 5.3.

The type of filtering used determines the bandwidth of a system that has less than the full audio range. Low-pass filters only allow frequencies below a selected cutoff frequency to pass through, blocking all higher frequencies. For example, a traditional voice grade telephone line has a cutoff frequency of 3,400 Hz. This means that if a musician plays the piano over a traditional telephone line, the listener at the other end won't hear the notes higher than G7# (3,322.4 Hz). In other cases, the provided audio bandwidth of a system is less than what is allowable. This is often true in the case of amplitude modulated (AM) broadcast radio stations, which are legally allowed to transmit 10 kHz of audio bandwidth. Even so, most stations don't use all of their allowable bandwidth, and the filters employed in AM receivers usually have a cutoff frequency of 7 kHz or less, intentionally limiting their audio fidelity. This is usually done to reduce costs, acknowledging that AM radio is mostly suited for broadcasting news and talk shows, and that frequency modulated (FM) radio stations, with their 15 kHz bandwidth and two-channel stereo sound, are better suited for broadcasting music.

Table 5.3 Typical bandwidths of audio systems

Audio source	Audio bandwidth (Hz)	
	Lower	Upper
Subwoofer	20	200
Single sideband radio communications	300	3,000
Traditional voice grade telephone	300	3,400
Wideband telephone	50	7,000
AM radio broadcasts	20	7,000
FM radio broadcasts	30	15,000
Stereo amplifier	20	20,000
CD player	20	20,000
Record player	20	20,000
Headphones	20	20,000
Full range loudspeaker	20	20,000

Most audio equipment measurements involve inputting a reference signal from the AF generator into the DUT, and then comparing the input signal to the output signal to measure the DUT's performance. Because distortion is undesirable when producing sound, a common example involves measuring the total harmonic distortion plus the noise (THD + N) of an audio amplifier. This is done by generating continuous sine wave signals, either at a single frequency or by stepping through a range of frequencies, that are passed through a line level input connector on the amplifier, such as an auxiliary or AUX port. Most AF generators are designed to easily output signals that are electrically compatible with an amplifier's line level inputs, because the characteristics of these signals are standardized, allowing audio components from different manufacturers to be interconnected.

A high-quality AF generator produces a nearly pure sine wave that, for all practical purposes, has no harmonic energy. However, as the signals pass through the DUT, the amplifier's non-linearity will cause harmonics of varying frequencies and amplitudes to be added to the input signal. Some noise is also added, sometimes by the AC power supply or by external interfering signals. The signals output by the amplifier are passed through a notch filter that's tuned to match the fundamental frequency of the input signal. The notch filter removes the fundamental frequency, leaving only a residual signal that consists of harmonics and noise. The THD + N metric is then calculated, usually as a percentage but sometimes in decibels (dB), as the amplitude of the residual divided by the amplitude of the entire signal. Manufacturers usually specify only the percentage value and the frequency or range of frequencies, for example, "0.02% at 1 kHz" or "0.05% at 20 Hz to 20 kHz," but they ideally should specify other measurement parameters, including the input signal amplitude, the load impedance, the amplifier gain, and the measurement bandwidth [3]. Audio analyzers, such as the one pictured in Figure 5.4, provide everything needed to perform THD + N and other types of audio measurements, by combining the AF generator with multiple measurement instruments and built-in software.

It's now time to move on and discuss RF measurements. As we've discussed, the audio spectrum is limited to 20 kHz, and the ability of the human ear to distinguish between frequencies is limited to perhaps 1 Hz in the best case. Both of these factors make AF measurements far less demanding than RF measurements. However, keep in mind that most of the concepts and measurement methods we'll discuss in Section 5.2, including phase comparisons and heterodyne methods, and most of the measurement instruments, including oscilloscopes and frequency counters, are also applicable to AF measurements [7, 15].

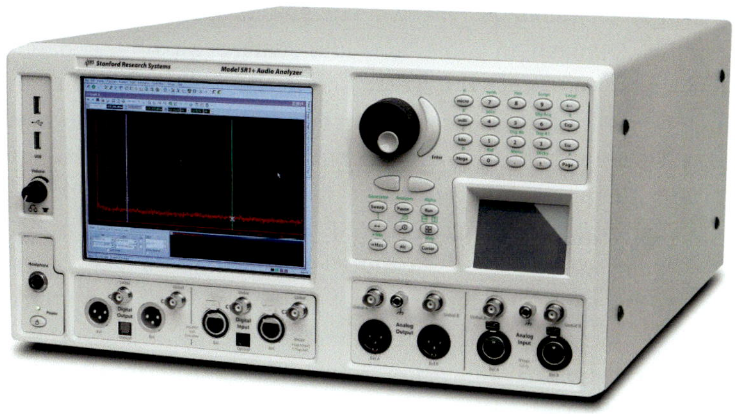

Figure 5.4 A combination audio frequency generator and analyzer (courtesy of Stanford Research Systems).

5.2 Radio Frequency Measurements

The ITU designates 12 numbered frequency bands in the RF spectrum, as listed in Table 5.4. The band number indicates the range for frequencies for the band, with a band numbered N extending from 0.3×10^N Hz to 3×10^N Hz [16]. Bands 4 through 11 are considered as the radio bands [1], and the bands numbered 1, 2, 3, and 12, denoted by the shaded rows in the table, are not allocated for radio communications [2]. The microwave portion of the RF spectrum is usually defined as bands 9 through 11 (UHF, SHF, and EHF), where the wavelengths are shorter than a meter but longer than a millimeter.

Table 5.4 Radio frequency bands

Band		Frequency nomenclature	Frequency range	Unit	Wavelength range	Unit
Name	Number					
ELF	1	Extremely low	3–30	Hz	100,000–10,000	km
SLF	2	Super low	30–300	Hz	10,000–1,000	km
ULF	3	Ultra low	300–3,000	Hz	1,000–100	km
VLF	4	Very low	3–30	kHz	100–10	km
LF	5	Low	30–300	kHz	10–1	km
MF	6	Medium	300–3,000	kHz	1–0.1	km
HF	7	High	3–30	MHz	100–10	m
VHF	8	Very high	30–300	MHz	10–1	m
UHF	9	Ultra high	300–3,000	MHz	1–0.1	m
SHF	10	Super high	3–30	GHz	10–1	cm
EHF	11	Extremely high	30–300	GHz	10–1	mm
THF	12	Tremendously high	300–3,000	GHz	1–0.1	mm

The calculation for wavelength (λ) was provided in Eq. (5.3), but RF wavelength can be approximated even more quickly as $\lambda = 300/f$, where λ is in meters and f is the frequency in megahertz. This works because the speed of light is very close to 300×10^6 meters per second and because 1 MHz = 1×10^6 cycles per second. It, of course, follows that a 1 MHz signal, which could be the carrier frequency of an AM broadcast station or a standard frequency produced by an oscillator, has a wavelength of 300 m.

For some applications and technologies, the wavelength of a signal is referred to more often than its frequency. This custom is followed in amateur radio, where all of the bands are known by their approximate wavelength, an example being the frequencies from 7.0 to 7.3 MHz that comprise the 40 m amateur band in the United States. It's also true for some microwave frequencies, such as the 21 cm hydrogen line discussed in Section 3.10, and especially true for lasers operating at optical frequencies above the RF spectrum. For example, the red-orange laser pointers used for talks and lectures often have their wavelength—which is typically between 630 nm and 650 nm—listed in fine print on their labels. Another long-established practice in consumer products is to conveniently hide the hard-to-remember frequencies from the user by using channel numbers. Some common examples are citizen's band radio, where 40 two-way communication channels occupy frequencies between 26.965 and 27.405 MHz (HF); over-the-air terrestrial television, where channels 2 to 13 occupy frequencies ranging from 54 to 216 MHz (VHF); and Sirius XM satellite radio, where more than 150 audio channels occupy frequencies ranging from 2320 to 2345 MHz (UHF).

Time and frequency metrologists encounter numerous RF frequencies, including the resonance frequencies of clocks and oscillators, the carrier frequencies of radio signals, the clock frequencies used in network communications, and, of course, the frequencies produced by every device being measured or calibrated. Table 5.5 lists some of the more common frequencies applicable to everyday time and frequency metrology. The table includes the RF band, the application, the frequency in hertz, the approximate wavelength in meters, and the period in nanoseconds.

Table 5.5 Some common frequencies applicable to time and frequency metrology

Band	Application	Frequency (Hz)	Wavelength (m)	Period (ns)
SLF	AC power	50	6,000,000	20,000,000
	AC power	60	5,000,000	16,667,000
ULF	IRIG-B time code carrier	1,000	300,000	1,000,000
LF	Quartz watch oscillator	32,768	9,152	30,518
	Radio time code carrier	60,000	5,000	16,667
	Standard frequency signals	100,000	3,000	10,000
MF	Standard frequency signals	1,000,000	300	1,000
	T1 communications channel	1,544,000	194	647
	E1 communications channel	2,048,000	146	488
HF	Standard frequency signals	5,000,000	60	200
	Standard frequency signals	10,000,000	30	100
VHF	Standard frequency signals	100,000,000	3	10
UHF	GPS L5 carrier	1,176,450,000	0.25	0.850
	GPS L2 carrier	1,227,600,000	0.24	0.815
	Hydrogen maser resonance	1,420,405,752	0.21	0.704
	GPS L1 carrier	1,575,420,000	0.19	0.635
	Rubidium clock resonance	6,834,682,611	0.04	0.146
	Cesium clock resonance	9,192,631,770	0.03	0.109

We described time interval measurements in Section 4.1, where a pulse from a DUT is compared to a pulse from a reference clock or time standard. The same process is followed in frequency measurements, but instead of a pulse, the DUT usually produces a sine wave that's compared to a sine wave from a reference oscillator or frequency standard. The most commonly measured frequency is 10 MHz, because most frequency standards provide 10 MHz outputs, and because most test instruments accept 10 MHz signals as their external time base.

Ideally, the measurement uncertainty of the frequency standard should be known with respect to the SI so that measurement traceability (Chapter 9) can be established. Of course, frequency measurements will only produce correct results if the frequency standard outperforms the DUT. A simple rule of thumb that protects against incorrect results is selecting a standard that is at least 10× more accurate than the DUT over the selected measurement interval. In other words, to determine if a DUT is accurate to 1×10^{-8} over a given interval, we need a frequency standard that's accurate to 1×10^{-9} or better over the same interval.

Getting back to the idea that understanding cycles is essential to understanding frequency measurements, let's examine one cycle of a sine wave (Figure 5.5). A sine wave completes one cycle, or one full oscillation, by moving from zero to its maximum value (peak), crossing through zero on the way to its minimum value (valley), and then moving back to zero. The period, T, is the time interval required to produce one cycle. The period is the reciprocal of the frequency of the sine wave, or $T = 1/f$, where T is expressed in seconds and f is expressed in hertz. The phase (ϕ) is the angle within the cycle that corresponds to a particular time, t. Phase can be measured in either radians or degrees, and a complete cycle consists of 2π radians or 360°.

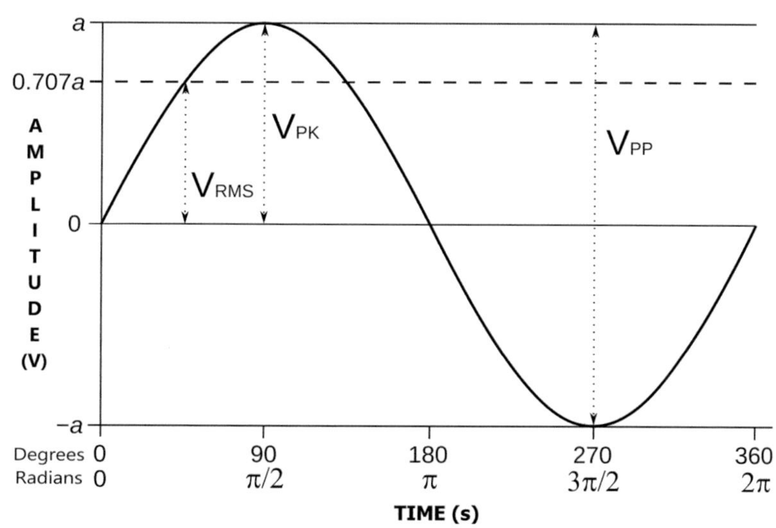

Figure 5.5 One cycle of a sine wave.

The amplitude of a sine wave is measured in volts (V), usually centered around a mean amplitude of 0 V, although a sine wave with a DC offset will have a mean amplitude that is offset from zero. The instantaneous points where the sine wave crosses its mean amplitude value, called zero crossings, occur twice during each cycle, at phase angles that are even multiples of π or 180°. The peak amplitude, V_{PK}, is the voltage difference between the mean amplitude and the maximum amplitude. The peak-to-peak amplitude, V_{PP}, is the voltage difference between the maximum amplitude and the minimum amplitude, so it can also be thought of as the peak-to-valley amplitude. The root mean square (RMS) amplitude, V_{RMS}, is obtained by multiplying V_{PK} by $\sqrt{2}/2$, or 0.707.

As we now know, and as we'll continue to be reminded of throughout the rest of this chapter and book, frequency is the reciprocal of period of a cycle, $f = 1/T$, where T is expressed in units of time. The reciprocal relationship of time and frequency means that we can measure frequency in either the time domain or frequency domain (Figure 5.6). A frequency measurement instrument can operate in either domain, and the measurements made in either domain are interchangeable and provide the same results. Time domain measurements are made with respect to time, using units that are either fractional parts or multiples of the second. Oscilloscopes, the topic of the next section, are perhaps the best examples of time domain measurement instruments. They measure the magnitude of an input signal versus time, and display a graph of the waveform, similar to the one shown in Figure 5.5.

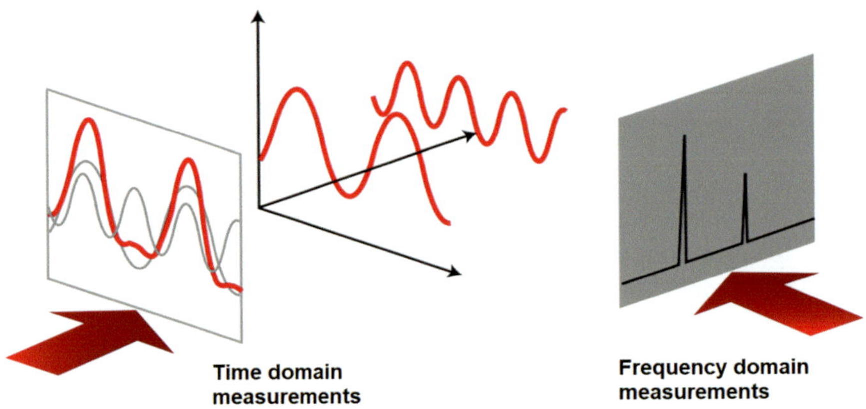

Figure 5.6 Time domain measurements compared to frequency domain measurements.

Frequency domain measurements are made with respect to frequency, using units that are either fractional parts or multiples of the hertz. A graph of a frequency domain measurement shows how a signal is distributed over a given bandwidth. A spectrum analyzer is perhaps the best example of a frequency domain measurement instrument, because it measures and displays the magnitude of an input signal versus frequency. The horizonal axis of a spectrum analyzer in is units of frequency, and the vertical axis is in units of power, typically dBm (decibels referenced to 1 mW). An example of a spectrum analyzer display is shown in Figure 5.7.

Figure 5.7 A spectrum analyzer display.

Spectrum analyzers separate a signal into its various frequency components by either sweeping across the frequency range, or through a process called Fourier analysis, named after the French mathematician Joseph Fourier (1768–1830). In his expansion theorem, Fourier proved that almost any periodic waveform, regardless of its shape or complexity, is composed of a series of sine waves with properly chosen amplitudes and phases. When the waveform is dissected into its sine wave components, the frequency of each sine wave is a harmonic of the fundamental frequency of the complete waveform. A perfect waveform would have just one frequency component, without any noise, distortion, or harmonics. If such a waveform existed, it would appear on a spectrum analyzer as a single line with zero bandwidth.

Keeping these basic concepts in mind, we'll describe several established methods for measuring the frequency of RF signals in the following sections. We'll start with one of the oldest and most illustrative methods, but not the most accurate; which is measuring frequency with an oscilloscope.

5.2.1 Measuring Frequency with Oscilloscopes

Oscilloscopes, previously discussed in Section 4.1.1 as an instrument for measuring time interval, can also be used to measure frequency in the time domain. Oscilloscopes allow us to view waveforms by graphing the amplitude of an input signal as a function of time. Their default configuration usually measures amplitude on the vertical or *y*-axis, and measures time on the horizontal or *x*-axis. This configuration is known as *Y–T*, or voltage-time mode. The most basic frequency measurement, one that works with even the oldest single-channel oscilloscopes, is a direct measurement that doesn't require a frequency standard. A direct measurement is performed by simply connecting the DUT signal to the oscilloscope, and then calculating the DUT's frequency from the waveform display.

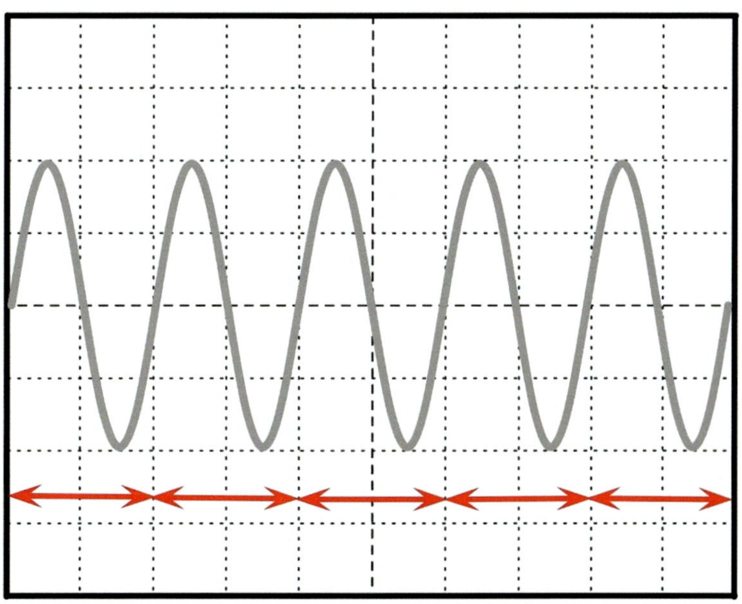

Figure 5.8 Direct frequency measurement with oscilloscope.

To illustrate this, Figure 5.8 shows a sine wave as it appears on an oscilloscope. The horizontal axis is marked with 10 equal time divisions. The time division setting, *TDS*, is user selectable. The next step is to count the number of time divisions needed to complete one cycle of the waveform, which we can call divisions per cycle, or *DPC*. The DUT frequency is then estimated as

$$f = \frac{1}{T} = \frac{1}{DPC \times TDS}. \tag{5.6}$$

In Figure 5.8, five cycles of the sine wave are shown on the oscilloscope display, indicated by the red arrows. Therefore, the period of the frequency is equal to two divisions of time, and $DPC = 2$. If $TDS = 1$ ms, then $f = 500$ Hz, if $TDS = 1$ μs, then $f = 500$ kHz, and so on [14, 17]. Of course, all modern oscilloscopes perform the simple calculation in Eq. (5.6) automatically, and provide a frequency readout on their display.

Using an oscilloscope to directly measure frequency is usually done when just a quick estimate of a DUT's frequency is needed, because the accuracy of this method is very limited. Even if an oscilloscope includes dedicated frequency counting hardware, which is not common, its accuracy will still be limited by its internal time base oscillator, which is usually no better than about 1×10^{-6}. For most oscilloscopes, the accuracy is further limited by its sampling rate, and the set or subset of samples that is used to estimate frequency. In some cases, a more accurate frequency estimate is obtained if the time divisions are set so that fewer cycles of the signal are displayed.

Just as the resolution of an oscilloscope's delta-time function is proportional to the time interval (Section 4.1.1), the resolution of its direct frequency measurements is proportional to the frequency. For example, the least significant digit of the frequency readout might be 0.1 Hz for a 10 kHz signal, and 100 Hz for 10 MHz signal, meaning that as the frequency increases by 1,000× the resolution gets proportionally coarser. The resolution, which is 1×10^{-5} in this example, is often a good indication of what the oscilloscope manufacturer considers as the best possible frequency measurement accuracy for the instrument, but the actual accuracy might be even worse than the resolution.

Better and more accurate methods of measuring frequency with an oscilloscope involve phase comparisons, where the phase of the DUT signal is compared to the phase of a signal from a frequency standard, rather than to the oscilloscope's time base oscillator. Because both signals are external to the oscilloscope, these measurements require an oscilloscope that can accept two input signals.

One of the oldest phase comparison methods was discovered long before the invention of the oscilloscope, and before RF measurements even existed. This method involves Lissajous patterns, which are named after Jules Antoine Lissajous (1822–1880), a French physicist who wanted to see the waves created by vibrations with his own eyes. For his doctoral thesis, Lissajous wrote about using sand patterns to study waveforms made by vibrating bars, which was a method pioneered by the German physicist and musician Ernst Chladni (1756–1827). Later, Lissajous performed experiments where he placed tuning forks in water, studying the ripples that they caused.

In 1855, Lissajous devised a way to study acoustic vibrations by using two tuning forks, each with a small mirror attached to one of its prongs, counterbalanced by a mass of equal weight attached to the other prong (Figure 5.9). The vibrating tuning forks were placed perpendicular to each other, and a light beam was aimed at one of the mirrors. The light beam was first reflected to the other mirror, and then to a large screen, where the two vibrations combined to form a two-dimensional pattern. By using tuning forks with slightly different frequencies, the patterns would appear in the form of rotating ellipses. Thanks to the persistence of vision in the human eye, which allows the eye and brain to retain a visual image of an event for perhaps 50 ms to 100 ms after it occurs, Lissajous had literally made it possible to "see" sound. He earned praise for his "beautiful experiments," and due in part to the high quality of his lectures and demonstrations, Lissajous patterns quickly became a standard topic in physics textbooks [18, 19].

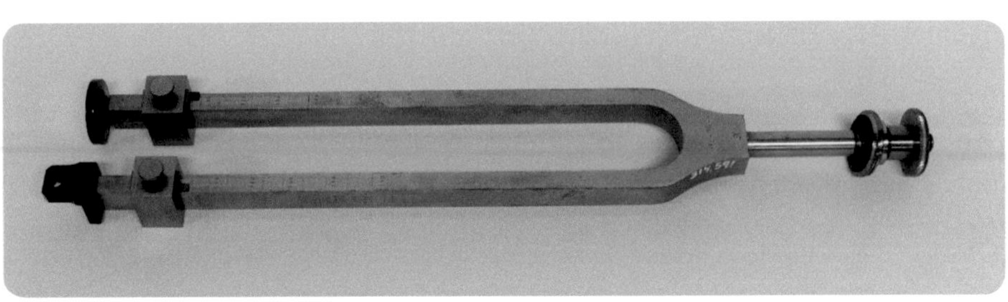

Figure 5.9 Tuning fork for a Lissajous apparatus (courtesy of National Museum of American History, Smithsonian Institution).

Lissajous patterns are generated on a two-channel oscilloscope by first configuring the instrument in *X-Y* mode. In this mode, the horizontal section of the oscillator, which is normally swept with the oscillator's internal time base oscillator in the standard *Y-T* mode, is now controlled by an external signal [14]. If we think of an oscilloscope as an instrument that draws waveforms, which is essentially what it is, the actual drawing is done by a "pen" that's generated by an electron beam in older cathode ray tube (CRT) oscilloscopes and by a digital signal in more modern instruments. When a sine wave is applied to the vertical input of an oscilloscope, the pen moves up and down as a function of the instantaneous applied voltage. This is how amplitude is measured. When a sine wave is applied to the horizontal input, the pen moves from side to side. If we simultaneously apply sine waves to both the vertical and horizonal inputs, the pen will draw a pattern that is a composite function of the instantaneous voltages on each input [20, 21]. It's basically the same thing that happens when a child plays with an "Etch A Sketch®" toy and turns both knobs at the same time.

Some examples of Lissajous patterns are shown in Figure 5.10. If the same sine wave, having the same frequency and amplitude and a 0° or 360° phase angle, is connected to both the vertical and horizontal inputs, the resulting pattern will be a stationary straight line tilted at a 45° angle with respect to the horizonal axis. If the two frequencies are nominally the same but have a relative offset, for example, two 10 MHz signals generated by different oscillators, the Lissajous pattern rotates due to the changing phase relationship of the signals. Ellipses are formed at phase angles of 45°, 135°, 225°, and 315°, and a circle is formed at 90° and 270°. When the two sine waves are 180° out of phase, the straight line seen at 0° and 360° is tilted in the opposite direction. If we measure the time required for a pattern to complete a full rotation, we can also make a coarse estimate of the frequency difference between the signals. More complex Lissajous patterns indicate the frequency ratio of the two sine waves, which can be helpful when comparing audio tones separated by one or more octaves [14, 20–23].

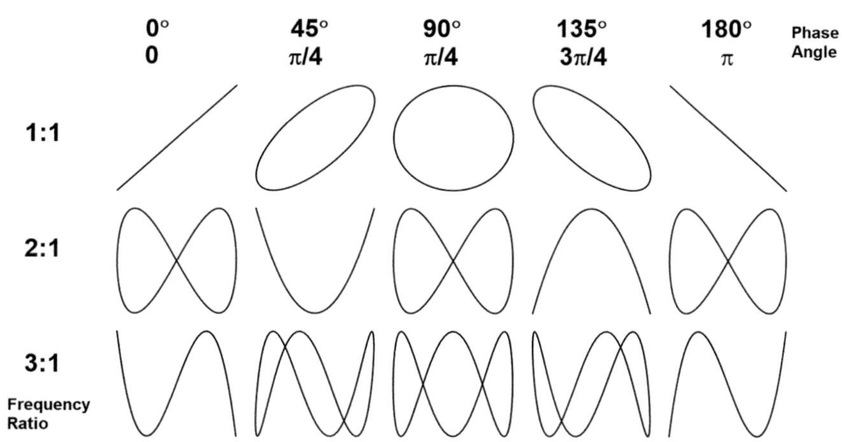

Figure 5.10 Examples of Lissajous patterns on an oscilloscope.

In a cultural context, Lissajous patterns are scientific icons, often found in artwork and in company logos, and sometimes even appearing as movie props, where an oscilloscope displaying a rotating pattern helps to establish the ambiance of a high-tech or laboratory environment. They are also useful teaching tools that are fun to play with, and have significant historical importance. However, there are many better measurement options, and Lissajous patterns are seldom if ever used for serious frequency measurements [14].

A simpler and more accurate phase comparison method involves simultaneously viewing two individual sine waves, and then comparing their relative motion. This method is sometimes called the oscilloscope pattern drift method [22]. It requires a two-channel oscilloscope, and a DUT that has the same nominal frequency as the reference. The nominal frequency, sometimes called the nameplate frequency, is the frequency that an oscillator is supposed to produce. This frequency is indicated by a label on the oscillator's output.

Let's consider an example where the DUT is a quartz oscillator, and the reference is a cesium frequency standard. Both devices have a nominal frequency of 10 MHz. The oscilloscope is triggered by connecting the cesium standard to channel 2, with the DUT signal connected to channel 1. The amplitude and position of both waveforms are then adjusted so that they overlap on the oscilloscope's display.

If we take a snapshot of the two sine waves at a given instant in time, t, we can record their phase difference, $\Delta\phi(t)$, which is $\phi_1(t) - \phi_2(t)$. If we measure phase angle in degrees rather than radians, $\Delta\phi(t)$ will always have a value between 0° and 360°, or ±180° because the two sine waves can only be ahead or behind each other by half of one cycle. If $\Delta\phi(t)$ = 0°, the two sine waves are in phase. If $\Delta\phi(t)$ equals –90° or +90°, the sine waves are in phase quadrature, with their peaks and zero crossings always separated by 90° (Figure 5.11).

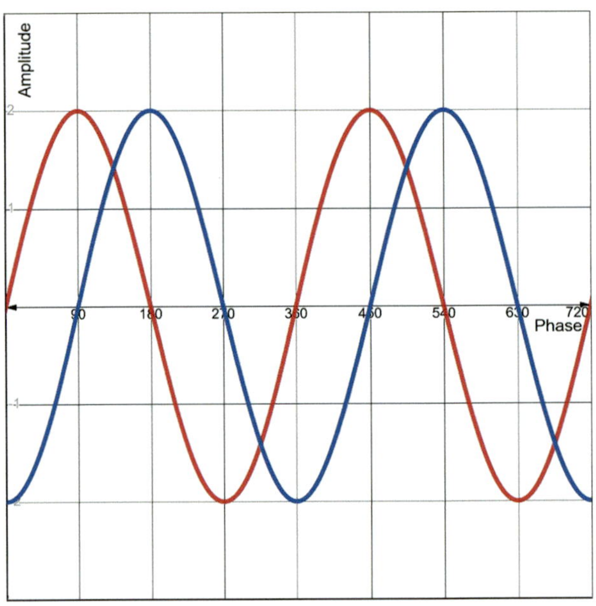

Figure 5.11 Two sine waves in phase quadrature.

If $\Delta\phi(t)$ = 180° (–180° and +180° are the same thing), the two sine waves are in phase opposition with their peaks and zero crossings always separated by 180° (Figure 5.12). When sine waves in phase opposition collide, they cause destructive interference, cancelling each other's energy because the sum of their amplitudes is zero. An example of this would be sound waves coming from two loudspeakers that are 180° out of phase, which results in some deadening of the sound. However, in this case the two sine waves are not physically interacting, they are just simultaneously displayed by the oscilloscope for comparison purposes.

In time and frequency metrology, the phase difference is almost always stated in time units, and time and phase are often used interchangeably. The conversion from an angular phase difference to a time difference is

$$\Delta t = \frac{\Delta\phi}{360 f}, \tag{5.7}$$

where Δt is the time difference in seconds, $\Delta\phi$ is the phase angle in degrees (use 2π instead of 360 for radians), and f is the frequency in hertz. Therefore, a 90° phase difference between 10 MHz signals corresponds to a time difference of 25 ns.

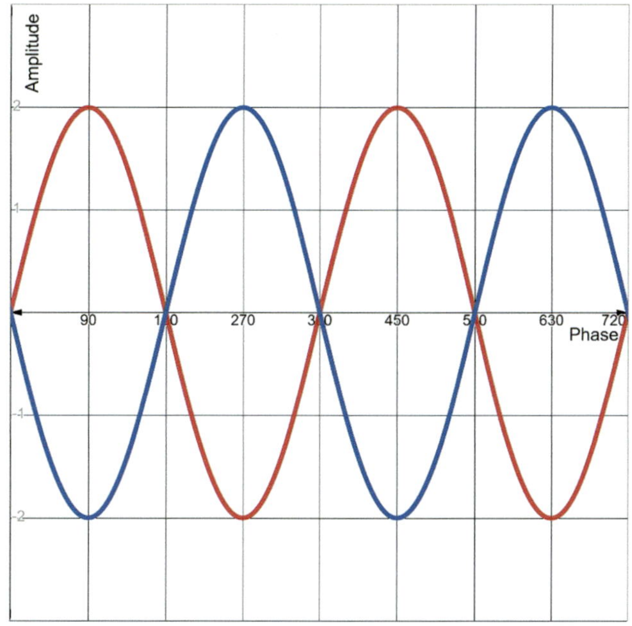

Figure 5.12 Two sine waves in phase opposition.

If the two sine waves being compared always had exactly the same frequency, they would both appear to be stationary on the oscilloscope display, just like our snapshots in Figures 5.11 and 5.12, and their phase relationship would never change. Of course, this is never the case in a real measurement, where the two sine waves always have a frequency difference that causes their phase relationship to constantly change. The reference sine wave will appear to be stationary on the oscilloscope display, and the DUT sine wave will move, with the direction of its motion indicating whether it is low or high in frequency with respect to the reference. In our example, where the DUT is a quartz oscillator and the reference is a cesium, the DUT sine wave will move rapidly (Figure 5.13).

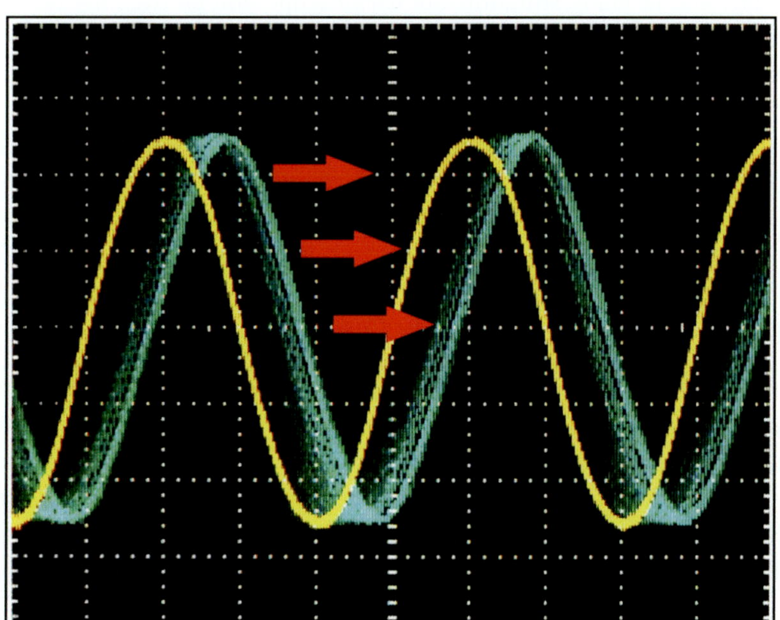

Figure 5.13 Phase comparison with an oscilloscope.

We can estimate the DUT's frequency difference by measuring the duration of a single cycle slip with a stopwatch. We could, if our reaction time was fast enough and our eyes were good enough, measure a partial cycle slip, such as a phase shift of 90° or 180°. However, measuring a full cycle slip is easier and provides more accurate results. The stopwatch is started at the instant when the two sine waves are exactly in phase, $\Delta\phi(t) = 0°$, which is the instant when they appear to be aligned on top of each other on the oscilloscope display. The stopwatch is allowed to run while the signals are out of phase, then stopped at the instant when a complete cycle slip has occurred and the sine waves appear to be exactly in phase again. The frequency difference between the DUT and the reference, f_{diff}, is estimated as

$$f_{\mathrm{diff}} = \frac{T}{t_{\mathrm{stop}}}, \tag{5.8}$$

where T is the period of the nominal frequency, in this case 100 ns for 10 MHz signals, and t_{stop} is the elapsed time indicated on the stopwatch, or the duration of one cycle slip. Therefore, if $t_{stop} = 1$ s, then $f_{diff} = 10^2$ ns/10^9 ns, or 1×10^{-7}. If the DUT is adjusted so that its sine wave moves more slowly, the frequency difference is reduced. For example, if the DUT is adjusted so that t_{stop} increases to 10 s, the frequency difference is reduced to 1×10^{-8}.

It's important to note that f_{diff} in Eq. (5.8) is unitless, which means that it's not expressed in units of hertz or any other unit. The use of unitless values is standard practice in frequency metrology because it allows us to easily compare the performance of oscillators that have different nominal frequencies. It's also standard practice to use scientific notation with unitless values, because the frequency differences measured by metrologists are usually very small numbers, especially when measuring atomic clocks. When we need to convert a unitless frequency difference to hertz, we can do so by simply multiplying the unitless value by the DUT's nominal frequency, f_{nom}. Referring to the previous example,

$$f_{\mathrm{nom}} \times f_{\mathrm{diff}} = (1 \times 10^{7}\,\mathrm{Hz})(1 \times 10^{-7}) = 1\,\mathrm{Hz}. \tag{5.9}$$

As we can see from these examples, oscilloscopes are indispensable for viewing waveforms and for demonstrating phase comparisons in a way that's easy to understand. They work well for simple frequency measurements, and in many situations an oscilloscope will be the only test instrument available. However, the shortcomings of even the pattern drift method quickly become obvious. For example, if we were to compare two oscillators with a frequency difference of 1×10^{-12}, a single cycle slip would take 10^5 s to complete, or more than a day. Measuring the duration of long cycle slips isn't practical, and it can be difficult to accurately measure the duration of short cycle slips or fractions of a cycle with a stopwatch. For this reason and for many others, oscilloscopes are best suited for frequency measurements that have low accuracy requirements. Frequency counters and time interval counters, the topics of the next two sections, are much better choices for most frequency measurements.

5.2.2 Measuring Frequency with Frequency Counters

Frequency counters (FCs) are the instruments most commonly used for frequency measurements. Accurate, easy to use, and with the ability to quickly display results, some FCs are single purpose instruments, especially those used to measure microwave frequencies. In most cases, however, measurements are made with a universal counter that's been configured to work as an FC. Universal counters can perform numerous types of measurements, including time interval (Sections 4.1.2 and 5.2.3), cycle counting (Section 4.1.4), period, frequency ratio, pulse width, duty cycle, and so on, but their most common application is measuring frequency.

The first FC, the Hewlett-Packard 524A, was developed by the American electrical engineer Alan Bagley (1924–2018) in 1950 [24]. Bagley was still a graduate student at Stanford University when he received a grant to work on nuclear-pulse-counting technology, with the objective of resolving two nuclear events that were separated by just 0.1 μs. The project was successful, resulting in a new product for Hewlett-Packard, and in a job offer for Bagley, who was hired after his graduation. Bagley was skilled at developing electronic counting circuits, and he soon developed a more general-purpose counter that could be used to measure the frequency of radio transmitters or to calibrate oscillators, and that could even serve as a frequency standard [25].

As shown in Figure 5.14, the 524A was huge compared to modern instruments, weighing 52 kg with dimensions of 71 × 55 × 36 cm. It used 88 vacuum tubes to measure and count frequencies up to 10 MHz, which was 50× higher than previously existing frequency meters. It's eight-digit display allowed counting frequencies up to 10 MHz with 1 Hz resolution. The six most significant digits were displayed by illuminating one number in a vertical bank of neon lamps, and the two least significant digits were read from analog meters.

Figure 5.14 The first commercial frequency counter, the Hewlett-Packard 524A (courtesy of www.hpmemoryproject.org).

The electronics technology needed to build fast counting circuits was brand new and state-of-the-art in 1950, but the internal operation of the 524A was straightforward, and the device was designed to be easy to use. The unknown sine wave frequency from the DUT was first converted to a square wave by a wide band squaring amplifier. The square wave was then applied to a "fast gate" circuit controlled by a 100 kHz quartz oscillator. This oscillator, and a series of frequency dividers, allowed the gate to be open for a period—called the gate time—of either 1 ms, 10 ms, 100 ms, 1 s, or 10 s. The operator selected the gate time by turning a knob on the front panel. Cycles of the DUT's signal were counted while the gate was open, and a reading was displayed when the gate was closed. The counting circuits were then reset to zero and the gate was reopened, which allowed the next measurement to begin.

If the selected gate time was 1 s, the counter reading showed the DUT frequency in units of hertz with no conversion necessary. Gate times other than 1 s required either multiplication or division to convert the counter's reading to hertz. For example, a 100 ms gate time required multiplying the reading by 10 to convert to hertz, and a 10 s gate time required dividing the reading by 10. The accuracy of the instrument was limited by the accuracy of the internal quartz oscillator that opened and closed the gate. It was also limited by a ±1 count quantization error that was unavoidable because the gate time wasn't synchronized to the DUT signal. This could cause a cycle to be missed or an extra cycle to be counted. Using the built-in oscillator, HP specified the accuracy of the 524A as 0.0005% (5×10^{-6}), ±1 count. If a better external frequency standard was available this accuracy would improve, but the ±1 count error would remain [24].

This was still remarkable performance at the time for an instrument that was so easy to operate, and there was an immediate demand for FCs, especially from the radio and defense industries. Frequency counters quickly became a major source of revenue and growth for Hewlett-Packard, and the success of the 524A and the improved instruments that followed led to HP's establishment of a frequency and time division that Bagley eventually managed. That division engineered a wide range of sophisticated products, including the cesium atomic clocks described in Section 3.6.

Frequency counters even played a part in the early technical education of Steve Jobs (1955–2011), the American inventor and businessman who co-founded Apple Computer. At the age of 12, Jobs famously called the home phone number of HP's co-founder Bill Hewlett (1913–2001). Jobs reportedly told Hewlett that he was building a frequency counter, and he asked him if HP had any spare parts they could share. Hewlett was impressed by the young man's enthusiasm and interest, and not only gave him the parts, but also a summer job in the plant where the counters were assembled [26].

The 524A, and all of the FCs that followed until about the mid-1970s, were direct or conventional counters. These devices were counters in the truest sense of the word, they simply counted cycles of the DUT signal for a gate time selected by the operator. The gate time wasn't synchronized with the DUT signal, and the gate was opened and closed by a time base circuit. For a 1 s gate time, the counter display was always in units of hertz, the resolution was always 1 Hz, and the only way to get finer resolution was to increase the gate time. A counter with more digits could count higher frequencies, but regardless of how many digits a conventional counter could display, it only displayed as many digits as were necessary. For example, if the DUT produced a 500 Hz signal, only three digits were needed, but seven digits were needed for a 5 MHz signal.

A better approach to frequency measurement was to first measure the period of the DUT signal, and to then invert the period to obtain the frequency. This method, called the period or reciprocal method, was again pioneered by Hewlett-Packard when they introduced their 5345A counter in 1974, and it was made possible by advances in high-speed electronics and the advent of the microprocessor [27, 28].

Reciprocal counters have many advantages when compared to conventional counters, and nearly all modern FCs implement the reciprocal method. In a reciprocal FC, the time base oscillator doesn't control the gate time. Instead, the count is triggered by the DUT signal, which opens two synchronized gates connected to two separate counters [29]. The input counter counts either a single cycle, or multiple cycles, of the DUT signal. Technically, only one cycle is needed, but for higher frequencies this could result in gate times that are too short and cause measurement errors. For this reason, most reciprocal counters automatically prescale the DUT signal so that the gate time is automatically set to meet the counter's specified resolution and still take measurements as fast as possible. Longer gate times can also be selected by the operator to produce measurements at a slower rate and get finer resolution [30]. The clock counter counts the cycles of the time base oscillator, which typically operates at 10 MHz. Because the input counter always counts whole cycles of the DUT signal, it doesn't have a ±1 count error. However, there is still a ±1 count error in the clock counter [28–33]. Correcting or reducing the quantization error from the clock counter involves using methods similar to those employed by TICs (Section 4.1.2).

Before getting to interpolation and other methods employed by the highest resolution reciprocal counters, let's look at how a basic reciprocal FC works (Figure 5.15). First, consider that the DUT signal has a nominal frequency DUT_f and a nominal period DUT_T. The time base oscillator has a nominal frequency TB_f and a nominal period TB_T. As always, the frequencies are expressed in hertz and the periods are expressed in seconds.

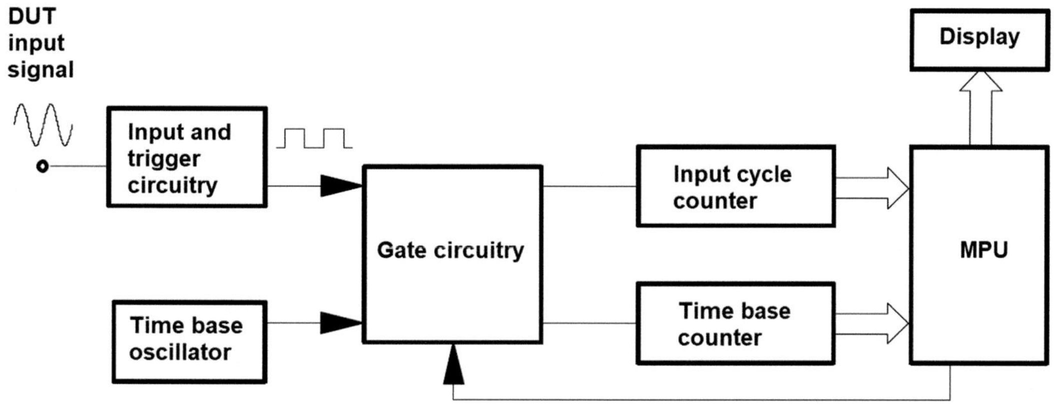

Figure 5.15 Simplified diagram of a basic reciprocal frequency counter.

The input and trigger circuitry converts the DUT signal from a sine wave to a square wave, and the two synchronized gates are opened by the rising edge of the signal and closed by the next rising edge. Therefore, the gate time is equal to the period DUT_T. During this period, the input counter counted just one cycle of the DUT frequency, so $INP_{cnt} = 1$. However, the clock counter counted up to CLK_{cnt}, a value determined by

$$CLK_{cnt} = \frac{DUT_T}{TB_T \pm 1} = \frac{TB_f \pm 1}{DUT_f}. \tag{5.10}$$

Therefore, for a 10 kHz DUT signal and a 10 MHz time base,

$$CLK_{cnt} = \frac{10^{-4}}{10^{-7} \pm 1} = \frac{10^7 \pm 1}{10^4} = 1000 \pm 0.0001, \tag{5.11}$$

where CLK_{cnt} is the result of a period measurement of the 10 kHz signal, expressed as a multiple of TB_T. The period measurement is then converted to a frequency measurement expressed in hertz,

$$f = \frac{INP_{cnt}}{(CLK_{cnt})(TB_T)} = \frac{1}{(1000 \pm 0.0001)(100)} = 10\text{ kHz} \pm 1\text{ mHz}. \tag{5.12}$$

Equation (5.12) is essentially the same equation as Eq. (5.1), which defines frequency as the reciprocal of time interval. That, of course, explains how a reciprocal FC got its name. If $INP_{cnt} > 1$, meaning that a longer gate time was implemented and more than one cycle of the DUT frequency was counted, then the denominator is adjusted accordingly by multiplying by INP_{cnt}.

In this example, the reciprocal counter is measuring a very low frequency of 10 kHz, and its 1 mHz resolution is 1,000× finer than the constant 1 Hz resolution of a conventional counter. This demonstrates the enormous advantage offered by a reciprocal counter when measuring low frequencies. Because the ±1 count error of the clock counter is a constant, a reciprocal counter with a 10 MHz time base and a 1 s gate time will provide eight digits of resolution, and be able to detect frequency differences of 1×10^{-7} with a single reading. The resolution will always be less than 1 Hz when measuring a frequency lower than the time base frequency. However, as the DUT frequency approaches the time base frequency, the resolution gets coarser, reaching 0.01 Hz at 100 kHz, and 0.1 Hz at 1 MHz. When the DUT signal has the same nominal frequency as the time base, a basic reciprocal counter has the same 1 Hz resolution as a conventional counter [28, 29].

Higher-resolution reciprocal FCs employ a variety of methods to display more digits, including multiplication of the time base frequency and the use of interpolation to remove the ±1 count error in the clock counter [32]. These methods are essentially the same as those employed by TICs to improve their single-shot resolution (Section 4.1.2). The only difference is that signals from two sources open and close the gate in a TIC. These signals, usually generated at a frequency of 1 Hz, are connected to the start and stop channels of the counter. In a reciprocal FC, the gate is opened and closed by the same signal, which is generated by the DUT at an arbitrary frequency. From there, the problem is the same—there are two unknown intervals that contribute resolution uncertainty to the measurements. The first unknown is the interval between the start trigger and the next time base cycle, and the second unknown is the interval between the last time base cycle and the stop trigger.

Because a ±1 count error introduces a resolution uncertainty that is equal to the period of the time base frequency, simply increasing the time base frequency will reduce the uncertainty. The original reciprocal FC, the Hewlett-Packard 5345A, didn't use interpolation at all, but instead multiplied the time base frequency from 10 MHz to 500 MHz. This improved resolution by 50×, and allowed a nine-digit display with no resolution uncertainty at a 1 s gate time [27]. More modern FC designs implement both time base multiplication and interpolation, with multiplication reducing the uncertainty of the clock counter's quantization error, and interpolation compensating for most of the remaining uncertainty.

Analog or digital interpolation methods are designed to measure the duration of partial time base cycles. Some FCs implement variations of the analog Nutt method [34] of charging and discharging capacitors that was previously described in Section 4.1.2. These FC designs employ two interpolation counters, one for the start trigger and one for the stop trigger, as shown in Figure 5.16 [29]. The start trigger counter measures the interval between the start trigger, which is the rising edge of the DUT signal that triggers the input counter, and the rising edge of the first cycle that is counted by the clock counter, or TB_1. The stop trigger counter measures the interval between the end of the last complete time base cycle and the stop trigger, or TB_2. The clock counter then returns a value corrected by interpolation, CLK_{cnti}, to remove the ±1 count error,

$$CLK_{\text{cnti}} = \frac{DUT_T}{TB_T + TB_1 - TB_2}. \tag{5.13}$$

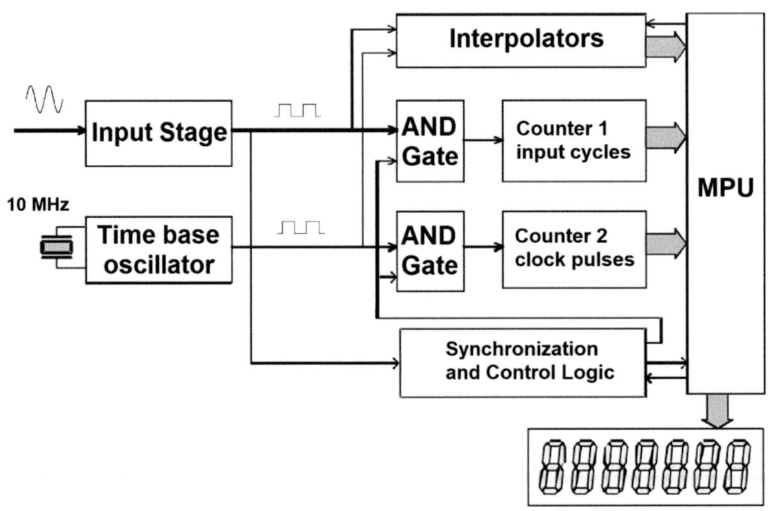

Figure 5.16 Simplified diagram of an interpolating reciprocal frequency counter [29].

Other frequency counters use digital interpolation, based on variations of the Vernier scale method also described in Section 4.1.2 [35]. One example of digital interpolation using a Vernier scale involves the use of two triggered Vernier oscillators (TVOs) that each run at the same frequency, which is slightly lower than the frequency of time base oscillator. For example, if 8-bit analog-to-digital conversion is used, the TVOs can be made to run at a frequency that has a ratio of 255/256 with respect to the time base. One TVO is synchronized to be in phase with the time base clock and is restarted when the start trigger arrives, and the other TVO repeats this process when the stop trigger arrives. Because their frequency differs from the time base, the TVOs are only in phase with the time base at a time of coincidence that repeats every 255 clock cycles. Referring again to Eq. (5.13), TB_1 is now the interval between the start trigger and the start coincidence, TB_2 is the interval between the stop trigger and the stop coincidence, and TB_T is the interval between the start and stop coincidence times. This removes the ±1 count uncertainty from the clock counter, and results in a 256× improvement in resolution [28, 30, 36].

The measurements made with a FC are directly referenced to the counter's time base oscillator. The internal time base oscillator on most FCs is a quartz oscillator, usually either a TCXO or an OCXO, but some FCs have internal rubidium oscillators. If the internal time base oscillator is used, its accuracy will only be known if the FC has been calibrated, and will typically be no better than about 1×10^{-8} for a quartz time base. Nearly all FCs can accept an external time base oscillator, typically with a nominal frequency of either 5 or 10 MHz, and to get the best results the best available frequency standard should be used as the external time base. Some devices now combine disciplined oscillators with FCs, and Figure 5.17 shows an instrument where a global navigation satellite system (GNSS) disciplined oscillator provides the time base for a built-in frequency synthesizer and counter.

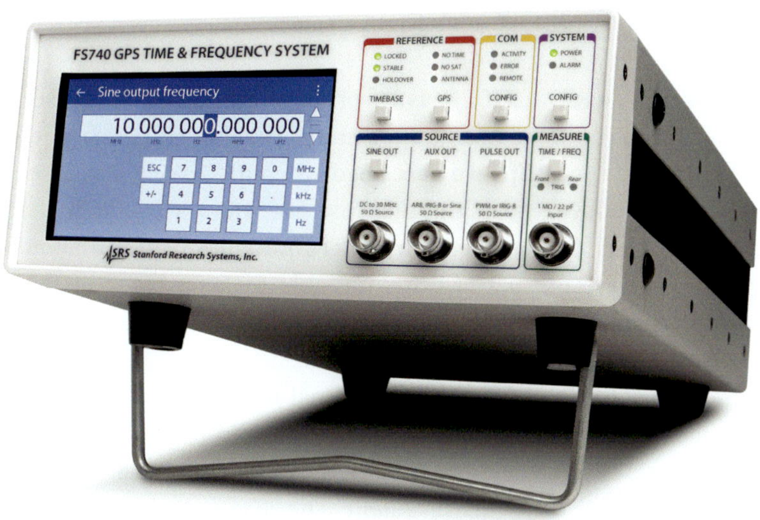

Figure 5.17 A GNSS disciplined oscillator with a built-in frequency synthesizer and frequency counter (courtesy of Stanford Research Systems).

The frequency difference between the DUT signal and the FC's time base oscillator is obtained as

$$f_{\text{diff}} = \frac{f_{\text{meas}} - f_{\text{nom}}}{f_{\text{nom}}} = \frac{\Delta f}{f}, \tag{5.14}$$

where f_{meas} is the reading obtained from the FC, and f_{nom} is the nominal frequency of the DUT. The smallest value of f_{diff} that can be detected with a single reading is determined by the number of digits on the FC's display. When measuring 10 MHz, a 10-digit FC can't detect frequency differences smaller than 1×10^{-9} without averaging. A 12-digit counter improves this situation by a factor of 100, and can detect frequency differences as small as 1×10^{-11} without averaging.

5.2.3 Measuring Frequency with Time Interval Counters

Frequency is commonly measured in the time domain with a TIC, an instrument that was previously described in Section 4.1.2 [22, 37, 38]. A frequency measurement performed with a TIC is a phase comparison that's conceptually identical to the oscilloscope phase comparisons described in Section 5.2.1, but potentially far more accurate. Like a two-channel oscilloscope, a TIC has inputs for two signals, in this case a start channel and a stop channel (Figure 5.18). The TIC usually starts counting when it receives a signal from the DUT and stops counting when it receives a signal from the reference. However, the order doesn't matter and the signals can be reversed, as long as the sign of the measurement result is reported correctly.

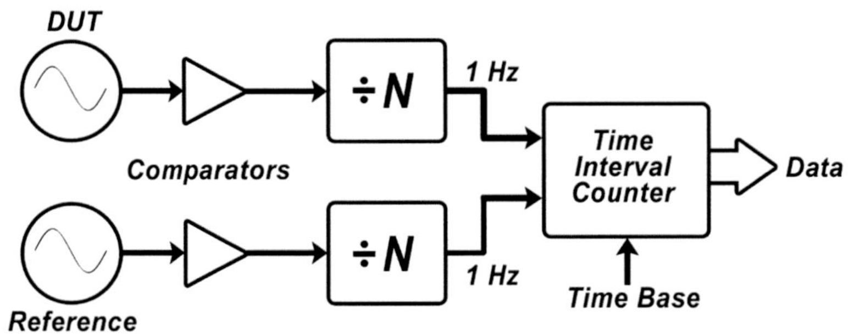

Figure 5.18 Simplified diagram of a TIC frequency measurement system.

When measuring frequency with a TIC, it isn't practical to directly measure standard frequency signals like 10 MHz. If we attempt that, the maximum counter reading will equal the period of the signal, or 100 ns in the case of 10 MHz. The cycle slips we described when discussing oscilloscope phase comparisons will cause "underflows" and "overflows," which in this case are counter readings that go from 100 ns back to 0 ns, or vice versa. If we graph data collected from TIC measurements of standard frequency signals, this cycle ambiguity creates a sawtooth pattern. It's possible to keep track of the cycles with computer software and to constantly correct for cycle ambiguity, but a better solution is to divide the standard frequency signals to lower frequencies before they are measured.

In most cases, the frequency is divided by N, the nominal frequency of the oscillator. This blocks $N - 1$ cycles of the nominal frequency, allowing just one cycle per second to pass through to the TIC. The ability to measure 1 pps (1 Hz) signals is a major advantage of a TIC, because it allows both time and frequency information to be obtained from the same measurement. The diagram in Figure 5.18 shows the frequency dividers as being external to the reference and DUT, but they are often built-in, and most atomic clocks and GNSS disciplined clocks have 1 pps outputs. The built-in dividers are usually programmable (sometimes called slewable), which allows their 1 pps output to be synchronized to agree with a time standard. If the devices being measured lack built-in frequency dividers, both fixed and programmable dividers are available as stand-alone circuits or instruments, and dividers have even been built-in to the TIC circuits [39]. A small circuit board that divides either a 1, 2.5, 5, or 10 MHz signal to 1 pps is shown in Figure 5.19.

As discussed in Section 4.1.2, it's important to make sure that the TIC is triggering at the correct voltage level on the input signal before beginning the measurements. This involves either carefully adjusting the trigger levels, or else using a TIC that has fixed trigger levels and that converts both input signals to an identical shape and amplitude prior to the measurement. If the purpose of the TIC measurement is to just measure frequency, and not both frequency and time, then the cable delays are unimportant and don't have to be measured.

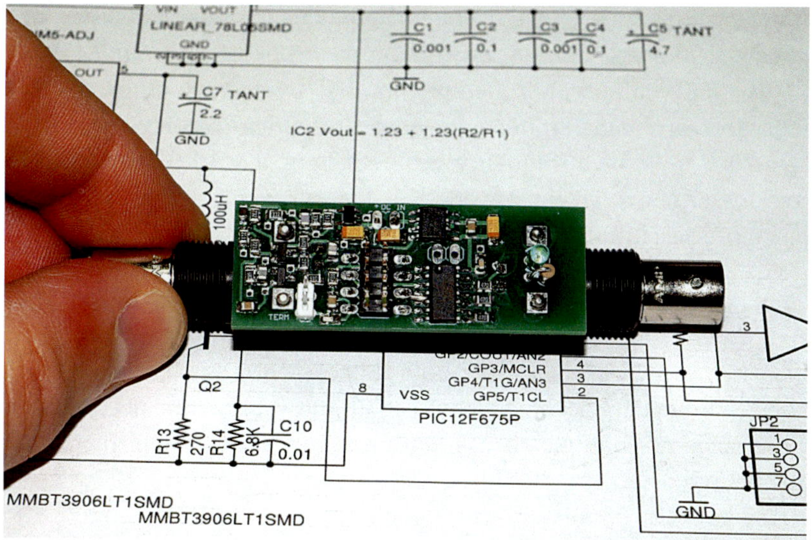

Figure 5.19 A frequency divider circuit (courtesy of TAPR).

Once the measurements begin, a new time interval reading is produced every second. As indicated by Eq. (5.14), a single reading from a FC shows us the frequency difference, because the reading can be subtracted from the known nominal frequency. However, in the case of a TIC measurement, at least two readings are needed to determine the frequency difference. A quick estimate of the DUT's frequency difference with respect to the reference is obtained as

$$f_{\text{diff}} = \frac{TI_2 - TI_1}{T} = \frac{\Delta t}{T}, \tag{5.15}$$

where TI_1 is the first measurement obtained from the TIC, TI_2 is the second reading obtained from the TIC, and T is the period between the two measurements.

If we follow the convention where the reference frequency is connected to the stop pulse, then the stop pulse will appear to be stationary and the start pulse from the DUT will move in the same fashion as the stationary and moving sine waves observed in the oscilloscope pattern drift method (Figure 5.13). As a result, the TIC readings will either get larger or smaller. For a series of TIC readings, i, where the readings always get larger, $TI_i > TI_{i-1}$, the DUT pulse is always arriving earlier than it did before. This means that the period of the DUT signal is shorter than the period of the reference signal, and because $f = 1/T$ its frequency is higher than the reference. For a series of readings that always get smaller, $TI_i < TI_{i-1}$, the opposite is true, the DUT pulse is always arriving later than it did before, because the period of its signal is longer and its frequency is lower than the reference.

Sign errors, the bane of all metrologists, are easy to make with TIC measurements. For example, if we reverse the order of the start and stop cables, or if we reverse the order in which we subtract the measurements, we also need to change the sign of the measurement result. However, if we avoid sign errors and keep all other variables equal, frequency domain measurements made with a FC will agree with time domain measurements made with a TIC, because

$$f_{\text{diff}} = \frac{\Delta t}{T} = \frac{\Delta f}{f}. \tag{5.16}$$

We discussed time interval resolution in detail in Section 4.1.2, but can reiterate here that the best TICs now have single shot resolution of less than 10 ps. A single shot resolution of 10 ps allows a TIC to detect frequency differences of 1×10^{-11} in 1 s, which is equivalent to the resolution of a 12-digit FC. A TIC with 1 ps resolution can, in theory, improve on this by an order of magnitude. However, a TIC's ability to detect very small frequency differences is often limited by factors other than resolution, including trigger errors, time base errors, and frequency divider noise. The technology is always improving, but as a general rule the ability to resolve frequency differences of 1×10^{-11} at 1 s can be thought of as the best-case performance for either a TIC or an FC. The measurement systems described in the next two sections are used when more resolution is needed.

5.2.4 Measuring Frequency with Heterodyne Systems

The use of FCs and TICs is widespread, and both instruments are excellent choices for measuring long-term accuracy and stability. However, higher-resolution systems are needed to measure short-term stability. Some of these systems continue to use counters, but employ heterodyne methods, rather than frequency dividers, to obtain the necessary low-frequency signals.

Heterodyning is the process of mixing two signals, at frequencies f_1 and f_2, to generate two new frequencies called heterodynes. The two heterodyne signals have frequencies equal to the sum, $f_1 + f_2$, and the difference, $f_1 - f_2$, of the mixed frequencies. The process is related to the same phenomena, the constructive and destructive interference of sine waves, that produces beat frequencies when tuning a piano (Section 5.1.1).

The presence of an audio beat frequency had been known to musicians for centuries, but the Canadian-American inventor Reginald Fessenden (1866–1932) was the first to apply frequency mixing to RF signals. Fessenden was classically trained, and he coined the word heterodyne based on his knowledge of the Greek language. Loosely translated, to heterodyne means to mix two difference forces, which in Fessenden's case meant mixing two sine waves of different frequencies. Fessenden was working towards the design of a radio receiver where the incoming radio signal was mixed with a continuous sine wave frequency from a stable oscillator. This would allow the generation of a lower frequency signal that the receiver could process [40]. Fessenden was granted a patent for the heterodyne principle in 1902 [41], but the stable continuous wave oscillator that he needed to build a heterodyne receiver didn't exist for at least another decade.

Superheterodyne receivers, an invention credited to the American electrical engineer Edwin Armstrong (1890–1954), became commercially available in 1924. These receivers included a tunable local oscillator (LO), making them more sophisticated than earlier heterodyne receivers with fixed-frequency oscillators, but the addition of the prefix "super" was still largely the result of 1920s advertising hype [42]. A century later, most radio receivers are still based on a superheterodyne design.

A frequency mixer, pictured in Figure 5.20, is a three-port circuit or device that converts frequencies by the process of heterodyning. The two input ports accept an RF signal at a frequency f_1 and a LO signal at a frequency f_2. The output port produces an intermediate frequency (IF) that can be either the sum or the difference of the two input frequencies, IF = $f_1 \pm f_2$. Because, most applications require only one of the heterodyne signals, the mixer suppresses the other, and in the case of the double balanced mixer (DBM) commonly used in frequency metrology, it also suppresses f_1 and f_2. As a result, the output of the mixer produces an IF signal that has either been up-converted (IF > RF) or down-converted (IF < RF). Down-conversion is practiced in conventional superheterodyne receivers such as AM/FM radios. For example, broadcast band receivers for AM (540 kHz to 1710 kHz) typically use a 455 kHz IF, and FM receivers (88 MHz to 108 MHz) typically use a 10.7 MHz IF [42, 43].

Down-conversion is also practiced in heterodyne systems designed for RF frequency measurements, but in these systems the IF, or beat frequency, is much lower. The DBM converts the DUT signal to a lower frequency, performing a function similar to the frequency dividers shown in Figure 5.18. However, when a frequency divider is used to divide the carrier, it also divides the frequency and phase fluctuations on the carrier by the same amount. This means that no measurement resolution is gained when dividers are used to convert to a lower frequency.

Figure 5.20 A double balanced frequency mixer (courtesy of Fairview Microwave).

In contrast to dividers, frequency mixers convert the carrier to a lower frequency without reducing the frequency or phase fluctuations [38]. Because frequency and phase information are preserved during the down-conversion, the resolution of the measurement system improves by the heterodyne factor, which is the ratio of the nominal frequency of the DUT to the beat frequency, $\frac{f_{\text{nom}}}{f_{\text{beat}}}$, or with respect to the mixer ports, $\frac{\text{RF}}{\text{IF}}$. Unlike divider systems that typically produce 1 Hz signals, the desired beat frequency produced by the mixer is usually not lower than 10 Hz, because the system has to have enough bandwidth to account for a wide range of possible frequency differences between the DUT and reference, including situations where a quartz oscillator is involved in the comparison.

Figure 5.21 shows a single-mixer heterodyne system for measuring frequency. The system includes a low-pass filter (LPF) and a DBM, in addition to a reciprocal frequency counter. The reference and DUT signals have the same nominal frequency, typically 10 MHz. Although not shown in the figure, the reference oscillator is connected to a frequency synthesizer that offsets its frequency from nominal, typically by either 100 Hz or 1 kHz. This signal is connected to the LO port on the mixer and the DUT signal is connected to the RF port. The DBM mixes the intentionally offset reference frequency with the DUT frequency, producing a beat frequency output by the IF port. The beat frequency signal is sent through a LPF to remove any carrier frequency harmonics, and is then measured by the FC. The frequency difference between the DUT and reference is determined by

$$f_{\text{diff}} = \frac{f_{\text{meas}} - f_{\text{beat}}}{f_{\text{nom}}} = \frac{\Delta f}{f}, \tag{5.17}$$

where f_{meas} is the FC reading, f_{beat} is the nominal beat frequency, and f_{nom} is the nominal frequency of the DUT. Note that this is almost the same equation as Eq. (5.14), except now the counter is measuring the beat frequency and not the carrier frequency, which makes Δf a smaller quantity, resulting in higher resolution [37, 38, 44, 45].

In theory, the improvement in resolution of a heterodyne system, when compared to a reciprocal FC, is the ratio of the nominal frequency to the beat frequency. This ratio, called the heterodyne factor, is 10^4 in our example for a 1 kHz beat frequency, 10^7 Hz/10^3 Hz, which means that the FC potentially gains four digits of usable resolution. For example, a 10-digit FC displaying hertz would show only two digits to the right of the decimal point when measuring 10 MHz (10 000 000. 00), but shows six digits to the right of the decimal point when measuring 1 kHz (1 000. 000 000).

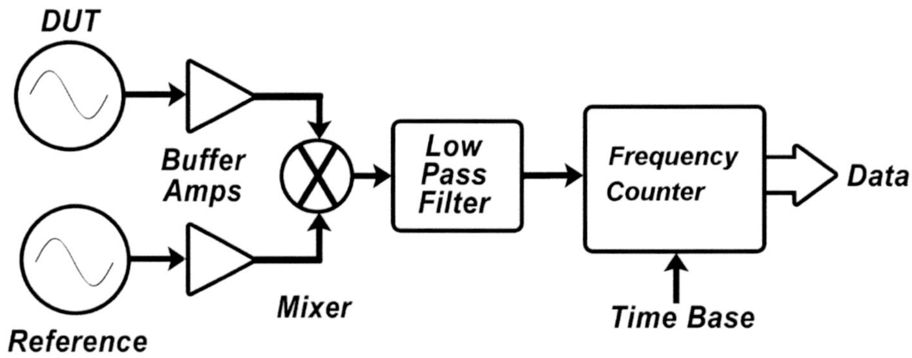

Figure 5.21 Simplified diagram of a single-mixer heterodyne system.

In practice, the full advantage of the heterodyne factor is not always realized, because the frequency synthesizer and other components introduce measurement noise, but a single-mixer heterodyne system can still improve upon the resolution of even a 12-digit FC. However, there are some disadvantages. Unlike a FC measurement, a heterodyne system requires some knowledge of the frequency difference of the DUT before the measurement begins, because its measurement bandwidth is limited. And unlike a TIC measurement that produces both frequency and time information, a single-mixer system measures frequency only, without producing time information [45].

The dual mixer time difference (DMTD) measurement system (Figure 5.22) combines the best features of the TIC and heterodyne methods. It performs high-resolution phase comparisons between a DUT and a reference frequency, using hardware that includes a transfer oscillator, two DBMs, two zero crossing detectors (ZCDs), and a TIC. The two DBMs are employed in parallel to down-convert two frequencies prior to comparing their phase.

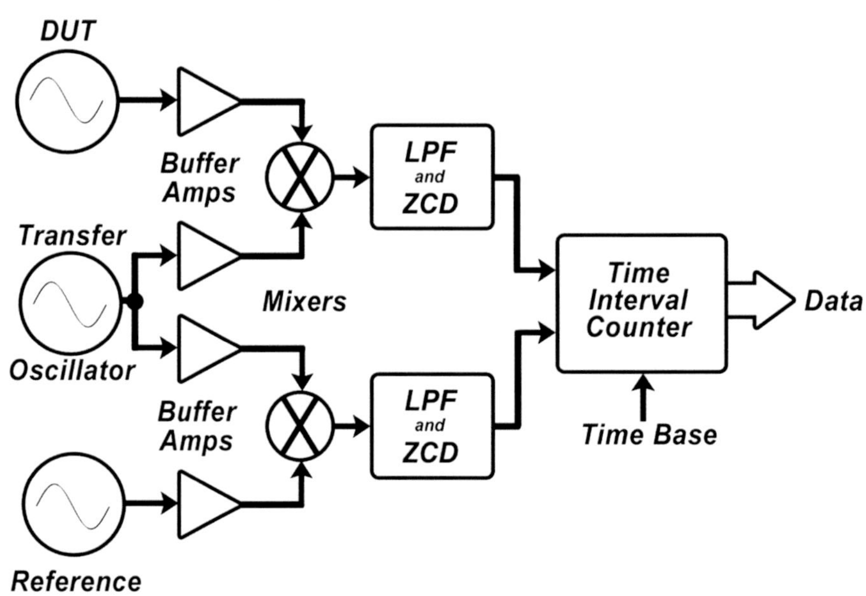

Figure 5.22 Simplified diagram of a dual mixer time difference measurement system.

The basic methodology of a DMTD measurement system was first implemented in an instrument called a phase meter, which was developed in 1966 by the American engineer Donald Maxwell (1913–2007), who was then with the General Electric Company in Syracuse, New York [46]. The DMTD method was refined in the mid-1970s, when systems were independently developed at NBS and at the National Research

Council (NRC) in Ottawa, Canada [44, 47]. These systems had immediate applications—a DMTD system was built under contract at NBS in 1975 to test the frequency stability of clocks that would fly on-board the first GPS satellites [48]. In the years that followed, DMTD systems became the preferred measurement system for comparing the frequencies of atomic clocks used in ensemble time scales (Section 4.4) at NBS/NIST, NRC, and other national laboratories [49], and commercial DMTD systems were made available as general-purpose frequency measurement instruments [50].

A DMTD system usually requires the DUT and the reference to have the same nominal frequency, which is typically 5 or 10 MHz, and some systems are designed to measure only one nominal frequency. Because DMTD systems are specifically designed to provide the high resolution necessary to measure very stable oscillators, their bandwidth is narrow, and the transfer oscillator's offset from the nominal frequency is small—often 10 Hz [48, 51], but just 1 Hz in some systems [49, 50]. The transfer oscillator is usually realized by a frequency synthesizer phase-locked to the reference oscillator [38]. Much like the common-view signal in a time transfer system (Section 4.3.3), the transfer oscillator doesn't need to be especially stable or accurate, because the noise it produces is common to both measurement channels and cancels when the phase difference is measured.

The transfer oscillator signal is heterodyned with both the reference oscillator signal and the DUT signal, producing two beat frequencies. The beat frequencies pass through a LPF that removes the harmonics, and through a ZCD that determines the zero crossing of each beat frequency cycle. The two ZCD outputs are connected to the start and stop channels of a TIC, which measures their phase and frequency difference as previously described in Section 5.2.3.

Measurement noise is contributed to a DMTD system by the transfer oscillator, by the TIC time base oscillator (Section 4.1.2), and by other system components. However, the amount of noise is reduced when the zero crossings are nearly coincident and the TIC reading is small. For this reason, a variable phase shifter is sometimes used to put the DUT signal nominally in phase with the reference before mixing the DUT signal with the transfer oscillator [47, 49]. As is the case in a conventional TIC frequency measurement, cycle slips are infrequent because of the low frequencies involved in the phase comparison. However, keeping track of the beat frequency cycles is still necessary to avoid cycle ambiguity (Section 5.2.3). Another solution is to time tag both the DUT and reference frequency zero crossings using a high-resolution external clock, and then recording the difference between time tags instead of recording a TIC reading [51].

The DMTD method typically provides the highest resolution of any analog frequency measurement system. The resolution is determined by the resolution of the phase comparison hardware (TIC or time tags) divided by the heterodyne factor. For example, if the phase comparison resolution is 100 ns, and the heterodyne factor is 10^6, based on a 10 MHz nominal frequency and a 10 Hz beat frequency, then the DMTD resolution can be estimated at $\frac{10^{-7}\,\text{s}}{10^6}$, or 0.1 ps. This is 100× more resolution than a 12-digit FC or a 10 ps TIC, and makes it possible to detect frequency differences of 1×10^{-13} in 1 s. Some DMTD systems improve this resolution by about a factor of two, and can detect frequency differences of about 5×10^{-14} in 1 s [50]. This is enough resolution for nearly all RF stability measurements, but even higher resolution can be obtained with digital sampling systems [52], the topic of the next section.

5.2.5 Measuring Frequency with Digital Sampling Systems

Digital sampling is the process of reducing a continuous analog waveform to a sequence of discrete samples. The individual samples are numbers assigned to the signal at a given point in time, $S(t)$, which can then be processed with general-purpose computers and software, or with dedicated digital signal processing (DSP) hardware. Figure 5.23 shows the digital sampling of a sine wave, with 12 samples taken every cycle. The sampling frequency, f_s, is expressed either in hertz or as samples per second. Therefore, if we assign an arbitrary nominal frequency of 1 MHz to the sample sine wave in Figure 5.23, f_s is equal to 12 MHz, or 12 MS/s.

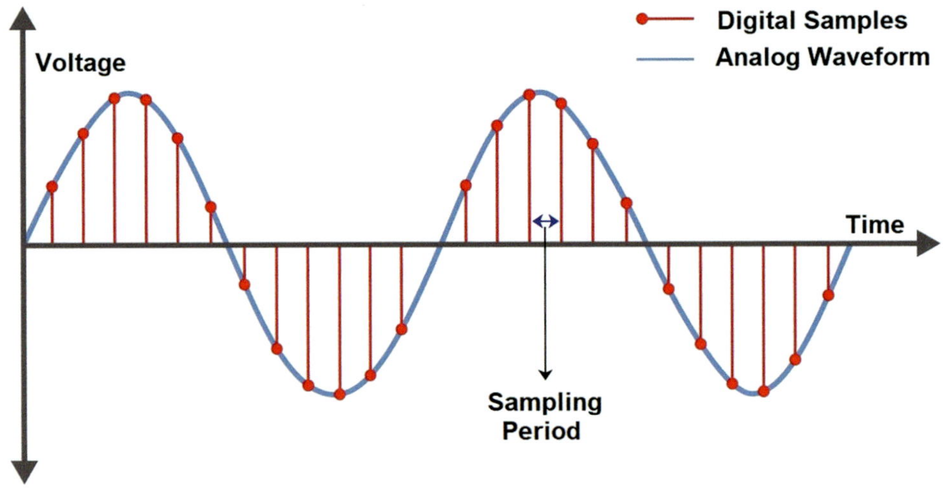

Figure 5.23 Digital sampling of a sine wave.

The basic principles of digital sampling were first presented by Harry Nyquist (1889–1976), a Swedish-American physicist and electrical engineer. Born in Sweden, Nyquist emigrated to the United States in 1907. After receiving his doctorate from Yale in 1917, he spent his entire career at AT&T and Bell Telephone Laboratories, retiring in 1954. His early theoretical work on determining the bandwidth requirements for transmitting information was summarized in a 1928 paper entitled "Certain Topics in Telegraph Transmission Theory" [53]. Nyquist determined that signals needed to be sampled at a rate that was at least twice as high as the maximum frequency component in the signal in order to obtain useful information. This means that at least two samples must be taken during every cycle. Therefore

$$f_s \geq 2f_{\max}, \tag{5.18}$$

where f_s is the Nyquist sampling rate, and $f_{\max}$ is the maximum frequency component of interest in the signal being sampled. It follows that the Nyquist frequency, f_N, is one half of the sampling rate, or

$$f_N = \frac{f_s}{2}. \tag{5.19}$$

Returning for a moment to audio frequency (Section 5.1), digital CD recordings provide an example of how the Nyquist sampling rate is applied. When a CD is recorded, analog sound waves are sampled at a rate of 44,100 S/s (44.1 kHz) for both stereo channels [54]. Because f_s = 44.1 kHz, it follows that f_N = 22.05 kHz, which is sufficient bandwidth to cover the approximate 20 kHz range of human hearing.

If an analog signal is undersampled, meaning that fewer than two samples per cycle are taken, it's not possible to unambiguously determine its frequency, because the various frequency components of the signal will overlap. This causes a situation where the components of the signal with frequencies higher than the Nyquist frequency will appear as lower frequency components. This effect is called aliasing, because it can disguise the frequency of a waveform reconstructed from samples, allowing it to assume an identity, or alias, that is not its own.

Aliasing due to undersampling can also cause distortion or artifacts when attempting to reconstruct a signal from its digital samples. Acquiring two samples per cycle is enough to determine the frequency of a waveform, but faster sampling rates are needed to fully reconstruct a waveform's amplitude and its other characteristics. The sine wave shown in Figure 5.23 is oversampled, because 12 samples per cycle is 6× the minimum Nyquist rate. This does a reasonably good job of replicating what the analog signal looks like, but more samples can provide an even better picture. Some amount of oversampling is desirable in frequency measurement systems because it avoids aliasing, increases resolution, and reduces noise. A challenge, of course, is determining the amount of oversampling that provides optimal performance without wasting resources.

Because digital sampling systems measure analog RF signals by utilizing digital software and hardware, they always include at least one analog-to-digital converter (ADC). An ADC is an electronic device or system that does exactly what its name implies, it converts an analog signal to a digital signal. The two most important specifications for an ADC are its sampling rate, f_s, and its resolution. Based on their sampling rates, ADCs are often categorized as being either fast or slow. Speed categories are relative and will likely always change as technology improves, but for the purpose of example, ADCs that can sample RF signals ≥10 MHz might be categorized as high-speed, and ADCs intended for sampling AF signals are categorized as low-speed. Regardless of its speed, an ADC is driven by a sampling clock that has to run at a frequency high enough to support its sampling rate. The sampling clock frequency is usually a multiple of f_s that can vary considerably depending upon the type and model of ADC. For example, an ADC capable of a 65 MHz sampling rate might require a 260 MHz sampling clock (4× multiple) to meet its specification.

The resolution of an ADC is expressed in bits, and indicates the number of discrete voltage differences, or levels, it can resolve from an analog signal. For example, a 16-bit ADC can resolve 65,536 discrete voltage levels (2^{16} = 65,536). However, due to jitter, or the variation in the placement of the sampling clock edges, one or more of the least significant bits might not be usable. Jitter becomes more of a concern as the ADC resolution increases, meaning that a noisy sampling clock that affects a 16-bit ADC might have no effect on a 10-bit ADC. It also becomes more of a concern when sampling at faster rates. Therefore, finding ways to reduce sampling clock jitter is one of the greater challenges faced by designers of high-speed and high-resolution digital sampling systems [55].

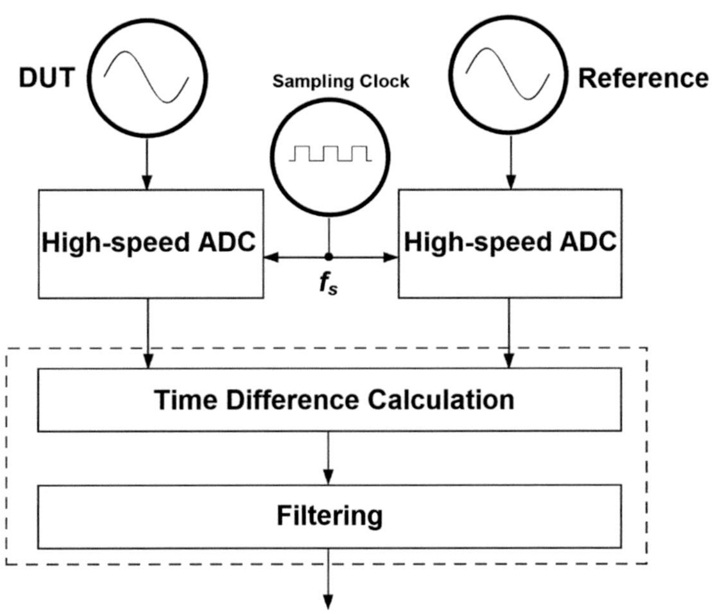

Figure 5.24 Simplified diagram of a direct sampling frequency measurement system.

Frequency measurements systems based on digital sampling have many different architectures, but they can be broadly grouped into two categories—systems that directly sample the incoming RF signals from the DUT and reference (Figure 5.24), and systems that sample IF signals obtained after down-converting the RF signals by heterodyning or another process (Figure 5.25). Systems that directly sample the RF signals can be thought of as pure digital designs, whereas those that sample the IF signal can be thought of as hybrid analog/digital designs. The inclusion of ADCs in both types of systems is a fundamental difference between digital sampling systems and the analog systems we've previously described. For example, the direct sampling system shown in Figure 5.24 replaces the two DBMs, the transfer oscillator, and the TIC in a DMTD system (Figure 5.22) with two ADCs, a sampling clock, and DSP software that performs the time difference calculations. Like a DMTD system, digital sampling systems produce both time and frequency information, and have the advantage of being able to compare oscillators that have different nominal frequencies.

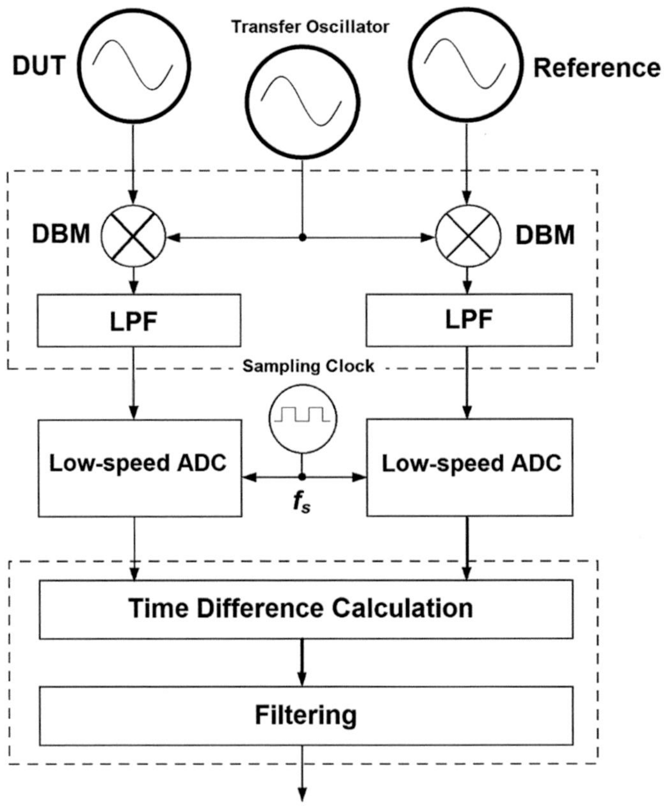

Figure 5.25 Simplified diagram of an IF sampling frequency measurement system.

Direct sampling systems use high-speed ADCs to simultaneously digitize the incoming signals from the DUT and the reference oscillator. To measure DUT signals at a nominal frequency of 10 MHz, the ADC sampling rate has to be at least 20 MHz, but preferably higher to support oversampling. This results in two voltage samples, $S_1(t)$ and $S_2(t)$ that are continuously output at a period of $T_s = 1/f_s$ and stored in a sequence. This process continues for a user-selectable sampling interval, which is typically 1 s, but which can be shorter. As we will see in Chapter 7, the selected sampling interval is equal to the shortest period for which frequency stability can be estimated, and is known as τ_0. During the sampling interval, the voltage samples are subtracted from each other to determine the time and frequency difference between the DUT and reference signals. After being filtered to remove any undesired high-frequency components, this measurement result is stored, and the measurement process is repeated [56].

The hybrid measurement system shown in Figure 5.25 works in nearly the same way as the direct sampling system, but here the DUT and reference signals are down-converted to an IF using the dual mixer method before being digitally sampled. Like a DMTD system, the IF sampling system requires a transfer oscillator and DBMs, but not a TIC. Because the IF signals being sampled are at a much lower frequency than the incoming RF signals, it allows the use of low-speed ADCs and has less stringent sampling clock requirements.

The first commercial digital sampling instrument for measuring phase noise and frequency was the Timing Solutions 5120A (Figure 5.26), introduced in 2005 and later manufactured by Symmetricom, Microsemi, and Microchip. The 5120A was designed as a self-contained instrument that didn't require a host computer, and it could measure sine waves with nominal frequencies ranging from 1 to 30 MHz. The DUT and reference signals were each digitized with high-speed ADCs that sampled at a 64 MHz rate before any other processing was performed. The digital signals were then mixed with a quadrature multiplier before their phase difference was computed. To cancel the noise from the sampling clock, the 5120A employed a cross correlation method that split both input signals into two paths so that parallel measurements could be made on both channels. This allowed the instrument to detect frequency differences of less than 5×10^{-15} in 1 s [52, 57], a resolution about 10× finer than a DMTD system.

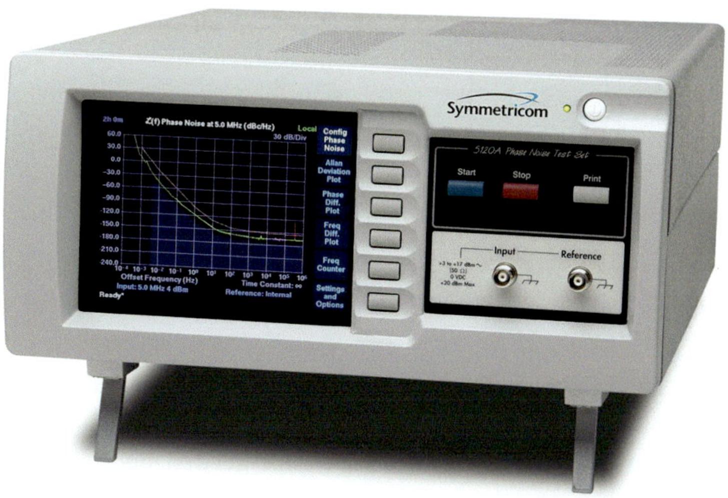

Figure 5.26 A direct digital sampling frequency measurement instrument (courtesy of Microchip).

Other commercial instruments with architectures similar to the 5120A typically require software that runs on a host computer, but they can provide comparable performance. For example, the Microchip 53100A Phase Noise Analyzer can also detect frequency differences of less than 5×10^{-15} in 1 s and can measure frequencies from 1 to 200 MHz [58]. A multi-channel measurement instrument developed by researchers in Europe has demonstrated the ability to detect frequency differences of less than 1×10^{-15} in 1 s, for input frequencies ranging from 5 to 400 MHz. This instrument utilizes a direct digital synthesizer (DDS) which has less phase noise than an ADC [59].

The architecture of a digital sampling frequency measurement system is similar to that of a software-defined radio (SDR). The term "software radio" was introduced by the electrical engineer Joseph Mitola (b. 1948) in 1991. Mitola was describing a then new type of radio architecture where all or most of the signal processing could be done digitally. This involved placing an ADC, in the case of a receiver, or a digital-to-analog converter (DAC), in the case of a transmitter, as close to the antenna as possible [60]. Because the signals they process are mostly digital, SDRs can implement components in software—including filters, modulators/demodulators, and encoders/decoders—that are implemented with analog hardware in conventional radios [61].

SDRs perform the same primary function as all other radios, which is receiving wireless signals through an antenna. However, unlike conventional radios, SDRs are programmable RF sampling instruments with DSP capabilities, which means that they can also be used to measure the frequency of oscillators. When their output data is analyzed with the appropriate software, general-purpose and relatively low-cost SDRs can produce high-resolution measurements, essentially equaling the performance of dedicated frequency measurement instruments [62, 63].

An SDR splits the incoming RF signal into two streams of time-series data. One is called the *I*, or in-phase, data stream because it has the same phase as the LO. The other is called the *Q*, or quadrature, data stream because its phase is shifted by 90° (Figure 5.11) with respect to the LO. Figure 5.27 shows an incoming RF signal received with an antenna and passing through a band pass filter (BPF) before being down-converted and split into *I* and *Q* signals using two mixers, a LO, and a 90° phase shifter. This simple example shows an analog front end, used in many SDR designs, where down-conversion is performed with analog methods before the *I* and *Q* signals are sampled with an ADC to form the data streams.

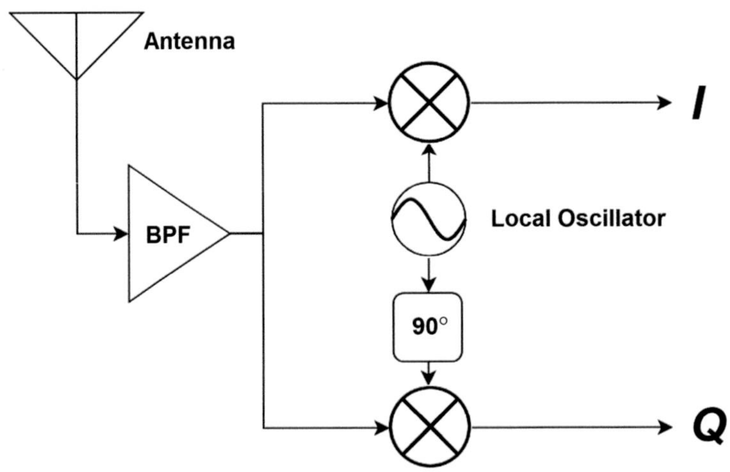

Figure 5.27 Down-converting a received RF signal into its *I* and *Q* components.

In a direct sampling receiver, the analog hardware isn't needed because the incoming RF signal is sampled with an ADC immediately after passing through the BPF. Digital down-conversion is then performed, using a numerically controlled oscillator (NCO) as the LO, to generate the *I* and *Q* data streams. An NCO is a digital signal generator that receives two inputs, a reference frequency signal and a digital command that specifies the output frequency. It then outputs a digitally sampled waveform, usually a sine wave, at the requested frequency.

The *I* and *Q* data streams are combined to form a two-dimensional *IQ* signal that represents how the RF carrier is modulated, allowing the reconstruction of its amplitude, its phase, and its frequency. The *IQ* data, in binary format, is output by an SDR to a host computer, and allows capturing all of the radio traffic within a given RF band with a fast Fourier transform (FFT) algorithm that converts the data from the time domain to the frequency domain [64, 65].

The SDR display pictured in Figure 5.28 is generated from processing *IQ* data. This example shows a 600 kHz segment of the spectrum of the AM broadcast band, with a center frequency of 1 MHz. The top portion of the display is a spectrum analyzer, graphing the amplitude of the received signals versus their frequencies. The continuously scrolling "waterfall" shows all of the RF signals present in the spectrum, or more properly, everything within the coverage range of the antenna. The signal bandwidths, limited to 10 kHz in the case of AM radio broadcasts, vary and change shape based on the type of audio being modulated, and the displayed colors indicate the relative signal strength. If an oscillator signal is connected to the RF input instead of a broadband antenna, the display wouldn't be nearly as interesting. A single spectral line would appear on both the spectrum analyzer and the waterfall, with an unvarying bandwidth that is near zero. However, the *IQ* data associated with the spectral line would include enough information to compute the phase and frequency difference of the DUT with respect to the reference.

Many low-cost SDRs have only two connectors, an RF input connector, which is normally intended for an antenna, and an output connector for data. These devices aren't suitable for frequency metrology because when a DUT is connected to the RF input, it can only be compared to the SDR's internal quartz oscillator. However, if the SDR has a reference clock input, a DUT connected to the antenna input can be compared to the reference clock in single-channel mode (Figure 5.29a). If the SDR has two RF inputs, one can be connected to the DUT and the other to the reference in dual-channel mode (Figure 5.29b). In both modes a NCO is locked to the sampling clock and digital down-converters (DDCs) produce the *I* and *Q* data streams. After decimation and filtering, the *IQ* data is formatted and sent over a network interface to the host computer for processing [62]. If two RF inputs are supported by the hardware, then a dual-channel mode that uses two channels of a single ADC can offer better performance than the single-channel mode due to noise cancellation. The noise introduced by the phase-locked sampling clock [66], as well as noise introduced by the ADC, is now common to both the DUT and reference input channels.

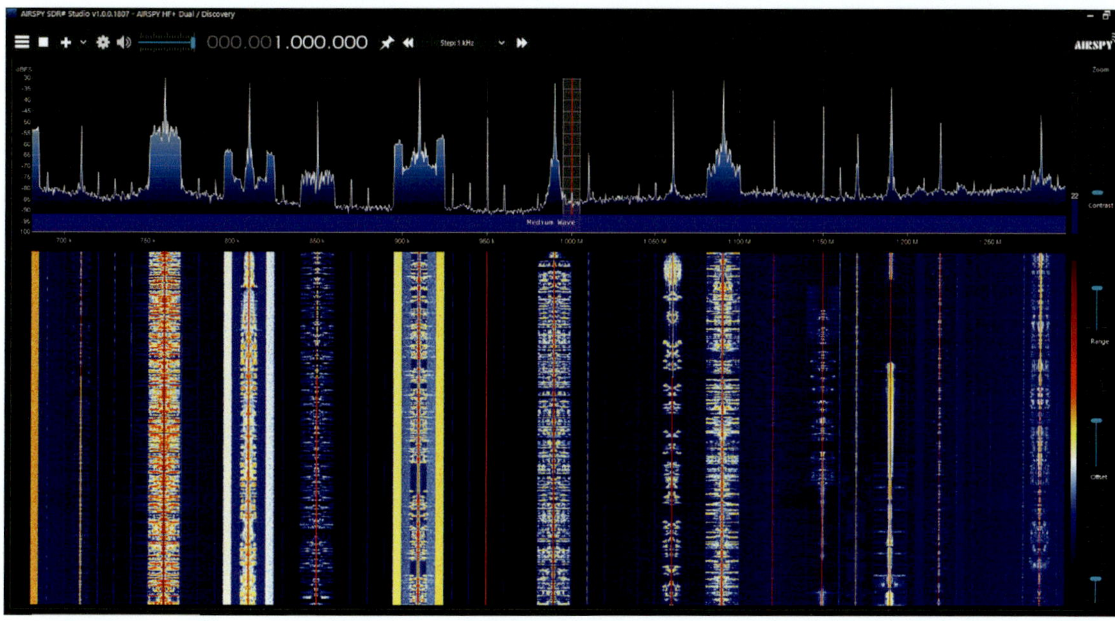

Figure 5.28 The spectrum and "waterfall" display generated by a SDR receiver.

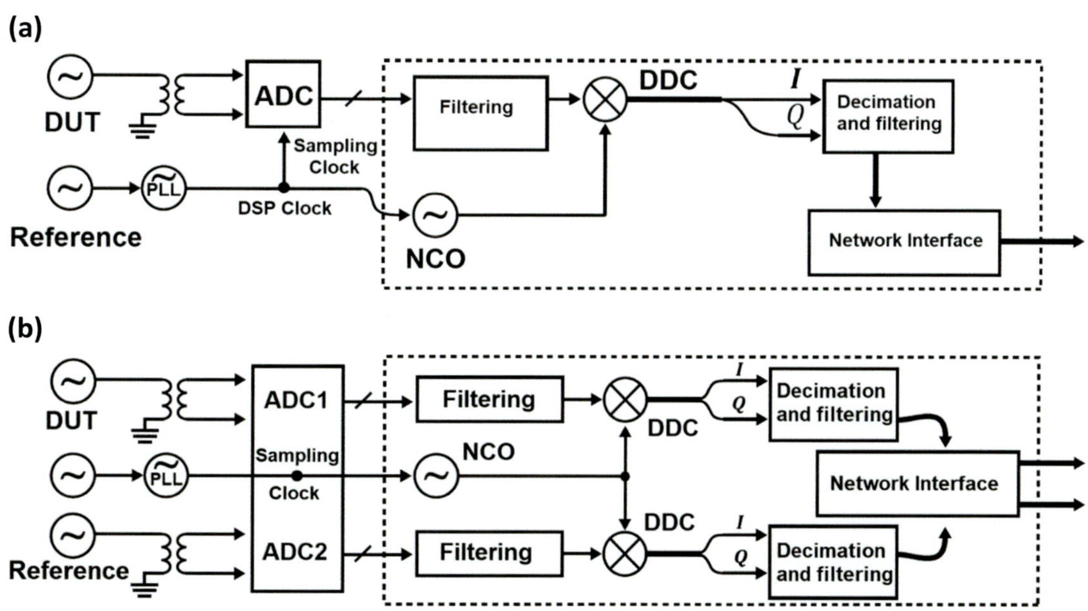

Figure 5.29 Simplified block diagrams of single-channel (a) and dual-channel (b) SDR frequency measurement systems [62].

Figure 5.30 shows an example of an SDR, the Ettus Research USRP N210, that has been utilized for frequency metrology. The instrument is customized by installing optional daughterboards that provide direct access to the ADC inputs for a range of nominal frequencies that cover the DUT signals of interest, from 1 to 250 MHz, for example. Measurements can be performed in single-channel mode, where the DUT is connected to an RF input and the reference is connected to the external clock input (REF IN), or in dual-channel mode, where the DUT and reference signals are connected to the two RF inputs (RF1 and RF2). The down-converted signals are sampled at a rate of 100 MHz with 14-bit ADC resolution. The resulting *IQ* data is streamed to a host computer through an Ethernet connection, but because 100 MS/s is more data than is needed—and perhaps more data than the link or computer can handle—the samples are decimated before being streamed and processed. This instrument and other similar SDRs can detect frequency differences smaller than 1×10^{-13} in 1 s, which is equivalent to the performance of the best DMTD systems [62, 63].

Figure 5.30 A software-defined radio suitable for frequency metrology (courtesy of Ettus Research).

5.3 Frequency Control

Until now, this chapter has focused on measuring the frequency difference between two oscillators. Another essential part of frequency metrology involves controlling the frequency of an oscillator to within an acceptable tolerance, by adjusting its frequency until it agrees with a reference. Without frequency control systems, clocks couldn't keep accurate time, telecommunication systems and electric power generation systems would fail, and frequency measurement instruments couldn't perform accurate measurements. Two primary methods of controlling frequency are the ubiquitous phase-locked loop (PLL) found in nearly all electronic circuits that require accurate frequency, and oscillators disciplined to agree with reference signals received by radio. Phase-locked loops and disciplined oscillators are discussed in the following sections.

5.3.1 Phase-Locked Loops

A PLL is a feedback control system that locks the frequency and phase of an output signal to the frequency and phase of a reference input signal. The reference and output signals can be either sinusoidal or digital [67]. A basic analog PLL (Figure 5.31) has three parts, a phase comparator or phase detector (PD), a loop filter, and a voltage-controlled oscillator (VCO). The name "phase-locked loop" is self-descriptive. The term "phase" refers to the relative phase difference between the input signal and the VCO signal, and "locked" means that the phase difference is kept constant. Because the phase difference doesn't change, it means that the input signal and the VCO signal have the same frequency. Finally, the term "loop" indicates that a PLL is a feedback loop that continuously measures and controls the VCO's output frequency.

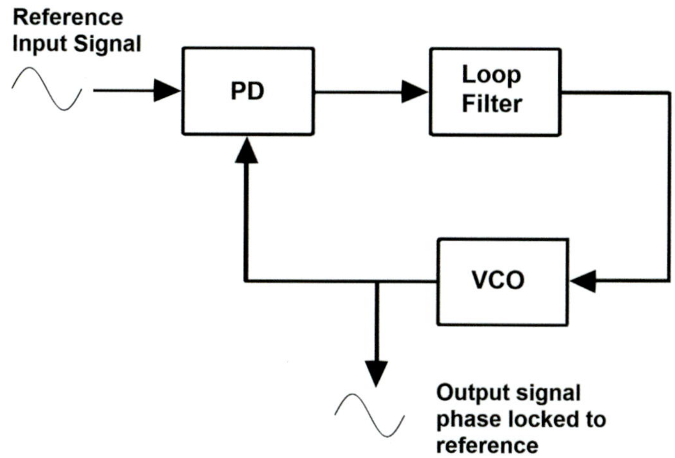

Figure 5.31 An analog phase-locked loop.

The VCO will oscillate at its free-running frequency, f_0, until the reference input signal is applied. The PD, which can be a frequency mixer, an analog multiplier, or a logic circuit, detects the phase difference between the input signal and the VCO. It outputs the phase difference, in the form of a control voltage, to a loop filter. The loop filter cleans up the control voltage by suppressing harmonics and noise before applying it to the VCO. For example, if a frequency mixer is used as the PD, the loop filter removes everything from the mixer output except the DC voltage that represents the phase difference between the signals. This control voltage changes the frequency of the VCO in a direction that reduces the phase difference between the VCO and the reference input signal.

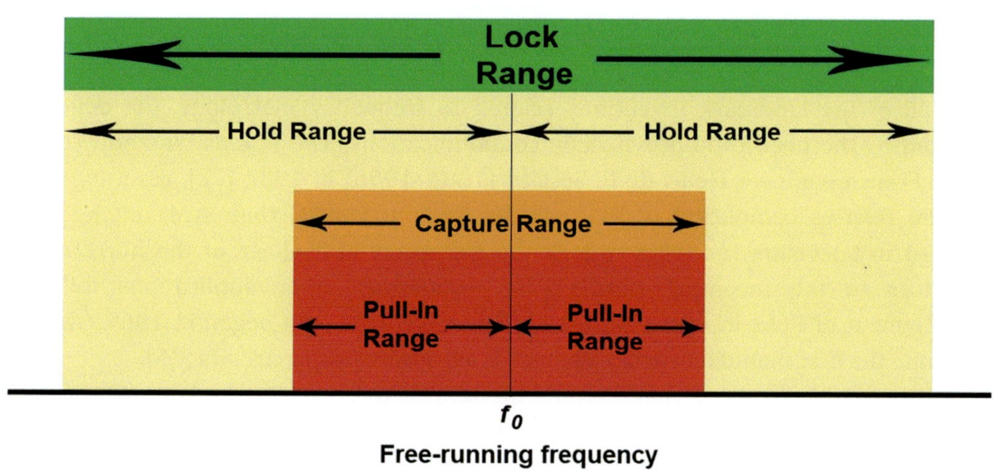

Figure 5.32 Lock, hold, capture, and pull-in ranges of a PLL.

As shown in Figure 5.32, the lock range—consisting of two hold ranges on either side of f_0—indicates the range of input frequencies that the PLL can track. The capture range, consisting of two pull-in ranges on either side of f_0 that are not always symmetrical, indicates the range of frequencies where the VCO can be tuned or adjusted so that it locks to the input signal. The lock and capture ranges determine the bandwidth of the PLL. Increasing the bandwidth can decrease the lock time but can also increase the noise, so a PLL might be configured to have a wide bandwidth during the initial lock process, and a narrow bandwidth after a lock condition occurs. When a PLL is locked, the phase and frequency of the VCO are equal to the phase and frequency of the input signal [67–70].

Frequency synthesizers, the circuits or instruments that can generate or synthesize a range of frequencies from a single reference frequency, are commonly implemented with PLL-based methods. An integer-*N* PLL is the simplest type of PLL synthesizer (Figure 5.33). It adds a programmable frequency divider to the feedback loop shown in Figure 5.31, and it performs frequency multiplication via a negative feedback mechanism. Instead of directly comparing the phase of the VCO to the reference frequency, f_{ref}, and locking the output frequency, f_{out}, to the reference, the frequency of the VCO signal is first divided by the integer N prior to the phase comparison. Therefore, a 100 MHz VCO can be phase locked to a 10 MHz reference signal if $N = 10$. Changing N, of course, changes the output frequency, but the integer-*N* method only allows the output frequency to be an integer multiple of the reference frequency, $f_{out} = f_{ref} \times N$, and it limits the resolution, or minimum step size of the synthesizer to be equal to f_{ref}. A fractional-*N* PLL removes some of these limitations by dynamically changing the value of N while the VCO is locked. For example, it divides a portion of the cycles by N and another portion by $N + 1$. When the dynamic switching is done with the correct proportions, the average division ratio equals N plus some arbitrary fraction. This improves the synthesizer's resolution because the minimum step size becomes a fractional portion of f_{ref} [69, 71].

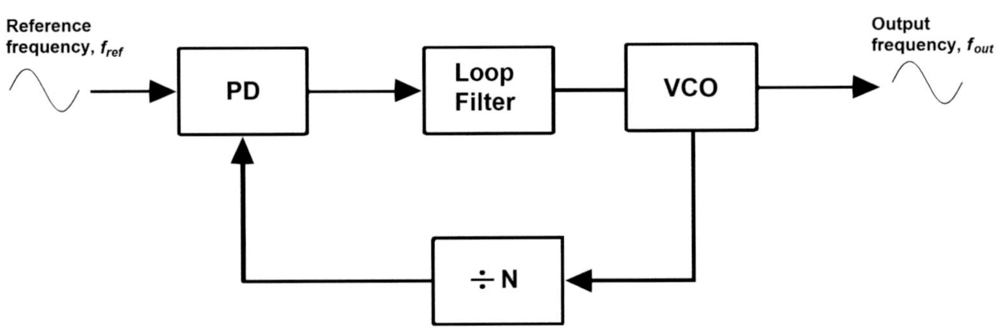

Figure 5.33 An integer-*N* PLL frequency synthesizer.

Like nearly all of the landmark inventions related to frequency metrology, the development of the PLL was motivated by the need to improve radio communications. The PLL as we know it today was first described by the French engineer Henri de Bellescize (1884–1966) in 1932 [72], resulting in a U. S. patent in 1935 [73]. The relative complexity of PLL circuits initially limited their use, but by the 1940s they were implemented in television receivers to lock the frequency and phase of the horizontal and vertical scanning oscillators to the incoming signals [74]. Numerous other applications followed, and the widespread deployment of PLLs in nearly all types of electronic circuits began in 1969 when the Signetics Corporation became the first manufacturer to produce PLLs as integrated circuits [75].

We noted at the end of Chapter 2 that it would be difficult to overstate the importance of quartz clocks and oscillators to time and frequency metrology, and the same thing could be said of PLLs. The majority of electronic circuits that include quartz oscillators also include PLLs, including radio receivers, where the LO is usually a PLL-based frequency synthesizer [42]. Nearly every electronic instrument described so far in this book utilizes PLL circuitry, including the atomic clocks described in Chapter 3, and the measurement instruments described in this chapter and in Chapter 4. For example, when a FC or a TIC is connected to an external time base, its internal oscillator is phase locked to the external oscillator.

5.3.2 Disciplined Oscillators and Clocks

Disciplined oscillators are instruments that lock the frequency of their local oscillator to an external reference frequency that's known to be accurate. The term was perhaps first used in 1971 to describe a system where a quartz oscillator was phase locked to an atomic oscillator using signals distributed by cables [76]. That usage is still technically correct, because the reference frequency for a disciplined oscillator can be distributed through any medium. However, the term is most often used to describe instruments that obtain their reference frequency from radio signals that are controlled to agree with UTC.

We described several time transfer systems in Section 4.3 that use radio signals to measure the time difference of a remote clock with respect to a reference clock, making it possible to synchronize the remote clock to agree with the reference. A similar process occurs when frequency is transferred. Radio signals can be used to measure the frequency difference of a remote oscillator with respect to a reference oscillator, making it possible to adjust the remote oscillator to agree with the reference. The practice of using radio signals as a reference for frequency measurements was established over a century ago. As we touched upon in Section 2.2, NBS radio station WWV began broadcasting carrier frequencies that were controlled by the United States frequency standard in 1923. This service made a frequency standard freely available to anyone with receiving equipment who was within the coverage area of the signals [77].

The standard frequency signals provided by WWV and other similar other radio stations are broadcast in the HF region of the radio spectrum. For example, the WWV broadcast frequencies have included 5, 10, and 15 MHz since 1935 [78]. These frequencies equal the nominal frequencies of oscillators used as frequency standards, so they conveniently allow oscillators to be calibrated using zero beat methods, where the output frequency of the oscillator is mixed with the RF carrier, and the oscillator is adjusted until the resulting beat frequency is as close to 0 as possible [79].

The standard carrier frequencies found in the HF spectrum provide an advantage when measuring frequency, but the fact that HF radio signals propagate mostly by skywave is a definite disadvantage. Skywave propagation means that HF radio waves radiate upward until they are refracted, or bent, by the ionosphere and reflected back to Earth. The HF signals are reflected from different ionospheric layers at different times, and the heights of the various layers constantly change, which in turn changes the length and delay of the radio path. The path delay changes cause Doppler frequency shifts that change the frequency of the received signal, sometimes by more than 1 Hz. These Doppler frequency shifts are often small enough to be considered insignificant when calibrating quartz oscillators, but as oscillator technology advanced, HF signals simply weren't stable enough to use as a frequency reference. When atomic oscillators were introduced in the 1950s (Chapter 3), most remote frequency calibrations were already being referenced to VLF or LF radio signals that propagate by groundwave.

Disciplined oscillators appeared shortly before the satellite age began, and the first units controlled their LO frequency with VLF radio signals, which was the best option then available. The invention of the disciplined oscillator can be credited to the American physicist John Alvin Pierce (1907–1996), who had a distinguished career as both a radio researcher and an inventor. Pierce worked at the Cruft Laboratory of Harvard University from 1934 to 1974, where he made essential contributions to the development of the Loran and Omega radio navigation systems (Section 15.3). In 1956, Pierce began a series of frequency measurements that were referenced to the 16 kHz signals transmitted by the British radio station GBR. His receiver in Cambridge, Massachusetts, was located some 5200 km from the transmitter in Rugby, England, but Pierce still found that he could measure frequency with an accuracy of about 1×10^{-9}. When publishing his results, Pierce included an elegant statement that summarized why disciplined oscillators would revolutionize frequency metrology in future years:

> *"There is an obvious relation between the measurement of the frequency of an oscillator and the automatic control of that frequency."* [80]

Pierce went on to describe two instruments already in use in his laboratory that locked the LO frequency to the received VLF signal. One instrument, which Pierce called a "frequency regulator," used two phase detectors operated in quadrature to control the LO frequency. The other more elaborate instrument (Figure 5.34) included a superheterodyne receiver and a mechanical servo system that adjusted the LO. These two instruments were likely the first examples of the now ubiquitous disciplined oscillator.

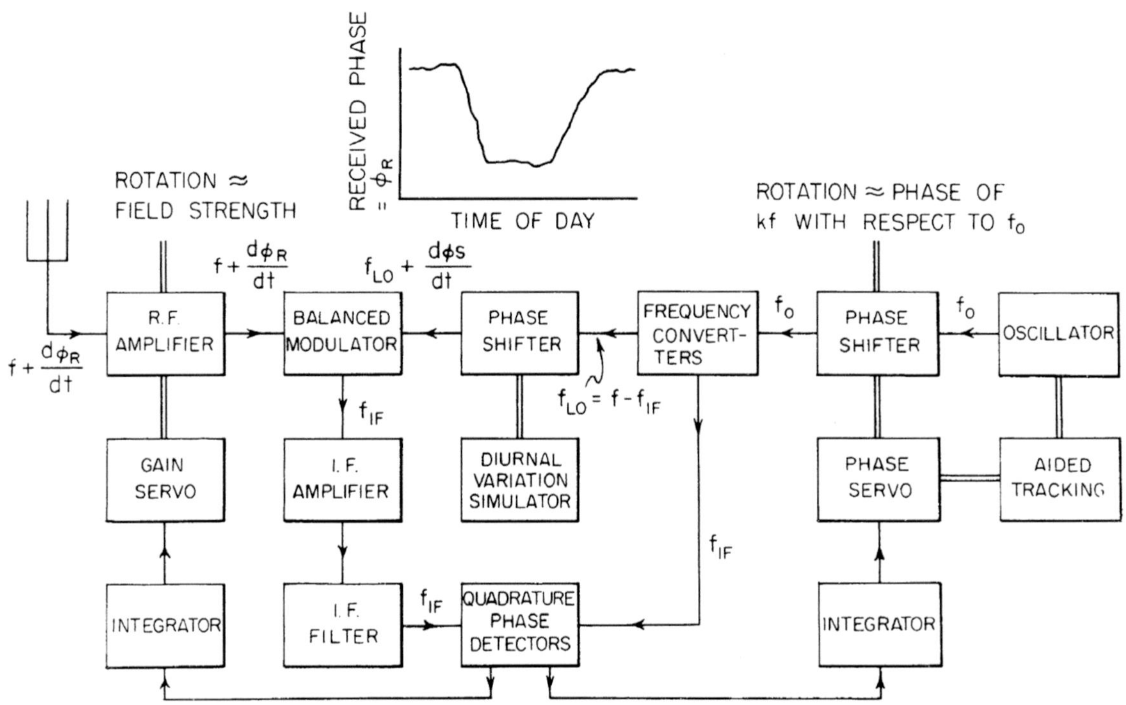

Figure 5.34 Pierce's early disciplined oscillator design based on VLF radio reception [80].

Disciplined oscillators referenced to LF signals, including the 100 kHz broadcasts from the Loran-C navigation system [81] and the 60 kHz broadcasts from NIST radio station WWVB [82], were commonly used as frequency measurement standards in the 1970s and 1980s. When these instruments were located close enough to the radio transmitter to receive groundwave signals—a range typically limited to about 2,000 km—they could control the frequency of a quartz oscillator so that it was accurate and stable to within parts in 10^{12} with WWVB, or to parts in 10^{13} with Loran-C, when averaged for one day. However, just as VLF/LF signals had replaced HF signals as the reference for high-accuracy frequency measurements in the 1950s, they were in turn replaced in the 1980s and 1990s by UHF signals transmitted from GPS satellites.

The development of GPS quickly revolutionized both time synchronization and frequency control. The first operational GPS satellite was launched into orbit in 1978, and the first receivers designed to recover time and frequency information from the atomic clocks on board the satellites appeared shortly afterwards [83, 84]. Commercial GPS disciplined oscillators (GPSDOs) became readily available in the mid-1980s [85]. From there, GPSDOs proliferated at an exponential rate compared to all previous types of disciplined oscillators, due to several major advantages. The first advantage was that their frequency was typically accurate and stable to within parts in 10^{13} when averaged for 24 hours, a level of performance that usually exceeded the best VLF/LF disciplined oscillators. Another advantage was their ability to work anywhere as Earth, unlike VLF/LF disciplined oscillators, which can only work reliably when they are located within the groundwave coverage area of a terrestrial transmitter. Finally, and perhaps most importantly, they could automatically synchronize a 1 pps signal to within less than 1 μs of UTC, which allowed them to also serve as GPS disciplined clocks (GPSDCs). Due to these capabilities, many existing technologies were enhanced by the mass production of GPS disciplined oscillators and clocks, and many new technologies were enabled.

The number of GPSDOs and GPSDCs now in operation likely exceeds one billion (10^9), with many deployed in critical infrastructure systems. Figure 5.35 shows an example of a small, low-cost GPSDO/ GPSDC with 1 pps and 10 MHz outputs. Because of their overwhelming importance to frequency and time control, much of Chapter 12 is devoted to discussing their performance characteristics and limitations. Other chapters discuss the importance of GPS to coordinating world time (Chapter 9), to navigation (Chapter 15), to frequency control in telecommunication and electric power generation systems (Chapters 16 and 17), and to time synchronization in financial markets (Chapter 18).

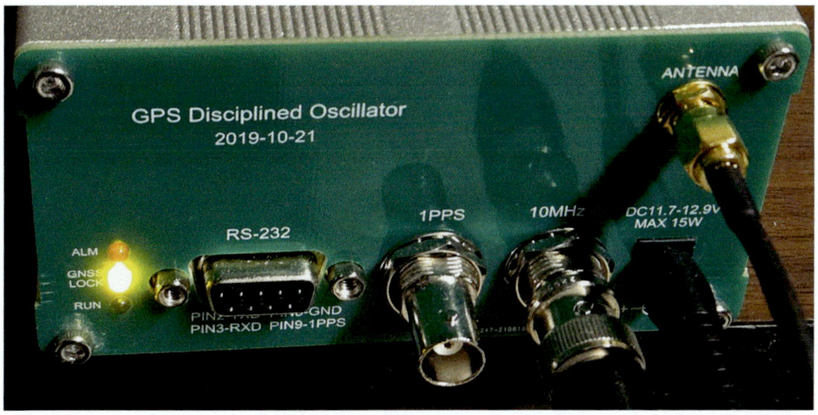

Figure 5.35 A small, low-cost GPSDO/GPSDC with 1 pps and 10 MHz outputs.

For now, we will conclude this chapter by discussing an element common to all disciplined oscillators, the control loop that adjusts and locks the disciplined oscillator's frequency to the received reference signals. A high percentage of disciplined oscillators utilize control loops based on proportional-integral-derivative (PID) controllers, which is the method most commonly implemented by process control systems [86]. A PID controller performs a straightforward function, it corrects the error, *e*, between a measured process variable, PV, and a desired set point, SP. Some common applications of PID controllers are thermostats, where SP is equal to the desired temperature of a room or building, and cruise control systems in automobiles, where SP is equal to the desired speed of the vehicle. In these applications, the PID controller continuously measures and adjusts the temperature or speed, keeping it as close to SP as possible. In a disciplined oscillator application,

$$PV = f_{\text{diff}}, \tag{5.20}$$

where f_{diff} is the frequency difference between the LO and the received reference signal. As we have reemphasized throughout this chapter, we can measure f_{diff} with either a frequency or phase comparison. Because the goal is locking the LO to the radio reference, it follows that SP = 0, which means that any non-zero frequency difference is considered to be an error, *e*, that requires correction.

We can think of a disciplined oscillator as an instrument that is always "steering" the LO frequency towards the reference frequency. The steering analogy is appropriate, because much of the theory of PID control was developed as part of a research and development effort aimed at automatically steering the USS *New Mexico*, a U. S. Navy battleship. The effort was led by the mathematician and engineer Nicolas Minorsky (1885–1970), who emigrated to the United States from Russia in 1918, after obtaining a doctorate in applied sciences and serving in the Russian Navy. Upon arriving in the United States, Minorsky first worked at the General Electric Company in New York, where he pursued the automatic steering problems that had interested him during his naval career. Minorsky based his analysis on the real world experience he'd acquired by watching helmsman steer ships. From these observations, he learned that accurate steering was based not only on the current error of the ship's steering, but also on the past steering errors, as well as the current rate of change of the errors. In 1922, Minorsky published the theoretical analysis for a three-term control system that we now know as the PID controller, the best known of his many contributions to the field of control theory [87].

As Minorsky had theorized, a PID controller algorithm involves three terms (Figure 5.36). As just discussed, the error, *e*, in a disciplined oscillator system will be any non-zero frequency difference, and each of the three terms can provide a proportional response to the error. The *P*-term is proportional to the error, based on the current measurement of f_{diff}. The *I*-term is proportional to the integral of the error, based on past measurements of f_{diff}. Finally, the *D*-term is proportional to the derivative of the error, based on its rate of change. The *P*-term is calculated as

$$P = K_p e(t), \tag{5.21}$$

where *P* is the proportional output of the controller, K_p is the proportional gain, *e* is the error, and *t* is the time of the error.

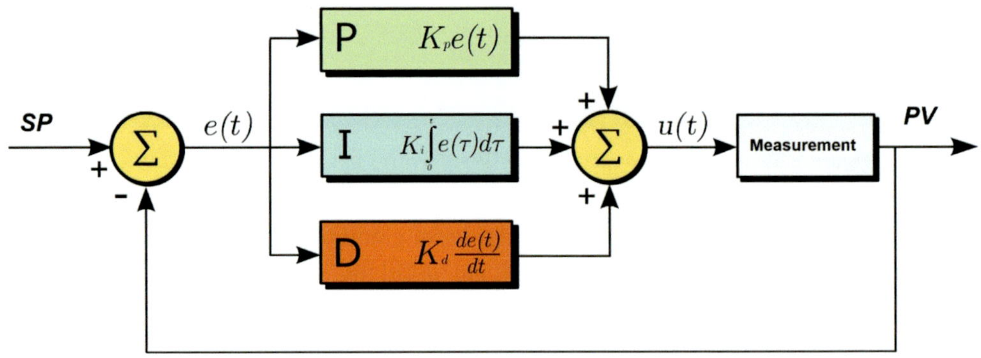

Figure 5.36 Diagram of a PID controller.

The *I*-term accounts for past errors that remain after the *P*-term is applied, and integrates these errors over time. By doing so, it corrects the accumulated frequency difference that wasn't previously corrected. During the locking process, the magnitude of *e* and the *P*-term will decrease proportionally, but the *I*-term will continue to grow to compensate for residual errors. When the LO is locked to its external reference, both *e* and the *P*-term will be near 0, and the controller will essentially output only the *I*-term, which will change by the small amounts necessary to maintain lock. The *I*-term is calculated as

$$I = K_i \int_0^t e(\tau)d\tau, \tag{5.22}$$

where *I* is the integral output of the controller, K_i is the integral gain, *t* is the instantaneous or present time, and *e* is the error. The inclusion of the dummy variable τ indicates that we are integrating an equation where time is a variable, and that the output of the equation will be a function of time.

The *D*-term is the rate of change of the error. It's calculated by determining the slope of the error over time (its first derivative with respect to time), and then multiplying this rate of change by the derivative gain,

$$D = K_d \frac{de(t)}{dt}, \tag{5.23}$$

where *D* is the derivative output of the controller, K_d is the derivative gain, *e* is the error, and *t* is the instantaneous time.

The PID controller outputs the sum of the three responses to the error as a time-dependent control variable, $u(t)$, calculated as

$$u(t) = P + I + D, \tag{5.24}$$

where P, I, and D are the terms defined in the previous three equations [86, 88].

The control variable, $u(t)$, is converted to a form that the LO can accept as a frequency correction. For example, the frequency correction could be sent in the form of a control voltage or in the form of a digital command. After the frequency correction is sent, a new measurement of f_{diff} is made, the P, I, and D terms are updated, the control variable is updated, and the control loop process repeats. The bandwidth of the control loop is usually matched to the full or partial tuning range of the LO, which limits the magnitude of an allowable frequency correction.

As shown in Eqs. (5.21) through (5.23), each control term includes a gain parameter, and PID controllers can be tuned by adjusting the gain parameters. Increasing the gain causes the controller to more aggressively respond to errors, but applying too much gain can cause the controller to overshoot and undershoot the set point by a wide margin. Some adaptive algorithms automatically adjust the gain to better respond to different conditions, for example, they may use different gain parameters during locked and unlocked situations. A small percentage of disciplined oscillator products allow users to change the gain parameters, but most products don't allow them to be changed, because the PID has already been customized by the manufacturer to work with the selected LO and reference receiver. The requirements for a locked condition—which is usually indicated by a front panel light or a machine-readable variable—are chosen by the designer of the disciplined oscillator. A locked condition usually means that f_{diff} has been below a predefined threshold for a predefined period.

The control loops and frequency steering algorithms implemented in disciplined oscillators are usually proprietary to the manufacturer, and some are far more sophisticated than a rudimentary PID controller. Disciplined oscillators can, for example, employ algorithms based on the Kalman filter estimators that are sometimes implemented in time scales (Section 4.4). A Kalman filter might even be used to filter the measurement data before PID processing [89]. Other algorithms measure quantities other than frequency, most commonly by reading a temperature sensor. This allows them to provide real-time compensation for temperature changes and other environmental effects that degrade the LO's performance [90].

The steering algorithm might also try to predict the magnitude and interval of the frequency corrections applied to the LO based on historical data. These predictions can be applied when the reference signals can't be received, allowing a disciplined oscillator to operate normally for a limited period without having a reference [91]. The reference signals can be unavailable for a variety of reasons—for example, the radio transmitter might be off the air, another RF source might be interfering with or "jamming" the reference signal, or the antenna cable might be cut or disconnected. The holdover capability of a disciplined oscillator is its level of performance when it's forced to operate without a reference signal. A combination of a stable LO and a refined holdover algorithm can sometimes keep a disciplined oscillator locked for multiple hours or even days, but not indefinitely. If a disciplined oscillator lacks a holdover algorithm, its holdover capability depends entirely on the stability of its LO. Therefore, if all other things are equal, a disciplined rubidium oscillator will have better holdover capability than a disciplined quartz oscillator.

We've now covered the basic methods of measuring time and frequency signals, and we can turn our attention to characterizing the performance of clocks and oscillators. We'll start by discussing what it means for a clock or oscillator to be accurate, the topic of Chapter 6.

References

1. International Telecommunications Union, "Volume 1: Articles," *ITU Radio Regulations*, 2020.

2. Federal Communications Commission, "FCC Online Table of Frequency Allocations," *47 C.F.R. § 2.106*, August 23, 2022.
3. B. Metzler, *Audio Measurement Handbook*, Audio Precision, 1993.
4. R. K. Cook and A. J. Bedard, Jr., "On the Measurement of Infrasound," *Geophysical Journal International*, vol. 26, no. 1–4, pp. 5–11, December 1971.
5. H. F. Olson, *Music, Physics, and Engineering*, 2nd edition, Dover Publications, 1967.
6. M. J. Moravcsik, *Musical Sound: An Introduction to the Physics of Music*, Paragon House Publishers, 1987.
7. T. D. Rossing, *The Science of Sound*, Addison-Wesley, 1982.
8. I. Asimov, *Understanding Physics, Volume I: Motion, Sound, and Heat*, Dorset Press, 1966.
9. International Standards Organization, "Acoustics – Standard tuning frequency (standard musical pitch)," *ISO 16-1975 (E)*, 1975.
10. W. F. Snyder and C. L. Bragaw, "Achievement in Radio: Seventy Years of Radio Science, Technology, Standards, and Measurement at the National Bureau of Standards," *National Bureau of Standards Special Publication* 555, October 1986.
11. R. C. Bickerton and G. S. Barr, "The origin of the tuning fork," *Journal of the Royal Society of Medicine*, vol. 80, no. 12, pp. 771–773, December 1987.
12. J. Jendzurski and N. G. Paulter, "Calibration of Speed Enforcement: Down-The-Road Radars," *Journal of Research of the National Institute of Standards and Technology*, vol. 114, no. 3, pp. 137–148, May-June 2009.
13. G. L. Harris, ed., "Calibration Procedures for Weights and Measures Laboratories," *National Institute of Standards and Technology Interagency Report 8250*, 83 p., May 2019.
14. J. J. Carr, *Elements of Electronic Instrumentation and Measurement*, Prentice-Hall, 3rd edition, 1996.
15. R. P. Turner, *Frequency and Its Measurement*, Howard W. Sams & Company, Inc., 1975.
16. International Telecommunications Union, "Nomenclature of the Frequency and Wavelength Bands Used in Telecommunications," *Recommendation ITU-R V.431.8*, August 2015.
17. P. Crozier, *Electronic Instruments and Measurements*, Breton Publishers, 1985.
18. E. Maor, *Trigonometric Delights,* Princeton University Press, 1998.
19. C. Taylor, *The Art and Science of Lecture Demonstration*, Taylor & Francis, New York, 1988.
20. J. Czech, *The Cathode Ray Oscilloscope: Circuitry and Practical Applications*, Interscience Publishers, New York, 1957.
21. J. T. Stanley and J. B. Milton, "Basic Laboratory Methods for Measurements or Comparison of Frequencies and Time Intervals," *National Bureau of Standards Report 10 744*, 131 p., June 1972.
22. G. Kamas and S. L. Howe, eds., "Time and Frequency User's Manual," *National Bureau of Standards Special Publication 559*, 256 p., November 1979.
23. A. S. Morris, *Measurement & Instrumentation Principles*, Butterworth-Heinemann, 2001.
24. A. S. Bagley, "The High-Speed Frequency Counter – A New Solution to Old Problems," *Hewlett-Packard Journal*, vol. 2, no. 5, pp. 1–4, January 1951.
25. C. House, "Oral History of Alan Bagley," *Computer History Museum Reference No. X5727.2010*, February 2010.
26. W. Isaacson, *Steve Jobs*, Simon & Schuster, 2011.
27. J. L. Sorden, "A New Generation in Frequency and Time Measurements," *Hewlett-Packard Journal*, vol. 25, no. 10, pp. 2–8, June 1974.
28. Hewlett-Packard Company, "Fundamentals of the Electronic Counters," *Hewlett-Packard Application Note 200*, 5965–7660E, 44 p., March 1997.
29. S. Johansson, "New Frequency Counting Principle Improves Resolution," *Proceedings of the 2005 IEEE International Frequency Control Symposium*, pp. 628–635, August 2005.
30. P. Boven, "Increasing the resolution of reciprocal frequency counters," *Proceedings of the 50th VHF Conference in Weinheim (UKW-Tagung Weinheim)*, 9 p., September 2005.

31. S. T. Dawkins, J. J. McFerran, and A. N. Luiten, "Considerations on the Measurement of the Stability of Oscillators with Frequency Counters," *IEEE Transactions on Ultrasonics, Ferroelectrics, and Frequency Control*, vol. 54, no. 5, pp. 918–925, May 2007.
32. E. Rubiola, M. Lenczner, P.-Y. Bourgeois, and F. Vernotte, "The Ω Counter, a Frequency Counter Based on the Linear Regression," *IEEE Transactions on Ultrasonics, Ferroelectrics, and Frequency Control*, vol. 63, no. 7, pp. 961–969, July 2016.
33. E. Rubiola, "On the Measurement of Frequency and of Its Sample Variance with High-Resolution Counters," *Review of Scientific Instruments*, vol. 76, no. 5, pp. 054703/1–6, May 2005.
34. R. Nutt, "Digital Time Intervalometer," *Review of Scientific Instruments*, vol. 39, no. 9, pp. 1342–1345, September 1968.
35. R. G. Baron, "The Vernier Time-Measuring Technique," *Proceedings of the Institute of Radio Engineers (IRE)*, vol. 45, no. 1, pp. 21–30, January 1957.
36. D. C. Chu, M. S. Allen, and A. S. Foster, "Universal Counter Resolves Picoseconds in Time Interval Measurement," *Hewlett-Packard Journal*, vol. 29, no. 12, pp. 2–11, August 1978.
37. D. A. Howe, D. W. Allan, and J. A. Barnes, "Properties of Signal Sources and Measurement Methods," *Proceedings of the 35th Annual Symposium on Frequency Control*, pp. 669–716, May 1981.
38. S. R. Stein, "Frequency and Time – Their Measurement and Characterization," Chapter 12 in vol. 2 of *Precision Frequency Control: Oscillators and Standards*, pp. 191–232, E. A. Gerber and A. Ballato, eds., Academic Press, 1985.
39. M. A. Lombardi, "Remote Frequency Calibrations: The NIST Frequency Measurement and Analysis Service," *National Institute of Standards and Technology Special Publication 250–229*, 90 p., June 2004.
40. H. G. J. Aitken, *The Continuous Wave: Technology and American Radio, 1900–1932*, Princeton University Press, 1985.
41. R. A. Fessenden, "Wireless Signaling," *United States Patent 706,740*, filed September 28, 1901, granted August 12, 1902.
42. J. J. Carr, *The Technician's Radio Receiver Handbook*, Butterworth-Heinemann, 2001.
43. S. J. Erst, *Receiving Systems Design*, Artech House, 1984.
44. D. W. Allan, "The Measurement of Frequency and Frequency Stability of Precision Oscillators," *National Bureau of Standards Technical Note 669*, 31 p., May 1975.
45. W. J. Riley, "Handbook of Frequency Stability Analysis," *National Institute of Standards and Technology Special Publication 1065*, 136 p., July 2008.
46. D. E. Maxwell, "A 5 to 50 MHz Direct-Reading Phase Meter with Hundredth-Degree Precision," *IEEE Transactions on Instrumentation and Measurement*, vol. IM-15, no. 4, pp. 304–310, December 1966.
47. D. W. Allan and H. Daams, "Picosecond Time Difference Measurement System," *Proceedings of the 29th Annual Symposium on Frequency Control*, pp. 404–411, May 1975.
48. D. W. Allan, "Report on NBS Dual Mixer Time Difference System (DMTD) built for Time Domain Measurements Associated with Phase 1 of GPS," *National Bureau of Standards Interagency Report 75–827*, 18 p., January 1976.
49. G. Brida, "High Resolution Frequency Stability Measurement System," *Review of Scientific Instruments*, vol. 73, no. 5, pp. 2171–2174, May 2002.
50. L. Sojdr, J. Cermak, and G. Brida, "Comparison of High-Precision Frequency-Stability Measurement Systems," *Proceedings of the 2003 Joint Conference of the IEEE International Frequency Control Symposium and the European Frequency and Time Forum (IFCS/EFTF)*, pp. 317–325, May 2003.
51. W. J. Riley, "A Small Dual Mixer Time Difference (DMTD) Clock Measuring System," *Hamilton Technical Services White Paper*, 25 p., May 2010.
52. S. R. Stein, J. Grove, J. Hein, R. A. Martin, B. Mesander, M. Phillips, P. Schweiger, J. Shima, and W. Solbrig, "Comparison of Heterodyne and Direct-Sampling Techniques for Phase-Difference Measurements," *Proceedings of the 2005 NCSLI International Workshop and Symposium*, 10 p., July 2005.
53. H. Nyquist, "Certain Topics in Telegraph Transmission Theory," *Transactions of the American Institute of Electrical Engineers*, vol. 47, no. 2, pp. 617–644, April 1928.

54. International Electrotechnical Commission (IEC), "Audio recording – compact disc digital audio system," *IEC International Standard 6098*, 1999.
55. R. Reeder, W. Green, and R. Shillito, "Analog-to-Digital Converter Clock Optimization: A Test Engineering Perspective," *Analog Dialogue*, vol. 42, no. 2, pp. 1–7, February 2008.
56. K. Mochizuki, M. Uchino, and T. Morikawa, "Frequency-Stability Measurement System Using High-Speed ADCs and Digital Signal Processing," *IEEE Transactions on Instrumentation and Measurement*, vol. 56, no. 5, pp. 1887–1893, October 2007.
57. J. Grove, J. Hein, J. Retta, P. Schweiger, W. Solbrig, and S. R. Stein, "Direct-Digital Phase-Noise Measurement," *Proceedings of the IEEE International Ultrasonics, Ferroelectrics and Frequency Control Joint 50th Anniversary Conference*, pp. 287–291, August 2004.
58. J. Miles, *Phase Station™ — 53100A Phase Noise Test Set User's Manual*, Jackson Labs, rev. 1.09, 245 p., December 2023.
59. C. E. Calosso, F. Vernotte, V. Giordano, C. Fluhr, B. Dubois, and E. Rubiola, "Frequency Stability Measurement of Cryogenic Sapphire Oscillators with a Multichannel Tracking DDS and the Two-Sample Covariance," *IEEE Transactions on Ultrasonics, Ferroelectrics, and Frequency Control*, vol. 66, no. 3, pp. 616–623, March 2019.
60. J. Mitola, "The Software Radio Architecture," *IEEE Communications Magazine*, vol. 33, no. 5, pp. 26–38, May 1995.
61. R. G. Machado and A. M. Wyglinski, "Software-Defined Radio: Bridging the Analog-Digital Divide," *Proceedings of the IEEE*, vol. 103, no. 3, pp. 409–423, March 2015.
62. J. A. Sherman and R. Jordens, "Oscillator Metrology with Software Defined Radio," *Review of Scientific Instruments*, vol. 87, no. 5, pp. 054711/1–11, May 2016.
63. C. Andrich, A. Ihlow, J. Bauer, N. Buester, and G. Del Galdo, "High-Precision Measurement of Sine and Pulse Reference Signals Using Software-Defined Radio," *IEEE Transactions on Instrumentation and Measurement*, vol. 67, no. 5, pp. 1132–1141, May 2018.
64. A. Barron, *Software Defined Radio*, Radio Society of Great Britain, 2nd edition, 2017.
65. T. F. Collins, R. Getz, D. Pu, and A. M. Wyglinski, *Software-Defined Radio for Engineers*, Artech House, 2018.
66. A. C. C. Olaya, C. E. Calosso, J-M. Friedt, S. Micalizio, and E. Rubiola, "Phase Noise and Frequency Stability of the Red-Pitaya Internal PLL," *IEEE Transactions on Ultrasonics, Ferroelectrics, and Frequency Control*, vol. 66, no. 2, pp. 412–416, February 2019.
67. D. K. Misra, *Radio-Frequency and Microwave Communication Circuits: Analysis and Design*, John Wiley & Sons, 2nd edition, 2004.
68. F. M. Gardner, *Phaselock Techniques*, John Wiley & Sons, 1966.
69. J. W. M. Rogers, C. Plett, and I. Marsland, *Radio Frequency System Architecture and Design*, Artech House, 2013.
70. H. W. Silver, "Experiment #68 – Phase Locked Loops, the Basics," *QST*, vol. 92, no. 9, pp. 71–72, September 2008.
71. C. Barrett, ed., "Fractional/Integer-N PLL Basics," *Texas Instruments Technical Brief SWRA029*, 55 p., 2010.
72. H. de Bellescize, "La réception synchrone," *L'Onde Électrique*, vol. 11, no. 126, pp. 230–240, June 1932.
73. H. de Bellescize, "Synchronizing System," *United States Patent 1,990,428*, filed September 29, 1932, granted February 5, 1935.
74. K. R. Wendt and G. L. Fredendall, "Automatic Frequency and Phase Control of Synchronization in Television Receivers," *Proceedings of the Institute of Radio Engineers (IRE)*, vol. 31, no. 1, pp. 7–15, January 1943.
75. A. Grebene and H. Camenzind, "Phase Locking as a New Approach for Tuned Integrated Circuits," *1969 IEEE International Solid-State Circuits Conference - Digest of Technical Papers*, pp. 100–101, February 1969.
76. R. Stone, "Disciplined Time and Frequency Oscillator," *Proceedings of the 3rd Annual Precise Time and Time Interval Meeting (PTTI)*, pp. 211–223, November 1971.
77. H. J. Walls, "The Standard-Frequency Set at WWV," *QST*, vol. 8, no. 3, pp. 9–12, October 1924.
78. G. K. Nelson, "A Century of WWV," *Journal of Research of the National Institute of Standards and Technology*, vol. 124, article no. 124025, pp. 1–31, September 2019.
79. R. J. Blume, "Fade-Canceling Zero Beat Indicator for Reception of Standard Radio-Frequencies," *Review of Scientific Instruments*, vol. 28, no. 9, pp. 703–708, September 1957.

80. J. A. Pierce, "Intercontinental Frequency Comparison by Very Low-Frequency Radio Transmission," *Proceedings of the Institute of Radio Engineers (IRE)*, vol. 45, no. 6, pp. 794–803, June 1957.
81. B. Bourke and B. Penrod, "An Analysis of a Microprocessor Controlled Disciplined Frequency Standard," *Proceedings of the 37th Annual Symposium on Frequency Control*, pp. 485–491, June 1983.
82. R. J. Hesselberth, T. P. Donaher, and J. E. Sandahl, "Disciplined Oscillator System with Frequency Control and Accumulated Time Control," *United States Patent 4,525,685*, filed May 31, 1983, granted June 25, 1985.
83. A. J. Van Dierendonck, Q. D. Hua, J. R. McLean, and A. R. Denz, "Time Transfer Using NAVSTAR GPS," *Proceedings of the 13th Annual Precise Time and Time Interval Meeting (PTTI)*, pp. 389–418, December 1981.
84. D. D. Davis, M. Weiss, A. Clements, and D. W. Allan, "Construction and Performance Characteristics of a Prototype NBS/GPS Receiver," *Proceedings of the 35th Annual Symposium on Frequency Control*, pp. 546–552, May 1981.
85. A. MacIntyre and S. R. Stein, "A Disciplined Rubidium Oscillator," *Proceedings of the 40th Annual Symposium on Frequency Control*, pp. 465–469, May 1986.
86. K. Åström and T. Hägglund, *PID Controllers: Theory, Design, and Tuning*, 2nd edition, Instrument Society of America, 1995.
87. S. Bennett, "Nicholas Minorsky and the Automatic Steering of Ships," *IEEE Control Systems Magazine*, vol. 4, no. 4, pp. 10–15, November 1984.
88. A. Visioli, *Practical PID Control*, Springer-Verlag, 2006.
89. R. Hardin and M. Yankowski, "Performance of a PID Phase Lock Loop with Kalman Filter Input data," *Proceedings of the 1992 IEEE International Frequency Control Symposium*, pp. 238–256, May 1992.
90. B. M. Penrod, "Adaptive Temperature Compensation of GPS Disciplined Quartz and Rubidium Oscillators," *Proceedings of the 1996 IEEE International Frequency Control Symposium*, pp. 980–987, June 1996.
91. Y. Shmaliy and L. Arceo-Miquel, "Efficient Predictive Estimator for Holdover in GPS-Based Clock Synchronization," *IEEE Transactions on Ultrasonics, Ferroelectrics, and Frequency Control*, vol. 55, no. 10, pp. 2131–2139, October 2008.

Chapter 6

Accuracy

Fast is fine, but accuracy is everything.

— Wyatt Earp

Accuracy is a word that we all know, a word that we associate with truth, correctness, or exactness. If we say that a reporter provided an accurate account of a news story, we're saying that we believe the reporter was essentially telling the truth. If we say that a book or a movie based on real-life events is accurate, it again means that we think it's essentially true. Being accurate is more or less synonymous with being correct, but accuracy typically involves providing more information than correctness. For instance, if we drove our car to work and someone asks us how we got there, we can be correct by simply stating that we got there "by car." Or, we can provide more information and say that we got there by driving a 2006 Toyota Avalon, with a Cassis Red Pearl exterior, a bad brake light, and 177,000 miles on the odometer. That might still be correct, and it's certainly a more accurate description of the vehicle that we drove to work, but it's probably too much information for that situation. However, there are other situations where we have to include all of the details needed for an accurate description, because simply being correct can cause confusion or misunderstandings.

Our most common conception of accuracy probably relates to exactness, which we can associate with sports where points are scored by a hitting a target. In the sport of basketball, for instance, the baskets at each end of the court are fixed targets, and common statistics like field goal accuracy (FGA) indicate the percentage of attempted shots that went into the basket. Statistics like FGA are easy to understand because they treat accuracy as a binary condition—a shot that goes in the basket is accurate, and a shot that doesn't go in is inaccurate. Many other sports have fixed targets, including the goalposts in football and the nets in soccer and hockey, and these sports have statistics similar to FGA.

The most iconic representation of accuracy involves "hitting the bullseye" of a circular fixed target, such as the target used in the game of darts. The bullseye—the smallest ring in a series of concentric rings—is located at the very center of the target. The amount of points players score depends on the accuracy of their shots or throws, and the closer they come to the bullseye the more points they score. Figure 6.1 shows a dartboard, where the dart stuck inside the bullseye is the result of a highly accurate throw, and where the two darts stuck outside the bullseye resulted from two less accurate throws.

As Time Goes By: A Comprehensive Guide to Time and Frequency Measurements
Michael A. Lombardi

ISBN 978-981-5352-28-3 (Hardcover), 978-1-042-01532-0 (eBook)
www.jennystanford.com

DOI: 10.1201/9781042015320-9

Figure 6.1 A dartboard, where an accurate throw has hit the bullseye.

Metrology is all about truth, correctness, and exactness, and it also has fixed targets that were established by internationally accepted definitions of units [1]. The *International Vocabulary of Metrology* (*VIM*) has changed its definition of accuracy slightly over the years, and other standards organizations have done the same and will likely continue to do so [2]. However, most definitions of accuracy are similar to the one in the 3rd edition of the *VIM* that first appeared in 2008. There, accuracy is defined as the "closeness of agreement between a measured quantity value and a true quantity value of a measurand" [3]. The "true quantity value of a measurand" is the fixed target that metrologists always aim for. In this book, the measurand is always time or frequency, and as we've already learned, both the second and hertz have defined "true quantity" values.

Some metrology publications strongly suggest that the term accuracy should only be used in a qualitative sense, and never used quantitatively, which prevents accuracy from being associated with numbers [4, 5]. Even the *VIM* reminds us, in an explanatory note to its definition of accuracy, that "The concept 'measurement accuracy' is not a quantity and is not given a numerical quantity value" [3, 6]. This suggests that it's OK to say that we made an accurate measurement, or that an instrument is accurate, but it's not OK to say that an oscillator is accurate to within 1×10^{-11}, or that a clock is accurate to within 1 μs. The thinking behind this line of reasoning is that quantitative statements should be reserved for discussions of measurement uncertainty, a subject area that begins by acknowledging that no measurement is perfect, and that attempts to quantify all of the doubts that we have about a measurement result (Chapter 8). Once we acknowledge that no measurement is perfect, our next thought might be that true measurement accuracy can't ever be realized, followed by the conclusion that attempts to quantify accuracy can't be valid and therefore shouldn't even be allowed.

This line of reasoning might make sense on a purely semantic level, but it fails at the common sense level. For instance, consider a golfing metrologist who comes home smiling after a great day on the course. The day was highlighted by a hole-in-one, where her tee shot landed about two meters from the pin and rolled directly into the cup. When asked why she's smiling, the golfer doesn't tell her spouse that she made a hole-in-one, but only says that she made "an accurate shot." When her spouse asks "how accurate?," the golfer responds that she's not allowed to quantify the accuracy of the shot, but she can qualitatively say that it was accurate, and she'd be more than happy to discuss the shot's uncertainty.

That conversation seems silly, and it would seem equally silly if metrologists were never allowed to quantify time and frequency accuracy with numbers. People unfamiliar with measurements and measurement uncertainty are still familiar with the concept of accuracy, and when they ask about the accuracy of an oscillator or clock, metrologists should be able to provide them with an answer. Therefore, in this chapter we'll discuss accuracy in both a qualitative and a quantitative sense, without losing sight of the fact that no measurement is perfect, and that every measurement made to determine accuracy has an associated uncertainty. With that in mind, the next two sections discuss the frequency accuracy of oscillators and the time accuracy of clocks. Time accuracy always depends on frequency accuracy, so we'll discuss frequency accuracy first.

6.1 Frequency Accuracy

We previously defined nominal frequency (Section 5.2.1) as the frequency that an oscillator is supposed to produce. This frequency is usually indicated by a label on the oscillator's output. For example, if an oscillator has a connector labeled "10 MHz," the oscillator's nominal frequency is 10 MHz.

The nominal value defines an ideal frequency with zero uncertainty, in accordance with its SI definition and the "true quantity value" that we discussed earlier. In a perfect world, only perfect signals with frequencies of exactly 10 MHz would come out of connectors with "10 MHz" labels, but in our imperfect world, signals always deviate from their nominal value by some amount. The idealized nominal frequency can't be realized, but it can still serve as our accuracy baseline. Therefore, the frequency accuracy of a DUT is the difference between its actual frequency and its nominal value.

The frequency accuracy of a DUT can only be determined by measurement, and the measurement can only be valid if the reference is more accurate than the DUT. This is mostly common sense, but it's essential to remember, because perhaps the biggest fear shared by all metrologists is accepting a false measurement result and allowing a product or instrument to pass a test that it should have failed. We noted in Section 5.2 that a simple rule of thumb that protects against incorrect results is selecting a reference that is at least 10× more accurate than the DUT. This 10:1 accuracy ratio has likely been a rule of thumb since the early part of the 20th century. However, it was perhaps first introduced as a formal metrology principle in 1950, when a United States military standard on gage inspection stated that "In general practice, gage tolerances should not exceed 10 percent of the tolerance of the part to be gaged" [7]. This became colloquially known as the "Gagemaker's Rule." These types of rules were later given the acronym TAR, which stands for test accuracy ratio.

The equipment and time required to support a 10:1 TAR isn't always available, and this necessarily led to the use of smaller accuracy ratios that could still reduce measurement risk to an acceptable level. The best known of these ratios is the 4:1 TAR that resulted from efforts in the 1950s to improve the measurement reliability of the U. S. Navy's guided missile programs. The 4:1 TAR subsequently became part of many military and industrial standard documents, and is still included in some standards documents today. The mathematical analysis that supported it was developed in 1955 by Jerry Hayes (1928–2022), a mechanical engineer at the United States Naval Ordnance Laboratory in Corona, California [8]. The 4:1 TAR provided a simple rule that metrologists could follow in the days before electronic calculators and computers, when most calculations were still performed with slide rules. As computing power increased and measurement requirements became more stringent, the use of TARs or similar test uncertainty ratios (TURs) became less common, and they were often replaced by a more comprehensive measurement uncertainty analysis.

It's clear that measurement uncertainty analysis, which we'll cover in Chapter 8, provides more information about the validity of a measurement than relying on a simple accuracy ratio. Even so, knowing something about the TAR is still mandatory when planning a measurement. For the purposes of example, Table 6.1 lists typical frequency accuracies, when averaged for a 1-day period, for five devices, including two quartz oscillators (TCXO and OCXO), two atomic oscillators (rubidium and cesium), and a GPSDO. It also lists a typical frequency accuracy for a GPSDO over a 10-day period. The green areas in the vertical columns indicate where the TAR between the reference and DUT is high enough to support a valid accuracy measurement. For example, a rubidium oscillator should be able to easily determine the accuracy of a TCXO or OCXO over a 1-day interval. A GPSDO should be able to determine the accuracy of a cesium over a 10-day interval, but not in one day, when the TAR is probably about 1:1. As noted earlier, selecting a reference with sufficient accuracy to measure the accuracy of a DUT is mostly just common sense, but situations where a DUT's accuracy is reported incorrectly are not unusual. When it happens, it's usually because either the reference or the measurement system wasn't capable of supporting the reported accuracy.

Table 6.1 Typical accuracy ratios for the reference and DUT in frequency metrology

Reference accuracy / DUT accuracy		TCXO	OCXO	Rubidium	Cesium	GPSDO (1 day)	GPSDO (10 days)
		$\pm 1 \times 10^{-7}$	$\pm 1 \times 10^{-8}$	$\pm 1 \times 10^{-11}$	$\pm 1 \times 10^{-13}$	$\pm 1 \times 10^{-13}$	$\pm 1 \times 10^{-14}$
TCXO	$\pm 1 \times 10^{-7}$	1:1	10:1	10^4:1	10^6:1	10^6:1	10^7:1
OCXO	$\pm 1 \times 10^{-8}$	—	1:1	10^3:1	10^5:1	10^5:1	10^6:1
Rubidium	$\pm 1 \times 10^{-11}$	—	—	—	10^2:1	10^2:1	10^3:1
Cesium	$\pm 1 \times 10^{-13}$	—	—	—	1:1	1:1	10:1
GPSDO (1 day)	$\pm 1 \times 10^{-13}$	—	—	—	1:1	1:1	10:1
GPSDO (10 days)	$\pm 1 \times 10^{-14}$	—	—	—	—	—	—

Throughout Chapter 5, we described numerous methods to measure the frequency difference, f_{diff}, between a DUT and a reference frequency. We noted that f_{diff} can be measured in either the measurement or time domain. Therefore, repeating Eq. (5.16), we know that

$$f_{\text{diff}} = \frac{\Delta t}{T} = \frac{\Delta f}{f}. \tag{6.1}$$

We can now state that frequency accuracy is essentially the same thing as f_{diff}, with a few caveats that we'll get to shortly. Frequency accuracy is also essentially the same thing as the other terms often used to describe the difference between two frequencies, including frequency offset, relative frequency, frequency bias, frequency error, fractional frequency difference, and so forth.

The first caveat involves the duration of a frequency accuracy measurement. In some of the Chapter 5 examples, we described methods of determining f_{diff} with a single measurement or just a few measurements. Measurements that quickly determine the frequency difference between two signals aren't generally thought of as accuracy measurements. In practice, frequency accuracy measurements are continuously made for multiple hours or days, and the resulting data makes it possible to estimate the accuracy of an oscillator at a given point in time, or over a specified interval.

The second caveat is that valid measurements of frequency accuracy must include an associated measurement uncertainty (Chapter 8). As metrologists know, reporting uncertainty is a requirement for valid measurements of all physical quantities, not just frequency. It's important to realize the distinction between a traceable frequency accuracy measurement with a known uncertainty, and a quick measurement of the frequency difference between two oscillators where the measurement uncertainty isn't a concern. The latter measurement might perfectly meet the needs of most students, technicians, scientists, and engineers, but it doesn't qualify as a valid measurement of frequency accuracy made for metrological purposes.

The third and final caveat is that valid measurements of frequency accuracy must be traceable to the SI. As we'll discuss in detail in Chapter 9, a measurement is only traceable if the reference can be linked to SI through an unbroken chain of calibrations. Let's briefly consider a situation where we are measuring the frequency difference of a TCXO relative to a rubidium oscillator. We know for a variety of reasons—including reading the oscillator specification sheets, and from the knowledge and experience gained from previous measurements—that the rubidium oscillator is more accurate than the TCXO by a factor of about 10^4. We also know that the rubidium oscillator isn't linked to the SI through an unbroken chain of calibrations. Our measurement correctly reports the difference between the TCXO and its nominal frequency, providing us with same result we would have gotten if we'd used a more accurate reference, because the uncertainty contributed by the rubidium oscillator is already too small to be discernible. However, it's still not a valid measurement of frequency accuracy, not because it's a relative measurement, but because it isn't a traceable measurement. Valid measurements of frequency accuracy can be relative measurements, in fact, they almost always are. However, because they are traceable, they are relative to the SI.

We just stated that almost all frequency accuracy measurements are relative measurements. That leads to an obvious question, what constitutes an absolute frequency accuracy measurement? The *VIM* doesn't define what an absolute measurement is, but in frequency metrology the term is generally reserved for comparisons of experimental atomic clocks with a cesium fountain primary frequency standard (Section 3.8). As of 2025, cesium fountain primary standards provide the best physical representation of the SI second, and will continue to do so for as long as the SI second is defined based on energy transitions of the ^{133}Cs atom (Section 3.4).

Absolute frequency accuracy measurements are relatively rare, usually only performed by scientists at national metrology laboratories. These scientists are usually engaged in measuring the clock transition of an atomic clock that can provide a secondary representation of the SI second, which in most current research efforts is an atomic clock based on optical, rather than microwave clock transitions (Chapter 13). An absolute frequency measurement of an optical clock will likely be either a direct comparison to a primary standard through a stable transmission medium such as a fiber optics network (Chapter 14), or an indirect comparison to a primary standard by contributing optical clock data to the BIPM for its calculation of International Atomic Time (TAI), and then comparing the results to the other primary standards that also contributed TAI data during the same period [9–11]. The generation of TAI is described in Section 9.4.3.

Like all measurements, absolute frequency accuracy measurements still have measurement uncertainties, and their magnitude depends on the measurement method and the uncertainties associated with the primary standards used as references. Comparing optical clocks directly to each other will usually produce results that have much smaller uncertainties than the comparisons made against cesium fountain primary standards. Even so, if the measurement doesn't involve a standard that represents the current SI definition of the second, it should be categorized as a relative measurement.

6.1.1 Frequency Accuracy Graphs

Graphs of frequency accuracy can have several different forms, but most use the standard Cartesian coordinate format (x, y), where the x-axis shows the elapsed time of the measurement, in units that can range from fractions of a second to months or years. For relatively low-accuracy measurements made in the frequency domain, it's common for the y-axis to indicate the difference between the measured and nominal frequency of the DUT, Δf, in units of hertz.

Figure 6.2 provides one example of this type of graph. Here, the DUT is not an oscillator, but instead is the carrier frequency from NIST radio station WWV. The carrier frequency was broadcast with a nominal frequency of 5 MHz, but was subjected to Doppler frequency shifts as it traveled from the transmitter to the receiver [12]. In this example, the receiver was located about 80 km from the station, and the signal was measured for one UTC day, a period of 24 hours. The reference frequency for the measurement was a GPSDO that provided an excellent TAR, which on average exceeded 10^4:1. The receiver was close enough to the transmitter for the signal to mostly propagate by groundwave during the period between sunset and sunrise, from about 0200 to 1400 UTC, and during the period after sunrise until dusk, from about 1500 to 2200 UTC. The DUT frequency was accurate to within ±0.1 Hz during these periods of groundwave reception. However, during the hours surrounding sunrise and sunset, the DUT frequency was much less accurate. This loss of accuracy was caused by the reception of signals that propagated mostly by skywave after being reflected by the ionosphere, which include the skywave reception of interfering signals from other radio stations. The largest Doppler frequency shift occurred at sunrise, between about 1400 and 1500 UTC, when the accuracy of the received carrier exceeded 0.7 Hz.

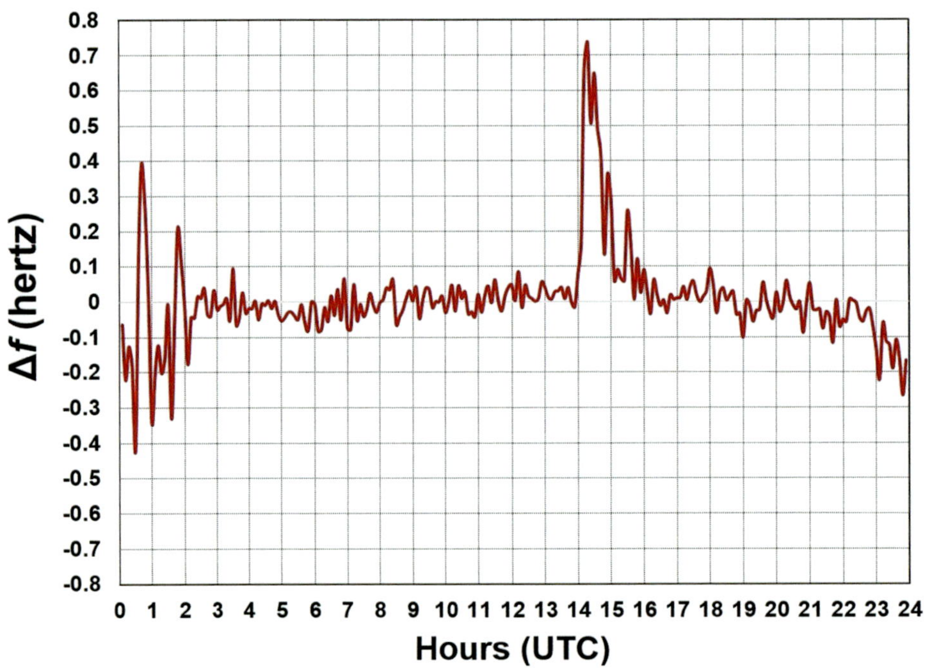

Figure 6.2 Frequency accuracy graph with the y-axis labeled with units of hertz.

A second type of frequency accuracy graph, typically reserved for high-accuracy measurements of primary standards made in the frequency domain, is shown in Figure 6.3. In this graph, the y-axis is labeled with the unitless frequency difference, $\frac{\Delta v}{v}$, equivalent to $\frac{\Delta f}{f}$. The graph shows a comparison of nine cesium fountain primary frequency standards that were compared to TAI for a period of nearly 10 years, and the x-axis is labeled with the Modified Julian Dates (MJD). Each data point represents a measurement of the frequency accuracy of a cesium fountain with respect to TAI, and the vertical bars attached to the data point represent the estimated measurement uncertainty. The minor divisions on the y-axis indicate frequency differences of 1×10^{-15}. The dashed horizontal line positioned at 0 on the y-axis represents the frequency of TAI [13].

For time domain measurements, the graphs that illustrate frequency accuracy are the result of a phase comparison, and they're called phase difference or time difference graphs. The *x*-axis divisions indicate the elapsed time of the measurement, the quantity T shown in Eq. 6.1. The *y*-axis divisions indicate the phase difference, $\Delta\phi$, between the DUT and the reference frequency. However, because the measurements are usually made with instruments that express their results in units of time and not in radians or degrees, the *y*-axis is nearly always labeled with units of time, providing the quantity Δt shown in Eq. (6.1).

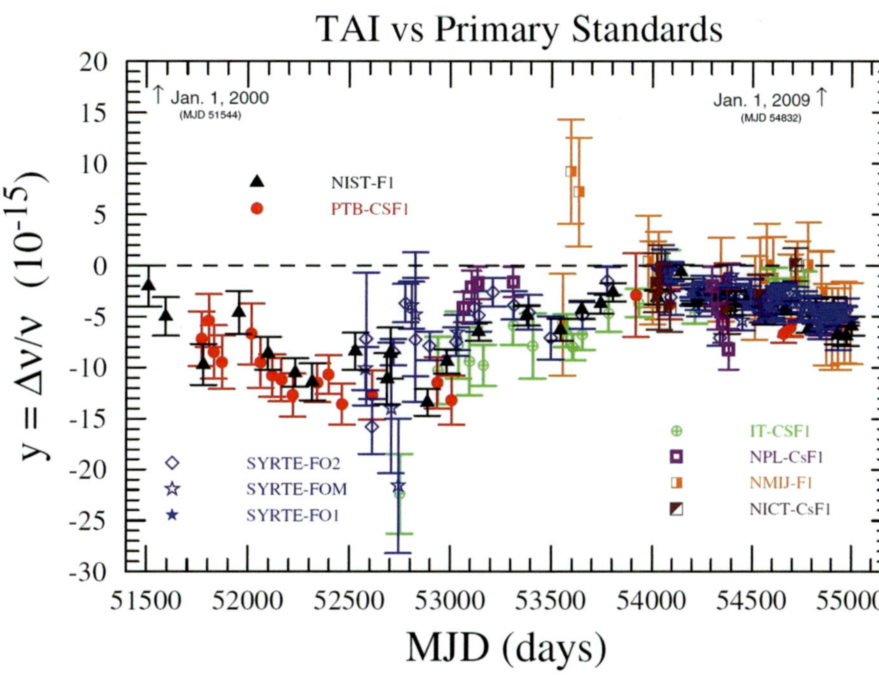

Figure 6.3 Frequency accuracy graph with the *y*-axis labeled with unitless values [13].

The frequency accuracy of the DUT for a given measurement interval is estimated as $\frac{\Delta t}{T}$. The easiest way to obtain Δt, which can often be estimated by simply looking at the graph, is to subtract the two endpoints of the *y*-series data that contains the time differences,

$$\Delta t = y(n) - y(1), \tag{6.2}$$

where n is the number of phase or time differences.

A better estimate of Δt can be obtained from the slope of the phase or time differences, which is the reason why metrologists often say that "frequency is determined by the slope of the phase." If we have an adequate TAR, the frequency difference between the DUT and the reference will be constant or nearly constant. This means that the accumulation of time differences will be a linear function, and therefore a linear least squares line can be fitted to the data to determine its slope. The slope, β_1, and *y*-intercept, β_0, for the least squares lines are calculated as

$$\hat{\beta}_1 = \frac{SS_{xy}}{SS_{xx}} \tag{6.3}$$

and

$$\hat{\beta}_0 = \bar{y} - \hat{\beta}_1 \bar{x} \tag{6.4}$$

where

$$SS_{xy} = \sum_{i=1}^{n}(x_i - \bar{x})(y_i - \bar{y}) = \sum_{i=1}^{n} x_i y_i - \frac{\left(\sum_{i=1}^{n} x_i\right)\left(\sum_{i=1}^{n} y_i\right)}{n} \tag{6.5}$$

and

$$SS_{xx} = \sum_{i=1}^{n}(x_i - \bar{x})^2 = \sum_{i=1}^{n} x_i^2 - \frac{\left(\sum_{i=1}^{n} x_i\right)^2}{n}, \tag{6.6}$$

where n is the number of phase or time difference values used in the estimate, and where β_1 is now substituted for Δt to determine frequency accuracy.

If an adequate TAR was maintained, the slope of the least squares line will be nearly identical to the slope of the actual data, because of the nearly constant frequency difference between the DUT and the reference. In fact, to get a good estimate of frequency accuracy, the measurement period T should be long enough to show this linear slope. A "quiet" time difference graph with an obvious linear correlation ensures that Δt is correctly indicating how the time difference of the DUT is changing, and is not dominated by noise produced by the measurement system or the reference frequency.

To estimate the "goodness of fit" of the least squares line to the time difference data, it's common to calculate r, the Pearson linear correlation coefficient. The possible values for r range from -1 to $+1$, with the sign of r simply indicating if the slope of the time difference data is negative or positive. If r were to equal 0, there would be no correlation between the x and y values, but when r is approximately equal to either -1 or $+1$, the x and y values have a strong linear correlation. When a strong correlation exists, all of the data points will lie near the least squares line, which provides confidence that the frequency accuracy is being correctly estimated. The Pearson correlation coefficient is calculated as

$$r = \frac{SS_{xy}}{\sqrt{SS_{xx} SS_{yy}}}. \tag{6.7}$$

Figure 6.4 shows a sample time difference graph of a free-running atomic oscillator that was compared to a GPSDO for a period of seven days, using a method where a TIC recorded the average time difference every hour. During this period, the total accumulated time difference, Δt, was nearly 900 ns, as indicated by both the actual time difference data and by the least squares line that was fitted to the data. If we subtract the two endpoints of the time difference data as shown in Eq. (6.2), the average frequency accuracy, f_{acc}, for the seven-day period is $+1.4 \times 10^{-12}$. By using the slope of the least squares line from Eq. (6.3), we get a more rigorous estimate but only a slightly different result, $f_{acc} = +1.5 \times 10^{-12}$. The graph shows that the time difference data are noisier than the least squares line, indicating that some noise is being contributed by the reference GPSDO. However, the strong linear correlation is obvious even without performing mathematical analysis ($r = +0.99$), so we can have confidence in this estimate of frequency accuracy.

In some cases, such as time domain comparisons between national frequency standards, obtaining a TAR much higher than 1:1 will be difficult; therefore, the slope of the time difference data will often be too small to be easily detected without very long measurement periods, because r will likely be near 0. Time difference graphs are still commonly used in these situations because they can show that two standards have equivalent frequency accuracy, providing the basis for international frequency and time coordination (Chapter 9).

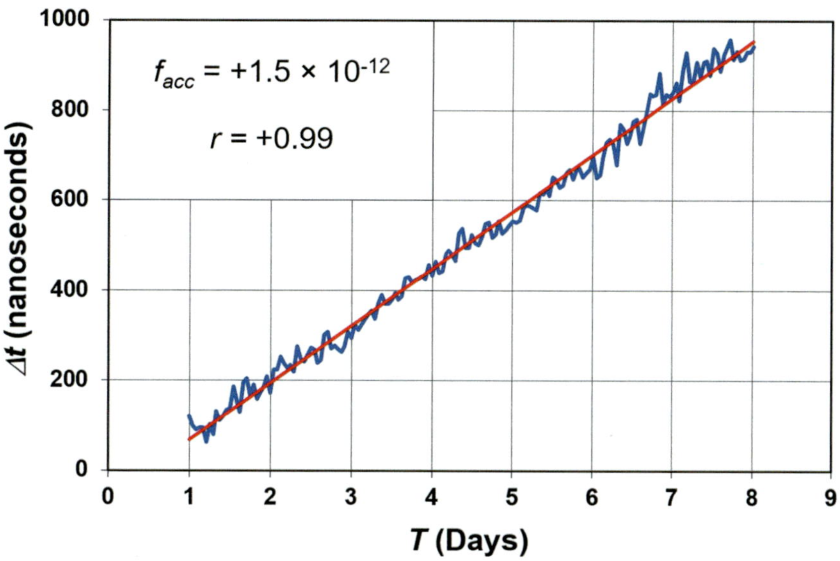

Figure 6.4 Frequency accuracy graph based on the slope of the time differences between an atomic oscillator and a GPSDO.

6.1.2 Frequency Accuracies of Oscillators

Oscillator datasheets nearly always include specifications for stability (Chapter 7), but they often don't include specifications for accuracy. The most notable exceptions are commercial cesium beam oscillators, which typically have accuracy specifications ranging from $\pm 1 \times 10^{-12}$ to $\pm 5 \times 10^{-13}$, and ironically, low-cost quartz oscillators, which sometimes provide specifications for the frequency accuracy that can be expected over a wide temperature range. The accuracy specifications for quartz oscillators are often provided in parts per million (ppm), which is equivalent to parts in 10^6. Quartz wristwatch manufacturers also sometimes provide an accuracy specification, translated for the consumer into time units. For example, a quartz wristwatch's accuracy might be specified as ±15 s per month, which is equivalent to frequency accuracy of about $\pm 5 \times 10^{-6}$.

The reason that accuracy specifications aren't usually provided is that the frequency accuracy of an oscillator depends on how well it is adjusted, and the limit of its accuracy during a given period is defined by its stability during that same period. It's not possible for an oscillator to have an accuracy better than its stability. For example, if the frequency of a rubidium oscillator is only adjusted once per year, its accuracy during the year between adjustments can be no better than its stability during that same year. Therefore, providing only a stability specification on a rubidium datasheet is considered sufficient because it establishes an accuracy limit, and the actual accuracy is determined by how often and how well an oscillator's frequency is adjusted.

Accuracy specifications are provided for cesium oscillators because the SI second is defined based on cesium resonance; therefore, cesium oscillators are intrinsically accurate frequency standards. A cesium oscillator that's working properly and that's operated within its specified temperature range should meet its accuracy specification at all times, so we can think of its specified accuracy as being essentially guaranteed. This means that if the frequency accuracy specification is $\pm 1 \times 10^{-12}$, we should be able to take a brand new cesium standard out of the box, plug it into an AC outlet, and get signals that are always accurate to within $\pm 1 \times 10^{-12}$ without performing any initial or periodic frequency adjustments. However, even if the specification easily meets the accuracy requirement for a given application, the cesium's output should still be measured to prove that it's working properly. Once measured, a cesium can be adjusted to exceed its specified accuracy, but again, its accuracy during a given period will be limited by its stability during that same period.

When discussing quartz and atomic oscillators in Chapters 2 and 3, we noted that the quality factor, *Q*, of an oscillator is related to both its accuracy and stability. The reason that *Q* relates to accuracy is that a high-*Q* resonator won't oscillate at all unless it is near its natural resonance frequency, making it potentially more accurate than a resonator allowed to operate across a wider range of frequencies [14]. The resonance width, which is mostly determined by the operating temperature range and allowable tuning range of a quartz oscillator, or by the atomic linewidth of an atomic oscillator, basically sets upper and lower boundaries on an oscillator's frequency accuracy. Table 6.2 summarizes the typical and best-case frequency accuracies for various types of quartz and atomic oscillators when averaged for a period of one day. It also lists the typical and best-case *Q* factors for each oscillator type.

Table 6.2 A summary of frequency accuracies and *Q* factors for various types of oscillators

	Frequency accuracy (1 day)		Quality factor, Q	
Oscillator type	**Typical**	**Best case**	**Typical**	**Best case**
XO (wristwatch)	$\pm 1 \times 10^{-5}$	$\pm 1 \times 10^{-6}$	10^4	10^5
TCXO	$\pm 5 \times 10^{-7}$	$\pm 1 \times 10^{-7}$	10^5	3×10^6
OCXO	$\pm 5 \times 10^{-8}$	$\pm 5 \times 10^{-10}$	10^6	3×10^6
Rubidium	$\pm 5 \times 10^{-9}$	$\pm 5 \times 10^{-12}$	10^7	3×10^7
Active hydrogen maser	$\pm 5 \times 10^{-12}$	$\pm 5 \times 10^{-14}$	10^9	10^9
Cesium beam	$\pm 5 \times 10^{-13}$	$\pm 2 \times 10^{-14}$	10^8	10^8
Cesium fountain	$\pm 1 \times 10^{-15}$	$\pm 1 \times 10^{-16}$	10^{10}	10^{10}

6.2 Time Accuracy

To achieve time accuracy, the fixed target that metrologists aim for is UTC, and a clock that is synchronous with UTC has "hit the bullseye." The time accuracy of a clock is equivalent to its time difference with respect to UTC. As discussed in Section 4.2, the time accuracy of a DUT is determined by measuring the time difference between the DUT's on-time marker (OTM) and the OTM of a trusted UTC reference. The OTMs for both the DUT and the reference are usually signals with a frequency of 1 Hz, generated in the form of 1 pps signals. To avoid ambiguity, the OTMs are labeled with a time code or time tag that provides time-of-day information. This information typically includes the UTC hour, minute, and second, and in the case of OTMs generated at frequencies higher than 1 Hz, the fractional part of the second.

Frequency accuracy requirements are usually specified as an average frequency over a given period, but time accuracy requirements are often specified as a threshold or hard limit that cannot be exceeded. For critical infrastructure systems, including telecommunication systems (Chapter 16) and systems that generate electrical power (Chapter 17), the time accuracy requirement is often specified as ±1 μs. This doesn't mean that a clock can meet the requirement by keeping time to within ±1 μs of UTC on average, but it instead means that a clock's time should never be allowed to differ by more than ±1 μs from UTC. For this reason, simple metrics such as the maximum time interval error (MTIE) are sometimes included in standards documents for telecommunication systems. These metrics indicate the maximum time error that a clock is allowed to have, which can be thought of as its worst-case time accuracy [15].

In order to comply with a time accuracy requirement, a clock usually has to be periodically synchronized to keep its time in agreement with UTC, or have its oscillator frequency continuously adjusted so that it maintains synchronization with UTC. The former situation is the way that radio-controlled clocks work (Chapter 10), and the latter situation is the way that disciplined oscillators and clocks work (Section 5.3.2

and Chapter 12). Without time or frequency adjustments, a clock's accuracy will continuously get worse, because free-running clocks always accumulate a time difference due to the frequency difference of their oscillator with respect to UTC. The relationship between the frequency difference of a clock's oscillator, and the time difference that a clock accumulates between adjustments, both with respect to UTC, is shown in Table 6.3. The last four rows of the table contain the time accuracy estimates that metrologists often commit to memory—clocks with oscillators accurate to about one part in 10^5, 10^8, 10^{11}, and 10^{14} per day, have respective time accuracies of 1 s, 1 ms, 1 μs, and 1 ns per day.

Table 6.3 The relationship between frequency accuracy and time accuracy

Frequency difference with respect to UTC	Measurement period	Accumulated time difference with respect to UTC
$\pm 1.00 \times 10^{-3}$	1 s	±1 ms
$\pm 1.00 \times 10^{-6}$	1 s	±1 μs
$\pm 1.00 \times 10^{-9}$	1 s	±1 ns
$\pm 2.78 \times 10^{-7}$	1 hour	±1 ms
$\pm 2.78 \times 10^{-10}$	1 hour	±1 μs
$\pm 2.78 \times 10^{-13}$	1 hour	±1 ns
$\pm 1.16 \times 10^{-5}$	1 day	±1 s
$\pm 1.16 \times 10^{-8}$	1 day	±1 ms
$\pm 1.16 \times 10^{-11}$	1 day	±1 μs
$\pm 1.16 \times 10^{-14}$	1 day	±1 ns

6.2.1 Time Accuracy Graphs

Time accuracy graphs are similar to the phase or time difference graphs used to estimate frequency accuracy, such as the graph shown in Figure 6.4. The *x*-axis indicates the period of the measurement, and the *y*-axis indicates the time difference between the DUT and the reference clock. However, there is one critical difference between a time domain graph that estimates frequency accuracy and a graph that estimates time accuracy. The measurements shown on-time domain graphs of frequency accuracy are obtained with phase comparisons between two frequencies that don't have to be, and usually aren't, synchronized to a UTC time source. Therefore, the measured time differences aren't relative to UTC, and can have a magnitude that's entirely arbitrary. For purposes of clarity, and to avoid labeling the *y*-axis with large numbers, it's common practice to normalize the measured time differences to 0 by subtracting the midpoint of the entire data set from each individual data point. In contrast, the magnitude of the time differences shown on the *y*-axis of a time accuracy graph are never arbitrary, and the time differences are never normalized to 0. Because the measurement is always referenced to a clock synchronized to UTC, the *y*-axis labels indicate the actual time difference between the DUT and UTC.

A graph of the time accuracy of the UTC(NIST) time scale (Section 4.4) compared to UTC is shown in Figure 6.5. The graph includes one data point every five days as reported by the BIPM, for the five-year period from 2019 to 2023. Only a few of the measurements, shown near the left side of the graph, fall outside of a range of ±3 ns. The BIPM reports a measurement uncertainty for the UTC(NIST) comparisons of 2.7 ns, similar to the range shown on the graph [16]. The exceptional accuracy of UTC(NIST) is possible because it consists of a group of extremely stable atomic clocks, and because its output is continuously steered to agree with UTC by making small adjustments as often as once per week. Therefore, the time variations of UTC(NIST) usually fluctuate on both sides of 0 on the *y*-axis, with 0 representing UTC, the fixed target that NIST metrologists are aiming for.

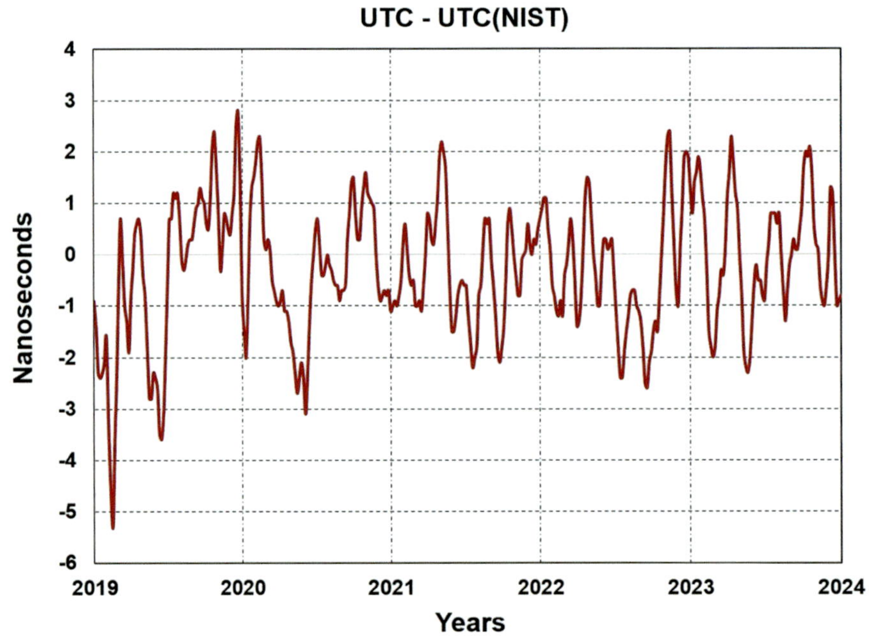

Figure 6.5 Time accuracy of UTC(NIST) compared to UTC for a five-year period.

Any time scale or clock that is continuously measured against UTC so that it has a quantifiable accuracy can be used as a time reference for other accuracy measurements. For example, GPS disciplined clocks (GPSDCs) are directly referenced to UTC(USNO), the time scale maintained by the United States Naval Observatory. However, their time accuracy is still limited by GPS signal propagation, and by signal delays in the GPSDC hardware that haven't been measured and removed, including delays in the GPS receiver, antenna, and antenna cable. We'll discuss the factors that contribute uncertainty to a GPSDC in detail in Chapter 12, but for now we can say that their signal delays can be estimated by comparing their output to UTC(NIST) or another time standard with quantifiable accuracy.

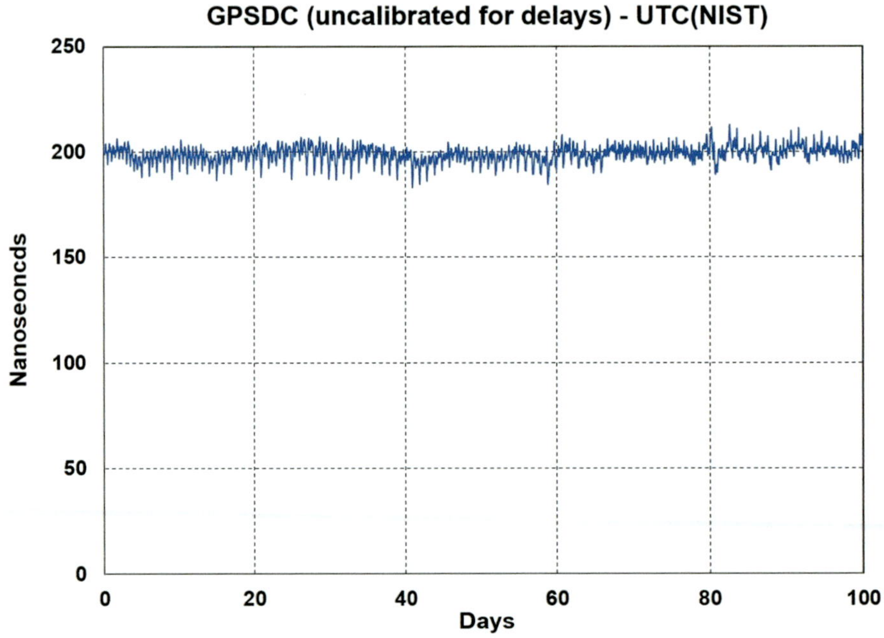

Figure 6.6 Time accuracy of a GPSDC that hasn't been calibrated for signal delays.

To illustrate this, Figure 6.6 shows the result of a 100-day comparison of a GPSDC to UTC(NIST). The time differences have no discernible slope for the entire 100-day period, indicating that there is essentially no frequency difference between the GPSDC and UTC(NIST) when averaged over a long period. This should be the case, because both clocks are continuously steered to agree with UTC. However, there is a constant 200 ns time difference, because this GPSDC has not been calibrated to remove its signal delays. If a delay constant of 200 ns was applied to this GPSDC, its output would "hit" the UTC target, and fluctuate on both sides of 0 on the *y*-axis, with variations of about ±10 ns.

After the delays of a GPSDC have been measured, and after delay compensation has been applied, their average time output can be within a few nanoseconds of UTC. This was demonstrated by measurements conducted at NIST in 2021 and 2022, when four GPSDCs were each calibrated against UTC(NIST) for a two-week period, with corrections applied to the calibration data to account for the small time differences between UTC, UTC(NIST), and UTC(USNO), which provides the time reference for GPS. After the final calibration results were applied to the four GPSDCs, they were then recalibrated using the same method for a one-year period, so that any residual biases from the original two-week calibrations could be removed. The one-year measurement revealed some small biases—there was an approximate 1 ns calibration bias in one GPSDC, and an approximate 3 ns calibration bias in the other three units. Figure 6.7 shows the time differences of the four GPSDCs compared to UTC for a six-week period after their calibration biases were corrected, with UTC again represented by 0 on the *y*-axis. All four units now agreed with UTC to within ±2.5 ns during the entire six-week post-calibration period [17].

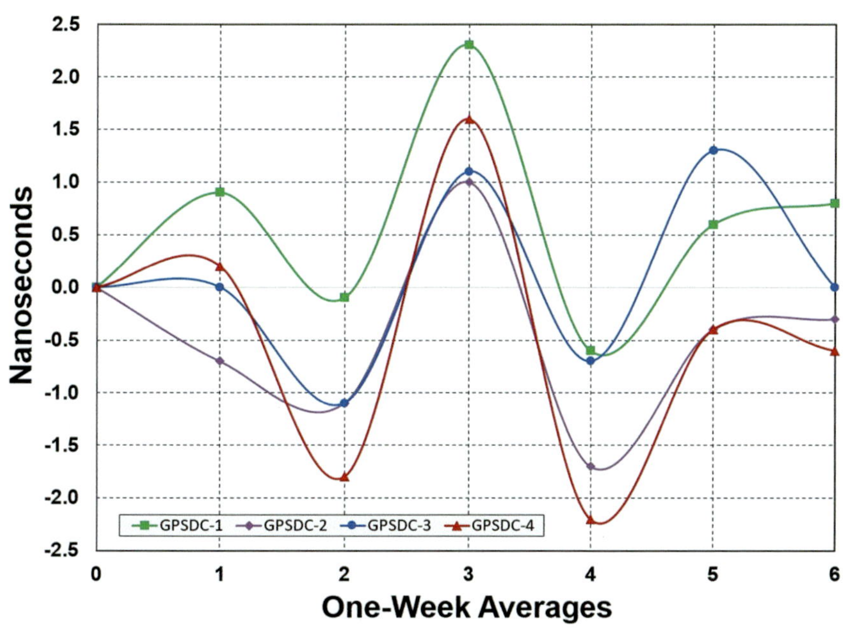

Figure 6.7 Time accuracy of four GPSDCs after a long-term signal delay calibration [17].

6.2.2 Time Accuracies of Clocks

Clocks based on quartz, rubidium, cesium, or hydrogen maser oscillators all have one thing in common—they can't recover accurate time by themselves. Their 1 pps output won't be aligned with UTC unless it is synchronized to a UTC reference. Prior to synchronization, the time accuracy of the 1 pps signal can be anywhere within a range of ±0.5 s. After synchronization, a clock will accumulate a time difference based on the frequency accuracy of its oscillator (Table 6.3), so keeping a clock synchronized to UTC requires periodic adjustments of either its frequency or its time.

Table 6.4 shows typical accumulated time differences for various clock types during a 1-day period. The small time differences shown in the last column for a GPSDC are marked with an asterisk. They account for the short-term "wiggles" shown in Figures 6.6 and 6.7, but long-term accumulated time differences aren't usually an issue with a GPSDC. The short-term time variations with respect to UTC are caused by various factors that we'll discuss in Chapter 12, but they'll average to near 0 over long periods because GPS is continuously steered to agree with UTC. Therefore, if the signal delays in the GPSDC hardware remain constant, a GPSDC will maintain a constant time accuracy with respect to UTC for an indefinite period.

Table 6.4 Accumulated time difference of various clock types during a 1-day period

	Frequency accuracy (1 day)		Accumulated time difference (1 day)	
Clock type	**Typical**	**Best case**	**Typical**	**Best case**
XO (wristwatch)	$\pm 1 \times 10^{-5}$	$\pm 1 \times 10^{-6}$	±1 s	±100 ms
TCXO	$\pm 5 \times 10^{-7}$	$\pm 1 \times 10^{-7}$	±50 ms	±10 ms
OCXO	$\pm 5 \times 10^{-8}$	$\pm 5 \times 10^{-10}$	±5 ms	±50 μs
Rubidium	$\pm 5 \times 10^{-9}$	$\pm 5 \times 10^{-12}$	±500 μs	±500 ns
Active hydrogen maser	$\pm 5 \times 10^{-12}$	$\pm 5 \times 10^{-14}$	±500 ns	±5 ns
Cesium beam	$\pm 5 \times 10^{-13}$	$\pm 2 \times 10^{-14}$	±50 ns	±2 ns
Cesium fountain	$\pm 1 \times 10^{-15}$	$\pm 1 \times 10^{-16}$	±100 ps	±10 ps
GPSDC	$\pm 1 \times 10^{-13}$	$\pm 2 \times 10^{-14}$	< 10 ns *	< 2 ns *

6.3 Measurement System Accuracy

As we discussed at the beginning of this chapter, the *VIM* defines accuracy as the "closeness of agreement between a measured quantity value and a true quantity value" [3]. It follows then that an accurate measurement system, when given a true quantity value as an input, should output a measured quantity value that closely agrees with the true value [18]. Therefore, we can test and validate the accuracy of a measurement system by using it to measure a known quantity.

We can test the accuracy of a time or frequency measurement system by using input signals that have a known time or frequency difference. For example, we can validate the accuracy of a TIC system used to measure frequency in the time domain (Section 5.2.3) by using any stable oscillator that generates both 10 MHz and 1 pps signals, a low-noise frequency synthesizer that can accept a 10 MHz input signal, and a low-noise frequency divider that can divide by 10^7. As shown in Figure 6.8, the 1 pps signal from the frequency standard is connected to the stop channel on the TIC. The 10 MHz signal from the frequency standard is connected to the external time base input of the frequency synthesizer, which is configured to produce an output signal of 10 MHz with a small frequency offset. The output of the frequency synthesizer is then input to the frequency divider, and the 1 pps output of the divider is connected to the start channel on the TIC.

The smallest known quantity that can be measured with this method depends on the resolution of the frequency synthesizer. If the synthesizer has 1 μHz resolution, applying a 1 μHz offset to the 10 MHz input signal results in a frequency difference of 1×10^{-13} between the signals connected to the TIC's start and stop channels. If the slope of the time difference values is estimated, and the data are graphed as shown in Figure 6.4, then an accurate measurement system should return a result of $f_{\text{diff}} = +1 \times 10^{-13}$. If the measurement system isn't capable of measuring that small of a frequency difference, the offset of the synthesizer can be increased. For example, we could increase the offset to 10 μHz to see if the system can measure a frequency difference of 1×10^{-12}.

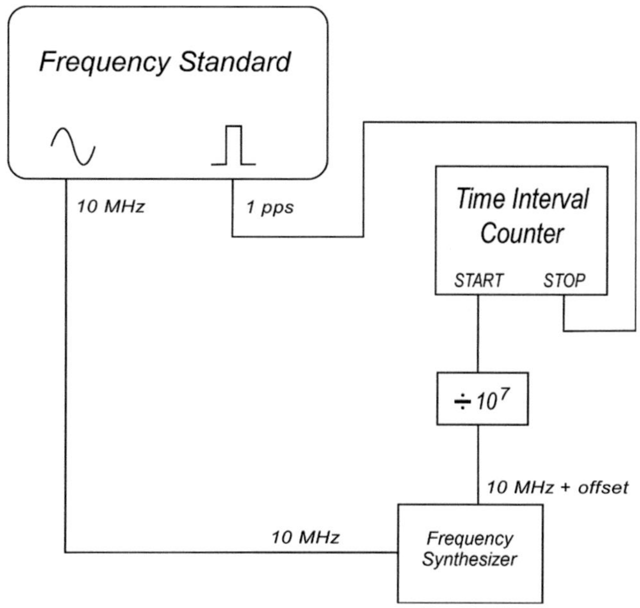

Figure 6.8 Testing a frequency measurement system by measuring a known quantity.

These types of tests can determine the usable accuracy of a measurement system and the amount of averaging that's needed to get the correct answer. They can also serve as diagnostic tests that prove that a measurement system is working properly. The same basic method can be used to test the accuracy of a time transfer system before deploying it to a remote location. For example, Figure 6.9 is a block diagram of a common-view GPS time transfer system (Section 4.3.3), where the two GPSDCs are compared to signals that have a known frequency difference due to the offset introduced by a synthesizer [19].

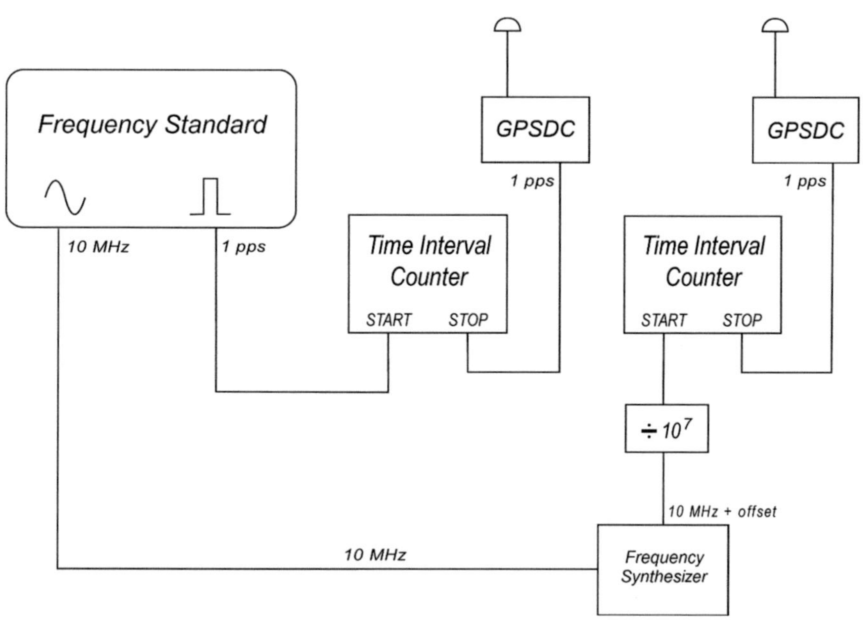

Figure 6.9 Testing a common-view GPS time transfer system by measuring a known quantity.

Accuracy is extremely important to metrologists, and it's one of the two characteristics that are most often used to describe clock and oscillator performance. Stability is the other characteristic, and it's the topic of Chapter 7.

References

1. Bureau International des Poids et Mesures, "The International System of Units (SI)," 9th edition, ver. 2.01, December 2022.
2. R. Buccianti, M. Cibien, L. Mari, and B. I. Rebaglia, "Accuracy, Trueness, and Precision: Considerations Based on the International Vocabulary of Metrology (VIM, 3rd edition) and Related Standards," *Proceedings of the XIX IMEKO World Congress: Fundamental and Applied Metrology*, pp. 1065–1069, September 2009.
3. Joint Committee for Guides in Metrology (JCGM), "International vocabulary of metrology – Basic and general concepts and associated terms (VIM)," *JCGM 200:2012*, 3rd edition, 2012.
4. B. N. Taylor and C. E. Kuyatt, "Guidelines for Evaluating and Expressing the Uncertainty of NIST Measurement Results," *National Institute of Standards and Technology Technical Note 1297*, 1994 edition, 24 p., September 1994.
5. S. Bell, "The Beginner's Guide to Uncertainty of Measurement," *National Physical Laboratory Good Practice Guide No. 11*, no. 2, 33 p., March 2001.
6. C. Müller-Schöll, "Accuracy – Review of the Concept and Proposal for a Revised Definition," *Acta IMEKO*, vol. 9, no. 5, pp. 414–418, December 2020.
7. United States Department of Defense, "Gage Inspection," *MIL-STD-120*, 198 p., December 1950.
8. J. L. Hayes, "Factors Affecting Measurement Reliability," *U. S. Naval Ordnance Laboratory Technical Memorandum No. 63–106*, 13 p., October 1955.
9. M. Pizzocaro, F. Bregolin, P. Barbieri, B. Rauf, F. Levi, and D. Calonico, "Absolute Frequency Measurement of the $^1S_0$–$^3P_0$ Transition of ^{171}Yb with a Link to International Atomic Time," *Metrologia*, vol. 57, pp. 035007/1–11, May 2020.
10. N. Nemitz, T. Gotoh, F. Nakagawa, H. Ito, Y. Hanado, T. Ido, and H. Hachisu, "Absolute Frequency of ^{87}Sr at 1.8×10^{-16} Uncertainty by Reference to Remote Primary Frequency Standards," *Metrologia*, vol. 58, pp. 025006/1–13, February 2021.
11. H. Leopardi, K. Beloy, T. Bothwell, S. M. Brewer, S. L. Bromley, J.-S. Chen, S. A. Diddams, R. J. Fasano, Y. S. Hassan, D. B. Hume, D. Kedar, C. J. Kennedy, D. R. Leibrandt, A. D. Ludlow, W. F. McGrew, W. R. Milner, D. Nicolodi, E. Oelker, T. E. Parker, J. M. Robinson, S. Romisch, J. A. Sherman, L. Sonderhouse, J. Yao, J. Ye, X. Zhang, and T. M. Fortier, "Measurement of the $^{27}Al^+$ and ^{87}Sr Absolute Optical Frequencies," *Metrologia*, vol. 58, pp. 015017/1–8, January 2021.
12. M. A. Lombardi, "Measuring the Frequency Accuracy and Stability of WWV and WWVH," *QST*, vol. 107, no. 3, pp. 33–37, March 2023.
13. T. E. Parker, "Long-Term Comparison of Caesium Fountain Primary Frequency Standards," *Metrologia*, vol. 47, no. 1, pp. 1–10, February 2010.
14. J. Jespersen and J. Fitz-Randolph, "From Sundials to Atomic Clocks: Understanding Time and Frequency," *National Institute of Standards and Technology Monograph 155*, 304 p., 1999 edition, March 1999.
15. S. Bregni, "Measurement of Maximum Time Interval Error for Telecommunications Clock Stability Characterization," *IEEE Transactions on Instrumentation and Measurement*, vol. 45, no. 5, pp. 900–906, October 1996.
16. The data and uncertainty information provided here is publicly available, obtained from the *Circular T* published monthly by the BIPM, from 1996 to the present. The current *Circular T*, and an archive of past publications, is available at: https://www.bipm.org/en/time-ftp/circular-t.
17. M. A. Lombardi, B. R. Patla, A. N. Novick, D. Matsakis, and S. J. Mitchell, "A Simple Method for Calibrating GPS-Disciplined Clocks via Direct Comparison to a UTC(k) time scale," *URSI Radio Science Bulletin*, no. 378, pp. 51–58, September 2021.
18. J. P. Bentley, *Principles of Measurement Systems*, Pearson Education Limited, 4th edition, 2005.
19. M. A. Lombardi and A. N. Novick, "Comparison of the One-Way and Common-View GPS Measurement Techniques Using a Known Frequency Offset," *Proceedings of the 34th Annual Precise Time and Time Interval Meeting (PTTI)*, pp. 39–51, December 2002.

Chapter 7

Stability

I'll stay a week or two
I'll stay the summer through
But I am telling you
I must be going

— Bert Kalmer and Harry Ruby, "Hello, I Must Be Going"

Stability is a characteristic of an object, situation, or condition that makes it unlikely to move or change. When something is stable, we can expect it to be almost the same in the future as it is in the present. Stability is related to solidity, staying power, and steadiness. For example, a house built on stable ground with a stable foundation has solidity. It has walls and ceilings that aren't likely to move or shift, and it can sometimes last for hundreds of years. A stable friendship, relationship, or marriage has staying power. It can survive the ups and downs of everyday life without breaking up or weakening, and it can sometimes last for an entire lifetime. A stable job can provide a steady income and peace of mind without the constant fear of not being able to pay the bills. Generally speaking, stability makes it easier for us to face the future without anxiety, and most of us would agree that stability leads to a higher quality of life.

It should come as no surprise, then, that stability also leads to a higher quality of oscillators and clocks. In fact, stability is an inherent measure of oscillator quality, because unlike accuracy, the stability of an oscillator can't be improved by adjustment. As we discussed in Chapter 6, the accuracy of a clock or oscillator during a given period can never be better than its stability during the same period. Therefore, stability is the foundation for accuracy. It tells us how accurate an oscillator or clock could possibly be, if we were somehow able to perfectly adjust its frequency or time.

When measuring frequency or time, stability indicates how well an oscillator or a clock maintains the same frequency or time difference with respect to the measurement reference over a given period. A frequency or time difference that "stays the same" during a given period is stable, regardless of whether the difference is large or small, or in other words, regardless of whether or not the oscillator or clock is accurate. In fact, stable oscillators and clocks might not be accurate, and devices that are temporarily accurate might not be stable. Confusing stability specifications with accuracy specifications is a common mistake, a mistake sometimes made by even experienced metrologists.

As Time Goes By: A Comprehensive Guide to Time and Frequency Measurements
Michael A. Lombardi

ISBN 978-981-5352-28-3 (Hardcover), 978-1-042-01532-0 (eBook)
www.jennystanford.com

DOI: 10.1201/9781042015320-10

To help us better understand the differences between stability and accuracy, Figure 7.1 provides four examples, each accompanied by two illustrations. The first illustration is based on the fixed target analogy described in Chapter 6, which in this case is a dartboard with its bullseye surrounded by concentric rings. The second illustration is a Cartesian *x*/*y* graph, where either frequency or time is the quantity indicated by the *y*-axis, and where 0 represents either the ideal nominal frequency of an oscillator or the time of arrival of the UTC second. The *x*-axis indicates the measurement period.

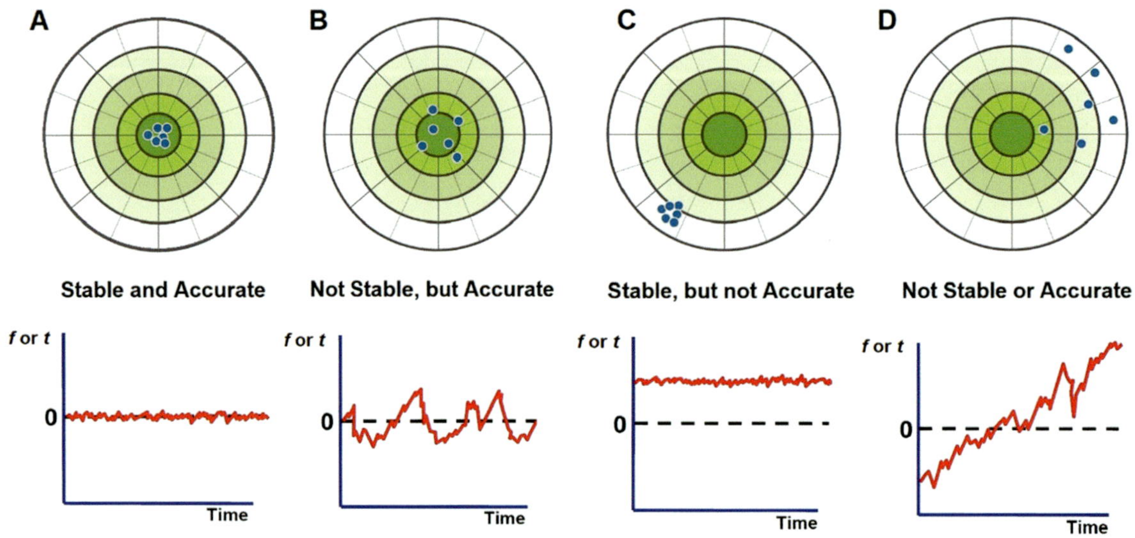

Figure 7.1 Examples of the differences between stability and accuracy.

In example A of Figure 7.1, the measurements are all inside the bullseye, and they form a straight line, aligned on 0, on the *y*-axis of the graph. We can think of example A as a representation of a primary standard such as a cesium fountain clock (Section 3.8). Because a primary standard directly realizes the SI second, it simultaneously achieves both accuracy and stability across all measurement periods.

In example B, the measurements fall either inside or just outside of the bullseye. When graphed, the measurements fluctuate on both sides of 0, but they always stay within a bounded range. This indicates a signal that sometimes is extremely accurate, and that is always accurate within its bounded range. However, the signal isn't stable during short periods. We can think of example B as a representation of the frequency of a GPSDO, or the time of a GPSDC whose signal delays have been calibrated and removed (Section 6.2.1). These devices aren't especially stable during short periods, but they are both accurate and stable when averaged for long periods, because they are steered to agree with UTC. Other types of oscillators and clocks disciplined to a UTC source, including those controlled by LF or HF radio signals, can also be represented by example B. In those cases, their short-term instabilities will be larger than they are with GPS devices due to the effects of terrestrial radio propagation. However, if we allow for a larger bounded range on the dartboard and a larger *y*-axis scale on the graph, the illustration shown in example B still works.

In example C, the measurements are tightly clustered in a space near the edge of the dartboard, far away from the bullseye, with very little variation between them. The graph shows measurements that form a straight line far away from 0. Both illustrations represent an inaccurate oscillator or clock that produces the wrong frequency or time, but that is always wrong by the same amount. We can think of example C as a representation of the frequency of a hydrogen maser (Section 3.10), which has exceptional stability, but outputs a frequency that might differ from its nominal value by a relatively large amount. For example, the frequency stability of a hydrogen maser for a given period might be 1×10^{-15}, but its frequency accuracy for the same interval might be 5,000× worse (5×10^{-12}). Because the maser in this example is so stable, it's frequency could be adjusted to agree with a primary standard or with UTC to obtain far better accuracy.

In example D, the measurements are scattered around the dartboard, with only one measurement coming near the bullseye. The graph shows measurements with a linear trend that's interrupted by some large short-term variations, and that briefly zig zags around 0 on the y-axis. Both illustrations represent an oscillator or clock that is neither stable or accurate over a long period, but that can still be temporarily accurate over a short period, briefly producing frequency very close to its nominal value, or time that agrees with UTC. We can think of example D as a low-quality quartz oscillator whose frequency is neither accurate nor stable. If a clock referenced to this oscillator is synchronized to UTC, it will only stay accurate for a short period before it starts to rapidly accumulate a time difference.

With these basic concepts in mind, we'll continue by describing the statistics used to estimate frequency and time stability, beginning with frequency.

7.1 Estimating Frequency Stability with Statistics

To estimate oscillator stability, we need statistics that can quantify the dispersion of its frequency during a given period. Going back to the dartboard analogy, an oscillator is stable if we repeatedly measure its frequency and the results are tightly clustered in a small area. If the measurement results are spread across a large area, the oscillator is unstable. As new measurements are added, the statistics used as metrics to estimate stability should update their estimates as a function of dispersion. A decreasing metric means that the dispersion is decreasing and the stability is getting better. An increasing metric means that the dispersion is increasing and the stability is getting worse.

For most types of measurements, dispersion is estimated with the variance (σ^2) and its square root, the standard deviation (σ), each expressed mathematically with the Greek letter sigma (σ). These classical statistics estimate dispersion as the numerical spread of a data set with respect to its mean value. The standard deviation is calculated as

$$\sigma = \sqrt{\frac{\Sigma_{i=1}^{N}(y_i - \mu)^2}{N}}, \tag{7.1}$$

where y_i is an individual data point in a data set with a population of N, and where μ is the mean value of all of the data points in the data set.

If a data set has a normal distribution, its data points are symmetrically distributed around their mean value. The normal distribution was first described by the German mathematician Carl Friedrich Gauss (1777–1855) in 1823, so it's also called the Gaussian distribution [1]. The probability density in a normal distribution is highest near the mean, which simply means that a data point is more likely to occur near the mean than far away from the mean. Normal distribution curves have the familiar bell-shape shown in Figure 7.2, where the highest density of data points occurs near the mean. About 68.27% of the data points fall within one standard deviation (1σ) of the mean, and about 95.45% fall within 2σ. Nearly all of the data points in a normal distribution—about 99.73%—will fall within 3σ of the mean. What we now call the standard deviation was originally called the "mean error" by Gauss. The term standard deviation was introduced in lectures presented in the 1890s by Karl Pearson (1857–1936), the British statistician whose name was previously given to the correlation coefficient described in Section 6.1.1.

The variance and standard deviation have countless applications in metrology and everyday life. They only work properly, however, when applied to stationary data where the mean value is relatively constant, because they estimate dispersion with respect to the mean. Obtaining a stationary set of measurements requires measurements that are time independent, which means that their properties aren't dependent on the time when the measurement was made. The measurement data can follow a cyclical pattern, but only if the pattern doesn't include a time-dependent trend that is continuously moving in the same direction. Stationary data is a white noise process, consisting of measurements that are time independent, which means that the measurements will look essentially the same regardless of when they were observed [2].

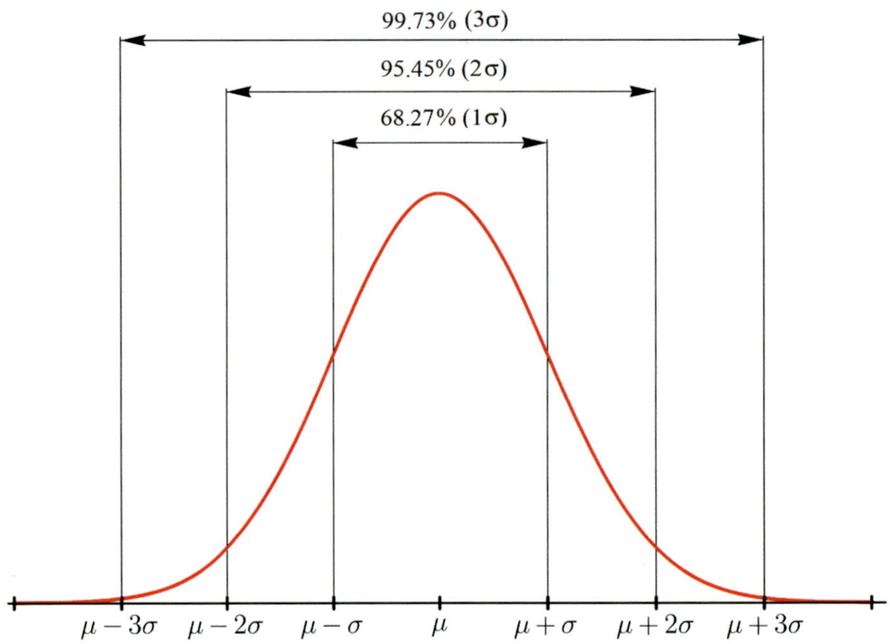

Figure 7.2 The normal distribution curve and the standard deviations from the mean value.

When classical statistics are applied to non-stationary measurement data, the results are misleading. A simple example involves the time-dependent measurements of the height of a tree. A tree continues to grow throughout its lifetime, so its height is a function of its age, although its growth rate gradually slows as the tree gets older [3]. To illustrate this, Figure 7.3 provides modeled data of a tree that has grown more than 20 m in the first 20 years of its life, with its height increasing each year. Because the tree is always growing, its mean height, and therefore the variance and standard deviation of its height, are essentially meaningless. In this example, its mean height was 6 m at age 10, and 12 m at age 20. If the tree was measured again in future years, its mean height would continue to increase. As long as the tree is alive and still growing, its mean height will never converge to any particular value.

Removing the time-dependent trend from non-stationary data can be accomplished with a two-sample variance, where the difference between two successive data points is taken, rather than taking the difference of each data point with respect to the mean. Therefore, if we begin with a non-stationary set of N data points, and then subtract each data point from the previous data point, we can produce a stationary data set with $N - 1$ data points. We can then estimate dispersion from the stationary data using the standard deviation. When we apply successive differencing to the tree height data shown in Figure 7.3, we can estimate the rate of dispersion of the tree's annual growth (Figure 7.4). In this example, the mean growth rate for the 20-year period is about 1.1 m per year, with a standard deviation of about 0.2 m.

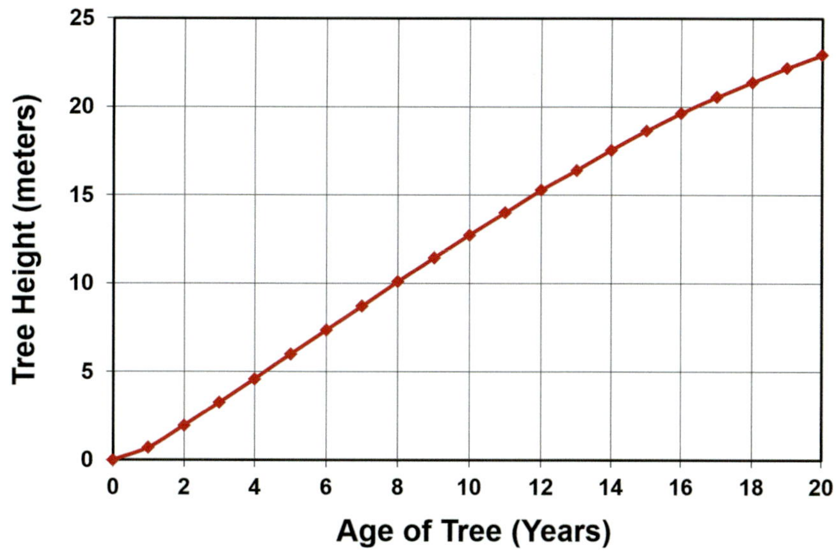

Figure 7.3 Modeled data of the height of a tree as function of time.

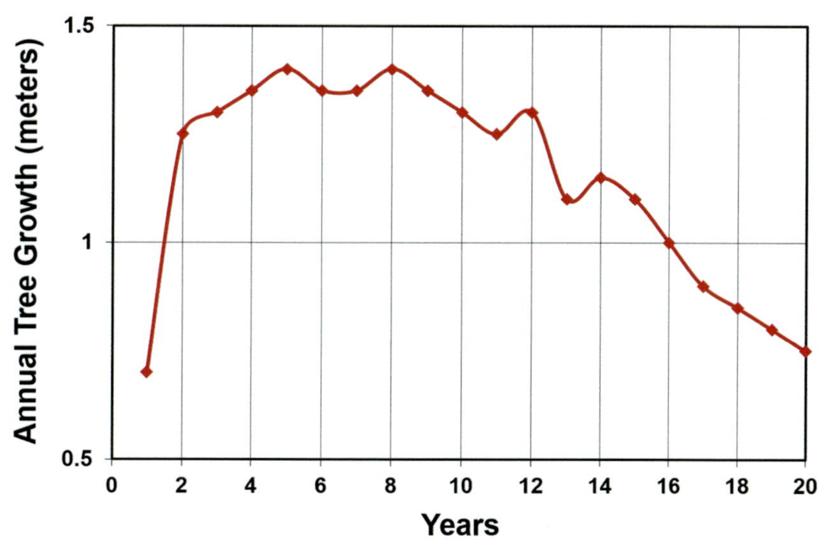

Figure 7.4 Estimating the growth rate of a tree by successive differencing of height data.

Using successive differencing to characterize dispersion was a method that likely was first applied to ballistics, the science related to projectiles and firearms, beginning in the late 19th century. The concept is usually credited to the French military officer and ballistician Emmanuel Vallier (1849–1921), and it proved invaluable in determining the accuracy of field artillery. As noted in the *Handbook of Ballistics*, published by Cranz and Becker in 1921:

> *"It might happen that during the tests some variable disturbance is at work, causing the mean point of impact to alter its position, or effecting a continuous change of the arithmetic mean. Thus the temperature might rise, or the bore become heated, or the velocity of the wind might change. In such a case the method of calculating from the arithmetic mean would give no results, and unless the whole series of observations is to be abandoned, recourse must be had to the method of successive differences. Vallier has pointed out that the calculation of the probable deviation from the successive differences is independent of any change in position of the mean point of impact."* [4]

The method of successive differences first appeared in the statistical literature in 1941, introduced by John von Neumann (1903–1957), a Hungarian-American mathematician and scientist whose work was again motivated by ballistics research. While working at the Institute for Advanced Study in Princeton, New Jersey, von Neumann published his mean square successive difference (MSSD) method, in collaboration with three researchers from the U. S. Army's Ballistic Research Laboratory [5]. That same year, von Neumann derived and published the distribution of the ratio of the MSSD to the standard variance [6]. Because the MSSD involved differencing a pair of measurements or observations, it was also known as the "pair variance" or the "two-sample variance." As von Neumann wrote:

> *"There are cases, however, where the standard deviation may be held constant, but the mean varies from one observation to the next. If no correction is made for such variation of the mean, and the standard deviation is computed from the data in the conventional way, then the estimated standard deviation will tend to be larger than the true population value. When the variation in the mean is gradual, so that a trend (which need not be linear) is shifting the mean of the population, a rather simple method of minimizing the effect of the trend on dispersion is to estimate standard deviation from differences."* [5]

Perhaps inspired by the work of Vallier, von Neumann, and others, frequency metrologists began heavily relying on non-classical statistics based on successive differencing in the 1960s, shortly after the introduction of commercial atomic clocks. It was an appropriate approach, because the data obtained from clock and oscillator measurements is usually non-stationary, with a mean and variance that never converges to any particular value, but instead changes whenever new measurements are added [7, 8].

To illustrate this, we can revisit the data from Figure 6.4 in the previous chapter, which showed the time differences from a comparison between an atomic oscillator and a GPSDO. This same data is shown in Figure 7.5, but the time differences now have their midpoint normalized to 0 for purposes of clarity. The range of the data series is ~895 ns, with a minimum value of −447.5 ns and a maximum value of +447.5 ns. At the end of the seven-day period, the mean is very close to the midpoint of 0, but prior to the end of the period the mean was a negative number, moving towards 0 with an obvious time dependency. For example, the mean was −370 ns after one day, −250 ns after three days, and −128 ns after five days. After the data were differenced using overlapping samples, x_2-x_1, x_3-x_2, x_4-x_3, and so on, the time-dependent trend was gone, and the mean converged to near 0 for each of the seven days.

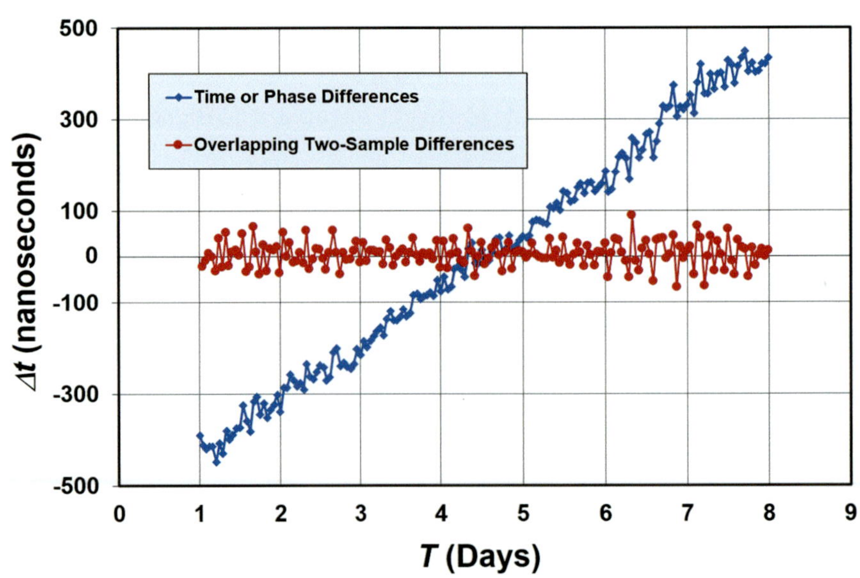

Figure 7.5 Removing the trend from non-stationary data by successive differencing.

The best known statistic in all of time and frequency metrology is called the Allan deviation (ADEV), and it employs the method shown in Figure 7.5 to remove the trend from non-stationary data. The Allan deviation is a two-sample deviation that's well suited for frequency metrology because it's convergent for most types of oscillator noise. In the years since its introduction, ADEV has become an internationally recognized metric for estimating and specifying frequency stability [9].

The Allan deviation is named after David Allan (b. 1936), the American physicist who first described the statistic and who applied it to the AT(NBS) time scale in the late 1960s. However, much of the foundational work was done by Allan's colleague, the American physicist James Barnes (1933–2002), who had begun working on ensemble time scales before Allan [10]. As was described in Section 4.4, Barnes led the development effort for the first ensemble time scale at NBS, named NBS-A, which consisted of four quartz clocks and one rubidium clock. That time scale was built in 1962, which was about five years before the redefinition of the SI second based on cesium resonance (Section 3.4), and about two years before Hewlett-Packard introduced the first successful commercial cesium beam clock (Section 3.6). Therefore, Barnes was originally focused on estimating quartz clock stability.

In 1964, Barnes and Allan collaborated on a paper describing the difficulties associated with estimating the frequency stability of quartz clocks, which have a very long-term random noise process with a $\frac{1}{f}$ dependency. They noted that this noise "extends to 1 cycle per year and probably even lower," and that it was non-convergent when estimated with traditional statistics. They stated that while the noise "is not in the realm of 'short-term' frequency fluctuations, it does have a very profound influence on their definition and measurement" [11]. They then proposed new statistical models that they felt could be helpful, and the work formed the basis for Allan's master's degree thesis at the University of Colorado [10, 12].

In February 1966, Barnes and Allan each published closely related papers in a special issue of the *Proceedings of the IEEE* that was dedicated to the topic of frequency stability. The papers were companion documents, and had already been published together in an NBS technical report that was released in September of the previous year [13]. By then, both scientists were focused on atomic clock stability. In his paper entitled "Atomic Timekeeping and the Statistics of Precision Signal Generators," Barnes described the first difference of phase measurements as being proportional to the average frequency for the measurement interval, and the second difference as being related to the drift of the oscillator frequency. Barnes also began considering the second-order differences as metrics for quantifying frequency stability, noting that they were convergent for oscillator noise as the sample size increased [14]. Allan's paper was called "Statistics of Atomic Frequency Standards," which was the same title as his recently completed master's degree thesis [15]. In this paper, Allan took the final steps towards producing a frequency stability metric based on the two-sample method [10]. While doing so, he provided experimental data obtained from measurements of ammonia masers, and from both cesium beam and rubidium clocks.

Barnes went on to become the first chief of the newly formed Time and Frequency division of NBS in September 1967, a position he held until 1982. Allan went on to lead the time scale efforts at NBS/NIST until retiring in 1992. In 1971, Barnes successfully led an Institute of Electrical and Electronics Engineers (IEEE) sub-committee that proposed using the statistic that we now call ADEV as a metric for estimating frequency stability [16]. This led to the establishment of ADEV as an IEEE standard for estimating frequency stability, a status it has now had for more than 50 years. From the 1970s to the present, the datasheets for all types of oscillators—both quartz and atomic—have included frequency stability specifications that are based on ADEV.

Defined as a time domain statistic, ADEV is essentially the standard deviation of the differences between successive frequency measurements over a given period. The period is expressed mathematically using the Greek letter tau (τ), and ADEV is expressed mathematically as $\sigma_y(\tau)$, where y denotes frequency. When applied to a series of phase or time measurements, the successive differencing is actually done twice. The first-order differences are time differences (Δt), and the second-order differences are frequency differences (Δf), as shown in Table 7.1 and as first described by Barnes [14].

Table 7.1 First and second-order differences obtained from a series of time measurements

Time measurements (ns) sampled at 1 s intervals (x_i)	First order differences (ns) ($\Delta t = x_{i+1} - x_i$)	Frequency measurements ($y_i = \frac{\Delta t}{1\ s}$)	Second-order differences ($\Delta f = y_{i+1} - y_i$)
331.44	—	—	—
332.51	1.07	1.07×10^{-9}	—
333.55	1.04	1.04×10^{-9}	-3×10^{-11}
334.60	1.05	1.05×10^{-9}	$+1 \times 10^{-11}$
335.67	1.07	1.07×10^{-9}	$+2 \times 10^{-11}$
336.72	1.05	1.05×10^{-9}	-2×10^{-11}
337.78	1.06	1.06×10^{-9}	$+1 \times 10^{-11}$
338.85	1.07	1.07×10^{-9}	$+1 \times 10^{-11}$
339.91	1.06	1.06×10^{-9}	-1×10^{-11}
340.96	1.05	1.05×10^{-9}	-1×10^{-11}

The basic ADEV equation [7, 9, 16, 18] for determining frequency stability is

$$\sigma_y(\tau) = \sqrt{\frac{1}{2}\langle \bar{y}_{i+1} - \bar{y}_i \rangle^2}, \tag{7.2}$$

where $\bar{y}_i$ is the average frequency difference of the DUT with respect to the reference, measured over the sampling or averaging period τ. We can also think of τ as the time interval between the beginnings of two successive measurements of the average frequency [16]. Because ADEV is a function of the measurement period, τ should always be defined when reporting or specifying stability.

The differencing of successive frequency measurements (shown in the rightmost column of Table 7.1) is an essential step when estimating stability, because it removes the trend contributed by the oscillator's inaccuracy, and converts non-stationary data to stationary data. We can see from Eq. (7.2) that ADEV squares the second-order differences, just as the standard deviation, Eq. (7.1), squares the differences between each measurement and the mean. The angle brackets $\langle\ \rangle$ in Eq. (7.2) imply that measurements are averaged for an infinite period, from $i = 1$ to $i = \infty$ (infinity). This is, of course, impossible in practice so ADEV is estimated from a finite data set as

$$\sigma_y(\tau) \cong \sqrt{\frac{1}{2(M-1)}\sum_{i=1}^{M-1}(\bar{y}_{i+1} - \bar{y}_i)^2}, \tag{7.3}$$

where M is the number of frequency measurements. ADEV can also be estimated from a series of time difference measurements, x_i to x_N, as

$$\sigma_y(\tau) \cong \sqrt{\frac{1}{2(N-2)\tau^2}\sum_{i=1}^{N-2}(x_{i+2} - 2x_{i+1} + x_i)^2}, \tag{7.4}$$

where N is the number of measurements, and where the measurements are again spaced by the interval τ [7, 9, 16–18].

As shown in Table 7.1, time difference measurements are first-order differences, and frequency measurements are second-order differences; therefore Eq. (7.4) processes one more measurement than Eq. (7.3), and $N = M + 1$. If we apply Eq. (7.4) to the x_i data series in the leftmost column of Table 7.1, the frequency stability is estimated as 1.17×10^{-11} at $\tau = 1$ s, a number about 100× smaller than the frequency accuracy of 1×10^{-9} over the same interval. This represents a common situation, because as previously noted, the accuracy of an oscillator can never be better than its stability, and is often orders of magnitude worse, especially in the case of free-running quartz and rubidium oscillators.

In Eqs. (7.3) and (7.4), ADEV is estimated using non-overlapping samples. This means that consecutive measurements never share samples, which makes them completely independent of each other. However, in nearly all cases—including most examples published in the scientific literature and in the examples found throughout this book—ADEV is calculated with overlapping samples. Measurements obtained from overlapping samples are not completely independent of each other, but by increasing the number of measurements and the associated degrees of freedom, the overlapping process decreases the width of the confidence interval for a stability estimate. The simplest approximation of a confidence interval, *CI*, without considering the type of oscillator noise, is

$$CI = \pm\frac{\sigma_y(\tau)}{\sqrt{N}}, \tag{7.5}$$

where N is the number of frequency measurements used to estimate ADEV [18].

To understand why more measurements are available with overlapping samples, consider that the fixed period at which measurements are taken, which is also the shortest period where we can estimate frequency stability, is called τ_0. For example, if we take samples every second, then $\tau_0 = 1$ s. The non-overlapping and overlapping versions of ADEV each provide identical frequency stability estimates at τ_0 because both statistics are working with the same number of samples. However, it's customary to average measurements over a longer period by using a positive integer, m, as an averaging factor (Figure 7.6), so that $\tau = m\tau_0$ [18]. As m increases, the ratio of the number of samples available to the overlapping and non-overlapping versions of ADEV increases at essentially the same rate, as shown in Table 7.2.

Table 7.2 A comparison of the number of frequency measurements for non-overlapping and overlapping versions of ADEV, calculated from time difference data, when N = 1,000

	Number of frequency measurements		
Averaging factor (m)	Overlapping samples ($N - 2m$)	Non-overlapping samples $\left(\frac{N}{m}-2\right)$	Ratio
1	998	998	1:1
2	996	498	2:1
4	992	248	4:1
8	984	123	8:1
16	968	61	~16:1
32	936	30	~29:1
64	872	14	~62:1
128	744	6	124:1

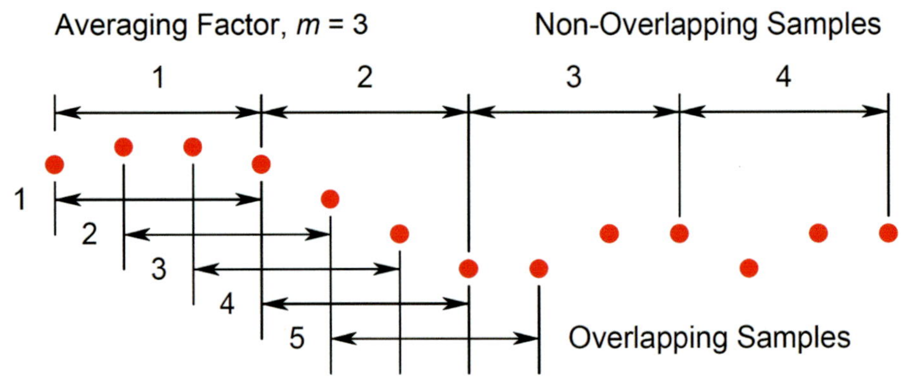

Figure 7.6 Overlapping and non-overlapping samples [18].

The equation for estimating ADEV from a series of frequency measurements with overlapping samples is

$$\sigma_y(\tau) \cong \sqrt{\frac{1}{2m^2(M-2m+1)} \sum_{i=1}^{M-2m+1} \left\{ \sum_{j=1}^{j+m-1} (\bar{y}_{j+m} - \bar{y}_j)^2 \right\}}, \tag{7.6}$$

and the equation for estimating ADEV from a series of time difference measurements with overlapping samples is

$$\sigma_y(\tau) \cong \sqrt{\frac{1}{2(N-2m)\tau^2} \sum_{i=1}^{N-2m} (x_{i+2m} - 2x_{i+m} + x_i)^2}, \tag{7.7}$$

where the averaging factor, m, has been added to Eqs. (7.3) and (7.4) [9, 18].

In 1981, Allan and Barnes published the Modified Allan deviation (Mod ADEV), expressed mathematically as Mod $\sigma_y(\tau)$. A two-sample deviation that always uses overlapping samples, Mod ADEV processes more frequency measurements at long averaging periods than non-overlapping ADEV, but slightly less than overlapping ADEV. Each frequency measurement is obtained with a triangular weighting function, instead of the uniform average used by the original versions of ADEV. The triangular weighting function has a period of 2τ and a height of $\frac{1}{\tau}$ which makes the calculation slightly more complex. The increased complexity was more of a concern in 1981 than it is today, and the authors noted that while it was easy to program, it took longer to compute than $\sigma_y(\tau)$. They then reassured the reader that this was "only of significance for the smaller computer and handheld calculator" [19].

The equation for estimating Mod ADEV [9, 17–19] from a series of frequency measurements is

$$\text{Mod}\ \sigma_y(\tau) = \sqrt{\frac{1}{2m^4(M-3m+2)} \sum_{j=1}^{M-3m+2} \left\{ \sum_{i=j}^{j+m-1} \left(\sum_{k=1}^{i+m-1} [\bar{y}_{k+m} - \bar{y}_k] \right) \right\}^2}, \tag{7.8}$$

where M is the number of frequency measurements. Mod ADEV can also be estimated from a series of $N = M + 1$ time difference measurements as

$$\text{Mod}\ \sigma_y(\tau) = \sqrt{\frac{1}{2m^2\tau^2(N-3m+1)} \sum_{j=1}^{N-3m+1} \left\{ \sum_{i=j}^{j+m-1} [x_{i+2m} - 2x_{i+m} + x_i] \right\}^2}. \tag{7.9}$$

To avoid confusion when comparing or specifying oscillators, it's important to state which statistic was used, because Mod ADEV generally provides lower frequency stability estimates than ADEV. However, Mod ADEV provides the same estimate of frequency stability for τ_0 as both the non-overlapping and overlapping versions of ADEV. A key benefit of Mod ADEV is its improved characterization of oscillator noise (Section 7.1.1). In addition, Mod ADEV provides the foundation for the Time deviation (TDEV), which we'll discuss in Section 7.2. As a result, it has been utilized as a standard estimator of both frequency and time stability in the telecommunications industry since the early 1990s [20].

Other statistics descended from ADEV have been developed to either improve the characterization of oscillator noise, or to improve long-term stability estimates without making additional measurements. The best known of these statistics are the Hadamard deviation, $H\sigma_y(\tau)$, and the Total deviation, $\sigma_{\text{total}}(\tau)$, each of which exists in several different forms [9, 18].

The Hadamard deviation was named in honor of the French mathematician Jacques Hadamard (1865–1963) and his Hadamard Transform, and was first described in 1971 [21]. The Hadamard deviation is a three-sample deviation that examines the second-order differences of the frequency measurements, or the third-order differences of the phase measurements. Unlike ADEV, it is insensitive to linear frequency drift, an advantage when characterizing rubidium clocks. Since the 1990s, it has been used to help characterize the clocks in the GPS satellite constellation, which have mostly been rubidium devices [22, 23].

The Total deviation was proposed in 1995 by the American physicist Dave Howe (b. 1949) as a way to improve the confidence of ADEV's stability estimates when τ is greater than 10% of the total measurement period. It forms a longer virtual data series by reflecting existing data and adding it to both ends of the existing data series. This increases the number of samples and the associated degrees of freedom at long averaging periods [24, 25].

As we have seen, ADEV and its related statistics were not new concepts even when they were first proposed in the 1960s, because they were based on the previously existing two-sample variance and deviation. However, by addressing the shortcomings of using classical statistics for frequency metrology, ADEV provided a standard, internationally accepted metric that made it easy to compare the stabilities of different types of oscillators. As we'll describe in the next section, ADEV also enabled two extremely useful aspects of frequency metrology, the ability to estimate frequency stability for any averaging period, and the ability to characterize oscillator noise.

7.1.1 Graphing Frequency Stability and Characterizing Oscillator Noise

Frequency stability graphs based on ADEV or related statistics use common base 10 logarithms to display data that extends across multiple orders of magnitude. They plot log $\sigma_y(\tau)$ on the y-axis to indicate unitless frequency stability, and log τ on the x-axis to indicate the averaging period. For this reason, frequency stability graphs are often colloquially referred to as "sigma-tau" graphs. On most ADEV graphs, the τ values are spaced using either the octave method or the decade method. For the octave method where $\tau_0 = 1$, the sequence is 1, 2, 4, 8, 16, 32, 64, and so on. For the decade method, octaves are still used, but they start over at all powers of 10, so for $\tau_0 = 1$, the sequence is 1, 2, 4, 10, 20, 40, 100, and so on. These methods save computational time, which was a significant issue when the statistics were introduced. However, computers are much more powerful now, so it's common to use the all-tau method, where ADEV is calculated for every possible value of τ.

Figures 7.7 and 7.8 provide sample graphs that estimate stability with overlapping ADEV and that provide confidence intervals that are indicated by the dashed lines. In both graphs, the confidence intervals get wider as τ increases, because the longer-term stability estimates are based on fewer measurements. Both graphs also show that the frequency stability improves as τ increases, until a noise floor is reached and averaging more measurements no longer improves the stability estimates.

Figure 7.7 graphs the frequency stability of a 32.768 kHz quartz wristwatch. To improve its performance, the wristwatch oscillator is regulated using a method called inhibition compensation, where the quartz oscillator is intentionally run at a frequency slightly higher than nominal, and to compensate, a small number of cycles are skipped at regular intervals with digital logic [26]. The graph uses the octave method, and τ ranges from 60 s to about three days. The stability is estimated as 6.7×10^{-8} at $\tau = 60$ s, and the noise floor is reached when τ is equal to about 1 hour.

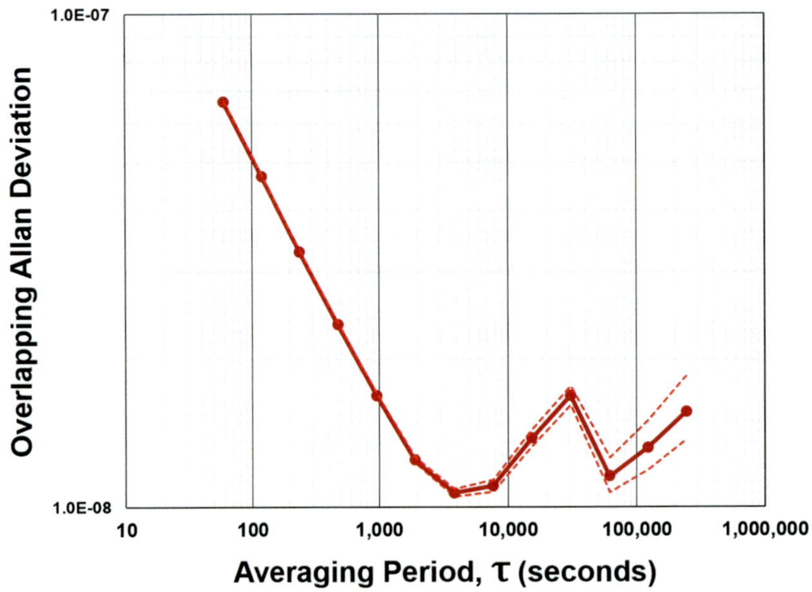

Figure 7.7 Quartz wristwatch oscillator stability (octave method with confidence intervals).

Figure 7.8 graphs the stability of a free-running rubidium with a nominal frequency of 10 MHz. The rubidium oscillator is free running. The decade method is used for this graph, and τ ranges from 1 s to 10,000 s. The stability is estimated as 3.3×10^{-12} at $\tau = 1$ s. The noise floor of about 6×10^{-13} is reached when $\tau = 1{,}000$ s.

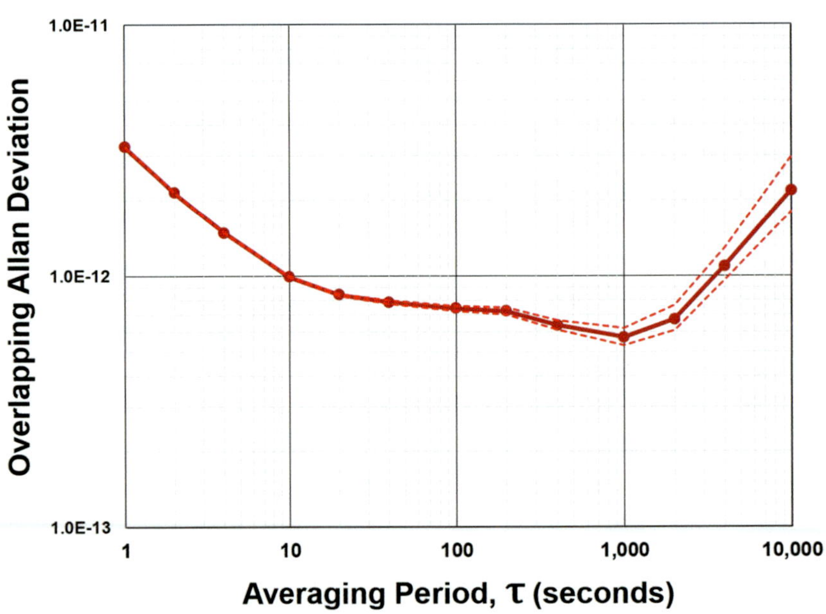

Figure 7.8 Rubidium oscillator stability (decade method with confidence intervals).

Figure 7.9 provides graphs of the accuracy (a) and stability (b) of three 5 MHz cesium oscillators as measured over a 60-day period. In (a), the accuracy of the three devices is shown to vary considerably, and cesium C is nearly two orders of magnitude more accurate than cesium A. However, the overlapping ADEV graph in (b) indicates that their stability is in relative agreement. This again demonstrates that accuracy is a function of how an oscillator's frequency has been adjusted. The all-tau method without confidence intervals is used for the graph, and τ ranges from 720 s to about two weeks. The measurement period isn't long enough for the cesium oscillators to reach a noise floor, but all three devices are stable to about 2×10^{-14} at $\tau = 10^5$ s, which is in agreement with the manufacturer's specification of 2.7×10^{-14} for the same period [27].

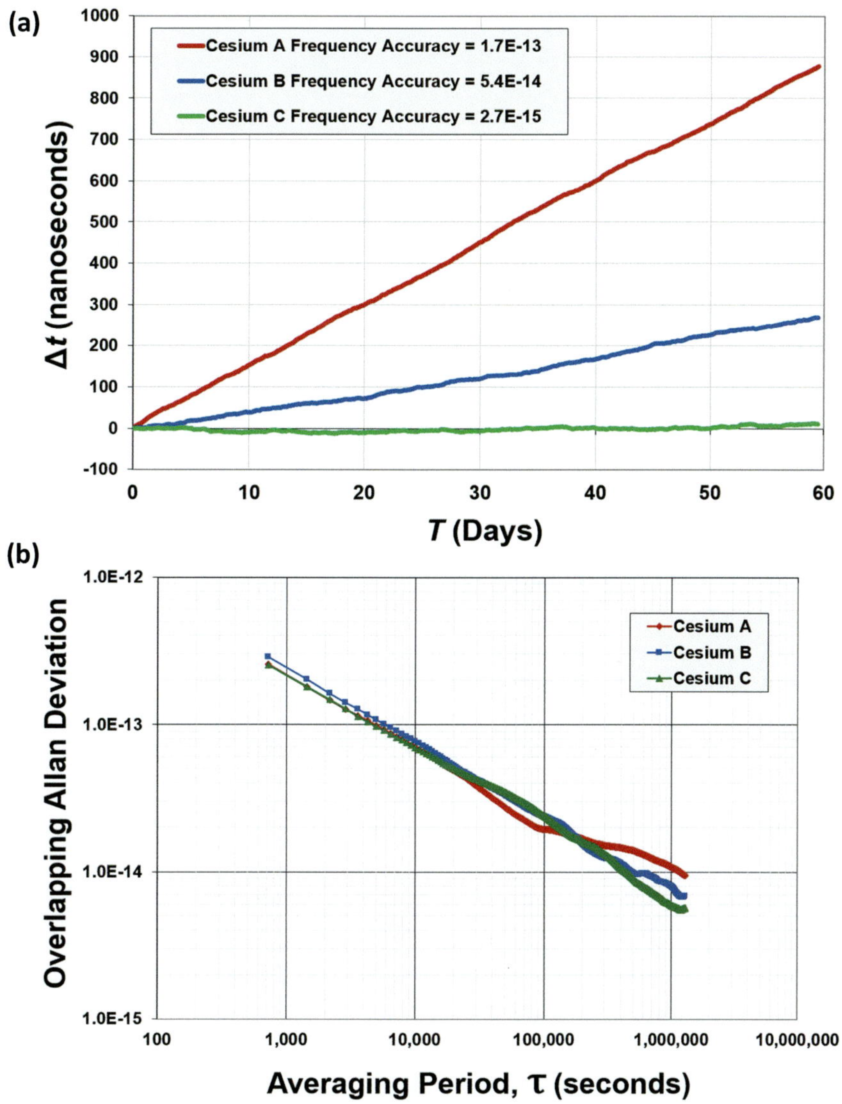

Figure 7.9 The accuracy (a) and stability (b) of three cesium oscillators measured for 60 days.

In addition to estimating frequency stability, ADEV can help identify and characterize the various types of oscillator noise. Five types of oscillator noise are commonly identified and discussed: white phase, flicker phase, white frequency, flicker frequency, and random walk frequency. Figure 7.10 illustrates how the noise type is identified from the slope of the curve on an ADEV graph. The slope is determined by raising τ to the power of $\frac{\mu}{2}$, where μ is the power law exponent for the time domain [18].

Oscillator phase noise is separated into two components, white phase and flicker phase. White phase noise consists of phase fluctuations that are evenly distributed across the frequency band of interest with no correlation in time. Because white noise has equal power at all frequencies within the band of interest, its power spectral density is flat, and it averages down at a rate proportional to the averaging period. Therefore, if a measurement is dominated by white phase noise, the stability improves by a factor of 10 when τ is increased by a factor of 10. White noise was given its name because it's somewhat analogous to white light, where every band of the visible spectrum is equally bright, and where the combined colors appear to be white.

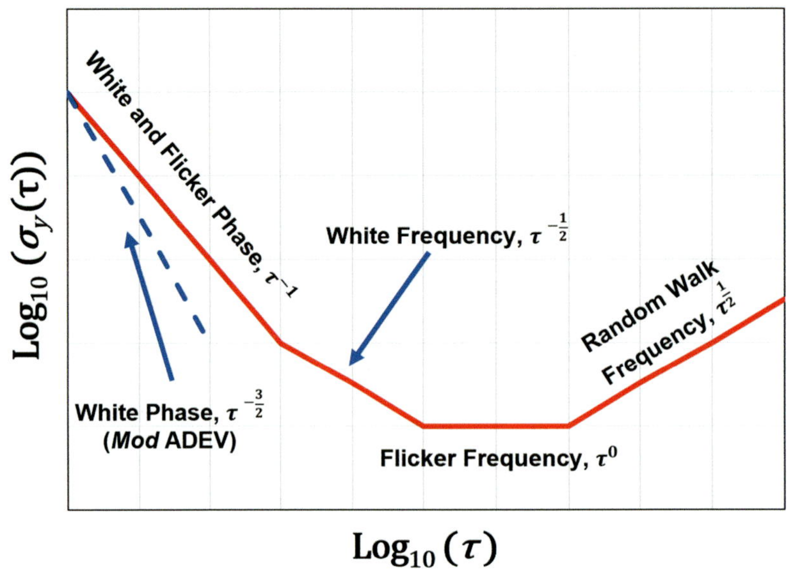

Figure 7.10 Characterizing the type of oscillator noise from the slope of the ADEV curve.

Flicker phase noise, also known as $\frac{1}{f}$ phase noise, isn't evenly distributed across the frequency band. Unlike white noise which has equal power at all frequencies, flicker noise has more power at lower frequencies because its power spectrum is inversely proportional to f. This means that as the frequency of the noise goes up, the intensity of the noise goes down. The power of the noise decreases at a rate of 3 dB per octave of frequency, which means that if the frequency of the noise doubles, the power of the noise is halved. This type of noise is also called pink noise, because visible light with this same power spectrum has a pink appearance.

Because it's a non-stationary random noise process that naturally occurs in a diverse group of systems, $\frac{1}{f}$ noise has long been a topic of discussion in electronic devices. It was first noticed as an excess of low-frequency noise in vacuum tubes [28]. The discussion in electronic devices ever since. The initial observations of noise in vacuum tubes are attributed to Walter Schottky (1886–1976), a German physicist who invented the screen grid vacuum tube in 1915 in an effort to make a more stable RF amplifier. Schottky was studying the current fluctuations in vacuum tubes in 1918 when he noticed the effect now called shot noise, which is white current noise that has the same power at all frequencies.

The discovery of $\frac{1}{f}$ noise can be attributed to the Swedish-American physicist John Bertrand Johnson (1887–1970). While working at Bell Telephone Laboratories in 1925, Johnson began studying Schottky's predictions of shot noise in vacuum tubes. Johnson's experiments demonstrated that the predictions held true at higher frequencies, but noted that "the effect is much larger in the lower range of frequencies than the theory predicts," and his measurements showed a dependency that was inversely proportional to frequency [29, 30]. The following year, Schottky used the word "flicker" to describe this effect, while acknowledging and attempting to confirm Johnson's work:

"If we had to do with emission of light instead of electrons, we would speak of a chaotic variation of light intensity taking place over the surface of the cathode, a phenomenon which we should describe by the word 'flicker'. If, as I shall endeavor to further confirm, Johnson's explanation of the phenomenon is the correct one, then we may use the analogy and call the new effect the 'flicker effect'." [31]

As shown in Figure 7.10, ADEV can't distinguish between white phase and flicker phase noise, because it identifies both as having a slope of τ^{-1}. However, Mod ADEV can make this distinction, identifying white phase noise as having a steeper slope of $\tau^{-3/2}$. For most measurements the distinction isn't critical, because both types of phase noise will average down rapidly as τ increases, but Mod ADEV indicates that white phase noise averages down faster than white flicker noise.

Frequency noise is also separated into white and flicker components. White frequency noise consists of frequency fluctuations that average down as τ increases, because like white phase noise, the noise is evenly distributed across the frequency band. However, the rate of improvement due to averaging is slower for frequency noise than phase noise. For white frequency noise, it's proportional to the square root of the averaging period, with the ADEV curve having a slope of $\tau^{-1/2}$.

Flicker frequency noise is $\frac{1}{f}$ frequency noise, and it doesn't average down at all. It represents the lowest and flattest part of the ADEV curve, where the slope is τ^{0}. This is the noise floor we referred to earlier when discussing Figures 7.7 and 7.8. Frequency metrologists often use "flicker" as an adjective when discussing the noise floor, and sometimes call it the "flicker floor," the "flicker frequency floor," or the "flicker noise floor." Regardless of what it's called, it represents the limit of the oscillator's stability, because flicker frequency noise has a low-frequency limit where it becomes independent of τ [30]. When this limit is reached, there is nothing more to be gained from averaging, so oscillator stability isn't usually estimated or specified for values of τ past this point. If more measurements are averaged, the stability estimates will get worse due to random walk frequency noise, the last of the five types of oscillator noise.

The name random walk is self-descriptive, because it describes frequency noise that causes an oscillator to eventually "walk" away from its stable frequency, by taking a series of successive random frequency steps. The difference between two steps is nearly constant, but the direction of the steps is random, and the steps might eventually return the frequency to the same place where the walk began. The slope for random walk noise is $\tau^{1/2}$, and its power spectrum is inversely proportional to f^2, which means that random walk noise intensifies at low frequencies, even more so than flicker noise.

Confusingly, random walk noise can also be called either red noise or Brown noise. Like white noise and pink noise, the name "red noise" originates from the visible light spectrum. Red light has the longest wavelength of all visible light (~700 nm), so the type of noise most prevalent at low frequencies with long wavelengths is called red noise. The term "Brown noise" is capitalized because it doesn't refer to the color brown, but instead was named in honor of the Scottish botanist Robert Brown (1773–1858). In 1827, Brown described the jittery, random motion of grains of pollen immersed in water that he'd observed while using a microscope. This became known as Brownian motion, perhaps the first example of what we now know as random walk noise. The term random walk was introduced by Pearson in 1905 [32].

The five types of oscillator noise are summarized in Table 7.3. It's useful to consider what the five noise types look like when viewed on a phase or time difference graph, because it illustrates the types of measurement data that can have their noise level reduced by averaging. As shown in Figure 7.11, white phase, flicker phase, and to a lesser extent, white frequency noise, are stationary noise types where averaging more measurements improves the results. However, even though ADEV has already removed the trend contributed by frequency inaccuracy from the measurements, flicker frequency and random walk frequency are still non-stationary noise types, which means that they don't converge to any particular value. All free-running oscillators exhibit non-stationary frequency noise when measured for a long enough period. This simply indicates that an oscillator's frequency accuracy can't stay the same forever. It eventually changes due to factors both internal and external to the oscillator.

Table 7.3 The five types of oscillator noise identified in frequency stability analysis

Noise type	Description	ADEV slope
White phase	Fluctuations in the phase of a signal that are evenly distributed across a given bandwidth. The stability improves at a rate proportional to the averaging period. Unlike ADEV, Mod ADEV can distinguish between white and flicker phase noise, and identifies white phase noise as having a slope of $\tau^{-3/2}$.	τ^{-1}
Flicker phase	Also known as $\frac{1}{f}$ phase noise. Unlike white phase noise, flicker phase noise is not evenly distributed across the frequency band. As the frequency goes up, the intensity of the noise goes down, and vice versa. For example, if the frequency is halved, the power of the noise doubles.	τ^{-1}
White frequency	Fluctuations in the frequency of a signal that have the same power at all frequencies across a given bandwidth. The stability is still improving, but at a rate proportional to the square root of the averaging period.	$\tau^{-1/2}$
Flicker frequency	Also known as $\frac{1}{f}$ frequency noise. Here, the oscillator has reached a noise floor that shows its best possible stability, and there nothing to be gained by more averaging. Oscillator specification sheets seldom include stability estimates past this point.	τ^{0}
Random walk frequency	Successive random steps in frequency. The difference between two steps is nearly constant, but the direction of the steps is random. The oscillator stability can have long periods of apparent up or down trends, with sudden and unpredictable changes in direction, perhaps eventually returning to where the walk process began.	$\tau^{1/2}$

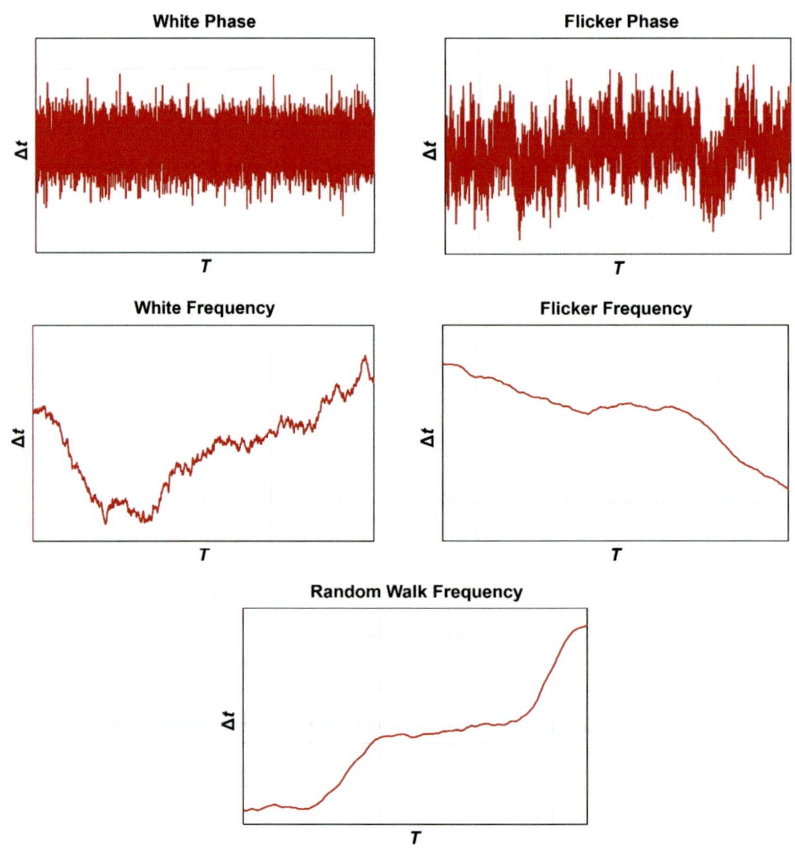

Figure 7.11 Oscillator noise, as it appears on a phase or time difference graph.

Nearly all of the examples provided in this book estimate frequency stability in the time domain, by utilizing ADEV or a related statistic. As previously noted, ADEV was defined as a time domain statistic, and it was designed to meet the needs of timekeeping, time transfer, time scales, frequency control, and the other applications that are the main focus of this book. However, as we've touched upon earlier in this section, oscillator noise is related to power spectral density (PSD), and frequency stability can be estimated in the frequency domain in terms of PSD. When this is done, stability is characterized as a function of f, which in this case is the Fourier or sideband frequency [18]. When discussing spectrum analyzers in Section 5.2, we noted that Fourier proved that periodic waveforms are composed of a series of sine waves, with each component having its own frequency. The frequencies of these components are known as Fourier frequencies. The random phase and frequency of an oscillator, and therefore its stability, can be modeled as

$$S_y(f) = h_\alpha f^\alpha, \tag{7.10}$$

where $S_y(f)$ is the PSD of the frequency fluctuations, y, on one side of the carrier frequency, h_α is the intensity or amplitude of the noise, f is the Fourier or sideband frequency, and α is the power law exponent for the frequency domain [9, 18, 33, 34].

As is the case with ADEV, the five basic types of oscillator noise in a frequency domain characterization are identified by the slope of the curve on a log-log graph. For ADEV, the slope is indicated by raising τ to the power of $\frac{\mu}{2}$ (Table 7.3). For PSD graphs, which are log-log graphs of $S_y(f)$ versus f, the slope is indicated by raising f to the power of α The exponents that determine the noise slopes for frequency and time domain graphs are summarized in Table 7.4.

As shown in Table 7.4 and Figure 7.12, the order of the noise types is reversed in the frequency domain when compared to the time domain, and the noise floor is now white frequency as opposed to flicker frequency noise. As we can glean from the discussion earlier in this section, random walk and flicker frequency noise intensify at low Fourier frequencies near the carrier, and along with the white frequency noise floor, are usually the noise types of most interest in the frequency domain. Equations for converting $S_y(f)$ to $\sigma_y(\tau)$ or Mod $\sigma_y(\tau)$ are provided in the *IEEE Standard 1139* [9] and elsewhere [18, 35].

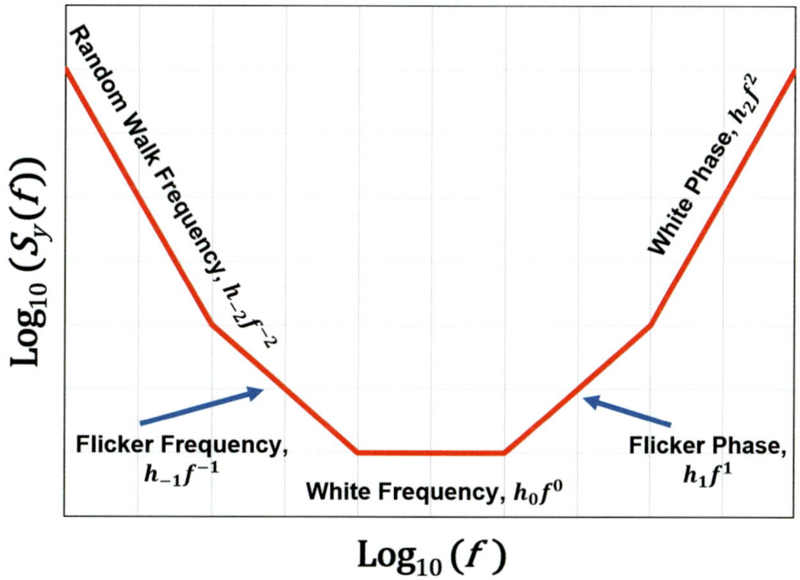

Figure 7.12 Characterizing the type of oscillator noise from the slope of the PSD curve.

Table 7.4 Slope characteristics of oscillator noise in the frequency and time domains

Oscillator noise type	Slope characteristics (exponents) of log-log graphs		
	Frequency domain	Time domain	
	$S_y(f)$	$\sigma_y(\tau)$	Mod $\sigma_y(\tau)$
	a	$\frac{\mu}{2}$	$\frac{\mu}{2}$
Random walk frequency	−2	1/2	1/2
Flicker frequency	−1	0	0
White frequency	0	−1/2	−1/2
Flicker phase	1	−1	−1
White phase	2	−1	−3/2

7.1.2 Frequency Stabilities of Various Types of Oscillators

It's common for metrologists to refer to the short-term, medium-term, and long-term stability of oscillators. Short-term stability generally refers to periods ranging from less than 1 s to 100 s. Medium-term stability generally refers to periods ranging from 100 s to one day, and long-term stability generally refers to periods of one day or longer. Quartz oscillators, especially OCXOs, usually have good short-term stability. However, they usually have poor medium-term stability and they always have poor long-term stability. Rubidium oscillators have good short and medium-term stability, and much better long-term stability than the best quartz oscillators, but their long-term stability is still quite poor when compared to a cesium beam or hydrogen maser. Table 7.5 summarizes the typical frequency stabilities for high-quality examples of the various oscillator types, with a graph provided in Figure 7.13. A GPS disciplined oscillator is also included for comparison purposes.

Table 7.5 Frequency stabilities of various types of oscillators

τ (seconds)	Typical frequency stability, $\sigma_y(\tau)$				
	OCXO [36]	Rubidium	GPSDO	Cesium beam [27]	Hydrogen maser [37]
1	5×10^{-13}	3×10^{-12}	5×10^{-12}	5×10^{-12}	8.5×10^{-14}
10	8×10^{-13}	1×10^{-12}	2×10^{-12}	3.5×10^{-12}	2×10^{-14}
100	3×10^{-12}	7×10^{-13}	1×10^{-12}	8.5×10^{-13}	5×10^{-15}
1 000	1×10^{-11}	6×10^{-13}	8×10^{-13}	2.7×10^{-13}	2×10^{-15}
10 000	3×10^{-11}	2×10^{-12}	3×10^{-13}	8.5×10^{-14}	1.5×10^{-15}
100 000	3×10^{-10}	4×10^{-12}	3×10^{-14}	2.7×10^{-14}	1×10^{-15}

For purposes of example, Figure 7.14 compares the stabilities of three signals; an external reference obtained from common-view GPS comparisons with UTC(NIST), a free-running rubidium oscillator, and the same rubidium oscillator when it's being disciplined by the external reference [38]. As shown in the graph, it takes about 30 minutes for the external reference stability to drop below the rubidium stability, so 30 minutes should be the approximate period of frequency corrections. In this example, there is a bump in the ADEV curve of the disciplined rubidium that results from the temporary instability introduced by the frequency correction. However, from that point forward the stability of the disciplined oscillator keeps getting better, and the stability of the free-running oscillator keeps getting worse.

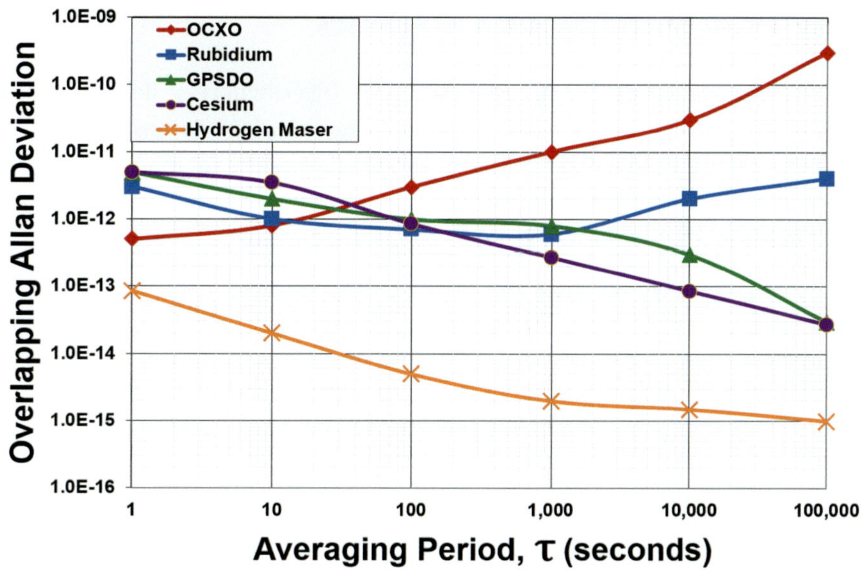

Figure 7.13 Frequency stabilities of various oscillator types.

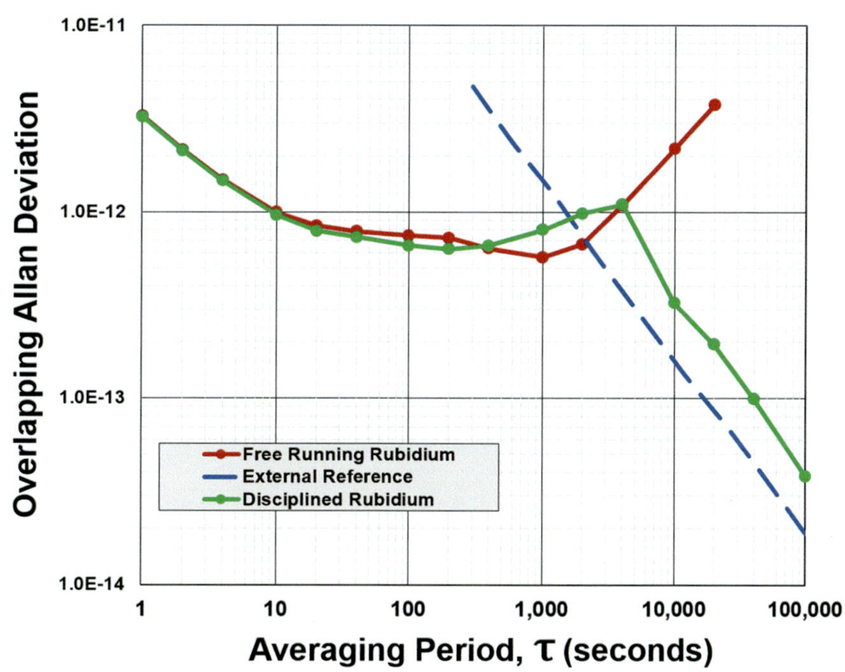

Figure 7.14 External reference stability compared to free-running and disciplined oscillators.

The specification sheets of GPSDOs and other disciplined oscillators always include ADEV values. It's important to know, however, that ADEV was designed and intended for use with free-running oscillators, and the noise characterization process that we've just described doesn't apply to disciplined oscillators. For example, the LO noise that affects the stability of a GPSDO is always being corrected by signals received from the satellites. This means that a GPSDO should never reach a flicker frequency noise floor, nor will it exhibit random walk frequency noise, as long as it continues to receive satellite corrections from the satellites. Over very long periods, such as weeks and months, a GPSDO should be more stable than any free-running oscillator. However, at τ = 1 day, only the best GPSDOs will be as stable as a cesium oscillator, and all GPSDOs will be less stable than a hydrogen maser.

7.2 Estimating Time Stability with Statistics

The Time deviation (TDEV) is expressed mathematically as $\sigma_x(\tau)$, where x denotes time, and is used to estimate the time stabilities of clocks and time transfer systems. TDEV is based on Mod ADEV, and like ADEV and Mod ADEV can be credited to David Allan, who was asked by members of the U. S. telecommunications industry to develop a metric for characterizing time stability in their networks [12]. Allan and his colleagues employed TDEV in their own work in 1990, by first using it to estimate the stability of NIST time signals transmitted over telephone lines [39]. Soon afterwards, TDEV was included in international telecommunication standard documents as a metric for specifying the stability of network time signals [20].

TDEV is defined [9] as

$$\sigma_x(\tau) = \frac{\tau}{\sqrt{3}} \text{Mod}\,\sigma_y(\tau), \tag{7.11}$$

where multiplying Mod $\sigma_y(\tau)$ by $\frac{\tau}{\sqrt{3}}$ normalizes TDEV so that it equals the classical standard deviation for white phase noise processes when $\tau = \tau_0$.

The presence of a reasonable amount of white phase noise is acceptable in a clock because it doesn't cause its time difference to continuously increase with respect to UTC. Instead, the time differences will both increase and decrease, and converge to a mean value with a magnitude that can be reduced by averaging [34, 39]. For example, the time distributed by terrestrial and satellite radio signals has day-to-day variations, known as diurnal effects, that are largely white phase noise processes. The magnitude of this noise is reduced by averaging, especially when τ is an even multiple of one day.

7.2.1 Graphing Time Stability and Characterizing Clock Noise

Time stability graphs based on TDEV plot log $\sigma_x(\tau)$ on the y-axis to indicate the time stability in units of seconds, and log τ on the x-axis to indicate the averaging period, also in seconds. A TDEV graph looks very similar to an ADEV graph, as both are log-log graphs that display data extending across multiple orders of magnitude. However, a key difference between ADEV and TDEV graphs is the typical direction of the curve. The curve on an ADEV graph of a free-running oscillator typically trends downward for a while until a noise floor is reached, but the curve of a TDEV graph for a free-running clock begins trending upwards nearly immediately.

This is illustrated in Figure 7.15, which is a time stability graph of the same free-running rubidium oscillator whose frequency stability was previously graphed in Figure 7.8. When used as a free-running clock, the rubidium is stable to $< 10^{-11}$ s (10 ps) if $\tau < 30$ s. However, the stability (and therefore the potential accuracy) of the free-running rubidium clock rapidly gets worse, exceeding 10^{-8} s (10 ns) when $\tau = 10{,}000$ s. A noise floor of 2 ps is quickly reached, when $\tau \le 2$ s.

The five types of oscillator noise summarized in Table 7.3 are the same five types of clock noise characterized by TDEV, but they are transposed by +1 [18]. This means that 1 is added to each of the exponents of the slope characteristics for Mod $\sigma_y(\tau)$ that were shown in Table 7.4. As a result, white phase noise is the only noise type with a negative exponent, and therefore the only noise type that TDEV can reduce by averaging. White phase noise is characterized as having a slope of $\tau^{1/2}$ on the TDEV curve, so it averages down at a rate proportional to $\sqrt{\tau}$. The noise floor is reached when the dominant noise type is flicker phase, which now has a slope of τ^0.

The other three types of clock noise were originally given the same names as their frequency noise counterparts [34], but because a time stability metric should ideally characterize noise as phase noise rather than frequency noise, they were later given different names. Therefore, random walk phase noise is synonymous with white frequency noise, flicker walk phase noise is synonymous with flicker frequency noise, and random run phase noise is synonymous with random walk frequency noise. Both the original and the newer names are listed in Figure 7.16, which illustrates how clock noise is characterized from the slope of the TDEV curve.

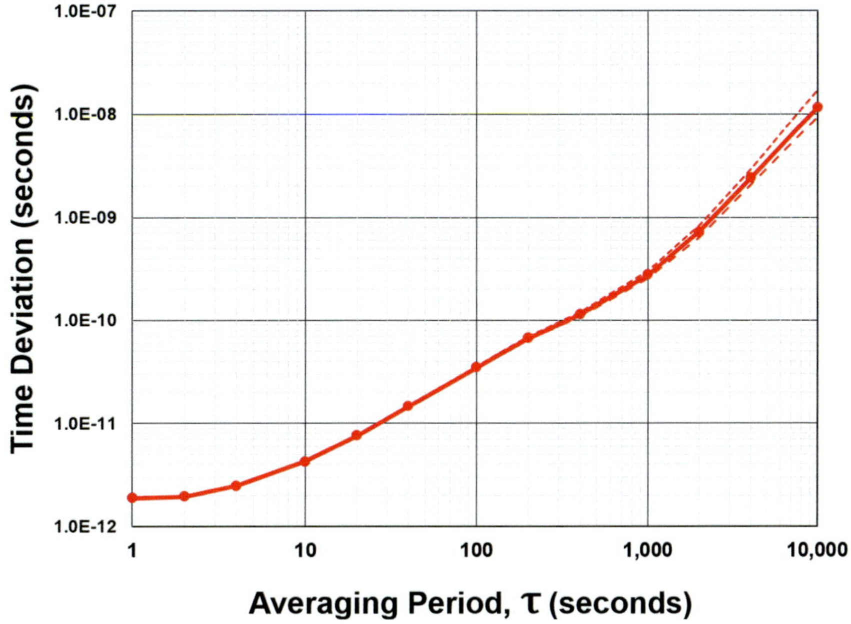

Figure 7.15 Time stability of a rubidium clock as estimated with TDEV.

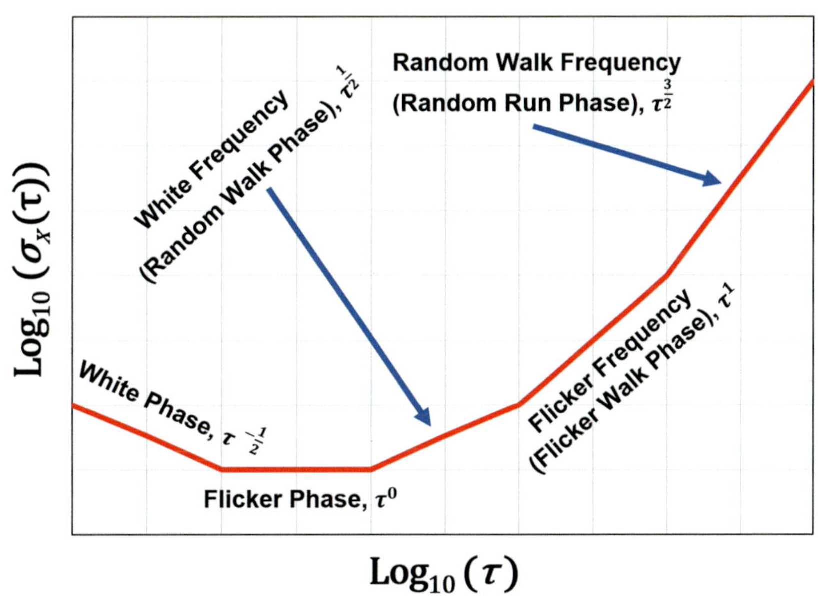

Figure 7.16 Characterizing the type of clock noise from the slope of the TDEV curve.

7.2.2 Time Stabilities of Various Types of Clocks

The time stabilities of various clock types are summarized in Table 7.6 and shown graphically in Figure 7.17. Both the table and graph have converted, $\sigma_x(\tau)$ to units of nanoseconds for ease of comparison. It's easy to see that there's a wide variation in time stability between the various types of free-running clocks. For example, the time stability, and therefore the potential accuracy, of a clock referenced to a free-running OCXO exceeds 10 µs at τ = 1 day, but a hydrogen maser is stable to about 50 ps over the same interval.

Table 7.6 Time stabilities of various types of clocks

	Typical time stability, $\sigma_x(\tau)$, in nanoseconds				
τ (seconds)	OCXO	Rubidium	GPSDC	Cesium beam	Hydrogen maser
1	0.0003	0.002	0.003	0.003	0.00005
10	0.005	0.006	0.01	0.02	0.0001
100	0.2	0.04	0.06	0.05	0.0003
1 000	5.8	0.35	0.46	0.16	0.001
10 000	173	11.5	1.7	0.5	0.009
100 000	17 300	23	1.7	1.6	0.06

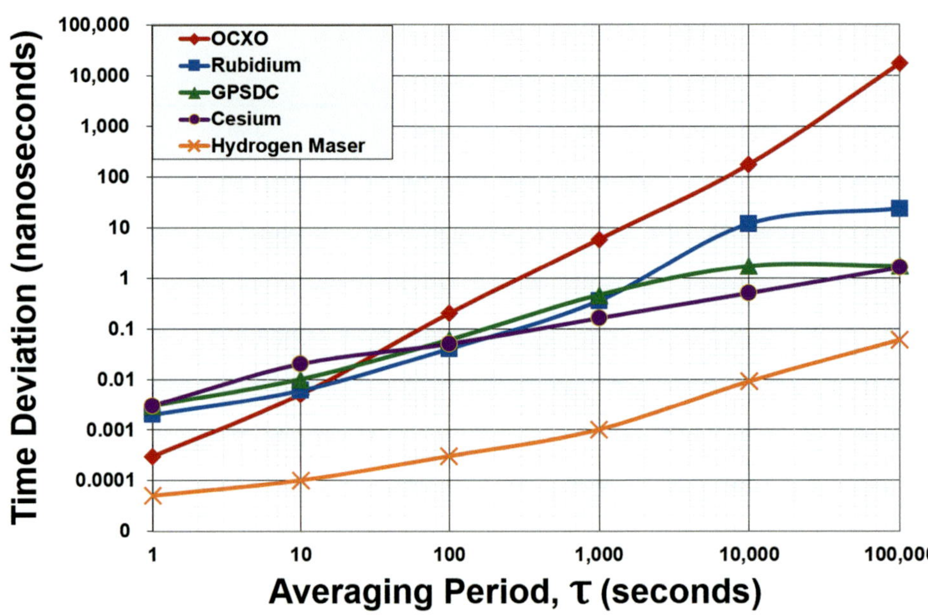

Figure 7.17 Time stabilities of various clock types.

For all five types of clocks, the TDEV curve has an upward trend with the time stability worsening as τ increases. For the free-running clocks, this upward trend will continue indefinitely. However, as we discussed at the beginning of this chapter, a clock disciplined to a UTC source will be accurate within a bounded range and will be stable when averaged for a sufficiently long period. Both things are true in the case of a GPSDC. The sample data provided in Table 7.6 was obtained from a GPSDC that had the same time stability for averaging periods of 10^4 s and 10^5 s, which flattened the end of the TDEV curve as shown in Figure 7.17.

If the time stability of a GPSDC is measured for longer periods, the bounded range of TDEV when applied to a disciplined clock becomes evident. This is illustrated in Figure 7.18, an all-tau TDEV graph that shows data from a GPSDC that was measured for several months, so that τ could be extended to 10^6 s with reasonable confidence. The TDEV curve in this example doesn't have an upward trend. The GPSDC time stability varies, but it stays within a fairly narrow range of 0.6 to 2.1 ns, for averaging periods ranging from one hour to more than 10 days.

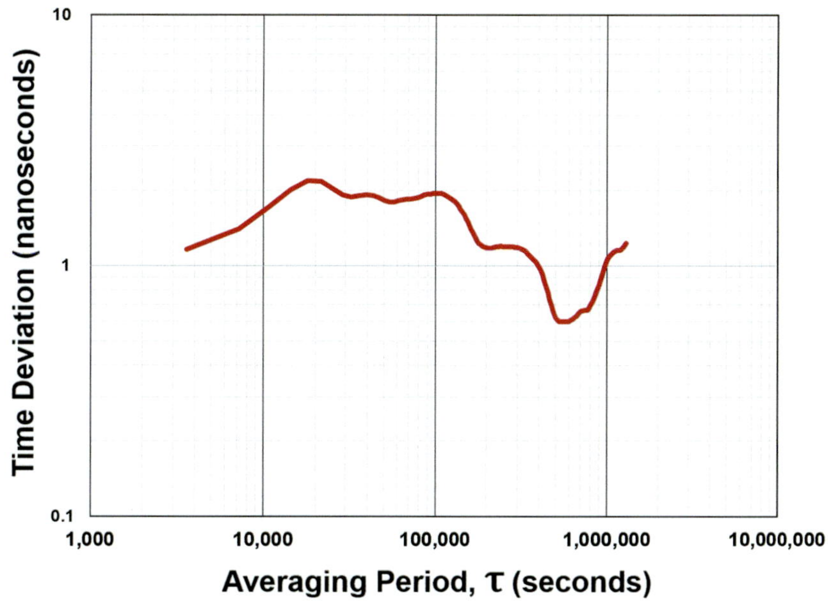

Figure 7.18 Long-term time stability of a GPS disciplined clock.

7.3 Limitations of Systems Measuring Short-Term Stability

The noise floor of a measurement system needs to be lower than the noise floor of the DUT that it measures, and not all frequency measurement systems have enough resolution to correctly measure the short-term stability of an oscillator. To illustrate this, Figure 7.19 compares measurements of the same OCXO obtained with four different frequency measurement systems that were each previously described in Chapter 5—a frequency counter (FC), a time interval counter (TIC), a single mixer heterodyne system, and a dual-mixer time difference (DMTD) system.

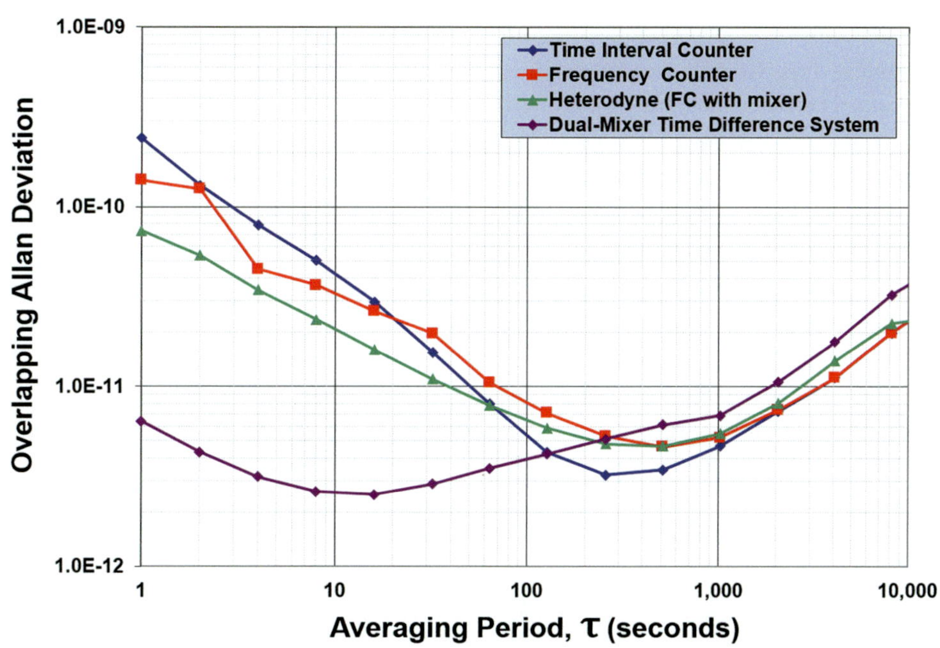

Figure 7.19 Stability estimates of the same OCXO when measured with different systems.

The same universal counter, with a hydrogen maser serving as its external time base, was used for the first three measurements. The counter had a 12-digit display and a time interval resolution of 150 ps. For the frequency counter method, the universal counter was configured as an FC and the OCXO frequency was directly measured. For the time interval method, the counter was configured as a TIC and the OCXO frequency was divided to 1 pps and compared to a 1 pps signal from the hydrogen maser reference. For the mixer method, the counter was configured as an FC, but the incoming OCXO signal was first heterodyned, using a double-balanced mixer, with a signal locked to the hydrogen maser that was offset in frequency by 1 kHz. Finally, a commercially available instrument was used for the DMTD measurement.

As shown in Figure 7.19, only the DMTD system could correctly show that the OCXO stability was about 7×10^{-12} at $\tau = 1$ s, reaching its noise floor of $< 3 \times 10^{-12}$ when $\tau \cong 10$ s. The other three systems had insufficient resolution, and measurement noise that exceeded the OCXO noise at short averaging periods, so they incorrectly indicated that the OCXO stability was between 7×10^{-11} and 1.5×10^{-10} at $\tau = 1$ s. These results were at least one order of magnitude worse than the DMTD measurement in all three cases. The three universal-counter-based systems also incorrectly indicated that the OCXO noise floor isn't reached until after several hundred seconds of averaging, when the DMTD system was correctly indicating that a random walk noise process had already begun. Even so, all four measurement systems produce similar results for averaging periods of several hundred seconds or longer. This demonstrates that the choice of a measurement system is seldom critical when measuring long-term stability, but it's always critical when measuring short-term stability. This is important to realize and remember, because confusing measurement system noise with DUT noise is a very common mistake.

We've now discussed both accuracy and stability, the two primary characteristics used to describe clock and oscillator performance. Both characteristics can help us quantify how much doubt that we have about a specific measurement result. Measurement uncertainty is the topic of Chapter 8.

References

1. C. F. Gauss, *Theoria Combinationis Observationum Erroribus Minimis Obnoxiae* (*Theory of the Combination of Observations Least Subject to Errors*), translated by G. W. Stewart, Society for Industrial and Applied Mathematics (SIAM), 1995.
2. R. J. Hyndman and G. Athanasopoulos, *Forecasting: Principles and Practice*, OTexts, 2nd print edition, June 2018.
3. Q. V. Cao, "Annual Tree Growth Predictions from Periodic Measurements," *Proceedings of the 12th Biennial Southern Silvicultural Research Conference*, General Technical Report SRS-71, pp. 212–215, February 2003.
4. C. Kranz and K. Becker, *Handbook of Ballistics, Volume 1—Exterior Ballistics*, Translated from the 2nd German edition, London: His Majesty's Stationary Office, 1921.
5. J. von Neumann, R. H. Kent, H. R. Bellinson, and B. I. Hart, "The Mean Square Successive Difference," *The Annals of Mathematical Statistics*, vol. 12, no. 2, pp. 153–162, June 1941.
6. J. von Neumann, "Distribution of the Ratio of the Mean Square Successive Difference to the Variance," *The Annals of Mathematical Statistics*, vol. 12, no. 4, pp. 367–395, December 1941.
7. D. W. Allan, "Should the Classical Variance Be Used as a Basic Measure in Standards Metrology?," *IEEE Transactions on Instrumentation and Measurement*, vol. IM-36, no. 2, pp. 646–654, June 1987.
8. J. Jespersen, "Introduction to the Time Domain Characterization of Frequency Standards," *Proceedings of the 23rd Annual Precise Time and Time Interval Meeting (PTTI)*, pp. 83–102, December 1991.
9. IEEE Ultrasonics, Ferroelectrics, and Frequency Control Committee, "IEEE Standard Definitions of Physical Quantities for Fundamental Frequency and Time Metrology — Random Instabilities," *IEEE Standard 1139–2022*, 2022.
10. D. B. Sullivan, "Time and Frequency Measurement at NIST: The First 100 Years," *Proceedings of the 2001 IEEE Frequency Control Symposium*, pp. 4–17, April 2001.
11. J. A. Barnes and D. W. Allan, "Effects of Long-Term Stability on the Definition and Measurement of Short-Term Stability," in *Proceedings of the IEEE-NASA Symposium on the Definition and Measurement of Short-Term Frequency Stability,* NASA SP-80, pp. 119–123, November 1964.

12. D. W. Allan and J. Levine, "A Historical Perspective on the Development of the Allan Variances and Their Strengths and Weaknesses," *IEEE Transactions on Ultrasonics, Ferroelectrics, and Frequency Control*, vol. 63, no. 4, pp. 513–519, April 2016.
13. J. A. Barnes and D. W. Allan, "Two Papers on the Statistics of Precision Frequency Generators," *National Bureau of Standards Report 8878*, September 1965.
14. J. A. Barnes, "Atomic Timekeeping and the Statistics of Precision Signal Generators," *Proceedings of the IEEE*, vol. 54, no. 2, pp. 207–220, February 1966.
15. D. W. Allan, "Statistics of Atomic Frequency Standards," *Proceedings of the IEEE*, vol. 54, no. 2, pp. 221–230, February 1966.
16. J. A. Barnes, A. R. Chi, L. S. Cutler, D. J. Healey, D. B. Leeson, T. E. McGunigal, J. A. Mullen, W. L. Smith, R. L. Sydnor, R. F. C. Vessot, and G. M. R. Winkler, "Characterization of Frequency Stability," *IEEE Transactions on Instrumentation and Measurement*, vol. IM-20, no. 2, pp. 105–120, May 1971.
17. D. W. Allan, "Time and Frequency (Time-Domain) Characterization, Estimation, and Prediction of Precision Clocks and Oscillators," *IEEE Transactions on Ultrasonics, Ferroelectrics, and Frequency Control*, vol. UFFC-34, no. 6, pp. 647–654, November 1987.
18. W. J. Riley, "Handbook of Frequency Stability Analysis," *National Institute of Standards and Technology Special Publication 1065*, 136 p., July 2008.
19. D. W. Allan and J. A. Barnes, "A Modified 'Allan Variance' with Increased Oscillator Characterization Ability," *Proceedings of the 35th Annual Symposium on Frequency Control*, pp. 470–475, May 1981.
20. S. Bregni, "Twenty-Five Years of Applications of the Modified Allan Variance in Telecommunications," in *IEEE Transactions on Ultrasonics, Ferroelectrics, and Frequency Control*, vol. 63, no. 4, pp. 520–530, April 2016.
21. R. A. Baugh, "Frequency Modulation Analysis with the Hadamard Variance," *Proceedings of the 25th Annual Symposium on Frequency Control,* pp. 222–225, April 1971.
22. S. T. Hutsell, W. G. Reid, J. D. Crum, H. S. Mobbs, and J. A. Buisson, "Operational Use of the Hadamard Variance in GPS," *Proceedings of the 28th Annual Precise Time and Time Interval Meeting (PTTI)*, pp. 201–214, December 1996.
23. D. A. Howe, R. L. Beard, C. A. Greenhall, F. Vernotte, W. J. Riley, and T. K. Peppler, "Enhancements to GPS Operations and Clock Evaluations Using a 'Total' Hadamard Deviation," *IEEE Transactions on Ultrasonics, Ferroelectrics, and Frequency Control*, vol. 52, no. 8, pp. 1253–1261, August 2005.
24. D. A. Howe, "An Extension of the Allan Variance with Increased Confidence at Long Term," *Proceedings of the 1995 IEEE International Frequency Control Symposium*, pp. 321–329, May 1995.
25. D. A. Howe, "The Total Deviation Approach to Long-Term Characterization of Frequency Stability," *IEEE Transactions on Ultrasonics, Ferroelectrics, and Frequency Control*, vol. 47, no. 5, pp. 1102–1110, September 2000.
26. M. Lombardi, "The Accuracy and Stability of Quartz Watches," *British Horological Journal*, vol. 150, no. 2. pp. 57–59, February 2008.
27. Microchip Technology Inc., "5071A Primary Frequency Standard Datasheet," October 2022.
28. M. S. Keshner, "1/*f* Noise," *Proceedings of the IEEE*, vol. 70, no. 3, pp. 212–218, March 1982.
29. J. B. Johnson, "The Schottky Effect in Low Frequency Circuits," *Physical Review*, vol. 26, no. 1, pp. 71–85, July 1925.
30. R. F. Voss, "1/*f* (Flicker) Noise: A Brief Review," *Proceedings of the 33rd Annual Symposium on Frequency Control*, pp. 40–46, May 1979.
31. W. Schottky, "Small-Shot Effect and Flicker Effect," *Physical Review*, vol. 28, no. 1, pp. 74–103, July 1926.
32. K. Pearson, "The Problem of the Random Walk," *Nature*, vol. 72, no. 1865, p. 294, July 1905.
33. D. W. Allan, "The Measurement of Frequency and Frequency Stability of Precision Oscillators," *National Bureau of Standards Technical Note 669*, 31 p., May 1975.
34. D. W. Allan, M. A. Weiss, and J. L. Jespersen, "A Frequency Domain View of Time-Domain Characterization of Clocks and Time and Frequency Distribution Systems," *Proceedings of the 45th Annual Symposium on Frequency Control*, pp. 667–678, May 1991.
35. E. Rubiola and F. Vernotte, "The Companion of Enrico's Chart for Phase Noise and Two-Sample Variances," *IEEE Transactions on Microwave Theory and Techniques*, vol. 71, no. 7, pp. 2996–3025, July 2023.

36. Microchip Technology Inc., "MX-043 Oven Controlled Crystal Oscillator Datasheet," September 2023.
37. Microchip Technology Inc., "MHM-2020 Active Hydrogen Maser Datasheet," October 2022.
38. M. A. Lombardi, "A NIST Disciplined Oscillator: Delivering UTC(NIST) to the Calibration Laboratory," *NCSLI Measure: The Journal of Measurement Science*, vol. 5. no. 4, pp. 46–54, December 2010.
39. D. W. Allan, D. D. Davis, J. Levine, M. A. Weiss, N. Hironaka, and D. Okayama, "New Inexpensive Frequency Calibration Service from NIST," *Proceedings of the 44th Annual Symposium on Frequency Control*, pp. 107–116, June 1990.

Chapter 8

Uncertainty

Would I lie to you?
Would I lie to you honey?
Now would I say something that wasn't true?

— Annie Lennox/David Allan Stewart, "Would I Lie to You?"

When properly practiced, metrology is a pure discipline. There are no ulterior motives or agendas other than finding the true value of a measurand. Because truth is the foundation of metrology, the biggest fear that metrologists face is not telling the truth by reporting a false measurement result. Even when it happens accidentally or unintentionally, reporting a false measurement result is equivalent to telling a lie, and lies violate everything that metrology stands for.

The truth-telling business can really get complicated, because the only true values metrologists accept are the unit definitions in the SI. The defined values of a quantity are known, but the true value of a measured quantity is unknowable in principle [1]. The *Guide to the Expression of Uncertainty in Measurement*, the international standard that metrologists call the *GUM*, reminds us that the true value of a quantity is "a value that would be obtained by a perfect measurement" but that "true values are by nature indeterminate" [2]. Because it's not possible to make a perfect measurement, metrologists do the only thing that they can do. They make the best measurements they can using the resources that they have available. To acknowledge that their measurements are imperfect, they then report not only their results, but also the amount of doubt that they have about their results.

Measurement uncertainty is the doubt about the true value of the measurand that still remains after a measurement is completed [3]. It is quantified and described through a process that analyzes the closeness of agreement between the measured quantity and its true quantity value established by the SI definition, the property that we defined as accuracy in Chapter 6. The uncertainty analysis also includes a probability distribution, like those described in our discussion of stability in Chapter 7. Therefore, a true uncertainty analysis quantitatively describes both the accuracy and stability of the device under test.

Before moving on, let's stop for a moment to consider the significance of acknowledging that uncertainty exists, and how meaningful it is when we can quantify the uncertainty. Acknowledging and quantifying

As Time Goes By: A Comprehensive Guide to Time and Frequency Measurements
Michael A. Lombardi

ISBN 978-981-5352-28-3 (Hardcover), 978-1-042-01532-0 (eBook)
www.jennystanford.com

DOI: 10.1201/9781042015320-11

uncertainty are two things that aren't generally practiced in our everyday lives. We are routinely presented with all kinds of written, spoken, and visual information, but none of the information we receive is accompanied by an uncertainty statement. For instance, we might be listening to someone who is telling us things that aren't absolute lies, but that are only about half true. That person—especially when they know they are distorting the truth—isn't going to state that their words have a 50% uncertainty. In another situation, we might be listening to someone who sincerely believes that they're telling the truth, but who still can't quantify the truthfulness of their words because they're simply repeating what they've heard elsewhere. Again, this person's words won't be accompanied by an uncertainty statement.

In sharp contrast to these examples, metrologists are trained to always acknowledge, look for, and quantify uncertainty. They are also trained to get as close to the truth as possible, by reducing the uncertainty as much as possible. In most cases, especially in industrial metrology, a conservative approach is favored that leans toward overestimating measurement uncertainties. However, when a new type of device is first introduced—such as a new type of atomic clock—it might not be possible to accurately measure its performance using previously existing standards and methods. For this reason, and also because of the exuberance that often occurs when scientists publish breakthrough measurement results, measurement uncertainties are sometimes underestimated. Uncertainties that are underestimated might remain undiscovered for years, but they are eventually found and corrected when better measurement tools become available.

If we compare most facets of everyday life to the way that good metrology is practiced, it's easy to see why metrologists have such a satisfying profession. The constant search for truth in measurements provides a sense of integrity that's not always obtainable in other career fields. Keeping these basic thoughts in mind, let's continue by exploring the background, history, and some of the basics of measurement uncertainty analysis.

8.1 History and Basics of Measurement Uncertainty Analysis

It seems likely that humans have an innate desire for the answers to their questions to be exact and unambiguous, because it's comforting to get answers that we can simply trust without further questioning. We know that this is usually true in the case of small children, and it's likely also true for most adults. In fact, even scientists are sometimes slow to embrace the concept of uncertainty. A traditional view of science mirrors everyday life, where uncertainty is often thought of as an undesirable state that should be avoided, and one of the easiest ways to avoid something is to not acknowledge that it exists.

The seeds for measurement uncertainty analysis were planted when classical statistics were developed and first applied by scientists. As we discussed in Section 7.1, Gauss described data that were normally distributed in 1823, and he demonstrated that the highest probability of distribution occurred near the mean value. What we now call measurement uncertainty analysis involves finding a probability distribution that has the true value of the measured quantity somewhere within its range, even though the true value itself is unknowable.

Long before uncertainty analysis was practiced or even conceptualized, it was known that multiple measurements of the same quantity produced different results. Therefore, the preferred practice was to always make several measurements, and to accept the mean value of a group of measurements as the measurement result. This provided more confidence than accepting a single measurement value.

Stopping there would have been convenient, but it was soon learned that even mean values are subject to errors based on the number of measurements that were used to obtain the mean. That observation was made in what can be considered as the first study of measurement uncertainty analysis, which was published in 1908 and entitled "The Probable Error of a Mean" [4]. The author of the study used the pseudonym "Student" instead of his real name, most likely because his employer didn't want their connection to the author to be known, because that would let their competitors know which statistical methods they were using. Surprisingly, about two decades passed before "Student" was identified as William Sealy Gossett (1876–1937), an English chemist, statistician, and brewer [5, 6]. Gossett worked for the Guinness Brewery from 1899 until his death. Most of those years were spent in Dublin, Ireland, but Gossett returned to England for the last part of his career, when he was promoted to the position of Head Brewer for the entire Guinness company.

Gossett was interested in the uncertainty of a measurement result when the number of samples was too small to have a normal distribution. We described the normal distribution curve in Chapter 7, where 68.27% of measurements fall within 1σ of the mean, where 95.45% fall within 2σ, and where 99.73% fall within 3σ (Figure 7.2). A perfect normal distribution curve requires the degrees of freedom, or the number of variables that are unconstrained and therefore free to vary, to be infinite. However, all data sets will start to converge to a normal distribution when the degrees of freedom are sufficiently high.

The degrees of freedom are equal to the number of measurements minus the number of constrained variables. For example, when n measurements are taken to determine a sample mean, the mean is a constrained variable, so the degrees of freedom are equal to $n - 1$. As was discussed in Chapter 7, when the degrees of freedom are increased by taking more measurements, the confidence interval narrows, which means that the mean measurement result has less uncertainty. However, in many situations it's not practical, and sometimes not even possible, to collect enough measurements to form a normal distribution curve.

Gossett addressed this problem by deriving tables of multipliers that could be used to determine confidence intervals equivalent to a normal distribution, even when the number of measurements was small. Because of the byline he used in his original paper [4], his landmark statistical work became known as Student's t distribution. Table 8.1 shows the multipliers that can be applied to a standard deviation obtained from measurements with a limited number of degrees of freedom, to obtain the normal distribution confidence intervals equivalent to 1σ, 2σ, and 3σ. As shown in the table, the standard deviation of a measurement with just five degrees of freedom would need to be multiplied by 1.11 to obtain a 1σ confidence interval, by 2.65 to obtain a 2σ confidence interval, and by 5.51 to obtain a 3σ confidence interval.

Table 8.1 Normal distribution multipliers for measurements with a limited number of degrees of freedom

Degrees of freedom	Probability		
	1σ (68.27%)	2σ (95.45%)	3σ (99.73%)
1	1.84	13.97	235.78
2	1.32	4.53	19.21
3	1.20	3.31	9.22
4	1.14	2.87	6.62
5	1.11	2.65	5.51
10	1.05	2.28	3.96
20	1.03	2.13	3.42
30	1.02	2.09	3.27
40	1.01	2.06	3.20
50	1.01	2.05	3.16
100	1.01	2.03	3.08
1000	1.00	2.00	3.01
∞	1.00	2.00	3.00

The central limit theorem is a key concept in probability theory, and it tells us that if we have a set of data points, x_i, with a mean (μ) and a standard deviation (σ), then the sample mean $\bar{x}$ of a random sample of size n will approach a normal distribution as the sample size gets larger. This is true regardless of how the x_i data are distributed. It's commonly taught in statistic courses and textbooks that the central limit theorem will hold if $n \geq 30$, which is a pragmatic approach to estimating when a data set will converge to a normal distribution [7]. This can be seen in Figure 8.1, where the probability density of a data sample with 30 degrees of freedom looks much like a normal distribution curve, and where t distributions with fewer degrees of freedom are still bell shaped and symmetric, but have a lower probability density near the mean and wider "tails." We can see from Table 8.1 that a measurement with 30 degrees of freedom still has about a 2% error with respect to a normal distribution for a 1σ confidence interval, increasing to about 9% for 3σ. A measurement with 1,000 degrees of freedom is equal to a normal distribution for 1σ and 2σ confidence intervals, and has an error of less than 1% for a 3σ confidence interval.

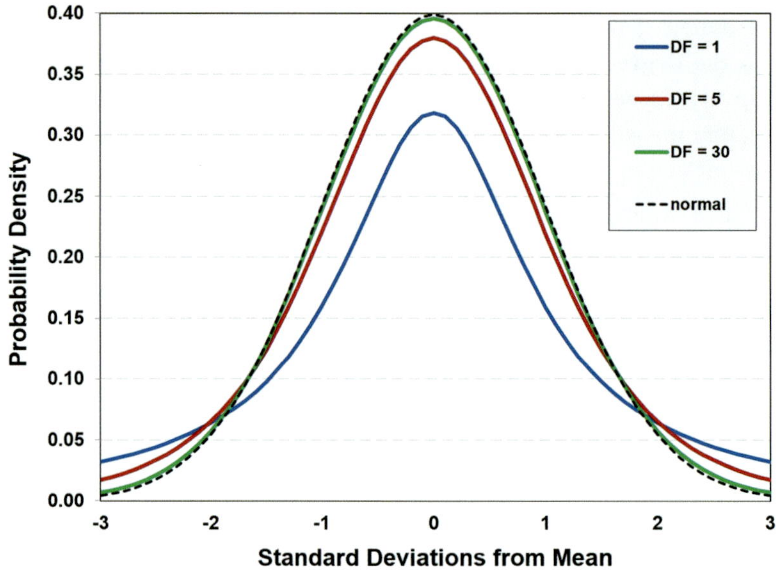

Figure 8.1 Distribution curves for varying numbers of degrees of freedom.

Gossett used the word "uncertainty" twice in the introduction to his 1908 paper [4], but the use of the term "measurement uncertainty" wasn't well established in either the scientific literature or in the vocabulary of metrologists until the 1960s. Before then, it was more common to simply focus on accuracy, and for laboratory manuals and textbooks to caution against making errors that would prevent measurements from being accurate. The American mathematician Churchill Eisenhart (1913–1994) can be credited for much of the shift that took place in the 1960s, when the scientific community became far more aware of the importance of measurement uncertainty analysis. Eisenhart was hired by the National Bureau of Standards (NBS) in 1946 to establish a statistical engineering laboratory. He served as the laboratory's chief until 1963, and remained at NBS as a senior research fellow until his retirement in 1983.

In the early 1960s, the statistical engineering laboratory of NBS was contracted by the United States Army to develop standard statistical procedures that the Army could implement for their research and development programs. This work was guided by Eisenhart, and it resulted in a series of five pamphlets that the Army distributed for its own use. Each pamphlet covered different aspects of measurement data analysis, and both the Army and NBS soon recognized that following these statistical procedures could benefit other groups both inside and outside the U. S. government. To make this happen, the entire series was released to the general public in 1963, in the form of an NBS handbook entitled *Experimental Statistics*. The last chapter of the handbook, Chapter 23, was written by Eisenhart and entitled "Expression of the

Uncertainties of Final Results" [8]. The short chapter outlined the basic principles of what we now know as modern measurement uncertainty analysis. However, despite the efforts of Eisenhart and his fellow NBS statistician John Mandel (1914–2007), who wrote extensively about uncertainty in his 1964 book *The Statistical Analysis of Experimental Data* [9], the general scientific community remained mostly unaware of the importance of assigning uncertainties to measurement results.

This finally started to change in 1968 when Eisenhart published a four-page paper in the journal *Science*. The paper had the same title as his handbook chapter, and it essentially contained the same material, but the material was now directed towards the general scientific community [10]. The paper emphasized the need for scientists to express the uncertainty of their measurement results with unambiguous and statistically valid methods, and to always include enough information to make the results believable when subjected to critical evaluation [11].

Eisenhart wrote that the accuracy required of a reported measurement value "depends primarily on its intended use," but that a "reported value whose accuracy is entirely unknown is worthless." He acknowledged that the actual error of a reported measurement result, or "the magnitude and sign of its deviation from the truth," was unknowable, but that the uncertainty of a reported value should always be indicated by "stating credible limits to its likely inaccuracy." Eisenhart conceded that no single form of stating uncertainty was "universally satisfactory," but he strongly recommended some "minimum standards of good practice." His recommendations included identifying and evaluating both the precision and the systematic errors of the measurement process. The precision referred to the closeness of successive independent measurements, made under the same conditions, to each other. The systematic errors referred to any tendencies or biases in the measurement process that caused it to measure something other than what was intended. The good news was that reasonable bounds could usually be placed on the systematic errors [10].

Eisenhart's methods slowly gained acceptance, but it was still difficult to compare measurement results without an international consensus on how to express measurement uncertainty. The International Committee for Weights and Measures (*Comité International des Poids et Mesures*, CIPM) recognized that this was a problem and started the long process that led to the *GUM* in 1977. During that year, the CIPM asked the BIPM to address the problem and to make a recommendation, and a detailed questionnaire was sent to 32 national metrology laboratories. Most of them responded by early 1979, and nearly all of the respondents agreed that it was important to develop an internationally accepted procedure for expressing measurement uncertainty. This method would combine all of the sources that contributed to the uncertainty of a measurement into one uncertainty value. However, their responses provided no apparent consensus on the method to be used.

To try and reach a consensus, the BIPM formed a working group comprised of staff members from national metrology laboratories. The working group succeeded in producing a brief outline for expressing measurement uncertainty that was approved by the CIPM as a recommendation in 1981. However, the brief outline was insufficient. A guidebook was needed that prescribed in detail how measurement uncertainty should be expressed, and that was written broadly enough to address the wide-ranging measurement requirements of industry and commerce. The CIPM asked the International Organization for Standardization (ISO) to develop this guidebook, and the ISO Technical Advisory Group on Metrology (TAG 4) was given the responsibility. In addition to ISO and the BIPM, five other standards organizations participated in TAG 4, representing the interests of electrical and electronic technologies, clinical chemistry, pure and applied chemistry, pure and applied physics, and legal metrology. Their work eventually resulted in the first publication of the *GUM* in 1993 [12]. The *GUM* was corrected for errors and reprinted in 1995, and the current version [2] is maintained by the Joint Committee for Guides in Metrology (JCGM). The JCGM membership includes the original seven members of TAG 4 and the International Laboratory Accreditation Cooperation (ILAC).

The terminology introduced by Eisenhart has evolved slightly over the years. For example, what Eisenhart called precision was long known to metrologists as either the random or statistical uncertainty, and what he called systematic error was known as systematic uncertainty. The informal use of the terms statistical and systematic uncertainty is still common, but since the first publication of the *GUM* the two types of uncertainties have been called Type A and Type B, with the type assignment depending upon the method of evaluation.

The following sections briefly describe the internationally accepted methods prescribed by the *GUM* for evaluating and reporting measurement uncertainties, and discuss how these methods can be applied to time and frequency measurements.

8.2 The *GUM* Method of Evaluating and Reporting Measurement Uncertainty

The underlying principle of the *GUM* is that a measurement quantity, even though it cannot be known exactly, is a stochastic variable with a probability function [13]. Stochastic is an adjective derived from an ancient Greek word (*στόχοσ*, pronounced *stókhos*) whose meaning is related to aiming, guessing, or conjecture. It follows that a stochastic variable has a distribution or pattern that can be analyzed and well described by the use of statistics, but its exact value cannot be determined.

The *GUM* states that a measurement result should be reported as

$$Y = y \pm U, \tag{8.1}$$

where Y is the measurand (the defined quantity being measured), y is a value obtained by measurement that is an estimate of the measurand, and U is the expanded measurement uncertainty. For example, in the case of a time interval measurement, Y is a defined time interval, and y is obtained from either a single time interval measurement, or by averaging a series of time interval measurements. For all types of measurements, the expanded uncertainty U defines the coverage or confidence interval. The true value of the measurand cannot be exactly determined, but is believed to lie somewhere within the coverage interval

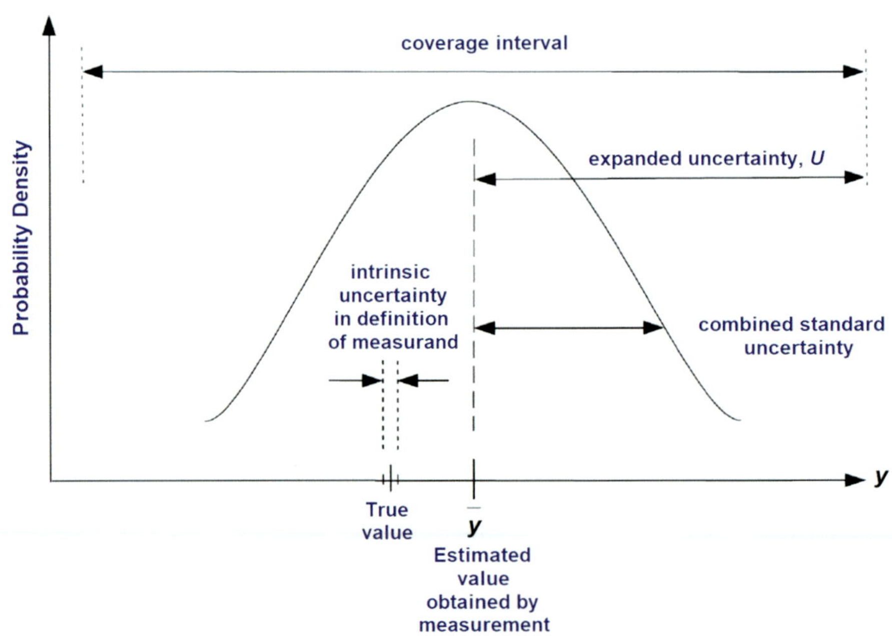

Figure 8.2 Estimating measurement uncertainty by establishing a coverage interval where the true value of the measurand is believed to be included.

(Figure 8.2). The *GUM* also tells us, as shown in the figure, that even the defined value of a measurand has intrinsic uncertainty in its definition The intrinsic uncertainty establishes the minimum uncertainty with which a measurand can be determined, and it can only be reduced if the measurand is "more completely defined" [2].

The use of a probability distribution was nothing new, as it was based on the landmark statistical work of Gauss, Pearson, Gossett, and others. However, the *GUM* used probability to add something new to the measurement uncertainty concept. It equated probability to the "degree of belief" that we can have in a measurement result, and by doing so it revolutionized the thinking and terminology of measurement [14]. With that in mind, let's look at the *GUM* method for determining the value of U.

8.2.1 Identifying the Contributors to Measurement Uncertainty

The *GUM* model is based on the mathematical relationships between the measurand Y and the input quantities X_i on which Y depends,

$$Y = f(X_1, X_2, \dots X_N), \tag{8.2}$$

where the function f should contain every quantity that "that can contribute a significant component of uncertainty to the result of the measurement" [2]. Therefore, the first step in determining U is to identify and list the components or sources that contribute uncertainty to the measurement.

The *GUM* includes a section entitled "Practical Considerations" which reminds readers that the guidance it provides "cannot substitute for critical thinking, intellectual honesty and professional skill." The section also states that the evaluation of uncertainty is "neither a routine task nor a purely mathematical one; it depends on detailed knowledge of the nature of the measurand and of the measurement" [2]. These words tell us that in order to properly identify each of the sources that contribute uncertainty to a measurement, a metrologist has to be familiar with all aspects of the measurement, including not only the measurand, but also the DUT, the reference, and the measurement instrumentation. Without this familiarity, they can easily underestimate the uncertainty by failing to account for one or more of its contributors.

There is, of course, a large body of information that metrologists can use to help identify the sources that contribute to measurement uncertainty. This information includes the results of previous measurements, measurement tutorials, and scientific literature that discusses the measurand. The information provided in the previous chapters of this book has touched upon some of the uncertainty contributors for various types of time and frequency measurements. For example, as described in Chapter 2, the frequency produced by a quartz oscillator is sensitive to variations in temperature. This is easy to infer because much of quartz oscillator research and development has involved looking for new methods and designs, including the TCXO and OCXO, that reduce the effects of temperature variations. Therefore, anyone who analyzes the uncertainty of a quartz clock or oscillator has to account for temperature variations.

From previous chapters of this book we can also infer that cable delays always contribute to the uncertainty of time interval measurements (Section 4.1.3), that human reaction time always contributes to the uncertainty of direct comparison measurements of stopwatches (Section 4.1.4), that signal path delays always contribute to the uncertainty of time transfer systems (Section 4.3), that the uncertainty of the time base oscillator always contributes to the uncertainty of frequency measurements made with oscilloscopes or frequency counters (Sections 5.2.1 and 5.2.2), and that the resolution of a measurement instrument can potentially contribute uncertainty to both time and frequency measurements, especially when only a small number of measurements are made (discussed throughout Chapters 4 and 5).

Although the *GUM* asks for every "significant component of uncertainty" to be identified, it doesn't define what "significant" means. Most metrologists would agree that it usually isn't necessary to include every possible source of uncertainty in an uncertainty analysis. Some sources introduce uncertainties that are considered to be so far "down in the noise" that their contributions to the total measurement uncertainty can be justly considered as too small to matter. However, there are some types of measurements where everything matters. One example would be an evaluation of a primary standard, such as a cesium fountain clock (Section 3.8). Another example is a measurement that tests whether a legal or technical measurement requirement is being met, such as a ±1 μs time accuracy requirement in a telecommunication network (Chapter 16). In this situation, ignoring a potential source of uncertainty might allow a test to be passed that should have been failed. This allows a false measurement result to be accepted, which in our example could cause a failure of the system whose operation depends on ±1 μs time accuracy.

Ultimately, it is up to the metrologist to decide what to include and what to leave out of an uncertainty analysis, and their decision is usually based on the way that the measurement result is expected to be used. After the uncertainty contributors have been identified, the next step is to evaluate each source of uncertainty with either the Type A or Type B method, as described in the next two sections.

8.2.2 The Type A Method of Evaluating Uncertainty

The Type A method evaluates a source of uncertainty by using statistical analysis. This method can be used if we have a series of repeated measurements consisting of n independent observations of X_i. This series of measurements is used to estimate $u(x_i)$, the standard uncertainty of the input quantity, which is typically done by calculating its standard deviation (σ), as described in Section 7.1. The standard uncertainty $u(x_i)$ has a probability distribution equivalent to one standard deviation.

The classical standard deviation (σ) is used to evaluate some uncertainty sources in time and frequency metrology, but as we discussed throughout Chapter 7, measurements of clocks and oscillators usually produce non-stationary data with a time-dependent trend. To remove this trend, statistics that employ successive differencing are typically used instead of the standard deviation, most notably the Allan deviation, $\sigma_y(\tau)$, the modified Allan deviation, Mod $\sigma_y(\tau)$, and the time deviation, $\sigma_y(\tau)$. In the context of the *GUM*, the various forms of the Allan deviation are Type A methods of evaluation that estimate the uncertainty contributed to a measurement by the frequency instability of an oscillator or by the time instability of a clock.

We previously noted how uncertainties evaluated with the Type A methods were once known as statistical or random uncertainties. Although few uncertainties are entirely random, a Type A evaluation is appropriate when the source of the uncertainty is difficult to control, and its boundaries can't be defined without statistical analysis. To illustrate this, consider a situation where a metrologist is evaluating the uncertainty contributed by the environmental temperature. They first need to know the probability distribution of the temperature during the period of the measurement, so they can determine the effect that temperature variations might have had on the DUT. If the measurement was performed in a laboratory where the temperature is carefully monitored and controlled, for instance, always kept to within ±1 °C of 23 °C, the distribution limits of the environmental temperature are already known and can be used without further statistical analysis. However, if the measurement was performed outdoors where the temperature is uncontrolled, a Type A evaluation might be needed to determine the temperature's probability distribution while the measurement was in progress.

8.2.3 The Type B Method of Evaluating Uncertainty

The Type B method is defined as a "method of evaluation of uncertainty by means other than the statistical analysis of a series of observations" [2]. That broad definition implies that any uncertainty evaluation that doesn't use the Type A method can be classified as using a Type B method, and that the two classifications cover all possible types of evaluation [3, 15].

Like the Type A method, the Type B method estimates the standard uncertainty of one of the input quantities X_i shown in Eq. (8.2), and the standard uncertainty is again equivalent to one standard deviation (σ). However, when the Type B method is used, statistical analysis isn't performed on a series of measurement data, usually because no measurement data are available. Instead, the quantity's standard uncertainty $u(x_i)$ is "evaluated by scientific judgement based on all of the available information on the possible variability of X_i". The information can come from previous measurement data (including data from calibration certificates), from manufacturer's specifications, from uncertainties assigned to reference data taken from handbooks, and from the metrologist's experience with the materials and instruments that were used to perform the measurement [2]. In many cases, a Type B evaluation is used to estimate and establish the limits of a known bias in a measurement process.

A Type B evaluation often starts with an uncertainty estimate that represents 100% of the probability distribution. This type of estimate is used when we assume that we have exact knowledge about the limits of a distribution, but no specific knowledge about the distribution of measured values between the limits. If only the limits are known, metrologists can apply a uniform probability distribution, which the *GUM* calls a rectangular distribution because of its shape (Figure 8.3). A rectangular distribution encompasses all possible outcomes of a measurement.

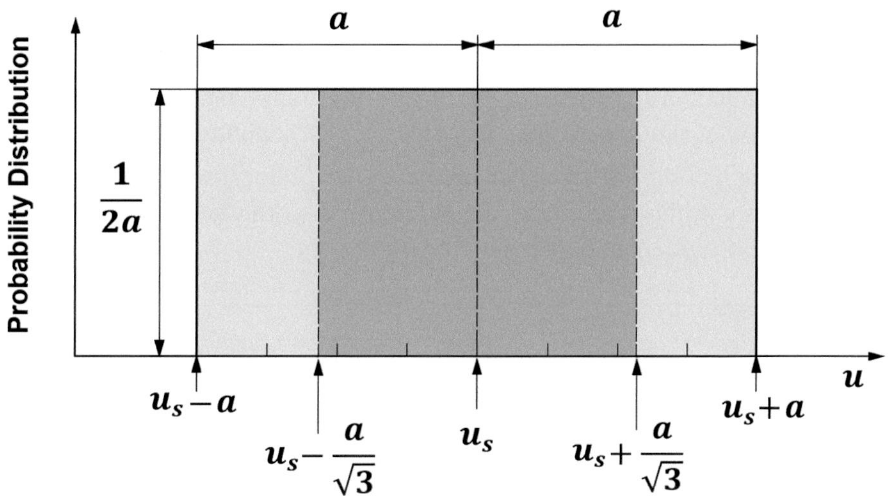

Figure 8.3 A rectangular probability distribution.

A rectangular distribution can be demonstrated by rolling a single die. A die only includes the numbers from 1 to 6, so we know in advance that 1 and 6 are the exact limits of the distribution, represented by the vertical lines on each side of the rectangle. We also know that there are only six possible outcomes for each roll, and that each outcome has the same $\frac{1}{6}$ probability. If we roll the die enough times and count the occurences of each outcome, we'll find that the totals for each outcome are essentially the same, represented by the horizontal line at the top of the rectangle.

As shown in Figure 8.3, for an uncertainty source u_s, we can define the limits of a rectangular distribution as $u_s - a$ to $u_s + a$, where a is the half-width of the distribution. The probability density is the same for all possible outcomes of u_s, equal to $\frac{1}{2a}$. We can convert an uncertainty with a rectangular distribution to a standard uncertainty with this simple equation,

$$u(x_i) = \sigma = \frac{a}{\sqrt{3}}. \tag{8.3}$$

The resolution of a measurement instrument, such as a stopwatch or a time interval counter, is a common example of an uncertainty that has a rectangular distribution. The uncertainty is found in the least significant digit of a measurement reading, which is rounded to the nearest integer value. For example, a digital counter with 1 ns resolution will round down to 49 ns for readings from 49.00 ns to 49.49 ns, and will round up to 50 ns for readings from 49.50 ns to 49.99 ns. Therefore, the half-width of the distribution is actually half of the resolution of the least significant digit, so for this example we should use $a = 0.5$ ns in Eq. (8.3). If we use $a = 1$ ns to specify the counter's full resolution, then the divisor in Eq. (8.3) should be $2\sqrt{3}$, which equals $\sqrt{12}$.

The normal and rectangular distributions are the most common distributions used in uncertainty analysis, but there are many other types of probability distributions that are used for special situations. We'll discuss just one more, the triangular distribution, which like the rectangular distribution was named because of its shape. The triangular distribution also encompasses all possible outcomes of a measurement because its limits are assumed to be exactly known. However, unlike the rectangular distribution, the triangular distribution has an estimated central value, because it is believed that a measurement sample is likely to be closer to the center of the distribution than it is to the limits.

Triangular distributions are appropriate for situations where an uncertainty is influenced by two or more factors that each have a rectangular distribution. Therefore, it can be demonstrated by rolling two dice. When we roll two dice, there are 11 possible outcomes—the numbers from 2 to 12—and 36 possible combinations. This means that some outcomes have higher probabilities than others, because they can result from multiple combinations. As shown in Figure 8.4, only one combination can result in an outcome of 2 or 12, so each of those outcomes has just a $\frac{1}{36}$ probability. Six combinations can result in rolling a 7, so that outcome has the highest probability, $\frac{6}{36}$ or $\frac{1}{6}$. Because the outcomes with the highest probabilities (6, 7, and 8) are in the center, and the outcomes with the lowest probabilities (2 and 12) are at the limits, the resulting distribution is shaped like a triangle.

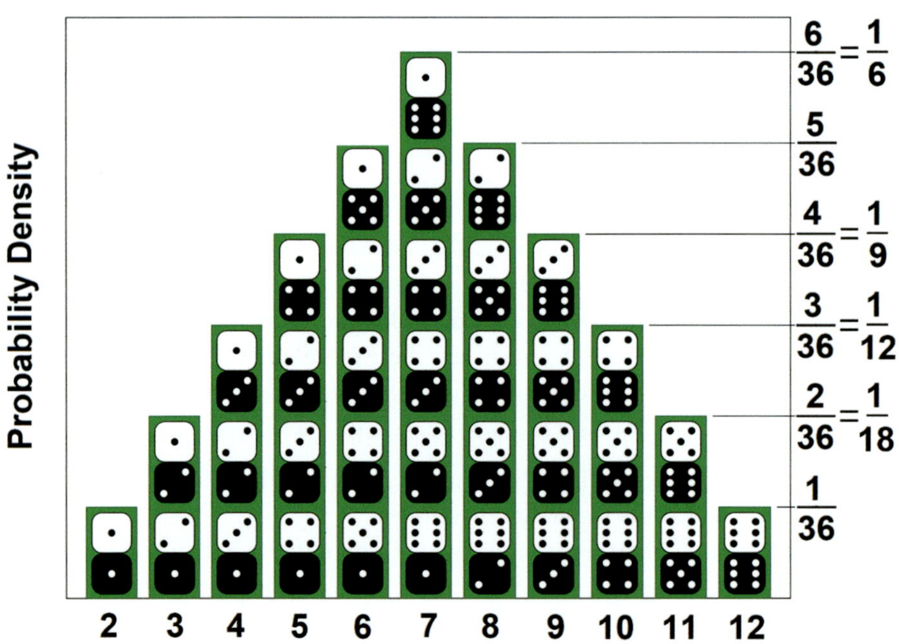

Figure 8.4 A triangular distribution formed by the possible outcomes of a roll of two dice.

Figure 8.5 shows that for an uncertainty source u_s, we can again define the limits of a triangular distribution as $u_s - a$ to $u_s + a$, where a is the half-width of the distribution, as we did with the rectangular distribution. However, the probability density is no longer flat, but now highest in the center of the distribution, where it equals $\frac{1}{a}$. We can convert an uncertainty with a triangular distribution to a standard uncertainty with this simple equation,

$$u(x_i) = \sigma = \frac{a}{\sqrt{6}}. \tag{8.4}$$

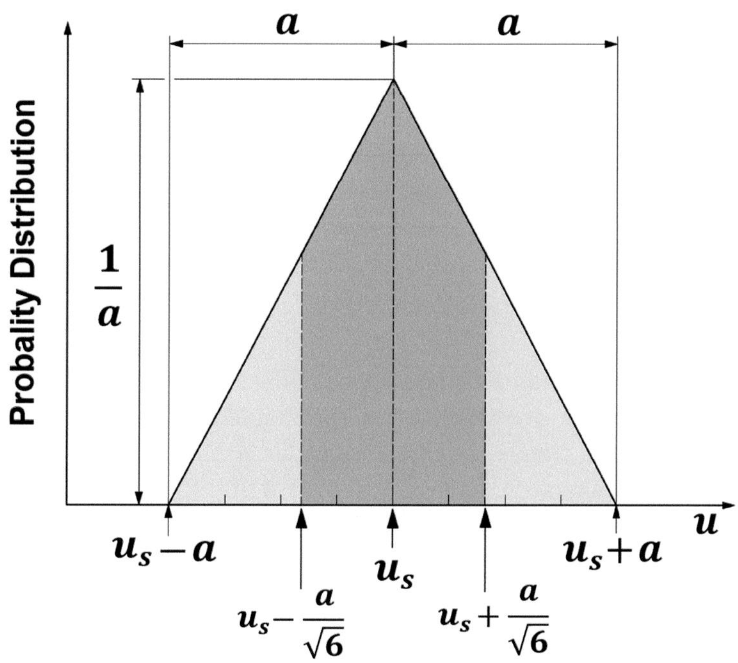

Figure 8.5 A triangular probability distribution.

8.2.4 Combining Measurement Uncertainties

The combined standard uncertainty of y is obtained by combining all of the previously evaluated sources of uncertainty, the input quantities X_i, into a single value [2]. As we can recall from Eq. (8.1), y is the estimated value obtained from the measurement. The combined standard uncertainty of y is expressed mathematically as $u_c(y)$.

Before combining the individual uncertainties into a single value, we have to decide if the sources of uncertainty are independent of each other, or if they are correlated. We previously described the Pearson linear correlation coefficient, r, in Section 6.1.1 when discussing frequency accuracy. The possible values for r range from −1 to +1. A value of 0 indicates no correlation between the variables, and values of −1 or +1 respectively indicate a 100% negative or a 100% positive correlation.

Even though it isn't difficult to determine the correlation coefficient for two variables, determining the correlation coefficients for a large number of input quantities X_i with respect to each other can involve a lot of math. In most cases, a metrologist won't have access to enough data to calculate all of the correlation coefficients, because some of the input quantities were evaluated with the Type B method. This means that some correlation coefficients will likely need to be estimated, which is easy enough when it's apparent that the correlation is near 0 or near 100%, but which can become mostly guesswork when the correlation lies somewhere in the middle.

For these reasons, the assumption is usually made that the contributors to measurement uncertainty are either completely correlated or completely independent. If we assume the contributors to uncertainty are completely correlated to each other, we can combine them by simply adding them together,

$$u_c(y) = u(x_1) + u(x_2) \ldots + u(x_N), \tag{8.5}$$

where $u_c(y)$ is the combined standard uncertainty and $u(x_i)$ is the standard uncertainty of one of the uncertainty contributors. Making the assumption that the sources of uncertainty are completely correlated is the most conservative approach to uncertainty analysis, because it leaves no risk that the combined uncertainty will be underestimated. However, it is seldom practiced because complete correlation of all of the uncertainty contributors is unlikely to occur. For example, if we make a measurement with a time interval counter, the uncertainties caused by the time base frequency, the trigger level settings, and the laboratory temperature are likely to be independent of each other with no correlation. Therefore, the most common approach to combining uncertainties, not just for time and frequency measurements but for all areas of metrology, is to use the root sum of squares (RSS) method. This method squares each of the uncertainty contributors, adds them together, and arrives at a combined standard uncertainty by taking the square root of the sum,

$$u_c(y) = \sqrt{u(x_1)^2 + u(x_2)^2 + \ldots u(x_N)^2}. \tag{8.6}$$

The RSS method nearly always produces a smaller combined uncertainty than adding the uncertainties together, with the largest difference occurring when the magnitude of all the uncertainty contributors is about the same. To provide a simple example, consider a time interval measurement with three uncertainty contributors, $u(x_i)$, that each have a value of 2 ns. If we simply add them together per Eq. (8.5), then $u_c(y) = 6$ ns. If we use the RSS method per Eq. (8.6), then $u_c(y) = \sqrt{12}$ – 3.5 ns. However, if one uncertainty source dominates all the others, the percentage difference between the two combining methods can be small. For instance, if we modify our example so that one source contributes 100 ns of uncertainty, with the other two contributions remaining at 2 ns, then $u_c(y)$ is equal to 104 ns when combining by addition and 100 ns when using the RSS method, a difference of just 4%.

Finally, there are situations where we know that two sources of uncertainty are correlated to each other, but that both sources are independent of a third uncertainty source. An example might be a time measurement that identifies both the variations in equipment delays and the variations in laboratory temperature as uncertainty sources, even though temperature is the likely cause of the equipment delay variation. In this situation, the RSS method can still be used, but the correlated uncertainties should be added together before they are squared,

$$u_c(y) = \sqrt{u(x_1)^2 + (u(x_2) + u(x_3))^2}. \tag{8.7}$$

8.2.5 Expanding and Reporting Measurement Uncertainties

Like each of its individual components, the combined standard uncertainty $u_c(y)$ has a coverage probability of 68.27%, equivalent to one standard deviation. This means that there is only about a two out of three probability that the true value of the measurand will be included in the distribution, and also means there is about a one in three probability that it won't be included. Therefore, the *GUM* recommends multiplying the standard uncertainty by a coverage factor to obtain an expanded uncertainty,

$$U = ku_c(y), \tag{8.8}$$

where U is the expanded uncertainty, k is the coverage factor, and $u_c(y)$ is the combined standard uncertainty. Technically speaking, multiplying by any value greater than one will convert a standard uncertainty into an expanded uncertainty, but k is usually in the range of 2 to 3, which means that the coverage probability provided by most uncertainty statements will exceed 95%. Table 8.2 shows the coverage probability of U for selected values of k. Once U has been determined, the final measurement result can be reported in the recommended form $Y = y \pm U$, that was previously given in Eq. (8.1).

Table 8.2 Coverage factors and coverage probability for an expanded uncertainty, *U*

Coverage factor, *k*	Coverage probability, %
1	68.27
1.28	80
1.65	90
1.96	95
2	95.45
2.33	98
2.58	99
3	99.73
4	99.99

The procedures we've just discussed for performing a measurement uncertainty analysis, based on the *GUM* method, are summarized in Table 8.3. The last two sections of this chapter provide examples of applying this method to several types of time and frequency measurements.

Table 8.3 A summary of procedures in a measurement uncertainty analysis based on the *GUM* method [2]

Procedure	Result
Obtain knowledge about the quantity to be measured (the measurand)	Knowledge of Y
Perform the measurement	Estimate of y
Identify the sources that contribute to the uncertainty of the estimate of y	$X_1, X_2, X_3, \ldots, X_N$
Evaluate each source of uncertainty, the input quantities X_i, with either the Type A or Type B method	Probability distributions of $X_1, X_2, \ldots, X_N$
Convert each source of uncertainty to a standard uncertainty	$u(x_1, x_2, \ldots, x_N)$
Combine the uncertainty sources into a single value	$u_c(y)$
Expand the combined uncertainty by multiplying by a coverage factor k	U
Report the measurement result	$Y = y \pm U$

8.3 Examples of Time Measurement Uncertainty Analysis

Measurement uncertainty analysis can be applied to all types of time measurements using the methods that were just described. We'll provide two examples here, one with a large uncertainty in units of milliseconds, and the other with a small uncertainty in units of nanoseconds. The first example estimates the uncertainty of a time interval measurement made with a digital stopwatch. The second example estimates the uncertainty of a time accuracy measurement where common-view observations of satellites were used to compare a free-running clock to UTC.

8.3.1 Measurement Uncertainty Analysis of a Stopwatch Calibration

For stopwatch calibrations, the measurand Y is the defined value of a reference time interval. The measurement of that interval results in the estimate y. For the purposes of example, we'll consider a measurement based on the direct comparison method described in Section 4.1.4, where the reference is provided by audio time signals broadcast by an HF radio station that's synchronized to UTC, which could be WWV or any of the other HF stations described in Section 10.3. The uncertainty analysis for this measurement is shown in Table 8.4.

Table 8.4 An example of a measurement uncertainty analysis for a stopwatch calibration

Source of uncertainty	Evaluation method	Distribution	σ (Type A) or a (Type B), ms	Divisor	Standard uncertainty $u(x_1)$, ms
Differential delay between start and stop times of DUT due to reaction time of metrologist	A	Normal	100	1	100
Differential delay between start and stop times of reference audio time signal due to radio propagation	B	Rectangular	1	$\sqrt{3}$	0.6
Resolution of DUT display	B	Rectangular	10	$\sqrt{12}$	2.9

Measurand, Y	3 hours (10,800 s)
Measurement estimate, y	10,800 s – 10,799.14 s = 0.86 s
Combined standard uncertainty	$u_c(y) = \sqrt{(100)^2 + (0.6)^2 + (2.9)^2} = 100.04 \text{ ms} = 0.1 \text{ s}$
Expanded uncertainty ($k = 2$)	$U = 2(0.1) = 0.2$ s
Measurement result expressed in time units	0.86 s ± 0.2 s
Measurement result expressed as a percentage	0.008% ± 0.002%

As summarized in the table, the DUT was measured by comparing the reading on its display to a reference time interval of three hours. The metrologist was required to start and stop the stopwatch when the reference audio time signals were heard, and essentially all of the measurement uncertainty was due to the metrologist's reaction time. Their reaction time had been previously evaluated with the Type A method, by taking the standard deviation of a series of measurements. The other two uncertainty sources were each evaluated with the Type B method, but their contributions to the combined uncertainty were insignificant. A typical uncertainty requirement for a stopwatch calibration is 0.01%, so the DUT in this example is within tolerance [16].

8.3.2 Measurement Uncertainty Analysis of the Time Accuracy of a Clock

As discussed in Chapter 6, the time accuracy of a clock is its time difference, Δt, with respect to UTC. As we'll discuss in Chapter 9, the arrival time of the UTC second is calculated by the BIPM from a weighted averaged of local UTC time scales distributed around the world. In this example, the measurand Y is the arrival time of the UTC second as defined by the BIPM. The value of y is an estimate of time accuracy. It's obtained by comparing the DUT to a reference time scale whose time difference is known with respect to the UTC calculations [17].

The reference time scale is usually operated by a national metrology laboratory and is known as UTC(k), where k represents the name of the laboratory. In this example, the measurement is based on common-view observations of GPS satellites (Section 4.3.3), with the satellites providing the comparison link between the DUT and UTC(k). Table 8.5 provides a sample uncertainty analysis, where we assume that y is equal to 100 ns, and that the uncertainty is reported for a measurement period of 24 hours.

Table 8.5 A sample measurement uncertainty analysis of the time accuracy of a clock, based on common-view observations of GPS satellites

Source of uncertainty	Evaluation method	Distribution	σ (Type A) or a (Type B), ns	Divisor	Standard uncertainty $u(x_1)$, ns
Inherent time stability of measurement method, $\sigma_x(\tau)$, for τ = 24 hours	A	Normal	0.5	1	0.3
Time offset due in bias in GPS antenna coordinates (assumes coordinates have been surveyed to within 1 minute)	B	Rectangular	3	$\sqrt{3}$	1.7
Time offset due to bias in calibration of measurement system delays	B	Rectangular	7	$\sqrt{3}$	4
Time offset due to bias in calibration of cable delays	B	Rectangular	1	$\sqrt{3}$	0.6
Resolution of time interval counter	B	Rectangular	0.1	$\sqrt{12}$	0.03
Time offset caused by delay changes in equipment or cables due to temperature	B	Rectangular	5	$\sqrt{3}$	3
Time offset caused by changes in GPS signal delays due to uncertainties in ionospheric corrections	B	Rectangular	3.5	$\sqrt{3}$	2
Time offset caused by changes in GPS signal delay due to multipath signal reflections	B	Rectangular	3.5	$\sqrt{3}$	2
Average UTC – UTC(k) time difference during measurement period, obtained from the BIPM	A	Normal	2	1	2

Measurand, Y	Defined arrival time of UTC second (calculated by BIPM)
Measurement estimate, y	Clock – UTC(k) = 100 ns
Combined standard uncertainty	$u_c(y) = \sqrt{(0.3)^2+(1.7)^2+(4)^2+(0.6)^2+(0.03)^2+(3)^2+(2)^2+(2)^2+(2)^2}$ = 5.9 ns
Expanded uncertainty (k = 2)	U = 2(5.9) = 11.8 ns
Measurement result (for 24 hours period)	100 ns ± 11.8 ns

The table lists nine sources of uncertainty. The first five, including the inherent stability of the measurement method, the systematic biases in the GPS antenna coordinates, measurement system delay calibrations, and cable delay calibrations; as well as the resolution of the time interval counter, were established as constants and don't change during the measurement. The last four uncertainty sources can change during the measurement, but the limits of their rectangular distribution are assumed to be wide enough to cover all possible changes. The nine uncertainty sources are assumed to be uncorrelated, so they're combined with the RSS method.

Finally, we should reiterate that the time accuracy was arbitrarily chosen to be 100 ns for the purposes of this example. Regardless of what the time accuracy is, it will change each time we report results if the DUT is a free-running clock, due to the frequency inaccuracy of the clock's oscillator, as discussed in Chapter 6. However, if the DUT is a clock disciplined to a source of UTC, its measured time accuracy should have good repeatability from day to day.

8.4 Examples of Frequency Measurement Uncertainty Analysis

Estimating the uncertainty of a frequency measurement is straightforward, due to the well-established statistical tools that estimate frequency stability, specifically the Allan deviation and its derivatives (Chapter 7). However, the uncertainty analysis for a frequency measurement requires more than just estimating an oscillator's stability, it also has to account for the oscillator's frequency accuracy. We'll consider two examples here, a simple analysis of uncertainty that can be applied to most frequency measurements, and the more complex analysis of the uncertainty of a cesium fountain primary frequency standard.

8.4.1 Measurement Uncertainty Analysis of the Frequency of an Oscillator

As discussed in Chapter 6, the frequency accuracy of a clock is its frequency difference, f_{diff}, with respect to its nominal frequency, which is labeled on the oscillator and defined by the SI. As we'll further discuss in Chapter 9, the best estimate of the SI frequency is the frequency of UTC, which is calculated by the BIPM from a weighted averaged of atomic oscillators distributed around the world.

In this example, the measurand Y is the nominal frequency, as defined by the SI and estimated by UTC. The value of y is an estimate of the average frequency accuracy, obtained by comparing the DUT, a rubidium oscillator, to the frequency reference, a GPSDO, for a period of 24 hours. As discussed in Section 6.1, the test accuracy ratio between the reference and DUT must be high enough, preferably at least 10:1, to support a valid measurement of the DUT's accuracy. In addition, the accuracy of the reference frequency must be known with respect to UTC. Both requirements are met in this example.

The measurement method is unspecified in this example, but in practice it can be either a frequency or a phase comparison, performed with any of the several methods described in Section 5.2 that have noise floors low enough to support this type of measurement. Table 8.6 provides a sample uncertainty analysis where the result is reported for a measurement period of 24 hours. The expanded uncertainty ($k = 2$) is essentially the DUT's Allan deviation for $\tau = 24$ hours, multiplied by 2.

Table 8.6 A sample measurement uncertainty analysis of the frequency accuracy of an oscillator

Source of uncertainty	Evaluation method	Distribution	σ (Type A) or a (Type B), unitless	Divisor	Standard uncertainty $u(x_1)$, unitless
Frequency accuracy of reference with respect to UTC during 24-hour period	A	Normal	1×10^{-13}	1	1×10^{-13}
Frequency stability of DUT, $\sigma_y(\tau)$, for $\tau = 24$ hours	A	Normal	4×10^{-12}	1	4×10^{-12}
Measurement noise floor, $\sigma_y(\tau)$, for $\tau = 24$ hours	A	Normal	1×10^{-15}	1	1×10^{-15}

Measurand, Y	Nominal frequency of DUT (defined by SI)
Measurement estimate, y	$\frac{f_{\text{meas}} - f_{\text{nom}}}{f_{\text{nom}}} = 2 \times 10^{-11}$
Combined standard uncertainty	$u_c(y) = \sqrt{(1\times10^{-13})^2 + (4\times10^{-12})^2 + (1\times10^{-15})^2} = 4 \times 10^{-12}$
Expanded uncertainty ($k = 2$)	$U = 2(4 \times 10^{-12}) = 8 \times 10^{-12} = 0.8 \times 10^{-11}$
Measurement result (for 24 hours period)	$2 \times 10^{-11} \pm 0.8 \times 10^{-11}$

8.4.2 Measurement Uncertainty Analysis of a Cesium Fountain Primary Frequency Standard

Cesium fountains are called primary frequency standards because they physically realize the SI second, which is defined based on energy transitions of the cesium atom, more accurately than any other device (Section 3.8). A primary standard can be thought of as a source of "absolute accuracy," but their accuracy still has limits that can only be established through measurement uncertainty analysis.

The uncertainty analysis of a primary standard is complex for several reasons. The first reason is that the usual practice of measuring a DUT by comparing it to a more accurate reference can't be followed, because by definition a more accurate reference doesn't exist. Another reason is that the analysis requires more than just general metrological knowledge—in the case of a cesium fountain, it also requires detailed knowledge of atomic physics. Many different physical effects can influence the measurement of the resonance frequency of a cesium atom, and each effect has to be accounted for in the uncertainty analysis. Finally, primary frequency standards aren't mass-produced instruments, and they aren't delivered with an instruction manual and a warranty. Instead, they are unique, "one-off" type instruments, designed by scientists as a laboratory apparatus that can be tweaked and modified whenever necessary, as part of a continuous effort to get better results and advance the state of the art. Therefore, a detailed understanding of how a primary standard works is a prerequisite for evaluating its uncertainty, and this means that the evaluation is usually performed by the scientists who designed, built, and tested the standard.

For the reasons just described, a thorough discussion of the uncertainty analysis of a cesium fountain would go far beyond the scope of this chapter, but we can still provide a simplified example. The first step usually involves comparing the frequency of the fountain to a hydrogen maser (Section 3.10). Masers have exceptional frequency stability even though they lack inherent accuracy, and the stability of a cesium fountain can be estimated by comparing its frequency to a maser for a period of about one month. The results can be evaluated with the Type A method, again by using the Allan deviation or one of its derivatives.

The next steps involve identifying and determining the magnitude of each of the physical effects that bias the fountain's frequency and cause it to be less accurate. A few systematic biases can be measured and evaluated with the Type A method, but most are evaluated with the Type B method. The magnitude of the bias is usually expressed as a unitless frequency difference that represents parts in 10^{16}. If the bias is larger than about 0.1×10^{-16} and is possible to correct, a correction is applied to the measurements to remove the bias. The corrections remove most, but not all, of the frequency bias. The part that remains is the uncertainty of the correction. An uncertainty for each correction, as well an uncertainty for each bias that was considered too small to correct, is estimated by examining its probability distribution. The correction uncertainties are then combined into a single value by use of the RSS method.

The three largest frequency biases in a cesium fountain are usually due to the Zeeman effect (Section 3.7), blackbody radiation (Section 3.4), and the gravitational redshift. Even though the magnetic fields in a fountain are usually weak and well-controlled, the Zeeman effect can bias the frequency by 1×10^{-13} or more. Blackbody radiation, which occurs when the temperature in the microwave cavity is not absolute zero (0 K), can bias the frequency by more than 1×10^{-14}. The gravitational redshift is a relativistic effect that biases the frequency by about 1.1×10^{-16} for every meter of altitude. This bias is small at low altitudes, but will exceed 1×10^{-13} if the clock is located at an altitude greater than 1,000 m. Fortunately, the corrections for each of these biases usually have uncertainties smaller than 1×10^{-16}.

There are two other frequency biases are harder to correct, even though their magnitude is comparatively small. The collisions between the atoms produce a frequency bias that's typically estimated as a few parts in 10^{15}, but can be difficult to correct to within 1×10^{-16}. The frequency bias due to distributed cavity phase (DCP) is even smaller, but it typically isn't corrected at all because it's difficult to model. The DCP is due to numerous factors that cause the atoms to change their spatial position as they pass upwards and downwards through the microwave cavity. It results in a Doppler frequency shift that usually exceeds 1×10^{-16}, making it the largest contributor to measurement uncertainty in some cesium fountain designs.

Table 8.7 provides a sample uncertainty analysis that includes the five systematic frequency biases that were just described, but it's important to note that many other biases are accounted for in published fountain evaluations [18–22]. The numbers in the table, including the final measurement result of $\pm 3.3 \times 10^{-16}$ with respect to the SI second, can be considered similar to the results from actual evaluations, but they don't represent the uncertainty of any particular cesium fountain. It's also important to note that the table only provides a combined standard uncertainty, $u_c(y)$ where $k = 1$, and not an expanded uncertainty, U. This follows the conventional practice for the evaluation of a primary frequency standard, as it avoids confusion when results are compared internationally.

Table 8.7 A sample measurement uncertainty analysis of a cesium fountain primary frequency standard

Source of uncertainty	Evaluation method	Frequency correction (parts in 10^{16})	Uncertainty of correction (parts in 10^{16})
Frequency instability	A	—	2
Zeeman effect	B	1000	0.4
Blackbody radiation	B	200	0.8
Gravitational redshift	B	1000	0.3
Collisional shift	B	50	1.5
Distributed cavity phase	B	0	2

Measurand, Y	SI second
Measurement estimate, y	Variable, but would be 0 by definition if it were possible to compensate for all measurement uncertainties
Combined standard uncertainty	$u_c(y) = \sqrt{(2)^2 + (0.4)^2 + (0.8)^2 + (0.3)^2 + (1.5)^2 + (2)^2}$ $= 3.3 \times 10^{-16}$
Measurement result	Realization of SI second within $\pm 3.3 \times 10^{-16}$

As we've described in this chapter, before metrologists can truthfully report a measurement result, they first need to quantify the amount of doubt that they have about the measurement. This is done by performing an uncertainty analysis. To complete the verification of a measurement result, metrologists also have to prove that the result can be traced back to the SI. Traceability is the subject of Chapter 9.

References

1. C. Ehrlich and R. Dybkaer, "Uncertainty of Error: The Error Dilemma," *NCSLI Measure: The Journal of Measurement Science*, vol. 6, no. 3, pp. 72–77, September 2011.
2. Joint Committee for Guides in Metrology, "Evaluation of Measurement Data—Guide to the Expression of Uncertainty in Measurement," *JCGM 100*, 2008.
3. A. Possolo, "Simple Guide for Evaluating and Expressing the Uncertainty of NIST Measurement Results," *NIST Technical Note 1900*, 103 p., October 2015.
4. Student, "The Probable Error of a Mean," *Biometrika*, vol. 6, no. 1, pp. 1–25, March 1908.

5. H. Hotelling, "British Statistics and Statisticians Today," *Journal of the American Statistical Association*, vol. 25, no. 170, pp. 186–190, June 1930.
6. W. J. Youden, *Experimentation and Measurement*, National Science Teachers Association Vistas of Science Series, Scholastic Book Services, 127 p., 1962. Also available as *NIST Special Publication 672*, reprinted May 1994.
7. C. H. Brase and C. P. Brase, *Understandable Statistics: Concepts and Methods*, 3rd ed., DC Heath and Company, 1987.
8. C. Eisenhart, "Expression of the Uncertainties of Final Results," Chapter 23 in "Experimental Statistics," *National Bureau of Standards Handbook 91*, pp. 23-1 to 23-6, August 1963.
9. J. Mandel, *The Statistical Analysis of Experimental Data*, John Wiley & Sons, 1964.
10. C. Eisenhart, "Expression of the Uncertainties of Final Results," *Science*, vol. 160, no. 3833, pp. 1201–1204, June 1968.
11. R. M. Lindstrom, "Believable Statements of Uncertainty and Believable Science," *Journal of Radioanalytical and Nuclear Chemistry*, vol. 311, no. 2, pp. 1019–1022, February 2017.
12. BIPM/IEC/IFCC/ISO/IUPAC/IUPAP/OIML, *Guide to the Expression of Uncertainty in Measurement*, 1st edition, Switzerland: International Organization for Standardization (ISO), 1993.
13. P. Howarth and F. Redgrave, *Metrology–In Short*, 3rd edition, European Association of National Metrology Institutes (EURAMET), 2008.
14. C. Ehrlich, "Terminological Aspects of the 'Guide to the Expression of Uncertainty in Measurement (*GUM*)'," *Metrologia*, vol. 51, no. 4, pp. S145–S154, July 2014.
15. A. Possolo and J. Meija, *Measurement Uncertainty: A Reintroduction*, 2nd edition, Sistema Interamericano de Metrologia (SIM), 2022.
16. J. C. Gust, R. M. Graham, and M. A. Lombardi, "Stopwatch and Timer Calibrations," *National Institute of Standards and Technology Special Publication 960-12*, 66 p., January 2009.
17. M. A. Lombardi and A. N. Novick, "Remote Time Calibrations via the NIST Time Measurement and Analysis Service," *NCSLI Measure: The Journal of Measurement Science*, vol. 1, no. 4, pp. 50–59, December 2006.
18. T. P. Heavner, E. A. Donley, F. Levi, G. Costanzo, T. E. Parker, J. H. Shirley, N. Ashby, S. Barlow, and S. R. Jefferts, "First Accuracy Evaluation of NIST-F2," *Metrologia*, vol. 51, no. 3, pp. 174–182, May 2014.
19. S. Weyers, V. Gerginov, M. Kazda, J. Rahm, B. Lipphardt, G. Dobrov, and K. Gibble, "Advances in the Accuracy, Stability, and Reliability of the PTB Primary Fountain Clocks," *Metrologia*, vol. 55, no. 6, pp. 789–805, October 2018.
20. S. Beattie, B. Jian, J. Alcock, M. Gertsvolf, R. Hendricks, K. Szymaniec, and K. Gibble, "First Accuracy Evaluation of the NRC-FCs2 Primary Frequency Standard," *Metrologia*, vol. 57, no. 3, 15 p., April 2020.
21. A. Takamizawa, S. Yanagimachi, and K. Hagimoto, "First Uncertainty Evaluation of the Cesium Fountain Primary Frequency Standard NMIJ-F2," *Metrologia*, vol. 59, no. 3, 15 p., April 2022.
22. X.-L. Wang, J. Ruan, D.-D. Liu, Y. Guan, J.-R. Shi, F. Yang, Y. Bai, H. Zhang, S.-C. Fan, W.-J. Wu, S.-H. Zhao, and S.-G. Zhang, "First Evaluation of the primary frequency standard NTSC-CsF2," *Metrologia*, vol. 60, no. 6, 15 p., November 2023.

Chapter 9

Traceability

This is the guy behind the guy behind the guy.

—Jerry (played by Joe Mantegna) in the movie "Things Change"

Traceability is a self-descriptive word, because it refers to the ability to trace something. That something can be either a quantity or a quality, and it can be traced by following the evidence left behind by actions or events that have already occurred. The evidence leads us to the source of what we are trying to trace. Examples of traceability are easy to find in everyday life—the small amount of snow that's still on the ground could be traced back to the snowfall that occurred last week. The footprints our dog made in that same snow can be traced back to the door where the dog enters and leaves the house. A river or stream can be traced back to the source of its water, which might be a lake or an underground spring. And while it might be less obvious, facial expressions, including smiles, frowns, and expressions of worry or surprise, can usually be traced to the events that just happened in someone's life.

If the trace leads back to a trusted source of information, traceability becomes more meaningful. We can verify information if we can trace it back to a trusted source. Consider, for instance, a news article written by a journalist. If the journalist interviewed trustworthy sources with a history of providing facts, then it should be easy to verify the information found in the article. However, if the journalist interviewed dubious sources with a history of engaging in rumors and speculation, verifying the information might prove to be impossible.

Different levels of trust can, of course, lead to different levels of verification. For instance, consider a situation where Phoebe is filing her income taxes. Before deciding whether she can take a specific tax deduction, she first needs to verify that the deduction is allowed. She could trace her decision about the deduction to her Uncle Arthur, a person whom Phoebe has always trusted. Uncle Arthur is a great guy, and whenever Phoebe has called him for financial advice in the past, he's always been happy to offer his opinions. However, after thinking about it for a while, Phoebe realizes that even though her uncle has many endearing qualities, he's definitely not an authoritative reference when it comes to taxes, because his own taxes have been audited several times. To avoid having her taxes audited, Phoebe decides to consult with a professional accountant, so she can trace her deduction decision to a more trusted reference than her uncle.

As Time Goes By: A Comprehensive Guide to Time and Frequency Measurements
Michael A. Lombardi

ISBN 978-981-5352-28-3 (Hardcover), 978-1-042-01532-0 (eBook)
www.jennystanford.com

DOI: 10.1201/9781042015320-12

Metrological traceability follows these same lines of reasoning—everything is based on an established hierarchy of trust. The source of traceability resides at the highest level of the hierarchy, and it's the authoritative reference that provides the basis for all of the measurements made at lower levels. The measurement uncertainty increases as we go down through the hierarchical levels and move further away from the source [1]. As shown in Figure 9.1, the measurement hierarchy is often depicted as a traceability pyramid.

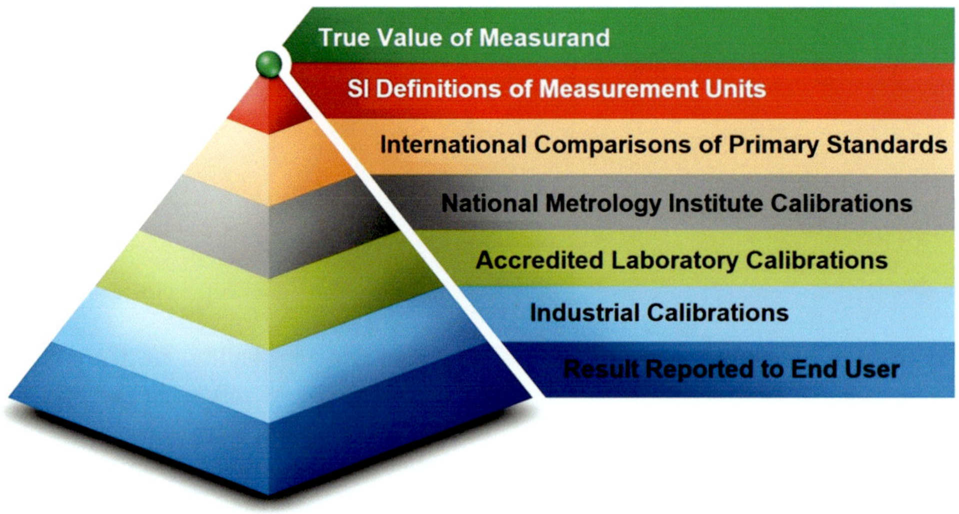

Figure 9.1 The Traceability Pyramid.

The apex of the traceability pyramid represents the true value of the measurand, which by international agreement is defined by the SI. The next level down represents the results of international comparisons between primary standards that provide physical realizations of the SI units. This level is followed by three levels of calibrations, the calibrations performed by national metrology institutions, the calibrations performed by accredited calibration laboratories, and the calibrations made in the various sectors of industry. The lowest level of the pyramid represents the traceable measurement results that are reported to end users.

Illustrations of the traceability pyramid, depicted in various ways, can be found hanging on the walls of calibration and standards laboratories all over the world. These depictions are useful because they nicely illustrate the hierarchical structure of metrology. However, as we are about to see, metrological traceability is not defined as a pyramid, but rather as a chain.

9.1 The Definition of Metrological Traceability

The *International Vocabulary of Metrology* (*VIM*) defines metrological traceability as

> *"the property of a measurement result whereby the result can be related to a reference through a documented unbroken chain of calibrations, each contributing to the measurement uncertainty"* [2]

Breaking the definition down into smaller parts is helpful, because it eliminates some of the common misconceptions about what metrological traceability really means [3]. The first thing the definition tells us is that traceability is "the property of a measurement result." This is important to know, because it's not unusual for the word "traceable" to be used as an adjective that modifies something other than a measurement. For example, phrases such as a "traceable time signal," "a traceable clock," "a traceable instrument," and "a traceable laboratory," are commonly heard but they are incorrect. It's true, of course, that time signals, clocks, instruments, and laboratories can all contribute to a traceable measurement. However, traceability is only a property of the measurement result itself.

The definition then tells us that in order for a measurement result to have the property of traceability, several conditions must be satisfied. The first condition is that we have to ensure that "the result can be related to a reference." That seems obvious, because all measurements have a reference, but the *VIM* stops short of telling us exactly what the reference should be. It's important to remember that even though the *VIM* defines the words and phrases used in metrology, it doesn't provide specific guidance on how to apply the definitions in a laboratory setting. Other guidance documents, however, specifically tell us that the reference should be the SI. Perhaps most notably, the international standard for testing and calibration laboratories, *ISO/IEC 17025*, states that laboratories should ensure that measurement results are traceable to the SI unless it is not technically possible [4]. This guidance acknowledges the fact that the SI is always the preferred reference. It's also the ultimate reference because, as stated elsewhere in this book, its unit definitions are accepted as truth and their measurement uncertainty is zero. It's technically possible to use the SI as the reference for all time and frequency measurements, so we can unequivocally state that time and frequency measurement results are only traceable when they are referenced to the SI.

Returning to the *VIM* definition, the next condition is that there must be "a documented unbroken chain of calibrations" that connects the measurement result to the reference. This phrase depicts traceability as a chain, and not as the pyramid shown in Figure 9.1. A traceability chain is analogous to a physical chain, with both consisting of a series of connected links. In a traceability chain, each link is a calibration that was conducted by collecting and analyzing measurements from a comparison between a DUT and a more accurate reference. The definition also implies that the links should be documented, and that the chain must be "unbroken," meaning that it should extend all the way from the SI to the final measurement result.

Finally, the last part of the *VIM* definition tells us that each of the calibration links in the traceability chain are "contributing to the measurement uncertainty." This reminds us that traceability and measurement uncertainty go hand in hand, and that measurement results with unknown uncertainties can't be traceable. In order to know the uncertainty of a measurement result, we have to know the uncertainty of every calibration link in the traceability chain. Once the uncertainty is known, its magnitude is unimportant—a measurement result can have a large uncertainty and still be traceable. If fact, a note to the *VIM* definition reminds us that "traceability of a measurement result does not ensure that the measurement uncertainty is adequate for a given purpose" [2].

Keeping the *VIM* definition of traceability in mind, we'll continue our discussion by describing why tracing measurements back to a standard unit has been, and always will be, an essential part of verifying a measurement result.

9.2 The Importance of Standard Measurement Units

We noted at the very beginning of this book that all measurements are comparisons. When humans first began measuring things, they needed references for their comparisons that were naturally occurring and convenient to use. The repeatable, naturally occurring astronomical events involving the Earth and the Sun provided a convenient comparison reference for time measurements. The variations in the astronomical time references, although considered large today, were so small that they went unnoticed for centuries.

Length measurements were also based on comparison references that had a natural basis and were convenient to use, but these references weren't repeatable and stable like the astronomical time references. The units and comparison references for the earliest length measurements were often based on human anatomy, which of course made them subject to large variations. The cubit was one of the first of these units, and the rods and sticks that measured cubits are often regarded as the first true measurement references.

The length of a cubit was considered equivalent to the length of a man's extended forearm, from his elbow to the tip of his middle finger.

Many ancient and medieval cultures based their length measurements on cubits, so the unit's exact origins are unclear. However, the cubit may have been first used in Egypt around 3000 BC, during the same approximate era when the Egyptians first began using shadow clocks. We do know that the Egyptians were already well versed in length measurements made with cubits when they built the Great Pyramid of Giza. Estimates of the Great Pyramid's age have historically varied by thousands of years, but based on radiocarbon dating measurements with a 95% confidence interval, it was likely completed sometime between 2620 and 2484 BC [5]. It was built to a height of 280 cubits, with sides that measured 440 cubits wide at their base.

As the construction of the Great Pyramid demonstrated, the cubit worked well as a measurement reference in a controlled environment, but it wasn't a true measurement standard. If we compare the cubit to the meter, it was about a half meter long. However, the cubit's length varied substantially based on the size of the human used as a comparison reference, and it wasn't always measured in the same way. For example, the common, or short, cubit in Egypt was divided into six palms, each with a width of four digits (fingers), making it 24 digits wide. This was equivalent to a length of about 45 cm. The royal cubit that the Egyptians used to build the Great Pyramid added a seventh palm, that presumably represented the palm of the Pharoah who ruled Egypt. This made a royal cubit rod 28 digits wide and about 52.4 cm long, which was much longer than a typical forearm (Figure 9.2). The cubit rods used by cultures in regions outside of Egypt ranged in length from about 44 cm to 64 cm [6, 7].

Figure 9.2 A replica of a royal cubit rod (courtesy of Queensland Museum, Australia).

Other length units based on human anatomy had similar issues that eventually required their values to be standardized. One example is the hand, a unit still used to measure the height of a horse. It was originally based on the breadth of the four extended fingers of a man's hand, which on average is about 3.5 inches (8.89 cm). To avoid confusion caused by variations in hand size, its length was standardized as 4 inches (10.16 cm) by the English Parliament in 1540.

A more common example is the foot, which has been a convenient unit of length from ancient times to the present, because it allows distance to be measured by counting steps. The history of using the foot as a measurement unit is exceptionally confusing, because much like the cubit, its value was influenced both by variations in the size of humans and by the various ways it's been measured. Depending on when, where, and how it was measured, the values assigned to the foot have ranged from about 11 to 14 inches (27.9 to 35.6 cm) [7]. In the United States, the only industrialized nation where the foot is still the favored unit for length, one foot equals 12 inches (30.48 cm), a value that was formally standardized in 1893 [8].

There are countless other examples of measurement units similar to the cubit, hand, and foot, that were defined based on arbitrary comparison references. The variation in these units made it impossible to compare weights and measures even across small geographic regions, let alone internationally. The lack of measurement standards made everything more difficult, including commerce and trade, and all forms of communications between regions. As travel became more common and society became less parochial, the problems caused by the lack of measurement standards became more detrimental. These problems were most apparent in Europe, where a large number of nations were located within a relatively small geographic area, and where each nation had its own languages and traditions.

By the latter part of the 18th century, it had become obvious in Europe that even though there would always be barriers caused by the different languages spoken in different nations, a common measurement language was needed. A common measurement language would only be possible if a uniform metric system was established. To make measurement comparisons easier, all of the metric system's units should be based on a common numbering system, which ideally should be the decimal system.

The first steps toward a metric system were taken in France and they were aimed at establishing a standard unit for length measurement. This unit, called the *mètre* in France, the metre in the United Kingdom, and the meter in the United States, can trace its name to a Greek verb that means to measure, count, or compare. Most 17th- and 18th-century scientists seemed to favor a definition of the meter that involved the swings of a pendulum. As we saw in Chapter 1, pendulums were first studied by Galileo in the late 16th century, and pendulum clocks became a timekeeping reference after they were invented by Huygens in the mid-17th century. Because the period of a pendulum's swing was a function of its length, time measurements were already related to length in 18th-century France, and it seemed logical to some scientists to relate length measurements back to time [9]. However, a different direction was chosen, and the meter was defined to be a tiny fraction of the circumference of the Earth.

The definition of the meter was agreed upon during the French Revolution, a period of social and economic reform in France that began in 1789. The responsibility for measurement reform was given to the French Academy of Sciences, which formed a committee led by the mathematician and physicist Jean-Charles de Borda (1733–1799). Borda's background was substantial; he was a former military engineer whose studies of projectile motion had gotten him elected to the Academy at a young age. He participated in the American Revolutionary War, and afterwards was in charge of several vessels in the French Navy. Borda was also a skilled instrument maker and a strong advocate for decimalization and a metric system [7]. His enthusiasm for decimalization even made him responsible for a period during the French Revolution—lasting from October 1793 to April 1795—where the French followed a system of decimal time. This system defined the day as a period of 10 hours, with an hour being the period of 100 minutes, and a minute the period of 100 seconds [7, 10]. That change was considered far too radical for the people to accept, even during a revolution, and the French went back to defining days, hours, and minutes with the traditional duodecimal and sexagesimal numbering systems that we described in Chapter 1.

In 1791, Borda's committee chose to define the meter as one ten-millionth (10^{-7}) of the distance from the north pole to the equator, to be measured along the Paris meridian, the circle of constant longitude that passed through Paris (Figure 9.3). Based on this definition, a quadrant of the Paris meridian would be 10,000 km long, and the polar circumference of the Earth would equal 40,000 km. Establishing the length of the meter meant that the quadrant had to be surveyed, and Borda was personally involved in the survey preparations, working on both the surveying methods and the instrumentation design. Surveying the entire quadrant would have been impossible, but it was possible to survey a segment of the quadrant that covered about 10% of its length. The segment chosen extended from Dunkirk, a city in northern France on the English Channel, to Barcelona on the Mediterranean coast of Spain. This segment was a good representation of the entire quadrant for at least two reasons; it was located near the middle of the quadrant, and both Dunkirk and Barcelona were at sea level [7, 9, 11, 12].

The meridian arc survey began in 1792, and it took more than six years to complete. It was an exceptionally ambitious undertaking, similar in some ways to the clockmaking pursuits of John Harrison, but perhaps even more difficult because it took place during the French Revolution. The survey was led by two French astronomers, Jean Baptiste Joseph Delambre (1749–1842) and Pierre Méchain (1744–1804). Delambre measured the southern portion of the meridian arc, starting in Barcelona, and Méchain measured the northern portion, starting in Dunkirk. Both surveyors endured many hardships, but the survey was especially difficult for Méchain. At one point, he was seriously injured, he was arrested when it was thought that his instruments were weapons, and his work was interrupted when war broke out between France and Spain in 1793. Méchain was also fanatical about accuracy, and he unfortunately believed that his methods and instruments could achieve any desired accuracy. When his measurements didn't produce the results he expected, he blamed himself for discrepancies that he thought were caused by his own errors, but which were instead what we would now identify as measurement uncertainties. When the survey was finally completed, Méchain went back to Paris and became director of the Paris Observatory, but the doubts he still had about his measurements eventually led him back to Spain, where he tried to improve upon his original results. He passed away there in 1804, after contracting yellow fever [12].

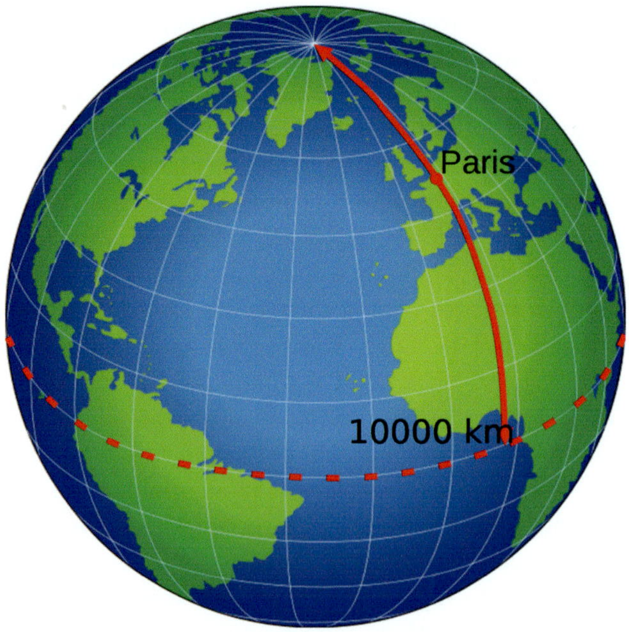

Figure 9.3 The original definition of the meter was one ten-millionth of the Paris meridian, which extended from the equator to the north pole.

Despite all the hardships, the data collected from the meridian arc survey were used to successfully determine the length of a meter. We now know that the measurement results underestimated the polar circumference of the Earth by about 8 km, an uncertainty of about 0.02%, which was a remarkable measurement for that era. The meter was officially defined by an artifact standard—a platinum bar that was placed in the French National Archives in 1799. Copies of the meter bar were made and distributed, but the meter wasn't accepted internationally as a measurement standard until 1875, as discussed in the next section.

9.3 The BIPM and the International System of Units (SI)

The kilogram, the standard unit of weight or mass, was also defined during the French Revolution. This definition, agreed upon in 1795, made the kilogram equal to the mass of one liter of liquid water, with the volume of a liter equal to one cubic decimeter (10 cm × 10 cm × 10 cm). An artifact standard for the kilogram was made out of platinum, and like the platinum bar that defined the meter, it was placed in the French National Archives in 1799. There was now a coherent system of units in place, where distance was defined in terms of meters, mass in terms of kilograms, and time in terms of seconds. This meter-kilogram-second system of units was called the MKS system, and although change happened slowly, it gradually gained acceptance both inside and outside of France during the first three-quarters of the 19th century.

At the beginning of March 1875—after many years of discussions regarding how measurement artifacts should be distributed and shared amongst nations, and six years after an International Meter Commission was formed in 1869—a conference convened at the French Foreign Ministry in Paris. Representatives from twenty nations, all members of the Meter Commission, accepted the invitation to attend the conference. After weeks of discussions and numerous drafts of proposals, representatives from 17 of the 20 nations signed the Meter Convention (*Convention du Mètre*) on May 20, 1875 [9]. To honor the signing of this historic treaty, May 20th is now celebrated annually as World Metrology Day.

The 17 original member states of the Meter Convention (listed alphabetically) were Argentina, Austria-Hungary, Belgium, Brazil, Denmark, France, Germany, Italy, the Ottoman Empire, Peru, Portugal, Russia, Spain, Sweden and Norway, Switzerland, the United States of America, and Venezuela. This group has grown to include 64 member states as of 2025, with another 36 nations represented as associate states.

The Meter Convention provided a foundation for an international system of units by creating the International Bureau of Weights and Measures (*Bureau International des Poids Et Mesures*, BIPM). The BIPM was established as an intergovernmental organization that would coordinate the new measurement system. The BIPM was immediately made responsible for the maintenance of the international prototypes of the meter and kilogram. It was also responsible for the future verifications of the prototypes, and for organizing periodic comparisons of the prototypes to the national standards of the member states.

It was already understood in 1875 that the system of units would be a formalization of the MKS system, but the BIPM's initial responsibilities were limited to just the meter and kilogram. This made sense, because the second was then based on the rotation of the Earth, and its determination remained in the hands of astronomical observatories [13]. It was also understood that the BIPM's responsibilities would include all future refinements in the definitions of measurement units, as well as all future expansion of the unit system.

The Meter Convention also determined how the BIPM would be governed, and it established the same three-bodied organizational structure that is still in use today. The BIPM is subordinate to the International Committee for Weights and Measures (*Comité International des Poids et Mesures*, CIPM), and the CIPM is in turn subordinate to the General Conference on Weights and Measures (*Conférence Générale des Poids et Mesures*, CGPM). The CIPM is an international advisory committee composed of 18 members, each belonging to a different state. Its role is to implement the metrological work that the member states have decided upon. The CGPM is a plenary, policy-making body that typically meets once every four years. It includes representatives from all of the member states, and a measurement unit can only be introduced or redefined if the CGPM approves. This three-bodied organizational structure can be loosely compared to a publicly traded company, where the BIPM is the company itself, the CIPM is the company's board of directors, and the CGPM is a meeting of the company's shareholders.

The BIPM is located in the Parc de Saint-Cloud, at Sèvres, in the suburbs southwest of Paris. Because the BIPM is an international organization that's considered to be both neutral and autonomous, its site has been classified as international territory since 1969. The site and its main building, the Pavillon de Breteuil, was granted to the BIPM by the French government in 1876. By then, the Pavillon de Breteuil was already more than two centuries old, having been built in 1672 during the reign of the French king Louis XIV (1638–1715), who had it constructed for his brother's family. It remained in that family until 1785, when it was acquired by the French queen Marie Antoinette (1755–1793). When the French Revolution began, the Pavillon de Breteuil became the property of the French state, and from then on it served a variety of purposes, including housing military personnel, high-ranking officials, and visiting foreign dignitaries. The building was badly damaged during the Franco-Prussian War of 1870–1871, and was in a sorry state of repair when it was given to the BIPM. The CIPM supervised its restoration, as well as the construction of a new building that was completed in 1878 to house the metrology laboratories [9, 14]. The entrance to the BIPM, with the Pavillon de Breteuil in the background, is shown in Figure 9.4.

Figure 9.4 The entrance to the International Bureau of Weights and Measures (BIPM).

One of the BIPM's first tasks was to produce two groups of artifact standards, one for the meter and the other for the kilogram. Because the artifact standards would serve as physical realizations of the quantities of the meter and kilogram, they had to be made as carefully as possible, from materials whose properties changed as little as possible. For both the meter and the kilogram, this material was chosen to be an alloy that was 90% platinum and 10% iridium. The BIPM then had to designate one of the meter artifacts as the international prototype meter (IPM), and one of the kilogram artifacts as the international prototype kilogram (IPK). The IPM and IPK would respectively serve as the international standards for length and mass. Once this was done, nearly identical copies of the IPM and IPK could be distributed to the member states of the Meter Convention, where they would serve as national standards [9, 13, 14].

The IPM was selected in 1889 from a group of 30 numbered meter bars that were delivered to the BIPM in 1886. The meter bars had an X-shaped cross section that provided maximum stiffness, so the length of the bar wouldn't change due to sagging or bending. The IPM was meter bar number 6 in the group of 30, and it was chosen after measurements determined that its length was closest to the meter bar placed in the French National Archives some 90 years earlier. The IPM and several copies were kept at the BIPM, but the remaining copies were distributed to the member states. For example, meter bars 21 and 27 were sent to the United States in 1890, with bar 27 serving as the U. S. national length standard until 1960 (Figure 9.5).

The IPK was chosen from a group of three cylinders, designated as KI, KII, and KIII, that were delivered to the BIPM in 1879. Each cylinder was about 39 mm in height and 39 mm in diameter, making them roughly the size of a golf ball. In 1889, KIII was designated as the IPK, again because measurements determined that its mass was closest to the kilogram placed in the French National Archives in 1799. Forty numbered copies of the IPK were delivered to the BIPM in 1884. As was the case with the meter bars, the IPK and several copies were kept at the BIPM, but the remaining copies were distributed to the member states. For example, copies K4 and K20 were sent to the United States in 1890, with K20 serving as the U. S. national mass standard until 2019 [9, 14]. The IPK, protected by three nested glass bell jars, is shown in Figure 9.6.

Figure 9.5 One end of meter bar 27, sent by the BIPM to the United States in 1890, where it served as the national standard for length measurements until 1960.

Figure 9.6 The international prototype kilogram (IPK) of the BIPM. The IPK served as the international standard for mass measurements from 1889 to 2019.

All mechanical units could be derived from the three base units of the MKS system, but the system had shortcoming when applied to electrical measurements, which were a huge topic of interest by the end of the 19th century, largely due to the work of James Clerk Maxwell (1831–1879). Maxwell was mentioned in Chapter 3 as the first to suggest an atomic clock, but he is best known for his formulation of electromagnetic theory, and his famed *Treatise on Electricity and Magnetism* was published about two years before the Meter Convention was signed. Maxwell's formulations were based on a variant of the MKS system, called the CGS system, where the centimeter was the unit of length, the gram was the unit of mass, and the second was the unit of time. The CGS system is still convenient to use for some areas of theoretical physics, but it simply wasn't suitable for practical electrical metrology.

To address these shortcomings, the Italian physicist and electrical engineer Giovanni Giorgi (1871–1950) proposed in 1901 that a fourth unit to be added to the MKS system, one chosen from the various units of electromagnetism. Giorgi's proposal was followed by several decades of discussion regarding which unit should be added, or even if an electrical unit should be added at all. Giorgi originally favored the ohm, the standard unit of electrical resistance, as the base unit, but the ampere, the standard unit of electrical current, was eventually chosen. It was confirmed as a base unit by the CGPM in 1948, forming an MKSA metric system that briefly preceded the SI. The CGPM approved the inclusion of two more base units in 1954—the kelvin, the base unit of temperature, and the candela, the base unit of luminous intensity [9, 13, 14].

In October 1960, almost 170 years after the meter was defined during the French Revolution, the CGPM formally gave the name International System of Units (*Système International d'Unités*, SI) to the metric system of units. The SI then included six base units, the three original MKS units plus the ampere, kelvin, and candela. The mole, used in chemistry as a measure of the amount of substance, became the seventh base unit in 1971. The unit names, unit symbols, and the quantities of the seven base units are listed in Table 9.1 [15].

Table 9.1 The seven base units of the International System (SI)

Unit name	Unit symbol	Quantity
meter	m	length
kilogram	kg	mass
second	s	time
ampere	A	electric current
kelvin	K	temperature
candela	cd	luminous intensity
mole	mole	amount of substance

Perhaps not surprisingly, the first two units of the SI, the meter and kilogram, were also the last units defined by artifact standards. However, at the same CGPM meeting where the SI was formally established in October 1960, the meter was redefined so that it was no longer an artifact standard. It was then redefined as a certain number of wavelengths (1,650,756.73) of the light released by a specific energy transition of the krypton-86 atom. The CGPM replaced this unwieldy definition in 1983, when the meter was redefined with respect to the fundamental physical constant of the speed of light. The meter is now defined as "the length of the path travelled by light in vacuum during a time interval with duration of 1/299 792 458 of a second" [15]. As we'll discuss in Section 13.2, this definition was made possible by optical frequency measurements made in the Time and Frequency division of NBS [16, 17].

The kilogram was the last artifact standard. The IPK defined the kilogram from 1889 until May 20, 2019. On that date, which was the 144th anniversary of the Meter Convention, the CGPM redefined not only the kilogram, but four of the other base units—the ampere, kelvin, candela, and mole. All were redefined in terms of fundamental physical constants, and all seven of the base units are now defined in terms of one or more of the seven physical constants listed in Table 9.2 [15]. Even so, the unit definitions in the new SI still closely approximate the original units, with the differences being insignificant for most practical purposes. For example, the new definition of the kilogram is based on the hyperfine transition frequency of a cesium atom, the speed of light, and the Planck constant, and it agrees within a few tens of micrograms, or a few parts in 10^8, with the old definition that was based on the mass of a liter of water [18]. The interdependencies of the SI base units, both to each other, and to the physical constants that define them, are shown in Figure 9.7.

Table 9.2 The seven defining physical constants of the SI

Defining constant	Symbol	Numerical value	Unit
hyperfine transition frequency of cesium	$\Delta\nu_{Cs}$	9 192 631 770	Hz
speed of light in vacuum	c	299 792 458	m s^{-1}
Planck constant	h	$6.626\ 070\ 15 \times 10^{-34}$	J s
elementary charge	e	$1.602\ 176\ 634 \times 10^{-19}$	C
Boltzmann constant	k	$1.380\ 649 \times 10^{-23}$	J K^{-1}
Avogadro constant	N_A	$6.022\ 140\ 76 \times 10^{23}$	mol^{-1}
luminous efficacy	K_{cd}	683	lm W^{-1}

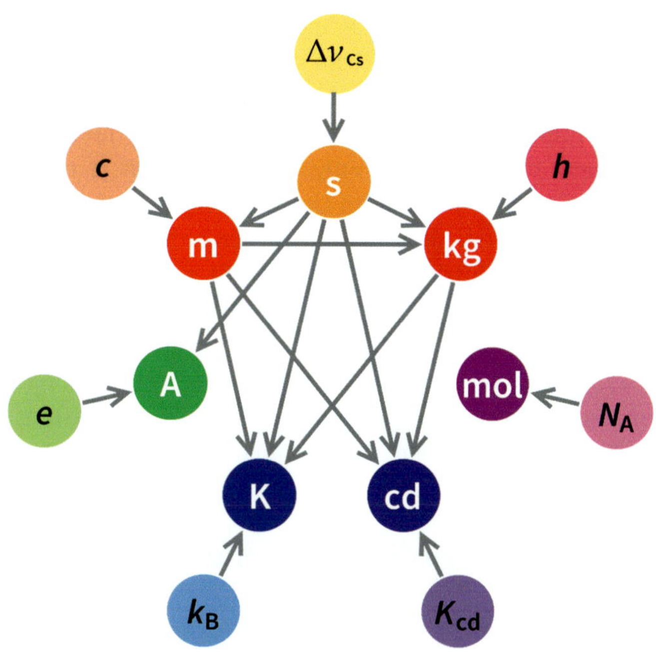

Figure 9.7 The relationships between the SI units and their defining physical constants (courtesy of Emilio Pisanty).

Only two SI units apply to time and frequency metrology, the second (s) and the hertz (Hz). The second is one of the seven base units, and as shown in Figure 9.7, it contributes to the definitions of five of the other base units. The mole is the only base unit that isn't tied to the second. As discussed in Section 3.4, the second was defined in 1967 as "the duration of 9 192 631 770 periods of the radiation corresponding to the transition between the two hyperfine levels of the ground state of the caesium 133 atom" [19]. The hertz, the standard unit of frequency, is one of the 22 coherent derived units that the SI provides with its own name and symbol. The SI defines each of the derived units as a product of a base unit, and the SI defines the hertz as s^{-1}. This is the same thing as $\frac{1}{s}$, and it reminds us once again that frequency is the reciprocal of time interval [15].

A traceability chain that extends all the way from the base unit definitions to an end user's measurement result is an abstract concept that can be hard to visualize. This is because the SI units aren't physical standards, but just defined values, and a DUT can't be directly compared to a defined value. To complete a traceability chain, there has to be a physical realization—an embodiment of the quantity being measured—before a real measurement can be made. Providing a physical realization of a quantity was, of course, the purpose of the artifact standards that once defined the meter and kilogram, and it's still the purpose of the primary standards used in all fields of metrology, including the cesium fountains that realize the SI second. Even so, measurements that directly compare a DUT to a primary standard are rare. The reasons are obvious—there are relatively few primary standards, and they reside at national metrology laboratories that are likely located far away from the DUT. If a comparison to a primary standard is done at all, it will likely be done only once every few years, which means that the measurement uncertainty of that particular link of the traceability chain won't be verified very often.

Fortunately, the issues that affect the establishment and verification of a traceability chain have been mostly overcome in time and frequency metrology, due to the international coordination of world timekeeping. This coordination effort includes all of the world's major timekeeping laboratories, and it results in Coordinated Universal Time (UTC), an atomic time scale based on the SI definition of the second, and the world's best approximation of the SI units for time and frequency. The next section describes how UTC is determined.

Before moving on to the next section, it's important to note that as this book is being written, it's common knowledge that the SI second will be redefined in the near future in terms of an optical clock transition (Chapter 13). A redefinition of an SI unit can only be approved by the CGPM, which meets every four years in Paris. A "roadmap" for the redefinition is already in place, and it indicates that the approval will most likely take place at either the 2030, 2034, or 2038 CGPM meeting. According to the criteria provided by the roadmap, a new definition will only be approved if it reduces the uncertainty of the physical realization of the second by 10× to 100× in the short term, and perhaps by even more than that in the long term. In addition, the new definition must ensure continuity with the current cesium definition, ensure the continuity and availability of the new SI second through UTC, enable a significant improvement in the quality of the UTC(*k*) time scales, and enable the distribution of the SI second to broad categories of users [20]. These criteria appear to be carefully thought out, and the redefinition should be seamless from a traceability perspective. Of course, the best part of a redefinition is the promise of smaller measurement uncertainties, which will not only benefit all existing technologies that rely on accurate time and frequency, but will also make new technologies possible.

9.4 The Determination of Coordinated Universal Time (UTC)

Before the SI transitioned from astronomical to atomic time, the most accurate measurements of time were made by astronomers. The process of establishing measurement traceability wasn't yet formalized, but astronomers understood that what we now call "the unbroken chain of calibrations" had to trace back to the rotational rate of the Earth. The rotational rate was accurately measured by measuring the angular position of stars more distant than the Sun. The clocks in astronomical observatories were synchronized and continuously adjusted based on the results of these observations.

The subject of time wasn't covered by the Meter Convention of 1875, even though the idea of establishing a worldwide or "universal" time standard had long been a topic of discussion. Because the angular position of the stars in the sky depended on the location of the observer on Earth, any attempt to establish a universal time would require a standard meridian to be chosen, so that all astronomical time measurements could be referred to the standard meridian. At the time of the Meter Convention, most ships at sea were already using the meridian that passed through the Royal Observatory in Greenwich, England (Figure 9.8) as the reference meridian on their charts and maps, but there was still no international consensus. The matter was finally settled by delegates from 25 nations who met at the International Meridian Conference held in Washington, DC, in October 1884. The delegates agreed that the Greenwich meridian would "be employed as a common zero of longitude and standard of time-reckoning throughout the whole world" [21]. As a result of this agreement, the Greenwich meridian became the international standard prime meridian with a longitude of 0°. The time referenced to the prime meridian was called Greenwich Mean Time (GMT).

Figure 9.8 The line of the prime meridian at the Royal Observatory in Greenwich, England (courtesy of Hans Bézard).

There was still ambiguity in the way that GMT was used. Astronomers used GMT to measure the duration of the astronomical day that began at noon, but everyone else used it for the civil day that began at midnight. There was hope expressed at the 1884 conference that both the astronomical and civil days would be standardized to always begin at midnight, but the inconsistent usage of GMT still persisted. Various scientific organizations recommended other designations for Greenwich Time, including Greenwich Civil Time (GCT) if the day began at midnight, and Greenwich Mean Astronomical Time (GMAT) if the day began at noon [22]. To try to stop the confusion, the International Astronomical Union (IAU) introduced the term Universal Time (UT) in 1928, noting that UT would "denote time measured from Greenwich Mean Midnight" and was "not ambiguous" [23]. Of course, the confusion didn't completely stop, as the term GMT is still used and perhaps always will be, but proper time measurements have been referenced to some form of UT since 1928.

Several different time scales referenced to the Earth's rotation have been designated as universal time, with a number added to the UT prefix. The UT0 time scale was the mean solar time of the prime meridian obtained from direct astronomical observations. As observatory clocks improved, especially when observatories began replacing their pendulum clocks with quartz clocks, it was found that the methods then used to measure the Earth's rotation angle were affected by polar motion. The small variations in position of the poles caused the longitude of the observing station to shift, which in turn contributed uncertainty to the time measurements.

These discoveries led to the UT1 time scale, which corrected for polar motion and made UT0 obsolete. The UT2 time scale applied corrections to UT1 for seasonal variations in the Earth's rotational speed. This sometimes had the effect of smoothing UT1, but because the seasonal variations are difficult to predict, UT2 often had no significant advantage over UT1 when measured over long intervals. For this reason, UT2 is now also considered obsolete, leaving UT1 as the principal universal time scale that's directly referenced to the Earth's rotation [21, 22, 24].

Until the 1920s, there were no clocks accurate and stable enough to definitively prove that the Earth's rotation had any uncertainty at all. As we saw in Chapters 1 and 2, this began to change with the development of the Shortt pendulum clock and the quartz clocks that followed, but the uncertainty of these clocks was still nearly as large as the uncertainty of the various versions of UT. That was no longer true after the invention of the cesium clock. By the late 1950s, cesium clocks could keep time that was not just slightly more accurate than UT, but already about 100× more accurate, a gap that has continued to widen ever since. This meant that astronomical time (UT) and atomic time (AT) will always diverge from each other, with their time difference becoming increasingly larger.

Keeping a continuous record of the time difference between UT and AT was important, and it required establishing a fixed origin, or epoch, for atomic time. When the USNO formed the first ensemble atomic time scale, known as A.1, in 1959 (Section 4.4), they defined this epoch in terms of the UT2 time scale. They declared that on January 1, 1958, at 0 hours, 0 minutes, and 0 seconds UT2, the time difference between atomic and astronomical time was 0 [25]. This convention was followed internationally, and January 1, 1958, became the epoch of a free-running, uninterrupted time scale that is based entirely on continuous measurements of atomic clocks. As we'll see in the next section, the free-running atomic scales that followed were given various names, and all of them provided huge advantages in frequency accuracy when compared to astronomical time scales.

Despite their accuracy advantage, there were many reasons why atomic time scales couldn't simply replace astronomical time scales, and the possibility of that happening wasn't seriously considered in the late 1950s. The first and foremost reason why it couldn't happen was based on world culture and tradition—from the beginning of recorded history, civil time had been based on the position of the Sun in the sky, and any sudden departure from that convention would simply be too radical for people to accept. Another reason why was that atomic clocks were a brand-new and unproven technology. Only a few existed, so access was limited, and the few that did exist were expensive to build, complex to operate, and not very reliable. In addition, the rotating Earth was a universal time reference, available for anyone to compare against, but individual atomic clocks were not. This meant that there would have to be ways of determining if the atomic clocks in various nations were being kept in agreement. Finally, the reason cited most often by scientists, both then and now, was that if people were forced to use atomic time, the difference between UT and AT would have to be kept small. For these reasons and others, it was decided to coordinate AT so that it was in relative agreement with UT.

The first efforts to coordinate atomic time internationally began on January 1, 1960, more than seven years before the SI second was redefined based on the cesium atom. On that date, five timing laboratories—NBS, the Naval Research Laboratory (NRL), and the USNO in the United States, and NPL and the Royal Greenwich Observatory (RGO) in the United Kingdom—began to coordinate the adjustments made to their atomic time scales [26]. These laboratories distributed time signals by radio, and their coordination efforts allowed them to synchronize their time transmissions to within a millisecond [27].

Laboratories in other countries gradually joined the atomic time coordination network, and the International Time Bureau (*Bureau International de l'Heure*, BIH) assumed the responsibility of coordinating atomic time internationally in 1961 [28]. The BIH, located in the Paris Observatory, was established in 1912 to help standardize the format of radio time signal broadcasts (Section 10.1). Since the 1920s, the BIH had been responsible for combining and publishing measurements of UT made by the world's leading observatories. These valuable experiences prepared the BIH for its next achievement, the development of a coordinated universal time scale based on atomic clocks.

9.4.1 Coordinating a Universal Time Scale Based on Atomic Clocks

Before the BIH began coordinating atomic time, they were already collecting data from distant atomic clocks via VLF phase comparisons, and they were using this data to calculate atomic time scales. Their efforts were concurrent with the USNO's efforts to produce A.1, but the BIH didn't publish atomic time scale data until 1961, when they were first authorized to do so. They followed the USNO's lead of using January 1, 1958, as the epoch for atomic time, and they named their atomic time scales using the convention A*n*, where *n* was the number of atomic clocks used in the average. By 1964, the BIH had changed this designation to AM, which was the mean value of all available atomic clocks, with each clock given equal weight. They also kept a separate, more stable time scale called A3 that was determined by averaging data from the three best clocks. The three best clocks were originally considered to be those located at NBS in the United States, at NPL in England, and at the *Laboratoire Suisse de Rechérches Horlogeres* (LSRH) in Switzerland. More atomic clocks were added to the group of three in 1966, but the name A3 was retained. The A3 time scale continued to evolve, and it eventually became International Atomic Time (*Temps Atomique International*, TAI), a name that was authorized by the CGPM in a series of decisions made from 1971 to 1975 [22, 28, 29].

The effort to internationally coordinate atomic time was first led by the Ukrainian-French astronomer Nicolas Stoyko (1894–1976), who began working at the BIH in 1924 and who served as its director from 1942 to 1965. Stoyko was one of the first scientists to definitively measure the variations in the Earth's rotation. From 1934 to 1937, he painstakingly tabulated data from pendulum clocks at observatories in the United States, France, and Germany, discovering that the same seasonal variations in the Earth's rotation could be measured from all locations, and that a day in January was about 2 ms longer than a day in June [22, 30]. These results were confirmed, with smaller uncertainties, with measurements made by Stoyko and other scientists using groups of quartz clocks during the 1940s.

Stoyko's atomic time scale work was done in close collaboration with his spouse, the astronomer Anna Stoyko (1914–1991), who had joined the BIH in 1947. They worked together on calculating the A3 time scale and its predecessors, and on estimating its frequency and time differences with respect to astronomical time. The Stoykos retired together in 1965, and the French astronomer Bernard Guinot (1925–2017) became the new BIH director. Guinot joined the Paris Observatory in 1952, and remained as the BIH director until 1985, overseeing much of the development of the UTC system as we know it today.

The BIH work on atomic time scales was based on the premise that civil time would continue to be based on the position of the Sun in the sky. This meant that there would always have to be two atomic time scales. One atomic time scale would be free running and uninterrupted. Its frequency would be far more accurate than the Earth's rotation, but its time would diverge from astronomical time. This was the A3 time scale that was later named TAI. The other atomic time scale would have its frequency, and occasionally its time, adjusted to agree with the Earth's rotation. The adjusted time scale was called Coordinated Universal Time (UTC).

The name "Coordinated Universal Time" had been used informally since the early 1960s [22, 29]. The name can't be definitively attributed to any one person, and as Guinot wrote, the time scale might have received its name "spontaneously" [28, 31]. The name choice was probably obvious to scientists because the words "Universal Time" paid homage to the astronomical time scales which were then the world's time reference, and because "Coordinated" referred to one of the first applications of atomic time scales, which was coordinating the transmission of radio time signals [27].

After at least several years of informal and common usage, the use of the name "Coordinated Universal Time" was formally approved by the IAU in August 1967, which was about two months before the

SI second was redefined. The meeting was held in Prague, Czechoslovakia, and was attended by a number of key figures in atomic timekeeping history, including Essen, Guinot, Markowitz, and Winkler [22, 26, 32].

It is often noticed that the acronym UTC doesn't match either the French spelling, *Temps Universel Coordoone*, or the English spelling of Coordinated Universal Time, so it is often thought that UTC was used instead of TUC or CUT to avoid favoring one language over another. That might have been part of the reasoning, but it also seems likely that beginning the acronym with "UT" was desirable because it was the rendering previously used with UT1 and UT2.

If we conveniently ignore the fact that the names TAI and UTC weren't formally approved as of 1960, we can say that beginning in 1960, the BIH made the UTC frequency approximately agree with the Earth's rotation rate by constantly applying an offset to the nominal frequency of TAI. They defined this offset at the beginning of the year, maintained it throughout the year, and changed it, if necessary, at the beginning of the following year. The offset was also applied to the nominal carrier frequencies of radio time signals, including those transmitted by NBS radio stations WWV and WWVH in the United States, because the carrier frequencies were used to distribute the national frequency standard [33]. This frequency correction method of generating UTC was practiced until 1971, with the corrections listed in Table 9.3. The magnitude of the corrections implies that the frequency accuracy of the Earth's rotation ranged from 1.3 to 3 parts in 10^8 during this period, which ensured that the time difference between UTC and astronomical time would be always be less than 1 s. However, because the Earth's rotation rate wasn't especially stable, there were always small residual time differences that couldn't be ignored, because the BIH was then committed to keeping UTC to within ±0.1 s of astronomical time. They accomplished this by occasionally making instantaneous time corrections in addition to the continuous frequency corrections. These time corrections, and the dates when they occurred, are listed in the rightmost column of the table [24, 34, 35].

Table 9.3 UTC frequency corrections from 1960 to 1971, before the advent of leap seconds

Year	UTC frequency correction, parts in 10^{10} $\frac{f_{UTC} - f_{AT}}{f_{AT}} \times 10^{-18}$	Implied frequency accuracy of Earth's rotation	UTC time corrections (s)
1960	−150	-1.5×10^{-8}	None
1961	−150	-1.5×10^{-8}	+0.05 (August 1)
1962	−130	-1.3×10^{-8}	None
1963	−130	-1.3×10^{-8}	−0.1 (November 1)
1964	−150	-1.5×10^{-8}	−0.1 (April 1) −0.1 (September 1)
1965	−150	-1.5×10^{-8}	−0.1 (January 1) −0.1 (March 1) −0.1 (July 1) −0.1 (September 1)
1966	−300	-3.0×10^{-8}	None
1967	−300	-3.0×10^{-8}	None
1968	−300	-3.0×10^{-8}	+0.1 (February 1)
1969	−300	-3.0×10^{-8}	None
1970	−300	-3.0×10^{-8}	None
1971	−300	-3.0×10^{-8}	None

As indicated in Table 9.3, the fixed UTC frequency offset was held constant at -300 parts in 10^{10} from 1966 to 1971. This offset was both larger and more constant that the offset of previous years, and was the result of a more accurate and stable determination of TAI, which had benefited from the use of the Loran-C radionavigation system. An acronym for long-range navigation, the Loran-C system was composed of stations that broadcast stable groundwave signals on a common carrier frequency of 100 kHz. These stable signals reduced the uncertainty of the atomic clock comparisons, which in turn gave the BIH better data to use for their TAI calculations. The uncertainty of the Loran-C path delay varied as a function of the distance between the atomic clock and the broadcast station, but was usually about 1 μs, as opposed to about 10 μs for clock comparisons based on VLF signals [22, 28, 29, 31]. However, there were a limited number of Loran-C stations, and each had a limited range of about 2,000 km, which essentially meant that only the laboratories located in North America and Europe could contribute their clock data to TAI and UTC. The invention of GPS removed these geographic restrictions in the 1980s, allowing UTC to become a true worldwide time scale.

The original UTC system was far more useful to frequency and time metrologists than its predecessors. It brought time and frequency metrology out of the exclusive realm of the astronomical observatory, and into a realm that was shared with calibration and standards laboratories, which was a huge benefit for both commercial and industrial measurements. Even so, the original UTC system still seemed peculiar. It was peculiar to frequency metrologists because it didn't take advantage of the frequency accuracy of atomic clocks, even though atomic clocks were by far the most accurate frequency source that the world had ever known. Instead, the atomic frequency was intentionally offset from its nominal value, making the frequency accuracy of UTC much worse than it could be. It was also peculiar to time metrologists because it inserted sub-second time steps into UTC at irregular intervals, a practice that didn't make sense in a world where we expect our clocks to update in even intervals of one second.

To address these peculiarities, the UTC system was revamped at the end of 1971, resulting in a new and improved time scale [36, 37]. A final sub-second time step of −0.1077580 s was made to UTC at the end of 1971, and in that same instant, the frequency corrections ended. The time step made TAI – UTC equal to exactly 10 s on January 1, 1972, at 0 hours, 0 minutes, and 0 seconds UTC. The end of the frequency corrections meant that the frequency of TAI and UTC would be exactly the same from that point forward. However, unlike TAI, UTC would still be synchronized to agree with the astronomical time scale UT1. This was accomplished by occasionally adding an integer second to the UTC time scale. The added second was called a leap second, because it was analogous to the leap day that we add to the calendar approximately once every four years.

The allowable tolerance for UT1 – UTC was originally set to ±0.7 s, but it was relaxed to ±0.9 s in 1974. This still ensured that atomic and astronomical time would always agree to within less than one second. The difference between the two time scales is called DUT1, where DUTI = UT1 – UTC, and the current value of DUT1 is transmitted with 0.1 s resolution by some of the time signal stations described in Chapter 10. Much finer resolution of 0.1 μs can be obtained from the International Earth Rotation Service (IERS), which produces daily DUT1 estimates [22].

The UTC system that was just described is still in use. As of 2025, a total of 27 leap seconds have been added to UTC, and TAI – UTC = 37 s. Table 9.4 lists the years when leap seconds were added. The last leap second occurred on December 31, 2016, making this the longest period in UTC history without a leap second, eight years and still counting. The previous record was seven years, from 1998 to 2005. Figure 9.9 shows the approximate time differences between TAI and UTC, beginning at the atomic time scale epoch in 1958, and continuing to 2024. The horizonal lines on the graph indicate when time adjustments were made to UTC.

The BIH ceased operations at the end of 1987, and its responsibilities were transferred to the IERS and the BIPM on January 1, 1988. The IERS was given the responsibility of determining UT1 and when leap seconds were needed, and the BIPM was given the responsibility of determining UTC [28]. This made the

BIPM not only the home of the SI, but also the home of UTC, which was both fitting and proper for the purposes of traceability. The next three sections describe how the BIPM calculates UTC, beginning with a discussion of the laboratories that contribute their atomic clock data to the calculations.

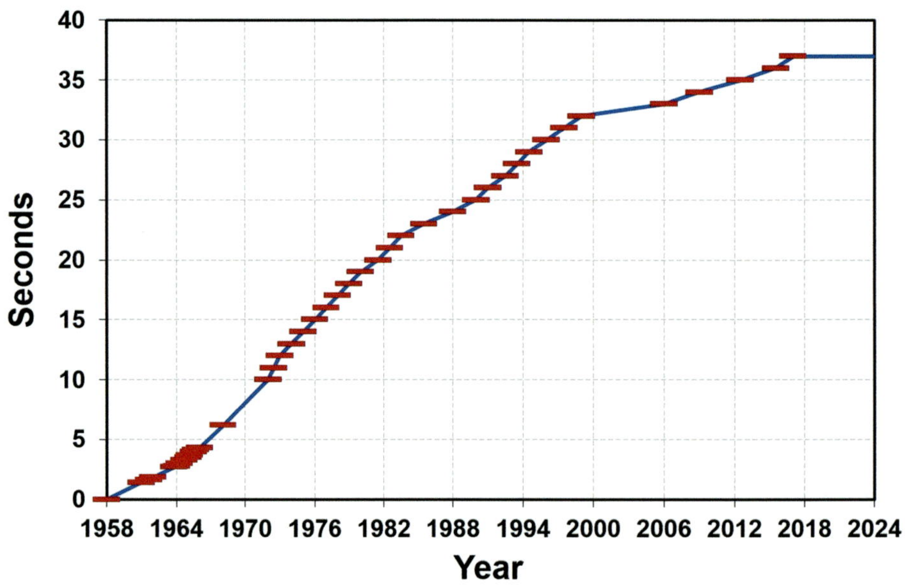

Figure 9.9 Approximate time differences between TAI and UTC from 1958 to 2024.

Table 9.4 Leap seconds added to UTC from 1972 to 2024

Leap second date	Leap second year	Total per date
June 30	1972, 1981, 1982, 1983, 1985, 1992, 1993, 1994, 1997, 2012, 2015	11
December 31	1972, 1973, 1974, 1975, 1976, 1977, 1978, 1979, 1987, 1989, 1990, 1995, 1998, 2005, 2008, 2016	16
Combined total	27	

9.4.2 The Laboratories that Contribute Atomic Clock Measurements to UTC

The UTC calculations are performed as part of a continuous international comparison of atomic clocks. This comparison has been designated by the CIPM's Consultative Committee for Time and Frequency (CCTF) as key comparison *CCTF-K001.UTC*. The BIPM organizes the key comparison, performs the UTC calculations, and publishes the results.

Every laboratory that participates in the key comparison contributes to UTC, and maintains a local time scale that approximates UTC. The BIPM calls these local time scales UTC(k), where k represents the acronym assigned to the laboratory. The key comparison benefits its participants in two ways. First, it allows them to establish traceability to UTC and to the SI, through measurement comparisons with their local UTC(k) time scale. Second, it allows the participants that distribute signals from their local UTC(k) time scale to establish a traceability chain—consisting of a documented, unbroken chain of calibrations with known measurement uncertainties—that extends from the SI second to the end user [39].

To participate in the key comparison, a laboratory has to be part of either a national metrology institute (NMI) or a designated institute (DI), and the NMI or DI has to reside in either a member or associate state of the CGPM. In most cases, the NMI has been assigned, either informally or through legislation, the responsibility of being the official timekeeper for their nation. However, in some nations, the official timekeeping responsibility has been assigned to a DI, or it's shared by the NMI and one or more DIs. The member states can fully participate in the CGPM activities, but the associate states only have observer status. However, both the member and associate states have agreed to participate in the CIPM mutual recognition arrangement (CIPM MRA), which means that all nations participating in the key comparison have agreed to accept each other's measurement results [40].

Participation in the key comparison also requires each laboratory to operate a time scale consisting of one or more atomic clocks, to operate time transfer equipment that outputs measurement data in the standard format requested by the CCTF, and to send their data to the BIPM on a continuous basis [39]. Table 9.5 lists the laboratories that contribute to UTC as of 2025.

Table 9.5 The laboratories, sorted alphabetically by country, that contribute to UTC as of 2025

UTC(*k*) acronym	Laboratory (English spelling)	City	Country
AGGO	Argentinian-German Geodetic Observatory	La Plata	Argentina
IGNA	National Geographic Institute	Buenos Aires	Argentina
INTI	National Institute of Industrial Technology	Buenos Aires	Argentina
ONBA	Naval Observatory of Buenos Aires	Buenos Aires	Argentina
AUS	National Measurement Institute	Sydney	Australia
BEV	Federal Office of Metrology and Surveying	Vienna	Austria
BY	Belarussian State Institute of Metrology	Minsk	Belarus
ORB	Royal Observatory of Belgium	Brussels	Belgium
SMD	Metrology Division of the Quality and Safety Department	Brussels	Belgium
IBM	Bolivian Institute of Metrology	La Paz	Bolivia
IMBH	Institute of Metrology of Bosnia and Herzegovina	Sarajevo	Bosnia and Herzegovina
INXE	National Institute for Metrology, Quality and Technology	Rio de Janeiro	Brazil
LRTE	Time and Space Reference Laboratory	Sao Carlos	Brazil
ONRJ	National Observatory of Rio de Janeiro	Rio de Janeiro	Brazil
BIM	Bulgarian Institute of Metrology	Sofiya	Bulgaria
NRC	National Research Council of Canada	Ottawa	Canada
BIRM	Beijing Institute of Radio Metrology and Measurement	Beijing	China
JATC	Joint Atomic Time Commission	Lintong	China
NIM	National Institute of Metrology	Beijing	China
NTSC	National Time Service Center of China	Lintong	China
INM	National Metrology Institute of Colombia	Bogota	Colombia
ICE	Costa Rican Metrology Laboratory	San José	Costa Rica
IFZG	State Office for Metrology	Zagreb	Croatia
TP	Institute of Photonics and Electronics, Czech Academy of Sciences (IPE/ASCR)	Praha	Czech Republic

UTC(k) acronym	Laboratory (English spelling)	City	Country
DFM	Danish National Metrology Institute	Hørsholm	Denmark
NIS	National Institute for Standards	Cairo	Egypt
MIKE	Center for Metrology	Espoo	Finland
CNES	National Center for Space Studies	Toulouse	France
OP	Paris Observatory	Paris	France
DLR	German Aerospace Center	Oberpfaffenhofen	Germany
DTAG	Deutsche Telekom AG	Frankfurt	Germany
IFAG	Federal Agency for Cartography and Geodesy	Wettzell	Germany
PTB	Physikalisch-Technische Bundesanstalt	Braunschweig	Germany
EIM	Hellenic Institute of Metrology	Thessaloniki	Greece
HKO	Hong Kong Observatory	Hong Kong	Hong Kong (China)
SCL	Standards and Calibration Laboratory	Hong Kong	Hong Kong (China)
BFKH	Government Office of Capital City of Budapest, Metrology Department	Budapest	Hungary
NPLI	National Physical Laboratory of India	New Delhi	India
IDN	Standardization Agency of Indonesia	Jakarta	Indonesia
NSAI	National Metrology Laboratory of Ireland	Dublin	Ireland
INPL	National Physical Laboratory of Israel	Jerusalem	Israel
CAO	Cagliari Astronomical Observatory	Cagliari	Italy
IT	National Institute of Metrological Research	Torino	Italy
NAO	National Astronomical Observatory	Mizusawa	Japan
NICT	National Institute of Information and Communications Technology	Tokyo	Japan
NMIJ	National Metrology Institute of Japan	Tsukuba	Japan
KZ	Kazakhstan Institute of Standardization and Metrology	Astana	Kazakhstan
LT	Center for Physical Sciences and Technology	Vilnius	Lithuania
LUX	Luxembourg Metrology Bureau	Belvaux	Luxembourg
BOM	Bureau of Metrology of Macedonia	Skopje	Macedonia
NMLS	National Metrology Institute of Malaysia	Sepang	Malaysia
CNM	National Metrology Center	Queretaro	Mexico
MASM	Mongolian Agency for Standardization and Metrology	Ulaanbaatar	Mongolia
MBM	Bureau of Metrology	Podgorica	Montenegro
ESA	European Space Agency	Noordwijk	Netherlands
VSL	National Metrology Institute of the Netherlands	Delft	Netherlands
MSL	Measurement Standards Laboratory	Lower Hutt	New Zealand
JV	Norwegian Metrology Service	Kjeller	Norway

UTC(*k*) acronym	Laboratory (English spelling)	City	Country
CNMP	National Metrology Center of Panama	Panama	Panama
INCP	National Quality Institute of Peru	Lima	Peru
AOS	Space Research Center of Polish Academy of Sciences	Borowiec	Poland
PL	Consortium of laboratories in Poland	Warsaw	Poland
IPQ	Portuguese Quality Institute	Caparica	Portugal
NIMB	National Institute of Metrology	Bucharest	Romania
SU	Russian Metrological Institute of Technical Physics and Radio Engineering	Moscow	Russian Federation
MTC	King Abdullah Center for Crescent Observations and Astronomy	Makkah	Saudi Arabia
SASO	Saudi Standards, Metrology and Quality Organization	Riyadh	Saudi Arabia
DMDM	Directorate of Measures and Precious Metals	Belgrade	Serbia
SG	National Metrology Center	Singapore	Singapore
SMU	Slovak Institute of Metrology	Bratislava	Slovakia
SIQ	Slovenian Institute of Quality and Metrology	Ljubljana	Slovenia
ZA	National Metrology Institute of South Africa	Pretoria	South Africa
KRIS	Korea Research Institute of Standards and Science	Daejeon	South Korea
ROA	Royal Institute and Observatory of the Armada	San Fernando	Spain
SL	Department of Measurement Units, Standards and Services	Colombo	Sri Lanka
SP	Research Institutes of Sweden	Boras	Sweden
CH	Federal Institute of Metrology	Bern	Switzerland
TL	Telecommunication Laboratories	Chung-Li	Taiwan
NIMT	National Institute of Metrology	Pathumthani	Thailand
DFNT	Metrology Laboratory of the Directorate-General for Communications and Informatics	Tunis	Tunisia
UME	National Metrology Institute	Gebze-Kocaeli	Turkey
UA	National Scientific Center, Institute of Metrology	Kharkiv	Ukraine
UAE	Emirates Metrology Institute	Abu Dhabi	United Arab Emirates
NPL	National Physical Laboratory	Teddington	United Kingdom
APL	Applied Physics Laboratory	Laurel	United States of America
NIST	National Institute of Standards and Technology	Boulder	United States of America
NRL	United States Naval Research Laboratory	Washington	United States of America
USNO	United States Naval Observatory	Washington	United States of America
UTE	National Administration of Power Plants and Electrical Transmissions	Montevideo	Uruguay
UZ	Uzbek National Institute of Metrology	Tashkent	Uzbekistan
VMI	Vietnam Metrology Institute	Ha Noi	Vietnam

The 91 laboratories listed in Table 9.5 are located in 68 nations. Each of the laboratories maintains a local UTC(k) time scale, where k is the acronym assigned to the laboratory. Some UTC(k) time scales have just a single cesium clock, but others are ensemble time scales with multiple atomic clocks (Section 4.4). More than 400 atomic clocks regularly contribute to UTC, and nearly all of them are either 5071A cesium beam clocks (Section 3.6) or active hydrogen masers (Section 3.10). Measurement data from all of the local UTC(k) time scales, including measurements of their individual clocks, are continuously sent to the BIPM. The clock data provides the inputs for the UTC time scale algorithms, as described in the next section.

9.4.3 The UTC Time Scale Algorithms

The principal time scale algorithm at the BIPM is called ALGOS. The algorithm was originally implemented by the BIH in 1973, and has since been updated whenever necessary to accommodate the ever-increasing number of atomic clocks that regularly contribute to UTC, and to also accommodate the continuous improvements made to time transfer systems [22, 24, 28, 38]. ALGOS processes measurement data from atomic clocks located at the laboratories listed in Table 9.5, and outputs a free atomic time scale, *Échelle Atomique Libre*, that's known by the acronym EAL.

The EAL time scale is based on a weighted average of the contributing clocks, and it was designed to be more stable in the long term than any of the individual clocks. The general equation for expressing EAL at a given time t is

$$\mathrm{EAL}(t) = \sum_{i=1}^{N} w_i [h_i(t) + h'_i(t)], \tag{9.1}$$

where N is the number of contributing atomic clocks, w_i is the relative weight assigned to the clock H_i, $h_i(t)$ is the measurement of clock H_i at time t, and $h'_i(t)$ is a correction applied to the measurement of clock H_i to ensure the continuity of the time scale if the weights of individual clocks change, or if N changes. Whenever necessary, the clock weights are adjusted by adaptive algorithms, which ensures that the sum of the clock weights is always equal to 1 [38, 41–43].

ALGOS is based on the simple premise that the most stable and predictable clocks are the best clocks. Therefore, EAL assigns the most weight to the clocks that exhibit the best stability. The stability of each contributing clock is estimated in a two-part process. First, the stability of the local UTC(k) time scale to which the clock belongs is estimated by comparing it to EAL. As we'll describe in the next section, the UTC(k) time scale data is obtained via a time transfer link and submitted to the BIPM in a standardized file format. If the UTC(k) time scale consists of just one clock, then the time scale and clock stability are, of course, the same. However, if the UTC(k) time scale consists of more than one clock, a separate clock file is submitted along with the time scale file, again using a standardized format. The clock file includes UTC(k) – Clock measurements for every individual clock in the local time scale. Once the stability of an individual clock is estimated with respect to its own local time scale, its stability can also be estimated with respect to EAL.

The stability estimate for each clock is updated every month, but the standard period of observation used to determine its weight is one year, with a filter ensuring that the most recent measurements receive more consideration than older measurements. A clock isn't allowed to contribute to the UTC calculation until its stability has been observed for a minimum period of five months. These long and cautious observation periods guard against disruptions, and they help to ensure the long-term stability of EAL and UTC [43, 44].

To prevent the most stable clocks from dominating EAL, a maximum clock weight of 0.7% was established in 1996. In January 2001, the weight limit was changed to $\frac{2}{N}$, where N was again the number of contributing atomic clocks [45]. This meant that if N exceeded 285, the maximum clock weight would be less than 0.7%. As the number of clocks continued to increase, the weight limit was changed to $\frac{2.2}{N}$ in July 2002, to allow the most stable clocks to make a larger contribution. That limit was in effect for more than 11 years, but it was increased to $\frac{4}{N}$ in January 2014, allowing clocks to have a 1% maximum clock weight if $N = 400$ [44].

Figure 9.10 shows the dramatic change that occurred in the EAL clock weights after January 1, 2014 (Modified Julian Date 56658) when hydrogen masers, the most stable atomic clocks, were allowed to increase their contributions. As of 2024, hydrogen masers account for more than 90% of EAL, which has improved the time scale's stability. Further changes to the clock weights, which could make EAL even more stable, are being evaluated [46].

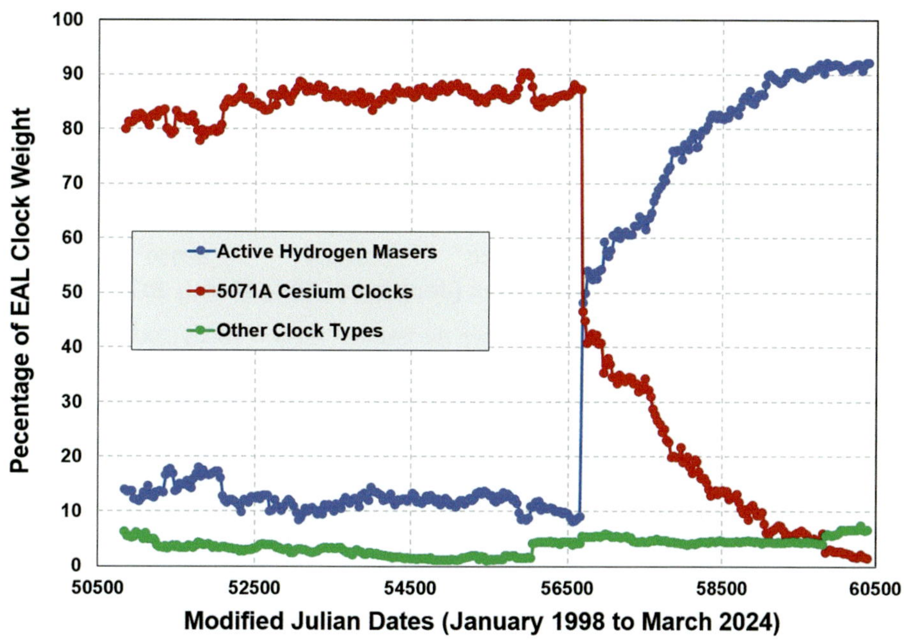

Figure 9.10 Percentage of EAL weight assigned to various atomic clock types (1998 to 2024).

Each month, the frequency of EAL is compared to measurements from a small number of primary frequency standards (PFS)—typically about 10—that submit measurement data to the BIPM. Most of the PFS are cesium fountain clocks (Section 3.8), but cesium beam clocks, rubidium fountain clocks, and optical clocks (Chapter 13) are sometimes included in the primary standard submissions. The PFS comparisons produce a frequency steering correction that's necessary to maintain agreement with the SI definition of the second. When this steering correction is applied to EAL, the resulting time scale is TAI. As previously described and shown in Figure 9.9, leap seconds are necessary to maintain agreement with astronomical time scales derived from the rotational rate of the Earth. When leap seconds are applied to TAI, the resulting time scale is UTC [22, 43].

Figure 9.11 is a graph of the unitless frequency differences of EAL versus TAI, $f_{EAL} - f_{TAI}$, for the period from January 2000 to March 2024. Beginning in late 2012, the frequency difference has remained nearly constant, about 6.5×10^{-13}, as indicated by the flat portion of the graph. During this period, the range of variation in the monthly steering corrections has been less than $\pm 1 \times 10^{-15}$. Figure 9.12 is a graph of the unitless frequency differences of the primary frequency standards versus TAI, $f_{PFS} - f_{TAI}$, for the same period. The flat portion of the graph that begins in late 2012 again shows that the frequency differences have variations of less than $\pm 1 \times 10^{-15}$. The close agreement demonstrated since 2012 is the result of the exceptional frequency stability of the hydrogen masers that now comprise most of EAL, and the exceptional frequency accuracy of the primary standards.

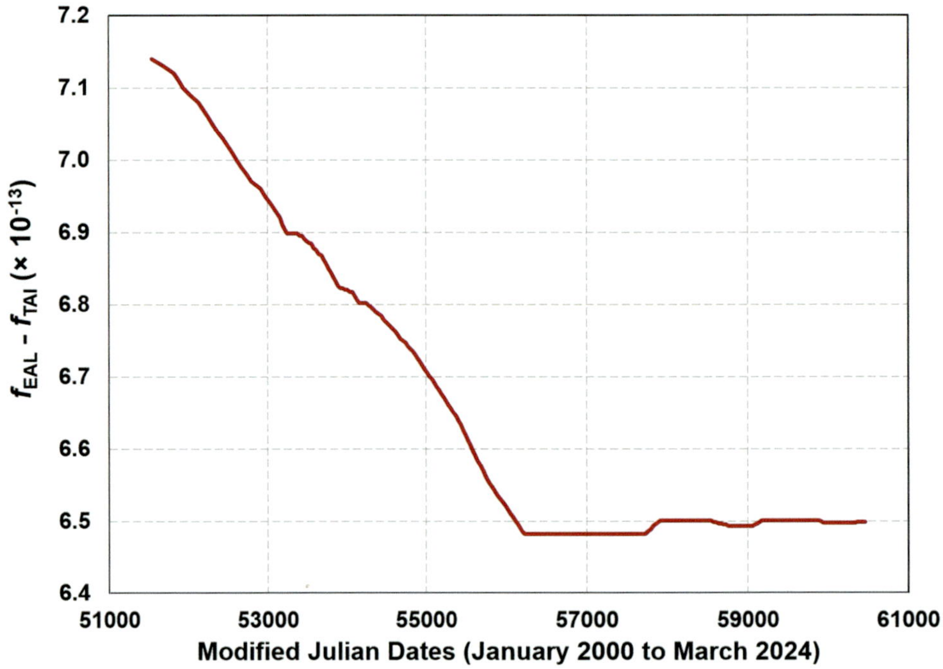

Figure 9.11 Frequency differences between the EAL and TAI time scales (2000 to 2024).

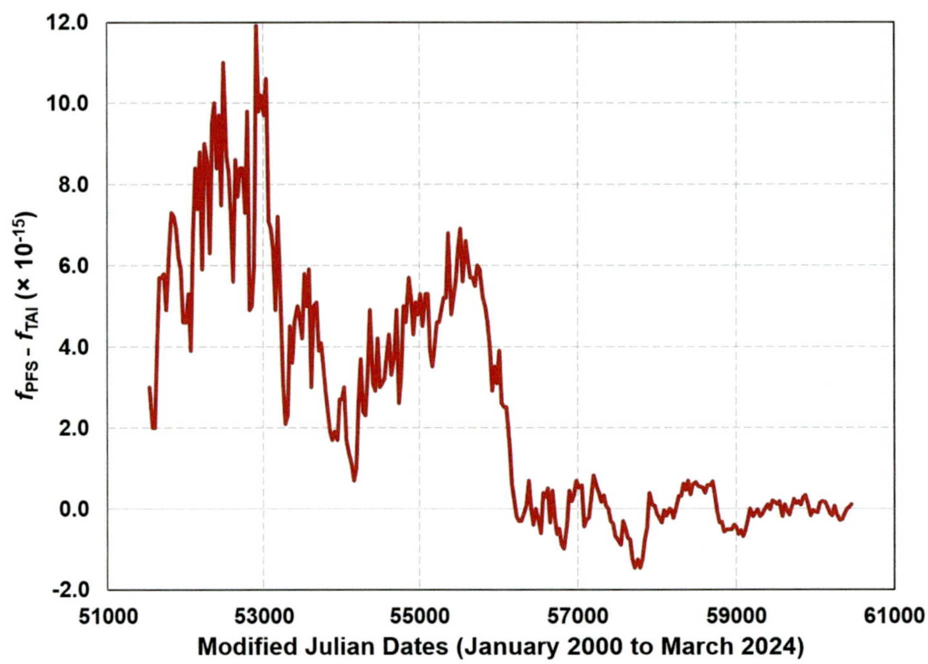

Figure 9.12 Frequency differences between the primary standards and TAI (2000 to 2024).

Figure 9.13 shows the relationship between EAL, TAI, and UTC. It's important to note that neither EAL nor TAI is distributed outside of the BIPM, but they are the "in-house" time scales that the BIPM needs to compute before UTC can be determined. The distributed time scale is UTC, which is the international standard for both time and frequency measurements. The next two sections describe the time transfer links that connect the local UTC(k) time scales to UTC, and explain how the BIPM reports the UTC(k) time differences and measurement uncertainties.

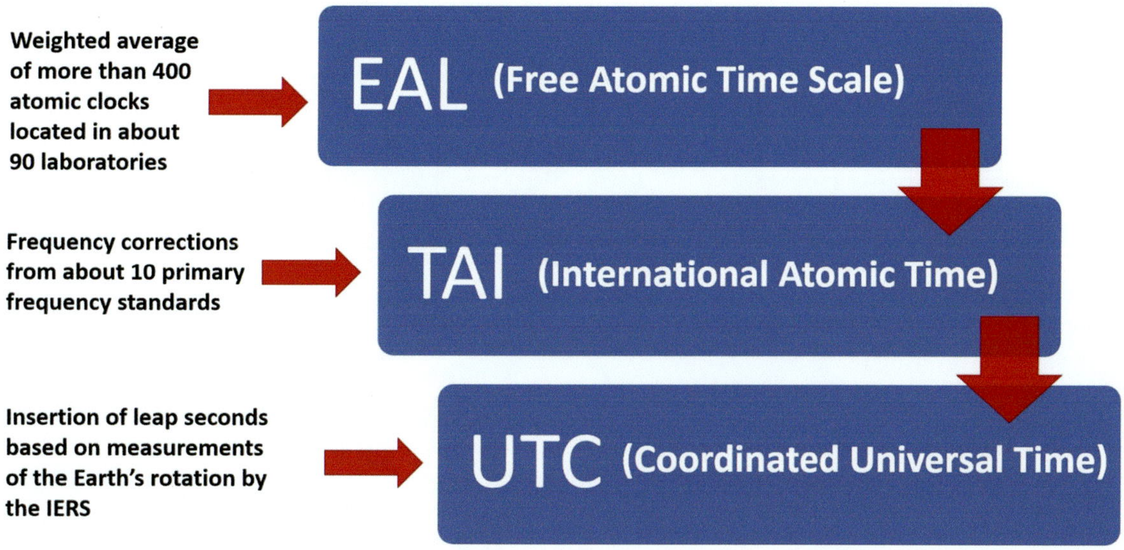

Figure 9.13 The relationship between EAL, TAI, and UTC.

9.4.4 Linking UTC(*k*) Time Scales to UTC

As we noted in Section 9.4.2, every UTC contributor is required to operate time transfer equipment that produces measurement data in the standard format requested by the CCTF and the BIPM. Each laboratory connects a 1 pps signal from their UTC(*k*) time scale, calibrated for delays, to their time transfer equipment. The time transfer equipment implements either a passive or active satellite link to UTC. The passive links only require the reception of satellite signals, but the active links require both the reception and transmission of signals that are relayed by satellites.

The passive links are based on the common-view time transfer method that was previously described in Section 4.3.3. The signals received by the UTC contributor are broadcast by a global navigation satellite system (GNSS). The parts of the signal used for the common-view comparison depend on the method used. It can include the spread spectrum codes modulated onto the carrier frequency, such as the coarse acquisition (C/A) and precision (P) codes broadcast by the GPS satellites, the phase of the received carrier frequency, or some combination of codes and carriers. The carrier has a much higher frequency than the chip rates of the codes, for example, the frequency of the GPS L1 carrier (1575.42 GHz) is more than 1500× higher than the C/A code chip rate (1.023 megachip/s). The carrier-phase-based methods of time transfer require more complex data processing, but they potentially have much higher resolution and lower uncertainties than the code-based methods [47].

The active time transfer links are based on the two-way satellite time and frequency transfer (TWSTFT) method that was previously described in Section 4.3.4. Both the received and transmitted signals pass through a transponder on a communications satellite that stays in a fixed location above the Earth's equator. The transponder forms a communications channel between the receiving and transmitting antennas [48].

Regardless of the time transfer method used, all of the UTC(*k*) time scales are compared to the Physikalisch-Technische Bundesanstalt (PTB), the NMI of Germany, which serves as the pivot laboratory and common reference for the BIPM key comparisons. To fulfill this role, PTB operates equipment that supports all of the time transfer methods that the BIPM accepts [43]. Figure 9.14 provides a map that shows the locations of the UTC contributors, with the inset showing the laboratories that are located in Europe. The map also indicates whether the laboratory operates only GNSS time transfer equipment, which is true in most cases, or both GNSS and TWSTFT equipment.

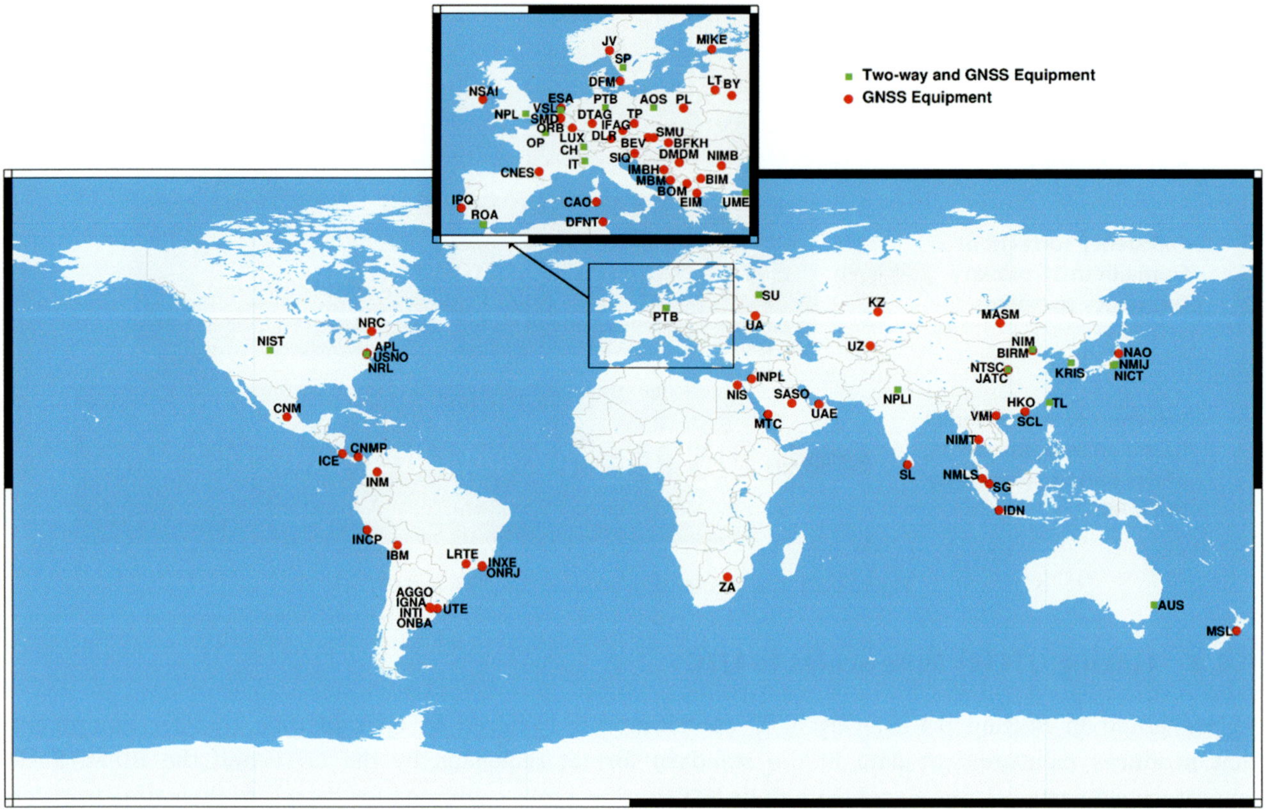

Figure 9.14 A map of the laboratories that contribute data to UTC via time transfer links (courtesy of the BIPM).

The first common-view GPS receivers used for international time comparisons were designed and built at NBS in the early 1980s, during a period when only a few GPS satellites were in orbit. The development effort was led by the electrical engineer Dick Davis (1932–2018), and it resulted in the production of what became known as the NBS-type receivers. These receivers were single channel, which meant that they could only track one GPS satellite at a time, and they were programmed to follow a tracking schedule. They received the C/A code broadcast by the satellite on the L1 band frequency of 1575.42 MHz, and they measured the time difference between the GPS signal and a local clock with a built-in TIC [49]. Two NBS receiver designs were manufactured and sold commercially by Allen Osborne Associates, who gave them the model names of TTR-5 and TTR-6. By 1984, the NBS-type receivers had enabled international common-view comparisons between the United States, Europe, and Japan, with uncertainties of about 10 ns for time, and about 1×10^{-13} for frequency, when averaged for several days. These uncertainties were more than 10× smaller than those obtained from previous comparisons using Loran-C signals [50].

The NBS-type receivers provided most of the UTC(k) links for more than a decade, and a few remained in use for several decades. However, the GPS single-channel method, which the BIPM called GPS SC, is no longer practiced. Beginning in the 1990s, the NBS-type receivers began to be replaced by multi-channel common-view receivers. These receivers were designed using small and inexpensive GPS receiver boards that were made for consumer applications. Like the NBS-type receivers, the multi-channel receivers were limited to the reception of the L1 band and to measurements of the C/A code, but they could simultaneously track eight or more satellites. Most of the receiver boards selected for time transfer applications were manufactured by Motorola, with the Motorola VP Oncore proving to be especially useful, because its stability was comparable to the NBS-type receivers [51].

The most prevalent common-view receiver built around the VP Oncore was the TTS-2, designed at the Polish Academy of Sciences (AOS) by the physicist Jerzy Nawrocki (b. 1952), in cooperation with the physicist Wlodzimierz Lewandowski (b. 1949) of the BIPM. The TTS-2, and the series of improved time transfer receivers that followed, are sold commercially by PikTime Systems in Poland. At the height of their usage, more than 30 laboratories used TTS-2 receivers to link their time scales to UTC. As of 2025, some laboratories continue to use the TTS-2 and other multi-channel, C/A code receivers as their primary link to UTC. This time transfer method is called GPS multi-channel (GPS MC).

The NBS-type receivers stored data in a format where the measurements of each GPS satellite were collected for 13 minutes (780 s), a period long enough to ensure that time transfer receiver had the most recent GPS navigation message. The navigation method includes the parameters necessary to correct for ionospheric delay, and requires 750 s to download [52]. The satellite tracks were recorded in 16-minute segments to allow for the constraints of the original single-channel receivers. The first two minutes of each 16-minute segment allowed the receiver to lock to the GPS signal before beginning the measurements. This was followed by 13 minutes of time interval measurements that determined the time difference between the local time scale and the satellite, and one minute after the measurements were completed that was reserved for data processing and preparing for the next segment.

The use of 16-minute segments also conveniently allowed the day, which consists of 1,440 minutes, to be evenly divided into 90 segments. However, the GPS satellite orbits are based on the sidereal day, which is the period of the Earth's rotation when referenced to distant stars that appear at a fixed location in the sky, rather than to the Sun. The period of a sidereal day is 23 hours, 56 minutes, and 4.09 s, or about four minutes shorter than the UTC or mean solar day, so only 89 full-length segments are recorded daily. The start time of the first segment varies, beginning four minutes earlier than the previous day for three consecutive days, followed by a day where the first segment begins 12 minutes later than the previous day. Therefore, the repeating sequence for the start time of the first segment—expressed in UTC hours and minutes—is 00:14, 00:10, 00:06, 00:02, 00:14, and so on.

The time transfer files are standard text files, with one line of the text devoted to each satellite tracked during the 16-minute segment. Therefore, only one line of text was needed per segment when the GPS SC method was utilized, but multiple lines per segment are needed for the GPS MC method. Each line contains the date and start time of the segment, the duration of the track, which is nominally 780 s, the elevation and azimuth angles of the satellite, and information about the measured time difference between the local time scale and the satellite, including the measurement corrections applied for ionospheric and tropospheric delays. Detailed information about GPS signals is provided in Chapter 12.

With some minor modifications, the NBS file format became the standard for the submission of GNSS time transfer data to the BIPM. After years of informal usage, the standard was formally published in 1994 [53], and is now called the Common GPS GLONASS Time Transfer Standard (CGGTTS). The standard was updated in 2015 to provide additional information, including extra data columns for ionospheric measurements made by dual-frequency receivers [54]. Figure 9.15 shows the CGGTTS file header and the satellite tracks for the first 16-minute segment of the day, as recorded by a single-frequency GPS receiver implementing the GPS MC method. The file header contains the Cartesian coordinates of the antenna (XYZ), and the calibrated time delays that were compensated, including the internal delay of the receiver (INT DLY), the delay introduced by the antenna cable (CAB DLY), and the delay introduced by the cable that connects the receiver to the UTC(k) time scale (REF DLY).

Beginning in the early 2000s, laboratories began using dual-frequency geodetic GPS receivers as their time transfer link to UTC, a development that can be credited to the physicist Pascale Defraigne (b. 1970) of the Royal Observatory of Belgium. Geodetic receivers are professional-grade instruments that receive both the L1 (1575.42 GHz) and L2 (1227.6 GHz) GPS carrier frequencies. They were designed for precise positioning and surveying applications where the antenna coordinates can be determined to within a few centimeters or less, but Defraigne discovered that they could also be used for time transfer. When used for time transfer, the receivers had to accept—either in their stock form or through modification—an external frequency and time reference from a local UTC(*k*) time scale. The acceptance of these signals made UTC(*k*) the receiver's local clock, which meant that the receiver's pseudorange measurements were then made by comparing the phase of the satellite carrier frequency to the phase of UTC(*k*).

```
GGTTS GPS DATA FORMAT VERSION = 01
REV DATE = 11/20/2013
RCVR = Generic Single-Frequency, C/A Code Time Transfer Receiver
CH = 12
IMS = 99999
LAB = NIST
X = -1288331.833 m
Y = -4721664.612 m
Z =  4078681.021 m
FRAME = WGS84
COMMENTS = Lab Code - 10002, UTC Code - 0010002
COMMENTS =
INT DLY = 25.5
CAB DLY = 119.8
REF DLY = 182.4
REF = UTC(NIST)
CKSUM = 07

PRN CL  MJD  STTIME TRKL ELV AZTH   REFSV      SRSV     REFGPS    SRGPS  DSG IOE MDTR SMDT MDIO SMDI CK
             hhmmss  s  .1dg .1dg   .1ns     .1ps/s     .1ns    .1ps/s .1ns     .1ns.1ps/s.1ns.1ps/s
 02 FF 56842 001400  780 807 2428  -5049146     -47         201     -112   14  65   68   -1  137   -5 E9
 04 FF 56842 001400  780 349 0500    -76293       0         113       50   31   4  116   22  195   19 81
 05 FF 56842 001400  780 198 1696   3610220     -26         180     -218   23  88  198  -67  338  -67 33
 06 FF 56842 001400  780 580 0460    -24402      37         132      -61   11  56   79    7  147    7 8A
 10 FF 56842 001400  780 410 1160   1325608       2         279     -156   22  28  101   -8  194  -20 CF
 12 FF 56842 001400  780 630 3045  -2159513     -43         157      -98   11  18   75   -4  145   -8 DE
 17 FF 56842 001400  780 178 0850   1062788      45         174       42   31  95  217   59  281   27 E7
 24 FF 56842 001400  780 230 2318    314713       1         101       -5   23  92  169   39  311   42 91
 25 FF 56842 001400  780 256 3150   -247955       0         111      -72   23 108  153  -37  240  -30 E0
```

Figure 9.15 Sample GPS time transfer data stored in the CGGTTS format.

The pseudorange measurements and other raw positioning data are stored by geodetic receivers in a standardized format called RINEX (Receiver INdependent EXchange). The RINEX files contain enough information to determine the time differences between the satellite and local time scale signals without the use of a TIC, and Defraigne wrote software to convert RINEX files to the BIPM's CGGTTS format [55]. Versions of her software are distributed by the BIPM and have led to the release of similar software by receiver manufacturers.

When single-channel receivers were in use, a UTC(*k*) time scale could only be compared to one satellite at a time, and comparisons between two time scales could only occur when the same satellite was scheduled to be simultaneously measured from both time scale sites. This meant that the time scale comparisons weren't continuous, and that only a small amount of measurement data could be collected. The multi-channel receivers greatly improved this situation, and eliminated the need for a tracking schedule, because at least one common satellite was usually visible at both time scale sites. However, if the baseline between the time scales was too long, there could still be periods when no satellites were in common view, which prevented any measurements from being made. There could also be periods where the only satellite in common view was at a low elevation angle at one or both sites, which meant that the corrections applied for its ionospheric and tropospheric delays would have large uncertainties that were less likely to cancel when the measurements were differenced.

For these and other reasons, the BIPM changed the way that they processed GPS time transfer data after most of the single-channel receivers had been replaced with multi-channel receivers. Beginning in September 2006, their data processing shifted from the classic common-view method to the all-in-view method. This processing change meant that the common-view comparison signal was no longer broadcast by an individual satellite, which was referenced to its own clock and influenced by its position in the sky. Instead, it was now an estimate of the GPS reference time that was being used to synchronize all of the satellites. The estimate was obtained by measuring signals from all of the satellites visible from the receiver's location. Measuring all visible satellites guarantees the inclusion of satellites at high elevation angles, with high signal-to-noise ratios, and with ionospheric and tropospheric delays that can be corrected with sub-nanosecond uncertainty. It also allows the continuous comparison of two time scales located anywhere on Earth, because it doesn't require any of the individual satellites to actually be in common-view [56].

The time transfer links implemented with dual-frequency receivers are more stable than the single-frequency links. Most of the improvement in stability is due to a reduction in the uncertainty contributed by ionospheric delays. The BIPM uses two methods to process dual-frequency receiver data, called GPS P3 and GPS PPP. Both methods can provide link uncertainties that are usually well below 5 ns (k = 1), which is roughly a factor of two smaller than the uncertainty of the GPS MC method when implemented with a single-frequency receiver over the same baseline. As of 2024, roughly 70% of the UTC(k) laboratories use GPS PPP as their primary GNSS method for linking to UTC, with the remaining 30% using either the GPS P3 or GPS MC method.

The GPS P3 method derives its name from the precision code (P code) that is modulated on both the L1 and L2 carrier frequencies (the C/A code is only modulated on L1). The GPS signals are delayed slightly as they pass through the ionosphere, but the delay is frequency dependent, meaning that signals at different frequencies are delayed by different amounts. Therefore, the delay can be estimated from pseudorange measurements of the P code on both L1 and L2 carriers. If the receiver's internal delays are known for both L1 and L2, a third pseudorange measurement that's mostly free of ionospheric delays can be calculated as

$$\mathrm{P3} = \frac{f_1^2}{f_1^2 - f_2^2}(\mathrm{P1} - RxDLY_1) - \frac{f_2^2}{f_1^2 - f_2^2}(\mathrm{P2} - RxDLY_2), \tag{9.2}$$

where f_1 and f_2 are the respective carrier frequencies of L1 and L2, P1 and P2 are the respective pseudorange measurements of the P code on L1 and L2, and $RxDLY_1$ and $RxDLY_2$ are the respective delays added to P1 and P2 by the receiver [57].

The most common GNSS link method, GPS PPP, is based on precise point positioning (PPP), a technique used by the geodetic community that has an uncertainty of just a few centimeters. The position estimates are calculated from measurements that the receiver performs on its own based on direct observations of the satellites, and from satellite orbit and clock data that's obtained from a network of tracking stations. These satellite orbit and clock data are used to correct the receiver measurements, which reduces the measurement uncertainty. The real-time receiver measurements are made with a single GNSS receiver and the collected data is stored as a RINEX file. Using the information obtained from direct observations, the ionospheric delay can be removed in the same fashion as the P3 method just described, and the phase angle of the carrier can be measured with respect to the reference clock. This makes it possible to refine the receiver's position estimate so that it has an uncertainty smaller than one wavelength of the carrier, which is about 19 cm for L1, and about 24 cm for L2. The techniques used are analogous to those implemented by time interval and frequency measurement systems (Chapters 4 and 5) to get more resolution, and the distance measurements translate directly to time measurements. The periods of the L1 and L2 carrier cycles are each less than 1 ns—about 635 ps and 815 ps—respectively.

The satellite orbit and clock data that corrects the receiver's measurements are obtained from the International GNSS Service (IGS), a network that has more than 500 tracking stations. Each station is equipped with a geodetic receiver that uploads its measurements to a data center for processing. The IGS began working with the BIPM in 1998 to develop a series of "clock products," in the form of data files distributed via the Internet that provide more accurate orbit and clock information than that obtained from direct observations of the satellites. These clock products are tabulated at 5-minute and 30-s intervals, and have been published via the Internet since 2004. They are provided in three forms: an ultra-rapid clock solution that's published four times daily, a rapid clock solution published about 17 hours after the end of each day, and a final clock solution, published weekly, within 13 days after the end of the solution week. The corrections to the receiver measurements can be made with several different software packages, some publicly available and some developed in-house [58, 59]. The BIPM applies IGS corrections not only to the GPS PPP links based on carrier-phase measurements, but to all of the GNSS links to UTC [43].

Every UTC(k) laboratories listed in Table 9.5 maintain a GNSS link to UTC. About 15% of these laboratories also maintain a TWSTFT link. The BIPM began accepting TWSTFT links for UTC contributions in 1999 [60], and these links provide even smaller uncertainties than the GNSS links, typically less than 2 ns (k = 1). However, because signals are transmitted as well as received, TWSTFT links require a larger investment in equipment. The required equipment includes a low-noise amplifier, a power amplifier, frequency up and down converters, a satellite modem that can both modulate and demodulate spread spectrum signals, a TIC, and a large satellite dish with a feed horn antenna that both transmits and receives signals. The antenna is pointed at a geostationary communication satellite, located in a fixed position above the equator. Figure 9.16 shows a dish antenna, 3.8 m in diameter, that transmits and receives UTC(NIST) signals for transatlantic comparisons with NMIs in Europe.

Figure 9.16 A dish antenna at NIST in Boulder, Colorado, that provides the TWSTFT link to the BIPM key comparison.

The time signals are transmitted in the form of pseudorandom noise (PRN) codes via binary phase shift keying (BPSK) modulation, with the BPSK sequence synchronized inside the modem to a 1 pps signal from the local UTC(k) time scale. Because different modem manufacturers use different PRN codes, all of the participating laboratories are required to use the same modem type to ensure compatibility [61]. The cost of satellite usage is proportional to the amount of bandwidth used by the signal, which is in turn determined by the chip rate of the PRN codes. The higher the chip rate, the more bandwidth is required. The chip rate is typically 1 megachip/s, but a higher chip rate of 2.5 megachip/s is sometimes used.

The modems used for time signals often lack a bandwidth-limiting filter, and typically require a bandwidth that's at least 1.5× higher than the chip rate to avoid interfering with neighboring signals on the satellite transponder.

The uplink frequencies for the time signal transmissions are assigned by the satellite operator. The BIPM links are implemented on the K_u band, which is primarily used for satellite communications. The uplink frequencies range from about 13.75 to 14.5 GHz, and signals are received on downlink frequencies that range from about 10.7 to 12.75 GHz. The transmitted signals are low power, nominally about 1 W. They reach the satellite after a nominal path delay of about 1/8 of a second, and the satellite transponder then serves as a "bent pipe" communications channel. After the uplink signal is detected by the transponder antenna, it first passes through a low-noise amplifier, followed by a frequency translator that converts the signal to its downlink frequency. The signal is then amplified and retransmitted back to Earth, again with a nominal path delay of about 1/8 of a second.

The TWSTFT comparisons between UTC(k) laboratories are performed according to a pre-determined schedule. The comparisons last for two minutes and are repeated every two hours. During each comparison, each laboratory transmits its own time signal while simultaneously receiving the signal from the other laboratory. The comparison results are submitted to the BIPM in a standardized format that's similar to the CGGTTS format used for GNSS links. If a laboratory submits both a GNSS RINEX file and TWSTFT data, their measurements can be processed and smoothed using a combination of GPS PPP and TWSTFT, a time transfer method that the BIPM calls TWGPPP.

The time transfer methods that link the UTC(k) time scales to UTC are summarized in Table 9.6. The table lists two methods that utilize GLONASS, the Russian satellite navigation system, that were not discussed in the text. These methods, known by the acronyms GLN MC and GPSGLN, have been approved by the BIPM but are seldom used. This of course can change, and it seems highly likely that other GNSS systems will be approved as UTC time transfer links in the future, including Galileo and BeiDou. More information about GLONASS, Galileo, and BeiDou is provided in Chapter 12.

Table 9.6 The satellite time transfer methods that link the UTC(k) time scales to UTC

Acronym	Description	Satellite type
GPS MC	GPS all-in-view, multi-channel, C/A code.	GNSS
GPS P3	GPS all-in-view, multi-channel, dual-frequency, P code.	GNSS
GPS PPP	GPS precise point positioning technique.	GNSS
GLN MC	GLONASS common-view, multi-channel, C/A code.	GNSS
GPSGLN	Combination of GPS MC and GLN MC.	GNSS
TWSTFT	Two-way satellite time and frequency transfer.	Communications
TWGPPP	Combined smoothing of TWSTFT and GPS PPP.	GNSS & Communications

9.4.5 Time Differences and Measurement Uncertainties of the UTC(k) Time Scales

As our discussion has shown, the BIPM calculates UTC using measurement data supplied by clocks and time scales that are located elsewhere. Therefore, UTC is a virtual time scale that doesn't produce physical signals. UTC is defined by its time differences with respect to the UTC(k) time scales, which do produce physical signals, and which serve as the reference for physical measurements. The UTC – UTC(k) time differences are published monthly by the BIPM in a document called *Circular T*, which is typically available around the 10th day of a month, and which covers the period of the previous month (Figure 9.17). This means that when the *Circular T* is published, the most recent time differences are already at least 10 days old, and the oldest time differences can be about 45 days old.

Bureau International des Poids et Mesures

CIRCULAR T 437

2024 JUNE 10, 16h UTC

UTC

ISSN 1143-1393

BUREAU INTERNATIONAL DES POIDS ET MESURES
THE INTERGOVERNMENTAL ORGANIZATION ESTABLISHED BY THE METRE CONVENTION
PAVILLON DE BRETEUIL F-92312 SEVRES CEDEX TEL. +33 1 45 07 70 70 tai@bipm.org

The contents of the sections of BIPM *Circular T* are fully described in the document " Explanatory supplement to BIPM Circular T " available at https://webtai.bipm.org/ftp/pub/tai/other-products/notes/explanatory_supplement_v0.7.pdf

1 - Difference between UTC and its local realizations UTC(k) and corresponding uncertainties. From 2017 January 1, 0h UTC, *TAI-UTC* = 37 s.

Date 2024 0h UTC			APR 29	MAY 4	MAY 9	MAY 14	MAY 19	MAY 24	MAY 29	Uncertainty/ns			Notes
		MJD	60429	60434	60439	60444	60449	60454	60459	u_A	u_B	u	
Laboratory *k*			[*UTC-UTC*(*k*)]/ns										
AGGO	(La Plata)	123	693.8	697.1	677.3	685.1	659.0	645.8	633.3	0.7	2.9	3.0	
AOS	(Borowiec)	123	-20.0	-20.9	-22.8	-24.7	-26.4	-27.5	-27.3	0.2	3.4	3.4	
APL	(Laurel)	123	-0.4	0.4	1.6	3.0	1.2	0.7	-0.2	0.2	NC	-	
AUS	(Sydney)	123	-243.6	-238.0	-233.9	-203.4	-203.0	-190.2	-173.6	0.2	2.9	2.9	
BEV	(Wien)	123	-20.6	-19.3	-12.9	-7.2	-4.8	0.5	-3.9	0.2	2.9	2.9	
BFKH	(Budapest)	123	10647.2	10696.1	10737.9	10794.9	10844.1	10886.7	10934.6	0.7	7.2	7.2	
BIM	(Sofiya)	123	404.0	451.7	504.2	556.2	605.6	655.7	709.9	0.2	2.9	2.9	
BIRM	(Beijing)	123	-1.6	-0.5	1.6	0.2	-1.5	-1.3	-0.1	0.2	3.3	3.3	
BY	(Minsk)	123	-2.0	-1.8	-1.7	-2.0	-1.5	-1.4	-1.2	1.5	3.1	3.4	
CAO	(Cagliari)	123	-12679.6	-12799.2	-12912.9	-13020.6	-13129.8	-13251.6	-13368.7	0.7	NC	-	
CH	(Bern-Wabern)	123	-1.8	-0.1	-0.4	0.1	-1.4	-2.9	-1.2	0.2	1.7	1.7	
CNES	(Toulouse)	123	2.3	-1.9	-7.7	-4.5	-1.7	-0.2	-1.0	0.2	2.9	2.9	
CNM	(Queretaro)	123	0.4	-2.0	-2.3	-6.7	-4.9	-2.5	-2.0	2.0	4.2	4.6	
CNMP	(Panama)	123	-10.4	-4.9	13.6	6.8	9.2	4.6	-6.9	0.5	5.4	5.4	
DFM	(Horsholm)	123	4.7	4.6	5.3	5.4	5.4	-1.6	-1.0	0.2	2.9	2.9	
DFNT	(Tunis)	123	-2488.4	-2588.4	-2689.8	-2786.8	-2879.2	-2977.8	-3067.1	0.7	NC	-	
DLR	(Oberpfaffenhofen)	123	-	-	-	-	-	-	-				
DMDM	(Belgrade)	123	1.1	10.1	15.9	13.1	-0.6	1.7	-4.3	0.2	3.8	3.8	
DTAG	(Frankfurt/M)	123	48.0	41.7	43.1	38.0	33.8	19.3	20.2	0.2	3.3	3.3	

Figure 9.17 A portion of Section 1 of the BIPM's monthly *Circular T* document.

Section 1 of the *Circular T* shows the UTC – UTC(k) time differences for every laboratory that contributes to UTC (listed in Table 9.5). The time differences are reported at 5-day intervals, for the Modified Julian Dates that end with either a "4" or a "9." These international time scale comparisons have been ongoing since the days of the BIH, and the BIPM has published the *Circular T* every month since March 1988. Because the time scale comparisons never stop, and because measurement results are continuously recorded and made available, the link in the traceability chain between the SI and the UTC(k) time scales is never broken.

A series of UTC – UTC(k) measurements can provide the time accuracy of UTC(k) over a given interval, and the same measurement series can also provide information about a time scale's frequency. For example, the frequency accuracy of a time scale over a given interval can be determined from the slope of the time difference measurements (Section 6.1.1), and the frequency stability of a time scale, where τ_0 = 5 days, can be determined with the Allan deviation (Section 7.1). The time differences between any two time scales can be obtained by simply taking the difference between their respective UTC – UTC(k) measurements.

Many UTC(k) laboratories now keep time that's continuously accurate to within a few nanoseconds of UTC. For purposes of example, Figure 9.18 graphs the time accuracy of the UTC(NIST), UTC(PTB), and UTC(USNO) time scales with respect to UTC over a three-year period (2021 to 2023), when all three time scales were usually within ±2 ns of UTC, and always within ±3 ns.

Since January 2005, the *Circular T* has included the uncertainties of the UTC – UTC(k) measurements in units of nanoseconds, with a resolution of 0.1 ns [62]. This information is provided in three columns, with two of the columns listing the uncertainties evaluated with the Type A and Type B methods, u_A and u_B, and a third column listing the combined standard uncertainty u. It's important to note that u is not an expanded uncertainty, therefore $k = 1$ and the confidence interval is 68.27%. The combined standard uncertainty is obtained with the RSS method as

$$u = \sqrt{u_A^2 + u_B^2}. \tag{9.3}$$

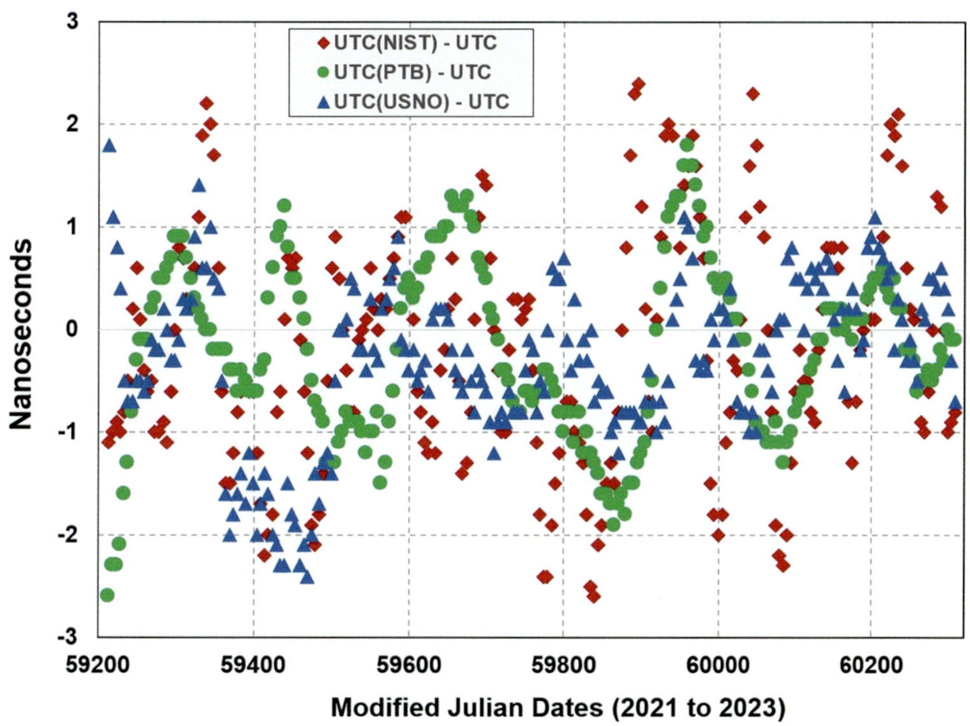

Figure 9.18 Time accuracy of the UTC(NIST), UTC(PTB), and UTC(USNO) time scales with respect to UTC (2021 to 2023).

The statistical uncertainty evaluated with the Type A method is the time stability of the link between UTC(k) and UTC(PTB), which is adjusted if necessary to account for the instability of UTC(PTB) with respect to UTC. Because PTB is the pivot laboratory for all of the links, its uncertainty is evaluated first using several methods, including a matrix of link comparisons involving multiple laboratories. This allows evaluating the correlations that appear when a single laboratory is involved in more than one link [62, 63]. The stability of the links has continuously improved due to improvements in both the UTC(k) time scales and in the time transfer methods, such as GPS PPP, that are now commonly practiced. Therefore, u_A has been reduced to 0.2 ns for many of the links, and is less than 1 ns for most of the links.

The systematic uncertainties evaluated with the Type B method are the uncertainties in the delay calibration of the time transfer equipment at the UTC(k) laboratory site. The delay calibrations are performed by both the BIPM and by the laboratories that the BIPM classifies as "Group 1," which is a designation given to a small number of laboratories that are deemed to have sufficient resources and calibration capabilities. The BIPM calibrates the time transfer equipment at the Group 1 laboratories, and the Group 1 laboratories calibrate the time transfer equipment at the other laboratories located within their regional metrology organizations (Section 9.5). The calibrations are expected to be done once every two years, and they're performed by sending a reference time transfer system with known delays to the laboratory under test [43]. The reference system is compared to the time transfer system under test for a recommended period of at least four days, and when the data and reference system are returned to either the BIPM or the Group 1 laboratory, the uncertainty analysis is performed.

The value of u_B is the uncertainty of the UTC(*k*) laboratory's most recent equipment calibration, plus an aging term that gets increasingly larger over time until the next calibration is performed. The aging term was introduced to acknowledge that the equipment delays in time transfer equipment always change over time, which is why periodic calibrations are necessary [64]. Under normal circumstances, u_B will always be larger than u_A and will dominate the link uncertainty. Due to the nature of the RSS method, u_B will often be essentially equal to the combined standard uncertainty u. The laboratories with the most stable and carefully calibrated equipment have reduced u_B to about 1.5 ns, and u_B is now less than 10 ns for all UTC(*k*) links.

9.4.6 Rapid UTC (UTCr)

The latency of UTC limits its usefulness for some time metrology applications. For example, the UTC(*k*) time scales are periodically adjusted to agree with UTC by applying frequency or time corrections to the instrument that controls the time scale's output signals, such as the auxiliary output generator used at NIST (Section 4.4). The magnitude of the corrections is usually determined by trying to predict the future performance of a UTC(*k*) time scale by analyzing its past performance. The *Circular T* provides the past performance data, but it's only updated once per month, which isn't often enough for metrologists who are trying to accurately estimate the magnitude of a time scale correction. To address this problem and make UTC more useful, the BIPM began computing a rapid realization of UTC, named UTCr, in January 2012. After being tested for 18 months, UTCr became an official BIPM product in July 2013 [65].

The UTC(*k*) – UTCr measurements are published every Wednesday. The data published on Wednesday are for the daily measurements made on Monday through Sunday of the previous week. Therefore, the UTCr latency is just 3 to 9 days, compared to 10 to 45 days for UTC, and the interval between measurement results is 1 day for UTCr, compared to 5 days for UTC. As of 2025, about 70% of the UTC(*k*) laboratories also contribute to UTCr.

The BIPM had to meet two objectives when designing UTCr. The first objective was to generate a new UTC time scale that was published more often and had less latency than UTC itself. The second objective was to make UTCr agree as closely as possible with UTC. The first objective was met by asking the UTC(*k*) laboratories to submit their time transfer data every day, and by simplifying some of the data processing procedures. For example, if a UTC(*k*) link is normally processed with either the GPS MC, GPS P3, or TWSTFT methods, the UTCr link is processed in the exact same way. However, if the UTC link is based on the GPS PPP method, it has to be processed in a different way for UTCr. This was necessary to avoid the delays caused by the latency in the IGS data, which GPS PPP depends on. The problem was solved by having UTCr replace the GPS PPP method with the GPS P3 method. This sped up the processing, but it slightly increased the Type A uncertainty of the link (typically by just a few tenths of a nanosecond). The TWGPPP links were also replaced by UTCr, by using either a TWSTFT or a GPS P3 link, depending upon which of the two options has the smaller Type B uncertainty.

To meet the second objective of keeping UTCr and UTC in close agreement, the BIPM developed a stability algorithm for UTCr that's very similar to the one used for UTC. Both time scales use a computation interval of about 30 days to update their clock stability estimates, and both determine the maximum clock weight as $\frac{2.5}{N}$. However, because N is smaller, the individual clocks contributing to UTCr have slightly more weight than they do in UTC. The time differences between the two time scales are minimized by making the UTCr algorithm always steer towards past values of UTC, rather than towards past values of UTCr. This prevents the two time scales from diverging from each other by more than a few nanoseconds, and ensures that the difference between them will be near zero when averaged over a long period.

UTCr is slightly less accurate than UTC, but still about 50% more accurate than the best UTC(*k*) time scales, so it's a sufficiently accurate reference to use for UTC(*k*) time scale adjustments, which can now be made weekly if necessary. Due to the smaller number of participating clocks, the UTCr frequency is slightly less stable than UTC, perhaps about 20% less stable when measured over a multi-year period. However, because both time scales are so exceptionally stable, a 20% difference sounds worse than it really is. For example, based on the performance of the primary frequency standards, a reasonable estimate of the multi-year stability of UTC is about 3×10^{-16}. If the UTCr stability is degraded by 20%, then the difference in stability between UTCr and UTC is just 0.6×10^{-16}. The time difference between the two time scales on a given date can be obtained from the published data by using a UTC(k) time scale as a "common-view" reference,

$$\mathrm{UTCr} - \mathrm{UTC} = [\mathrm{UTCr} - \mathrm{UTC}(k)] - [\mathrm{UTC} - \mathrm{UTC}(k)]. \tag{9.4}$$

This difference will almost always be smaller than the combined standard uncertainty of the UTC(k) time transfer link that's provided on the *Circular T* [43, 65].

Even though its performance is nearly equivalent, it's still important to remember that UTCr depends entirely on the existence of UTC. Therefore, UTCr can't replace UTC, but it can help to make UTC more useful. The two time scales fulfill different objectives—UTC provides the best possible realization of the SI second and is the ultimate source of traceability, and UTCr provides additional benefits to metrologists by reducing the latency.

9.5 The Role of Regional Metrology Organizations

To better support the work of the BIPM and traceability in general, the world's NMIs cooperate with each other via regional metrology organizations (RMOs). There are six RMOs, each covering a large geographic area, that fulfill several important tasks in international metrology. For example, the RMOs peer review the quality systems maintained by the NMIs. They also assess the calibration and measurement capabilities, known as the CMCs, of each NMI. The CMCs are reviewed every five years and published in the BIPM's key comparison database [40]. The geographic regions covered by the six RMOs are mapped in Figure 9.19, and their acronyms, names, and years of formation are listed in Table 9.7.

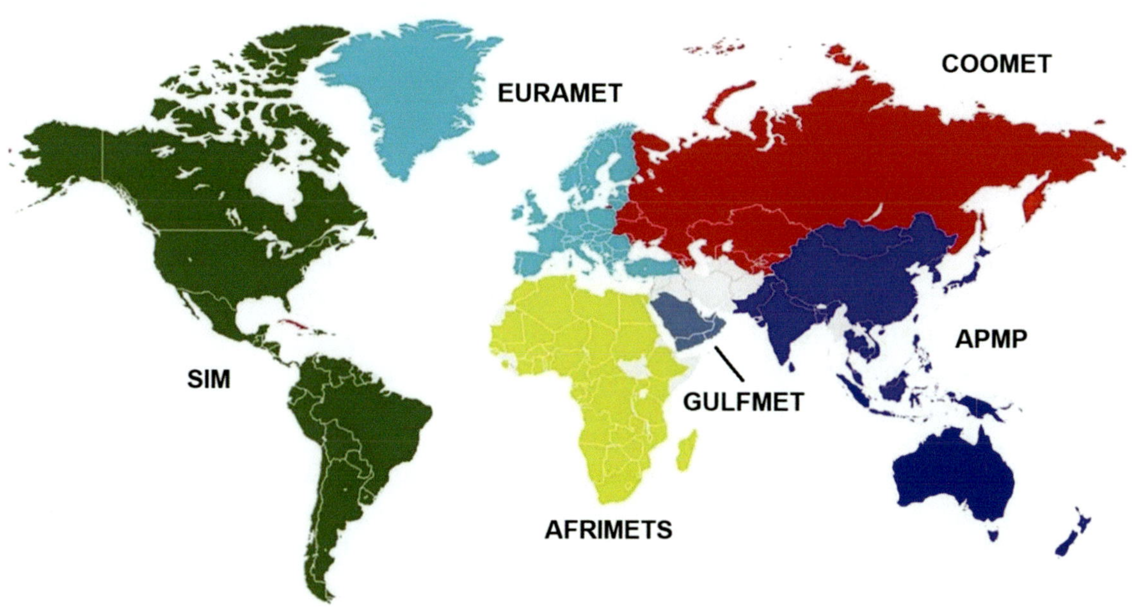

Figure 9.19 A map of the world's regional metrology organizations.

Table 9.7 The world's regional metrology organizations, sorted alphabetically by acronym

RMO acronym	RMO name (English spelling)	Year of formation
AFRIMETS	Intra-Africa Metrology System	2006
APMP	Asia Pacific Metrology Program	1977
COOMET	Euro-Asian Cooperation of National Metrological Institutions	1991
EURAMET	European Association of National Metrology Institutes	1987
GULFMET	Gulf Association for Metrology	2015
SIM	Inter-American Metrology System	1979

In addition to their work with quality systems, the RMOs also support the BIPM key comparisons by organizing supplementary comparisons. In time metrology, the RMOs support the key comparisons by helping the BIPM calibrate the time transfer equipment at UTC(k) laboratories, a function performed by the Group 1 laboratories as described in Section 9.4.4. They also organize and conduct supplementary comparisons that allow some laboratories to establish traceability to the SI without participating in the key comparison. According to CCTF guidelines, laboratories that are unable to participate in the key comparison can establish traceability via comparisons with another NMI or DI that does participate. This, of course, requires the comparison procedures and measurement uncertainties to be appropriately documented, and then validated by the RMO [43, 66]. In addition to allowing a traceability chain to be initially established, a supplementary comparison can also provide a backup or alternative traceability chain for laboratories whose participation in the key comparison has been temporarily interrupted [67].

A prime example of an RMO supplementary comparison is the SIM Time Network (SIMTN), which began operation in 2005, and has since been extended to compare the time scales operated in 26 nations, as shown in Figure 9.20. The network is operated by the Inter-American Metrology System (*Sistema Interamericano de Metrologia*, SIM), which covers the largest land area of any RMO, extending through North, Central, and South America and the Caribbean Islands. The SIMTN comparisons are made using time transfer links that are similar to the BIPM's GPS MC method, but have two significant differences; the common-view data is uploaded every 10 minutes and instantly processed by Internet servers in Canada, Mexico, and the United States, and there is no pivot laboratory [68]. This method allows the results of $\frac{N^2-N}{2}$ bilateral time comparisons, where N is the number of laboratories submitting time transfer data, to be published via the Internet every 10 minutes.

The SIMTN includes laboratories that participate in the BIPM key comparison, but it also includes other laboratories that currently lack the resources needed to participate. For example, the time scales at some SIM laboratories are based on a single rubidium clock, rather than a cesium clock. These laboratories plan to join the key comparison when the necessary resources are available. In the meantime, they can still establish traceability to the SI, at reduced levels of uncertainty, by making bilateral comparisons with UTC(k) laboratories via the SIMTN.

The measurements collected by the SIMTN enabled the development of a supplementary time scale called SIM Time (SIMT). Work began on SIMT in 2008, and it was first generated as an operational time scale in January 2010. The development effort for SIMT was led by the physicist José Mauricio López-Romero (b. 1961) and his colleagues at the National Metrology Center (*Centro Nacional de Metrología*, CENAM) in Mexico. The SIMT time scale algorithms are based entirely on the time difference measurements submitted to the SIMTN, without the use of any individual clock files or primary standards data.

The individual clocks in SIMT are defined as the individual time scales of the SIMTN participants, but only the time scales that include at least one cesium clock are allowed to have weight in the SIMT computation. The eligible clocks, which typically number from about 10 to 12, are weighted according to their frequency stability in much the same fashion as the clocks in UTC. This streamlined method allows SIMT to publish new results every hour. The fast SIMT updates supplement the far less frequent updates made by UTC and UTCr, at the expense of a small decrease in stability and accuracy. For example, measurements made over a period of about 18 months have estimated that the frequency stability of SIMT is about 1×10^{-14} at $\tau = 5$ days, when directly compared to UTC. The instability with respect to UTC is understandably due to SIMT's smaller number of clocks, its lack of contributions from primary frequency standards, and its less stable time transfer links. During this same measurement period, SIMT was consistently accurate to within ±10 ns of UTC, an accuracy that was mostly limited by Type B uncertainties in the SIMTN time transfer equipment [69].

Figure 9.20 The locations of the laboratories in the SIM Time Network.

9.6 The Use of Coordinated Universal Time (UTC) as a Measurement Reference

Because UTC provides the best approximation of the SI second, it provides an authoritative reference for all measurements of time, frequency, and time interval. As discussed in Chapter 6, the accuracy of a clock is determined by measuring its time difference with respect to UTC. The accuracy of an oscillator is determined by measuring its frequency difference with respect to its nominal frequency, and the best estimate of the nominal frequency is obtained by comparison to the frequency of UTC. Time interval measurements (Section 4.1) are referenced to the duration of the SI second as approximated by UTC. Traceability to the SI can be established for each of these measurements if an unbroken chain of comparisons can link the measurement result back to UTC, and if the uncertainty of each comparison link is documented.

UTC is also the reference for the time-of-day and date notation used in everyday timekeeping. When used by itself, UTC can serve as an unambiguous time-of-day and date reference, which is especially useful when time tagging any type of measurement data. When a time zone offset is applied to UTC, it can also serve as the local time-of-day and date reference for any given time zone. Both applications of UTC are described in the next section.

9.6.1 The Use of UTC as a Time-of-Day and Date Reference

As discussed in Section 8.4, the Greenwich meridian officially became the prime meridian at the International Meridian Conference in 1884, and the time referenced to the prime meridian became a standard known as Greenwich Mean Time (GMT). A driving force behind this development was the Scottish-Canadian inventor and railroad engineer Sanford Fleming (1827–1915), who is often credited as being the inventor of standard time. Some historians consider that credit to be overly generous, but even if Fleming wasn't the inventor of standard time, he was certainly its strongest advocate.

Fleming emigrated from Scotland to Canada when he was 18, and he soon gained a well-earned reputation as a hard worker who had the ability to draw people together and to present and discuss progressive ideas. Shortly after arriving in Canada, Fleming was apprenticed as a surveyor, and in 1849, he co-founded the Royal Canadian Institute, which started out as a professional institute for surveyors and engineers but eventually became a prestigious general scientific society. Fleming began his innovative, productive, and often combative career as a railroad engineer around 1853. After retiring from the railroads in 1880, he spent the rest of his life pursuing his many passions. Chief among those passions was his goal of establishing a standardized system of world time.

Fleming was largely responsible for the convening of the 1884 International Meridian Conference. He represented Canada at the conference, and he advocated not only for a prime meridian but also for a worldwide standard for time zones. Fleming proposed dividing the world into 24 time zones, the same number used to divide the day into hours. Twelve time zones would be located both east and west of the prime meridian, each having an equal width of 15°, or one hour, of longitude [70].

When the Meridian Conference was convened, Fleming had been publicly speaking and writing about the need for a standardized time system for about eight years, and had probably been thinking about it much longer than that, due to his career as a railroad engineer [71]. He had likely been interested in a 24-hour clock system for quite a while, because the 12-hour system based on a.m. and p.m. (abbreviated from the Latin *ante meridiem* and *post meridiem*) uses the same hour numbers during the daytime and nighttime, and this sometimes caused confusion on railroad schedules. According to a widely circulated story, first told by Fleming himself [72], his advocacy for standard time began in July 1876 when he was traveling in Ireland and was expecting to catch a train at 5:35 p.m., which was the time printed on the schedule. Unfortunately, the schedule had a typographical error—p.m. was mistakenly substituted for a.m.—and Fleming had to stay overnight and catch his train in the morning.

The story was probably embellished to prove a point [71], but 1876 seems to be the year when Fleming began to show great enthusiasm not just for a 24-hour clock, but for a uniform time system for the entire world [21]. He began presenting his ideas in public, and expressed them in writing in a pamphlet called *Terrestrial Time: A Memoir*, which was published in March 1878. Fleming's initial vision of Terrestrial Time wasn't referenced to any meridian. Instead, he conceptualized time as being referenced to the center of the Earth, to "clearly bring out the idea, that it is equally related to every point on the surface of the globe." As shown in Figure 9.21, Fleming visualized the Earth as a clock, where the hour hand pointed to each of the 24 divisions as it became noon. The hour hand revolved from east to west, moving at the same speed as the Earth on its axis, and it always pointed towards the Sun. The 24 divisions were designated by the letters of the alphabet, with the letters J and Z omitted, and Fleming even suggested that clock faces should be modified to use letters instead of numbers [72].

Fleming's initial work was mostly focused on time only, with little mention of time's link to longitude, but soon afterwards he became more aware of the importance of establishing a prime meridian as a common reference for both time and geography [71]. He then began making a strong and persuasive case that the prime meridian should be 180° from Greenwich, which is the meridian now recognized as the International Date Line. Fleming preferred that meridian because it could be thought of as neutral rather than national. During the years before and after the International Meridian Conference, Fleming also gave different names to his proposed timekeeping systems, dropping the use of "Terrestrial Time" in favor of "Cosmopolitan Time," "Cosmic Time," and eventually the term "Universal Time" that we use today [21, 73]. Fleming eventually accepted that the Greenwich meridian should be designated as the prime meridian, but he continued to think about time in a universal sense. It seems clear that he never liked the term Greenwich Mean Time due to its nationalistic implications.

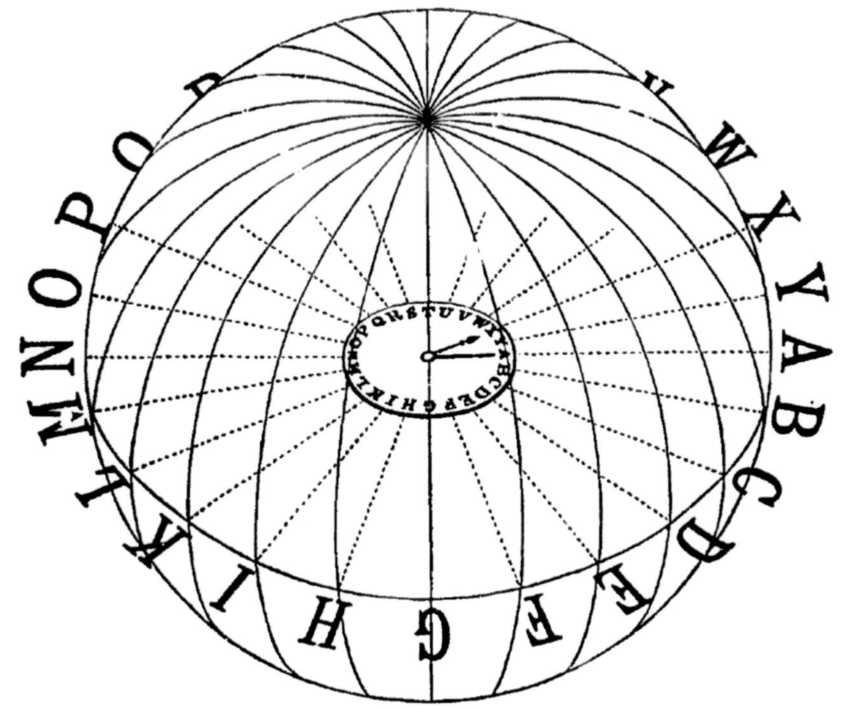

Figure 9.21 Fleming's original concept of world time zones referenced to the center of the Earth [72].

Fleming wasn't the first person to propose the use of time zones—in fact, the United States had a standardized time zone system in place for its railroad system by 1883. This system was largely based on the ideas of Charles Dowd (1825–1904), an educator and school administrator who felt that rail travel was unreliable and confusing due to the lack of a standardized time system. Dowd wrote in 1870 that the United States should be divided into four time zones—called Washington, first hour, second hour, and third hour—that were bounded by meridians separated by 15° of longitude [74]. Dowd referenced his proposed time zone system to the meridian passing through Washington, DC, which is located at about 77° west longitude with respect to Greenwich. However, in 1872, he modified his system by moving all of the time zone sections about 2° east, so that they were aligned with the Greenwich meridian and thus separated from Greenwich Time by an integer number of hours. A further modified version of Dowd's proposal was accepted as a standard by the U. S. railroad system in November 1883, and the four time zones were given the names that they still have today—Eastern, Central, Mountain, and Pacific [21, 71].

Unlike Fleming, Dowd never extended his time zone ideas outside of the United States. However, Fleming wasn't the first to hypothesize about an international time zone system. This had previously been done in 1858 in an obscure book entitled *Miranda!*, published in London by the Italian mathematician Guiseppe Barilli (1812–1894), who wrote using the pseudonym of Quirico Filopanti [75]. Similar to Fleming's later proposals, Barilli described an international time zone system based on 24 divisions, each with a longitudinal width of 15°, and each referenced to a meridian that passed through Rome. Only 500 copies of the book are believed to have been printed, however, and it seems to have had no impact on world timekeeping during Barilli's lifetime. It is not known if Fleming was even aware of it.

Fleming worked tirelessly to get his own time zone proposals accepted, but as he learned throughout his career, achieving an international consensus was essentially impossible. To say that acceptance was gradual would be an understatement, because the system of 24 one-hour time zones circling the Earth, all referenced to the prime meridian that passes through Greenwich, required over a century of legislative actions made by individual nations. A few nations have still opted for a system where the time is offset by an hour referenced to the prime meridian, plus some fraction of an hour, which can be either 30 or 45 minutes [71]. In addition, the time zone boundaries that pass through nations don't follow a strict longitudinal line, as shown by the map of world time zones in Figure 9.22. For example, in the United States, the time zone boundaries are maintained by the Department of Transportation, and they usually follow state boundaries, county boundaries, or rivers. In many nations, the time zones boundaries follow the national border, even if the nation spans more than one time zone. China, for instance, has just one time zone, even though its geographic area spans five of the one-hour longitude segments.

The time zone letter designations proposed by Fleming remain in common usage, most notably in radio and military communications. The 12 time zones to the east of the prime meridian are given the letters A through M, with the letter J skipped. The zones extend eastward in 15° increments of longitude, where A is the time zone centered on 15° east longitude (UTC + 1 hour) and M is the time zone extending from 172.5° to 180° east longitude (UTC + 12 hours), covering half of the 15° segment that surrounds the International Date Line.

The 12 time zones to the west of the prime meridian are given the letters N through Y, extending westward in 15° increments, where N is centered on 15° west longitude (UTC – 1 hour), and where Y extends from 172.5° to 180° west longitude (UTC – 12 hours), covering the other half of the segment surrounding the International Date Line. The time zone centered on the prime meridian at 0° longitude is called Z, or Zulu (pronounced ZOO LOO) when spoken in the standard phonetic alphabet accepted by several organizations, including the ITU, the International Civil Aviation Organization (ICAO), and the North Atlantic Treaty Organization (NATO). Therefore, even though it has no official usage in time metrology, "Zulu" can be considered equivalent to UTC when used for time-of-day notation.

The hourly offset between local time and UTC is affected not only by the time zone offset, but also by the practice followed in some nations of advancing the time by one hour during the summer months, so that sunset occurs at a later hour than usual. The period when the hour is advanced is called daylight saving time (DST) in the United States and Canada, and summer time (ST) in the European Union and other countries. The rules regarding DST and ST are established by both local and national laws, and are always subject to change, based on the current economic concerns and public opinion. The use of DST can create confusion when looking at a time zone map, because the number of hours between local time and UTC is no longer equal to the number of 15° time zone divisions between the local longitude and the prime meridian. For example, Boulder, Colorado, is located at about 105° west longitude, or seven time zones from the prime meridian. When Boulder is on Mountain Standard Time (MST), the local time is UTC – 7 hours, which agrees with the time zone offset. However, when DST is in effect and Boulder is on Mountain Daylight Time (MDT), the local time is UTC – 6 hours.

To avoid the ambiguity introduced by time zone offsets and by DST, the international standard for date and time representation, *ISO 8601*, states that local time should always be expressed with respect to UTC [76]. For example, if the local time in Boulder, Colorado, is 17:37:57 when MST is in effect, it should be expressed as 17:37:57-07, to indicate that the local time is offset by –7 hours from UTC. In most cases, expressing the offset in hours is all that is required. The only exceptions to this occur in the few nations where the UTC offset also includes a fractional part of the hour, requiring both hours and minutes to be included in the offset information. Seconds are never included in the offset information, because they are always the same for both local time and UTC.

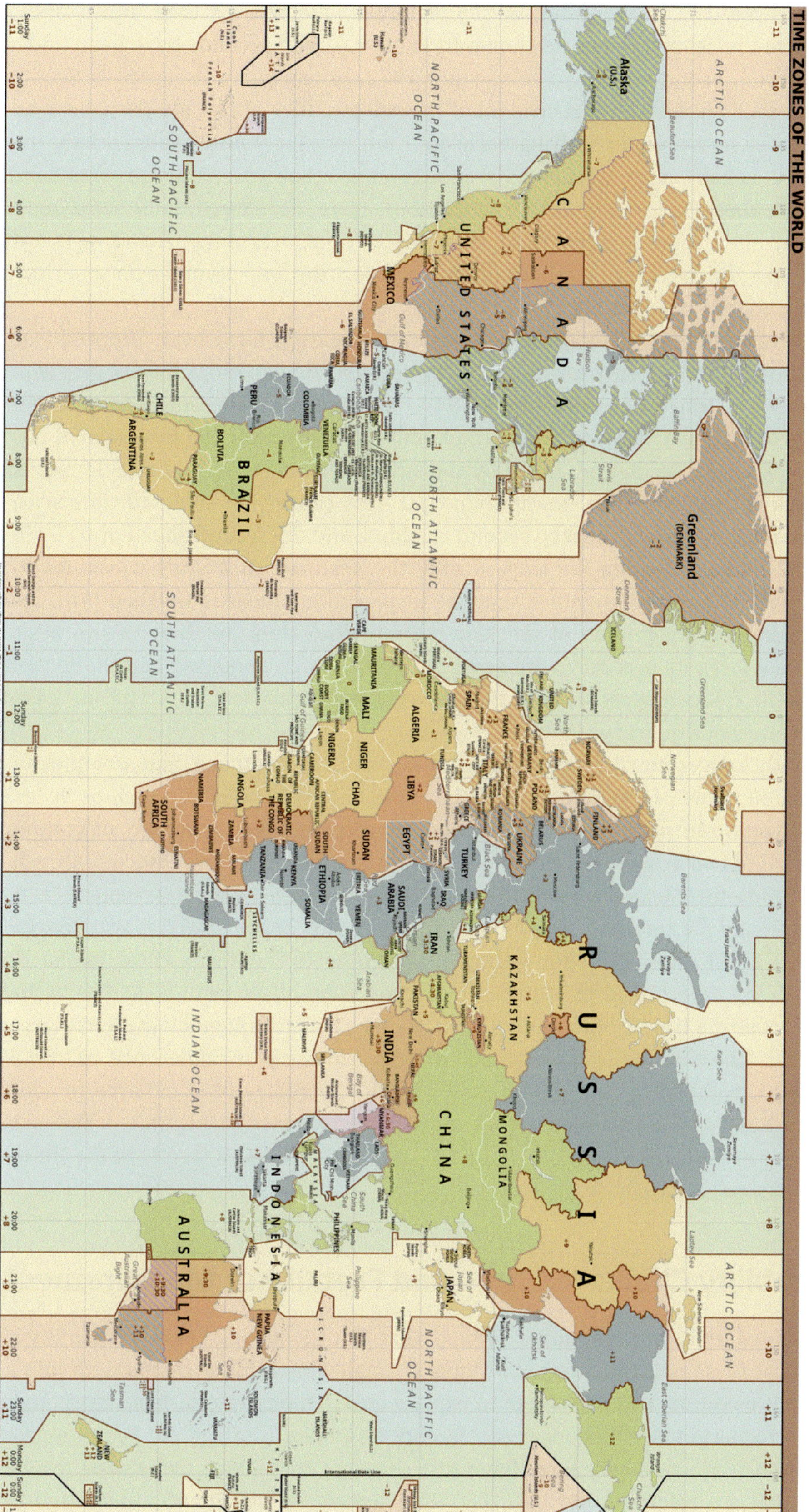

Figure 9.22 Map of world time zones.

Unlike local time, UTC time-of-day is completely unaffected by time zones and DST, but UTC is periodically affected by leap seconds. As shown in Table 9.4, leap seconds have been added to UTC at a rate of approximately one second every two years since 1972. However, as discussed in Section 3.5, the practice of inserting leap seconds into UTC is scheduled to end by 2035 [77]. This decision was reached in 2022 because the number of systems requiring accurate time had increased exponentially in recent years, and because leap seconds had caused the failure of some systems that weren't designed to handle them properly. To guard against these failures, some organizations have developed their own methods of handling leap seconds that unfortunately differ from standard UTC conventions. Instead of inserting a one-second time step at the end of a day, these methods gradually insert the leap second over a period of multiple hours, a process called leap second smearing.

At least several different methods of leap second smearing have been proposed or implemented by various organizations, and most are similar to the method documented by Google, which bears a strong resemblance to the pre-leap second practices of UTC (Section 9.4.1). The Google leap second smear begins 12 hours before the occurrence of a leap second and ends 12 hours after the leap second has occurred. During the 24-hour period, every second produced by the smeared time scale is 11.6 μs longer than the SI second, making the smeared time scale run slower than UTC. This causes the smeared time scale to be 0.5 second behind UTC in the instant before the leap second is added, and 0.5 second ahead of UTC after the leap second occurs. In the 12 hours following the leap second, the smeared time scale continues to slow down until it again agrees with UTC, which signifies that the leap second smear is finished [78]. Leap second smears succeed in reducing the magnitude of the leap second time step, but they create a situation where the smeared time scale loses its accuracy with respect to UTC for an extended period. For example, the Google method reduces the frequency accuracy of the smeared time scale to 1.16×10^{-5} for 24 hours, which is the approximate accuracy of a quartz wristwatch.

In short, both conventional leap seconds and smearing methods add complexity to systems that require a continuous time-of-day reference. This complexity will be eliminated, or at least pushed further down the road, when the practice of leap seconds is discontinued. However, discontinuing leap seconds also means that DUT1 will grow to become much larger than the current ±0.9 second threshold.

We can conclude this section by noting that government legislation has made UTC the official time-of-day reference in many nations, and that the responsibility for keeping UTC is usually given to laboratories that maintain a UTC(k) time scale. For example, according to federal statutes in the United States, local time-of-day is always determined by subtracting a time zone offset from UTC. The boundaries of the nine time zones that cover the United States and its territories are determined by the Secretary of Transportation. UTC is defined as "the time scale maintained through the General Conference of Weights and Measures and interpreted or modified for the United States by the Secretary of Commerce in coordination with the Secretary of the Navy" [79]. This means that the UTC reference for the United States is jointly provided by NIST, an agency of the Department of Commerce, and by the USNO, a facility of the United States Navy that is part of the Department of Defense.

9.7 Traceability Chains for Time and Frequency Measurements

There isn't a defined limit on the number of links that can be included in a traceability chain. Even so, it's better for practical reasons if the number of links is small. Traceability chains with just a small number of links are easier to maintain, and are easier to explain and demonstrate to laboratory auditors or assessors, who often have to decide whether legal traceability has been established. Fortunately, most traceability chains in time and frequency metrology have just three or four links, due to the availability of a variety of time signals that are referenced to UTC. The time signals include those transmitted by terrestrial radio stations, by network time servers, and by GNSS satellites. Each of these signals can link a measurement to UTC and the SI, via a UTC(k) time scale [3].

Figure 9.23 provides a diagram of a generic, four-link traceability chain that links the measurement of a DUT to UTC and the SI. This diagram can be used to model a measurement that involves network or radio time signals that are referenced to a UTC(*k*) time scale. The actual establishment of legal traceability requires evidence that the chain is unbroken, which is provided by documenting the measurement uncertainty of links A through D.

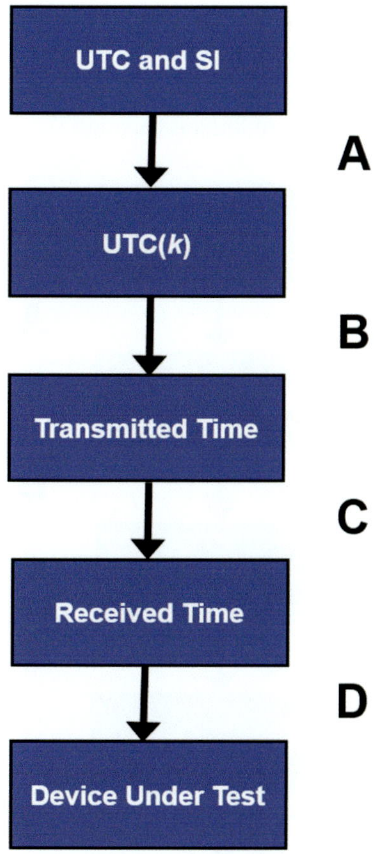

Figure 9.23 A generic four-link traceability chain for a measurement that involves the direct reception of radio or network time signals that are referenced to UTC.

In Figure 9.23, link A of the traceability chain connects UTC to UTC(*k*). The difference between the two time scales, UTC – UTC(*k*), can be expressed in terms of either time or frequency. The value of UTC – UTC(*k*), as well as its measurement uncertainty, can be conveniently obtained from measurement results published in the *Circular T*. This small uncertainty is insignificant for most types of measurements.

Link B of the traceability chain connects the UTC(*k*) time scale to the transmitted time signal, which we can call $UTC(k)_{TX}$. The transmitted time signal is often generated by a clock that's controlled by UTC(*k*) through a time transfer link, but remotely located—such as the clock at a radio station site or the clock in a network data center. The difference between UTC(*k*) and the transmitted signals, or $UTC(k) - UTC(k)_{TX}$, can be expressed in terms of either time or frequency. In most cases, the UTC(*k*) laboratory that controls the time signal broadcasts has already compensated for most of this difference, so the remaining measurement uncertainty is often insignificant. Even so, documentation about the uncertainty of link B should be made available by the UTC(*k*) laboratory.

Link C of the traceability chain connects the transmitted time signal to the received time signal via either a wireless radio path, or a wired network path. The difference between the transmitted and received signals, or $UTC(k)_{TX} - UTC(k)_{RX}$, can be expressed in terms of either time or frequency. Determining the uncertainty of link C requires knowledge of the uncertainty of the path delay between the transmitter and receiver, as well as knowledge of the uncertainties introduced by the receiving equipment.

Link D of the traceability chain connects the received time signals to the measurement of the DUT. The difference between received time signals and the signals produced by the DUT, or $UTC(k)_{RX}$ – DUT, can be expressed in terms of either time or frequency. The uncertainty of link D is the uncertainty of the measurement method used to compare the two signals.

Figure 9.24 is a slightly modified version of the diagram in Figure 9.23. This version can be used to model a measurement that involves GPS signals that are referenced to the UTC(USNO) time scale. The actual establishment of legal traceability again requires evidence that the chain is unbroken, which is provided by documenting the measurement uncertainty of links A through D.

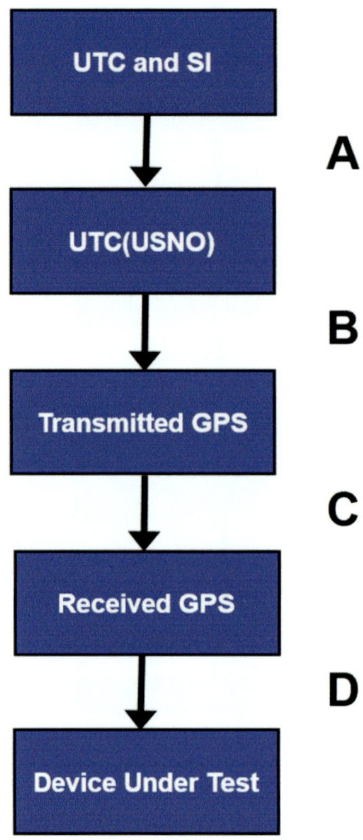

Figure 9.24 A generic four-link traceability chain for a measurement that involves the direct reception of GPS signals that are referenced to UTC(USNO).

In Figure 9.24, link A of the traceability chain connects UTC to UTC(USNO), because UTC(USNO) is the GPS time reference. The difference between the two time scales, UTC – UTC(USNO), can be expressed in terms of either time or frequency. The value of UTC – UTC(USNO), as well as its measurement uncertainty, can be conveniently obtained from measurement results published in the *Circular T*, and will be insignificant for most types of measurements.

Link B of the traceability chain connects the UTC(USNO) time scale to the predicted value of UTC(USNO) that's transmitted by GPS, which we can call $UTC(USNO)_{GPS}$. The UTC(USNO) – $UTC(USNO)_{GPS}$ difference has historically been less than ±1 ns on average (Section 12.4) and is insignificant for most measurements. Documentation about the uncertainty of link B can be obtained from the USNO, or obtained with slightly larger uncertainties, typically about ±3 ns, from the *Circular T* [3].

Link C of the traceability chain connects the transmitted GPS signals to the received GPS signals via a wireless radio path. The difference between the transmitted and received signals, or UTC $(USNO)_{GPSTX}$ – UTC

(USNO)$_{GPSRX}$, can be expressed in terms of either time or frequency. The factors that determine the uncertainty of link C are discussed in detail in Section 12.7.

Link D of the traceability chain connects the received GPS signals to the measurement of the DUT. The difference between the received GPS signals and the signals produced by the DUT, or UTC(USNO)$_{GPSRX}$ – DUT, can be expressed in terms of either time or frequency. The uncertainty of link D is the uncertainty of the measurement method used to compare the two signals.

The traceability chains illustrated in Figures 9.23 and 9.24 involve the use of time signals that are freely available to anyone who has the proper receiving equipment. Figure 9.25 provides a diagram of an even simpler traceability chain, a generic three-link chain that can be used to model a measurement that involves a paid subscription to a remote calibration service provided by a UTC(*k*) laboratory. In this example, the service provider will monitor and report the uncertainty of links A and B, but the service client is still responsible for determining the uncertainty of link C.

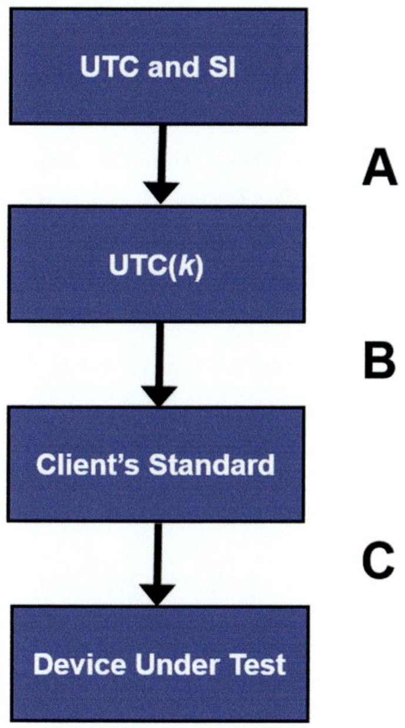

Figure 9.25 A generic three-link traceability chain for a measurement that involves a paid subscription to a remote calibration service operated by a UTC(*k*) laboratory.

In Figure 9.25, link A of the traceability chain is identical to Figure 9.23, connecting UTC to UTC(*k*), with the necessary measurement data obtained from the *Circular T*, and accounted for by the service provider.

Link B of the traceability chain directly connects the UTC(*k*) time scale to the client's standard, via a dedicated time transfer link that's included as part of the paid subscription. This link can be either a common-view link via navigation satellites, a two-way link via a communication satellite, or a network link through an optical fiber. The service provider documents the link B uncertainty in the calibration reports that it provides to the client.

Link C of the traceability chain connects the client's standard to the measurement of the DUT, and is equivalent to link D in the previous two examples. The uncertainty of link C is the uncertainty of the measurement method used to compare the two signals.

It's important to note that even when traceability has been established, the measurement uncertainties might still be too large to meet the industrial or regulatory requirements of a specific application. For example, a traceable measurement with an uncertainty of 100 μs won't satisfy the requirements of a telecommunications network that requires 1 μs uncertainty. This is mostly just common sense, but there is still the occasional misconception that traceability by itself will satisfy all measurement requirements.

In the next three chapters, we'll take a detailed look at the freely available radio, network, and satellite navigation signals that can be used to establish link C in the measurement models illustrated in Figures 9.23 and 9.24. These signals help complete the unbroken chain that connects end-user measurements to UTC and the SI. We'll focus not only on the time signals, but also on the types of clocks that the time signals control. Chapters 10 and 11 discuss the clocks controlled by radio time signals and by network time signals. Chapter 12 discusses the clocks controlled by satellite navigation signals, which provide essential support to critical infrastructure systems that depend on high-accuracy time.

References

1. E. L. Crow, "An Analysis of the Accumulated Error in a Hierarchy of Calibrations," *IRE Transactions on Instrumentation*, vol. I–9, no. 2, pp. 105–114, September 1960.
2. Joint Committee for Guides in Metrology (JCGM), "International Vocabulary of Metrology—Basic and General Concepts and Associated Terms (VIM)," *JCGM 200:2012*, 3rd edition, 2012.
3. D. Matsakis, J. Levine, and M. A. Lombardi, "Metrological and legal traceability of time signals," *Proceedings of the 49*th *Annual Precise Time and Time Interval Meeting (PTTI)*, pp. 59–71, January 2018.
4. International Organization for Standardization (ISO), "General Requirements for the Competence of Testing and Calibration Laboratories," *ISO/IEC 17025:2017*, 2017.
5. M. W. Dee, C. B. Ramsey, A. J. Shortland, T. F. G. Higham, and J. M. Rowland, "Reanalysis of the Chronological Discrepancies Obtained by the Old and Middle Kingdom Monuments Project," *Radiocarbon*, vol. 51, no. 3, pp. 1061–1070, 2009.
6. M. H. Stone, "The Cubit: A History and Measurement Commentary," *Journal of Anthropology*, vol. 2014, no. 489757, 11 p., January 2014.
7. H. A. Klein, *The World of Measurements*, New York: Simon and Schuster, 1974.
8. L. V. Judson, "Weights and Measurements of the United States: A Brief History," *National Bureau of Standards Special Publication 447*, March 1976.
9. T. Quinn, *From Artefacts to Atoms: The BIPM and the Search for Ultimate Measurement Standards*, Oxford University Press, 2012.
10. J. E. Barnett, *Time's Pendulum*, Plenum Press, 1998.
11. A. Robinson, *The Story of Measurement*, London: Thames & Hudson, 2007.
12. K. Adler, *The Measure of All Things*, Simon & Schuster, 2001.
13. J. Kovalevsky and T. J. Quinn, "The International System of Units (SI)," *C. R. Physique*, vol. 5, no. 8, pp. 799–811, October 2004.
14. C. H. Page and P. Vigoureux, "The International Bureau of Weights and Measures 1875–1975," *National Bureau of Standards Special Publication 440*, May 1975.
15. Bureau International des Poids et Mesures, "The International System of Units (SI)," 9th edition, ver. 2.01, December 2022.
16. K. M. Evenson, J. S. Wells, F. R. Petersen, B. L. Danielson, and G. W. Day, "Speed of Light from Direct Frequency and Wavelength Measurements of the Methane-Stabilized Laser," *Physical Review Letters*, vol. 29, no. 19, pp. 1346–1349, November 1972.
17. D. A. Jennings, R. E. Drullinger, K. M. Evenson, C. R. Pollock, and J. S. Wells, "The Continuity of the Meter: The Redefinition of the Meter and the Speed of Light," *Journal of Research of the National Bureau of Standards*, vol. 92, no. 1, pp. 11–16, January-February 1987.
18. H. Fang, F. Bielsa, S. Li, A. Kiss, and M. Stock, "The BIPM Kibble Balance for Realizing the Kilogram Definition," *Metrologia*, vol. 57, no. 4, pp. 1–10, July 2020.
19. Resolution 1 of the *13th Conference Generale des Poids et Mesures (CGPM)*, 1967.
20. N. Dimarcq, M. Gertsvolf, G. Mileti, S. Bize, C. W. Oates, E. Peik, D. Calonico, T. Ido, P. Tavella, F. Meynadier, G. Petit, G. Panfilo, J. Bartholomew, P. Defraigne, E. A. Donley, P. O. Hedekvist, I. Sesia, M. Wouters, P. Dubé, F. Fang, F. Levi,

J. Lodewyck, H. S. Margolis, D. Newell, S. Slyusarev, S. Weyers, J.-P. Uzan, M. Yasuda, D.-H. Yu, C. Rieck, H. Schnatz, Y. Hanado, M. Fujieda, P.-E. Pottie, J. Hanssen, A. Malimon, and N. Ashby, "Roadmap towards the redefinition of the second," *Metrologia*, vol. 61, no. 1, pp. 1–19, January 2024.

21. D. Howse, *Greenwich Time and the Longitude*, London: Philip Wilson Publishers, Ltd., 1997.
22. D. D. McCarthy and P. K. Seidelmann, *Time: From Earth Rotation to Atomic Physics*, Cambridge University Press, 2nd edition, 2018.
23. International Astronomical Union (IAU), "Commission 4. (Ephemerides)," *Transactions of the IAU*, vol. 3, p. 224, July 1928.
24. P. Kartaschoff, *Frequency and Time*, Academic Press, 1978.
25. W. Markowitz, "The Atomic Time Scale," *IRE Transactions on Instrumentation*, vol. I-11, no. 3, pp. 239–242, December 1962.
26. R. A. Nelson, D. D. McCarthy, S. Malys, J. Levine, B. Guinot, H. F. Fliegel, R. L. Beard, and T. R. Bartholomew, "The Leap Second: Its History and Possible Future," *Metrologia*, vol. 38, no. 6, pp. 509–529, December 2001.
27. "Co-Ordination of Time and Frequency Transmissions," *Journal of the IEE*, vol. 6, no. 65, p. 268, May 1960.
28. B. Guinot, "History of the Bureau International de l'Heure," *Polar Motion: Historical and Scientific Problems, IAU Colloquium 178*, vol. 208, pp. 175–184, 2000.
29. B. Guinot and E. F. Arias, "Atomic Time-Keeping from 1955 to the Present," *Metrologia*, vol. 42, no. 3, pp. S20–S30, June 2005.
30. R. Rigby, "Timekeeper for the World," *The Rotarian*, vol. 103, no. 3, pp. 22–25, September 1963.
31. C. Audoin and B. Guinot, *The Measurement of Time: Time, Frequency and the Atomic Clock*, Cambridge University Press, 2001.
32. International Astronomical Union (IAU), "31. Commission de L'Heure," *Transactions of the IAU*, vol. 13, no. 2, pp. 173–182, August 1967.
33. National Bureau of Standards, "National Standards of Time and Frequency in the United States," *Proceedings of the Institute of Radio Engineers (IRE)*, vol. 48, no. 1, pp. 105–106, January 1960.
34. P. K. Seidelmann, *Explanatory Supplement to the Astronomical Almanac*, Mill Valley, California: University Science Books, 1992.
35. Bureau International de l'Heure (BIH), *Annual Report for 1972*, 1973.
36. J. McA. Steele, "The Improved UTC System to be Introduced on 1 January 1972," *International Hydrographic Review*, vol. XLIX, no. 2, pp. 137–148, July 1972.
37. A. R. Chi and H. S. Fosque, "A Step in Time: Changes in Standard-Frequency and Time-Signal Broadcasts—January 1, 1972," *NASA Technical Note D-7065*, 31 p., January 1973.
38. P. Tavella and C. Thomas, "Comparative Study of Time Scale Algorithms," *Metrologia*, vol. 28, no. 2, pp. 57–63, 1991.
39. Consultative Committee for Time and Frequency (CCTF), "Participation in the Ongoing Key Comparison in Time, CCTF-K001.UTC," *CCTF WGMRA Guideline 7*, April 2015.
40. O. Werhahn, D. A. Olson, C. Kuanbayev, and A. Henson, "The CIPM MRA—Success and Performance," *Metrologia*, vol. 60, no. 4, 9 p., July 2023.
41. G. Panfilo and E. F. Arias, "Algorithms for International Atomic Time," *IEEE Transactions on Ultrasonics, Ferroelectrics, and Frequency Control*, vol. 57, no. 1, pp. 140–150, January 2010.
42. E. F. Arias, G. Panfilo, and G. Petit, "Timescales at the BIPM," *Metrologia*, vol. 48, no. 4, pp. S145–S153, July 2011.
43. G. Panfilo and F. Arias, "The Coordinated Universal Time (UTC)," *Metrologia*, vol. 56, no. 4, 26 p., June 2019.
44. G. Panfilo, A. Harmegnies, and L. Tisserand, "A New Weighting Procedure for UTC," *Metrologia*, vol. 51, no. 3, pp. 285–292, May 2014.
45. G. Petit, "Towards an Optimal Weighting Scheme for TAI Computation," *Metrologia*, vol. 40, no. 3, pp. S252–S256, June 2003.
46. J. Milton and G. Panfilo, "The Maximum Weight in UTC: Proposal for a New Approach," *IEEE Transactions on Ultrasonics, Ferroelectrics, and Frequency Control*, vol. 70, no. 4, pp. 336–343, April 2023.

47. T. E. Parker and D. Matsakis, "Time and Frequency Dissemination: Advances in GPS Transfer Techniques," *GPS World*, vol. 15, no. 11, pp. 32–38, November 2004.
48. D. Kirchner, "Two Way Time Transfer Via Communication Satellites," *Proceedings of the IEEE*, vol. 79, no. 7, pp. 983–990, July 1991.
49. D. D. Davis, M. Weiss, A. Clements, and D. W. Allan, "Construction and Performance Characteristics of a Prototype NBS/GPS Receiver," *Proceedings of the 35th Annual Symposium on Frequency Control*, pp. 546–552, May 1981.
50. D. W. Allan, D. D. Davis, M. Weiss, A. Clements, B. Guinot, M. Granveaud, K. Dorenwendt, B. Fischer, P. Hetzel, S. Aoki, M.-K. Fujimoto, L. Charron, and N Ashby, "Accurate of International Time and Frequency Comparisons Via Global Positioning System Satellites in Common-View," *IEEE Transactions on Instrumentation and Measurement*, vol. IM-34, no. 2, pp. 118–125, May 1985.
51. J. Nawrocki, W. Lewandowski, and J. Azoubib, "Time Transfer with GPS Multi-Channel Motorola Oncore Receiver using CCDS Standards," *Proceedings of the 29th Annual Precise Time and Time Interval Meeting (PTTI)*, pp. 319–328, December 1997.
52. M. A. Weiss and D. W. Allan, "An NBS Calibration Procedure for Providing Time and Frequency at a Remote Site by Weighting and Smoothing of GPS Common View Data," *IEEE Transactions on Instrumentation and Measurement*, vol. IM-36, no. 2, pp. 572–578, June 1987.
53. D. W. Allan and C. Thomas, "Technical Directives for Standardization of GPS Time Receiver Software," *Metrologia*, vol. 31, no. 1, pp. 69–79, 1994.
54. P. Defraigne and G. Petit, "CGGTTS-Version 2E: An Extended Standard for GNSS Time Transfer," *Metrologia*, vol. 52, no. 6, 22 p., October 2015.
55. P. Defraigne and G. Petit, "Time Transfer to TAI Using Geodetic Receivers," *Metrologia*, vol. 40, no. 4, pp. 184–188, June 2003.
56. G. Petit and Z. Jiang, "GPS All in View Time Transfer for TAI Computation," *Metrologia*, vol. 45, no. 1, pp. 35–45, 2008.
57. V. Zhang and Z. Li, "Measured Ionospheric Delay Corrections for Code-Based GPS Time Transfer," *NCSLI Measure: The Journal of Measurement Science*, vol. 10, no. 3, pp. 66–71, September 2015.
58. F. Lahaye, D. Orgiazzi, P. Tavella, and G. Cerrotto, "GPS Time Transfer: Using Precise Point Positioning for Clock Comparisons," *GPS World*, vol. 17, no. 11, pp. 44–49, November 2006.
59. P. Defraigne, N. Guyennon, and C. Bruyninx, "GPS Time and Frequency Transfer: PPP and Phase-Only Analysis," *International Journal of Navigation and Observation*, vol. 2008, article 175468, 7 p., July 2008.
60. Bureau International des Poids Et Mesures (BIPM), "Annual Report of the BIPM Time Section," vol. 13, 100 p., 2000.
61. Z. Jiang, V. Zhang, T. E. Parker, G. Petit, Y.-H. Huang, D. Piester, and J. Achkar, "Improving Two-Way Satellite Time and Frequency Transfer with Redundant Links for UTC Generation," *Metrologia*, vol 56, no. 2, 15 p., February 2019.
62. W. Lewandowski, D. Matsakis, G. Panfilo, and P. Tavella, "The Evaluation of Uncertainties in [UTC – UTC(k)]," *Metrologia*, vol. 43, no. 3, pp. 278–286, May 2006.
63. W. Lewandowski, D. Matsakis, G. Panfilo, and P. Tavella, "Analysis of Correlations, and Link and Equipment Noise in the Uncertainties of [UTC – UTC(k)]," *IEEE Transactions on Ultrasonics, Ferroelectrics, and Frequency Control*, vol. 55, no. 4, pp. 750–760, April 2008.
64. G. Petit and G. Panfilo, "Discussion on the Calculation of the Uncertainties of [UTC – UTC(k)]," *Metrologia*, vol. 61, no. 4, 8 p., July 2024.
65. G. Petit, F. Arias, A. Harmegnies, G. Panfilo, and L. Tisserand, "UTCr: A Rapid Realization of UTC," *Metrologia*, vol 51, no. 1, pp. 33–39, 2014.
66. Consultative Committee for Time and Frequency (CCTF), "CCTF Criteria for Obtaining Traceability in Time and Frequency," *CCTF WGMRA Guideline 9*, June 2017.
67. R. Solis, "Traceability to UTC Using the SIM Time Network Bilateral Comparisons," *Proceedings of the 2018 Conference on Precision Electromagnetic Measurements (CPEM)*, 2 p., July 2018.
68. M. A. Lombardi, A. N. Novick, J. M. Lopez-Romero, F. Jimenez, E. de Carlos-Lopez, J. S. Boulanger, R. Pelletier, R. de Carvalho, R. Solis, H. Sanchez, C. A. Quevedo, G. Pascoe, D. Perez, E. Bances, L. Trigo, V. Masi, H. Postigo, A. Questelles, and A. Gittens, "The SIM Time Network," *Journal of Research of the National Institute of Standards and Technology*, vol. 116, no. 2, pp. 557–572, March-April 2011.

69. J. M. Lopez-Romero, M. Lombardi, E. de Carlos-Lopez, N. Diaz-Munoz, C. A. Ortiz, R. de Carvalho, and R. Solis, "Comparison of the Multinational SIM Time Scale to UTC and UTCr," *IEEE Transactions on Instrumentation and Measurement*, vol. 69, no. 7, pp. 4445–4452, July 2020.
70. M. M. Thomson, *The Beginning of the Long Dash: A History of Timekeeping in Canada*, University of Toronto Press, 1978.
71. I. R. Bartky, *One Time Fits All: The Campaigns for Global Uniformity*, Stanford University Press, 2007.
72. S. Fleming, *Terrestrial Time: A Memoir*, 37 p., London: Edwin S. Boot, Printer, March 1878.
73. S. Fleming, "Time Reckoning for the Twentieth Century," reprinted in *Annual Report of the Smithsonian Institution for the Year Ending June 30, 1886*, pp. 345–366, Washington: Government Printing Office, 1889.
74. C. F. Dowd, *System of National Time for Railroads*, Albany, New York: Weed, Parsons and Company, 1870.
75. Q. Filopanti, *Miranda! A Book of Wonders Hitherto Unheeded*, London: James Morgan, 1858.
76. International Organization for Standardization (ISO), "Date and Time—Representations for Information Interchange—Part 1: Basic rules," *ISO 8601-1:2019*, 2019.
77. C. Palmer, "Global Timekeepers Push Pause on the Leap Second," *Engineering*, vol 27, pp. 6–7, August 2023.
78. "Leap Smear," *Google Developers Guide* (https://developers.google.com/time/smear), October 2023.
79. 15 *United States Code* § 261 (2007).

PART III

Time Signals

Chapter 10

Clocks Controlled by Radio Time Signals

The whole world takes its time from Greenwich, but Greenwich, they say, takes its time from Admiral Boom.

— Bert Michaels (played by Dick Van Dyke) in the movie "Mary Poppins"

Frustrating as it is to time metrologists, a group of independent clocks can't keep the exact same time. We can prove this by synchronizing a group of wristwatches and then checking them again a month later. By then, the watches will disagree with each other, probably by at least a few seconds, and perhaps by much more. If we try the same experiment with a group of cesium clocks, the time differences will be nanoseconds instead of seconds, but again the clocks won't be able to keep the exact same time.

The British horologist Frank Hope-Jones (1867–1950) recognized these problems well over a century ago, and he became an advocate for collectivism, rather than individualism in clocks. The clocks of that era were typically self-wound, synchronized from different sources, and had mechanical oscillators with poor accuracy and stability. Therefore, none of them could even come close to keeping the same time as the others. Hope-Jones wrote in 1900 that

> *"Time-keeping is essentially one of those things that can be done much better by a Municipality, or a Company than by the individual. What is every one's business is nobody's. There is not an argument in favour of the public supply of such commodities as electricity, gas, or water, which does not apply with equal or even greater force to the distribution of time. Anything like correct measurement of time is a delicate operation, so difficult indeed that absolute accuracy is impossible even with the aid of carefully compensated seconds pendulums in our observatories. How then can we expect to do it for ourselves in every room and in every office, with clocks in which the length of the pendulum and the quality of the works are sacrificed to the ornaments of the cases?"* [1]

Hope-Jones was well aware that there was more to timekeeping than simply designing more accurate clocks. He noted that to make clocks keep uniform time, "it will not suffice merely to improve the standard and quality of our ordinary clocks; we must strike at the root of the evil—their *independence*" [1]. No matter how perfect their oscillators, all clocks would need to be periodically synchronized to a common reference before they would lose their independence. It would, of course, be best if this synchronization

As Time Goes By: A Comprehensive Guide to Time and Frequency Measurements
Michael A. Lombardi

ISBN 978-981-5352-28-3 (Hardcover), 978-1-042-01532-0 (eBook)
www.jennystanford.com

DOI: 10.1201/9781042015320-14

were automatic. By 1910, Hope-Jones had already devoted years to solving this problem—he was then working on the design of an "ideal synchronizer which both corrects the rate and sets the hands," thus forcing a group of electrically connected clocks to always display the same time [2].

Hope-Jones has since been proven right because the "standard and quality of our ordinary clocks" has improved substantially, but the problems of clock independence still exist. The quartz wristwatch is one obvious example of ordinary clocks getting better, because even the lowest-priced models typically keep time accurate to within several minutes per year, which is roughly the same amount of time that their mechanical predecessors gained or lost per week. Because oscillator technology is always improving, the next advances in ordinary clocks should be even more impressive. We might one day be able to buy low-cost clocks containing miniature atomic oscillators that would gain or lose just a small fraction of a second over the course of a century [3]. Even then, the problem of clock independence will still exist at some level unless all clocks are periodically synchronized to a common reference to ensure their accuracy.

The importance of keeping clocks synchronized to a common reference is readily evident when we look at the stringent timing requirements of the telecommunication, electric power, and financial industries (Chapters 16 through 18), and the consequences of not meeting these requirements. It's also readily evident in everyday life. A study published in 1994 checked the accuracy of 152 different clocks used to time stamp events in the emergency department of a large hospital. Many of the clocks tested were watches worn by medical personnel. It was found that 55% of the clocks were in error by more than a minute, and that only 3% were accurate to within 1 s. The slowest clock was more than 12 minutes slow, and the fastest was about seven minutes fast. This meant that two clocks in the same hospital could vary by nearly 20 minutes, causing potentially life-threatening problems [4]. The consequences of having the wrong time are not always that serious, but clocks that disagree can always cause problems.

The clock independence problem is still with us, but a tremendous amount of effort has gone into solving the problem, and a tremendous amount of progress has been made. The progress is the result of the wide-scale distribution of accurate time signals, which has allowed clocks to be synchronized no matter where they are located. The first attempts at a solution occurred even before the birth of Hope-Jones, when time signals were first distributed via telegraph lines in 1852 by the Royal Observatory in England [5]. However, telegraphic time signals could only be sent to places that were connected to the cumbersome land lines used by telegraphs. As we'll see in this chapter, the ideal medium for distributing time signals would prove to be wireless telegraphy—later known as radio—which eventually allowed time signals to be sent everywhere [6, 7].

10.1 The Origins of Time Signal Stations and Radio-Controlled Clocks

We can think of radio-controlled clocks (RCCs) as less sophisticated versions of the disciplined oscillators that were described in Section 5.3.2. An RCC serves only as a time standard, and not as a frequency standard. It periodically synchronizes the clock's time to a radio reference signal, but it doesn't adjust the oscillator that produces the clock's frequency. In contrast, a disciplined oscillator adjusts the frequency of its LO to agree with the radio reference signal—typically with a combination of hardware and software that functions as a PLL. Nearly all disciplined oscillators also provide a synchronized clock, which allows them to serve as a reference for both frequency and time.

The idea of an RCC is nearly as old as radio itself. In fact, almost immediately after it was known that information could be sent by wireless transmissions, early radio scientists began looking for ways to send time signals from one location to another. The first radio time signals were decoded by humans who used the information to manually synchronize clocks, and the automatic synchronization of clocks occurred years later after the technology advanced.

Most historians cite the experiments of the Italian inventor Guglielmo Marconi (1874–1937) as the first radio transmissions. In 1895—when he was just 21 years old—Marconi used a spark-gap transmitter and a vertical-wire antenna to successfully send radio signals over a distance of more than 3 km. By 1899, Marconi had established radio communications between France and England across the English Channel, and in 1901 he sent the first radio signals across the Atlantic Ocean [8]. In the midst of these historic experiments—before any radio stations existed, and before the public even believed Marconi's results—a proposal was made to use the new wireless medium for the transmission of time signals. The proposal was made by Howard Grubb (1844–1931), an Irish optical instrument designer who had worked with astronomical observatories for many years and who had developed a strong interest in timekeeping. Grubb gave a talk to the Royal Dublin Society in November 1898, where he introduced the concept of a radio-controlled clock [9]. The following year, Grubb's comments were published in the Society's proceedings:

> *"There is something very beautiful in this action of the 'Marconi' wave. In a city supplied with this apparatus we should be conscious as we hear each hour strike that above us and around us, swiftly and silently, this electrical wave is passing, conscientiously doing its work, and setting each clock in each establishment absolutely right, without any physical connection whatsoever between the central distributing clock, and those which it keeps correct by means of this mysterious electrical wave.*
>
> *We might go even still further, and although I do not put it forward as a proposition likely to be carried out in any way, except as an experiment, yet it undoubtedly would be perfectly possible to carry an apparatus in one's pocket, and have our watches automatically set by this electrical wave as we walk about the streets."* [10]

With these words, Grubb envisioned the automatic RCC about three-quarters of a century before it finally appeared. Grubb and others realized from the beginning that radio was a nearly perfect medium for delivering time signals. With radio, the correct time could be almost simultaneously delivered to every clock that resides within the coverage area of a time transmitter, with virtually no limit on the number of clocks that could potentially be synchronized. Because radio waves travel at the speed of light, the time delay between a terrestrial transmitter and receiver is so small that it can be completely ignored when accuracy to within a tenth of a second is sufficient. The United States Navy was lured by these potential benefits, and began broadcasting radio time signals in telegraphic code just five years after Grubb's prophetic suggestions.

10.1.1 The First Time Signals Broadcast by the United States Navy

Timekeeping is an essential part of military operations, and the United States Navy was serious about timekeeping long before the invention of radio. Like the Royal Observatory in England, the United States Naval Observatory (USNO) in Washington, DC, had been providing time over telegraph lines for several decades before radio was invented. The Navy telegraphic time service began in August 1865, just a few months after the end of the American Civil War. Time signals were sent from the USNO thrice daily—at 7 a.m., noon, and 6 p.m.—to the fire-alarm and police stations of Washington, DC.

The noon time signals were also sent to the Western Union Telegraph Company for redistribution, marking the beginning of an elaborate time transfer system with a long history. By May 1877, Western Union was distributing USNO time to every U. S. city with a population of more than 20,000 [11]. Due to this wide area time distribution network, Western Union became known to many as the nation's "official timekeeper." Figure 10.1 shows a Western Union display clock, photographed in about 1935. The placards surrounding the clock identify Western Union as "The Nation's Timekeeper" and advertise their time service as costing five cents per day—a cost that included renting a clock manufactured by the Self Winding Clock Company of New York. The USNO/Western Union time service lasted for about a century, continuing until the late 1960s [12].

Figure 10.1 A Western Union time service clock, synchronized over telegraph lines (c. 1935).

Despite the success of this partnership, the Navy's primary timekeeping concern wasn't providing time to Western Union. They were most concerned with providing accurate time to ships at sea, and it wasn't possible, of course, to run a telegraph line to a ship. The best they could initially do was to drop large time balls—with the drops synchronized to telegraphic time signals—from locations near the shore. Other clocks could be manually synchronized to the drop of the time ball, but these systems were only useful to sailors on waters within sight of the time ball. Time balls existed before telegraph time services, and the USNO had first installed one in 1845, located atop a building on a bluff above the Potomac River. Each day at noon, the time ball was dropped or thrown down by a person who received a verbal or hand signal from a person standing in front of the observatory's clock. In 1877, Western Union attracted considerable attention by erecting a time ball on top of its New York City headquarters that dropped daily at noon, except on Sundays. The drop of the Western Union time ball was synchronized to USNO time sent by telegraph from Washington.

In 1885, the time ball atop the USNO building was relocated to the State, War, and Navy Department building. The new location was adjacent to the White House, the home of the president of the United States, which signified the importance of accurate time to the government. By then, the time balls were no longer manually dropped, but instead were automatically dropped when the time signal was received by an apparatus that employed electrical circuits and a magnet [13].

Radio signals finally solved the problem of sending time signals to ships at sea. Clocks located at Navy yards in the coastal regions of the United States were synchronized to the USNO clock in Washington via telegraph lines, and the time was then retransmitted as a radio signal. The Navy's first experiments of sending radio time signals to ships at sea began in September 1903, when short-range, low-powered signals were transmitted from the Navesink Highlands in New Jersey. The first official time signal transmissions began at the radio station of the Boston Navy Yard in Massachusetts on August 9, 1904 [14]. That same year, the USNO superintendent Colby Mitchell Chester (1844–1932) included these prophetic comments while thanking the telegraph companies for their cooperation in distributing time:

> *"Perhaps with the development of wireless telegraphy it is not too much to expect that the day will come when not only every land station but every ship at sea will receive the daily noon signal from our standard clock, in which practically all of the uncertainties of longitude determinations at sea will vanish. When that day comes (and it now seems to be entirely practicable) this clock may be heard in every land and on every ocean of the globe, and be available daily both for standard time and navigation everywhere."* [15]

As John Harrison had proven in the 18th century, having an accurate shipboard clock was critical for navigation (Section 1.6). During Harrison's era, the accuracy of a ship's clock could only be measured before and after a voyage. During the voyage, the best that could be done was to apply an estimated daily time offset. The Navy's radio time service changed everything—by receiving reference time signals while still at sea, sailors could repeatedly check the accuracy of shipboard clocks and make adjustments whenever necessary [16].

After their initial efforts in 1903, the Navy continued to invest in a network of time signal stations that remained in operation until World War II. The number of stations increased to 19 by 1909, with each having a reliable range of about 100 miles (161 km). Soon afterwards, however, advances in radio technology increased the power and range of the stations and their numbers were reduced [11]. By 1916, a smaller but improved network of nine high-power transmitters—listed in Table 10.1—sent USNO time signals to the Atlantic and Pacific Oceans, the Great Lakes, and the Gulf of Mexico [17].

Table 10.1 The United States Navy time signal stations that were operational in 1916

Call Sign	Location	Frequency (kHz)	Wavelength (m)	Region
NAA	Arlington, Virginia	120	2,500	Atlantic
NAR	Key West, Florida	300	1,000	Atlantic
NAT	New Orleans, Louisiana	300	1,000	Gulf of Mexico
NAJ	Great Lakes, Illinois	198	1,512	Great Lakes
NPE	North Head, Washington	150	2,000	Pacific
NPG	Mare Island, California	120	2,500	Pacific
NPK	Point Arguello, California	400	750	Pacific
NPL	San Diego, California	150	2,000	Pacific
NPW	Eureka, California	214	1,400	Pacific

The first of the Navy's high-power radio stations, identified by the call sign NAA, began daily time signal broadcasts on March 31, 1913, from Arlington, Virginia [11]. NAA was built as part of the Navy's effort to establish a worldwide communication network, and the station wasn't limited to broadcasting time signals. The broadcast facilities included a 100 kW rotary spark-gap transmitter and three large transmission towers of ornamental steel construction. The largest tower was 183 m tall, about 14 m taller than the nearby Washington Monument, and the other two towers reached a height of 137 m (Figure 10.2). The center of the towers formed an isosceles triangle where the base and height were each about 107 m in length [18].

Time signals from NAA were broadcast on a wavelength of 2,500 m (~120 kHz), with other frequencies used later. The signals were sent daily at noon and 10 p.m. and they had a range of about 3,000 miles (4,828 km). However, due to the presence of their other stations, the USNO only expected NAA to cover "the country east of the Rocky Mountains and the greater part of the North Atlantic Ocean." The mean daily error for the transmitted signals for 1916 was 30 ms, with a 20 ms delay introduced by the direct telegraph wire connecting the station to the USNO clock on the other side of the Potomac River [17]. A 1913 article praising the new time signals stated that only "one-twentieth of a second is lost in transmission between the Arlington station and San Francisco" [19].

Figure 10.2 The three NAA transmission towers at Arlington, Virginia (1913).

The time signals sent by NAA and the other Navy stations were designed to be easy to decode. The transmissions lasted for five minutes, beginning at 11:55 a.m. for the noon time signal. During the first four minutes, a dot with a duration of 350 ms was sent once per second, but the 29th dot and the last five dots of each minute were omitted. The pattern differed slightly during the last minute before noon, when the 29th dot and the last 10 dots of the minute were omitted. The omission of dots at the end of minutes allowed those listening to the signal to better hear the ticks of the clocks that they were synchronizing. The transmission ended with a long dash of 1 s in duration that was sent exactly at noon, which served as the OTM used for clock synchronization [16, 20].

The Arlington towers became a permanent part of radio history in October 1915, when they were used by engineers from the American Telephone and Telegraph Company (AT&T) to send the first voice messages across the Atlantic Ocean. The messages were received by fellow AT&T engineers who were listening at the Eiffel Tower in Paris. Because the radio signals were only sent from west to east, the confirmation that the messages from Arlington were heard was cabled back across the Atlantic [21].

The Navy decommissioned NAA in the late 1930s. The radio towers were considered a hazard to aircraft approaching the new Washington National Airport, and were dismantled in 1941. The NAA call sign is still in use at this writing—it's been assigned since 1961 to a VLF station in Cutler, Maine that the Navy uses to communicate with submarines.

10.1.2 The First Time Signals Broadcast from the Eiffel Tower

Radio time signal experiments began in Europe shortly after the U. S. Navy began their time signal broadcasts. By 1910, the experiments in France led to the introduction of the best known of all of the early time signal stations. The station was identified by the call sign FL and operated from the Eiffel Tower in Paris. The tower was named after the civil engineer Gustave Eiffel (1832–1923), whose company designed and built the tower for the 1889 Paris Exposition, which was held to celebrate the 100th anniversary of the French Revolution. The Eiffel Tower was built to a height of 312 meters, which made it the tallest structure in the world. It continued to hold that distinction until 1930, when it was finally surpassed by the Chrysler Building in New York City.

The Eiffel Tower is now one of the world's most iconic and most visited structures, but many Parisians initially felt it was unsightly. They had opposed its initial construction, and after it was built, they wanted it to be removed. It was therefore scheduled for demolition in 1910, when the 20-year land use permit that had been granted to Eiffel would be returned to the city of Paris. Fortunately, radio was invented during that 20-year period, and the tower was spared because it had proven to be useful as an antenna mast.

The first radio telegraphy stations at the Eiffel Tower were installed under the direction of Gustave-Auguste Ferrié (1868–1932), a pioneering radio scientist and an officer in the French Army. Ferrié had specialized in the army's telegraph service, and he was present in 1898 when Marconi conducted radio telegraphy experiments between Wimereux, France and Dover, England. He was immediately won over by the new technology and was appointed to a committee to explore radio telegraphy for military communications between France and England, which led to his founding of the army's radio telegraphy service in 1900. In 1903, Eiffel contacted Ferrié and offered him the use of the tower as an antenna mast for an experimental radio station. The two men joined forces and succeeded in getting the tower designated as a station of the French radio service in 1904. From there, the tower was used for numerous radio experiments and for military radio communications, becoming an essential part of the French defense system. As a result, Eiffel's original 20-year permit was extended for another 70 years, saving the tower from demolition [22, 23].

The antenna for the French Army's original radio telegraphy station at the Eiffel Tower was a single wire that was suspended from an insulator attached to the tower's top floor. The radio transmitter and other equipment were originally housed in small buildings near the tower's base (Figure 10.3). By 1908, the antenna system had been expanded to four wires, each 380 m in length. The Paris city council, concerned that the noise from the radio transmitter and high-voltage conductors were a risk to both the aesthetics of the park and public safety, arranged for an excavation project to move the radio station underground. The underground station was connected to a fan-shaped, six-wire antenna system moored to a metal gallows at the top of the tower after passing through insulators. The wires, 425 m in length and 5 mm in diameter, were anchored at their other end to decorative pylons on the ground so that they would go unnoticed, with the insulated pull-off cables descending into a shaft to the underground station [22]. The construction of the underground station was delayed for several months when surging flood waters from the Seine River inundated the site and damaged the radio equipment in January 1910. Figure 10.4 is a drawing of the tower's six-wire antenna station after the underground station was completed.

Knowing that the radio signals from the Eiffel Tower could cover a large region, the French Bureau of Longitudes began urging the French Army to broadcast time signals in May 1908. By broadcasting time signals, FL could allow ships at sea to synchronize their clocks and accurately determine longitude, providing the same service in Europe that NAA and the other Navy stations were already providing in the United States. The Ministry of War agreed, and after funding was approved and construction was completed, the first time signals from the Eiffel Tower were broadcast on May 23, 1910 [5, 23, 24]. The reference clock was located at the nearby Paris Observatory and was connected to the station by telegraph wires.

Figure 10.3 Buildings housing the French Army's radio telegraphy equipment at the base of the Eiffel Tower (c. 1908).

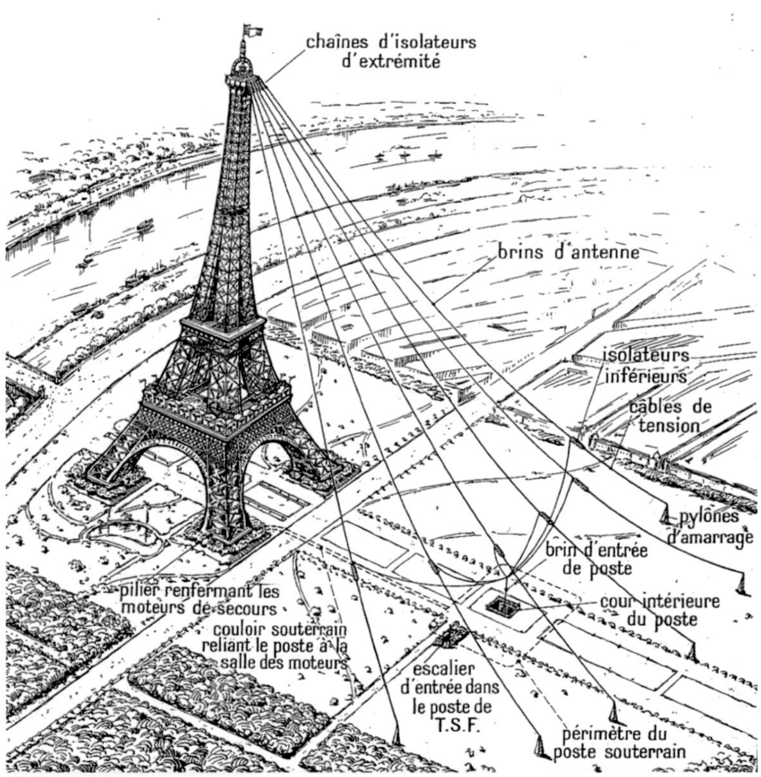

Figure 10.4 Drawing of the six-wire antenna system and underground radio telegraphy station at the Eiffel Tower (c. 1911).

The time signals were originally sent at midnight, 12:02 a.m., and 12:04 a.m. Paris Time, on a wavelength of 2 km (~150 kHz). The original radius of the signal was estimated as nearly 3,000 km, a coverage area that essentially doubled after repairs from the flood damage were completed [25]. Then, the radiated power was about 50 kW and the signals spanned the Atlantic Ocean, reaching the United States and Canada [26]. Legal time in Paris was shifted so that its minutes and seconds agreed with Greenwich Mean Time in March 1911, and FL began broadcasting GMT on July 1, 1911 [24].

There were countless benefits to having accurate time available, and users other than mariners soon began decoding the time signals from Paris and elsewhere—including business people, clock makers, jewelers, military personnel, railroad planners, and radio hobbyists. As the numbers of stations and listeners grew, it was soon found that transmitted time signals could differ from each other by several seconds. To improve this situation, a meeting was scheduled at the Paris Observatory in October 1912, with the objective of "effecting practical unification of radio time signals, and to prepare plans for an international time service to suit the needs of all." This meeting, attended by delegates from 16 nations, resulted in the formation of the International Time Bureau (*Bureau International de l'Heure*, BIH), an organization that was responsible for the international coordination of time until that task was given to the BIPM in 1988 [5].

The meeting at the Paris Observatory also proposed the formation of an international time service, provided by a network of stations that would begin sending time signals in an identical format on July 1, 1913. All signals would be referenced to GMT, and to avoid errors, it was recommended that the telegraphic time codes be sent automatically and not by hand. The Eiffel Tower station, then assigned to a wavelength of 2,500 m (~120 kHz), was chosen as the center of the new international time service [27]. Eight other stations were also selected, including NAA and another U. S. station, KPH in San Francisco, as well as stations located in Brazil, Eritrea, Germany, the Philippines, Somalia, and Sudan [26]. The number of time signal stations continued grow, and by 1920 there were 33 stations located in 17 different countries that were broadcasting time signals referenced to GMT [28].

The time signal broadcasts from the Eiffel Tower survived World War I and continued well into the 1920s [29]. The tower's innumerable contributions to radio have continued to the present day—it still serves as a mast for radio and television antennas.

10.1.3 The First Radio-Controlled Clocks

The standardization of radio time signals led to the first commercial products that could be loosely classified as RCCs. These products weren't actually radio controlled but instead were just receiving sets designed for listening to the time signals from FL, NAA, and other stations. The sets included signal charts that explained the telegraphic code, allowing listeners who decoded the signals to manually synchronize a clock. One such product was the Horophone, fittingly designed by Hope-Jones, and offered for sale by the Synchronome Company in London. An avid amateur radio enthusiast in addition to being a horologist, Hope-Jones had cofounded Synchronome in 1895 and was the company's managing director when the Horophone was introduced in June 1913. The Horophone was designed to be wall mounted like the telephones of that era and included a receiver with a glass-enclosed crystal detector, a connection for an antenna, headphones for listening, and a hook where the listener could hang their pocket watch [30, 31].

An early version of the Horophone, alongside an advertisement, is shown in Figure 10.5. The advertisement announced the upcoming "International Service of Time Signals" that would occur on or after July 1, and included a circular diagram with the schedule and format of the time signals from FL in Paris, and from the German time signal station located in Norddeich, which then had the call sign KND. Several other companies sold similar time signal receivers, including Marconi's Wireless Telegraph Company, which introduced their product in 1914 [32].

The first clocks that were actually controlled by radio utilized synchronization methods that were only semi-automatic. This was the best that could be done given the state of the technology, because the telegraphic time codes only contained an OTM and a sequence of signals that counted down to the OTM, which wasn't enough information for a clock to set itself. Therefore, an operator first needed to set the hands of the clock close to the correct time. When the time signal was received, the clock would align its hands with the OTM and begin running, adjusting its hands as necessary when further OTMs were received to keep the time accurate. These types of clocks were elaborate electromechanical machines that were difficult to build, and they were mostly built for research purposes and specialized applications. Few were ever sold commercially.

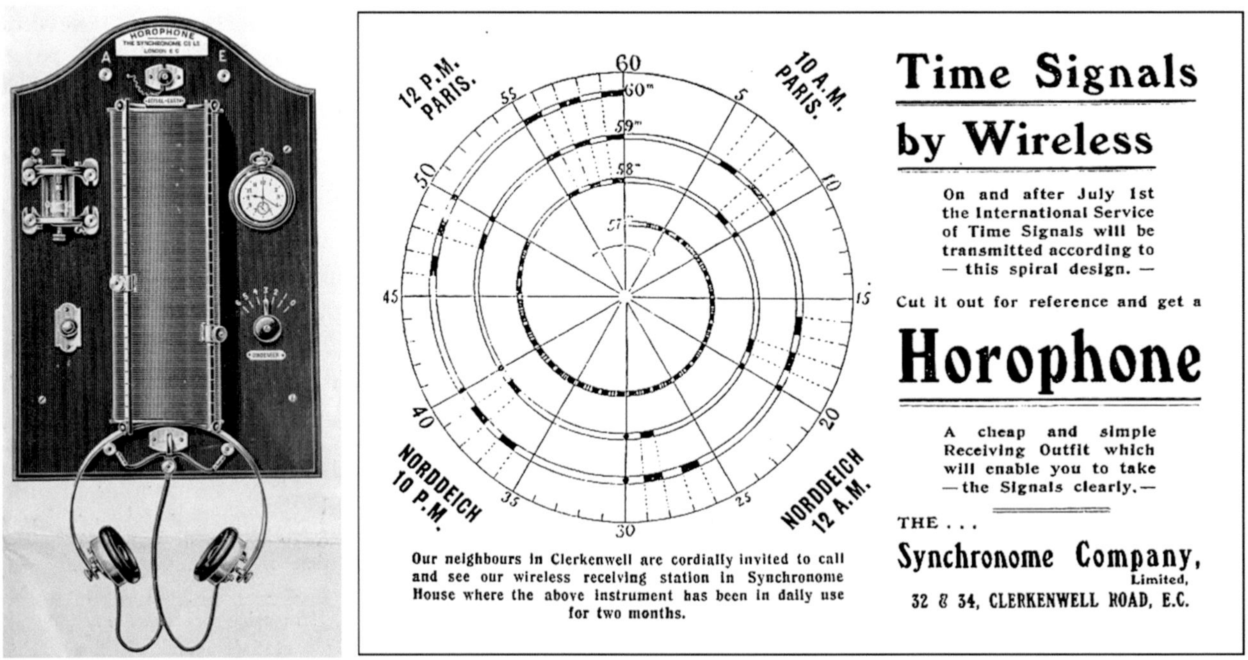

Figure 10.5 The Horophone time signal receiver, alongside an advertisement (1913).

If a single person could be credited with the invention of the RCC, the British clockmaker Alfred Ball (1866–1932) would likely be the strongest candidate. Ball was apprenticed as a youth to a watch and clockmaker, and he quickly became interested in electrical clocks. He reputedly built an electric clock as early as 1887, and by 1898 he had obtained an electric clock patent. By then, he was the owner of a small battery company in Plymouth, England, that supplied dry cells to Gent & Company in Leicester, which was one of the largest industrial clockmakers of that era. In 1904, Gent hired Ball and purchased his battery company, simultaneously adding an expert electrical horologist to their staff and dry cell manufacturing to their business [33]. Shortly after Ball's arrival, Gent began producing electric clock systems that Ball either invented or co-invented, where a master clock could synchronize the dials of other clocks. The best known of his inventions was the "Pul-syn-etic" system, which became a registered trademark of Gent, and a line of products that was marketed with great success [34].

Ball became interested in controlling a clock by radio shortly after the international time service began in 1913. He quickly declared himself as an advocate of the "wireless control, or rather supervision, of public clocks," proposing a system in 1915 where a clock located at a town hall, synchronized to the time signal from Paris, could in turn synchronize eight of the town's principal clocks by radio [35]. The idea apparently led to a working design, and during a March 1924 lecture given to the British Horological Institute, Ball's friend and fellow horologist Herbert White announced that Ball had designed a system to "cause the Paris Time signal to automatically correct turret clocks," but "the time was not ripe for publication." Based on this, it seems likely that changes made to the format of the Paris Time signals in 1923 had caused Ball's clocks to stop working.

Annoyed by these changes, Ball began using newly available time signals, referenced to the Royal Observatory in London and broadcast by the British Broadcasting Corporation (BBC), for his future wireless clock designs. The BBC signals, commonly known as the "six pips," were first broadcast in February 1924 and are still being broadcast more than a century later. Their format is much simpler than telegraphic code, consisting only of a series of six tones or "pips" that occur on each of the five seconds leading up to the hour, and on the hour itself [36].

Ball added his new wireless clock invention to the "Pul-syn-etic" clock system, and used the system to control eight electric clocks at his home in Leicester, synchronized to reference time signals broadcast from the BBC station in Daventry, about 50 km away. A diagram of one of Ball's RCC designs is shown in Figure 10.6. A radio and antenna tuned to the BBC station in Daventry produced pulses generated by the six pips, which triggered an electromagnet (A) that moved a 12-tooth ratchet wheel shown in the top left of the diagram. When the sixth pip was received and the ratchet wheel had made one-half turn, it energized a synchronizing magnet (M) that corrected the master clock shown on the right side of the diagram. There is no record that an RCC was ever sold commercially by Gent & Company, but prototype models with clock faces marked "Auto Controlled by Wireless from Daventry" were constructed [36]. It was reported in 1928 that the RCC at Ball's home "has operated without any attention whatever for the last six months, and the maximum error recorded has not exceeded one second fast or slow" [37].

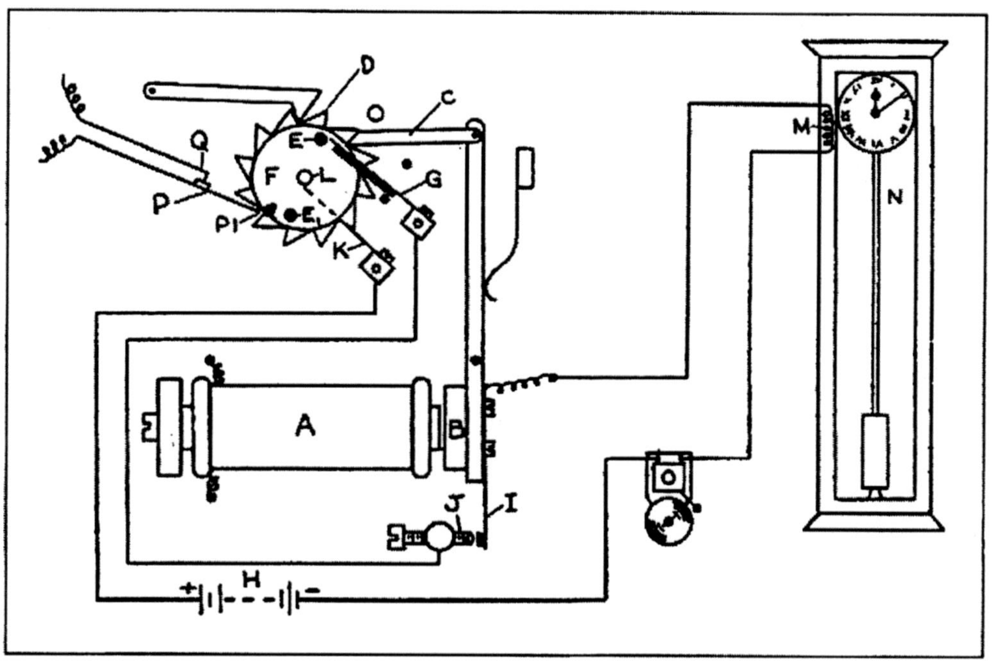

Figure 10.6 A diagram of one of Ball's RCC designs, controlled by the six-pip BBC signals.

Ball worked tirelessly on other RCC designs for the rest of his life, but the technology needed to fulfill his ambitions simply wasn't available. The fully automatic, accurate, and low-cost RCCs that we now take for granted didn't appear until about a half century after Ball's death. Their emergence required the development of several new technologies. For example, making RCCs that were fully automatic required time signal stations to replace their simple telegraphic time codes with digital time codes that contained enough information for clocks to set themselves. Digital time codes weren't available until the 1960s. Making RCCs accurate required replacing the pendulum and quartz clocks at time signal stations with atomic clocks, and reliable atomic clocks weren't available until the 1960s and 1970s. There also had to be improved ways of synchronizing the station clocks. The early time signal stations were synchronized via telegraph lines, but far more sophisticated time transfer methods, such as common-view satellite observations (Section 4.3.3), were needed to take advantage of atomic clock accuracy. These high-accuracy time transfer methods weren't practiced until the 1980s. Finally, making low-cost RCCs required microprocessors and software to decode the digital time codes. This technology didn't appear until the 1970s, and it took several more years before it had advanced enough to be used in battery-powered devices.

Once the necessary technologies were in place, the number of RCCs grew exponentially, and a high percentage of the clocks we now rely upon in everyday life are automatically controlled by radio. The remaining sections in this chapter describe the time signals currently distributed by radio—including signals from terrestrial radio stations, geostationary satellites, and mobile phone networks—and the types of clocks that these signals control. The time signals broadcast by satellite navigation systems are discussed separately in Chapter 12.

10.2 Time Signals from LF Terrestrial Radio Stations

Most of the recent advances in radio technology have involved microwave frequencies, so the broadcasting of LF time signals can be thought of as retro technology. The low-cost, single-function RCCs designed for the consumer market receive time signals on frequencies ranging from 40 to 77.5 kHz, which are even lower than the frequencies once used by NAA and FL.

Continuing to use technology that's more than a century old might seem counterintuitive in today's world, but LF radio is still an attractive medium for time distribution. Time signals require very little bandwidth, so they make good use of the LF spectrum, where bandwidth is in short supply. The LF band is essentially unknown to radio listeners because, with few exceptions, there isn't enough bandwidth available to broadcast audio. This mostly limits the use of the LF band to utility stations with low bandwidth requirements, including navigation beacons, weather stations, and time signal stations. For time signal stations, the bandwidth restrictions aren't even an issue. Only about 10 Hz of bandwidth is needed to transmit a time code, which typically consists of just 60 bits of information sent at the very slow speed of 1 bit per second (bps).

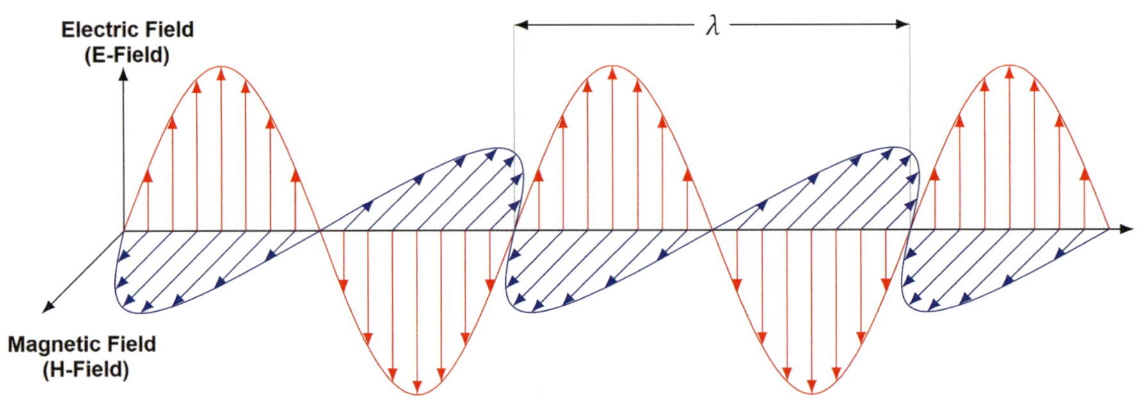

Figure 10.7 A vertically polarized electromagnetic wave.

Ideally, time signals should have stable path delays, cover a large geographic area, and be easy to receive with low-cost equipment. For the most part, LF radio checks all of these boxes. The path delays are stable because their long wavelengths give LF signals the ability to propagate as groundwaves that travel far beyond the horizon. To make this happen, the transmitted signals are vertically polarized, which means that the antenna's electric field (E-field) is perpendicular to its magnetic field (H-field) and to the Earth's surface (Figure 10.7). In other words, the E-field is made to oscillate up and down instead of sideways.

As the signal travels, its wavefront tilts forward at the top but angles downward at the bottom, so that it "drags" along the surface and follows the curvature of the Earth. Vertical polarization is necessary when transmitting LF groundwave signals because the height of the antenna above ground is much shorter than the wavelength of the signal. If the antenna's E-field was horizontally polarized, or parallel with respect to the Earth's surface, the ground would heavily attenuate every wave cycle, and the signal would soon be too weak to be useful [38].

Groundwave propagation makes LF time signals more stable than signals that follow an indirect route between the transmitter and receiver, such as the high-frequency (HF) time signals that we'll discuss in Section 10.3. Because groundwave signals stay close to the Earth's surface, their path delays can be accurately estimated by dividing the distance between the transmitting and receiving antennas by the speed of light constant (Section 4.3.1). Stable path delays also make it possible to recover an accurate and stable frequency that can be used to discipline a local oscillator (Section 5.3.2).

The large coverage area of LF time signals is enabled by a combination of groundwave and skywave propagation. The distance that a signal can travel by groundwave is a function of frequency; as the frequency goes up, so does the attenuation. Consider, for example, a low-cost RCC that is attempting to read an LF time code so it can synchronize its clock. It should be successful—based on typical product specifications and assuming little or no local RF interference—if the received signal strength is at least 40 dB(μV/m). If all other conditions were equal, and two 1 kW signals were broadcast from a vertical antenna, one at 60 kHz and the other at 10 MHz, the 60 kHz groundwave would travel about 10× further. Therefore, it can provide the signal strength that the RCC needs across a distance of about 1,000 km, compared to about 100 km for the 10 MHz signal [38, 39].

Even though the LF groundwave signals are strong, at a sufficient distance from the transmitter their strength is exceeded by skywave signals, and skywaves are used to synchronize RCCs located outside of the groundwave coverage area [40]. Skywave is most often associated with MF and HF radio signals, but it also affects LF propagation. Skywave propagation occurs when radio signals interact with the ionosphere, which is the portion of the upper atmosphere that extends from about 50 km to 650 km above the Earth's surface. In most parts of the atmosphere, the gas molecules are electrically neutral, but the ionosphere is formed when the solar radiation becomes so intense that the gas molecules split, or ionize. This leaves molecules that are missing an electron—the positively charged ions that give the ionosphere its name—and free electrons. When the free electrons reach a sufficient density, they begin to affect the propagation of radio signals. Some radio signals pass through the ionosphere, some are absorbed, and some have their paths bent so that they move back towards the Earth, an effect called refraction. The refracted signals can be thought of as being reflected or bounced from the sky back to Earth.

During the daytime, LF signals are reflected by the D layer, the part of the ionosphere closest to Earth, which extends from about 50 km to 90 km above the Earth's surface. This allows the signals to propagate within an Earth-ionosphere waveguide that's located in the space between the ground and the D layer boundary. Shortly after the sun sets, the D layer essentially vanishes, and the LF signals reach higher altitudes before being reflected back to Earth by the upper ionospheric layers (Figure 10.8). The wider waveguide, coupled with lower electron densities and less signal absorption, allows skywaves to travel further and with more intensity at night than they do in the daytime. For these reasons, most LF RCCs only try to synchronize at night, usually after the sun has set at both the transmitter and receiver sites [41]. Of course, the indirect skywave path is less stable than a direct groundwave path, so the goals of having both a stable path and a large coverage area are mutually exclusive. For low-cost RCC applications, the large coverage area is usually considered to be much more important.

In addition to having stable paths and a large coverage area, LF signals are also easy to receive. They can even be received inside buildings, which is another reason why the medium is attractive for time distribution. Their long wavelengths allow LF signals to pass through walls, buildings, and other obstacles, and to reach the antennas of RCCs located indoors. This is a major advantage, because reliable reception of other time signals, including HF signals and signals broadcast from satellites, is often difficult to achieve without an outdoor antenna.

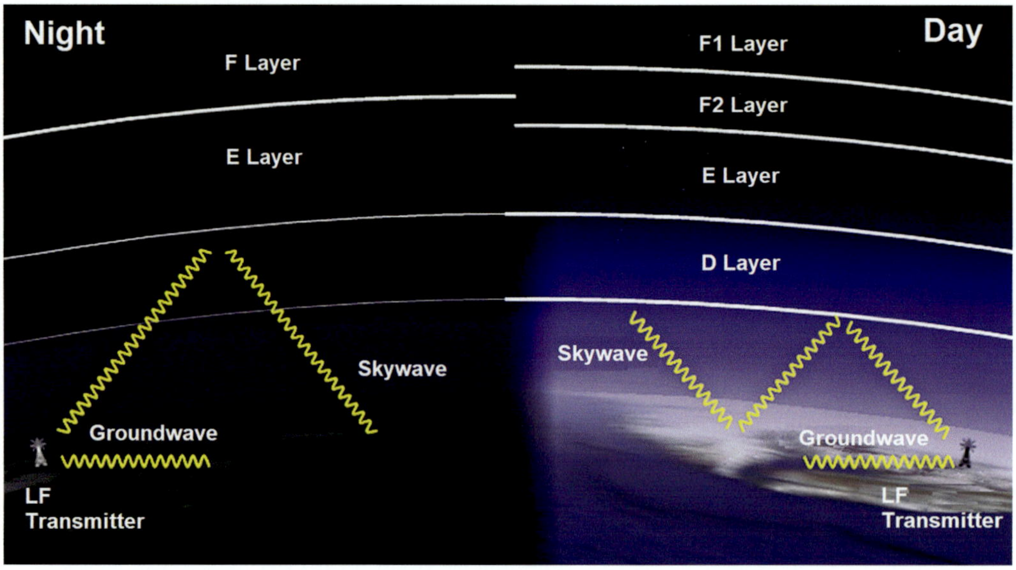

Figure 10.8 The daytime and nighttime ionospheric layers and their effect on LF signals.

Table 10.2 lists the LF time signal stations most often used for RCC applications, with each station described in the following sections. These stations broadcast digital UTC time codes that are synchronized to the UTC(*k*) time scales in their respective countries.

Table 10.2 The LF time signal stations most commonly used for RCC applications

Call Sign	Location	Frequency (kHz)	Radiated power (kW)	First year on air	First time code
WWVB	Ft. Collins, Colorado, USA	60	70	1963	1965
DCF77	Mainflingen, Germany	77.5	30	1959	1973
MSF	Anthorn, Cumbria, England	60	15	1950	1974
JJY	Mt. Ohtakadoya, Fukushima, Japan	40	13	1999	1999
	Mt. Hagane, Fukuoka, Japan	60	23	2001	2001
BPC	Shangqiu, Henan Province, China	68.5	50	2007	2007

10.2.1 Radio Station WWVB (United States)

Radio station WWVB is operated by the National Institute of Standards and Technology (NIST), which was called the National Bureau of Standards (NBS) when the station began operation on July 4, 1963. The station is located near Fort Collins, Colorado, about 80 km northeast of the NIST laboratories in Boulder, where it shares a 380-acre (1.5×10^6 m^2) site with the HF station WWV.

WWVB broadcasts time signals on a carrier frequency of 60 kHz, with a radiated output power of about 70 kW. The groundwave signals have a usable range that exceeds 2,000 km, and signals that propagate via both groundwave and skywave allow WWVB to reach the entire continental United States during both the daytime and nighttime hours. At night, the skywave signals can reach Alaska, Hawaii, Central America, and most of the populated regions of Canada.

The design and construction of WWVB can be largely credited to William Wilbur Brown (1889–1980), an American engineer whose career began during the formative days of radio. Before coming to NBS, Brown had already completed a distinguished career at the General Electric Company, where he worked from 1913 until his retirement in 1956 [42]. While working on a high-power HF transmission system in New Jersey, Brown participated in the transmission of President Woodrow Wilson's reply to Germany's World War I peace proposal in October 1918, a historic moment in radio communications. He later spent several decades as both a managing and project engineer in GE's transmitter engineering division, where his accomplishments included the design and installation of high-power VLF and LF transmission systems for the U. S. Navy.

After retiring from GE, Brown began a new career as a consultant. He arrived at NBS in 1958, where he utilized his vast experience to design WWVB's transmission system and to plan the station's construction. The WWVB call sign shares the first three letters as its HF predecessors WWV and WWVH (Section 10.3.1), and the letter "B" is commonly thought to refer to the station's proximity to Boulder. Ironically, the call letters include all of Brown's initials, which is perhaps more than just a coincidence.

Figure 10.9 provides a block diagram that shows the main components of WWVB [43]. The components include two antennas, three LF transmitters, the broadcast control equipment, an ensemble time scale that provides the station with its reference clock, and equipment that generates digital time codes by modulating the carrier frequency.

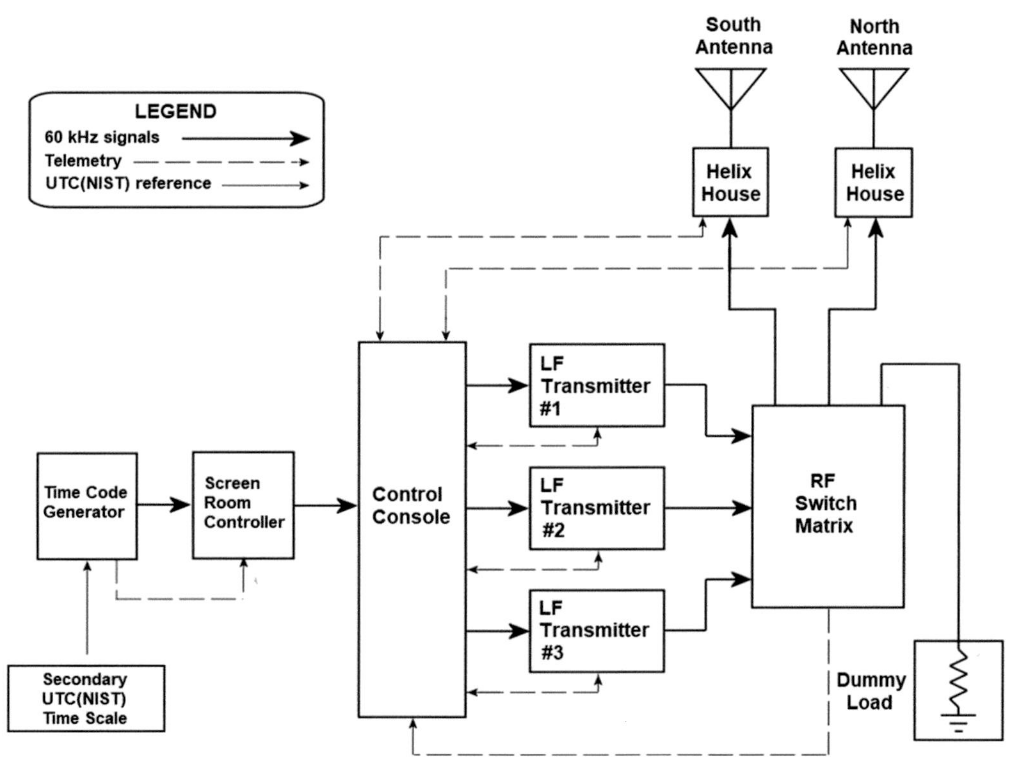

Figure 10.9 Block diagram of radio station WWVB.

The most significant feature of WWVB is its large antenna array, which covers a ground area of about 30 acres (121,406 m^2). The array includes two antenna systems, situated to the northwest and southeast of the WWVB building, and known as the north and south antennas. Each antenna system consists of four towers, a top-loaded vertical monopole antenna, a tuning system, and a ground plane. The four towers are arranged in a diamond pattern, as shown in the station site map in Figure 10.10 and in the aerial photograph in Figure 10.11. The towers are about 122 m tall, electrically grounded at their bases, and anchored by insulated guy cables. The distance between the down leads of the north and south antennas is 866 m.

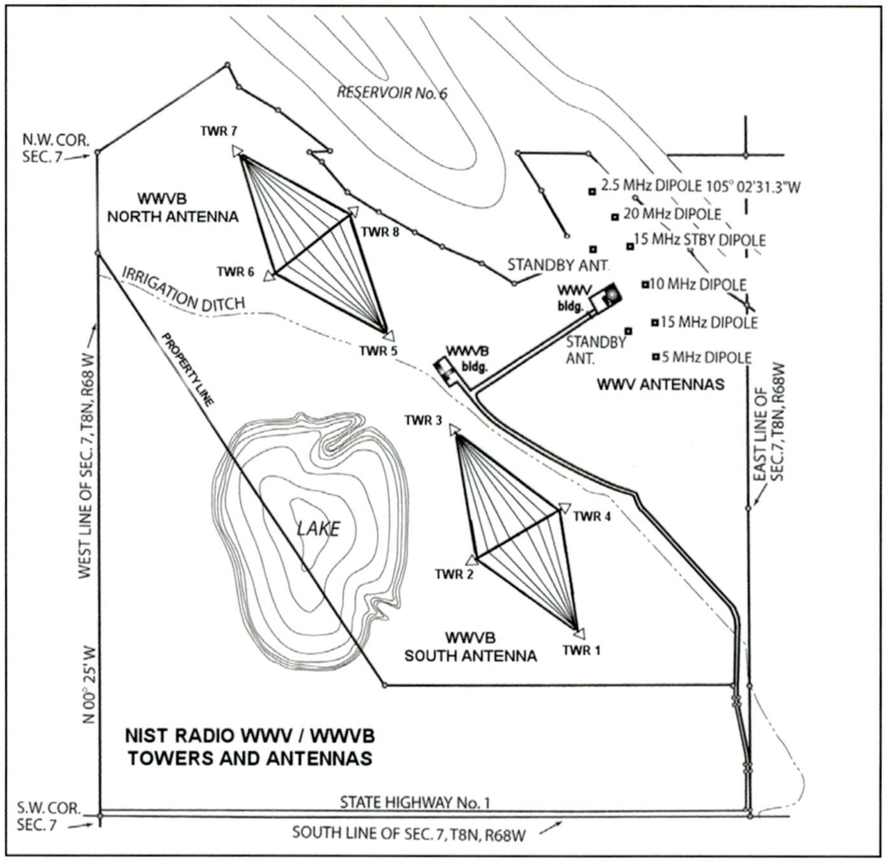

Figure 10.10 Site map of WWVB and WWV.

Figure 10.11 Aerial photograph of the WWVB site and antenna array.

The purpose of the four towers is to support the cables that form a top-loaded vertical monopole antenna. The antenna has two main parts, a capacitance hat and a down lead. The capacitance hat consists of aluminum cables that are arranged horizontally in a diamond shape and supported by a tower at each corner of the diamond. The capacitance hat isn't directly fastened to the towers, but instead is connected to insulated steel cables that extend to concrete counterweights near the tower's base. This electrically isolates the capacitance hat and allows it to "float" between the towers, with the tension from the counterweights always keeping it nearly parallel to the ground.

The down lead is the antenna's radiating element. It is kept nearly vertical by a tensioning cable and counterweight, and it descends from the center of the capacitance hat into a building that houses the tuning system, where it is connected to the transmitter. Even though the down lead is suspended as high above the ground as possible, it is still only 86 m long, which is short compared to the long wavelengths being transmitted. To achieve maximum efficiency, the length of the radiating element of a transmitting antenna should be a quarter-wavelength long, or 1,250 m in the case of 60 kHz, which is far too large to be practical to build. Top loading the down lead with the capacitance hat is a compromise solution that adds electrical "length" and makes the antenna reasonably efficient [43].

The two buildings that house the tuning systems for the north and south antennas are informally called helix houses. Each antenna is tuned by matching its approximate 1 Ω impedance to the 50 Ω impedance of the transmitters [44]. The impedance match is accomplished with a large inductor with a fixed horizontal winding, and a smaller, built-in variable inductor (Figure 10.12). The variable inductor is needed because the antenna impedance is always changing due to weather conditions such as wind or ice, so it requires constant tuning. When an impedance change is detected, an auto-tuning system rotates the variable inductor in the direction necessary to maintain an optimal impedance match. The helix house is also where each antenna connects to its ground plane. The ground plane consists of 300 underground wires, each about 400 m in length, that extend from the helix house in a radial pattern. The ground plane improves the antenna's efficiency by minimizing the ground loss of the transmitted signals.

Figure 10.12 The WWVB antenna tuning system inside the helix house.

The three WWVB transmitters are located inside the main station building. Two transmitters are always on the air, and the third serves as a backup or standby. The coaxial transmission lines that run from the transmitters to the north and south antennas are more than 400 m in length. Unless maintenance is being performed on one of the antennas, the station operates with two transmitters, each with its own antenna, that simultaneously broadcast the same information. The station's efficiency, or the percentage of transmitter output power converted to radiated RF output power, is only about 55% on average when broadcasting from a single antenna, but about 69% in dual antenna mode. Both transmitters typically output about 51 kW, resulting in a radiated output power of about 70 kW.

The station's ensemble time scale and three time code generators (TCGs) reside in an RF shielded room. The time scale includes four cesium clocks and is a secondary representation of the primary UTC(NIST) time scale in Boulder. The station time scale is continuously compared to the primary time scale with time transfer methods that include common-view observations of GPS satellites, and its frequency is periodically adjusted so that it always keeps time within ±25 ns of UTC [45]. The time scale provides the reference signals for three TCGs, including one that is on the air and two that are in standby mode. All three TCGs are continuously monitored by a control system that intercompares their time codes. If a TCG is found to be in error by more than 1 μs with respect to the other two, or if its output signal goes away, it is immediately taken off line. If the defective TCG was on the air, the control system immediately connects another TCG to the transmitters. A TCG is shown in Figure 10.13.

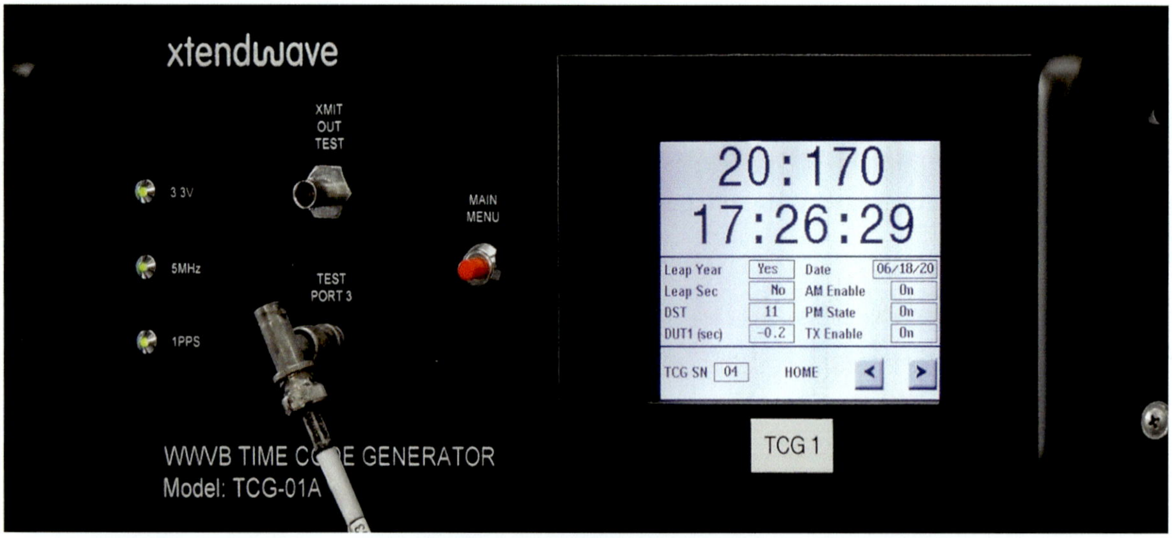

Figure 10.13 A WWVB time code generator.

Two unique digital time codes, both synchronized to UTC(NIST), are produced by modulating both the amplitude and phase of the 60 kHz carrier frequency. The AM time code, which has been broadcast since 1965 [46] and is now called the legacy time code, is decoded by nearly all WWVB RCCs. The PM time code has been broadcast since 2012, but is decoded by only a small percentage of RCCs as of 2025. Both time codes provide the UTC time and date, as well as advance notice of leap seconds and daylight saving time (DST). The legacy AM code also provides a DUT1 correction and a leap year indicator.

The legacy time code uses a simple form of modulation called amplitude shift keying (ASK), where digital data is sent by changing the amplitude of the carrier wave. A common form of ASK is on-off keying, most notably used in the transmission of Morse code, where information is sent by switching the carrier wave on and off. The ASK modulation used by WWVB is similar to on-off keying, but the carrier wave is never switched completely off, because its frequency is locked to UTC(NIST) and can be used by receivers as a frequency standard. Instead, WWVB drops the carrier power by 17 dB at the start of each UTC second.

This power drop causes an amplitude shift that's large enough for receivers to easily detect. In addition, because the amplitude shift occurs every second, it can easily be distinguished from other amplitude shifts caused when propagation conditions change or when the receiver's antenna is moved [47].

The drop in carrier power serves two purposes. The first purpose is to signal the start of a time code bit. A bit is sent every second, and the period that the carrier power is held low determines whether the bit has a value of 0 or 1. A 0 bit is sent when the carrier power is held low for 0.2 s, with full power restored for the next 0.8 s. A 1 bit is sent when the carrier power is held low for 0.5 s, with full power restored for the next 0.5 s. A frame marker is sent every 10 s, by holding the carrier power low for 0.8 s and then returning to full power for the next 0.2 s. The bits are encoded in BCD format, where multiple bits represent one decimal digit. The complete time code includes 60 bits and requires one minute to send, as shown in Figure 10.14 [43].

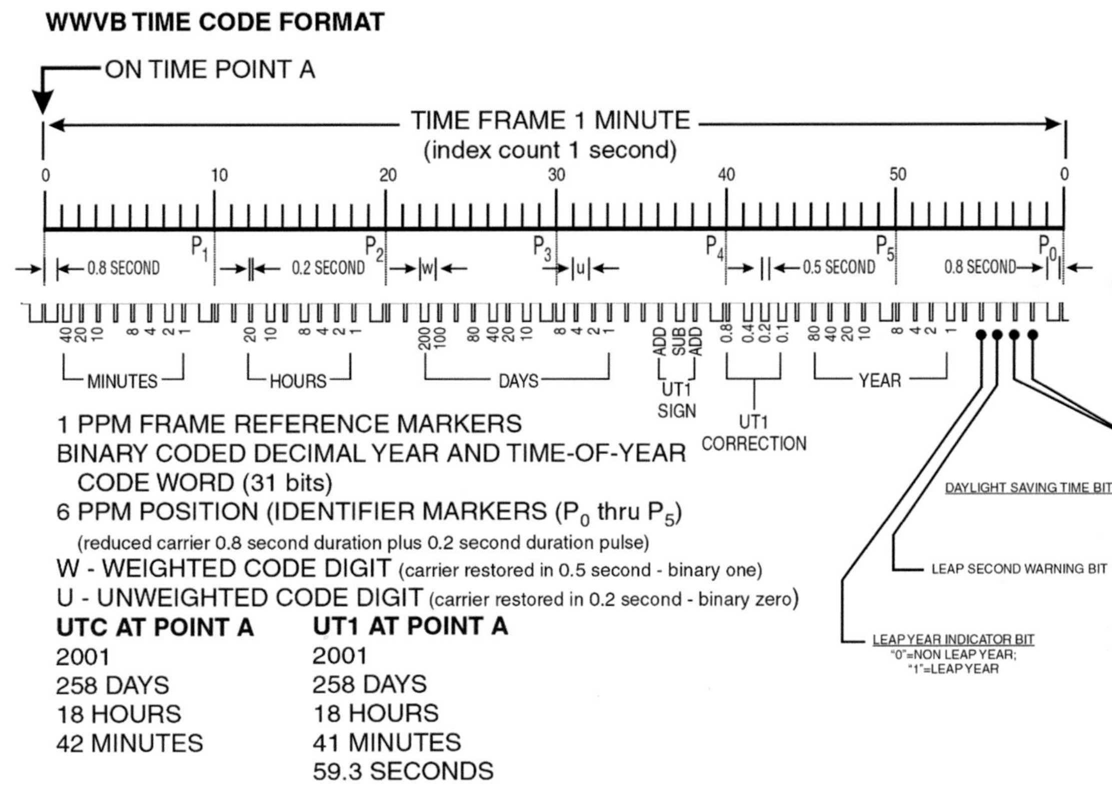

Figure 10.14 The legacy WWVB time code format.

The second purpose for the drop in carrier power is to mark the beginning of the current UTC second, providing a new OTM every second that RCCs can use to synchronize. Figure 10.15 shows the 17 dB drop in voltage as displayed on an oscilloscope. The green trace was produced by a 1 pps signal from the station's time scale, and the yellow trace was produced by a signal output from the TCG. The OTM is indicated by the first cycle of reduced carrier power. As shown, it aligns with UTC(NIST) to within the 16.67 μs period of 60 kHz.

The OTM is much easier to detect before the signal is transmitted than it is when the signal is received. For the received signal, the power drop occurs over a period of multiple cycles. This happens because the bandwidth of the transmitting antenna is much narrower than the carrier frequency, causing the carrier amplitude to slowly decay, even though the power drop was instantaneous. When loaded by the transmitter, the dual antenna system at WWVB has a bandwidth of roughly 1 kHz. The decay time is related to the period of the antenna bandwidth, which is about 1 ms, meaning that the full drop in carrier amplitude takes about 60 cycles to complete (60 × 16.67 μs = 1 ms). Receivers can, of course, detect a decrease in amplitude before the full power drop is complete, but reliable detection of the first cycle of reduced amplitude is difficult. Therefore, cycle ambiguity often limits the uncertainty of time signals received from WWVB to about plus or minus three cycles, or ±50 μs [43].

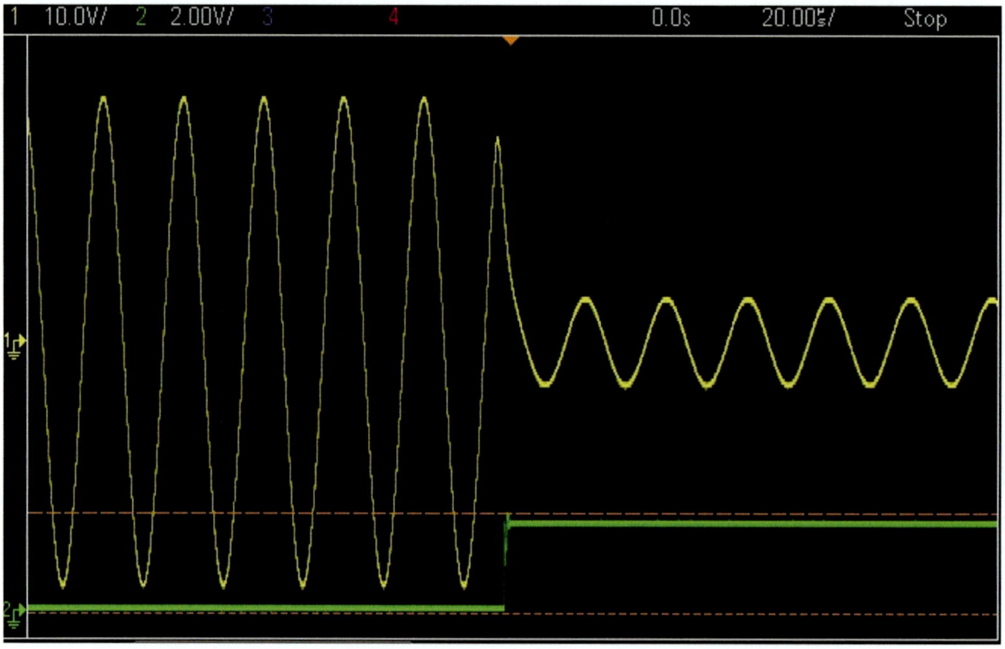

Figure 10.15 The WWVB on-time marker, coincident with the UTC second.

The PM time code was added to WWVB in 2012 in an effort to make decoding the time easier, by making it easier to distinguish between a 0 bit and a 1 bit. The PM code can potentially benefit RCCs located in areas where the signal strength is low, or where the level of local RF interference is high. It also provides RCCs with the ability to detect errors in the received time code, which can make time code reception more reliable. The PM code sends data in a completely different format than the AM code, but the data are still transmitted at a 1 bps rate, and a complete time code still has 60 bits and requires one minute to send.

The PM code transmits bits by binary phase shift keying (BPSK) of the carrier frequency. A 1 bit is sent by inverting the normal phase of the carrier, a 180° phase shift, for 1 s. A 0 bit is sent by maintaining the normal carrier phase for 1 s. The phase transition always occurs 0.1 s after the OTM that marks the beginning of the UTC second. Therefore, the phase transition always occurs when the carrier amplitude is 17 dB below full power. The information sent by the PM time code is summarized in Table 10.3 [48].

Table 10.3 The phase-modulated WWVB time code format

Function	Description	Bit length	Bit numbers
Frame synchronization	The time code is framed by a fixed pattern of bits, known to the RCC, that are transmitted in the first 13 s and last second of each minute. An RCC can synchronize by correlating this bit pattern to the information received by radio.	14	1–13, 60
Error code correction (ECC)	The five ECC bits are appended to the 26-bit minute counter to form a 31-bit Hamming code that allows correcting one bit error and detecting up to two bit errors.	5	14–18

Function	Description	Bit length	Bit numbers
Minute counter	The minute counter uses 26 bits to indicate the number of minutes that have elapsed since 00:00 UTC on January 1, 2000. An RCC can extract the month, day, year, day of year, and the hour and minute from this counter. The least significant bit of the minute counter is sent twice, making a total of 27 bits.	27	19–29, 31–39, 41–47
Reserved bits	Three bits are normally unused and reserved for special purposes.	3	30, 40, 50
DST status and pending leap second information	Indicates if DST is currently in effect, and if it starts or ends today. Also indicates if a leap second is scheduled at the end of the current month.	5	48–49, 51–53
Pending DST information	Indicates the date and time of the next DST transition, providing notification months in advance based on the current DST rules.	6	54–59

10.2.2 Radio Station DCF77 (Germany)

Radio station DCF77 is operated by the Physikalisch-Technische Bundesanstalt (PTB), the NMI of the Federal Republic of Germany, and the same institution that serves as the pivot laboratory for the BIPM key comparisons that generate UTC (Section 9.4.4). The station is located on the Mainflingen transmission site, about 25 km southeast of the PTB laboratories in Frankfurt (Figure 10.16). The Mainflingen site was established in 1949, and has since been home to a number of LF and MF radio stations, but DCF77 is now its best-known occupant.

Figure 10.16 Aerial photograph of the Mainflingen transmission site where DCF77 is located.

The official start date for the DCF77 time and frequency transmissions was January 1, 1959, but the first transmission attempts began in 1956. The station's call sign is self-descriptive: The "D" refers to Deutschland, the "C" was a code for LF transmissions, the "F" refers to the station's proximity to Frankfurt, and the "77" refers to its carrier frequency of 77.5 kHz.

DCF77 uses a 50 kW transmitter and a vertical, omnidirectional antenna with a height of 150 m. The antenna is top-loaded to increase its electrical length, in a similar fashion as the WWVB antenna. A standby antenna and backup transmitter are always available to keep down time to a minimum. The efficiency of the transmission system exceeds 60%, keeping the station's radiated output power between 30 kW and 35 kW. This power level and the station's convenient location in central Europe allow the DCF77 time signals to reach most of the European continent, as shown in Figure 10.17. The outer ring in the figure shows the approximate limit of the usable skywave coverage, which has been estimated to be about 1,900 km during the day and about 2,100 km at night [49].

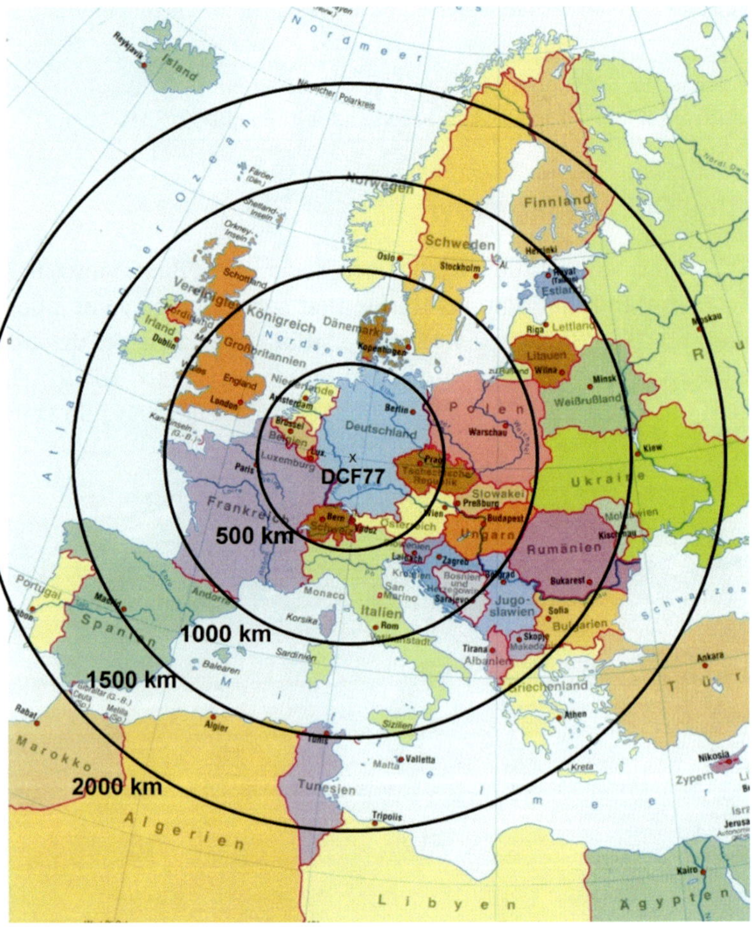

Figure 10.17 The coverage area of radio station DCF77 (courtesy of PTB).

Digital time codes, synchronized to UTC(PTB), are transmitted in both AM and PM formats. The AM time code was added in June 1973, and the parts of the code containing time-of-day and date information have since remained unchanged. The power is dropped by about 16.5 dB to mark the beginning of each second, except for the last second of the minute, second 59, when the carrier power is not reduced. A 0 bit is sent by holding the carrier power low for 0.1 s, with full power restored for the next 0.9 s. A 1 bit is sent by holding the carrier power low for 0.2 s, with full power restored for the next 0.8 s. A full time code, sent at a rate of 1 bps, requires 60 s to transmit.

The beginning of a new minute is marked by always sending a 0 bit in second 0. Seconds 1 through 14 are used for public safety or weather information, provided by other organizations, that PTB broadcasts as a public service. Second 15 indicates if DCF77 is operating using its backup transmission system. Seconds 16 through 19 contain notices about Central European Summer Time (CEST) and current status information, as well as leap second notices. A 1 bit is always sent in second 20 to indicate the start of the encoded time. Seconds 21 through 58 transmit the minute, hour, day of month, day of week, month number, and a two-digit year number, all encoded in BCD format, as well as several parity bits (Figure 10.18). The coded information always pertains to the minute that follows [49, 50, 51].

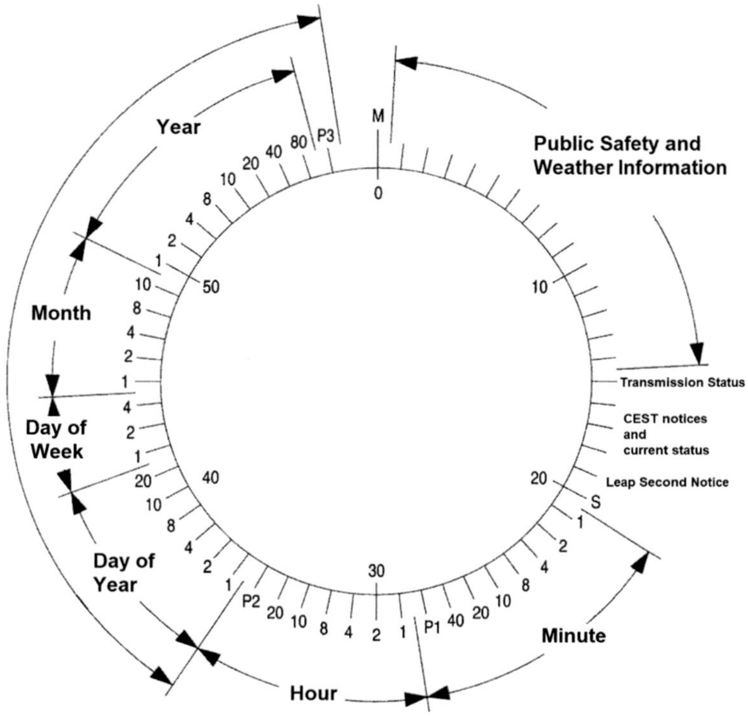

Figure 10.18 The DCF77 time code format.

The PM time code was added DCF77 in June 1983 [49]. It allows receivers to record the time of the arrival of the OTM with much higher resolution than the AM time code, keeping the uncertainty of the received time within one-half of the period of the carrier frequency, or within ±6.5 μs. The PM time code is also less sensitive to interference than the AM code. Despite these advantages, the use of the PM time code is not common, limited mostly to higher-cost industrial RCC products, while the low-cost RCCs found in homes and offices continue to use the AM time code.

The PM codes use a type of modulation called pseudorandom phase shift keying (PRPSK). After each bit is sent by the AM code, the same information is immediately resent as a pseudorandom, direct-sequence spread spectrum code that consists of 512 chips. A normal chip sequence represents a 0 bit, and an inverted chip sequence represents a 1 bit. The chip sequence is modulated onto the carrier by shifting the phase by about ±15.6°. The phase states are equally distributed in the chip sequence, so the average phase of the carrier is unchanged. The chip sequence is sent 0.2 s after the power drop that marks the beginning of the second, which is after the AM bit has been sent and the carrier has returned to full power. Each of the 512 chips lasts for 120 carrier cycles (~1.55 ms); therefore the full sequence takes 793 ms to send, ending 7 ms before the start of the next second. Both time codes are sent at a rate of 1 bps, and their data content is identical for seconds 15 to 58 of each minute, when the time-of-day and date information is sent (Figure 10.16). However, the PM code identifies the end and beginning of a minute differently, and the public safety and weather information sent by the AM code is excluded. The PM code sends a 0 bit in second 59, even though the carrier power is not dropped. It begins a new minute by sending a 1 bit in seconds 0 through 9, and a 0 bit in seconds 10 through 14 [50, 52].

10.2.3 Radio Station MSF (United Kingdom)

Radio station MSF is operated by the National Physical Laboratory (NPL), the NMI of the United Kingdom, which is located in Teddington, a suburb of London. The station began operations at the Rugby transmission site, located about 125 km northwest of London. The Rugby station was opened by the British government in 1926, who decided that the wireless services it offered would be state run by the Post Office. Standard frequency broadcasts from Rugby using the call sign MSF began in 1950, the result of a collaboration between NPL, who supplied the frequency standard, and the Post Office engineering department, who provided the use of their transmitters. The MSF signals were broadcast on both the HF (2.5, 5, and 10 MHz) and LF (60 kHz) bands.

The collaboration was encouraged by the atomic clock developer Louis Essen (Section 3.3), who had a strong interest in frequency measurements, and who noted in his memoirs that "we added a frequency of 60 kilohertz to cover Europe without ionospheric reflections" [53]. After Essen's new cesium clock began operating at NPL in 1955, the HF and LF signals from MSF were involved in the common-view measurements, made between NPL and the USNO, that helped define the atomic second (Sections 3.3 and 3.4).

The 60 kHz broadcasts from MSF at Rugby became a continuously operated NPL service in 1966, and the digital time code was added in September 1974. The Rugby transmission site was closed in 2007, and MSF was relocated to Anthorn, a village in Cumbria, England, near the Scottish border, which is about 430 km northwest of the NPL laboratories. The Anthorn station was formally inaugurated on April 1, 2007, and the Rugby station was turned off.

MSF uses a vertical monopole antenna to broadcast its digital time code on a 60 kHz carrier frequency. The station's radiated output power is about 15 kW. The MSF signals have a usable range of about 1,500 km, covering all of the UK and most of northern and western Europe, and overlapping the DCF77 signal coverage in many areas.

The MSF time code is synchronized to UTC(NPL), and sent using simple on-off carrier modulation, with the digits encoded in BCD format (Table 10.4). The first second of each minute, second 0, is marked by turning the carrier off for 0.5 s, then back on for the remaining 0.5 s. Seconds 1 through 59 always begin with the carrier off for at least 0.1 s, and always end with the carrier on for at least 0.7 s. During seconds 1 through 59, two bits (A and B) are sent in the intervals from 0.1 s to 0.2 s, and from 0.2 s to 0.3 s. The A and B bits are signaled with negative polarity, a 0 bit is sent with the carrier on, and a 1 bit is sent with the carrier off. The A bit is always set to 0 in seconds 1 through 16, which are used to transmit a DUT1 correction. The B bit is always set to 0 when the time-of-day and date information is sent in seconds 17 to 51.

Table 10.4 The MSF time code format

	Bits				Bits		
Second	**A**	**B**	**Description**	**Second**	**A**	**B**	**Description**
0	0	0	Minute marker	30	20	0	Day of month (1 to 31)
1	0	+0.1	Positive DUT1 correction. If DUT1 ≥ assigned weight, then bit B is set to 1. If DUT1 is negative or 0, all B bits are set to 0.	31	10	0	
2	0	+0.2		32	8	0	
3	0	+0.3		33	4	0	
4	0	+0.4		34	2	0	
5	0	+0.5		35	1	0	
6	0	+0.6		36	4	0	Day of week Sunday = 0 Saturday = 6
7	0	+0.7		37	2	0	
8	0	+0.8		38	1	0	

	Bits				Bits		
Second	**A**	**B**	**Description**	**Second**	**A**	**B**	**Description**
9	0	-0.1	Negative DUT1 correction. If DUT1 ≤ assigned weight, then bit B is set to 1. If DUT1 is positive or 0, all B bits are set to 0.	39	20	0	Hour (0 to 23)
10	0	-0.2		40	10	0	
11	0	-0.3		41	8	0	
12	0	-0.4		42	4	0	
13	0	-0.5		43	2	0	
14	0	-0.6		44	1	0	
15	0	-0.7		45	40	0	Minute (0 to 59)
16	0	-0.8		46	20	0	
17	80	0	Year (0 to 99)	47	10	0	
18	40	0		48	8	0	
19	20	0		49	4	0	
20	10	0		50	2	0	
21	8	0		51	1	0	
22	4	0		52	0	0	Both bits always set to 0
23	2	0		53	1	STW	Bit B is BST warning
24	1	0		54	1	P1	Parity bit (year)
25	10	0	Month (1 to 12)	55	1	P2	Parity bit (month/day)
26	8	0		56	1	P3	Parity bit (day of week)
27	4	0		57	1	P4	Parity bit (hour/minute)
28	2	0		58	1	BST	B =1 if BST is in effect
29	1	0		59	0	0	Both bits always set to 0

Bit A is used to transmit a sequence of "01111110" in seconds 52 to 59. This sequence doesn't appear elsewhere in the time code, so it uniquely marks the end of a minute. Bit B in second 53 is set to 1 to provide a one hour advance notice of British Summer Time (BST), and bit B of second 58 is set to 1 when BST is currently in effect [54].

10.2.4 Radio Station JJY (Japan)

Radio station JJY is operated by the National Institute of Information and Communications Technology (NICT), the NMI responsible for the official time of Japan. The NICT laboratories are located in Tokyo, and JJY broadcasts time signals on two frequencies from two mountaintop locations. The 40 kHz transmitter is near the summit of Mt. Ohtakadoya, located in the Fukushima Prefecture about 210 km northeast of Tokyo. The 60 kHz transmitter is near the summit of Mt. Hagane, located in the Fukuoka Prefecture about 870 km southwest of Tokyo (Figure 10.19).

The JJY call sign was assigned for many years to an HF time signal station that began operations in 1940. The HF station ceased operations in March 2001, less than two years after the more widely used 40 kHz signal from Mt. Ohtakadoya went on the air in June 1999. The 60 kHz broadcasts from Mt. Hagane began in October 2001, and the two LF transmitters are now jointly identified as JJY, which is now both a call sign and a registered trademark of NICT. Both stations distribute UTC(NICT), the national time standard of Japan.

Figure 10.19 The JJY transmitters at Mt. Ohtakadoya, 40 kHz, (a) and Mt. Hagane, 60 kHz (b).

The two stations each have 50 kW transmitters that are connected via an impedance matching system to a top-loaded vertical monopole antenna. The antennas have an umbrella-type design where a number of radial wires slope downward from the top of the mast. The 40 kHz antenna is taller, with a height of 250 m as opposed to 200 m for the 60 kHz antenna. The 60 kHz transmission system is nearly twice as efficient, producing about 23 kW of radiated output power as opposed to about 13 kW for the 40 kHz station [51, 55]. Both stations provide strong groundwave signals to all parts of Japan [40], and their skywave signals reach many ocean areas and parts of eastern Asia.

The same AM time code is broadcast from both sites, using ASK modulation that sends bits encoded in BCD format by switching the output power between 100% and 10%. Unlike WWVB, DCF77, and MSF, the JJY signal is always at full power at the start of each second. A 0 bit is sent by keeping the power high for 0.8 s, then holding it low for 0.2 s. A 1 bit is sent by keeping the power high for 0.5 s, then holding it low for 0.5 s. Frame markers are sent every 10 s by keeping the power high for 0.2, and then holding it low for 0.8 s. After each bit or frame marker is sent, the switch back to full carrier power is the OTM that signals the start of the second. A full time code, sent at 1 bps, requires 60 s to transmit.

The JJY time code format is listed in Table 10.5. During minutes 15 and 45, the last part of the time code differs from what's shown in the table. During those two minutes, JJY sends its call sign twice using Morse code, replacing the usual year information in seconds 40 to 48. Following the call sign transmission, status bits are sent during seconds 50 to 55 instead of the usual day-of-week and leap second information. The status bits are used to provide advance notice of scheduled service interruptions, and are set to 0 when no interruptions are planned [51, 55].

Table 10.5 The JJY time code format

Second	Bit	Description	Second	Bit	Description
0	M	Minute marker	30	8	Day of year, continued
1	40	Minute (0 to 59)	31	4	
2	20		32	2	
3	10		33	1	
4	0		34	0	Unused, always 0
5	8		35	0	
6	4		36	PA1	Parity bit for hour
7	2		37	PA2	Parity bit for minute
8	1		38	0	Unused, but available if Japan implements DST
9	P1	Frame Marker	39	P4	Frame Marker
10	0	Unused, always 0	40	0	Unused, but available if Japan implements DST
11	0		41	80	Year (0 to 99)
12	20	Hour (0 to 23)	42	40	
13	10		43	20	
14	0		44	10	
15	8		45	8	
16	4		46	4	
17	2		47	2	
18	1		48	1	
19	P2	Frame Marker	49	P5	Frame Marker
20	0	Unused, always 0	50	4	Day of week Sunday = 0 Saturday = 6
21	0		51	2	
22	200	Day of year 1 = January 1 365 = December 31 366 = December 31, leap year	52	1	
23	100		53	LS1	Leap Second Notice
24	0		54	LS2	
25	80		55	0	Unused, always 0
26	40		56	0	
27	20		57	0	
28	10		58	0	
29	P3		59	P0	Frame Marker

10.2.5 Radio Station BPC (China)

Radio station BPC is operated by the National Time Service Center (NTSC) of the Chinese Academy of Sciences, the NMI responsible for the official time of China. The NTSC laboratories are located in Xi'an, the capital city of Shaanxi Province, and BPC is located near Shangqiu, in Henan Province, about 700 km east of the laboratories. The station began broadcasting time signals on a carrier frequency of 68.5 kHz in 2007.

BPC utilizes a single tower umbrella-type antenna and a 100 kW transmitter. The radiated output power, assuming 50% efficiency, is near 50 kW. The usable range of the signal is estimated as 1,000 km for groundwave, and 3,000 km for skywave [51]. The station broadcasts an AM time code for 20 hours a day, with a four hour interruption from 2100 to 0100 UTC, during the early morning hours in China. It also broadcasts an experimental spread spectrum time code, using a system that might eventually include multiple transmitters [56], but as of 2025 those signals are not offered as a free service.

The AM time code is similar in some ways to those sent by other LF stations, but different in others. Similar to the other stations, BPC uses ASK modulation to send bits encoded in BCD format by switching the output power between 100% and 10%, with the carrier power dropped at the beginning of each second to serve as an OTM. However, unlike the other stations, the time code repeats every 20 s, and is sent at a rate of 2 bps, by transmitting both a most significant and least significant bit (MSB and LSB). Table 10.6(a) lists the method of transmitting the MSB and LSB, and Table 10.6(b) lists the format of the 20 s time code. The encoded time-of-day information is the Chinese Standard Time, using a 12-hour format, at the beginning of the 20 s time code frame. The beginning of the frame is marked by not reducing the carrier power.

Table 10.6 The BPC bit transmission method (a) and time code format (b)

(a)

Power drop (ms)	MSB	LSB
0	Start of time code	
100	0	0
200	0	1
300	1	0
400	1	1

(b)

	Bits				Bits		
Second	MSB	LSB	Description	Second	MSB	LSB	Description
0	—		Start of time code	10	HR	P1	Hour (0 = AM, 1 = PM) Parity, seconds 1 to 9
1	40	20	Second (0, 20, or 40)	11	0	16	Day of month (1 to 31)
2	0	0	Unused	12	8	4	
3	8	4	Hour (0 to 11)	13	2	1	
4	2	1		14	8	4	Month (1 to 12)
5	32	16	Minute (0 to 59)	15	2	1	
6	8	4		16	32	16	Year (0 to 99)
7	2	1		17	8	4	
8	0	4	Day of week 1 = Monday 7 = Sunday	18	2	1	
9	2	1		19	64	P2	Parity, seconds 11 to 18

10.2.6 LF Radio-Controlled Clocks and Wristwatches

Designing LF RCCs that automatically synchronized to the correct time, without any human assistance other than turning on the power, first became possible when WWVB, DCF77, and MSF added digital time codes to their broadcasts during the period from 1965 to 1974 (Table 10.2). The RCCs introduced during this period were mostly benchtop or rack-mounted instruments designed for scientific or industrial applications. They usually required an outdoor antenna, and before the advent of the microprocessor they required a substantial amount of electronic circuitry, which made them expensive to manufacture. The first such product was probably the Develco 3391, which was introduced in 1969 as a time standard for the electric power industry [57]. It synchronized to WWVB, which was then the only LF station providing a digital time code.

While waiting for more affordable products to appear, some radio hobbyists built their own RCCs. This trend started in 1973, when the American magazine *Radio Electronics* published a circuit design for a clock that synchronized to WWVB [58]. Circuit designs for clocks controlled by the European stations DCF77 and MSF were respectively published in 1974 [59] and 1975 [60]. However, few people outside of the technical community were aware of the industrial and hobbyist clocks, and RCCs remained mostly unknown to the general public for another decade.

The public awareness of RCCs skyrocketed after products intended for home use were demonstrated in 1985. The first of these clocks, the Kundo Space timer wall clock (Figure 10.20), and the solar-powered Junghans RCS-1, were both manufactured in Germany and both synchronized to signals from DCF77. Junghans, a German watch and clock manufacturer founded in 1861, was also the first company to produce a radio-controlled wristwatch. Their innovative wristwatch, which Junghans named the MEGA 1 (Figure 10.21), was introduced in 1990. It received DCF77 with an antenna located inside its leather wrist strap, overcoming the problem of a metal case [61, 62].

Figure 10.20 Kundo Space timer, the first radio-controlled wall clock (1985).

The early consumer-oriented RCC models were sold only in Europe and synchronized to either DCF77 or MSF. This was because the Asian stations JJY and BPC had yet to be built (Table 10.2), and because WWVB could originally output just 13 kW, which wasn't enough power to make RCCs work throughout the United States. This situation improved when WWVB increased its power output to 50 kW by 1999 [63], and to 70 kW soon afterwards. The WWVB power upgrade—coupled with the debut of the new stations in Japan and China—soon led to the annual production of many millions of RCCs, sold throughout Europe, the United States, and Asia. Many products can now receive signals from multiple stations, by taking advantage of their similar frequencies and time code formats. For example, the Casio wristwatch shown in Figure 10.22 utilizes their "Multi Band 6" technology, and can synchronize to all six of the stations listed in Table 10.2.

Figure 10.21 Junghans MEGA 1, the first radio-controlled wristwatch (1990).

Figure 10.22 A "Multi Band 6" radio-controlled wristwatch (c. 2010, courtesy of Casio).

The design details of LF RCCs vary, based on their price point and number of features, but they generally include just six main elements: a quartz oscillator, an antenna, a miniature radio receiver, a processing unit, a display, and a power supply. The oscillator is the same type found in any quartz wristwatch or clock, with a nominal frequency of 32.768 kHz and a typical accuracy of about 5×10^{-6}. This means the clock will either gain or lose about 0.5 s per day if it can't receive the radio time signal [41].

Designing small antennas that could effectively receive a long-wavelength signal was the biggest challenge faced by RCC developers. Fortunately, an RCC receiver has an easier job than the receiver inside of a disciplined oscillator, because it doesn't have to continuously receive signals and track the phase of the carrier. Instead, it only needs to pull in enough signal—during the usual one minute transmission of a time code frame—to separate the time code from the noise. The receiver only tries to obtain time codes at night, when the signal is the strongest. If it can decode just one time code every 24 hours, the RCC can always maintain sub-second accuracy.

The antenna is usually a ferrite rod wrapped with a coil of thin wire—the same type of antenna that's commonly found in an AM radio. The coil wrapped around the ferrite rod is tuned to resonance with a capacitor. If the RCC is designed to receive more than one frequency, it still usually has just one antenna, but the capacitance is varied as necessary. The ferrite rod directs the magnetic component of the incoming radio signal through the coil, which makes the antenna far more sensitive than it would have been if the coil were used on its own. This works well because ferrite, a ceramic material that's made by mixing iron oxide with other metals, is attracted to magnets and conducts magnetic fields. The downside is that a ferrite rod antenna is highly directional. It works much better if the ferrite rod is positioned at a right angle (broadside) with respect to the direction of the transmitter. Because ferrite rod antennas are so dependent on their orientation, they don't work well if the antenna is moving while trying to decode the signal, which can be a problem with radio-controlled wristwatches.

The antenna resides inside the clock's case and has to be small enough to fit. Therefore, its length varies depending on the clock type. Wall clocks typically have the longest antennas, with lengths ranging from 60 mm to 200 mm, and an antenna length of 60 mm is typical for alarm or desk clocks. In a wristwatch, space is at a premium, so the antenna length is necessarily limited to about 25 mm, and is often much less. As a general rule, the longer antennas provide better reception. The clock case, or at least the part of the case that's near the antenna, is made of plastic or another non-conductive material to avoid blocking signals. A ferrite rod antenna with dimensions of 60 mm × 10 mm, attached to a receiver module, is shown in Figure 10.23.

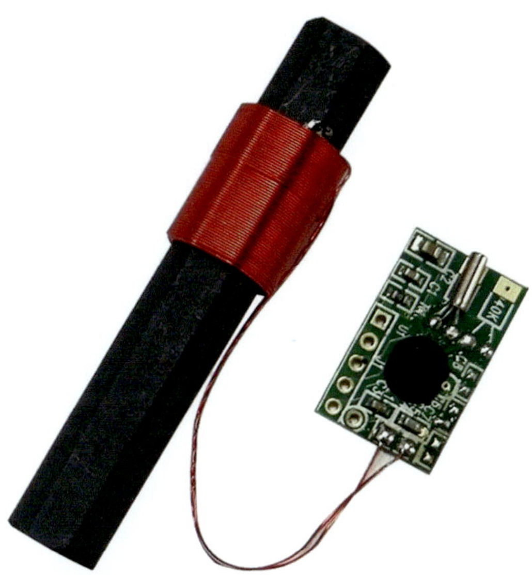

Figure 10.23 A ferrite rod antenna and receiver module for an LF radio-controlled clock.

The miniature radio receiver has most of its functions performed by a single integrated circuit. It's designed to cover a range of frequencies, typically 40 kHz to 80 kHz, which is just wide enough to receive all of the available time signal stations. The receiver usually also has a preamplifier with slow automatic gain control, which makes the amplitude changes used to modulate the time code easier to detect. The time signal station of interest is selected by using a small quartz crystal oscillator with a nominal frequency that matches the station's carrier frequency, 60 kHz, for example, in the case of WWVB, MSF, and JJY. The crystal serves as a filter so that the receiver is fixed tuned to a single frequency, ignoring other incoming frequencies. This simple design can be categorized as a tuned radio frequency (TRF) receiver, and dates back to the early days of radio, predating the superheterodyne receiver. It works well enough for RCCs, which can be fixed tuned to any of the LF time signal stations if just four crystal oscillators—with nominal frequencies of 40, 60, 68.5, and 77.5 kHz—are added to the receiver circuitry.

There are numerous methods that the receiver can use to detect and demodulate the incoming time signal [64]. The methods vary in complexity, reliability, and synchronization accuracy, but their primary function is to convert the incoming radio signal to a continuous stream of bits. These bits are sent to a processing unit, where software extracts the time-of-day and date information from the time code. Because most time signal stations transmit UTC (BPC's transmission of Chinese Standard Time is an exception), the software also converts UTC to local time by applying a correction obtained from a time zone setting that was previously selected by the user. Once this process is completed, the processing unit synchronizes the clock's display.

The RCC display can be either analog or digital. Digital displays usually show hours, minutes, and seconds in the HH:MM:SS format, with the display resolution limited to 1 s. An analog display requires a stepper motor to move the hands of the clock, and also requires the hands to be perfectly aligned to take advantage of the accuracy.

Some RCCs, such as clock radios, use AC power, but most run off batteries. The battery life is nearly equivalent to that of a regular quartz clock because the receiver is only turned on when the clock attempts to synchronize. Some RCCs synchronize only once per day—others perhaps as many as six times—but in all cases the receiver is idle for more than 99% of the day.

The advertisements for LF RCCs often make sensational claims about their accuracy, and manufacturers sometimes even label them as atomic clocks, which, of course, they are not. They obviously benefit from atomic time keeping because they synchronize to signals referenced to UTC, but they weren't designed for nanosecond, microsecond, or even millisecond-level accuracy. They were instead designed to provide consumers with clocks that never need to be manually set, and that should always be within ±0.5 s of UTC, or accurate to the nearest second.

Even so, anyone interested in time metrology will likely be curious about the true accuracy of an RCC. To satisfy that curiosity, we first need to remember that an RCC is a simple example of a one-way time transfer system where no attempt has been made to measure or compensate for path delay. For a groundwave path, the path delay is roughly 3.3 ms per 1,000 km (Section 4.3.1). If we know the length of the path and the type of propagation, we can make a good estimate of the path delay. Or, we can just make a very conservative and safe assumption that should apply to all LF RCCs, regardless of where they are located, and state that the path delay will never exceed 20 ms. That assumption should always be true, even if the station were 5,000 km away, a distance that's far beyond the reliable range of any of the LF stations. If we add another conservatively estimated delay of 10 ms to allow for the clock to process the radio signals and synchronize the display, we can still state with confidence that all RCCs should be accurate to within 30 ms of UTC at the time of synchronization, with some achieving much better accuracy.

We mentioned at the beginning of Section 10.1 that an RCC doesn't adjust the frequency of the clock's oscillator. For this reason, its accuracy at the time of synchronization is the best accuracy it can achieve. From that point forward, its time difference with respect to UTC will continue to increase until the next synchronization occurs. The graph in Figure 10.24 provides an example of typical accuracy, by showing measurements of the time differences between UTC(NIST) and a wristwatch controlled by WWVB. The wristwatch synchronizes five times daily—at local midnight and at 1, 2, 3, and 4 a.m. The midnight synchronization occurs after the quartz oscillator has free run for about 20 hours, and a daily correction of about 0.45 s is needed to compensate for the accumulated time difference. The other four synchronizations, indicated by the jagged lines at the top of the graph, occurred after the oscillator had free run for just one hour, so the clock corrections were just slightly larger than 0.02 s. These measurements indicate that the frequency accuracy of the oscillator is about 6×10^{-6}. The time difference with respect to UTC varies considerably during the day, but this particular wristwatch meets the typical RCC requirement of always being within ±0.5 s of UTC [65].

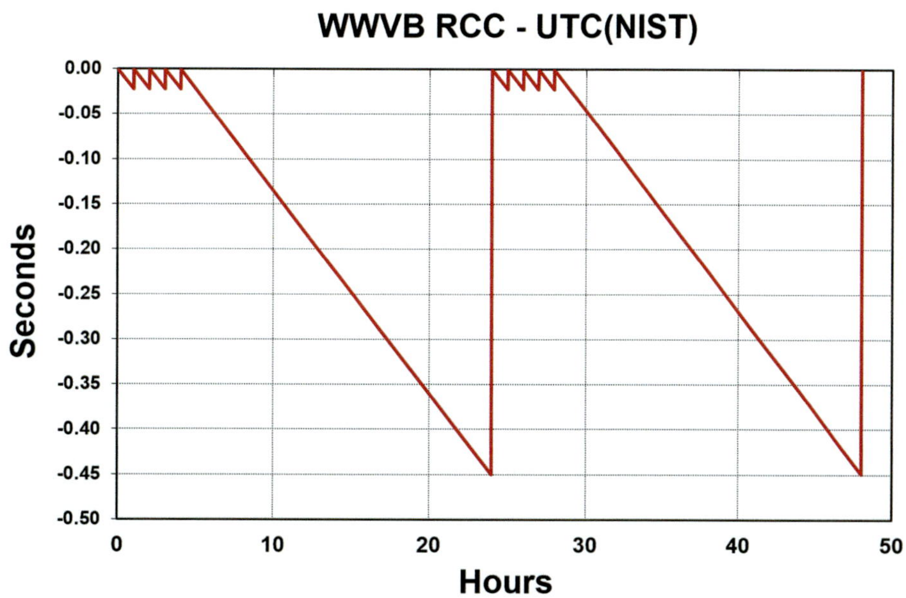

Figure 10.24 Accuracy of an LF radio-controlled clock compared to its UTC reference source.

10.3 Time Signals from HF Terrestrial Radio Stations

High-frequency (HF) radio, commonly known as shortwave radio, has long been associated with time and frequency metrology. When conditions are optimal, HF signals can provide worldwide coverage. The signals also have enough bandwidth, typically at least 6 kHz, to transmit audio with fidelity similar to commercial AM broadcast band stations. The audio includes ticks broadcast at 1 s intervals and accurate voice announcements of time that provide listeners with a "talking clock." The HF time signal stations are best known for their audio time signals, but they are also known for providing standard frequency carrier signals that can be used as references for frequency measurements at moderate accuracy levels (Section 5.3.2). In addition, most stations broadcast digital time codes, but their use is limited because nearly all low-cost RCCs receive the more reliable LF signals just described in Section 10.2. We'll begin this section by describing some general characteristics of HF radio signals, followed by descriptions of the available time signal stations.

The HF band occupies the spectrum from 3 to 30 MHz. However, 2.5 MHz—which is technically in the MF band—is always grouped with the HF frequencies for the purposes of time signal broadcasts. As radio amateurs and shortwave listeners know, the HF spectrum is a virtual wonderland because of its skywave signal propagation. When the right frequency is selected and the conditions are optimal, an HF signal with just a few watts of power can reach any location on Earth. The range of skywave signals is only limited by the furthest possible distance between two points on the Earth's surface—called the antipodal distance—which is about 20,000 km, one-half of the Earth's circumference.

We previously described in Section 10.2 how LF signals can reflect from the ionosphere's lowest layer, the D layer, in the daytime, and from higher layers at night when the D layer disappears (Figure 10.8). In contrast, HF signals always pass through the D layer without being reflected, and the first reflections occur in the E layer, at heights ranging from about 90 km to about 125 km above the Earth's surface. However, the E layer has relatively low levels of ionization content during the daytime, and it greatly weakens at night, so the F layers above the E layer are the most important ionospheric regions for HF propagation.

During the daytime, there are two distinct F layers: a lower layer called F1 and an upper layer called F2. The altitudes of these layers are determined by numerous, constantly changing variables related to the time of day, time of year, geographic location, and solar activity. As a very general approximation, F1 and F2 might have altitudes of 200 km and 300 km, respectively, during the winter, increasing to 300 km and 400 km, respectively, during the summer. At night, F1 and F2 combine into a single F layer with an altitude of about 250 km to 300 km. Due to the disappearance of the D layer and the combining of F1 and F2 into a single layer, just two ionospheric layers are present at night (E and F), but four are present during the daytime (D, E, F1, and F2), as shown in Figure 10.25.

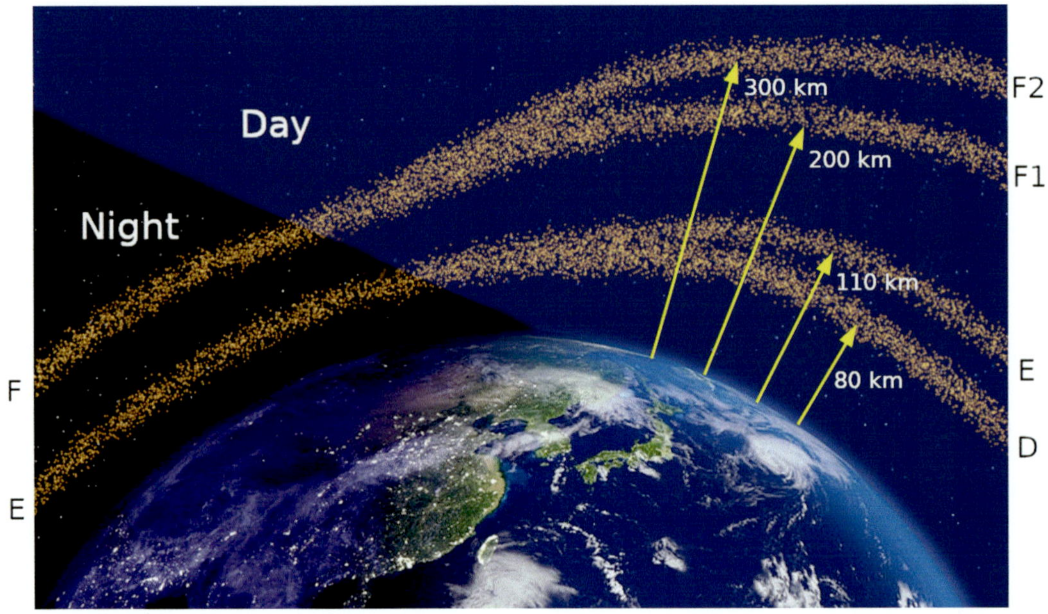

Figure 10.25 The daytime and nighttime layers of the ionosphere and their approximate height above the Earth's surface (credit: Carlos Molina).

Figure 10.26 is a simplified diagram showing how HF signals propagate. Some of the signal is groundwave, not affected by the ionosphere, but usually not traveling far enough to be useful, typically only reaching receivers within about 100 km of the transmitter [39]. Nearly all of the useful propagation is skywave, caused by refraction from the ionosphere. The signals that propagate at a steep enough angle pass through the ionosphere into space, but the remaining signals are reflected back to Earth. The skip distance is the distance between the transmitter and the point where the skywave signals return to Earth. The skip distance is related to the ionosphere's height at the point of reflection, and as a very general rule the maximum skip distance for E layer reflections is about 2,000 km. It's about twice that, or 4,000 km, for F layer reflections. Signals can reflect from the ionosphere multiple times, with each hop carrying the signal further. The skip zone—or the "dead zone" where the signals can't be received—begins just beyond the reach of the groundwave signals and ends where the first skywave signals return to Earth [66, 67].

When HF signals are used as a frequency or time reference, the variations in path delay caused by skywave propagation will obviously dominate the measurement uncertainty. When the height of the ionosphere is increasing at the point of reflection, the signal path gets longer and the received frequency decreases. Conversely, if the height of the ionosphere is decreasing, the signal path gets shorter and the received frequency increases. Therefore, the skywave affects the frequency accuracy of an HF signal by creating a Doppler frequency shift [68]. It also affects the time accuracy, because the constant variations in path delay can add uncertainties that can exceed 5 ms if the signal makes several "hops" off the ionosphere. The effects of skywave propagation aren't easy for a time metrologist to estimate or model, but because frequency is the reciprocal of time interval, and because the meter is defined with respect to the speed of light, everything is related. Time-of-flight (TOF) measurements of an HF signal can be used to calculate its Doppler frequency, and vice versa. Either the TOF or the Doppler frequency measurements can be used to estimate the height of the ionosphere [69].

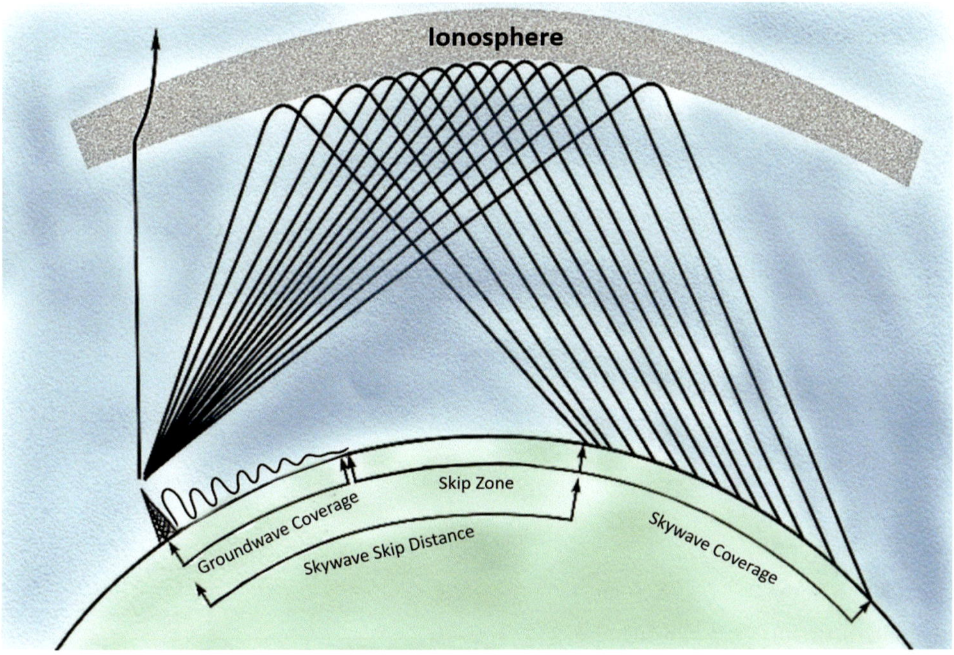

Figure 10.26 A simplified diagram showing the groundwave and skywave coverage areas of HF signals, and the skip zone, where no signals are received.

Despite the potentially large measurement uncertainties, HF radio is still a perfectly acceptable medium for synchronizing clocks at moderate accuracy levels. The main reason that HF signals are seldom used for RCC applications is not their inaccuracy, but instead their unreliable reception, which is due to all of the factors that influence propagation. The skip distance and the size of the skip zone can dramatically change based on current ionospheric conditions. Unless the receiver is located within the groundwave coverage of the transmitter, there is no single HF frequency that will always be usable, and a frequency that provided reliable reception for weeks or months may suddenly not work at all.

The number of HF time signal stations in operation declined by at least 50% from 1990 until 2025, partially because of the limitations in accuracy and reception reliability just described. However, the remaining HF stations continue to provide valuable services not available elsewhere. For example, the audio time signals of HF stations can be heard with any general coverage shortwave receiver, and they require only the human ear to decode. The digital time signals, although seldom used by consumer RCC products, continue to serve industrial applications. Most stations provide the same information on multiple frequencies so that at least one frequency should be usable at all times. This redundancy benefits users and makes worldwide coverage possible. Finally, scientists who study the ionosphere greatly benefit from the continuous transmissions of HF signals that are traceable to the SI, because this allows them to accurately measure the Doppler frequency shifts [69]. Table 10.7 provides a list of HF time signal stations, with brief descriptions provided in the following sections.

Table 10.7 HF time signal stations

Call sign	Location	Language	Frequencies (MHz)	Output power (kW)	First year of service
WWV	Ft. Collins, Colorado, USA	English	2.5	2.5	1923
			5	10	
			10	10	
			15	10	
			20	2.5	
			25	2.5	
WWVH	Kauai, Hawaii, USA	English	2.5	5	1948
			5	10	
			10	10	
			15	10	
CHU	Ottawa, Canada	English French	3.33	3	1929
			7.85	5	
			14.67	3	
BPM	Pucheng County, Shaanxi, China	Chinese	2.5	1	1980
			5	10	
			10	10	
			15	10	
HLA	Daejeon, South Korea	Korean	5	2	1984
PPE	Rio de Janeiro, Brazil	Portuguese	10	1	2008

10.3.1 Radio Stations WWV and WWVH (United States)

Radio stations WWV and WWVH are both operated by NIST, and WWV has been operational for more than a century, making it one of the longest-running scientific services in the United States. WWV began operations in 1923 by broadcasting unmodulated carrier signals that were used as a frequency reference. These signals were a key part of the early U. S. efforts to help the first commercial broadcast stations measure their transmission frequencies, so they could avoid interference with other stations (Section 2.2). WWV was only a frequency service until 1937, when 1 pps signals were added to the broadcasts as a time interval reference. Even then, the pulses weren't in phase with any time reference, because the Navy was designated as the official radio time broadcaster for the United States.

The Navy essentially stopped broadcasting time signals during World War II, and in 1944 they authorized the synchronization of the WWV time signals to the USNO reference clock. As a result, WWV became a time signal station shortly after the war ended, and began broadcasting a telegraphic time code in October 1945. Voice announcements of time, the station's best-known feature, began on January 1, 1950 [43, 70]. In 1960, WWV became the first radio station to broadcast a digital time code [71]. WWV was located at several different sites in the Washington, DC area before moving to Fort Collins, Colorado, to share the WWVB site in 1966. Prior to that move, it was joined on the air by its sister station, WWVH, which began operating in 1948 and added voice announcements of time in 1964. The Hawaiian Island of Maui was the original site of WWVH, but the station was moved to Kauai in 1971.

WWV and WWVH both broadcast time signals on 2.5, 5, 10, and 15 MHz, and WWV also broadcasts on 20 and 25 MHz, so a combined total of 10 transmitters are always on the air. Both stations output 10 kW on 5, 10, and 15 MHz, with less power radiated on the other frequencies (Table 10.7). The

WWV signals are referenced to the same secondary UTC(NIST) time scale as WWVB (Section 10.2.1). The WWVH signals are referenced to cesium clocks located on the Kauai site that are disciplined by the primary UTC(NIST) time scale in Boulder, Colorado via common-view GPS signals.

When listening to a shared frequency such as 10 MHz, it's sometimes possible to hear the audio from both stations. Listeners can easily tell the difference, however, because a man's voice announces the time on WWV and a woman's voice announces the time on WWVH. The WWVH announcement begins 15 s before the top of the minute and lasts for about 7 s, followed by the WWV announcement at 7 s before the top of the minute. Therefore, the two voice announcements never overlap. The top of the minute is marked with a 1,000 Hz tone from WWV and a 1,200 Hz tone from WWVH, each with a duration of 0.8 s. During the minutes, short pulses are transmitted at 1 s intervals that sound similar to the ticking of a clock. Both stations mark the top of the hour with a 1,500 Hz tone, also of 0.8 s duration [43]. The broadcast formats for the WWV and WWVH audio time signals are shown in Figure 10.27.

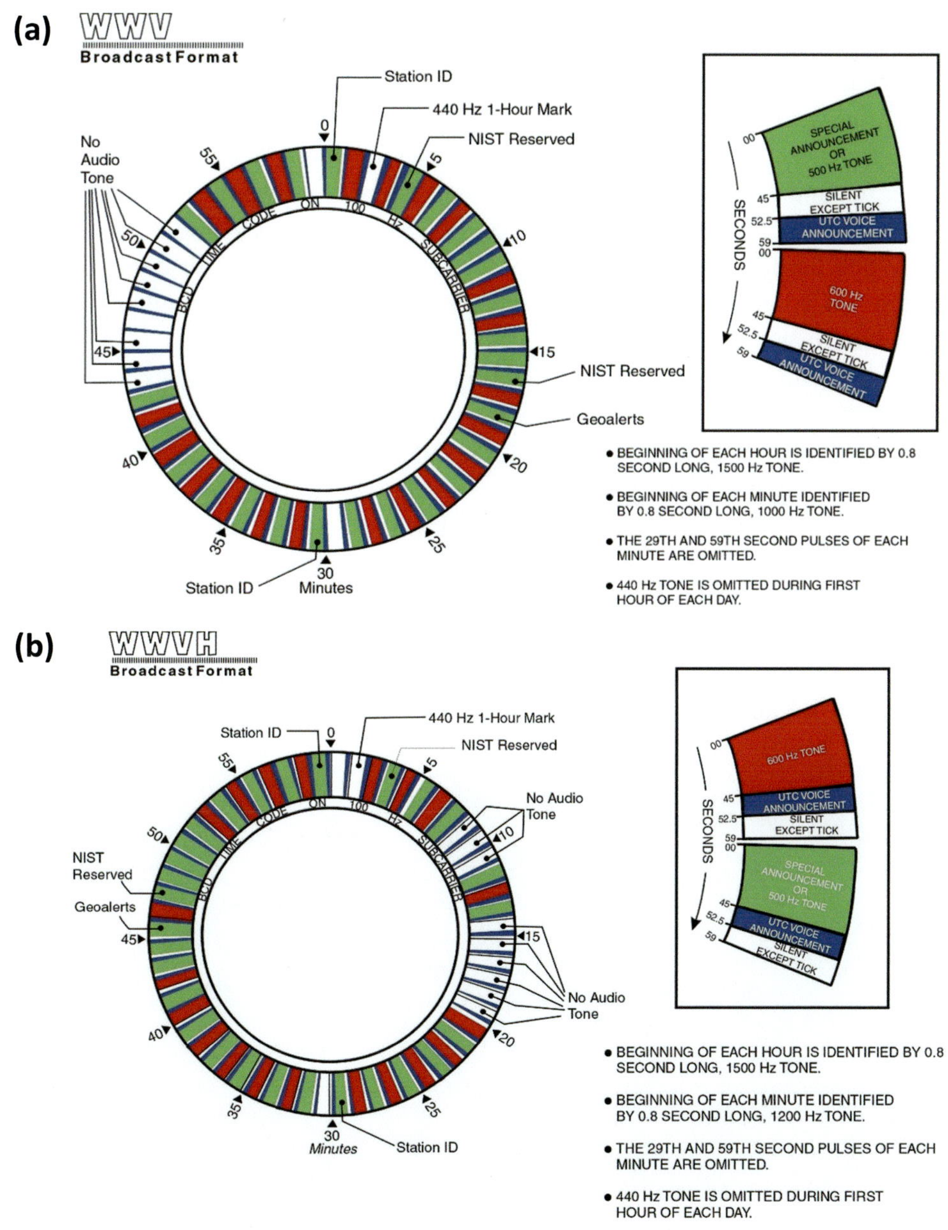

Figure 10.27 Broadcast formats for (a) WWV and (b) WWVH.

The iconic voice announcements of time are well known to most radio amateurs and shortwave radio listeners and have followed the same format since January 1, 1974. The time of the upcoming minute is announced with these words:

"At the tone, X hour(s), Y minute(s), Coordinated Universal Time."

From 1966 until the end of 1973, the time was announced as Greenwich Mean Time. Prior to that, the voices heard on WWV and WWVH respectively announced the local time for the Eastern and Hawaiian time zones. The voices were always, of course, pre-recorded.

The recordings used to announce the time on WWV, beginning in perhaps 1952 and continuing until January 1990, were provided by Don Elliott Heald (1922–2009), a prominent radio and television announcer whose career spanned more than 50 years. New voice recordings were made when the station began using digital storage for the audio, and the voice was changed to that of John Doyle (1942–2023), who, like Heald, was an established professional announcer. Doyle's voice has continued to announce the time on WWV to the present day, with the exception of a one-year period from August 1991 to August 1992, when the voice of Eric Smith filled in due to equipment problems.

The voice of Jane Barbe (1928–2003) has announced the time on WWVH from 1971 to the present, with the exception of a one-month period in 1991 when the voice of Johanna Stahl filled in due to equipment problems. Barbe was an accomplished voice actress and singer, but she was best known as a time announcer. Beginning in the 1960s, Barbe recorded announcements of time, temperature, weather, and other information for the Bell system and other telephone companies. Her voice was heard by millions of listeners, and she became nationally known as the "Time Lady."

In addition to the audio time signals, both stations transmit a digital time code that can be used to automatically synchronize an RCC. The time code is sent in BCD format on a 100 Hz subcarrier at a rate of 1 bps. Several companies once offered WWV RCCs, but the complexities associated with HF reception, along with the antenna requirements, prevented their widespread use. The most common example of a WWV RCC was the Heath GC-1000 Most Accurate Clock (Figure 10.28), introduced in 1983. Available as either a kit or an assembled product, it included a removable telescopic antenna that could be used indoors, but an outdoor antenna was recommended. It scanned for the time code on 5, 10, and 15 MHz, before selecting the strongest signal to decode, and included a loudspeaker for listening to the audio. Radio hobbyists and scientists [72] quickly recognized its utility, but during the 1990s the GC-1000 and similar products were superseded by the lower cost and more reliable clocks controlled by WWVB.

Figure 10.28 The Heath GC-1000 "Most Accurate Clock" that synchronized to WWV (1983).

10.3.2 Radio Station CHU (Canada)

Radio station CHU is operated by the Institute for National Measurement Standards of the National Research Council (NRC), the NMI and official timekeeper of Canada. The NRC laboratories are located in the Canadian capital of Ottawa, and CHU resides about 20 km to the southwest. The station is remotely controlled from the NRC laboratories and it broadcasts time signals referenced to UTC(NRC).

CHU has been located at its present site since 1947, but it began operations at the Dominion Observatory, an astronomical observatory in Ottawa that was once responsible for Canada's official time. Experimental time signals were sent from the Dominion Observatory beginning in 1923, and daily time transmission began in 1929 using the call sign VE9OB. The call sign was changed to CHU in 1938, and by then time signals had been broadcast 24 hours a day for several years on three carrier frequencies. By the end of 1938, the three frequencies assigned to CHU were 3.33 MHz, 7.335 MHz, and 14.67 MHz [73]. Two of the three frequencies are still used today, but the 7.335 MHz signal was changed to 7.85 MHz in 2009, when the ITU reallocated the frequencies from 7.30 to 7.35 MHz from fixed service to general broadcast use.

The choice of the CHU frequencies might seem odd, because the other HF time signal stations (Table 10.7) broadcast on carrier frequencies that include 5 MHz, 10 MHz, or both, matching the nominal frequencies of oscillators used as frequency standards. The CHU frequencies were chosen for practical reasons; they were just outside of the 80, 40, and 20 meter bands that were allocated to licensed radio amateurs [74]. This was beneficial to amateur radio operators for many years, because most of their HF receiving equipment wasn't "general coverage" until the 1980s, which meant that it couldn't receive all of the HF frequencies from 3 to 30 MHz. Instead, their equipment was limited to reception of the frequencies allocated to amateurs, with some additional bandwidth slightly below and above the amateur band that allowed CHU to be received. As a result, CHU could often be heard with receivers that couldn't tune in to other time signal stations.

CHU outputs 5 kW on 7.85 MHz, and 3 kW on 3.33 MHz and 14.67 MHz. The same information is sent on all three frequencies. A voice announcement of UTC, provided in both French and English, is broadcast during the last 10 s of each minute. The French announcement occurs first during odd-numbered minutes, and the English announcement occurs first during even-numbered minutes. The voice announcement refers to the time of the 1 kHz tone that follows, which is normally 0.5 s in duration, but is omitted during the 29th second of a minute. The top of the hour is marked by 1 kHz tone that lasts for a full second, followed by 9 s of silence.

CHU also broadcasts a digital time code in a format that NRC first used to distribute time to computers via telephone lines, beginning in the 1970s [73, 74]. The time code is sent during seconds 31 through 39 of every minute by audio frequency shift keying (FSK), using a mark tone of 2,225 Hz, a space tone of 2,025 Hz, and a data rate of 300 baud. This format follows the Bell 103 standard that was defined by AT&T in 1963 for the first commercial 300 baud computer modem. Therefore, computers with a Bell 103 compatible modem can decode the time from the audio signal. The information encoded in the FSK data stream includes the UTC time, the year and day of year, a leap second warning, a DUT1 correction, the current difference between the TAI and UTC time scales, and a Canadian daylight time indicator.

10.3.3 Other HF Time Signal Stations: BPM (China), HLA (Korea), and PPE (Brazil)

Three other HF time signal stations share the standard carrier frequencies that were first used by WWV and WWVH (Table 10.7). Due to the complex characteristics of skywave propagation, these stations are sometimes heard along with, or instead of, the other HF time signal stations that transmit on the same frequency.

The most powerful of the three stations is BPM, which is operated by NTSC, the same Chinese institution that operates the LF station BPC (Section 10.2.5). BPM is located in Pucheng County, in Shaanxi Province of China, about 100 km northwest of the NTSC laboratories. Since 1980, the station has broadcast time signals referenced to UTC(NTSC) on frequencies of 2.5, 5, 10, and 15 MHz. The 2.5 MHz transmitter outputs just 1 kW, but the 5, 10, and 15 MHz transmitters radiate 10 kW, matching the power output of WWV and WWVH. The 5 and 10 MHz signals are broadcast 24 hours per day, but the 2.5 MHz signal is only broadcast for 17.5 hours (0730 to 0100 UTC), and the 15 MHz signal for just eight hours (0100 to 0900 UTC).

The BPM signals are unusual because audio signals referenced to UT1 are transmitted for eight minutes each hour (Table 10.8). When UT1 is being transmitted, this creates an odd click-beep effect when BPM is simultaneously received with WWV or WWVH, because the OTMs for UT1 and UTC occur at different times. BPM also broadcasts a digital time code in BCD format that includes the UTC minute, hour, month, day of month, and year, as well as a DUT1 correction and a leap second indicator [51].

Table 10.8 The BPM transmission schedule

	Time within the hour (MM:SS)		
Time slot	**Start**	**Stop**	**Data content**
1	59:00	00:00	Station identification (Morse code and voice)
2	00:00	10:00	UTC time signal
3	10:00	15:00	Carrier frequency only (no modulation)
4	15:00	25:00	UTC time signal
5	25:00	29:00	UT1 time signal
6	29:00	30:00	Station identification (Morse code and voice)
7	30:00	40:00	UTC time signal
8	40:00	45:00	Carrier frequency only (no modulation)
9	45:00	55:00	UTC time signal
10	55:00	59:00	UT1 time signal

Radio station HLA is operated from the campus of the Korean Research Institute of Standards and Science (KRISS) in Daejeon, South Korea. KRISS is the NMI responsible for Korean standard time, and the HLA signals are referenced to the national time scale, UTC(KRIS). HLA went on the air on a part-time schedule in 1984, with continuous 24 hour broadcasts beginning in 2011. The UTC signals are transmitted on a carrier frequency of 5 MHz with an output power of 2 kW. The seconds are marked with short audio tones, with the tones omitted on the 29th and 59th seconds. Longer tones of 0.8 s in duration mark the minute and hour, with respective frequencies of 1,800 Hz and 1,500 Hz. The audio also transmits a DUT1 correction by using double tones, and a BCD time code is continuously broadcast on a 100 Hz subcarrier.

Radio station PPE is operated from the campus of the National Observatory in Rio de Janeiro, Brazil. The National Observatory is responsible for the official time of Brazil and the PPE time signals are referenced to the UTC(ONRJ) time scale. The PPE signals have been transmitted since November 2008, on a carrier frequency of 10 MHz with an output power of 1 kW. The signals are mainly intended for local reception in Brazil, but are sometimes heard in other countries. A woman's voice announces Brazilian time (UTC – 3 hours) in Portuguese every 10 s, with audio tones marking the seconds and minutes. Unlike the other HF time signal stations, PPE doesn't transmit a digital time code.

10.4 Time Signals from VHF Terrestrial Radio Stations (FM Broadcast Band)

The terrestrial time signals sent in the VHF (30 to 300 MHz) and UHF (300 to 3,000 MHz) radio bands don't propagate by groundwave or skywave, but instead are known as line-of-sight signals. These signals can't travel beyond the radio horizon, so they have the advantage of small and stable path delays, but the disadvantage of a limited coverage area. In addition, these signals are not always referenced to a UTC source, nor are they usually monitored by a national timing laboratory. Therefore, no guaranteed accuracy can be assumed and traceability to the SI is usually difficult to establish. Despite these limitations, VHF and UHF time signals are still very useful, and they provide time synchronization for large numbers of consumer products that have a built-in clock. This section describes the freely available radio time codes distributed by public and commercial radio in the FM broadcast band, and Section 10.5 describes time codes distributed by terrestrial VHF and UHF television stations.

The FM broadcast band is well known to all radio listeners as a medium for delivering news and entertainment. In most countries, the FM broadcast band covers the frequencies from either 87.5 or 88 MHz to 108 MHz, but Japan utilizes a narrower band (76 to 95 MHz) and Brazil a wider band (76 to 108 MHz). FM radio uses frequency modulation, which adds information to a radio signal by varying the instantaneous frequency of its carrier wave. It was invented by Edwin Armstrong (1890–1954), whose many other inventions include the superheterodyne receiver (Section 5.2.4). Armstrong began experimental broadcasts of FM signals from the Empire State Building in New York City in 1934. Two-channel audio was added to the FM format in the early 1960s, and it gave FM radio its best-known feature, which is high-fidelity stereo sound.

Some 30 years after stereo was added, it became possible to also transmit time codes and other types of information on FM radio by using a protocol best known as the Radio Data System (RDS), but officially called the Radio Broadcast Data System (RBDS) in the United States [75, 76]. This system has gained wide acceptance in both Europe and the United States, where nearly all FM radio stations now have RDS capability. FM stations began transmitting RDS information in the early 1990s, and the first receivers that could display the encoded information as text appeared shortly afterwards. The received information is used to identify the station and to describe the programming content that's currently being broadcast. It also includes the current date and time-of-day, which can be used to synchronize a clock.

To understand how the RDS works, consider that the bandwidth allocation for an FM station is 200 kHz, so it allows audio and data subcarriers with bandwidths totaling up to 100 kHz. This wide bandwidth provides FM radio with its ability to deliver high-fidelity audio, by providing a single channel for monaural audio, and two channels for stereo, with each channel having 15 kHz of bandwidth. These channels, along with a stereo pilot tone, occupy about 53 kHz of bandwidth. To avoid interfering with the audio signals, the RDS data is transmitted on a 57 kHz subcarrier. This works well, because the RDS sidebands use slightly more than 2 kHz of bandwidth, which is narrow enough to avoid interfering with the right stereo channel (Figure 10.29).

The RDS information is sent by BPSK modulation of the 57 kHz subcarrier. A total of 48 subcarrier cycles are used to send one bit, which results in a data rate of 1,187.5 bps. The data are sent in 26-bit blocks consisting of 16 data bits and 10 error correction bits, with four blocks forming a 104-bit group. Group 4A, transmitted once per minute and known as the clock time and date (CT) group, contains the encoded time-of-day and date information (Figure 10.30). The encoded information refers to the time and date of the next minute, and the end of the CT group transmission occurs within ±0.1 s of the transition to the next minute.

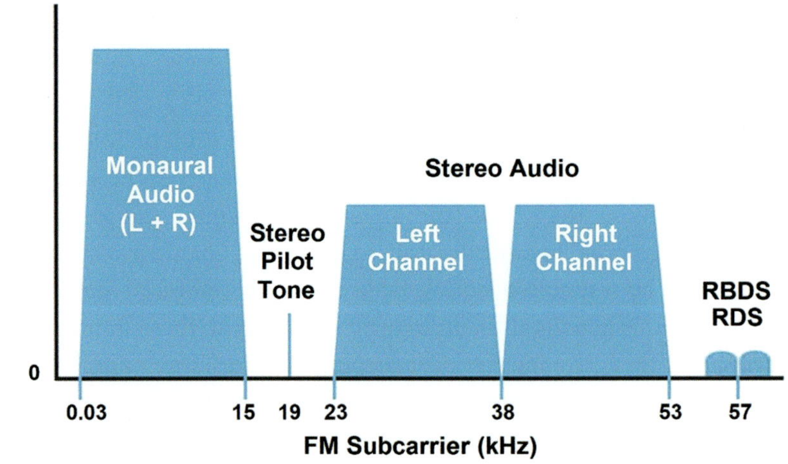

Figure 10.29 The FM audio subcarriers and the RDBS/RDS subcarrier.

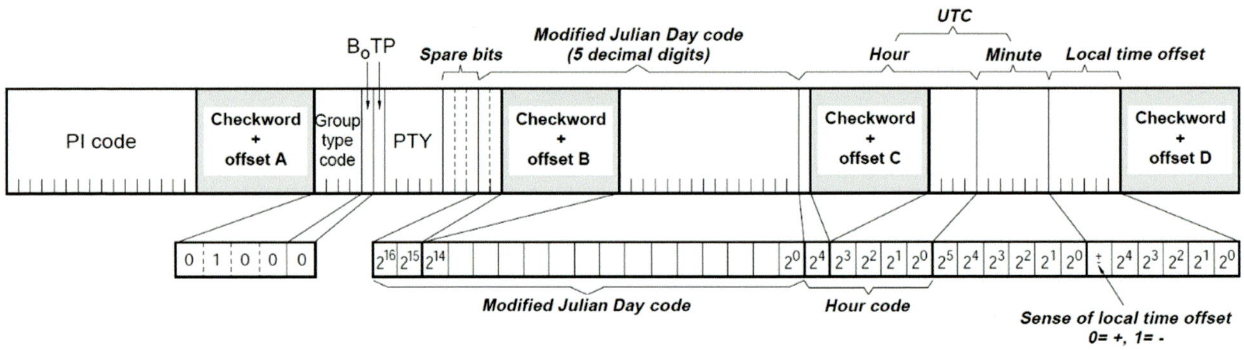

Figure 10.30 The RDS time code transmitted by FM radio broadcast stations.

Like the other RDS groups, Block 1 of the CT group begins with a 16-bit program identification (PI) code that contains a country code, information about the coverage area, and a program reference number. Block 2 is mostly reserved for other uses, but it contains the most significant bits of a 17-bit Modified Julian Day (MJD) number. The remaining MJD bits are included in Block 3, along with the MSB for the UTC hour (0 to 23). Block 4 includes the remaining bits for the UTC hour, the UTC minute (0 to 59), and the local time zone offset from UTC, with a range of −15.5 to +15.5 hours [75, 76].

The abundance of FM radio stations and receivers makes the RDS well suited for the large-scale distribution of time signals. For example, there are more than 10,000 FM radio stations in the United States, and RDS-compatible receivers are found in nearly all homes and automobiles. The RDS time code is often used to synchronize the internal clocks of these receivers, and is sometimes used to synchronize the main dashboard clock on automobiles. According to the RDS standard, FM stations should only transmit the CT group if the time is known to be accurate with respect to UTC, which implies accuracy to within ±0.1 s [75, 76].

Many FM stations synchronize the RDS clock with a GPS disciplined clock, so the accuracy requirement is usually easily met, but it's important to remember that keeping accurate time is not the first priority for station operators. Stations don't always adhere to the RDS standard, and CT groups are often broadcast with incorrect time. Because the time is obtained from the FM station that's currently being tuned in, a significant time step can occur when the receiver is tuned to a different station. Even so, the potential for high-accuracy timekeeping exists due to the stable line-of-sight VHF signals, and the RDS has been successfully used to calibrate clocks, located on both mobile and stationary platforms, in large-scale city-wide sensor networks [77].

10.5 Time Signals from VHF/UHF Terrestrial Television Stations

Television (TV) can be broadly defined as the medium used to transmit both moving images and sound. Defining what constitutes a TV signal is difficult, however, because television has undergone a vast transformation since the first TVs were offered to the public in the 1930s. For many years, it was possible to define TV signals as the freely available analog audio and video signals broadcast in the VHF spectrum from 54 to 216 MHz (channels 2 through 13), and in the UHF spectrum from 470 to 890 MHz (channels 14 through 83). This definition is now outdated, because most viewers in the United States and elsewhere now obtain their TV signals from cable systems, Internet streaming services, or from signals broadcast from communication satellites, with the access requiring a paid subscription.

In this section, we'll limit our discussion to the time signals broadcast on the original VHF/UHF frequencies by terrestrial television stations, the signals now known as free over-the-air (OTA) TV. However, the free OTA signals are no longer analog. The worldwide transition from analog to digital television (DTV) began around the year 2000. It was essentially completed in the United States by 2009, and is now essentially complete in all countries. We'll start by providing some historical information about the use of TV signals for time synchronization before describing the DTV time signal formats.

Television signals first captured the attention of time and frequency metrologists in the 1960s, during a period when digital time codes broadcast by radio were still extremely rare. Metrologists were attracted to TV signals because of their stable line-of-sight paths, and because their 6 MHz bandwidth allowed them to carry several types of information that could potentially serve as measurement references. They were also attracted by the huge number of installed televisions. Watching television had become one of America's favorite pastimes, and about 95% of U. S. households had a TV set by 1969.

The researchers at NBS in Boulder, Colorado, quickly saw this as a huge opportunity, and they envisioned a system where UTC(NBS) could be distributed to nearly every U. S. household by adding an RCC to their TV. At least several things would need to happen before this vision could become a reality. A time code would have to be designed that could be added to the broadcasts without causing any interference to the existing signals. A time code receiver would also need to be designed that could be added to an existing TV, and that could decode and display the time [78]. In addition, atomic clocks and encoding equipment synchronized to UTC(NBS) would need to be installed at the major television network centers. From there, the networks would need to distribute the time using their microwave relay systems to their network affiliates [79].

The vision was vigorously pursued for several years, and NBS successfully developed an experimental system, called *TvTime*, by cooperating with local television stations. The NBS researchers designed a time code that was hidden inside the video signal, and a time code receiver that could display the decoded time on an existing TV. Most of the design work was done by Dick Davis (previously mentioned in Chapters 4 and 9).

Analog television then used a video format consisting of 525 horizontal lines, a standard defined by the National Television System Committee (NTSC) in 1941. Images were "painted" by moving an electron beam across and down a cathode ray tube (CRT), the "picture tube" that served as the TV's screen. The beam painted every other line as it moved down the screen, alternating on each pass between even-numbered and odd-numbered lines. The entire 525 lines were painted 30 times per second in two fields of 262.5 lines, but the first 21 lines of each field, called the vertical blanking interval (VBI), weren't part of the visible picture. The experimental NBS time code was broadcast on line 17 of the VBI by use of BPSK modulation. It consisted of two 32-bit numbers that were sent on the first and third transmissions of line 17 after the start of each second. The time code contained the UTC hour, minute, and second, as well as the fractional part of the second with 1 μs resolution. The other 58 transmissions of line 17 contained a 2 MHz sine wave that phase locked a quartz oscillator inside the time code receiver [78].

If the NBS vision had been fulfilled, RCCs based on TV signals would likely have appeared in households more than a decade before the first clocks controlled by LF signals (Section 10.2.6). As fate would have it, it didn't happen, but the NBS efforts led to an important new feature that was eventually added to all TVs. This feature, called closed captioning, allows a text transcription of the audio portion of a TV program to be displayed on screen. Closed captioning resulted from successful nationwide tests of the *TvTime* system that were conducted from the American Broadcasting Corporation (ABC) network center in New York City in 1971. While successfully demonstrating their system's ability to send a time code, NBS also demonstrated their system's ability to send text messages to other ABC affiliates located across the country.

Despite the impressive results, the management at both NBS and ABC had concerns about the long-term expense and reliability issues involved with the continuous distribution of accurate time, but ABC quickly recognized the utility of using the VBI to send text. For years, organizations representing an audience of about 20 million hearing-impaired persons had been asking the TV networks to provide captioning for their programs, and the technology developed by NBS provided the networks with a solution [80]. From that point forward, ABC and the other TV networks were more interested in captioning technology than time distribution, and NBS turned its attention to other forms of time distribution, including signals relayed by satellites (Section 10.6). In recognition of its efforts, NBS, along with ABC and the Public Broadcasting System (PBS), received an Emmy award for its work on closed captioning technology in 1980.

Closed captioning eventually was standardized as part of the Extended Data Service (XDS), a service transmitted in line 21 of the VBI. This service also eventually included a time code containing UTC, the date, DST information, and a time zone offset [81]. By mutual agreement within the TV industry, the XDS time code was only inserted by PBS stations. In the days before DTV and digital recording, the PBS time code was useful for synchronizing the clocks of analog videocassette recorders (VCRs), whose time displays would infamously blink until their clocks were set. The Sony Corporation collaborated with PBS to develop a system where compatible VCRs would scan the channels for a time code, and then synchronize their clocks when the first time code was found. This synchronization system and the XDS itself were made obsolete by the advent of DTV.

The Advanced Televisions Systems Committee (ATSC) developed the DTV standards that have now completely replaced analog television in the United States, Canada, Mexico, and at least six other countries, with similar DTV standards now in use throughout the rest of the world. The original DTV standard in the United States, now called ATSC 1.0, was approved by the Federal Communications Commission (FCC) in 1996 and remains the predominant DTV standard as of 2025. When broadcast as terrestrial OTA TV, the DTV signals use the same VHF/UHF frequencies once used by analog TV, and each channel still has 6 MHz of bandwidth. However, DTV provides many advantages over analog TV, with the most significant advantage being higher resolution video. In addition, the transport layer includes a Program and System Information Protocol (PSIP) that carries both system and program information, similar to the RDS of FM radio (Section 10.5). The system information allows viewers to navigate and have access to the channels, and the program information helps viewers decide what to watch by providing a title and description [82].

The PSIP also includes a system time table (STT) that is transmitted once per second, and the STT includes a digital time code. The ATSC PSIP standard [83] stipulates that the time code as transmitted must be accurate to within ±1 s, and this should be easy to achieve because the STT generator is usually referenced to a GPS disciplined clock. However, signal delays are introduced by the STT generator, the PSIP generator, the various hardware used in the transport stream, and in the broadcast transmitter itself; therefore, the accuracy of a given DTV time code will depend on the uncertainty of the delay compensation provided by the station. The time-of-day information contained in the ATSC 1.0 broadcasts [82, 83], summarized in Table 10.9, essentially follows the format used by GPS. A DTV receiver with an RCC synchronized to the time code is shown in Figure 10.31. As is the case with other types of RCCs, users of DTV clocks have to select their time zone before UTC can be converted to local time.

Table 10.9 Time-of-day information included in the ATSC 1.0 broadcast

Function	Description	Bit length
system_time	The number of seconds that have occurred from the beginning of the GPS epoch (January 6, 1980) to the present. It can be converted to UTC date and time-of-day by applying the **GPS_UTC_offset.**	32
GPS_UTC_offset	The current time difference between GPS and UTC. This value reflects the number of leap seconds that have occurred from the beginning of the GPS epoch (January 6, 1980) to the present, and will change if an additional leap second is added or subtracted.	8
daylight_saving	Indicates whether or not DST is in effect and signals the day and hour for transitions into or out of DST.	16

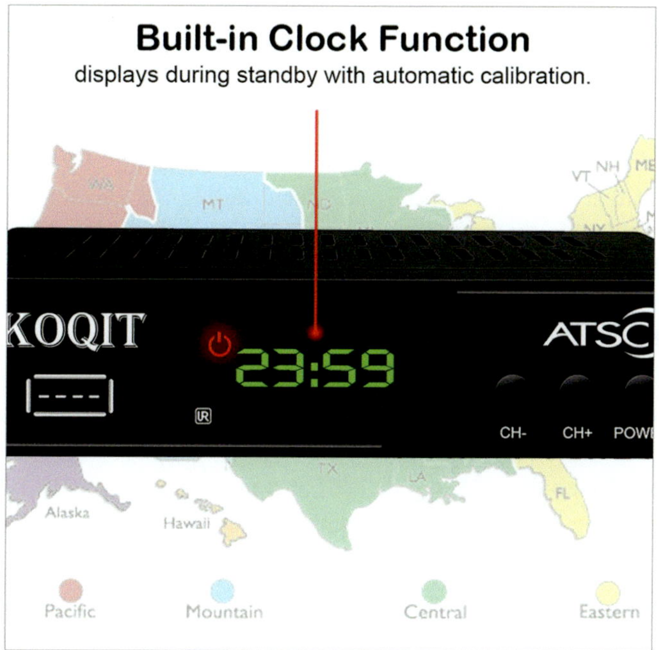

Figure 10.31 An RCC integrated into a DTV receiver (courtesy of Koqit).

The ATSC 3.0 standard, branded as "NextGen TV" and approved by the FCC for voluntary use in November 2017, has different synchronization requirements than ATSC 1.0 and transmits time in a different format. ATSC 3.0 requires its physical layer to be synchronized to a common UTC reference clock, so GPS disciplined clocks are installed at every transmitter site. Therefore, the ATSC 3.0 physical layer time scale, used for all media synchronization, runs at the same frequency as the GPS time scale [84].

Receivers can synchronize to UTC from an ATSC 3.0 OTA broadcast by obtaining time code information from both the physical layer and the ***SystemTime*** element structure delivered in Extensible Markup Language (XML) format by the Low Level Signaling (LLS) layer. The physical layer delivers a 32-bit field containing the number of seconds since 00:00:00 TAI, on January 1, 1970, which is the same epoch used by the Precision Time Protocol (PTP) described in Section 11.2.5. An additional 10-bit field contains fractional seconds in units of milliseconds, with two optional 10-bit fields containing fractional seconds in units of microseconds and nanoseconds, respectively. The conversion from TAI to UTC requires the information contained in the ***SystemTime*** element structure, which is delivered at least once every 5 s, and summarized in Table 10.10 [85].

Table 10.10 Time-of-day information included in the ATSC 3.0 ***SystemTime*** element structure

Name of element	Description	XML data type
@currentUtcOffset	The current offset in whole seconds between TAI and UTC.	short
@ptpPrepend	Includes the upper 16 bits signifying the number of seconds since the PTP epoch of January 1, 1970. This element will be set to 0 until the 32-bit field in the physical layer rolls over, an event that will occur about 136 years after the epoch.	unsignedShort
@leap59	Set to true if the last minute of the current UTC day will have 59 s (a negative leap second). Otherwise set to false.	Boolean
@leap61	Set to true if the last minute of the current UTC day will have 61 s (a positive leap second). Otherwise set to false.	Boolean
@utcLocalOffset	The offset between the time zone of the transmitter location and UTC.	duration
@dsStatus	Set to true if DST is in effect at the transmitter location, set to false if DST is not in effect.	Boolean
@dsDayOfMonth	If a transition to or from DST is scheduled during the current month, this element indicates the day of the month when the transition will occur.	unsignedByte
@dsHour	If a transition to or from DST is scheduled during the current month, this element indicates the hour (in local time at the transmitter site) when the transition will occur.	unsignedByte

The DTV time codes have considerable potential for RCC applications, but their use has been mostly limited to synchronizing the clocks found in TV equipment and systems. However, time metrologists sometimes utilize VHF/UHF TV transmissions as common-view signals (Section 4.3.3) for comparisons of atomic clocks. Common-view TV measurements can achieve sub-microsecond uncertainties, as first demonstrated in the 1960s with analog TV signals [86] and more recently with DTV signals [87, 88]. When this method is implemented, the clocks being compared must each be located within the coverage area of the TV transmitter.

10.6 Time Signals from Geostationary Satellites

Satellites with a geostationary Earth orbit, called GEO satellites, have an orbit period of one sidereal day—23 hours, 56 minutes, and 4.09 s—which is in synch with the period of the Earth's rotation when measured with respect to distant stars. The geosynchronous orbit, coupled with an orbit inclination angle of 0° with respect to the Earth's equatorial plane, makes the satellite stay in a fixed position above the equator. Therefore, when a geostationary satellite is observed from a fixed location on Earth, it will always be located in the same position in the sky. These features immediately attracted the attention of time metrologists when geostationary satellites first appeared in the 1960s, because they allowed line-of-sight time signals to be broadcast from a fixed position high above the Earth's surface. Their direct line-of-sight path provides better stability than terrestrial skywave signals, and their high vantage point provides a larger coverage area than terrestrial groundwave signals.

Geostationary satellites have an orbit radius of 42,164 km with respect to the center of the Earth, and fly at an altitude of 35,786 km above the Earth's equator. Signals broadcast from this altitude reach about one-third of the Earth's surface. To illustrate this, Figure 10.32 shows the coverage area of *EchoStar 105*, a satellite located above the equator at 105° west longitude. From this position, *EchoStar 105* can provide television coverage to all 50 U. S. states.

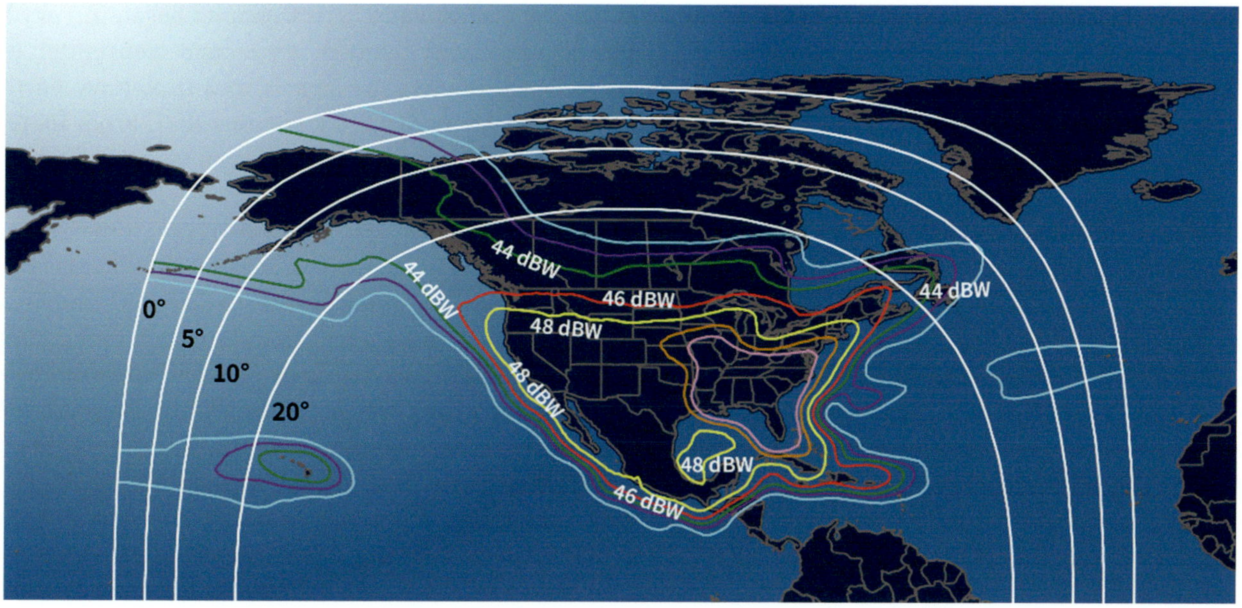

Figure 10.32 Coverage area of the *EchoStar 105* satellite (courtesy of EchoStar).

Because their coverage area is so large, the path delay between the satellite transmitter and a receiver on Earth can vary over a range of about 20 ms, with the delay a function of the receiver's location. The minimum possible path delay is about 119.4 ms. A delay that short would require the receiver to be located at mean sea level on the equator, and for the satellite to be located directly overhead, with an elevation angle of 90° and an altitude of 35,786 km. The maximum possible delay is about 139.3 ms. A delay that long would require the receiver to be located at the extreme edge of the coverage area, where the satellite is barely visible above the horizon, with an elevation angle near 0°. In this situation, the distance to the satellite is about 41,756 km. This path delay always varies at the sub-millisecond level—even when the receiving antenna never moves—because geostationary satellites aren't perfectly stationary. They move about their fixed position by repeatedly tracing a pattern in the sky, called an analemma, that has the shape of an asymmetrical figure-eight. They return to the same position after a period of one sidereal day, but their movement during the day causes the path delay to vary within a range of about 0.2 ms. This delay can be compensated if the satellite broadcasts its current position in addition to the current time.

The concept of a geostationary satellite was introduced by the British science fiction writer Arthur C. Clarke (1917–2008), who is known as the "godfather" of satellite communications [89]. Clarke's lifelong interest in space and space travel inspired him to join the British Interplanetary Society while he was still a teenager, and while still in his twenties, he predicted and explained geostationary satellites in a letter sent to the editors of *Wireless World* in 1945:

> *"An 'artificial satellite' at the correct distance from the earth would make one revolution every 24 hours; i.e., it would remain stationary above the same spot and would be within optical range of nearly half the earth's surface. Three repeater stations, 120° apart in the correct orbit, could give television and microwave coverage to the entire planet."* [90]

Clarke noted in his letter that the possibility of such a satellite was "perhaps half a century ahead," but in October 1945 he published a longer article in *Wireless World* that described the basic principles of a communications satellite with a remarkable amount of detail that was later proven to be mostly accurate [91]. In 1964, less than two decades after Clarke's suggestions, the first geostationary satellite, *Syncom 3*, was launched from Cape Kennedy, Florida. Within two months of this launch, the Olympics began in Tokyo, Japan, and Clarke's prediction of worldwide television coverage was fulfilled. An X-band transmitter in Tokyo sent video coverage of the Olympics to *Syncom 3*, which relayed the signals to a U. S. Navy Earth station in Point Mugu, California, so they could be broadcast by terrestrial TV stations [89]. From there, satellite technology has continued to rapidly advance, and geostationary satellites remain a critical part of most long-distance telecommunication systems.

As we discussed in Chapter 9, some of the national laboratories that operate UTC(*k*) time scales implement the two-way satellite time transfer method (Section 4.3.4), so they rely on GEO satellites as the "relay" that provides their primary link to UTC. Their use as a two-way relay system makes GEO satellites an essential part of time and frequency metrology, but their use as a one-way time transfer system isn't especially common. In the remainder of this section, we'll discuss two types of one-way GEO satellite time signals that can and have served RCC applications—the signals broadcast by UTC(*k*) laboratories, and the signals broadcast by subscription-based radio and television satellite systems. The time signals from GEO satellites that are part of regional navigation satellite systems are discussed in Section 12.9.

NBS was the first UTC(*k*) laboratory to broadcast time codes from geostationary satellites. These broadcasts began as an experiment in August 1971, and they involved rebroadcasting the WWV time signals—including the voice announcements and the digital time code—from the *ATS-3* satellite. The satellite time signal station, known as WWVS, operated on the VHF frequency of 135.625 MHz. Time signals were broadcast for two 15-minute periods on Monday through Friday of each week. This continued for two years, but WWVS never became an NBS service [92].

A long-running NBS satellite time code service began shortly afterwards, enabled by the launch of a National Oceanic and Atmospheric Administration (NOAA) satellite that was designed to monitor and forecast the weather. The satellite, called Synchronous Meteorological Satellite 1 (*SMS-1*), was launched in May 1974, and NOAA allowed NBS to include a digital time code and satellite position data in the satellite uplink message. The future NOAA weather satellites were renamed GOES (Geostationary Operational Environmental Satellites), controlled from an uplink station located in Wallops Island, Virginia. With assistance from NOAA, NBS was continuously broadcasting a time code and satellite position data from a single GOES satellite by the end of 1975, and from two GOES satellites by 1977. The two satellites were respectively located at 135° and 75° west longitude, and their signals were broadcast on two UHF carrier frequencies near 468.8 MHz. A 100 bps data stream was modulated on to the carriers, with eight of every 100 bits assigned to NBS. This allowed a 240-bit time and position code to be sent every 30 s [93].

The GOES time code originated from a clock and TCG at Wallops Island that was remotely synchronized to UTC(NBS) in Boulder. The path delay included the delay of the uplink signal from Wallops Island to the satellite, and the downlink signal from the satellite to a receiver on Earth. Depending on the receiver's location, the path delay could range from 244 ms to 276 ms. To match the midpoint of this range, the TCG was always advanced by 260 ms. This didn't stop the time code from arriving early or late, but it did guarantee that the received time would be accurate to within ±16 ms, even when the actual path delay was unknown.

The first RCCs that could synchronize to satellite signals were designed and built at NBS in 1974, in support of the GOES time code service. The first of these devices worked only in "uncorrected" mode, accepting the time code as received with its associated uncertainty of ±16 ms. Later versions were designed to automatically compensate for the path delay, which was something that no previous RCC could do, and NBS published the circuit diagrams and assembly language source code to encourage commercial companies to build them.

The original NBS design used the 4-bit Intel 4004, the first mass-produced microprocessor, to perform the necessary calculations. The location of the satellite uplink transmitter was fixed and known, so after a user specified the GOES receiver's antenna coordinates—which was done by adjusting thumbwheel switches—the receiver could decode the satellite's position data and calculate the path delay. The result of the path delay measurement was applied to the received time code, which compensated for the error in the 260 ms time code advance [94]. This delay compensation method kept GOES clocks accurate to within ±100 μs of UTC, an accuracy that was mostly limited by the uncertainty of the receiver and satellite position data.

Figure 10.33 shows a satellite-controlled clock built at NBS in 1976, and a promotional satellite clock built that same year to commemorate the United States bicentennial. An estimated 10,000 GOES satellite clocks were sold commercially by several manufacturers before the time code service was discontinued at the end of 2004 [93]. After the time code service was discontinued, the GOES satellites continued to fulfill their primary mission, which is providing NOAA with a continuous stream of environmental data for weather forecasting.

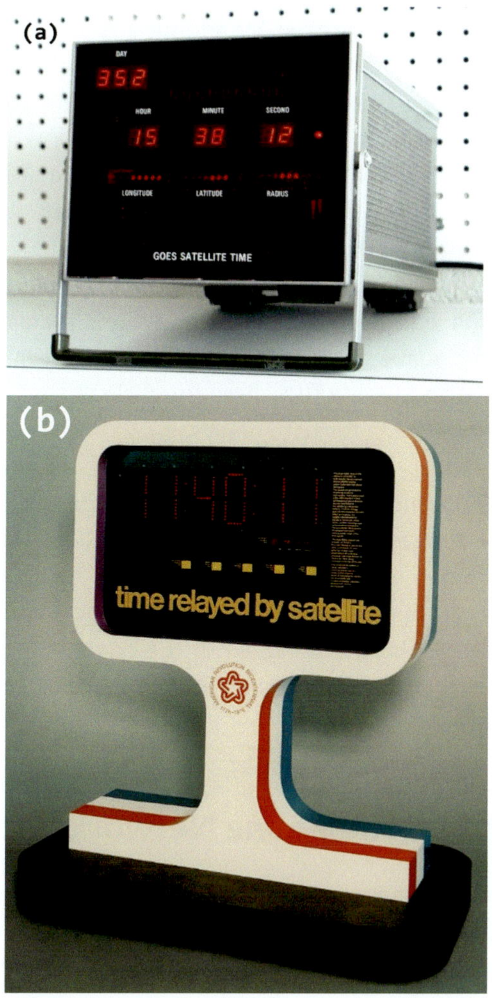

Figure 10.33 A satellite-controlled clock built at NBS in 1976 (a) and a promotional clock built that same year to commemorate the United States bicentennial (b).

A number of other UTC(k) timing laboratories have since experimented with broadcasting time codes from GEO satellites, but their efforts usually haven't led to the establishment of a service. Besides GOES, the most successful GEO satellite time code service was operated by the National Physical Laboratory of India (NPLI) from 1987 until 2006, with the time signals broadcast by the Indian National Satellite System (INSAT). A digital time code synchronized to UTC(NPLI) was originally broadcast from a multipurpose communications satellite named INSAT-1B. This satellite, and the INSAT-1D satellite that replaced it in 1990, were located directly above the equator at 74° east longitude. The frequency-modulated time code was transmitted from a dedicated satellite transponder providing 160 kHz of bandwidth on a frequency of 2599.675 MHz. The time code was sent at a rate of 1 byte per second, and a 14-byte sequence containing UTC information and satellite position data was sent four times every minute [95].

Much of the design work for the NPLI/INSAT time service can be credited to the physicist and time metrologist Amitava Sen Gupta (b. 1953), who was a frequent collaborator with NBS/NIST. Sen Gupta simplified and improved the accuracy of the algorithms used to compute the coordinates of the INSAT satellites from the Keplerian orbital elements provided by the control facility [96]. These algorithms were shared with NIST and integrated into the GOES time code service beginning in 1992 [93]. The INSAT service improved upon the accuracy of the GOES service, partially because the orbital element data for the satellite was updated more frequently, and the received signals were typically accurate to within ±20 μs of UTC(NPLI). This accuracy was further improved by adding a differential correction to the broadcast in 1996. The differential correction was obtained by receiving the time code at the transmitter site, and measuring the time difference between the transmitted and received signals. Users who subtracted the differential correction from their own measurement data could perform a common-view comparison with NPLI, making the INSAT service accurate to within ±1 μs of UTC(NPLI) at all receiving sites in India [97].

Since the demise of the GOES and INSAT services, subscription-based digital television and radio services that are broadcast from GEO satellites have become common, and these services almost always include a time code that can be used to synchronize the receiver's clock. However, the time codes often can't be received unless a paid subscription has been activated, and their format isn't always documented by a published standard, two factors that make them unsuitable for general-purpose RCC applications. Even so, the time codes are usually synchronized with a GPS disciplined clock, and should be accurate to well within ±1 s of UTC, even if there is no compensation for path delays.

Digital satellite television services commonly broadcast on microwave frequencies ranging from 10.7 GHz to 12.7 GHz, but other frequencies are sometimes used. Time codes are embedded in the content broadcast by the various satellite television services, and can include the ATSC time codes described in Section 10.5, or the time code included in the international Digital Video Broadcasting (DVB) standard for digital television. The DVB time code includes the UTC hour, minute, and second, as well as the MJD and a local time offset [98]. Time codes are also embedded in the content broadcast by Satellite Digital Audio Radio Service (SDARS) providers, such as Sirius XM radio in the United States, which occupies the frequencies from 2320 MHz to 2345 MHz.

10.7 Time Signals from Mobile Phone Networks

According to ITU estimates, 97.9% of the world's population was covered by digital phone networks as of 2023, and the nearly nine billion mobile phone subscriptions exceeded the world's population [99]. These staggering numbers indicate that mobile phone signals are by far the most widely used terrestrial-based time signals. The first-generation mobile phone networks, now known as 1G, were analog, but the second, third, fourth, and fifth-generation networks that followed are digital. These networks, respectively known as 2G, 3G, 4G, and 5G, require time synchronized to within ±10 μs or better and are discussed in Section 16.3. This accuracy requirement can usually only be met by installing GNSS disciplined clocks at the mobile phone base stations, with the base stations serving as "repeaters" that retransmit the time signals received from satellites.

Mobile phone signals, transmitted in numerous microwave frequency bands, can preserve much of the frequency accuracy of GNSS signals, and also much of the time accuracy, although some path delay is added as the line-of-sight signal travels from the base station to the phone. However, there are a huge number of base stations, with one or more located within a kilometer of the receiver in most populated areas, which means that the additional path delay is often just a few microseconds. The widespread availability of these signals has prompted some manufacturers to offer clock products that are referenced to mobile phone signals, and these products can provide performance nearly equal to a GNSS disciplined clock without requiring an outdoor antenna [100].

Many devices have their clocks synchronized by mobile phone signals—including wristwatches and computer tablets—but the main beneficiary of these time signals is the ubiquitous smartphone. A smartphone performs an endless variety of functions in addition to being a telephone. For example, a smartphone is a text messaging device, a camera, a video and audio recorder, a navigation device, a music player, a video game machine, a book reader, and a television. Its potential uses are limited only by our imagination because a smartphone includes a programmable, networked computer. This vast array of functions makes it easy to forget that a smartphone is also a radio-controlled clock.

Most smartphones synchronize their clocks to the system time of the mobile phone network. This is often done through a mechanism called Network Identity and Time Zone (NITZ), which allows networks to identify themselves to mobile devices. In addition to an ID, the network also sends time and date information to the mobile device, including the year, month, day, hour, minute, second, DST information, and the local time zone's offset from UTC. This information is sent to a smartphone when it registers on the network, when it relocates to a different time zone, when the transition to or from DST occurs, and whenever the network changes its identity. It can also be sent at any time during the signaling connection between the mobile network and the smartphone [101]. However, no specific rules enforce how often a NITZ update is sent, so the synchronization interval is variable, and there might be extended periods when updates aren't sent at all, as well as periods when a smartphone is outside the range of any network. During these periods, a smartphone's clock accumulates a time error, with the magnitude depending on the frequency accuracy of the clock's oscillator. Despite these limitations, a smartphone's clock is usually accurate to within ±1 s of UTC.

Figure 10.34 A mobile phone application showing the difference between the network system time and the time obtained from an Internet NTP server (courtesy of Timo Partl).

Smartphones can potentially synchronize to sources other than the mobile phone network. For example, all smartphones have internal GNSS receivers that can also supply time information. They also have wireless network capability that allows them to access the Internet and obtain the time from a Network Time Protocol (NTP) server (Section 11.2.4). Software applications are available that compare a smartphone's clock to the time obtained from GNSS or NTP. A screen display from one of these apps is shown in Figure 10.34. In this example, the time difference between the smartphone clock and a more accurate NTP time server is about 0.15 s. However, the time supplied by the mobile network is accurate enough for most users, and the root access to a smartphone's operating system is usually blocked, which prevents the use of GNSS or NTP as a synchronization source.

The smartphone represents the most significant step forward in the evolution of RCCs, with billions of devices in daily use. For many of us, perhaps most of us, the smartphone has replaced the wristwatch as the clock that we rely upon the most. By allowing us "to carry an apparatus in one's pocket" that has its time set by an "electrical wave as we walk about the streets," the smartphone has completely fulfilled the vision that Howard Grubb had more than a century ago [10].

As we've seen in this chapter, the mass production of RCCs has revolutionized everyday timekeeping, but accurate clocks aren't always controlled by radio. Countless numbers of accurate clocks are controlled by time signals sent over wired networks, as we'll discuss in Chapter 11.

References

1. F. Hope-Jones, "Electrical Time-Service," *Journal of the Institution of Electrical Engineers*, vol. 29, no. 142, pp. 119–140, January 1900.
2. F. Hope-Jones, "Modern Electric Time Service," *Journal of the Institution of Electrical Engineers*, vol. 45, no. 202, pp. 49–115, June 1910.
3. R. Wynands, "The Atomic Wrist-Watch," *Nature*, vol. 429, no. 6991, pp. 509–510, June 2004.
4. W. H. Cordell, M. L. Olinger, P. A. Kozak, and A. W. Nyhuis, "Does Anybody Really Know What Time It Is? Does Anybody Really Care?," *Annals of Emergency Medicine*, vol. 25, no. 5, pp. 1032–1036, May 1994.
5. D. Howse, *Greenwich Time and the Longitude*, London: Philip Wilson Publishers, Ltd., 1997.
6. M. A. Lombardi, "Radio Controlled Clocks," *Proceedings of the 2003 NCSLI International Workshop and Symposium*, 18 p., August 2003.
7. M. A. Lombardi, "The Evolution of Time Measurement, Part 5: Radio Controlled Clocks," *IEEE Instrumentation & Measurement Magazine*, vol. 15, no. 2, pp. 49–55, April 2012.
8. G. Weightman, *Signor Marconi's Magic Box*, Da Capo Press, 2003.
9. A. V. Simcock, "Sir Howard Grubb's Proposals for Radio Control of Clocks and Watches, 1898–1899," *Radio Time: The Radio-Controlled Clock and Watch Magazine*, vol. 4, no. 1, no. 10, pp. 18–22, Autumn 1992.
10. H. Grubb, "Proposal for the Utilization of the 'Marconi' System of Wireless Telegraphy for the Control of Public and Other Clocks," *Scientific Proceedings for the Royal Dublin Society*, vol. X, part I, no. 7, pp. 46–49, 1899.
11. G. A. Weber, *The Naval Observatory: Its History, Activities, and Organization*, Baltimore: The Johns Hopkins Press, 1926.
12. J. A. Bloore, "Self Winding Clock Co. of New York: History - 1886–1960s," *Clockdoc.org White Paper*, 19 p., September 2020.
13. A. McCrossen, "Time Balls: Marking Modern Times in Urban America, 1877–1922," *Material History Review*, vol. 52, pp. 4–15, Fall 2000.
14. L. S. Howeth, *History of Communications-Electronics in the United States Navy*, Bureau of Ships and Office of Naval History, U. S. Government Printing Office, 1963.
15. S. J. Dick, *Sky and Ocean Joined, The U.S. Naval Observatory 1830–2000*, Cambridge University Press, 2003.
16. W. H. G. Bullard, "The Naval Radio Service: It's Development, Public Service, and Commercial Work," *Proceedings of the Institute of Radio Engineers*, vol. 3, no. 1, pp. 7–28, March 1915.

17. Department of the Navy, *Annual Report of the Naval Observatory for the Fiscal Year 1916*, Washington: Government Printing Office, 1916.
18. S. C. Hooper, "Naval Communications—Radio Washington," *Proceedings of the Institute of Radio Engineers*, vol. 17, no. 9, pp. 1595–1620, September 1929.
19. A. H. Orme, "Regulating 10,000 Clocks by Wireless," *Technical World Magazine*, vol. XX, no. 2, pp. 232–233, October 1913.
20. H. C. Roters and H. L. Paulding, "Radio Electric Clock System," *Proceedings of the Institute of Radio Engineers*, vol. 18, no. 9, pp. 1537–1559, September 1930.
21. "By Wireless Phone from Arlington to Paris," *The Electrical Experimenter*, vol. 3, no. 8, pp. 393–448, December 1915.
22. T. Simpson and B. Liles, "The History of Wireless Telegraphy at the Eiffel Tower," *IEEE Antennas and Propagation Magazine*, vol. 65, no. 2, pp. 130–135, April 2023.
23. P. Gallison, *Einstein's Clocks, Poincare's Maps*, W. W. Norton & Company, 2003.
24. I. R. Bartky, *One Time Fits All: The Campaigns for Global Uniformity*, Stanford University Press, 2007.
25. F. Honore, "Wireless Time Signaling to Sea from the Eiffel Tower," *Scientific American*, vol. 102, no. 25, pp. 497–498, June 18, 1910.
26. Paris Bureau of Longitudes, *Wireless Time Signals: Radio-Telegraphic Time and Weather Signals Transmitted from the Eiffel Tower, and Their Reception*, English Translation, E. & F. N. Spon, Lmt., 1915.
27. G.-A. Ferrié, "Application of Wireless Telegraphy to Time Signals," *The Wireless World*, vol. 1, no. 5, pp. 297–300, August 1913.
28. "Time Signalling Stations," *The Wireless World*, vol. 8, no. 8, pp. 273–274, July 10, 1920.
29. W. G. W. Mitchell, "Time Signals: The Transmissions from Eiffel Tower," *The Wireless World and Radio Review*, vol. 10, no. 18, pp. 545–550, July 29, 1922.
30. The Synchronome Company Limited, *Untitled Brochure and Instructions for the Horophone*, 8 p., November 1913.
31. R. Kennedy, *The Book of Electrical Installations*, vol. 3, Caxton Publishing Company, 1915.
32. "A Time Signal Receiver," *The Wireless World*, vol. 1, no. 11, pp. 686–687, February 1914.
33. D. Boullin, "Achievements of Gent & Co., Part 1," *NAWCC Watch & Clock Bulletin*, vol. 60/1, no. 431, pp. 52–68, January-February 2018.
34. H. R. Langman and A. Ball, *Electrical Horology*, London: The Technical Press Ltd., 1935.
35. A. E. Ball, "Proposed Wireless Control of Public Clocks," *The Wireless World*, vol. 3, no. 25, pp. 15–20, April 1915.
36. D. J. Boullin, "The Amazing Mr. Ball," *Radio Time: The Radio-Controlled Clock and Watch Magazine*, vol. 15, no. 1, no. 43, pp. 9–38, Summer 2005.
37. "Clock Synchronising by Wireless," *The Wireless World*, vol. 23, no. 3, p. 70, July 18, 1928.
38. P. N. Saveskie, *Radio Propagation Handbook*, Tab Books Inc., 1980.
39. International Telecommunications Union, "Ground-Wave Propagation Prediction Method for Frequencies between 10 kHz and 30 MHz," *Recommendation ITU-R P.368-10*, August 2022.
40. N. Wakai, N. Kurihara, A. Otsuka, K. Inamura, and Y. Takahashi, "Wintertime survey of LF field strengths in Japan," *Radio Science*, vol. 41, no. RS5S13, 7 p., April 2006.
41. M. A. Lombardi, A. N. Novick, J. P. Lowe, M. J. Deutch, G. K. Nelson, D. S. Sutton, W. C. Yates, and D. W. Hanson, "WWVB Radio Controlled Clocks: Recommended Practices for Manufacturers and Consumers," *National Institute of Standards and Technology Special Publication 960-14*, 51 p., August 2009.
42. M. A. Lombardi and G. K. Nelson, "WWVB: A Half Century of Delivering Accurate Frequency and Time by Radio," *Journal of Research of the National Institute of Standards and Technology*, vol. 119, pp. 25–54, March 2014.
43. G. K. Nelson, M. A. Lombardi, and D. T. Okayama, NIST "Time and Frequency Radio Stations: WWV, WWVH, and WWVB," *National Institute of Standards and Technology Special Publication 250-67*, 161 p., January 2005.
44. P. M. Hansen and D. Gish, "WWVB Post Repair: Antenna Performance Measurement," Space and Naval Warfare Systems Center (SSC San Diego), *Technical Report 1834*, December 2003.
45. J. Levine, "Realizing UTC(NIST) at a Remote Location," *Metrologia*, vol. 45, no. 6, pp. S23–S33, December 2008.

46. "WWVB adds Time Code to Broadcast," *NBS Technical News Bulletin*, vol. 50, pp. 15–16, January 1966.
47. J. Lowe and K. Allen, "Increasing the Modulation Depth of the WWVB Time Code to Improve the Performance of Radio Controlled Clocks," *Proceedings of the 2006 IEEE Frequency Control Symposium*, pp. 615–621, June 2006.
48. Y. Liang, O. Eliezer, D. Rajan, and J. Lowe, "WWVB Time Signal Broadcast: An Enhanced Broadcast Format and Multi-Mode Receiver," *IEEE Communications Magazine*, vol. 52, pp. 210–217, May 2014.
49. A. Bauch, P. Hetzel, and D. Piester, 2010, "Time and Frequency Dissemination with DCF77: From 1959 to 2009 and Beyond," *PTB-Mitteilungen*, vol. 119, no. 3, pp. 3–26, 2009.
50. D. Piester, A. Bauch, J. Becker, and A Hoppman, "Time and Frequency Broadcast with DCF77," *Proceedings of the 43rd Annual Precise Time and Time Interval Meeting (PTTI)*, pp. 185–196, November 2011.
51. International Telecommunications Union, "Protection Criteria for Systems in the Standard Frequency and Time Signal Services," *Report ITU-R TF.2487-0*, September 2021.
52. P. Hetzel, "Time Dissemination via the LF Transmitter DCF77 Using a Pseudo-Random Phase-Shift Keying of the Carrier," *Proceedings of the 1988 European Time and Frequency Forum (EFTF)*, pp. 351–364, March 1988.
53. National Physical Laboratory and R. Essen, *The Memoirs of Louis Essen - Father of Atomic Time*, Her Majesty's Stationary Office (HMSO), 121 p., 2015.
54. "NPL Time and Frequency Services: MSF 60 kHz Time and Date Code," Her Majesty's Stationary Office (HMSO), 2 p., 2019.
55. "Time and Frequency Transmission Facilities," *National Institute of Information and Communications Technology (NICT) Brochure*, 8 p., June 2017.
56. P. Feng, Z. Zhang, G. Wu, Y. Bai, and X. Ding, "Low Frequency Time-Code Time Service Method Based on Frequency Sharing," *Proceedings of the SPIE*, Eighth International Symposium on Precision Engineering Measurement and Instrumentation, vol. 8759, pp. 875909/1–8, January 2013.
57. B. J. Laveau, "A New Time Standard," *Leeds & Northrop Technical Journal*, no. 7, pp. 12–17, Fall 1969.
58. D. Lancaster, "Experiment with WWVB," *Radio Electronics*, vol. 44, no. 8, pp. 48–51, August 1973, concluded in vol. 44, no. 9, pp. 98 & 101, September 1973.
59. P. Hetzel and L. Rohbeck, "Digitale Anzeige der vom Sender DCF77 Verbreiteten Amtlichen Zeit," *Funkschau*, vol. 46, pp. 727–730, 1974 (in German).
60. D. A. Bateman, "Time by Radio: A Digital Clock Synchronized to a Broadcast Atomic Time Standard," *Wireless World*, vol. 81, no. 1474, pp. 277–282, June 1975.
61. D. J. Boullin, 'Domestic Radio-Controlled Clocks - Part 1,' *Horological Journal*, vol. 130, no. 12, pp. 7–8, 21–22, June 1988, and "Part 2," vol. 130, no. 13, pp. 8–9, 23, July 1988.
62. D. J. Boullin, "The First Mass-Produced Radio Controlled Clocks: 1986–1989" & "Radio Controlled Wristwatches," *Radio Time: The Radio-Controlled Clock and Watch Magazine*, vol. 17, no. 3, no. 51, pp. 4–16 & 39–48, October 2009.
63. M. Deutch, W. Hanson, G. Nelson, C. Snider, D. Sutton, W. Yates, P. Hansen, and B. Hopkins, "WWVB Improvements: New Power from an Old Timer," *Proceedings of the 31st Annual Precise Time and Time Interval Meeting (PTTI)*, pp. 523–535, December 1999.
64. D. Engeler, "Performance Analysis and Receiver Architectures of DCF77 Radio-Controlled Clocks," *IEEE Transactions on Ultrasonics, Ferroelectrics, and Frequency Control*, vol. 59, no. 5, pp. 869–884, May 2012.
65. M. A. Lombardi, "How Accurate is a Radio Controlled Clock?," *The Horological Journal*, vol. 152, pp. 108–111, March 2010.
66. G. Jacobs and T. J. Cohen, *The Shortwave Propagation Handbook*, Cowan Publishing Corp., 1979.
67. I. Poole, "Radio Waves and the Ionosphere," *QST*, vol. 83, no. 11, pp. 62–64, November 1999.
68. M. A. Lombardi, "Measuring the Frequency Accuracy and Stability of WWV and WWVH," *QST*, vol. 107, no. 3, pp. 33–37, March 2023.
69. K. Collins, S. Cerwin, P. J. Erickson, D. Joshi, N. Frissell, and J. Huba, "Methods for Estimation of Ionospheric Layer Height Characteristics from Doppler Frequency and Time of Flight Measurements on HF Skywave Signals," *EGUsphere*, preprint, 24 p., May 2022.

70. G. K. Nelson, "A Century of WWV," *Journal of Research of the National Institute of Standards and Technology*, vol. 124, article no. 124025, pp. 1–31, September 2019.
71. "Experimental Timing Code Added to WWV Broadcasts," *NBS Technical News Bulletin*, vol. 44, pp. 114–115, July 1960.
72. M. V. Tollefson and R. H. Bloomer, Jr., "More Accurate Time from the Heathkit Most Accurate Clock," *Publications of the Astronomical Society of the Pacific*, vol. 98, pp. 520–523, May 1986.
73. M. M. Thomson, *The Beginning of the Long Dash: A History of Timekeeping in Canada*, University of Toronto Press, 1978.
74. C. C. Costain, H. Daams, J.-S. Boulanger, and R. J. Douglas, "Time Dissemination from the National Research Center of Canada," *Proceedings of the 37*th *Annual Symposium on Frequency Control*, pp. 78–80, June 1983.
75. International Electrotechnical Commission, "Specification of the Radio Data System (RDS) for VHF/FM Sound Broadcasting in the Frequency Range from 87.5 MHz to 108.0 MHz," *IEC 62016*, 169 p., 2015.
76. National Radio Systems Committee, "Specification of the Radio Broadcast Data System (RBDS)," *NRSC-4-B*, 36 p., April 2011.
77. L. Li, L. Sun, G. Xing, W. Huangfu, R. Zhou, and H. Zhu, "ROCS: Exploiting FM Radio Data System for Clock Calibration in Sensor Networks," *IEEE Transactions on Mobile Computing*, vol. 14, no. 10, pp. 2130–2144, October 2015.
78. D. D. Davis, J. L. Jespersen, and G. Kamas, "The Use of Television Signals for Time and Frequency Dissemination," *Proceedings of the IEEE*, vol. 58, no. 6, pp. 931–933, June 1970.
79. D. D. Davis, B. E. Blair, and J. F. Barnaba, "Long-Term Continental U. S. Timing System via Television Networks," *IEEE Spectrum*, vol. 8, no. 8, pp. 41–52, August 1971.
80. S. L. Danielson and D. A. Howe, "Use of the Television Vertical Interval to Broadcast Time for Everyone and Program Captions for the Deaf," *Communications Society*, vol. 10, no. 5, pp. 3–6, November 1973.
81. Consumer Technology Association, "Line 21 Data Services," *ANSI/CTA-608-E S-2019*, 128 p., April 2008.
82. B. J. Lechner, R. Chernock, M. K. Eyer, A. Goldberg, and M. S. Goldman, "The ATSC Transport Layer, Including Program and System Information Protocol (PSIP)," *Proceedings of the IEEE*, vol. 94, no. 1, pp. 77–101, January 2006.
83. Advanced Television Systems Committee, "ATSC Standard: Program and System Information Protocol for Terrestrial Broadcast and Cable," *ATSC A/65:2013*, 142 p., August 2013.
84. Advanced Television Systems Committee, "ATSC Standard: ATSC 3.0 System," *ATSC A/300:2024-04*, 31 p., April 2024.
85. Advanced Television Systems Committee, "ATSC Standard: Signaling, Delivery, Synchronization, and Error Protection," *ATSC A/331:2024-04*, 237 p., April 2024.
86. J. Tolman, V. Ptacek, A. Soucek, and R. Stecher, "Microsecond Clock Synchronization by Means of TV Synchronizing Pulses," *IEEE Transactions on Instrumentation and Measurement*, vol. IM-16, no. 3, pp. 247–254, September 1967.
87. D. Boehm, J. White, S. Mitchell, and E. Powers, "Clock Comparison Using Digital Television Signals," *Proceedings of the 41st Annual Precise Time and Time Interval Meeting (PTTI)*, pp. 319–326, November 2009.
88. T. I. Mondal, J. A. Sherman, and D. A. Howe, "Time Transfer Performance of the Broadcast Positioning System™ (BPS™)," *Proceedings of the International Technical Meeting of the Institute of Navigation*, pp. 88–97, January 2025.
89. D. J. Whalen, *The Origins of Satellite Communications: 1945–1965*, Smithsonian Institution Press, 2002.
90. A. C. Clarke, "V2 for Ionosphere Research? in 'Letters to the Editor'," *Wireless World*, vol. 51, no. 2, p. 58, February 1945.
91. A. C. Clarke, "Extra-Terrestrial Relays: Can Rocket Stations Give World-wide Radio Coverage?," *Wireless World*, vol. 51, no. 10, pp. 305–308, October 1945.
92. D. W. Hanson and W. F. Hamilton, "Satellite Broadcasting of WWV Signals," *IEEE Transactions on Aerospace and Electronic Systems*, vol. AES-10, no. 5, pp. 562–573, September 1974.
93. M. A. Lombardi and D. W. Hanson, "The GOES Time Code Service, 1974–2004: A Retrospective," *Journal of Research of the National Institute of Standards and Technology*, vol. 110, no. 2, pp. 79–96, March-April 2005.

94. J. V. Cateora, D. D. Davis, and D. W. Hanson, "A Satellite-Controlled Digital Clock," *NBS Technical Note 681*, 46 p., June 1976, and "Automatic Path Delay Corrections to GOES Satellite Time Broadcasts," *NBS Technical Note 1003*, 52 p., February 1978.

95. A. Sen Gupta, A. K. Hanjura, P. Banerjee, and B. S. Mathur, "Standard Time and Frequency Signal Dissemination Service via Indian Domestic Satellite INSAT-1B," *Proceedings of the 19th Annual Precise Time and Time Interval Meeting (PTTI)*, pp. 381–398, December 1987.

96. A. Sen Gupta, "A Program to Compute the Coordinates of a Geostationary Satellite for Time and Frequency Broadcast," *IEEE Transactions on Instrumentation and Measurement*, vol. 42, no. 2, pp. 480–489, April 1993.

97. A. Sen Gupta and B. S. Mathur, "Standard Time and Frequency Signal Broadcast Via INSAT—Accuracy Improvements Using Differential Mode," *IEEE Transactions on Instrumentation and Measurement*, vol. 46, no. 2, pp. 212–215, April 1997.

98. European Telecommunication Standards Institute, "Digital Video Broadcasting (DVB); Specification for Service Information (SI) in DVB systems," *DVB Document A038*, 197 p., June 2019.

99. International Telecommunication Union, "ITU DataHub," (https://datahub.itu.int), last accessed in 2024.

100. M. D. Korreng, "UTC Time Transfer for High Frequency Trading Using IS-95 CDMA Base Station Transmissions and the IEEE-1588 Precision Time Protocol," *Proceedings of the 42nd Annual Precise Time and Time Interval Meeting (PTTI)*, pp. 359–368, November 2010.

101. European Telecommunication Standards Institute, "Digital cellular telecommunications system (Phase 2+) (GSM); Universal Mobile Telecommunications System (UMTS); LTE; Network Identity and Time Zone (NITZ); Service description; Stage 1," *ETSC TS 122 042*, ver. 17.0.0, 10 p., April 2022.

Chapter 11

Clocks Controlled by Network Time Signals

"... when the Rabbit actually took a watch out of its waistcoat-pocket, and looked at it, and then hurried on, Alice started to her feet, for it flashed across her mind that she had never before seen a rabbit with either a waistcoat-pocket or a watch ..."

— Lewis Carroll, "Alice's Adventures in Wonderland"

As described in Chapter 10, time signals were sent via telegraph wires about half a century before they were sent by radio. When radio was invented, its first application was the transmission of telegraphic signals, and it was originally called wireless telegraphy. That usage continued without exception for at least a decade after Marconi's early experiments, and it took even longer before "radio" became a common word. This all makes sense, because the wired telegraph and the wireless radio were both invented with the same purpose in mind, which was allowing people to quickly communicate with each other over long distances, without the difficulties and delays involved when messages were physically carried from one place to another.

The primary figure in the development of telegraphy was Samuel Morse (1791–1872), an American inventor and painter. Morse is best known for helping to develop the telegraphic code that bears his name, but he was also heavily involved in the development of the electromechanical devices that were needed to send messages over wires, and in the commercialization of the telegraph industry. Morse had a grand vision of the telegraph's ability to connect people. He predicted in a letter written in 1838 that covering the United States with telegraph wires could provide "with the speed of thought, a knowledge of all that is occurring throughout the land, making, in fact, one neighborhood of the whole country" [1]. Even though his vision seems like a precursor of the modern Internet, it seems unlikely that Morse, nor anyone else, ever thought of the telegraph as something that could be used in homes as a person-to-person form of communication. There were some direct connections between organizations, such as the long-running time signal collaboration between Western Union and the USNO (Section 10.1), but most messages were sent from telegraph office to telegraph office. A person would write out their message and then deliver it to their local telegraph office to be sent. The receiving telegraph office would then deliver the message to the recipient.

As Time Goes By: A Comprehensive Guide to Time and Frequency Measurements
Michael A. Lombardi

ISBN 978-981-5352-28-3 (Hardcover), 978-1-042-01532-0 (eBook)
www.jennystanford.com

DOI: 10.1201/9781042015320-15

In contrast, the inventor of the telephone was thinking about person-to-person communication from the very beginning. When the Scottish-American scientist Alexander Graham Bell (1847–1922) first demonstrated the telephone in 1876, the telegraph industry was well established, with about 8,500 offices located in the United States alone. These offices were connected by over 300,000 km of wire, and they sent more than 30 million telegrams a year. Even so, Bell was thinking about a much larger communications network where a telephone was installed inside every home. Just two years after demonstrating his invention, Bell wrote to investors that the telephone "renders it possible to connect every man's house, office or manufactuary with a Central Station so as to give him the benefit of direct telephonic communication with his neighbors at a cost not greater than that incurred for gas or water." Bell added that just as cities had networks of gas and water pipes, that "cables of telephonic wires could be laid underground or suspended overhead." He predicted that these wires would allow "direct communication between any two places in the city" and that "future wires will unite the head offices of Telephone Companies in different cities," making it possible for people located far apart from each other to "communicate by word of mouth" [2].

It took some years, of course, for the telephone industry to develop and grow, but Bell's predictions all came true during his lifetime. More than 20% of U. S. households had telephones by 1910, and the number increased to nearly 40% by 1920. The first coast-to-coast call, from New York to San Francisco, was made by Bell himself in 1915 [2]. As we might guess, one of the first uses of the telephone was placing a call to ask for the correct time.

11.1 Voice Announcements of Time by Telephone

In the early days of telephones, there was no such thing as direct dialing. Calls had to be placed by first picking up the phone and talking to a telephone operator. The caller then asked the operator to connect them to the person they wanted to speak with. This was originally done by giving the operator the person's name or address, but as the number of telephone subscribers increased, it was done by providing a phone number, which originally consisted of a combination of letters and digits. The operator would then physically connect the caller's phone line to the line of the person they wanted to reach by plugging patch cords into a switchboard. A long-distance call was an even more complicated process; it had to go through multiple switchboards and involve multiple operators, and it could take more than 10 minutes for the call to be placed.

Needless to say, telephone operators had a very demanding job. They were required by the phone company to always be polite, exact, and helpful, even when the volume of calls was high and the pace was frantic. Their job was made even harder by callers who would pick up the phone and ask them for the correct time. The helpful operators would, of course, oblige, but the volume of time requests eventually became overwhelming. An announcement published by the New England Telephone Company in 1918 stated that 100,000 calls asking for the time of day were being made daily, and that the practice was being immediately discontinued. This was necessary so that the work force could be "relieved of unnecessary burdens" and could direct their energies "toward the furnishing of effective telephone service, and to that alone" [3]. A similar decision was made in Canada in 1922, when the Bell Telephone Company had their operators stop giving time to the public [4].

It wasn't efficient, nor was it fair, to ask telephone operators to provide the time on demand. However, it was clear that people liked being able to get accurate time through their telephones. In the years before people had access to smartphones, GPS, and the Internet, the telephone was often the only way for them to get accurate time. By making occasional phone calls, people could periodically set the clocks and watches in their homes and keep their lives on schedule. Therefore, the telephone companies soon began offering dedicated numbers that provided a time-of-day service. These numbers were answered by operators who had a single repetitive task—they sat in front of an accurate clock, and they announced the time at a regular interval. Figure 11.1 shows a 1929 advertisement for a time-of-day service offered by the Michigan Bell Telephone Company, where the time was announced every 15 s.

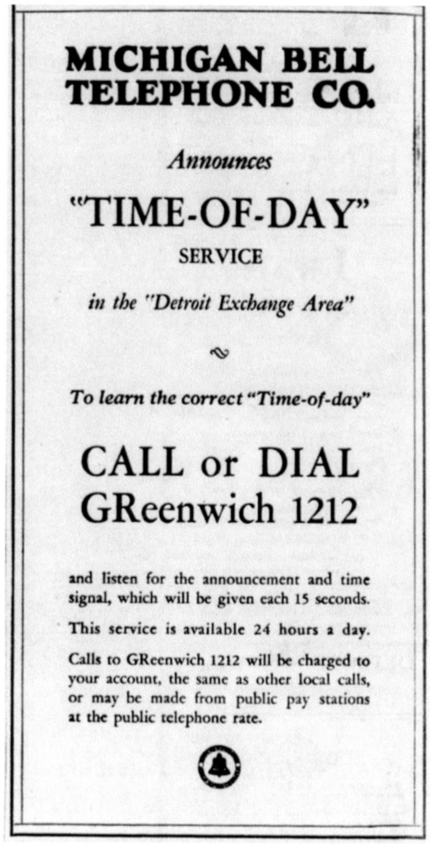

Figure 11.1 A 1929 newspaper advertisement for a telephone-time-of-day service.

Beginning in the 1930s, timing laboratories began offering more accurate telephone time-of-day services where the time wasn't tediously provided by a human operator, but instead automatically provided by a speaking clock. The first speaking clock service began at the Paris Observatory in February 1933, initiated by Ernest Esclangon (1876–1954), a French astronomer and mathematician. Esclangon was appointed director of the observatory in 1929, and he wanted to free himself and other staff members from the responsibility of answering calls from citizens asking for the exact time. The constant deluge of calls not only put stress on the staff, but it also prevented the telephone from being used for other business. Therefore, Esclangon had the French clockmaking company Brillé build a clock that could automatically announce the time during any hour of the day or night. He then worked with the local phone company to connect the speaking clock to telephone lines. On its first day of service, 140,000 calls were made to the speaking clock, although only a small percentage of the calls were connected. More phone lines were added, and the service quickly began handling about 250,000 calls per month [5].

The speaking clock, pictured in Figure 11.2, utilized an electromechanical control circuit to reference the rotational rate of a motor-driven cylinder to the Paris Observatory's pendulum clock. The cylinder contained 90 strips of pre-recorded "talking film," separated into three sections—one for the hour, one for the minutes and seconds, and one for the other words used in the time announcements. The sound from the strips in each section was read and amplified in sequence by three photoelectric cells [5]. The original voice recordings for the time announcements were provided by the French radio host and actor

Marcel Laporte (1891–1971), whose voice continued to be used until 1965. The service continued, using more advanced speaking clock systems and other voice announcers, until July 2022, when the volume of calls had dropped to just a few thousand per month. During the almost 90-year period of the Paris Observatory's speaking clock service, speaking telephone clocks became common. Numerous telephone time-of-day services were offered by national timing laboratories, local telephone exchanges, and by private companies that provided other messages along with the time announcements, including advertising and weather forecasts.

Figure 11.2 The original speaking telephone clock (courtesy of the Paris Observatory).

Telephone time-of-day services continued to increase in popularity until about the late 1970s. In 1974, which was near the peak of telephone time service usage, about four billion (4×10^9) telephone calls were made in the United States to obtain time-of-day or weather information, and it was estimated that about half of those calls were made just to obtain the time. About 1,000 speaking clock machines were then in operation. Most of the machines were synchronized to NBS radio station WWV, and most were located in telephone company central offices [6].

The drop in calls to telephone time-of-day services likely began when the ownership of quartz clocks and watches became common. Quartz clocks didn't need to be set as often as the clocks that preceded them, not only because they were more accurate, but also because they were battery-powered and seldom stopped. This differed from mechanical clocks, which needed to be resynchronized whenever someone forgot to wind them. The drop in calls was even more dramatic when other reference sources of accurate time became widely available. For example, when mobile phones began to replace the telephones connected to land lines, it wasn't necessary to make a call to get the time, because the time was always provided by the mobile phone's display.

Even though many other sources of accurate time are now available, some telephone time-of-day services are still operational as of 2025, and Table 11.1 lists the United States phone numbers of the NIST and USNO services. The time transmitted by these services is usually accurate to less than 1 ms, but the accuracy of the time received by the caller depends on the delays introduced by the telephone circuit that was used to make the call. For example, callers in the continental United States using ordinary land lines can usually expect signal delays of less than 30 ms. For wireless mobile phone networks or Voice over Internet Protocol (VOIP) networks, the delays can be larger, but should be less than 150 ms to meet ITU guidelines for avoiding audio distortion [7]. In the rare situations where long-distance calls are relayed by geostationary satellites, the delay will exceed 250 ms.

Table 11.1 Examples of telephone time-of-day numbers that are referenced to UTC

Agency	Description	Telephone Number
NIST	UTC announced every minute, audio from WWV radio broadcast.	(303) 499–7111
	UTC announced every minute, audio from WWVH radio broadcast.	(808) 335–4363
USNO	Announcements of both UTC and Eastern Time are provided every 15 s.	(202) 762–1401 (202) 762–1069
	Announcements of both UTC and Mountain Time are provided every 15 s.	(719) 567–6742

11.2 Synchronization and Control of Computer Clocks

The title of this chapter makes references to clocks that are controlled by network time signals. The telephone time-of-day services that we've just described were used for decades to manually synchronize countless numbers of clocks, but they didn't establish a formal control system. The first large group of clocks to be controlled by network time signals were the clocks found inside personal computers (PCs). Beginning in the 1980s, PCs, known in a network sense as clients, began to synchronize their clocks with accurate time codes distributed by servers. The network time codes allowed computer clocks to be controlled in much the same fashion as a radio-controlled clock or a disciplined oscillator.

Early PCs, including the Apple II and the original IBM PC, didn't include clock hardware as standard equipment. Rudimentary timekeeping functions were provided by the operating system (OS), which kept time by counting the interrupts generated by a programmable interrupt timer (PIT). However, the OS clock couldn't keep time when the PC was turned off, so it needed to be reset each time the computer was turned back on. This created a demand for optional battery-backed real-time clock (RTC) boards that could be plugged into a PC's expansion slots. The RTC kept time while the computer was turned off, and the OS clock was synchronized to the RTC when the computer was turned back on.

An RTC became standard equipment in IBM-compatible PCs in 1984, when the IBM PC/AT was introduced. The original RTC was based on the Motorola 146818 and had a resolution of 1 s. The PIT was based on the Intel 8254 and had a finer, but still coarse resolution of about 54.9 ms. It's interesting to note that the PIT resolution was directly related to the 3.579545 MHz color burst frequency first used by color televisions (Section 2.4). To allow the generation of color video, the main PC oscillator ran at 4× the color burst frequency, which was 14.31818 MHz. This frequency was fed to the 8254, which ran one of its channels at 1.193182 MHz, equal to 1/3rd of the color burst frequency. Each cycle of the 8254 frequency channel decremented a 16-bit programmable counter, which produced an interrupt each time it rolled over, resulting in an interrupt frequency of about 18.207 Hz. The 54.9 ms resolution of the OS clock was the period of the interrupt frequency.

As the speed of microprocessors increased, the interrupts were generated at faster rates. This led to a corresponding increase in PC clock resolution. For example, an OS kernel that handles 1,000 interrupts per second can update its time variables every millisecond. Modern microprocessors usually have a built-in clock cycle counter that can interpolate the interval between interrupts, allowing some OS clocks to have 1 μs, 100 ns, or even 1 ns resolution. However, the least significant digits of the time messages output by a computer clock often contain no usable information, because the RTC and PIT frequencies are usually derived from inexpensive quartz oscillators that aren't stable enough to support the clock's resolution. If the temperature of these oscillators is closely regulated, they might be able to maintain frequency accurate to $\pm 1 \times 10^{-6}$, allowing the clock to keep time that's accurate to within about ±0.1 s per day [8]. However, computer circuits are often subjected to large temperature variations, making it common for free-running computer clocks to gain or lose multiple seconds per day, which is why frequent synchronization is necessary.

The epochs for the OS time scales vary. For example, iOS, the OS used by Apple's mobile devices, has a time scale epoch of January 1, 2001, but Microsoft Windows uses January 1, 1601, a date that goes all the way back to the first century after the Gregorian calendar was established. The best-known epoch for a computer OS is midnight UTC on January 1, 1970. That's the epoch for Unix time, which originated with the Unix OS, but has since been inherited by many software applications. Unix time is often encoded as a signed 32-bit integer, which only allows a range of about ±68 years with respect to the epoch. Therefore, Unix time will roll over and go back to 0 on January 19, 2038, causing a potentially huge problem for systems that aren't updated prior to the rollover. Other OS clocks—including later variations of Unix—use more bits to encode the time and will avoid rollover problems for the foreseeable future.

11.2.1 Network Asymmetry

We previously described the first principle of time transfer—the accuracy of time transferred from a reference clock to a remote clock can be no better than the uncertainty of the path delay measurement (Section 4.3.1). This principle holds true when computer clocks are synchronized over networks. Computer clocks can be coarsely synchronized with one-way time transfer methods, where the network delay is ignored because there is no easy way to measure it, but most synchronization protocols are based on the loop-back time transfer method described in Section 4.3.2.

All loop-back methods measure the round-trip delay (*RTD*), which is the sum of two network delays—the delay from client-to-server and the delay from server-to-client (Figure 11.3). The *RTD* is then divided by two to obtain the one-way path delay. If the network connecting the server and client had perfect symmetry, the $\frac{RTD}{2}$ method would measure the one-way path delay with zero uncertainty.

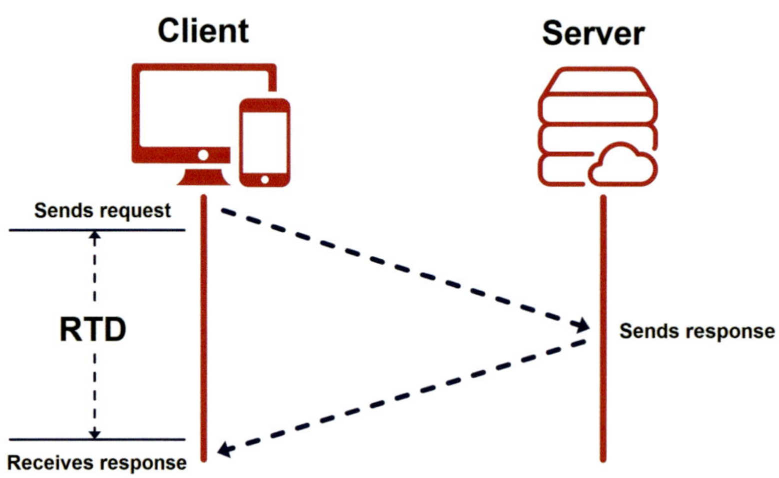

Figure 11.3 Measuring round-trip delay between a client and server.

Symmetry, derived from the Greek word *symmetros*, can be defined as a balance in dimensions, proportions, or measures. For example, a symmetrical object has equal dimensions on both sides. This often makes the object more pleasing to the eye, and symmetry is often equated with beauty and harmony in both art and nature. In timekeeping networks, symmetry can be equated to low measurement uncertainty, because if the network paths for signals traveling in both directions are balanced, the client-to-server and server-to-client path delays will be equal.

Unfortunately, real networks can never achieve perfect symmetry. Instead, real networks are asymmetric, which means that path delays from client-to-server and server-to-client aren't equal, and that the $\frac{RTD}{2}$ method of measuring one-path delay will always result in some measurement uncertainty. Network asymmetry is usually caused by one signal path being longer than the other. The difference in the path lengths can be fixed and permanent, introduced by the way the network was configured. Or, it can be variable and temporary, introduced by the way that the signals are currently being routed. Network asymmetry can even occur when the two signal paths have equal lengths. An example would be a network where the signals traveling in opposite directions pass through different types of cables or hardware, with different velocities of propagation (Section 4.1.3). Another example would be the common situation where network traffic is heavier in one direction, causing the signals traveling in that direction to be subjected to greater delays.

In all cases, the time uncertainty contributed by network asymmetry will be greater than 0% of the *RTD*, but will be theoretically limited to 50% of the *RTD*. Because the time uncertainty is always a percentage of the *RTD*, network connections with smaller *RTD*s will usually have smaller uncertainties. This means that a synchronization performed via a LAN usually has less uncertainty than a synchronization performed via a WAN. However, even if the server and client both reside on the largest WAN, the public Internet, the percentage difference in delays between the client-to-server and server-to-client paths is typically less than 5%. This perhaps sounds small, but it can cause the uncertainty to exceed 1 ms if the *RTD* is sufficiently large. The time uncertainty introduced by a given percentage of differential delay can be calculated as

$$U_{\mathrm{NA}} = \pm \frac{RTD}{\frac{400}{PDD} + 2}, \tag{11.1}$$

where U_{NA} is the time uncertainty due to network asymmetry, *RTD* is the round-trip delay, and *PDD* is the estimated percentage of differential delay. If one-path delay is 5% greater than the other, so that $PDD = 5$, then U_{NA} will be about ±1.2 ms for an *RTD* of 100 ms. In this example, dividing by two estimates the one-way path delay as 50 ms, but the actual one-way delay is either 48.8 ms or 51.2 ms, reflecting the 5% difference.

The actual amount of network asymmetry can't be determined from a series of measurements made between a client and server clock, even if an infinite number of round-trip measurements were made [9]. The only way to truly know the difference in delay between the two one-way paths would be to calibrate each path separately, using an independent time transfer system with sufficiently low uncertainty.

Asymmetry is usually the dominant source of uncertainty when synchronizing clocks over networks. It's so dominant that in most cases we can only be certain of one thing—when the client-server time difference exceeds 50% of the *RTD,* we know that at least part of the time difference is due to a client or server clock offset. In other cases, the presence of a clock offset can't be easily proven, because the entire client-server time difference could possibly be due to network asymmetry.

With these basic concepts in mind, we'll continue by describing some time codes and protocols commonly used to synchronize computer clocks, including methods implemented via ordinary telephone lines, WANs, and LANs.

11.2.2 Telephone Time Codes

The Internet became publicly available in the early 1990s, but most PCs didn't have access until at least a few years later. In the pre-Internet era, computer clocks were synchronized by connecting to time servers via ordinary telephone lines. The connection was made with a telephone modem, the same devices that were then used to connect PCs to larger mini and mainframe computers, and that were later used to establish low-speed Internet connections.

An early leader in the distribution of time codes by telephone was the NRC in Canada, which began distributing time codes compatible with the Bell 103 modem standard in the 1970s [4]. This same time code format is still broadcast by radio station CHU (Section 10.3.2). Beginning in the mid-1980s, NRC began offering an improved telephone time code service using time servers designed by Leitch Video, a manufacturer of broadcast television equipment. The Leitch servers phase locked an OXCO to a 5 or 10 MHz signal from an atomic frequency standard, and synchronized their transmitted time codes to within ±20 μs of a 1 pps input signal from UTC(NRC). The servers produced time codes in two formats—the SMPTE time code used by the television industry (Section 4.2.1), and the NRC time code, which was transmitted as ASCII (American Standard Code for Information Interchange) characters. The servers also supported a loop-back echo test for the NRC time code, which potentially allowed client computers to measure the *RTD*, and to apply a one-way delay correction to the received time [10].

A number of other NMIs launched telephone time code services referenced to UTC(*k*) time scales during the 1980s and 1990s. One example is the NIST Automated Computer Service (ACTS), which began operation in 1988. The ACTS servers, which were designed in-house at NIST, sent a shortened version of its ASCII time codes every two seconds to clients who connected at 300 baud, and a full ASCII time code every second to clients who connected at speeds of 1200 baud or faster.

The full ACTS time code was defined as a string of 50 ASCII characters that looks like this when displayed on a computer screen:

60399 24-03-30 00:31:56 50 0 +.0 045.0 UTC(NIST) *

From left to right, the time code contains the MJD, the UTC date and time, a DST indicator, a leap second indicator, a DUT1 correction, the OTM advance (in milliseconds), a label indicating that the time was referenced to UTC(NIST), and an OTM.

The two most interesting things about the ACTS time code are the OTM advance and the OTM itself, because both are elements of the loop-back method of time transfer. In the sample time code just provided, the OTM advance is 45.0 ms and the OTM is an asterisk (*). The OTM advance indicates that the server has sent the time code 45 ms early. When ACTS was designed, 45 ms was selected as the default value for the OTM advance based on experiments that NIST conducted with 1200 baud modems. This allowed 8 ms for the transmission of the OTM, 7 ms for the path delay to a typical user in the United States, and 30 ms for the differential delay inside the modem. The default OTM advance was just a rough estimate that didn't adapt well to all situations. For example, the path delay could be smaller or larger than 7 ms. In addition, when modem speeds increased to 9600 baud, the OTM transmission time dropped to 1 ms, but the differential delays in modems became less stable and often exceeded 30 ms. Therefore, the uncertainty of the time received by the client was improved by the default OTM advance, but the uncertainty could still be large, and it was often more than 10 ms.

This situation was improved by the loop-back method, which the ACTS servers implemented if the client software returned the OTM at least three consecutive times. The server marked the time, T_1, when the OTM was sent, and the time, T_2, when the OTM came back, estimating the *RTD* as $T_2 - T_1$. If three consecutive readings showed that the *RTD* was stable, the default OTM advance was changed to a value of $\frac{RTD}{2}$ and the OTM was changed from an asterisk to a pound sign (#). When this occurred, the uncertainty of the OTM received by the client was still limited by network asymmetry, but was likely reduced to about ±1 ms [11–13].

Telephone time codes work well and are still accessed by a small number of client computers that need to be physically isolated from unsecured networks like the public Internet. However, they are mostly obsolete, and have been replaced by time codes distributed via the Internet, as described in the following sections.

11.2.3 The First Internet Time Code Protocols—Daytime and Time

The Internet—along with GPS—ranks as one of the world's two most important time distribution systems. The time transmitted by GPS is far more accurate, and a very high percentage of Internet time servers are synchronized with GPSDCs. However, due to the billions of networked devices that make periodic time requests, the Internet is likely responsible for more direct clock synchronizations than GPS. Before we describe two of the first Internet time code services, the Daytime and Time protocols, it's worthwhile to take a brief look at how the Internet was developed and how it works.

The beginnings of the Internet trace back to 1969, when the Advanced Research Projects Agency (ARPA)—an obscure research arm of the United States DoD—began funding an academic research project to build a WAN to connect computers. The network built as the result of this project was originally called the ARPANET, but is known today as the Internet.

The Internet was designed to meet a few basic requirements. The network was packet-switched, which meant that the chunks of data exchanged by network computers, known as packets, weren't required to always travel over the same path. The computers were each identified by unique network addresses, but the packets they sent and received were allowed to travel over any usable path between the sender and receiver. Packet switching was an extremely important requirement, especially for a communications network funded by the DoD. If a network path was destroyed or made unavailable by an attack, the packets that normally traveled along that path could be rerouted, reaching their destination via a different path.

Another basic design requirement was to allow all different kinds of computer systems to communicate with each other. We now take this capability for granted, but the networks that predated the ARPANET were highly proprietary, and often only worked with equipment made by a single manufacturer. A proprietary network wasn't suitable for the DoD, which owned many incompatible computers and networks, and needed a way to connect them all together. Finally, the Internet was required to be a long-haul network that allowed computers separated by long distances to exchange data and communicate. At first the networked computers were scattered across just a few locations in the United States, but they were eventually scattered all over the world [14].

Beginning in late 1969, and continuing to the present day, every aspect of the inner workings of the Internet has been extensively documented by freely available publications called RFCs, an acronym for Request for Comments. The original RFCs were typewritten documents circulated amongst ARPA researchers working on a particular aspect of the network, but today they are formally published by the Internet Engineering Task Force (IETF). A simplified model of the four basic layers of the Internet's architecture, each well documented by the RFCs, is shown in Figure 11.4.

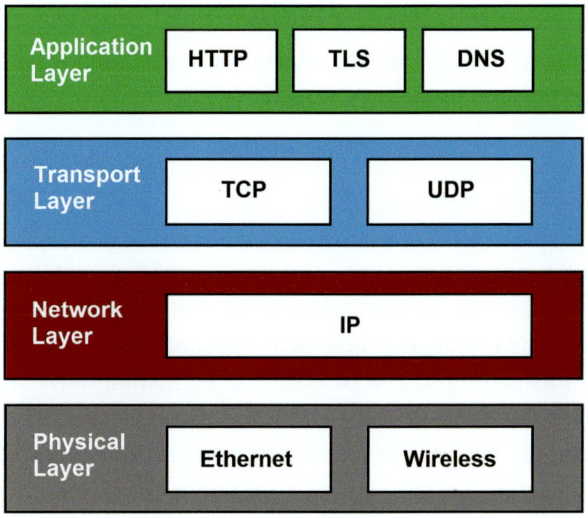

Figure 11.4 The four basic layers of the Internet's architecture.

The bottom layer of the Internet's architecture is called the link or physical layer. It includes the physical transmission medium that the Internet uses to send and receive signals. For example, the physical layer can include wired signals sent via Ethernet connections, wireless signals sent via wireless local area network (WLAN) connections, or as is usually the case, a combination of multiple types of physical links.

The network layer is where the Internet Protocol (IP) is implemented. This protocol assigns a globally unique address—called an IP address—to every host on the Internet, and handles the intricacies of directing the packets sent between senders and receivers. The transport layer's job is ensuring the reliable end-to-end transport of the packets, so it includes mechanisms for both error detection and error recovery. The two main protocols utilized by the transport layer are the Transmission Control Protocol (TCP) and the User Datagram Protocol (UDP). Both protocols use the IP layer as a common base, so they are respectively called TCP/IP and UDP/IP. They differ from each other in a key respect—TCP is connection-oriented and UDP is connectionless. TCP is much more reliable because it negotiates and establishes a connection between the sender and receiver before any data are actually sent. It obtains an acknowledgement when data are successfully transmitted, and it resends missing or erroneous data. UDP is faster because it doesn't go through the complex process of negotiating communication channels or data paths before it sends data. Instead, it sends data without establishing a prior connection with the receiver, and without checking afterwards to see if the data were actually received. This makes UDP unsuitable for any type of guaranteed delivery service [14, 15].

The application layer is the top layer of the Internet's architecture, and it's where the software applications that provide Internet services are located. The applications run on servers that continuously "listen" for user requests for specific services. These services are uniquely identified by the port numbers that they've been assigned. The port numbers have 16 bits, so they can have values ranging from 0 to 65,535. However, the best-known and most commonly requested Internet services usually have port numbers in the range from 0 to 1,024. The best-known example is the Hypertext Transfer Protocol (HTTP) which provides the foundation for the World Wide Web (WWW). Requests for unencrypted HTTP connections are made through port 80, but requests for HTTPS connections, meaning those encrypted with Transport Layer Security (TLS), are made through port 443. Another example is the Internet's domain name system (DNS), which translates the pure text names that identify Internet web sites to their numerical IP address. Requests to DNS servers use port 53 for an unencrypted connection, or port 853 for a connection encrypted with TLS.

The American computer scientist Jonathan Postel (1943–1998) was a central figure in the development of the Internet's architecture, including its first time code protocols. Postel began working on ARPANET protocols in late 1969 when he was hired as a postgraduate research engineer at the University of California in Los Angeles. He held numerous other positions throughout his career, while at the same time continuously serving—from 1969 until his death in 1998—as the editor of the RFC documents that define how the Internet works. In the RFC document that originally defined TCP, he wrote the words now known as Postel's law:

> *"TCP implementations should follow a general principle of robustness: be conservative in what you do, be liberal in what you accept from others."* [16]

Postel was indicating that network messages should always be sent in a technically correct format, but should always be accepted as long as the meaning of the message was clear. His words succinctly captured the inner workings and the essence of the entire Internet.

In 1983, Postel brought timekeeping to the Internet by publishing two documents—*RFC 867* and *RFC 868*—that defined the Daytime and Time protocols. The Daytime protocol can be implemented as either a TCP or a UDP service. In either case, the server listens for a request on port 13 and returns a time code when a request is received. The RFC indicated that the Daytime code didn't have to follow any particular syntax, but it recommended the use of ASCII characters, and stated that it should fit on one line. Because it was sent as ASCII characters, the Daytime code could be read by humans when it appeared on a computer screen, just like the NRC and ACTS telephone time codes that were described in the previous section. Two examples were given in the RFC—each containing the hour, minute, second, month, day, and year, and each indicating the local time zone [17]:

Tuesday, February 22, 1982 17:37:43-PST

02 FEB 82 07:59:01 PST

The Time protocol can also be implemented as either a TCP or a UDP service. In either case, the server listens for a request on port 37. When a request is received, a time code is sent in the form of a 32-bit binary number that represents the number of seconds since midnight on January 1, 1900 GMT—which was the reference time scale in effect at that epoch. The client software converts the 32-bit number to the UTC time and date, which can be converted to local time if the time zone is known. The unsigned 32-bit number has a range of about 136 years, so the time code will roll over and go back to 0 on February 7, 2036 [18]. This rollover will occur about two years before the Unix system time rollover mentioned in Section 11.2.

The Daytime and Time protocols are easy to implement and use, but they share several shortcomings. Both are limited to a resolution of 1 s, and both are examples of the simplest form of one-way time transfer, where no attempt is made to either measure or compensate for the path delay. Therefore, the best possible accuracy of the Daytime and Time protocols will be equal to the unknown path delay between the server and the client. In addition, both protocols lack the ability to provide an advance warning of leap seconds. This is usually handled by having the server adjust its clock at the instant when the leap second occurs, and by resynchronizing client clocks immediately afterwards.

Despite their shortcomings, the Daytime and Time protocols can still meet the requirements of some applications, and they are still supported by some Internet time servers, perhaps most notably by the servers that form the NIST Internet Time Service (ITS). The ITS was developed by the American physicist Judah Levine (b. 1940) and began operations in 1993 [19]. It responds to Daytime, Time, and Network Time Protocol (NTP) time requests. However, nearly all of the time requests made via the Internet—including nearly all of the approximately one million (10^6) time requests that the ITS handles every second—are NTP requests. The next section provides an overview of NTP, which is the primary method used to synchronize and control clocks via the Internet.

11.2.4 The Network Time Protocol (NTP)

Many thousands of NTP servers reside on the Internet, some publicly accessible and others having restricted access. These servers are operated by the national laboratories that maintain UTC(k) time scales, by universities, by small private companies and individuals, and by large corporations including Amazon, Apple, Google, and Microsoft. Client software for NTP is built-in to the OS of nearly every computer. If it isn't built-in, open source software is freely available, which makes it easy to add NTP capability to any networked computer. Countless other networked devices synchronize their clocks via NTP, often in a way that is completely transparent to the user. These devices include printers, sensors, medical instruments, appliances, media streamers, and standard networking hardware such as routers and switches. As previously mentioned, the NIST ITS handles about 10^6 NTP requests per second, or more than 86 billion requests per day, and the total number of time requests sent daily to all NTP servers is likely in the trillions.

NTP was invented by David Mills (1938–2024), an American computer scientist and engineer. Mills began working at the Communications Satellite Corporation in 1977, where he was involved in DoD-funded projects related to the ARPANET, and where he soon focused his work on the synchronization of networked computer clocks. According to Mills, NTP evolved from earlier work [20, 21], specifically from the time stamp messages in the Internet Control Message Protocol (ICMP) defined in *RFC 792* [22] and from the previously described Time protocol [18]. But Mills went far beyond all previous computer timekeeping efforts. He was concerned about accuracy, and his protocol included a time code that had more than enough resolution to support any type of OS clock or computer clock hardware. It also implemented the loop-back method to measure and compensate for network path delay.

Mills published the first NTP RFC—now referred to as version 0—in 1985. At this writing, Version 4 is still the most recent implementation, published as *RFC 5905* in June 2010 [24]. Beginning with version 1, all NTP implementations have been based on a hierarchical measurement system that's similar to the traceability pyramid shown in Figure 9.1. In the case of NTP, the clocks residing in the various levels of the measurement hierarchy are defined by their stratum numbers—with the term "stratum" borrowed from geology, where it refers to an individual layer in a formation of rock or sediment.

As shown in Figure 11.5, a stratum 1 server is directly connected to an internal or external UTC reference clock. The reference clock can be a UTC(*k*) time scale, a RCC referenced to a UTC(*k*) time scale, or a GNSS disciplined clock. The servers at lower levels have no direct connection to a UTC reference clock. Instead, they synchronize to another NTP server which can be in the same stratum, but is preferably in a higher stratum. For example, a stratum 2 server synchronizes its clock by connecting to one or more stratum 1 servers, and a stratum 3 server synchronizes its clock by connecting to one or more stratum 2 servers [23, 24]. The computers that operate below stratum 3 can have time uncertainties of tens of milliseconds or larger, and are typically not used to synchronize other clocks. Synchronization to a stratum 1 server is preferred, but even a stratum 1 server can occasionally transmit the wrong time, due to a variety of possible failure modes [25]. To safeguard against these failure modes, some software is designed to request the time from multiple NTP servers, which allows crosschecking the server accuracy before synchronizing the client's clock.

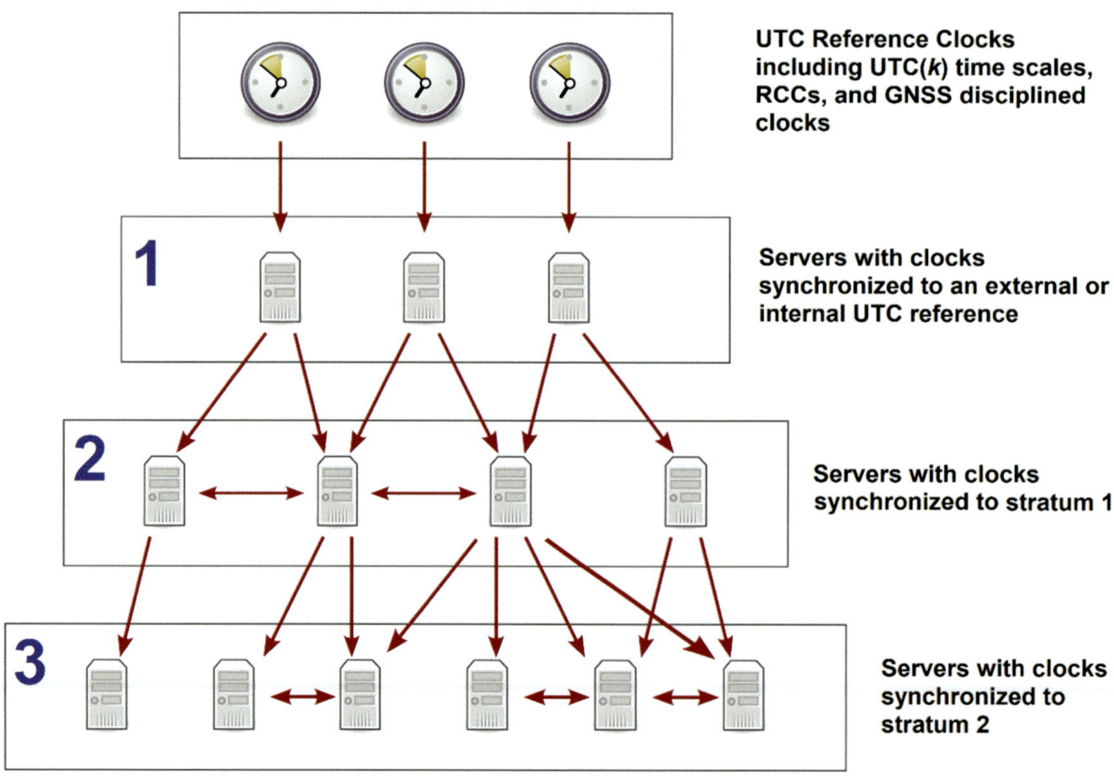

Figure 11.5 The stratum levels of NTP clocks.

The billions of Internet devices whose clocks are synchronized run either a subset of NTP—called the Simple Network Time Protocol (SNTP)—or the full implementation. The devices that connect to just one upstream server, and that have no dependent downstream clients, run SNTP. The full NTP implementation is usually intended for secondary servers, such as the stratum 2 servers in Figure 11.5, that connect to multiple upstream servers and that have multiple downstream servers or clients. However, some endpoint clients seeking the best possible accuracy also run the full implementation by routinely connecting to multiple servers. The full implementation disciplines the system clock's oscillator frequency so that it always keeps time accurate to within a certain threshold, with ±1 ms being a typical goal. This may seem like modest accuracy, but the inexpensive quartz oscillators found in most computer systems have poor frequency accuracy and stability. Therefore, maintaining even millisecond accuracy requires the client software to implement frequency control methods similar to those described in Section 5.3. To get the best results, the client software has to be configured to access an appropriate number of independent servers.

Figure 11.6 is a diagram of the clock disciplining control loop that's implemented in software. The control loop operates as a hybrid of a PLL and a frequency-locked loop (FLL). The PLL disciplining works best over short intervals when the noise is dominated by the instability of the network. The FLL disciplining works best over longer intervals when the noise is dominated by the instability of the clock's oscillator. An algorithm that compares and combines the time codes requested from multiple servers generates θ_r, which serves as the reference clock. The system clock, θ_c, is modeled as a variable frequency oscillator (VFO). A phase detector outputs the phase difference, V_d, between the reference and system clocks. A series of phase differences are then filtered by an algorithm that sends only the best estimates, V_s, to the loop filter. Inside the loop filter, the estimates are converted to predictions of both phase, x, and frequency, y, and these predictions are combined using PLL and FLL techniques to produce V_c, which is the signal used to adjust the clock. The clock adjustments are applied gradually and continuously, which achieves the goal of keeping the clock accurate without having to introduce large time steps [23, 24].

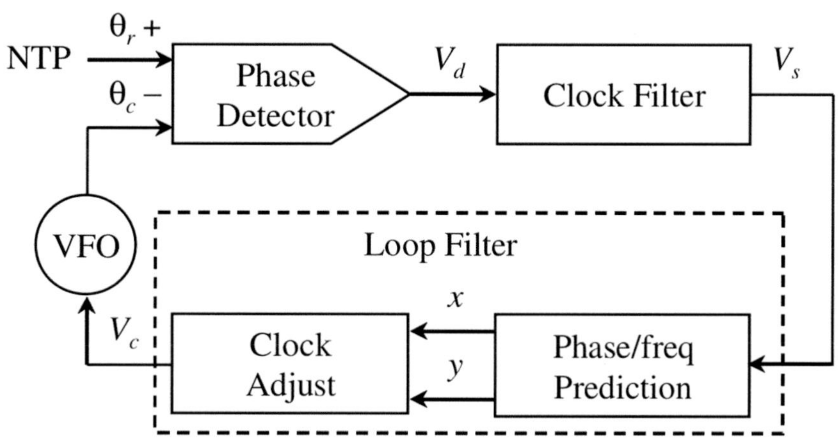

Figure 11.6 The clock disciplining feedback loop implemented by NTP [23].

NTP is always implemented as a UDP service, and the server listens for client requests on port 123. Table 11.2 describes the required fields in an NTP request and response packet, which has a minimum size of 48 bytes. All participating clients and servers exchange packets in this format, regardless of whether they operate in SNTP or full NTP mode. The packet sent by the client can ignore most of the fields, but at the very least it should declare itself as a client by setting the Mode field to 3. It should also record the time of its request by sending the originate time stamp.

Table 11.2 The 48 required bytes in the NTP response message

Byte Number	Field	Size (bits)	Description
1	Leap Indicator	2	00—no leap second warning 01—last minute of the day has 61 s 10—last minute of the day has 59 s 11—alarm condition (clock not synchronized)
	Version Number	3	The version number of the NTP implementation
	Mode	3	3—client 4—server 5—broadcast
2	Stratum	8	0—unspecified 1—reference clock (UTC(k), RCC, or GPSDC) 2 to 15—secondary clock (synchronized via NTP)
3	Poll exponent	8	An 8-bit unsigned integer that indicates the maximum interval between message in a client-server connection where the client clock is being disciplined. The value is expressed in seconds to the nearest power of 2. A typical default value might be 10, where 2^{10} s = 1024 s.
4	Precision exponent	8	An 8-bit signed integer that indicates the resolution of the system clock, in seconds, to the nearest power of 2. For example, a value of -20 indicates a resolution of 953.7 ns (2^{-20} s).
5 to 8	Root delay	32	A 32-bit unsigned number indicating the round-trip delay to the reference clock. The first 16 bits indicate whole seconds, and the last 16 bits indicate the fractional part of a second. The resolution is about 15.3 μs (2^{-16} s), but the root delay can be set to 0.
9 to 12	Root dispersion	32	A 32-bit unsigned number indicating the maximum time offset relative to the reference clock. The first 16 bits indicate whole seconds, and the last 16 bits indicate the fractional part of a second. The resolution is about 15.3 μs (2^{-16} s).
13 to 16	Reference Identifier	32	A four-character ASCII string that identifies the server's reference clock. For example, "GPS" refers to a GPSDC, WWVB refers to a WWVB RCC, NIST refers to UTC(NIST), and so on.
17 to 24	Reference time stamp	64	The local time when the system clock was last set or corrected.
25 to 32	Originate time stamp	64	The time, T_1, when the client sent its request to the server. This time stamp is recorded by the client's clock, and sent to the server as part of the request.
33 to 40	Receive time stamp	64	The time, T_2, when the client's request arrived at the server.
41 to 48	Transmit time stamp	64	The time, $T_{3,}$ when the server sent its response to the client.

The server's response to a client's request includes all of the information shown in Table 11.2, including time stamps sent in the 64-bit NTP time stamp format. In this format, the 64-bit number is unsigned, which means that it's always a positive number. The first 32 bits indicate the number of whole seconds that have elapsed since the NTP epoch of January 1, 1900, and the last 32 bits indicate the elapsed part of the second in progress. The first 32 bits are identical to the Time protocol and they are referenced to the same epoch, so they have the same range of 136 years. This means that the 64-bit NTP time stamps will roll over and go back to 0 on February 7, 2036. To prevent this, version 4 of NTP supports 128-bit time stamps [24]. This solves the rollover problem forever, because 2^{64} s is about 585 billion years. The potential resolution of the 64-bit time stamps is 2^{-32} s, about 233 ps, which should be more than enough resolution for any potential use of NTP. However, the 128-bit time stamps increase the potential resolution to 2^{-64} s, or about 54 zeptoseconds.

The NTP time difference measurement uses the originate, receive, and transmit time stamps described in Table 11.2. The time difference, *TD*, between the client and server clocks is estimated as

$$TD = \frac{((T_2 - T_1) + (T_3 - T_4))}{2}, \tag{11.2}$$

where T_1 is the time when the client made its request (originate time stamp), T_2 is when the client request was received by the server (receive time stamp), T_3 is when the server transmitted its response back to the client (transmit time stamp), and T_4 is when the packet sent by the server is received by the client. Note that the originate time stamp containing T_1 was sent by the client to the server, and the server simply echoes it back. Note also that T_4 is not included in any packet transmission and is only known to the client. Finally, note that the division by two in Eq. (11.2) is based on the standard loop-back method assumption that the one-way network path delay is equal to one-half of the *RTD*.

By using the same four time stamps provided in Eq. (11.2), the client can calculate the *RTD* to the server as

$$RTD = (T_4 - T_1) = (T_3 - T_2), \tag{11.3}$$

where the time required for the server to process the request, $(T_3 - T_2)$, is subtracted from the estimate of round-trip delay, $(T_4 - T_1)$, as measured by the client's clock. This prevents variations in the server's processing time from impacting the *RTD*. For most NTP servers, the request processing time is less than 100 µs. However, it can sometimes be larger, especially if the number of client requests meets or exceeds the server's capacity.

For stratum 1 servers—where the reference clock is expected to be accurate and stable—the uncertainty of an NTP time difference measurement will almost always be dominated by network asymmetry. This is especially true for measurements made on the public Internet. We'll provide several examples, using TD and *RTD* measurements made at NIST in Boulder, Colorado. These measurements were performed by making client requests from a computer with an internal clock that was directly connected by coaxial cable to a 1 pps output from UTC(NIST), with the cable delay measured and compensated to within ±1 ns.

In each of these measurements, the client sent a request to the server every 10 s. At the end of every 10-minute segment, four values were recorded. Two of these values were averages: the average time difference, TD_{AVG}, and the average round-trip delay, RTD_{AVG}, each obtained by averaging the results of 60 client requests. The other two values were obtained from just one of the 60 client requests, the one where the *RTD* was the smallest. When the *RTD* is at a minimum, the time difference value is called TD_{MIN}, and the round-trip delay value is called RTD_{MIN}. Even though TD_{MIN} and RTD_{MIN} were obtained using just 1/60th of the available data, they are assumed to have smaller uncertainties than the averaged values. This is because smaller *RTD* values are likely to be less influenced by network asymmetry. This assumption isn't always true, because it's possible for a network path with a large *RTD* to be more symmetric than a path with a small *RTD*. However, the assumption is true in most cases.

Figure 11.7 shows the results of a 48-hour TD (a) and *RTD* (b) comparison between the UTC(NIST) client and an NTP server referenced to UTC(NPLI) in New Delhi, India. This is a long, transoceanic network path, because Boulder and New Delhi are separated by about 12,400 km. Despite this distance, the TD measurements show that TD_{AVG} is usually between 0 and 4 ms, with a mean value of about 2.3 ms, and one large outlier of about 20 ms. The outlier is explained by the *RTD* measurements, which show an increase of about 40 ms that occurred at the same time. The change in the *RTD* was likely due to a rerouting of network packets, but it apparently lasted for less than 10 minutes, because it doesn't appear in the RTD_{MIN} data.

The values of TD_{MIN} are more accurate and stable than the TD_{AVG} values, always staying within a range of about 1 ms, with a mean value of just 44 µs. This supports the assumption that measurements made over network paths with smaller delays will have smaller uncertainties. It also tells us that even long-distance networks can sometimes be symmetric.

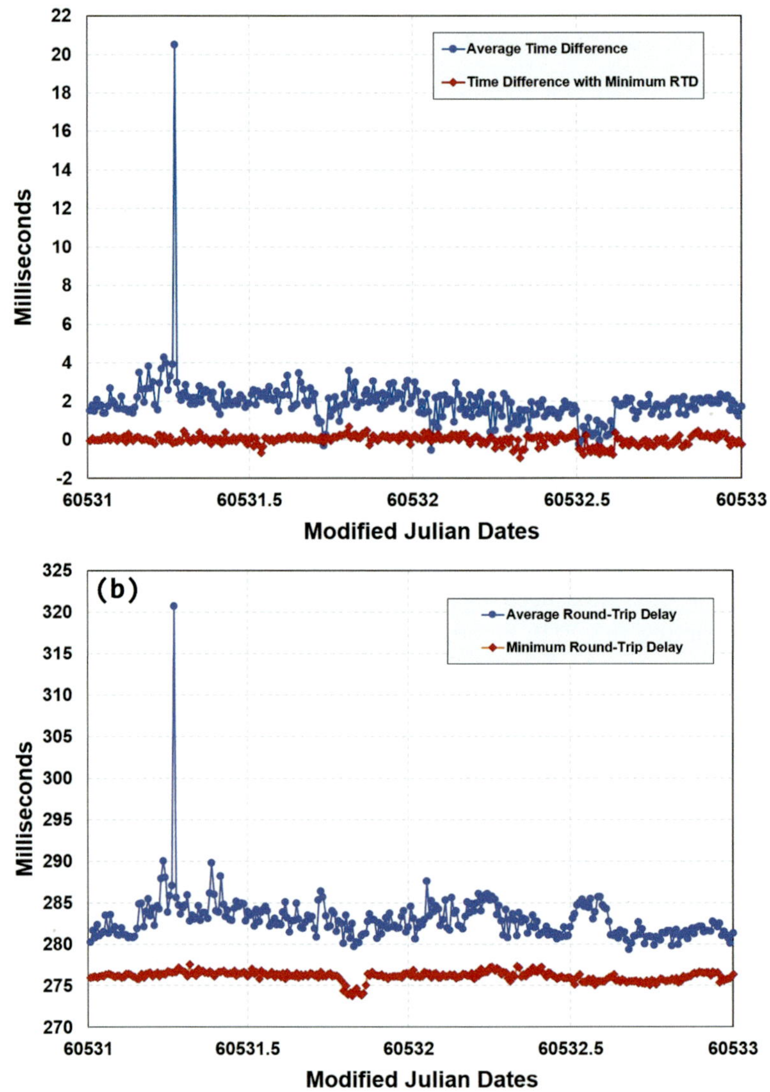

Figure 11.7 NTP measurements of time difference (a) and round-trip delay (b) made between UTC(NIST) and UTC(NPLI).

The examples in Figure 11.7 involve a comparison of two stratum 1 clocks that are known to be accurate and stable, so we can make some estimates about the amount of network asymmetry based on a TD measurement. For example, if TD_{AVG} = 3 ms, then the differential delay between the network paths to and from the server would equal 6 ms. If RTD_{AVG} = 280 ms, we can infer that one path had a delay of 137 ms, and the other a delay of 143 ms, which is larger by 4.4%. However, it's important to reiterate that the NTP loop-back measurements can't reveal which network path has the larger delay. As mentioned earlier in this chapter, determining the source of differential delay in an asymmetric network would require calibrating the two one-way paths individually by using another time transfer system. Any independent path delay calibrations would have to be done repeatedly, because the Internet's speed and routing are constantly changing, and the path delays change accordingly.

Figure 11.8 provides a more typical example by showing the results of a 48-hour TD (a) and *RTD* (b) comparison between the UTC(NIST) client and an NTP server in New Jersey. The server is referenced to a GPSDC that's accurate to within ±1 μs. In this example, averaging the NTP measurements has the effect of smoothing the data, and the TD_{AVG} values are more stable than the TD_{MIN} values. However, the mean results are similar, about -1.1 ms for TD_{MIN} and about -1.2 ms for TD_{AVG}, implying that the differential delay is about 2.2 ms. The one outlier in the TD_{MIN} data of 0.7 ms can be attributed to a temporary 1.4 ms change in the *RTD*. Because RTD_{AVG} was about 41 ms, we can infer that one path had a delay of 19.4 ms, and the other a delay of 21.6 ms, which is larger by about 11.3%. In this case, the network path from Boulder to New Jersey had more asymmetry, percentage wise, than the network path from Boulder to New Delhi. This illustrates the vagaries of using the Internet as a time transfer medium, and it also illustrates why achieving an uncertainty within ±1 ms is not always possible with NTP, even when the network path is relatively short.

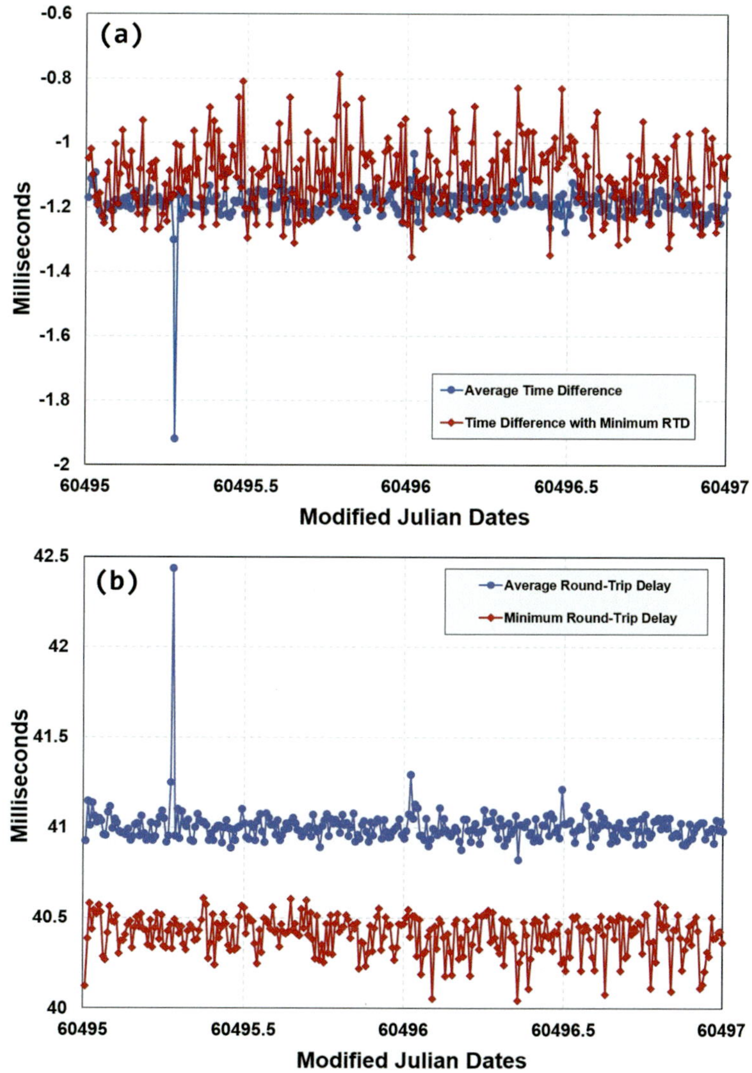

Figure 11.8 NTP measurements of time difference (a) and round-trip delay (b) made between UTC(NIST) and an NTP time server in New Jersey.

NTP can potentially measure time differences of less than a microsecond when operated on a well-controlled LAN. However, asymmetry will always contribute some measurement uncertainty, even if the LAN is as simple as an Ethernet crossover cable that directly connects two computers. In that example, uncertainty is contributed by the differential delay between the two network interface cards, and also by instabilities in both the client and server software. These uncertainties are usually too small to notice for NTP measurements performed on the public Internet, but they can be the dominant sources of uncertainty for measurements performed on a LAN [26].

To illustrate this, Figure 11.9 shows the results of a 48-hour TD (a) and *RTD* (b) comparison between a client and server that are collocated in a data center, and both referenced to a UTC(NIST) disciplined clock. In this example, averaging the NTP measurements again has the effect of smoothing the data, and the TD_{AVG} values are more stable than the TD_{MIN} values. The mean results are on opposite sides of zero, about +10 µs for TD_{MIN} and about –15 µs for TD_{AVG}. For TD_{AVG} this implies that the differential delay is about 30 µs. Because RTD_{AVG} is equal to 290 µs, we can infer that one path had a delay of 130 µs, and the other a delay of 160 µs, which is larger by about 23%. This example indicates that the percentage of differential delay can be especially large when the *RTD* is small. However, a small *RTD* is still advantageous because it reduces the theoretical limit of the uncertainty, which in this case is just ±145 µs, equal to 50% of the 290 µs *RTD*.

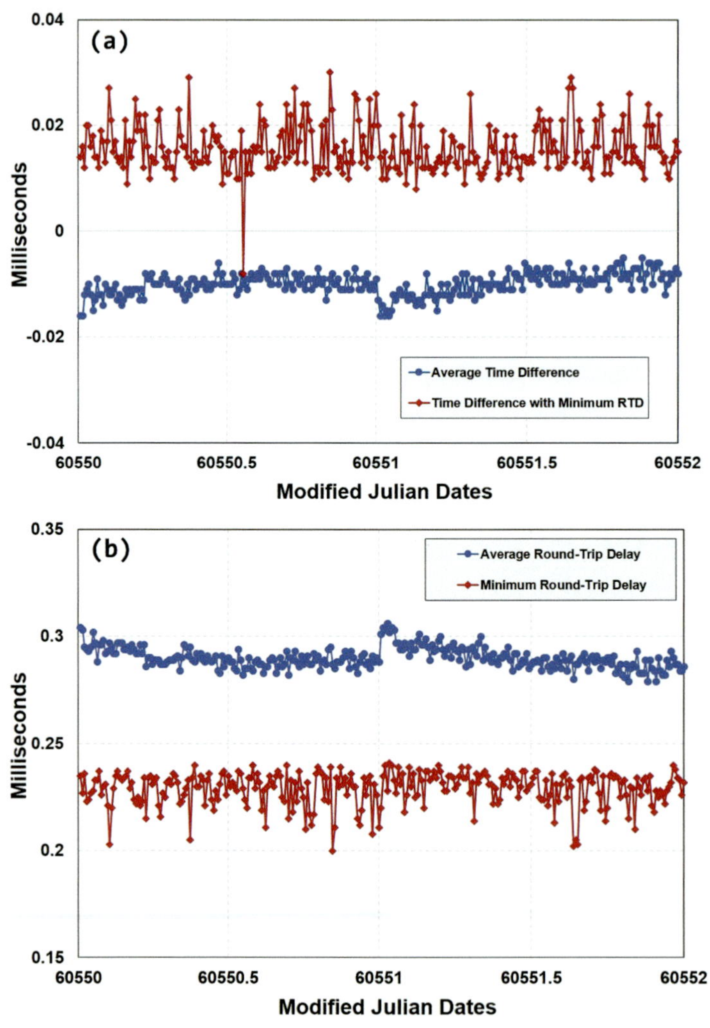

Figure 11.9 NTP measurements of time difference (a) and round-trip delay (b) made between a server and client that are both referenced to a UTC(NIST) disciplined clock.

In David Mill's own words, the "ultimate goal of NTP is to synchronize the clocks in all participating computers to the order of less than a millisecond or two relative to UTC" [23]. NTP achieves that goal in nearly all situations, which is remarkable considering the size and complexity of the Internet, and the fact that NTP can be implemented without any special hardware, making use of the ordinary computer clocks and network hardware that are already in place. The next section describes the Precision Time Protocol (PTP), which can achieve much smaller uncertainties than NTP by using hardware that's specifically designed for time transfer.

11.2.5 The Precision Time Protocol (PTP)

The clock synchronization protocol called PTP is defined by the *IEEE 1588* standard. The standard was first published in 2002 (v1), and revised in 2008 (v2). The most recent implementation at this writing is informally called PTP v2.1, and was published in 2019 [27]. From the beginning, PTP was designed to fill a niche not met by either NTP or GPS. Its purpose was supporting networked measurement, control, and communication systems where each component needed an RTC synchronized to a common time scale, so that all of the system's data could be successfully correlated [28]. These systems required much better accuracy than NTP could provide, and it often wasn't possible—or economically feasible—to install a GPSDC at every network node [29].

The *IEEE 1588* standardization effort was led by John Eidson (b. 1937), an American electrical engineer and inventor. Eidson's background was perfectly suited for this effort—he had vast experience in measurement and control systems, and was well aware of the importance of accurate time. After obtaining a doctorate from Stanford University, Eidson worked at Varian Associates before joining Hewlett-Packard in 1972. He remained at HP and its spinoff company Agilent Technologies until 2009, working on numerous measurement instrumentation and clock synchronization projects.

PTP is a packet-based system that was originally designed to operate over a LAN where the physical layer is Ethernet, and where the packets are multicast, or simultaneously transmitted to a group of network devices. However, the protocol can also be implemented in UDP unicast mode, as a direct transmission from one network point to another, with the sender and receiver each having their own network address.

A PTP network is based on a hierarchy of primary and secondary clocks. There are four clock types in the hierarchy—grandmaster clocks (GMC), boundary clocks (BC), transparent clocks (TC), and ordinary clocks (OC). The GMC serves as a primary time reference, and for purposes of redundancy or scalability, there can be more than one GMC in a PTP network. The secondary clocks—the BCs, TCs, and OCs—are GMC followers. Their role is to ensure that the time kept by a GMC is distributed throughout the network [27]. The GMC can be integrated into a network hardware device, such as a LAN switch, or it can be a standalone instrument. In most cases, the GMC is referenced to a GPSDC, but it can be referenced to another high-accuracy clock, such as a cesium clock or a UTC(k) time scale.

A boundary clock has multiple network ports. One or more ports receive time from the GMC, and one or more ports transmit time so that other segments of the network can be synchronized. Therefore, a BC simultaneously plays the roles of a follower and master clock—it synchronizes to a GMC through an upstream interface, and it synchronizes other clocks through a downstream interface. This capability can be built-in to what otherwise looks like standard networking hardware. For example, a router or Ethernet switch might be capable of operating as a BC.

Transparent clock capability is built-in to PTP-compatible network hardware, such as switches or routers, that have more than one network port [30]. A TC differs from a BC because it doesn't synchronize its own clock. Instead, it first measures the delay between an incoming and outgoing PTP message, and then updates the fields in the outgoing message to account for this delay. By compensating for the delay added by the network hardware, a TC lives up to its name by being invisible or transparent to the other PTP clocks.

The ordinary clocks are endpoint devices, and have just one network port. That port can either transmit or receive time—for example, a GMC can technically be thought of as an OC—but the term is usually reserved for a clock that only receives time. Some examples of OCs that only receive time are the clocks in sensors, measurement instruments, and client computers. Figure 11.10 shows a simple PTP network configuration that includes the four types of clocks. The ports labelled *M* indicate that a clock is operating as a master, and the ports labelled *F* indicate that the clock is operating as a follower. Note that a BC operates in both modes, and that a TC technically operates in neither mode, although a TC is perhaps best thought of as a passive follower.

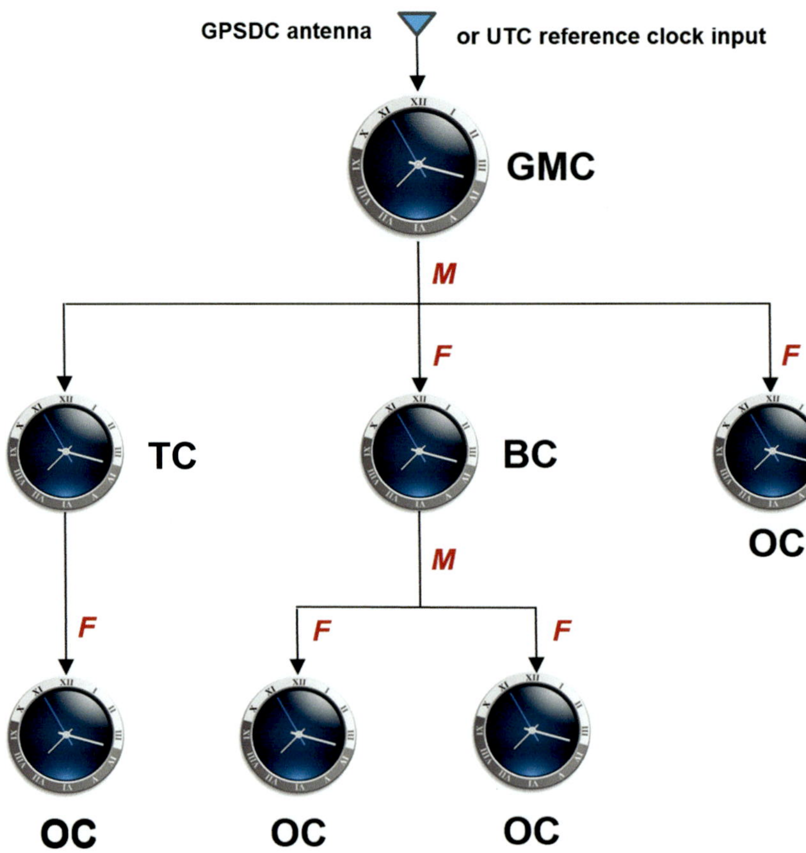

Figure 11.10 A simple PTP network configuration that includes all four clock types.

The *IEEE 1588* standard is well established and extensively documented, so the PTP equipment from different manufacturers should be compatible and able to coexist on the same network. However, some PTP products were designed to perform just one function and other equipment was designed to be as versatile as possible. This versatility can include the ability to support multiple versions of PTP, as well as the ability to operate in "hybrid" mode—serving, for example, as either a GMC or a BC, or even simultaneously handling both roles. The large number of options allow for a large number of potential configurations, and makes it possible to scale a PTP network to almost any size. When more than one reference clock is available, a PTP clock that operates as a follower executes an algorithm that selects the best synchronization source by comparing the expected performance of all of the available clocks. The comparison includes examining the accuracy, stability, and traceability of the available reference clocks, as well as their logical distance from the follower, with a shorter logical distance meaning that there are fewer BCs in the network path.

The PTP method for measuring the path delay between two PTP ports—for example, between a reference clock and a BC or OC—is very similar to the method used by NTP. Like NTP, PTP measures the RTD with the loop-back method, and it again estimates the one-way path delay as $\frac{RTD}{2}$. Both protocols are packet-based, designed to transfer time information via data packets containing time stamps. The latency of the packets depends on the network, and the format of the packets themselves doesn't provide either protocol with an inherent accuracy advantage [31].

Despite their similarities, there are several key differences that make PTP more accurate than NTP. One difference is that a PTP network measures the network delay far more often than an NTP network. Even when an NTP clock is being disciplined, the network delay is typically measured just several times per hour, but PTP networks are typically configured to measure the network delay at least once every 2 s, and sometimes as often as 128 times per second. Therefore, PTP uses more bandwidth than NTP, but it measures the network delays with less uncertainty. Another difference is that at least some of the NTP time stamps are obtained from software routines that read client computer clocks like those described in Section 11.2. The client clock oscillators aren't especially stable to begin with, and the latency introduced by the computer's OS can make the software-generated time stamps even less stable. In contrast, PTP time stamps are usually generated with dedicated hardware clocks that have better accuracy and stability. Finally, the intervals measured between time stamps are typically much shorter with PTP than NTP, because PTP is usually implemented on a LAN, and NTP is usually implemented on a WAN. When PTP is implemented on a WAN, the intervals between time stamps are still relatively short, because the various segments of the network are separated by boundary clocks. Each of these advantages allows PTP networks to be less affected by asymmetry, and to support more accurate timekeeping than NTP networks. As described in the previous section, NTP uncertainties are expressed in milliseconds, or in the best case tens of microseconds, but PTP uncertainties are nearly always expressed in units of microseconds or nanoseconds [32, 33].

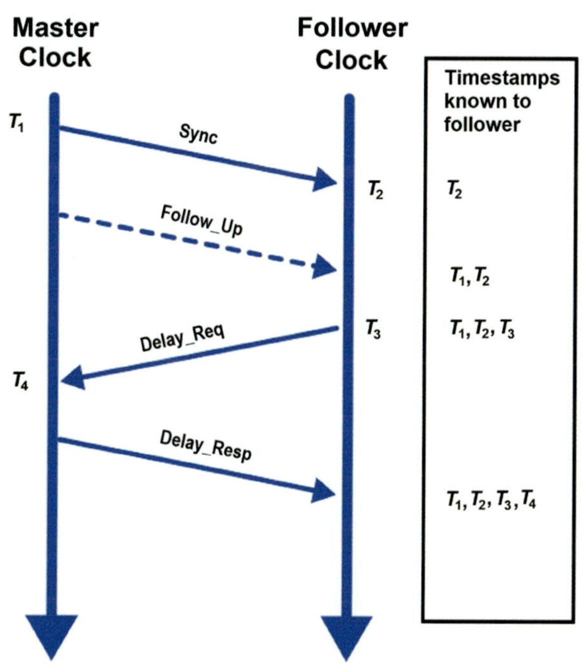

Figure 11.11 The PTP delay request-response path delay measurement.

The *IEEE 1588* standard calls the method that measures the network delay between two PTP ports the delay request-response mechanism. This mechanism, illustrated by the diagram in Figure 11.11, is always initiated by the master clock, and it involves the exchange of messages containing time stamps. The epoch for PTP time is the same as Unix time—midnight UTC on January 1, 1970. The PTP time scale is equivalent to TAI, but the GMC provides the current time difference between UTC and TAI so that UTC can be easily computed. The time stamps in PTP v2 and v2.1 generally have 96 bits. An unsigned 48-bit integer indicates the number of whole seconds that have elapsed since the PTP epoch, and an unsigned 32-bit integer indicates the fractional part of the second in units of nanoseconds. The nanosecond field overflows when it reaches 10^9 ns, and the second field then updates. If supported by the hardware, a 16-bit unsigned integer provides sub-nanosecond resolution in units of 2^{-16} ns, which is equivalent to about 15.3 fs [27].

The master clock sends *Sync* messages to the follower clock at the requested rate, which again is usually at least once every 2 s, and perhaps as often as 128 times per second. When the follower receives the *Sync* message, it records the time T_2. The *Sync* message contains an estimate of the time, T_1, when it was sent, but some of the generic hardware and software stacks used to implement PTP lack the ability to add an accurate time stamp to a message that they are in the process of generating. If this is the case, a *Follow_Up* message is sent immediately after the *Sync* message. The *Follow_Up* message contains a more accurate and higher resolution estimate of T_1, and when the message is received, the follower clock has knowledge of both T_1 and T_2.

The follower clock then sends a *Delay_Req* message to the master clock, recording the time of transmission as T_3. The master clock responds by sending a *Delay_Resp* message back to the follower at the time T_4. When the follower clock receives the *Delay_Resp* message, it now has knowledge of all four time stamps needed to measure the one-way network path delay, and can use this information to correct the time received from the master clock. The one-way delay (OWD) from the master clock to the follower is estimated as

$$OWD = \frac{(T_2 - T_1) + (T_4 - T_3)}{2} = \frac{(T_2 - T_3) + (T_4 - T_1)}{2}. \quad (11.4)$$

Due to the advantages of hardware-generated time stamps and the use of BCs and TCs that can correct their time stamps for latency, PTP can potentially eliminate nearly all of the asymmetry contributed by computer and network hardware, so that the remaining asymmetry is contributed by the medium itself [27]. However, in many cases, not all of the hardware in a PTP network was designed to correct for latency, especially when PTP is implemented on a WAN such as the public Internet. For example, a network router or switch that isn't designed for PTP applications can cause packets traveling in opposite directions to have asymmetric delays. Even when asymmetry exists, a PTP link can still do an excellent job of controlling the frequency of remote oscillators. This is because a GMC can essentially preserve the stability of its reference clock—including the stability of a GPSDC or a cesium clock—when transferring frequency across the network [34].

When implemented on a WAN, the accuracy of time transferred via a PTP link is usually much worse than the accuracy that the link's stability could support. For example, a PTP link might have nanosecond or even sub-nanosecond stability, but might still have a fixed 50 μs time offset due to the asymmetry in one or more network devices. The presence of network asymmetry often prevents long-haul PTP from meeting industrial time accuracy requirements, such as the common ±1 μs requirements in the telecommunication and electric power industries (Chapters 16 and 17).

Repeated measurements of a PTP link have often shown that its time offset can be essentially constant for periods of weeks or months. This suggests that the offset can be measured and compensated. However, the time offset can still change whenever the network hardware is reset, and the change can make any previous measurement invalid. For these reasons, meeting a stringent time accuracy requirement with PTP usually requires a network where every hardware and software component was designed to mitigate asymmetry [35]. This is possible on a LAN but usually isn't possible on a WAN. Therefore, sending time accurate to ±1 μs with PTP via a WAN will likely require periodic link calibrations made with an independent time transfer method, such as common-view GPS (Section 4.3.3).

PTP still supports the measurement and control systems that it was originally designed for, but its use has expanded and evolved since it was introduced in 2002. As we'll discuss in Chapters 16 through 18, PTP provides accurate time—usually over LANs—to telecommunication networks, electric power stations, and to the computers that time stamp stock market transactions. More recently, PTP has been implemented via WLANs and optical transport networks (OTNs). It also now synchronizes a more diverse group of clocks, including the clocks found in robotics systems and Internet of Things (IoT) devices [36]. The next section describes an extension of the *IEEE 1588* standard that was designed to improve upon PTP's accuracy—the protocol called White Rabbit.

11.2.6 White Rabbit (WR)

The White Rabbit (WR) protocol is named after the watch-carrying rabbit in Lewis Carroll's *Alice's Adventures in Wonderland*, a humorous nod to a fictional character who was very much concerned about time, even though he was perennially late. The protocol was developed as an open-source software and hardware project by the European Organization for Nuclear Research (*Conseil Européen pour la Recherche Nucléaire*, CERN) in Geneva, Switzerland. The WR project began at CERN in 2008 [37], the same year that PTP v2 was defined by the *IEEE 1588–2008* standard. It resulted in a synchronization protocol that cleverly combined PTP with Synchronous Ethernet (SyncE), a method for transferring frequency over the Ethernet's physical layer [38]. The *IEEE 1588–2019* standard that defines PTP v2.1 added a "High Accuracy Default PTP Profile" for sub-nanosecond synchronization, a profile that's entirely based on the open source implementation of WR [27, 39].

The original purpose of WR was to improve upon the time synchronization accuracy of processes in CERN's particle accelerator network, which was then based on a system of unidirectional links, with no provisions for measuring the network's delays. The original WR specifications called for a system that could transfer time with an accuracy of 1 ns and a resolution of less than 50 ps. The system was also designed to service more than 1,000 nodes, to cover distances of about 10 km, and to be able to transfer time to all nodes with a guaranteed upper bound in network latency. A main goal of the project was to meet these specifications by using existing standards, and then extending them as necessary. The existing standards chosen were Ethernet, which provided WR with its physical layer, and PTP, which provided WR with a packet-based method of time synchronization [37].

A WR network is similar to a PTP network, but it distributes accurate frequency through the physical layer, in addition to distributing packets containing accurate time stamps. In a standard Ethernet system, the transmit clocks on the Ethernet devices are referenced to free-running oscillators that are only required to be accurate to within ±100 parts per million (ppm), equivalent to $\pm 1 \times 10^{-4}$ [40]. In a SyncE system, the Ethernet clocks are phase locked to a primary reference oscillator, as shown in Figure 11.12. The SyncE standard only requires a frequency accuracy of ±4.6 ppm [38, 41], but in a WR network, the frequency accuracy is many orders of magnitude better, equivalent to the accuracy of the primary reference oscillator, which is usually a GPSDO or a cesium. The use of SyncE means that the entire physical layer of a WR network maintains the same frequency, a process called layer 1 or L1 syntonization.

The WR time information is transmitted in packets on the network or transport layer, in the same fashion as PTP. However, there are two key elements that distinguish WR from conventional PTP. The first element involves the close cooperation between PTP synchronization and L1 syntonization. The second element involves the enhanced timekeeping accuracy and resolution that is obtained through phase detection [42].

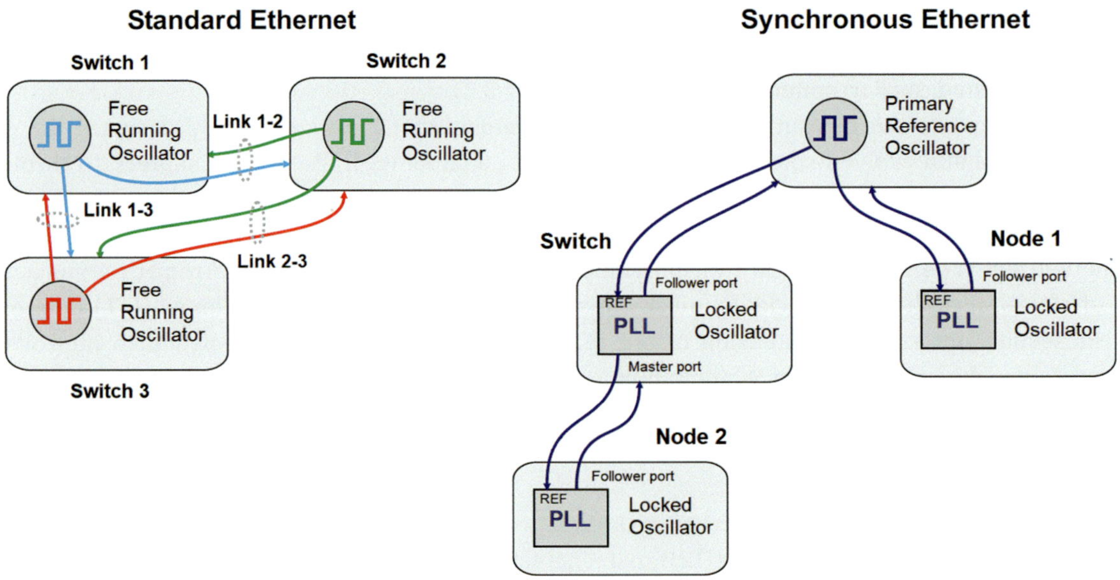

Figure 11.12 Standard versus Synchronous Ethernet (SyncE).

Figure 11.13 is a simplified diagram that illustrates the relationship between PTP synchronization and L1 syntonization in a WR network. As shown in the top half of the diagram, there are two WR nodes. Node A has a port operating in the master state, and node B has a port operating in the follower state. Both nodes are implementing the WR/PTP clock model. The local clock in this model is synchronized to a reference time scale obtained from a GMC, and a time counter inside the clock increments at each rising edge of the local PTP clock signal. As shown in the bottom half of the diagram, the L1 Tx clock signal is transmitted over the medium, and the L1 Rx clock is recovered from data received from the medium.

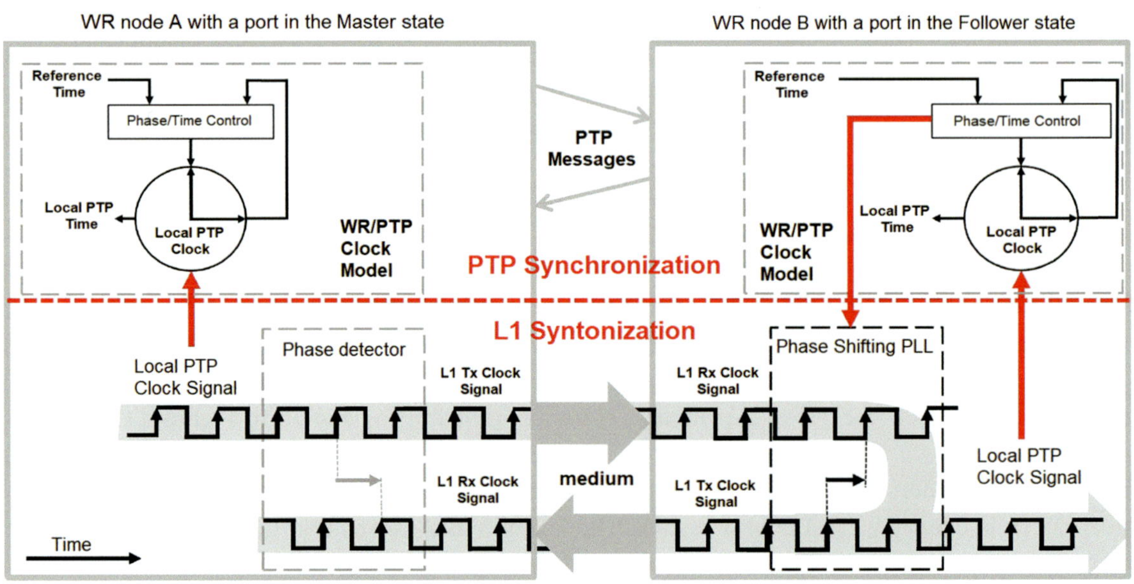

Figure 11.13 PTP synchronization and L1 syntonization in a WR network [42].

As shown in Figure 11.13, the two WR nodes A and B are directly connected by a network link. Because node A is configured as the master, the L1 Tx clock signal that it transmits is syntonized to its local PTP clock. Node B is configured as the follower, and its local PTP clock is syntonized to the L1 Rx clock signal received from node A. As a result, the PTP clocks in both nodes are syntonized to the same frequency. Once syntonization is achieved, the PTP time measurements made with respect to the GMC are used to synchronize the local PTP clock in node B. The initial synchronization is performed by adjusting the value of the time counter and correspondingly adjusting the phase of the local PTP clock.

Once synchronization is achieved, the time counter of node B is no longer adjusted. Instead, the synchronization of node B is maintained by a PLL that keeps the local PTP clock syntonized to the recovered L1 Rx clock signal. The PLL uses a digital DMTD system (Section 5.2.4) as a phase detector, and it maintains a constant phase offset by using a proportional-integral (PI) controller that constantly steers the phase towards a desired setpoint. These techniques give WR the ability to support sub-nanosecond time synchronization, and to provide time tags with a resolution of about 1 ps [42]. Commercial systems are now available that support both PTP and White Rabbit.

11.3 Web Clocks

Web clocks are time-of-day display clocks that are viewed with ordinary web browsing software. The clocks are hosted by a web site, and accessed by making either an HTTP or HTTPS request. In many cases, web clocks are just a small part of a web site's content, and are presented alongside a variety of other information. For example, the current time of day can be obtained by typing the word "time" into Internet search engines such as Google, Yahoo, or Bing, and small continuously running web clocks can be found in various places, including the BIPM web site at *bipm.org*. Other web sites are exclusively designed as web clocks, and can often display the time in multiple time zones. These sites can be hosted by either commercial companies or by NMIs that maintain UTC(k) time scales, such as the *time.gov* web site (Figure 11.14) that's provided as a service of NIST and the USNO. In most cases, the time displayed by a web clock is referenced to the web server's clock, which is typically synchronized and controlled via NTP.

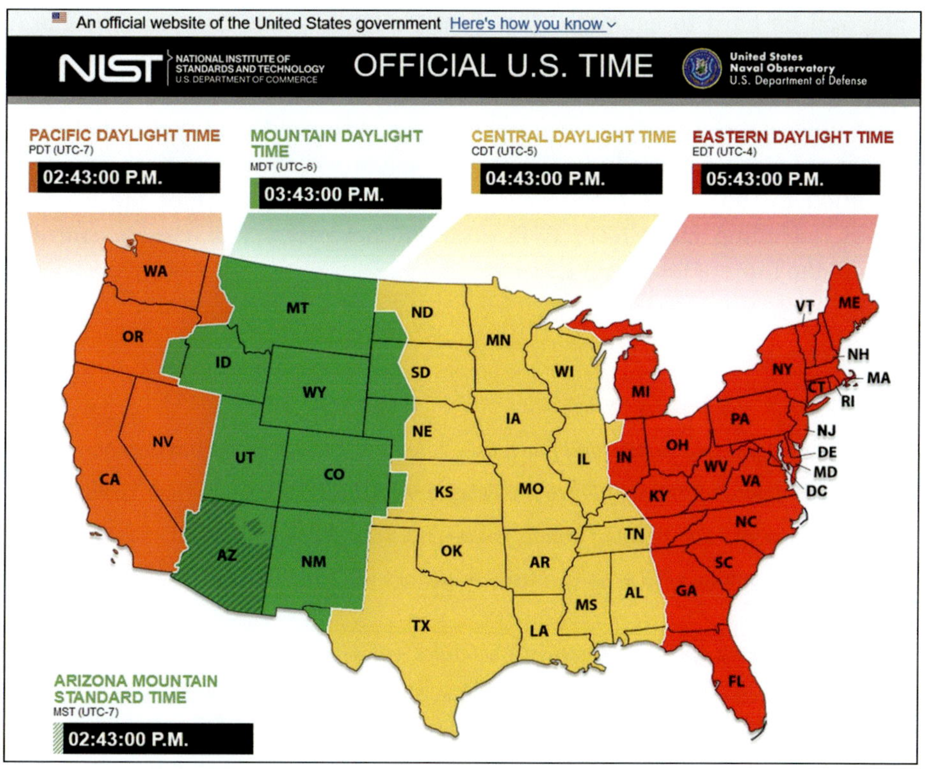

Figure 11.14 A portion of the *time.gov* web clock display.

The client software that's run by the web browser receives a time code from the server when the web site is first loaded, and again whenever it is reloaded. Some web clocks are programmed to refresh periodically, perhaps once every hour, so that they can receive a time code update. There is no standard format for the time code, which is sent from the server to the client as a data structure defined by the software developer, often using a data interchange format such as JavaScript Object Notation (JSON).

In between time code updates, web clocks keep time using their local system clock, and dynamically update the display by using a markup language such as HTML5 (Hypertext Markup Language 5) and a programming language such as JavaScript. The programming language provides the math functions needed to perform basic calculations, such as converting UTC to local time, compensating for the path delay to the server, and comparing the local client clock to the server clock.

If the client software records the times when it requested (T_1) and received (T_4) the time code, and if the server provides the time when the request was received (T_2) along with the time code (T_3), then the path delay can be compensated using the standard NTP method given in Eq. (11.2). When this is done, it becomes a simple matter for the web clock to display the time difference between the local system clock and the time server. This time difference is sometimes displayed with 1 ms resolution, which is a much finer resolution than a web clock's typical accuracy.

Web clocks are usually accurate enough for everyday purposes. They make it easy to check the current time, and they're a convenient reference for manually setting a wristwatch or clock. In many cases, they are more useful than an RCC, especially to travelers and to people communicating across long distances, because they can simultaneously display the time in multiple time zones.

However, web clocks have some obvious limitations. They can't synchronize a computer or smartphone clock, because they lack root access to the OS. They also aren't suitable for applications that require traceability, because their measurement uncertainty is difficult to estimate. Even when a web clock compensates for path delay, it's still usually referenced to a web server clock whose uncertainty is unknown. Like an RCC, a web clock keeps time using its local system clock during the period between synchronizations, so it can accumulate a large time error if the browser isn't periodically refreshed. Finally, the delay introduced by the time display can be hard to estimate. The display's update rate is limited by the frame rate of a computer or smartphone's video hardware, which is typically 60 Hz or higher, but the stability of the update rate usually depends on how well the software was written.

We'll continue to discuss clocks that are measured, synchronized, and controlled with network time signals in subsequent chapters, especially in Chapter 14 when we discuss optical time transfer. For now, we'll turn our attention to the clocks that provide the reference for most of the world's high-accuracy timekeeping systems, including most NTP, PTP, and WR networks. The clocks controlled by satellite navigation systems are the topic of Chapter 12.

References

1. E. L. Morse, *Samuel F. B. Morse: His Letters and Journals—Volume II*, Houghton Mifflin Company, 1914.
2. I. Lebow, *Information Highways & Byways: From the Telegraph to the 21*st *Century*, New York: IEEE Press, 1995.
3. W. R. Driver, "Announcement," *Cambridge Tribune*, vol. 41, no. 15, p. 12, June 8, 1918.
4. M. M. Thomson, *The Beginning of the Long Dash: A History of Timekeeping in Canada*, University of Toronto Press, 1978.
5. A. Selwyn, "The Speaking Clock of Paris," *Practical Mechanics*, vol. 1, no. 7, pp. 317–318, April 1934.
6. A. S. Risley, "The National Measurement System for Time and Frequency," *National Bureau of Standards Special Publication 445-1*, 72 p., June 1976.
7. International Telecommunications Union (ITU), "One-way Transmission Time," *ITU Recommendation G.114*, May 2003.

8. H. Marouani and M. R. Dagenais, "Internal Clock Drift Estimation in Computer Clusters," *Journal of Computer Systems, Networks, and Communications*, article no. 583162, 7 p., 2008.
9. N. Freris, S. R. Graham, and P. R. Kumar, "Fundamental Limits on Synchronizing Clocks Over Networks," *IEEE Transactions on Automatic Control*, vol. 56, no. 6, pp. 1352–1364, June 2011.
10. D. Jackson and R. J. Douglas, "A Telephone-Based Time Dissemination System," *Proceedings of the 18th Annual Precise Time and Time Interval Meeting (PTTI)*, pp. 541–553, December 1986.
11. J. Levine, M. Weiss, D. D. Davis, D. W. Allan, and D. B. Sullivan, "The NIST Automated Computer Time Service," *Journal of Research of the National Institute of Standards and Technology*, vol. 94, no. 5, pp. 311–321, September-October 1989.
12. J. Levine, M. A. Lombardi, L. Nelson, and V. Z. Zhang, "Technical Reference Manual for NIST Automated Computer Time Service (ACTS)," *National Institute of Standards and Technology Interagency Report 6611*, 89 p., July 2001.
13. J. Levine, "Timing in telecommunications networks," *Metrologia*, vol. 48, no. 4, pp. S203–S212, July 2011.
14. J. L. Carrell, L. A. Chappell, E. Tittle, and J. Pyles, *Guide to TCP/IP*, 4th ed. Boston: Course Technology, 2013.
15. F. Wilder, *A Guide to the TCP/IP Protocol Suite*, 2nd ed., Boston: Artech House, 1998.
16. J. Postel, "DoD Standard Transmission Control Protocol," *RFC 761*, 88 p., January 1980.
17. J. Postel, "Daytime Protocol," *Network Working Group RFC 867*, 2 p., May 1983.
18. J. Postel and K. Harrenstien, "Time Protocol," *Network Working Group RFC 868*, 2 p., May 1983.
19. J. Levine, "The NIST Internet Time Service," *Proceedings of the 25th Annual Precise Time and Time Interval Meeting (PTTI)*, pp. 505–513, November 1993.
20. D. L. Mills, "Precision Synchronization of Computer Network Clocks," *ACM SIGCOMM Computer Communication Review*, vol. 24, no. 2, pp. 28–43, April 1994.
21. D. L. Mills, "A Brief History of NTP Time: Memoirs of an Internet Timekeeper," *ACM SIGCOMM Computer Communication Review*, vol. 33, no. 2, pp. 9–21, April 2003.
22. J. Postel, "Internet Control Message Protocol," *Network Working Group RFC 792*, 21 p., September 1981.
23. D. L. Mills, *Computer Network Time Synchronization: The Network Time Protocol*, Taylor & Francis, 2006.
24. D. Mills, J. Martin, J. Burbank, and W. Kasch, "Network Time Protocol Version 4: Protocol and Algorithms Specification," *Internet Engineering Task Force RFC 5905*, 110 p., June 2010.
25. S. Sommars, 2017, "Challenges in Time Transfer Using the Network Time Protocol (NTP)," *Proceedings of the 48*th *Annual Precise Time and Time Interval Meeting (PTTI)*, pp. 271–290, January 2017.
26. A. N. Novick and M. A. Lombardi, "Practical Limitations of NTP Time Transfer," *Proceedings of the 2105 Joint Conference of the IEEE International Frequency Control Symposium and the European Frequency and Time Forum (IFCS/EFTF)*, pp. 570–574, April 2015.
27. IEEE, "Precision Clock Synchronization Protocol for Networked Measurement and Control Systems," *IEEE Standard 1588–2019*, November 2019.
28. J. C. Eidson and K. Lee, "Sharing a Common Sense of Time," *IEEE Instrumentation & Measurement Magazine*, vol. 6, no. 1, pp. 26–32, March 2003.
29. J. C. Eidson, *Measurement, Control and Communication Using IEEE 1588*, Springer-Verlag, 2006.
30. J. Han and D.-K. Jeong, "A Practical Implementation of IEEE 1588–2008 Transparent Clock for Distributed Measurement and Control Systems," *IEEE Transactions on Instrumentation and Measurement*, vol. 59, no. 2, pp. 433–439, February 2010.
31. L. Cosart, "Studying Network Timing with Precision Packet Delay Measurements," *Proceedings of the 40th Annual Precise Time and Time Interval Meeting (PTTI)*, pp. 165–186, December 2009.
32. G. Giorgi and C. Narduzzi, "Performance Analysis of Kalman-Filter-Based Clock Synchronization in IEEE 1588 Networks," *IEEE Transactions on Instrumentation and Measurement*, vol. 60, no. 8, pp. 2902–2909, August 2011.
33. P. Ferrari, A. Flammini, S. Rinaldi, A. Bondavalli, and F. Brancati, "Experimental Characterization of Uncertainty Sources in a Software-Only Synchronization System," *IEEE Transactions on Instrumentation and Measurement*, vol. 61, no. 5, pp. 1512–1521, May 2012.

34. K.-Y. Tu, C.-S. Liao, and S.-Y. Lin, "Remote Frequency Control via IEEE 1588," *IEEE Transactions on Instrumentation and Measurement*, vol. 58, no. 4, pp. 1263–1267, April 2009.
35. R. Exel, "Mitigation of Asymmetric Link Delays in IEEE 1588 Clock Synchronization Systems," *IEEE Communications Letters*, vol. 18, no. 3, pp. 507–510, March 2014.
36. Z. Idrees, J. Granados, Y. Sun, S. Latif, L. Gong, Z. Zou, and L. Zheng, "IEEE 1588 for Clock Synchronization in Industrial IoT and Related Applications: A Review on Contributing Technologies, Protocols and Enhancement Methodologies," *IEEE Access*, vol. 8, pp. 155660–155678, August 2020.
37. J. Serrano, M. Cattin, E. Gousiou, E. van der Bij, T. Wlostowski, G. Daniluk, and M. Lipinski, "The White Rabbit Project," *Proceedings of the 2nd International Beam Instrumentation Conference*, pp. 936–942, September 2013.
38. J.-L. Ferrant, M. Gilson, S. Jobert, M. Mayer, M. Ouellette, L. Montini, S. Rodrigues, and S. Ruffini, "Synchronous Ethernet: A Method to Transport Synchronization," *IEEE Communications Magazine*, vol. 46, no. 9, pp. 126–134, September 2008.
39. F. Girela-López, J. López-Jiménez, M. Jiménez- López, R. Rodríguez, E. Ros, and J. Díaz, "IEEE 1588 High Accuracy Default Profile: Applications and Challenges," *IEEE Access*, vol. 8, pp. 45211–45220, March 2020.
40. IEEE Computer Society, "IEEE Standard for Ethernet," *IEEE Standard 802.3-2022*, May 2022.
41. International Telecommunications Union, "Timing Characteristics of Synchronous Equipment Slave Clock," *Recommendation ITU-T G.8262/Y.1362*, November 2018.
42. M. Rizzi, M. Lipinski, T. Wlostowski, J. Serrano, G. Daniluk, P. Ferrari, and S. Rinaldi, "White Rabbit clock characteristics," *Proceedings of the 2016 IEEE International Symposium on Precision Clock Synchronization for Measurement, Control, and Communication*, 6 p., September 2016.

Chapter 12

Clocks Controlled by Satellite Navigation Systems

Time flies over us, but leaves its shadow behind.

— Nathaniel Hawthorne, "The Marble Faun – Volume II"

Most of the world's high-accuracy clocks are controlled by satellite navigation systems, primarily by global navigation satellite systems (GNSS). A GNSS is a constellation of satellites in medium Earth orbit (MEO) that distributes positioning, navigation, and timing (PNT) information by transmitting radio signals. These signals reach all areas of the Earth, and the number of GNSS receivers now exceeds the world's human population. As we touched upon in Section 10.7, these receivers are integrated into billions of smartphones, and they are also found in countless numbers of electronic systems and transport vehicles—regardless of whether the vehicle travels via land, water, or air.

Satellite navigation receivers have permanently changed the world with their amazing ability to tell us where we are and to lead us to our destination, so most of the attention that they attract is understandably focused on positioning and navigation. It's important to remember, however, that the time information provided by GNSS—the "T" in PNT—is the foundation for its positioning and navigation capabilities. The ability to determine position within an accuracy of one meter or better was made possible by atomic clocks that keep time accurate to within a few nanoseconds or less. The satellites carry atomic clocks and broadcast time signals as they fly over the Earth. As a result, satellite navigation systems do more than just navigate, they are also the world's best systems for high-accuracy time distribution.

The original GNSS, and by far the most prominent, is the Global Positioning System (GPS) that's owned and operated by the United States government. Because GPS is so important to world timekeeping, most of this chapter is about GPS. However, the other major GNSS constellations—GLONASS, Galileo, and BeiDou—are each described in Section 12.8. In addition, this chapter covers regional navigation satellite systems (RNSS) in Section 12.9, and low Earth orbit (LEO) satellite systems in Section 12.10.

The GPS sections of this chapter will focus on its ability to control clocks and oscillators by distributing time signals that are synchronized to UTC. We'll begin by briefly discussing the origins of GPS in Section 12.1, including the first efforts to use satellites for navigation and to launch atomic clocks into space. That section is followed by descriptions of the GPS satellite constellation (Section 12.2), the GPS

As Time Goes By: A Comprehensive Guide to Time and Frequency Measurements
Michael A. Lombardi

ISBN 978-981-5352-28-3 (Hardcover), 978-1-042-01532-0 (eBook)
www.jennystanford.com

DOI: 10.1201/9781042015320-16

signal format (Section 12.3), and the relationship between GPS time and UTC (12.4). GPS disciplined oscillators and clocks—called GPSDOs and GPSDCs—are the focus of Section 12.5, and the measurement uncertainties of those devices are discussed in detail in Sections 12.6 and 12.7.

The concepts described in the GPS sections—especially those pertaining to measurement uncertainty—also apply to the other GNSS, RNSS, and LEO satellite constellations that are described later. We'll conclude this chapter with Section 12.11, which discusses the vulnerability of satellite navigation signals to malicious and accidental interference.

12.1 The Origins of GPS

The origins of GPS can be traced back to the start of the Space Age, which began in October 1957 when the Soviet Union launched *Sputnik 1*, the first artificial satellite. The first American satellite, *Explorer 1*, was launched just four months later. The first satellites were mostly used for solar and atmospheric studies, but the attention quickly turned to using satellites for telecommunications. As was described in Section 10.6, the launch of the first geostationary communications satellite in 1964 immediately attracted the attention of time metrologists. The attention led to the development of radio-controlled clocks that received signals broadcast from satellites. It also led to the development of the TWSTFT method (Section 4.3.4) that uses geostationary satellites to link the major UTC(k) time scales to UTC (Section 9.4.4).

The first satellite-based navigation system was the Navy Navigation Satellite System (NNSS), commonly called the Transit system. This system was designed and developed by the Applied Physics Laboratory (APL) of Johns Hopkins University in Laurel, Maryland, and was operated by the United States Navy from 1964 to 1996. It was based on a LEO satellite constellation where the satellites flew at an altitude of about 1,100 km, circling the Earth every 107 minutes. Their 90° inclination angle with respect to the Earth's equator formed a polar orbit, and the satellites flew directly over the north and south poles during each revolution around the Earth. This configuration allowed just five satellites to provide a reasonable amount of global coverage.

The Transit satellites transmitted on two frequencies—150 MHz and 400 MHz—so that their frequency-dependent ionospheric delays could potentially be corrected. This same technique was later used by GPS. Only one satellite was usually observable at one time, but a Transit receiver could obtain a latitude and longitude position fix that was usually accurate to within less than 100 m during a single satellite flyover. Even so, the Transit system wasn't a real-time positioning system. A usable satellite flyover was only guaranteed to occur once every four hours, although at mid-latitude locations it would likely occur about twice that often [1, 2].

The Transit system took advantage of atomic clock accuracy, but like the geostationary time code services described in Section 10.6, the atomic clocks weren't carried by the satellites, but were instead located at ground stations. The satellites carried a quartz oscillator that had a nominal frequency of 5 MHz. The quartz oscillator signals passed through a group of frequency multipliers and dividers, and provided the reference frequency for the satellite's clock and for the radio transmissions. The 150 MHz and 400 MHz broadcasts were received by four satellite ground stations—located in California, Hawaii, Maine, and Minnesota—that measured the Doppler frequency of the received signals with respect to local atomic clocks that were referenced to UTC(USNO). The results of these measurements were used to compute frequency and time corrections that were periodically transmitted back to the satellites [2].

These methods allowed a system that was designed for navigation to also broadcast accurate time signals. In addition, because the rapidly moving satellites were broadcasting both time and position data, it allowed receivers on Earth to continuously compensate for path delay. Measurements conducted at NBS over an eight-month period in 1979 showed that the average accuracy of Transit satellite signals received in Boulder, Colorado, was within ±25 μs, with a standard uncertainty of about 10 μs [3].

More accurate clocks were put into space by the TIMATION system developed by the United States Naval Research Laboratory (NRL). The name TIMATION was coined by combining the words time and navigation, and the system's principal designer was the American physicist Roger Easton (1921–2014), who was a staff member at NRL from 1943 until his retirement in 1980. A main objective of the program was to make accurate clocks part of the payload of Earth orbiting navigation satellites. The presence of accurate clocks in space would allow a receiver on Earth to perform a passive ranging measurement, a measurement of the path delay between a clock in space and a clock on Earth, and to convert the path delay to distance [4]. Passive ranging measurements would not only advance the science of time transfer, but they would also demonstrate that three-dimensional navigation—the determination of latitude, longitude, and altitude—was achievable anywhere on the Earth's surface.

Two experimental LEO satellites, TIMATION I and II, were launched in 1967 and 1969, respectively. They flew at an altitude of about 925 km with a 70° inclination angle, circling the Earth once every 103 minutes. The satellites carried stable quartz clocks designed for use in space, and were tracked by several experimental ground stations that were built and operated by NRL. The range measurements were performed using a method called side tone ranging, where the signals transmitted by the satellites were not modulated, but instead were continuous wave (CW) subcarrier signals that were spaced by a desired set of tone frequencies. The ground station receiver combined two CW signals to obtain a beat frequency, and then compared the phase of the beat frequency to a reference signal generated by the receiver. The phase difference allowed the determination of both the path delay and distance between the satellite and receiver clocks [4, 5].

While the Transit and TIMATION programs were ongoing, the United States Air Force (USAF) was working on its own satellite program to support the navigation of aircraft. This classified effort was ambiguously named program 621B. It was started in October 1963 as a collaboration between the Space Division of the USAF and the Aerospace Corporation, which was founded in El Segundo, California, in 1960 to advance ballistic missiles and space systems that were essential to U. S. national security [6].

The president of the Aerospace Corporation was Ivan Getting (1912–2003), an American physicist and electrical engineer, who had extensive experience in missile detection, guidance, and control systems. In the 1940s, Getting's group at the Radiation Laboratory of MIT worked on automatic-tracking microwave radar systems that were deployed in World War II. Beginning in the 1950s, Getting led a project at the Raytheon Corporation to develop a mobile ballistic missile guidance system called Mosaic. These systems were all based on terrestrial radio transmitters, but he quickly made the transition to satellites. After founding the Aerospace Corporation, Getting became a passionate advocate for the establishment of a global positioning system where the radio transmitters were carried by satellites.

The studies produced by the Aerospace Corporation in support of program 621B were focused on developing a three-dimensional global positioning system. This required enough satellites to be launched so that four were always observable from a receiver's location. The various studies all recommended systems that had three features in common. First, the system had to accurately determine the positions of the satellites. Borrowing from the concepts of celestial navigation, this allowed satellites with known positions to essentially replace the stars. Second, the light from the stars had to be replaced with radio waves that penetrated the ionosphere and clouds, allowing the system to work in all types of weather. Finally, the system had to measure relative time, by comparing the time of arrival (TOA) at the receiver to the satellite time, which meant that all of the satellite clocks had to be synchronized to a common standard. Meeting the third requirement meant that the satellites would have to carry atomic clocks that were continuously monitored and corrected by ground stations [6].

In November 1972, the American engineer and Air Force Colonel Bradford Parkinson (b. 1935), was appointed as the director of program 621B. Parkinson had an exceptionally strong background in navigation, guidance, and control systems. He had graduated from the U. S. Naval Academy before being commissioned in the USAF, and had gone on to obtain a master's degree from MIT and a doctorate in Astronautics from Stanford University. Parkinson quickly assembled and assumed leadership of a team that combined the best aspects of the Transit system, the TIMATION program, and program 621B. These efforts prompted the United States DoD to establish a Joint Program Office (JPO) with the goal of developing and deploying a single system that could meet the positioning and navigation needs of all agencies involved in the defense of the United States. The USAF was designated as the lead agency, but all branches of the U. S. military—including the Navy, Army, and Marine Corps—were involved in the JPO effort, and they worked together towards their common goal [7, 8].

Due to the work of Parkinson, Getting, Easton, and countless others, the DoD approved a proposal in December 1973 for a passive one-way ranging system of 24 MEO satellites with a 12-hour orbital period, where each satellite would carry atomic clocks. The system was originally named Navstar, but because that name was later associated with the names of the individual satellites, it soon became known as Navstar GPS. In common usage, the name was shortened to just GPS, an acronym for Global Positioning System.

Under the umbrella of the cooperative JPO effort, the third TIMATION satellite developed by NRL was renamed *Navigation Technology Satellite 1* (*NTS-1*). Launched in July 1974, *NTS-1* was the first satellite to carry atomic clocks, and its payload included two rubidium clocks that were manufactured by the German company Efratom Electronics. Measurements of the *NTS-1* signals made at NRL and the USNO in 1976 indicated that the received time accuracy was less than 1 μs, and accuracies of tens of microseconds were measured in England and Australia [9].

The next NRL satellite, named *NTS-2*, was the first satellite that was completely designed and built using JPO funding, and also the first to include all of the features that would be later be employed by the Navstar satellites (Figure 12.1). It was launched in June 1977 with two cesium clocks on board that had been designed and constructed for space use by Frequency and Time Systems in Danvers, Massachusetts. The clocks had a frequency stability of about 2×10^{-13} at $\tau = 1$ day. Because *NTS-2* was fully compatible with the GPS ground control segment being developed by the USAF, it was used to evaluate the ground stations, software, and systems for the Navstar satellites, which were then being developed by Rockwell International [5, 8]. The first Navstar satellite was launched in February 1978, and three additional Navstar satellites were launched before the end of that year.

Figure 12.1 The prototype for the Navstar GPS satellites, *NTS-2*, which launched in 1977.

GPS was officially made available for civilian use in 1983, which immediately led to a commercial market for GPS products. The signals were freely available to civilians before 1983, but they were provided on a "use at your own risk" basis. The commercial market grew exponentially after GPS was declared a fully operational system, an event that didn't occur until July 17, 1995, more than 21 years after the original Navstar proposal was approved by the DoD [8]. Even then, there were still concerns about making the full accuracy of GPS available to the general public.

For national security reasons, a feature called Selective Availability (SA) intentionally degraded the accuracy of the civilian GPS signals by inserting a continuous sequence of errors that were much larger than the system's normal uncertainties. The use of SA limited the unaveraged timing uncertainty of GPS to 340 ns ($k = 2$), which was then roughly 5× larger than the uncertainty of the non-degraded time signals. The signals that provided the full accuracy of GPS were encrypted and could only be accessed with special receivers that were designed to accept a decryption key. The "keyed" receivers were only made available to the U. S. military and other authorized personnel. This situation continued until May 2, 2000, when a Presidential Order deactivated SA and made the full accuracy of GPS available to civilians [10, 11].

Time metrologists were immediately attracted to GPS, and the first commercial GPS receiver used as a clock was delivered to the USNO in 1979. The receiver was developed by Stanford Telecommunications in Sunnyvale, California, and was part of a system that included a large rack of equipment, with the receiver software running on a standalone computer that sat outside of the rack [12, 13]. Like the other GPS receivers first used for timekeeping—including the NBS common-view receivers previously described in Section 9.4.4—it was a single-channel unit that could track only one satellite at a time. Only one satellite was needed to receive the time, but a navigation solution required signals from four satellites, and continuous time signal reception also required multiple satellites. The early single-channel receivers, which included the first GPS handheld units, obtained a position fix by using a sequencing method. They tracked and measured a single satellite for a period long enough to obtain the necessary data, then switched to the next satellite.

Due to a strong demand for new products, GPS technology rapidly evolved, and the receivers quickly became smaller and faster, with the ability to simultaneously track multiple satellites. The first multi-channel receivers still relied on sequencing methods—using three channels, for example, to switch between eight satellites. Receivers with a dedicated tracking channel for every satellite first appeared in the early 1990s. By 1993, Motorola had introduced a GPS receiver module with dimensions of just 3.5 × 3.6 × 1.0 cm that had six dedicated tracking channels. The six-channel receivers were quickly replaced by eight-channel units, and receivers that could simultaneously track 12 satellites—the maximum number likely to be visible at a given location—were common by the end of the 1990s [14].

The control of GPS has also evolved. The GPS master control station was originally located at Vandenberg Air Force Base in California, but was moved in 1987 to Schriever Space Force Base (formerly known as Falcon Air Force Base and Schriever Air Force Base), located near Colorado Springs, Colorado. The 2nd Space Operations Squadron (2SOPS) is headquartered at Schriever and is responsible for the command and control of the GPS satellites. Originally part of the USAF, 2SOPS is now part of the United States Space Force (USSF), a new branch of the United States military that was organized under the Department of the Air Force in December 2019.

12.2 The GPS Satellite Constellation

The GPS constellation was designed to operate with a minimum of 24 operational MEO satellites, with the full constellation consisting of 31 satellites. As of 2025, a full constellation of 31 satellites has been fully operational for well over a decade, and additional satellites are also available as in-orbit spares. The satellites are divided into six orbital planes, with each plane containing at least four satellites. The orbital planes are tilted with respect to the Earth's equator, with an inclination angle of 55° (Figure 12.2). The satellites fly at a nominal height of 20,200 km above the Earth's surface. That height, when added to the Earth's radius, results in a nominal orbital radius of 26,560 km. Their orbital period equals one half of the period of a sidereal day, which is 11 hours, 58 minutes, and 2 s [10, 15].

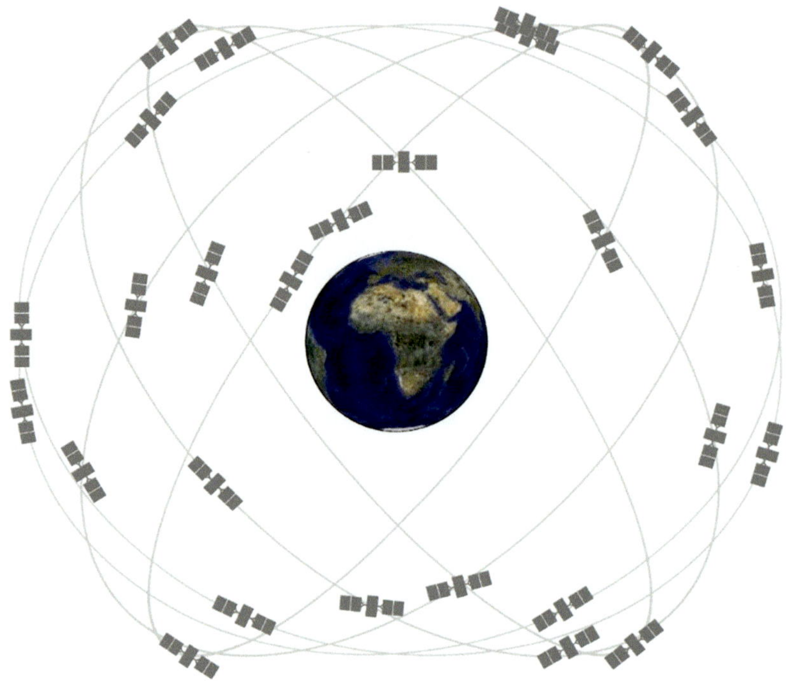

Figure 12.2 The GPS satellite constellation.

In addition to the master control station at Schriever Space Force Base, the GPS ground segment includes an international network of control, monitoring, and tracking stations that generate and send orbital and clock corrections to the satellites. The corrections ensure that each satellite's clocks are continuously synchronized to UTC(USNO).

To meet the demanding GPS specifications for positioning and navigation, the satellite clocks have to be exceptionally stable. To better understand this, consider a typical situation where clock corrections are only received once during each orbital period, and where the maximum acceptable contribution from the satellite clocks to the positioning uncertainty is one meter. The signals travel at a speed of about 3.3 ns per meter, so the satellite clocks have to be stable enough to keep time accurate to within 3.3 ns for about 12 hours without being adjusted. Maintaining this level of accuracy requires frequency stability of less than 7.6×10^{-14} at τ = 12 hours. The performance of the satellite clocks is continuously monitored and analyzed at NRL to ensure that they meet their accuracy and stability requirements [16].

The GPS clocks run at a nominal frequency of 10.23 MHz, but their frequency has to be compensated for relativistic effects. To understand why, consider that the atomic definitions of the second are based on stationary clocks located on the geoid. The geoid represents the hypothetical shape of the Earth, which coincides with the mean sea level and its imaginary extension both under and over the Earth's land areas. According to Einstein's special and general theories of relativity, a moving clock will run slower and lose time relative to a stationary ground clock, and a clock located above the geoid will run faster than a clock on the ground. Both of Einstein's theories—which were published in 1905 and 1915, respectively—have been confirmed by all of the experimental tests that have since been made, and the invention of GPS was essentially a physical realization of Einstein's view of space and time [17].

The GPS satellites are rapidly moving, traveling at a velocity of about 3.9 km/s. In accordance with the theory of special relativity, this causes the atomic clocks on board the satellites to lose about 7 μs per day with respect to an atomic clock on Earth. The satellites are also located high above the geoid, where the gravitational field is much weaker than it is on the Earth's surface. In accordance with the theory of general relativity, this causes the satellite's atomic clocks to gain about 45 μs per day with respect to an atomic clock on Earth. Therefore, a GPS clock uncompensated for relativistic effects would gain about 38 μs per day.

To compensate for the effects of relativity, the 10.23 MHz frequency of the satellite clocks is set about 0.0046 Hz lower than nominal by applying a frequency offset of about -4.46×10^{-10}. This intentional offset is sometimes called the "factory offset" by GPS clock manufacturers. If the eccentricity of the GPS orbits was 0—which would be the case if the orbits were truly circular—the factory frequency offset would be sufficient and no further corrections would be necessary. However, the eccentricity of a GPS orbit can be as high as 0.02, which can cause a satellite clock to be ahead or behind by as much as 45 ns. This additional relativistic effect is compensated by the receiver's firmware, which computes a time correction based on the orbital parameters transmitted by the satellites [10, 11, 17, 18].

Rockwell International produced three generations of GPS satellites between 1978 and 1997—known as Blocks I, II, and IIA—that carried two rubidium and two cesium clocks. Those satellites are no longer operational, and the majority of the GPS satellites launched since 1997 have only carried rubidium clocks. The only exception has been the Block IIF satellites manufactured by Boeing from 2010 to 2016, which carried a single cesium clock in addition to two rubidium clocks. Figure 12.3 is an artist's depiction of a Block III satellite manufactured by Lockheed Martin. The final launch of a Block III satellite is scheduled to occur in 2026, when they will be succeeded by Block IIIF.

Figure 12.3 Artist's depiction of an orbiting GPS Block III satellite (courtesy of the USAF).

As of 2025, the average operational life span of the GPS satellites now retired from the constellation is greater than 11 years, and the average age of the satellites still in service is greater than 13 years. Both figures have greatly exceeded the specified life expectancies of the satellites. Some satellites have provided exceptionally long service—for example, one of the Block IIR satellites launched in 1997 was still operational in 2025. These longevity figures are impressive, but a continuous launch schedule for new satellites will always be needed to replenish the constellation. Table 12.1 lists the production generations of the GPS satellites, organized by their block numbers.

Table 12.1 The production generations, organized by block number, of the GPS satellites

Block number	Launch period	Number of launches	Satellites operational as of 2025	Manufacturer	Atomic clocks on board
I	1978–1985	11	0	Rockwell International	Rubidium (2) Cesium (2)
II	1989–1990	9	0	Rockwell International	Rubidium (2) Cesium (2)
IIA	1990–1997	19	0	Rockwell International	Rubidium (2) Cesium (2)
IIR	1997–2004	13	6	Lockheed Martin	Rubidium (3)
IIR-M	2005–2009	8	7	Lockheed Martin	Rubidium (3)
IIF	2010–2016	12	12	Boeing	Rubidium (2) Cesium (1)
III	2018–2026	6	6	Lockheed Martin	Rubidium (3) Extra Slot (1)
IIIF	2027–2037	0	0	Lockheed Martin	To be determined

12.3 The GPS Signal Format

The GPS satellites broadcast on three carrier frequencies, L1, L2, and L5, that are available to civilians for use in PNT applications. Two other carrier frequencies, L3 and L4, are used for scientific and military applications and have restricted access. The "L" refers to the satellite L-band, which is typically defined as having two main sections, a lower L-band ranging from 1164 MHz to 1300 MHz where L2 and L5 reside, and an upper L-band ranging from 1559 MHz to 1610 MHz where L1 resides. The three carrier frequencies are each obtained by multiplying the 10.23 MHz nominal clock frequency—multiplying by 154 produces the L1 frequency of 1575.42 MHz, multiplying by 120 produces the L2 frequency of 1227.6 MHz, and multiplying by 115 produces the L5 frequency of 1176.45 MHz.

The GPS signal format is based on a channel access method called code-division multiple access (CDMA), which allows multiple transmitters to simultaneously send data over a single communications channel. The CDMA method was later implemented in mobile telephone networks (Section 16.3), but it wasn't well known before being included in the GPS architecture. Its inclusion can be credited to the American engineer James Spilker (1933–2019), who co-founded Stanford Telecommunications in 1973, and who was one of the principal designers of the GPS signal format. Spilker and his colleagues found that CDMA was perfectly suited for making passive ranging measurements from multiple satellites [8].

CDMA employs spread spectrum communication techniques, where information that could be sent using a narrow-bandwidth signal is deliberately spread over a wider bandwidth. This makes the signals more secure, more resistant to interference and noise, and easier to detect even when transmitted at very low power levels. In the case of GPS, the power level of the signals received on Earth is exceptionally small, typically about –160 dBW, which is equal to 1×10^{-16} W. The received power level is so small that it's below the thermal noise floor of the atmosphere, but due to the spread spectrum method, the GPS signals are still easily detectable.

CDMA also provides a code-division scheme that allows multiple satellites, each identified by its own unique PRN code, to simultaneously send information to a receiver while using the same carrier frequency. The GPS PRN codes are numbered from 1 to 32, but the number of operational satellites is intentionally limited to 31, so there is always at least one code that's not in use. PRN is an acronym for pseudorandom noise, and the codes got their name because they appear to be random noise, even though they are neither random nor noise. Instead, they are predefined sequences of rectangular pulses, known as chips, that are spread across the available bandwidth and that represent binary ones and zeros. The codes are continuously sent, and they always repeat after a predefined period. Because they are spread across the spectrum, the codes are essentially invisible to all conventional radio receivers. They can only be detected by receivers that can generate replicas of the predefined PRN codes and then match them to the incoming signal using correlation techniques, which is exactly how a GPS receiver works.

The PRN codes used by GPS—as well as the PRN codes used for TWSTFT (Section 9.4.3)—are also called Gold codes, named after the American mathematician Robert Gold who invented the method at Magnavox Research Laboratories in 1967 [19]. Gold codes are generated by first selecting a pair of code sequences with low cross-correlation, known as *m*-sequences. The selected *m*-sequences both have the same length of $2^n - 1$, where n is the size of the linear feedback shift register that was used to generate the *m*-sequence. A set of $2^n - 1$ Gold code sequences is then generated by taking the exclusive ORs (XORs) of the two sequences in all relative positions. Each of the generated codes will have low cross-correlation with respect to the other codes in the set; the maximum cross-correlation is $2^{(n+2)/2} + 1$ for even n and $2^{(n+1)/2} + 1$ for odd n. Due to the nearly orthogonal nature of Gold codes, a GPS receiver can easily determine which satellite is broadcasting an incoming signal, even during non-optimal conditions where RF noise or interference flips some of the code bits.

As illustrated in Figure 12.4, several types of predefined and repeating PRN codes are included in the L1, L2, and L5 signals transmitted by the GPS satellites. The vast majority of GPS products, including most disciplined oscillators and clocks, only receive the L1 signals, and the most widely received code is the coarse acquisition (C/A) code that is only available on L1. The C/A code repeats every millisecond and transmits 1,023 chips per millisecond (1.023 megachips/s), so the code frequency is 1.023 MHz.

Both L1 and L2 also include a precision code that only repeats once every seven days. The precision code is transmitted at a rate of 10,230 chips per millisecond (10.23 megachips/s), a code frequency of 10.23 MHz that matches the nominal GPS clock frequency. The public, unencrypted version of the precision code is known simply as the P code, and the encrypted version reserved for military use is called the P(Y) code. The satellites also broadcast an M-code reserved for military use on L1 and L2 with a code frequency of 0.5115 MHz. Like the P(Y) code, the details of the M-code are classified and not publicly available [10, 11, 18].

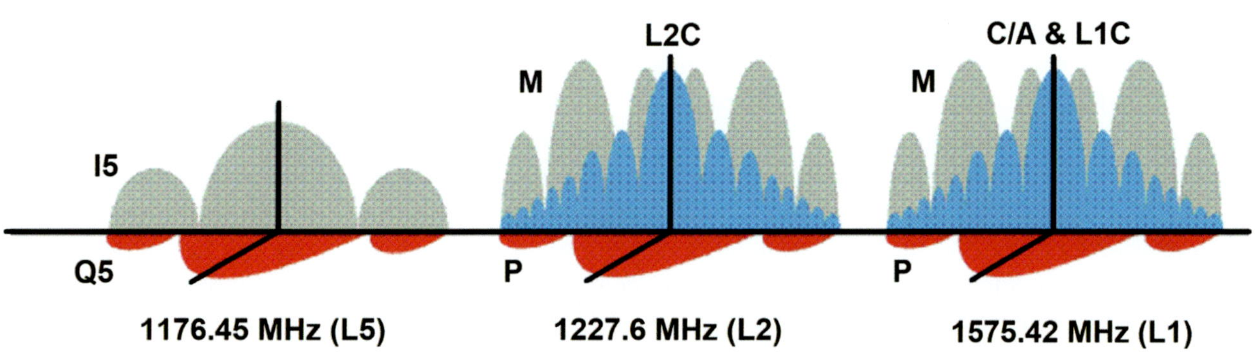

Figure 12.4 Simplified diagram of the L1, L2, and L5 frequency spectrum.

The L1 C/A code and the L2 P code have been available since GPS began, so they are known as the legacy signals. The L1 C/A code was the first GPS signal intended for commercial and civilian use, so it's known as the legacy civil signal. As part of an ongoing GPS modernization program, three new civil signals have since been added—the L5, L2C, and L1C signals shown in Figure 12.4. These signals have added new capabilities to GPS, but the legacy civil signals still support most of the huge base of existing equipment. The L5 signals were first broadcast in April 2014, and that same month the L2C codes were added to the L2 carrier. The L1C codes were added to the L1 carrier in March 2019.

When the new civil signals were introduced, they could only be broadcast by the recently launched satellites that were equipped with upgraded transmitters. The number of satellites having L5, L2C, and L1C capability has since increased each time a new satellite is launched, and well over half of the satellites are broadcasting the L2C and L5 signals as of 2025. Even so, we are still some years away from the day when all 31 satellites in the constellation can broadcast all four of the civil signals.

The L5 signals were designed for safety-of-life applications such as aircraft guidance, and are broadcast in a part of the lower L-band that's exclusively reserved for aviation safety services. Two PRN codes are transmitted on L5, an in-phase code called I5, and a quadrature phase code called Q5. The two codes are identical, differing only in their initial state, and both repeat every millisecond. The I5 and Q5 codes transmit 10,230 chips per millisecond, so their 10.23 MHz code frequency is the same as the P code and GPS clock frequencies.

The L2C signals contain two distinct PRN code sequences; one called the civil-moderate code (L2CM) and the other called the civil-long code (L2CL). The L2CM code is 10,230 chips long and repeats every 20 ms. The L2CL code is 767,250 chips long and repeats every 1.5 s. Each signal is transmitted at 511,500 chips per second, but they are multiplexed together to form a 1.023 megachips/s signal that matches the L1 C/A code frequency. The L2CM signal is modulated with a navigation message, but L2CL doesn't contain any modulated data. The long, dataless L2CM sequence provides L2C with a correlation strength that's about 250× higher than the L1 C/A code.

The L1C signals were the result of a collaborative effort between the United States and Europe to develop a common civil signal that would allow interoperability between GPS and Galileo. The L1C PRN codes are 10,230 chips long and repeat every 10 ms, so their code frequency of 1.023 MHz again matches the frequency of the L1 C/A code.

Three components—the RF carrier, the PRN codes, and a binary navigation message—are combined to form the GPS signals. Figure 12.5 is a simplified diagram that shows how the legacy signals are combined on L1 and L2. The PRN codes are combined with the navigation data using modulo-2 addition, where an XOR function is used to sum two bits. If the code chip and the data bit are the same—both 1s or both 0s—the resulting bit is 0, but if the two are different, the resulting bit is 1. The composite binary signal is then impressed on the carrier with BPSK modulation, where a 0 bit leaves the carrier unchanged, and where a 1 bit multiplies the carrier by −1, which shifts the carrier phase by 180°. Therefore, the carrier phase is shifted by 180° each time a bit transition from 0 to 1, or from 1 to 0, occurs [10].

The legacy navigation message is broadcast on L1 and L2 at a rate of 50 bps. As shown in Figure 12.6, the navigation message includes 25 frames of information. Each frame has five subframes, each subframe has 10 words, and each word has 30 bits, so the entire navigation message consists of 37,500 bits ($25 \times 5 \times 10 \times 30 = 37{,}500$). Each bit has a period of 20 ms, so it takes 750 s to send the entire message.

The first three subframes of every frame contain information specific to the satellite that's broadcasting the message. Subframe 1 contains information about the satellite's clock, the satellite's health, and other status information. Subframes 2 and 3 contain the ephemeris data for the satellite, which is the orbital information needed to compute the satellite's position. The information in subframes 1 through 3 is repeated every 30 s, and is usually updated every two hours.

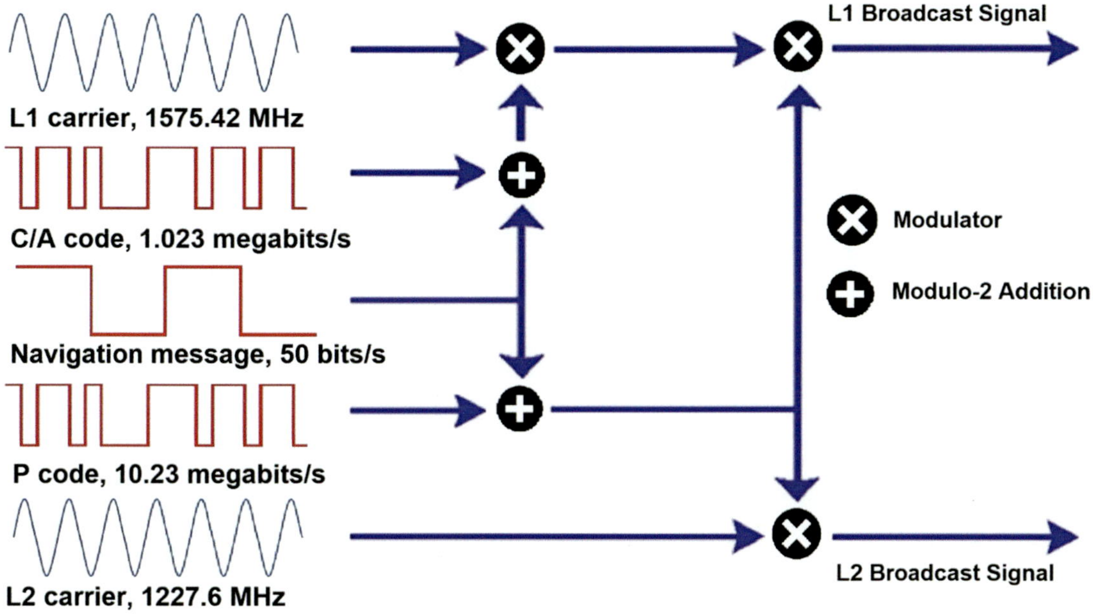

Figure 12.5 GPS signal modulation.

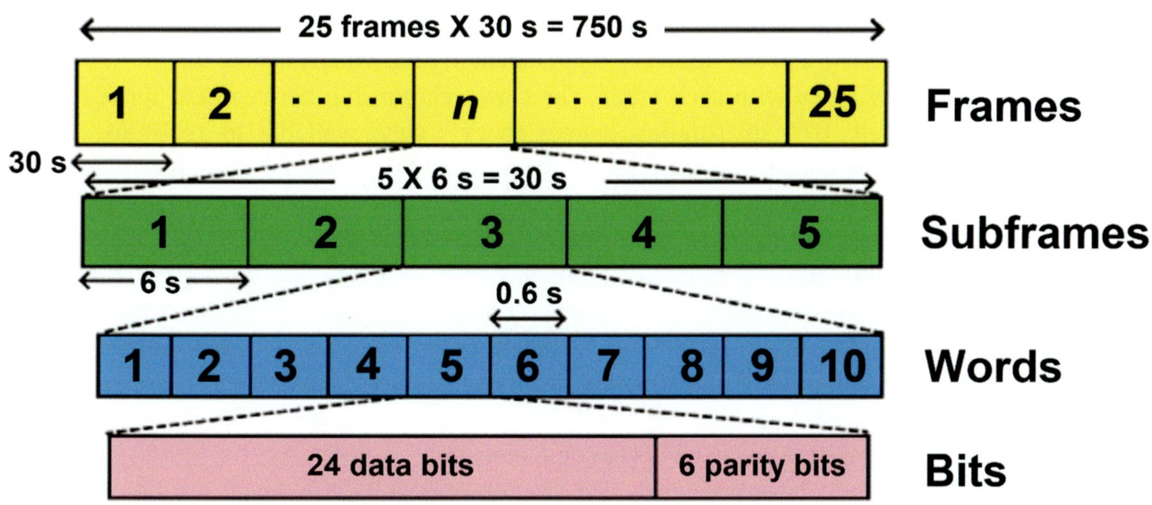

Figure 12.6 The data structure of the legacy navigation message.

Subframes 4 and 5 contain the GPS almanac. All satellites broadcast the same almanac, which consists of 25 pages. One page is broadcast in every frame, so receiving the entire almanac requires receiving the entire navigation message. The almanac contains the rough coordinates of all of the GPS satellites. When a GPS receiver is first turned on, it downloads the almanac from the first satellite it finds, and then uses the almanac data to find the other visible satellites. The almanac also contains information essential to the GPSDCs that we'll describe later in this chapter, including the parameters for the UTC correction (Section 12.4) and the ionospheric delay model (Section 12.7.5).

Table 12.2 summarizes the characteristics of the GPS L1, L2, and L5 signals, including their minimum RF power as received on Earth. As shown in the table, the frequency allocations for each band slightly exceed their minimum bandwidth requirement of 20.46 MHz, which is equal to 2× the maximum PRN code frequency of 10.23 MHz.

Table 12.2 Characteristics of the GPS L1, L2, and L5 signals

Band	Center frequency (MHz)	Available bandwidth (MHz)	PRN code name	PRN code length (chips)	PRN code period	PRN code frequency (MHz)	Minimum received RF power (dBW)	Data rate (bps)
L1	1575.42	24	C/A	1,023	1 ms	1.023	−158.5	50
			P	6.19×10^{12}	7 days	10.230	−161.5	50
			L1C	10,230	10 ms	1.023	−157.0	50
			M	—	—	5.115	−158.0	—
L2	1227.6	22	P	6.19×10^{12}	7 days	10.230	−161.5	50
			L2CM	10,230	20 ms	0.5115	−158.5	50
			L2CL	767,250	1.5 s	0.5115	−158.5	—
			M	—	—	5.115	−158.0	—
L5	1176.45	25	I5/Q5	10,230	1 ms	10.230	−157.0	50

12.4 GPS Time and Its Relationship to UTC(USNO) and UTC

GPS time (GPST) is referenced to UTC(USNO). The UTC(USNO) time scale is accurate to within a few nanoseconds of UTC (Figure 9.18), and this accuracy is passed on to GPS, which ensures that UTC, UTC(USNO), and GPST will always closely agree with each other. The close relationship between the three time scales makes GPS a highly accurate UTC distribution source that's freely available in real-time. When the measurement uncertainties are documented, a traceability chain to the SI can be established via the direct reception of GPS signals (Figure 9.24).

The GPST time code consists of just two numbers—the number of weeks that have elapsed since the GPS epoch of January 6, 1980, and the number of seconds that have elapsed during the current week. The week number is included in subframe 1 of the navigation message. The time of week (TOW) indicator that counts seconds is included in every subframe, and updates every 6 s. Unfortunately, the legacy navigation message broadcasts the week number as a 10-bit code, which limits its maximum value to 1,023 ($2^{10} - 1$). When the week number reaches a value of 1,023, it rolls back to zero and rolls the date back by about 19.7 years, which has caused some receivers to completely fail.

Week number rollover events have previously occurred on August 21, 1999, and April 6, 2019, and the next one is scheduled to occur on November 20, 2038. Some GPS receivers were only designed to handle one or two rollovers after the issuance of their firmware, and those devices have either already failed or will eventually fail if their firmware isn't updated. However, most receivers are now designed to handle all week rollovers without interruption. To increase the interval between rollovers by a factor of eight, a 13-bit week number was added to a civil navigation (CNAV) message that's now broadcast on L1C, L2C, and L5. The receivers that access the 13-bit CNAV message will only experience a week rollover once every 157.5 years, with the first rollover occurring in the year 2137.

Unlike UTC, GPST ignores leap seconds. However, GPST can be converted to UTC by applying two corrections, one large and one small, that are broadcast by the satellites. The large correction, expressed in integer seconds, compensates for the number of leap seconds that have been added to UTC since the GPS epoch. The small correction, expressed in nanoseconds, compensates for the most recent measured time difference between GPST and UTC(USNO). Nearly all GPS receivers apply both corrections by default, so disabling the corrections is often not even possible. Both corrections are included in page 18, subframe 4, of the GPS almanac [18]. Because the GPS signals are referenced to UTC(USNO), and because

UTC(USNO) is in turn referenced to UTC, a GPSDC that applies both corrections can provide a very close approximation of UTC in real-time [20].

The UTC offset correction, Δt_{UTC}, is computed as

$$\Delta t_{\mathrm{UTC}} = \Delta t_{\mathrm{LS}} + A_0 + A_1(t_{\mathrm{E}} - t_{\mathrm{ot}} + 604800(WN - WN_t)), \tag{12.1}$$

where

Δt_{LS} is the number of leap seconds added to UTC since the GPS epoch,

A_0 is the constant UTC offset parameter expressed in seconds,

A_1 is a unitless frequency offset value that allows the correction of the time error accumulated since the UTC reference time, t_{ot}, when A_0 was last updated,

t_{E} is GPST (the time being converted to UTC),

604800 is a constant that equals the number of seconds in one week,

WN is the GPS week number, and

WN_t is the UTC reference week number.

The first part of Eq. (12.1), $\Delta t_{\mathrm{LS}} + A_0$, takes care of most of the UTC correction. The Δt_{LS} term, expressed in integer seconds, is the leap second correction. The A_0 term is the small, nanosecond part of the correction, equal to the difference between the GPST and UTC(USNO) second markers. This correction is broadcast in units of seconds, but it typically has a value of less than 1×10^{-8} s, or less than 10 ns.

The remaining part of Eq. (12.1) fine-tunes the UTC output of a GPSDC by applying a drift correction provided by A_1. The drift correction is in the form of a unitless frequency offset value, and it accounts for the period between the current time and the time when A_0 was last updated. The magnitude of this correction is usually less than 1 ns, because the GPS broadcast normally updates A_0 more than once per day, and because the drift correction is usually about 1 ns for a whole day [21, 22].

The UTC(USNO) prediction broadcast by GPS is based on an extrapolation of the observed difference between GPST and UTC(USNO) from the start to the end of the previous day. Figure 12.7 shows the daily differences for the year 2021 between the UTC(USNO) predictions broadcast by GPS and the actual UTC(USNO) time scale [22]. The daily values all fell within a range of ±2.1 ns. Their mean value was essentially zero, −0.1 ns, and their standard deviation was 0.7 ns, which indicates that the expanded uncertainty (k = 2) of the GPS signal as transmitted in space is about ±1.4 ns with respect to UTC(USNO). The BIPM's *Circular T* document now publishes daily estimates of the time differences between UTC and the UTC predictions broadcast by GPS, GLONASS, Galileo, and BeiDou. This documentation can help complete a traceability chain to the SI for time measurements referenced to any of these four GNSS broadcasts [23].

Continuous measurements of the UTC − UTC(USNO) time difference made over the 15-year period from 2009 to 2023—obtained from the BIPM's *Circular T*—show that the two time scales always agreed to well within ±10 ns (Figure 12.8). The time difference has progressively gotten smaller, and was kept within ±2 ns during the most recent years shown on the graph. The average time difference during the 15-year period was 0.5 ns. The k = 1 uncertainty reported by the BIPM during the comparison period was as small as 1.0 ns and never exceeded 4.1 ns, indicating that the expanded uncertainty (k = 2) always stayed within a range from 2.0 to 8.2 ns. The UTC(USNO) time scale has a negligible amount of frequency inaccuracy with respect to UTC, as indicated by the gentle slope of the blue linear least squares line. Using the methods described in Section 6.1, the average frequency accuracy of UTC(USNO) during the 15-year period is estimated as -1.6×10^{-17}.

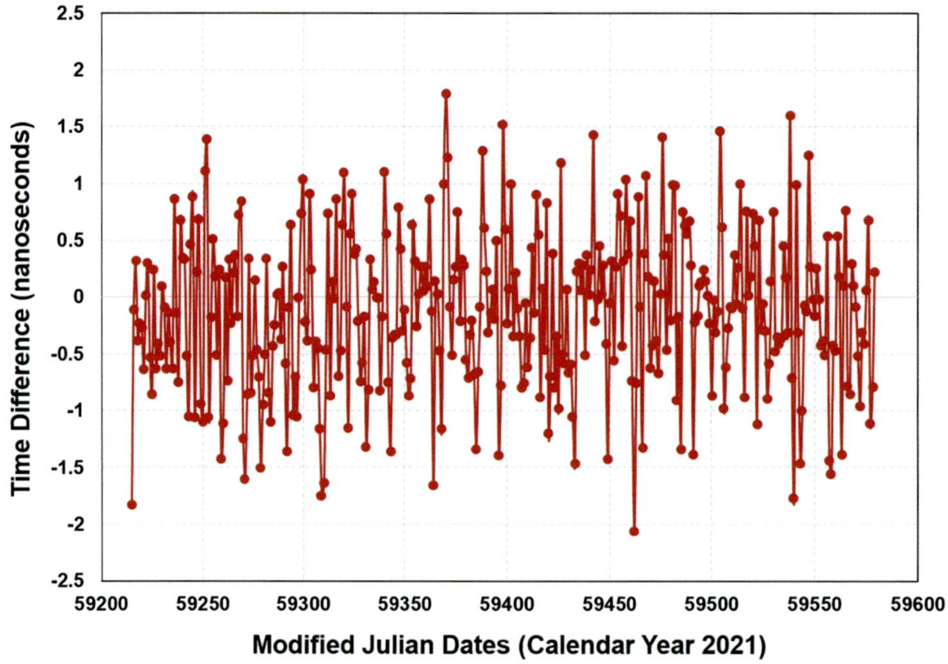

Figure 12.7 The daily time differences between the UTC(USNO) prediction broadcast by GPS and the actual UTC(USNO) time scale for the year 2021 [22].

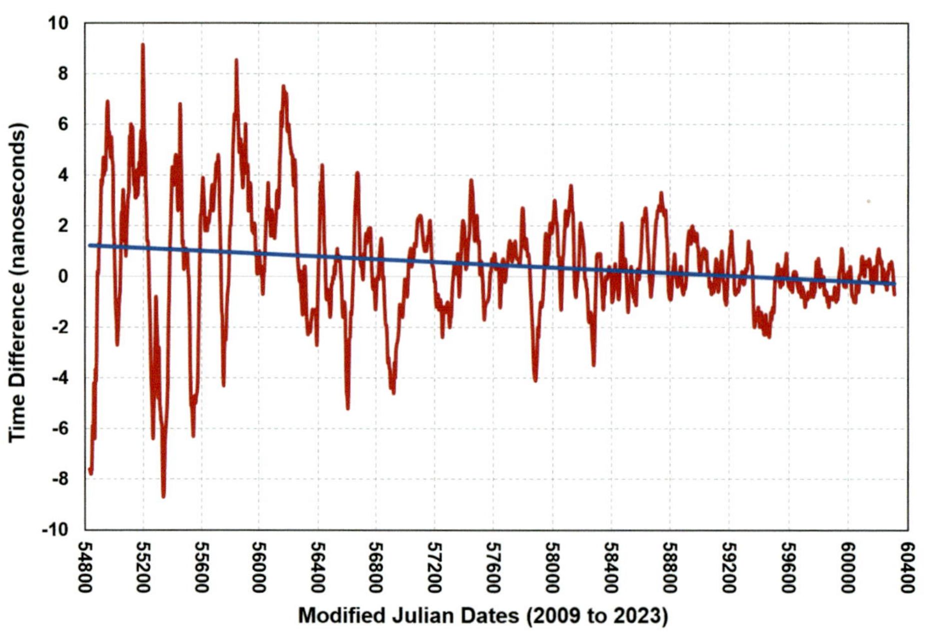

Figure 12.8 The time difference between UTC and UTC(USNO) during the 15-year period from 2009 to 2023.

As discussed in Section 9.6.1, the UTC(USNO) and UTC(NIST) time scales are both recognized as official sources of time in the United States. The two time scales are continuously compared using TWSTFT and GPS-based methods, and are kept in very close agreement [24]. The UTC(USNO) – UTC(NIST) time differences for the 15-year period from 2009 to 2023, again obtained from the *Circular T*, are shown in Figure 12.9. During this period, the time differences have never exceeded a range of ±20 ns, and have remained within ±4 ns in recent years. The average UTC(USNO) – UTC(NIST) time difference was 0.9 ns, and the average frequency difference was -2.1×10^{-17}.

The 1 pps signal from a GPSDC that's been calibrated for equipment delays is an accurate source of UTC(USNO)—a source we can call UTC(USNO) via GPS. This means that when a GPSDC is directly compared to UTC(NIST), the resulting time differences should closely agree with the UTC(USNO) - UTC(NIST) time differences shown in Figure 12.9. To illustrate this, Figure 12.10 overlays the results of a GPSDC – UTC(NIST) direct comparison onto the Figure 12.9 data. The calibrated GPSDC was based on a L1 C/A code receiver and a quartz LO, and it ran continuously at NIST in Boulder, Colorado, during the entire 15-year period.

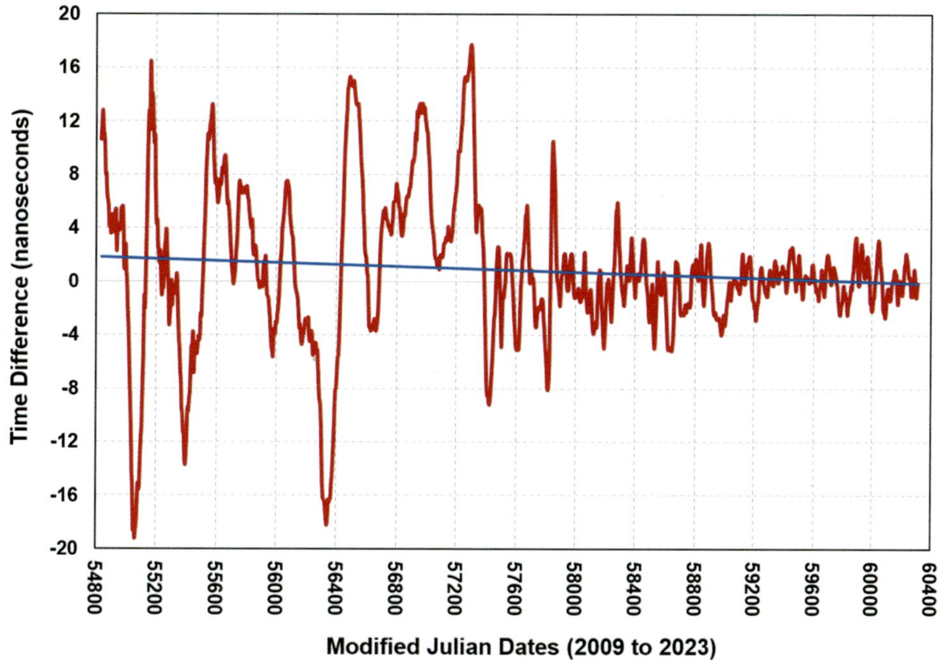

Figure 12.9 The time difference between UTC(USNO) and UTC(NIST) during the 15-year period from 2009 to 2023.

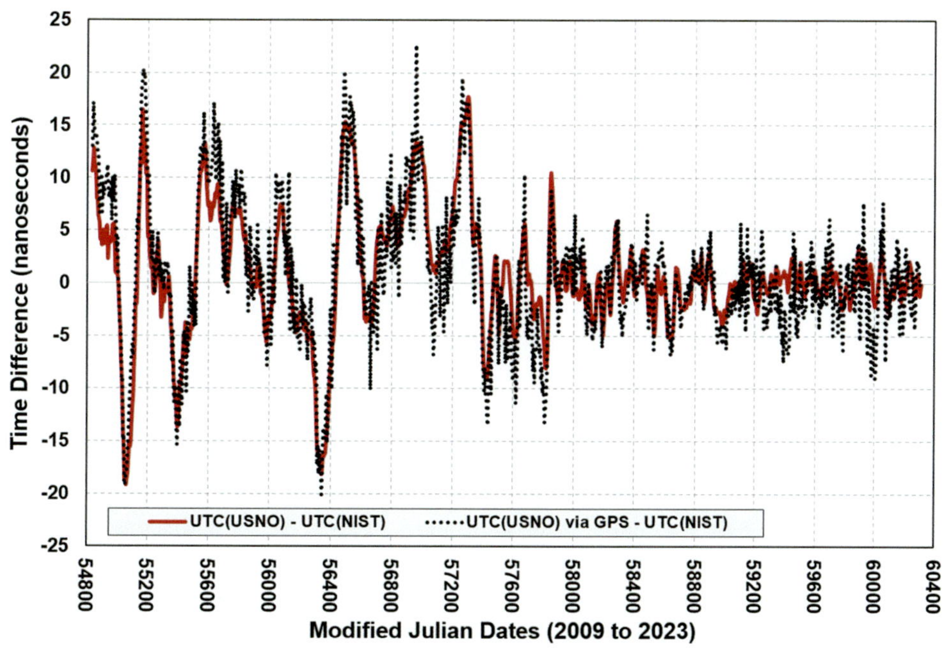

Figure 12.10 The time difference between UTC(USNO) via GPS and UTC(NIST), compared to the *Circular T* data, during the 15-year period from 2009 to 2023.

To match the interval of the *Circular T*, only one data point is graphed every five days, with the GPSDC data point obtained from a 24-hour average. The GPSDC measurement has larger outliers, mostly due to uncertainties in the ionospheric delay model and the receiver's susceptibility to temperature variations. However, the phase structure of the GPSDC measurement closely matches the *Circular T* results, and both measurements produced the same average time difference of 0.9 ns. The GPSDC's average frequency difference of -5.8×10^{-17} was slightly larger than the *Circular T* average due to the increase in measurement noise, but only differed by a few parts in 10^{17}. These very long-term measurements verify that the 1 pps output of a delay-calibrated GPSDC is essentially equivalent to UTC(USNO).

Table 12.3 shows the average time and frequency differences between UTC and UTC(USNO), and between UTC(USNO) and UTC(NIST), each obtained from *Circular T* data. It also shows the time and frequency difference between UTC(USNO) via GPS and UTC(NIST), as obtained from the direct comparison measurement. All differences are obtained from continuous measurements made over a 15-year period. The frequency difference estimates were obtained from Figures 12.8 through 12.10 as $\frac{\Delta t}{T}$, with Δt obtained by taking the slope of a linear least squares line (Section 6.1). The frequency differences are provided as parts in 10^{17}, which indicates that the frequency of a GPSDC—when measured over a sufficiently long period—is about 1,000× more accurate than the frequency of a free-running cesium beam clock. The frequency accuracy of a free-running cesium beam clock is usually limited to a few parts in 10^{14}, because unlike a GPSDC, their frequency is not being corrected to agree with UTC.

Table 12.3 Time and frequency differences between UTC, UTC(USNO), UTC(USNO) via GPS, and UTC(NIST), when averaged over a 15-year period

Time scale comparison	Time difference (ns)	Frequency difference ($\times 10^{-17}$)	Data source
UTC – UTC(USNO) from *Circular T*	0.4	−1.7	Figure 12.8
UTC(USNO) – UTC(NIST) from *Circular T*	0.9	−2.1	Figure 12.9
UTC(USNO) via GPS – UTC(NIST)	0.9	−5.8	Figure 12.10

12.5 GPS Disciplined Oscillators and Clocks

As was described in Section 5.3.2, a disciplined oscillator includes a LO, a receiver and antenna that receives signals transmitted by a UTC reference source, and a frequency or phase comparator. The comparator measures the frequency or phase difference between the LO and the incoming reference signal, and it converts this difference to a correction that is periodically applied to the LO by a control loop. This process is continuously repeated to keep the LO locked to the reference signal. A disciplined clock adds a frequency divider circuit that divides the disciplined oscillator frequency down to 1 pps. It also adds the hardware and software needed to synchronize the 1 pps signal to UTC.

These same basic methods are implemented in GPS disciplined oscillators and clocks, and the same device usually performs both the frequency and time functions. A GPSDO never requires the frequency of its LO to be adjusted. However, depending upon its performance requirements, a GPSDC might require calibrations and adjustments that compensate for signal and equipment delays. As we'll discuss in Section 12.7, delay compensation can reduce the uncertainty and improve the accuracy of a GPSDC.

A quartz oscillator—typically either a TCXO or an OCXO—serves as the LO in most GPS disciplined devices. However, the higher-cost devices often use either a conventional rubidium oscillator, or a rubidium or cesium CSAC, as their LO. The control loop parameters are designed to match the LO's stability characteristics. The period of the control loop can be fixed or variable, but an adaptive control loop is desirable because it allows corrections to be applied more frequently when needed. For example, reducing the control loop period when the device is first turned on can speed up the locking process. When the device is locked, the optimal period between corrections can be estimated by comparing the stability of the free-running LO to the stability of the received GPS signals, and determining the period where the LO instability first exceeds the GPS instability. The graph previously shown in Figure 7.14 is an example of this type of comparison.

The receiver in most GPS-disciplined clocks and oscillators is a single-frequency unit that receives the L1 signals on 1575.42 MHz. Only a small percentage of GPSDO/DC devices use multi-frequency receivers. The receiver usually consists of just a small circuit board or a single IC, and the same receiver model can often be found in the products offered by multiple manufacturers. Not all GPS receivers will work in a GPSDO/DC design, because the OEM products designed for positioning and navigation often lack some of the necessary features. The receivers suitable for GPSDO/DC designs always include a 1 pps output, and in some cases include a programmable frequency synthesizer. In addition, the receiver's firmware was usually written specifically for time and frequency applications. Figure 12.11 is a photograph of the interior of a typical GPSDO/DC, showing the receiver board and associated circuitry, and the 10 MHz OXCO that serves as the local oscillator. The connectors provide a computer interface, outputs for 1 pps and 10 MHz, and an input for the antenna.

Figure 12.11 The interior of a typical GPSDO/DC device showing the receiver board, the local oscillator, and the input and output connectors.

There are several types of GPS antennas. The simplest types are called patch antennas, because they are printed on a "patch" of dielectric material, and mounted on a larger piece of metal that serves as a ground plane. The enclosure of a patch antenna can be quite small—a typical size might be 35 × 35 × 15 mm—and they are often designed to affix to the indoor-side of a window, where they can obtain a partial view of the sky that's sufficient for receiving a few satellites. However, for best results, a GPS antenna should be located outdoors and high enough off the ground so that its view of the sky is unobstructed on all sides. The most common outdoor GPS antennas are also patch antennas, but the patch is covered by a plastic radome that protects it from the weather, as shown in Figure 12.12. Underneath the radome, the patch is mounted on a relatively large ground plane that blocks some of the signals that reflect from surfaces below the antenna.

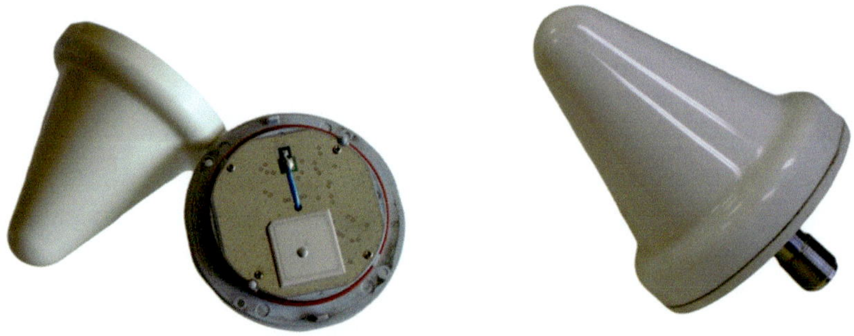

Figure 12.12 A GPS patch antenna with its radome removed (left) and attached (right).

Several other types of antennas—including the quadrifilar helix, choke ring, and pinwheel—can improve upon the performance of a patch antenna. These antennas can do a better job of blocking the reception of unwanted signals—specifically the multipath signals that don't travel directly from the satellite to the antenna, but instead reach the antenna after being reflected by surrounding objects.

The quadrifilar helix antenna, pictured in Figure 12.13, gets its name from its helical shape, which is similar to a corkscrew or a spiral staircase, and because the helix includes four conductors that are symmetrically spaced around a common axis. The conductors can be made of wire, but they are typically made from copper strips or tubing. The quadrifilar helix is well suited for GPS reception because its omnidirectional pattern allows it to receive signals coming from satellites in all directions. It's also specifically designed to receive GPS and other satellite signals that are right-hand circularly polarized (RHCP), meaning that the electric field of the incoming signal rotates clockwise as it moves forward. A reflected signal will reverse its polarization, becoming left-hand circularly polarized (LHCP), and a quadrifilar helix is able to reject most of the LHCP signals [25, 26].

Figure 12.13 A GPS quadrifilar helix antenna with its radome removed (left) and attached (right).

The choke ring antenna was invented at NASA's Jet Propulsion Laboratory (JPL) in the late 1980s. It was designed to improve the performance of omnidirectional antennas that were susceptible to positioning errors caused by multipath signals. It did so by shifting the antenna's gain pattern upward, which prevented the reception of signals reflected from objects at low elevation angles [27]. A choke ring antenna consists of a ground plane, a receiving element, and a series of concentric rings made of conductive material that act as shorted transmission lines, preventing the unwanted signals from reaching the receiving element (Figure 12.14).

Figure 12.14 A GPS choke ring antenna.

The optimal diameter of the ground plane is 1.5× the wavelength of the GPS signal, which is about 28.5 cm in the case of the 19 cm wavelength of L1. However, some choke ring antennas are constructed with smaller-diameter ground planes to reduce their size and weight, and others have larger-diameter ground planes designed to accommodate the lower frequency L2 and L5 signals. The receiving element, which is usually a patch antenna, is located in the center of the ground plane. It's surrounded by concentric rings, with a typical design having three to five rings. The depth of each ring is usually one-quarter of the wavelength of the GPS frequency, which is about 4.75 cm in the case of L1, but the depth can vary to accommodate the other carrier frequencies.

The pinwheel antenna was invented by the NovAtel Corporation of Canada in the late 1990s, and was specifically designed to be both smaller and lighter than a choke ring, with most of the reduction in size and weight due to the elimination of the large metal ground plane [28]. The term "pinwheel antenna" is a registered trademark of NovAtel, but due to their appearance, pinwheel antennas are sometimes colloquially referred to as "flying saucers" (Figure 12.15). The case has a diameter of about 18 cm, and it houses a phased array antenna that consists of multiple spiral slots that are optimized to only receive RHCP signals. An array of slot rings located outside the perimeter of spiral slots further reduces the multipath problem, by blocking LHCP signals from low elevation angles that reach the antenna's edges. Pinwheel antennas have a very stable phase center, which is an advantage in both navigation and timing applications. Their ability to reject multipath and other unwanted signals often compares favorably to a choke ring, even though pinwheel antennas are smaller and typically cost less.

Figure 12.15 A GPS pinwheel antenna.

The GPS antennas that were just described are nearly always designed as active antennas, which means that they have a built-in low noise amplifier (LNA) that typically adds about 25 dB to 40 dB of gain to the incoming signals. The LNA typically requires 5 V DC, which is supplied by the receiver through the antenna cable. The gain provided by the LNA allows the antenna to compensate for the attenuation in the antenna cable, which can be large at the GPS carrier frequencies. As a general rule, the antenna gain should exceed the cable attenuation by at least 6 dB to 10 dB. For example, the attenuation at the L1 frequency of 1575.42 MHz is about 20 dB for 30 m of RG-58 coaxial cable, so in this situation an antenna with 30 dB of gain will probably work, but an antenna with 25 dB of gain probably won't.

Cable runs longer than about 30 m nearly always require lower-loss cable than RG-58, and if the cable is sufficiently long, in-line amplifiers might need to be installed to boost the signal. Very long cable runs of hundreds of meters are sometimes required in city environments, because the antenna might be located on the roof of a tall building, many floors above the receiver's location. In this case, in-line amplifiers can be used to make the receiver work, but replacing the cable with an optical fiber is often a better solution. However, when optical fiber is used, signal converters are needed at both ends of the fiber. The RF signal received by the antenna is converted to an optical signal for transmission through the fiber, and the optical signal is converted back to RF before reaching the GPS receiver.

Figure 12.16 is an example of a simple GPSDO design. The frequency or phase difference between the GPS signal and the LO signal is measured using a phase comparator, which can be based on any of the methods described in Chapters 4 and 5. The LO typically has a nominal frequency of 10 MHz, so its signal is divided to a lower frequency prior to the phase comparison—often to 1 pps. The output of the phase comparator is read by a microprocessor unit (MPU) whose firmware executes a control loop, which is often some variation of a PID controller (Section 5.3.2). The control loop keeps the LO locked to GPS by continually issuing frequency or phase corrections. In this example, the LO is a VCO, so the corrections are issued by varying the control voltage. The loop period is usually less than an hour, even when the LO is a rubidium. If the LO is an inexpensive quartz oscillator with little or no temperature compensation, the loop period could be a few seconds or less.

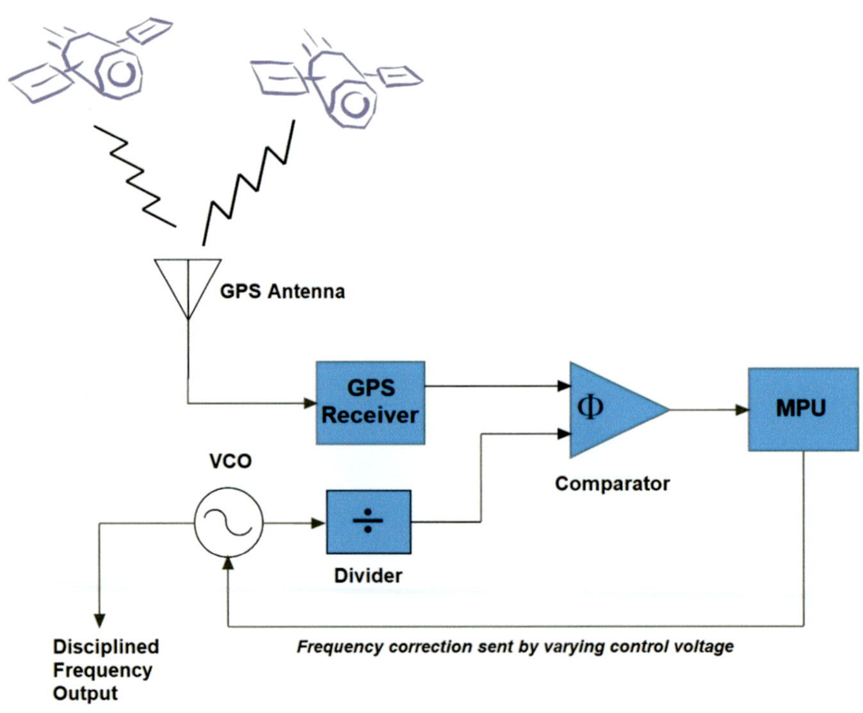

Figure 12.16 A GPSDO where the frequency is adjusted by varying the VCO control voltage.

Figure 12.17 shows an improved design that sends frequency corrections to a direct digital synthesizer. This design has at least two advantages. It doesn't disturb the stability of the free-running LO, because the corrections are applied to the output of the synthesizer. It also allows very small frequency corrections to be issued. For example, a DDS with 1 μHz resolution can apply corrections as small as 1×10^{-13} when the LO has a nominal frequency of 10 MHz.

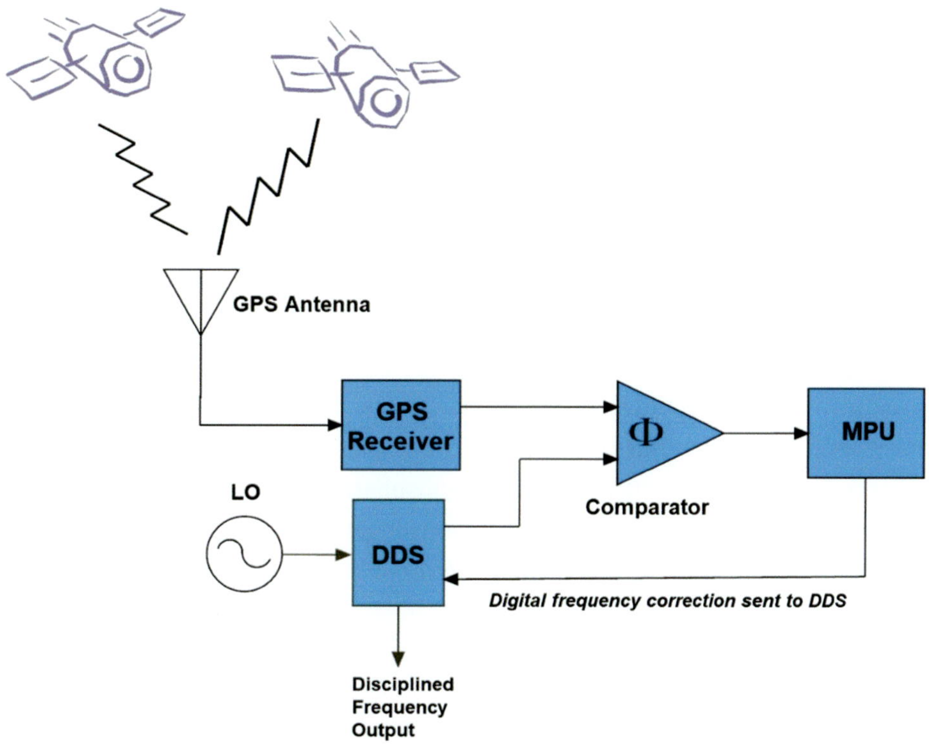

Figure 12.17 A GPSDO where the frequency is adjusted by correcting the output of a DDS.

A well-designed GPSDO outputs signals that are smooth and stable by avoiding large phase fluctuations. This means that the control loop has to ignore the short-term phase fluctuations of GPS signals and to rely mostly on the LO for its short-term stability. However, the control loop has to always track the GPS phase to get its accuracy, so it also needs to be able to quickly respond to situations when the LO frequency is disturbed, to prevent the device from losing lock. An optimally designed control loop will ensure that the disciplined frequency is always accurate and stable in both the short and the long term [29].

In a GPSDC design, a signal from the locked LO is divided to 1 pps, and the 1 pps signal is synchronized to the UTC second. Very accurate synchronization is possible, because the path delay from the satellites to the receiver can be potentially measured and compensated with an uncertainty of just a few nanoseconds. As noted in Section 12.2, the nominal height of the GPS satellites is 20,200 km above the Earth's surface, so the path delay from a satellite that is directly overhead is roughly 67 ms. This delay can increase to as large as 86 ms for a satellite near the horizon. In either case, compensating for the path delay to within one nanosecond requires measuring the path delay with an uncertainty of less than 2×10^{-8}.

A GPS receiver can routinely measure the path delay from the satellites with very small uncertainties, mostly because the satellite signals have the major advantage of coming from the sky. Unlike terrestrial HF and LF signals that propagate within a waveguide bounded by Earth's surface and the ionosphere (Chapter 10), satellite signals pass through the ionosphere on their way down to the receiver. The satellite path is longer, but it's usually unobstructed, and this allows the delay to be measured with less uncertainty than the delay from a terrestrial radio station.

Compensating for path delay is also easier with GPS because the positions of both the satellites and the receiver are accurately known. The position and velocity of each satellite can be estimated by using orbital data from the GPS broadcast. This allows the receiver, through a series of range measurements performed with multiple satellites, to calculate its own position. The time difference between the satellite and receiver clocks is then converted to distance by multiplying by the speed of light constant. Due to the factors that contribute to measurement uncertainty, the measured range isn't exactly equal to the true geometric range, and for this reason it's called the pseudorange. The pseudorange, p, is calculated as

$$p = \rho + c\,(\Delta t_r - \Delta t_s) + d_{ion} + d_{trap} + \varepsilon_{mp} + \varepsilon_{m}, \qquad (12.2)$$

where

ρ is the true geometric range,

c is the speed of light constant,

Δt_r and Δt_s are the time offsets of the receiver and satellite clocks with respect to GPST,

d_{ion} is the distance/delay added to the signal path by the ionosphere,

d_{trop} is the distance/delay added to the path by the troposphere,

ε_{mp} is the distance/delay added to the path by multipath signal reflections,

and ε_{rn} is the distance/delay added to the path by the receiver and antenna, and by other factors that add noise to the pseudorange measurement [10, 15].

Section 12.7 explores the factors that contribute to ionospheric, tropospheric, multipath, and receiver/antenna delays, and by doing so provides a method for evaluating the time uncertainty of a GPSDC. However, we'll first describe a method for evaluating the frequency uncertainty of a GPSDO.

12.6 Evaluating the Frequency Uncertainty of a GPSDO

A GPSDO typically outputs a sine wave that has a nominal frequency of 10 MHz. An ADEV graph, obtained from measurements comparing the 10 MHz output of a GPSDO to UTC(NIST), is shown in Figure 12.18. The graph shows that this GPSDO was stable to better than 1×10^{-11} at all averaging periods, when τ_0 was equal to 1 minute. The short-term stability is due to the presence of a stable LO, which in this case was an OCXO, and the long-term stability is due to the frequency corrections made with respect to GPS. The frequency stability at $\tau = 1$ day was 1.4×10^{-13}, which is typical performance for an OCXO-based device.

As previously discussed in Section 7.1.2, an ADEV graph for a disciplined oscillator typically includes a bump at the period where the external reference stability drops below the LO stability. The presence of the bump indicates the approximate period of the control loop that makes the frequency corrections. After the frequency corrections begin, there is a period where the GPSDO stability stays about the same or gets slightly worse, which in Figure 12.18 is the period when τ is between 100 s and 1,000 s. The stability rapidly improves from that point forward. A close inspection of this graph—and of most GPSDO stability graphs—will reveal smaller bumps that appear at even multiples of the correction period.

As we touched upon in Section 7.1.2, ADEV was designed to estimate the frequency stability of free-running oscillators, and not disciplined oscillators. The flicker frequency region of an ADEV graph will never be reached with a GPSDO because its frequency is continuously being corrected to agree with UTC. In our example, the GPSDO stability dropped below 1×10^{-13} after a period of about two days, and continued to improve afterwards. This improvement will continue indefinitely if more measurements are performed.

There will be periods when the rate of improvement will slow down, and other periods where the stability will temporarily get worse due to a variety of factors—including inexact frequency corrections or very-low-frequency noise sources caused by seasonal variations. However, a GPSDO should never reach a true flicker frequency noise floor. This means, for example, that if the frequency of a cesium clock is measured with respect to a GPSDO over a sufficiently long period, it should make little difference which make or model of GPSDO is used as the reference. The stability of all GPSDOs should be about the same when measured for periods of multiple weeks or months, provided that they always stay locked to the GPS signals.

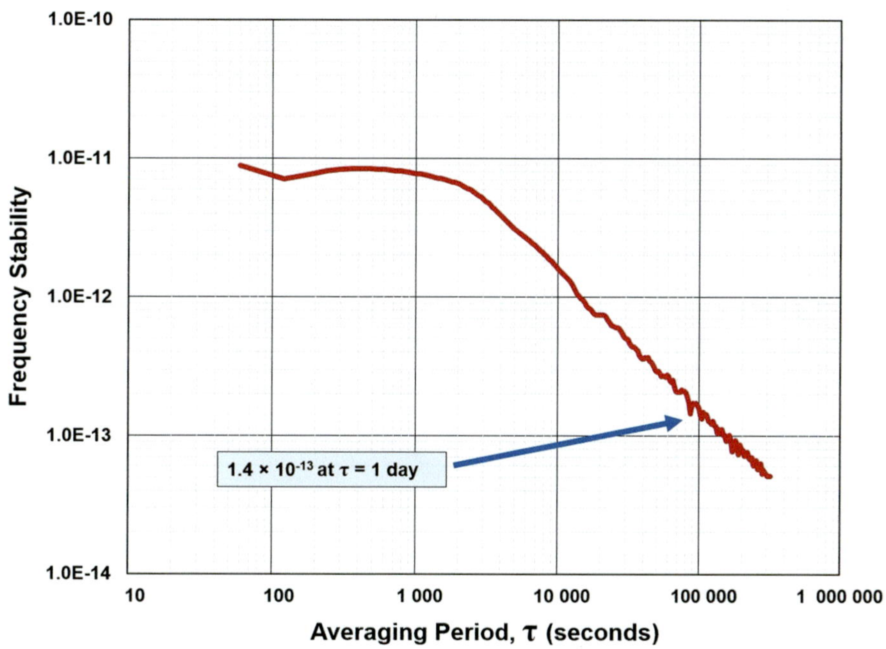

Figure 12.18 The frequency stability of a typical OCXO-based GPSDO.

There are, however, substantial differences between the frequency stabilities of GPSDOs produced by different manufacturers when they are measured for periods of one day or less. The short-term stability of a GPSDO, at intervals shorter than the interval between corrections, should be identical to the short-term stability of its free-running LO. Its medium-term stability, at intervals longer than the correction interval but shorter than one day, is design dependent and influenced by many factors. These factors include the quality of the receiver and antenna, the stability of the LO, the resolution of the comparator, the correction method, and the correction uncertainty.

Table 12.4 shows the measured frequency stabilities of ten GPSDOs that were compared to UTC(NIST). Values are shown where τ is equal to 1 s (when data were available), one minute, one hour, and one day. The ten devices were each produced by different manufacturers, and they each only received the L1 frequency. It's important to remember that these ten devices represent just a tiny sample of the many available GPSDO models, and that numerous devices might perform better or worse than those listed in the table.

As indicated in Table 12.4, six of the measured GPSDOs were stable to within parts in 10^{12} at $\tau = 1$ s. All of the devices reached a stability of at least 1×10^{-11} at $\tau = 1$ minute, and of at least 6×10^{-12} at $\tau = 1$ hour [29]. When comparing GPSDO performance, a good metric to use is their frequency stability at $\tau = 1$ day. A stability of 1×10^{-13} or better at $\tau = 1$ day typically indicates an instrument of high quality, and six of the 10 tested devices either met or exceeded this specification. A few came close to matching the stability of the best commercially available cesium beam clocks at $\tau = 1$ day, which, according to their manufacturer's specification, is about 0.3×10^{-13}. The cesium clock specification for stability at $\tau = 1$ day can potentially be improved upon by a multi-frequency GPSDO, although few such devices are commercially available as of 2025.

Table 12.4 Frequency stability of various GPSDOs at averaging periods of one day or less

GPSDO ID code	Local oscillator type	Frequency stability estimated with the overlapping Allan deviation, $\sigma_y(\tau)$			
		1 s	1 minute	1 hour	1 day
SYX	Rubidium	NA	1×10^{-11}	7×10^{-13}	0.5×10^{-13}
ET6	Rubidium	NA	1×10^{-12}	4×10^{-13}	0.6×10^{-13}
ERT	Rubidium	6×10^{-12}	1×10^{-12}	6×10^{-13}	0.7×10^{-13}
AR1	Quartz	6×10^{-12}	1×10^{-11}	1×10^{-12}	0.9×10^{-13}
BR8	Quartz	5×10^{-10}	9×10^{-12}	4×10^{-12}	1×10^{-13}
GD3	Quartz	NA	6×10^{-12}	3×10^{-12}	1×10^{-13}
PT1	Quartz	4×10^{-12}	1×10^{-11}	4×10^{-12}	3×10^{-13}
HPZ	Quartz	1×10^{-12}	2×10^{-12}	1×10^{-12}	4×10^{-13}
TS6	Rubidium	6×10^{-12}	8×10^{-13}	6×10^{-12}	4×10^{-13}
FL9	Rubidium	7×10^{-12}	3×10^{-12}	3×10^{-13}	7×10^{-13}

Unlike an uncertainty analysis of a GPSDC—the process we'll describe in Section 12.7—an uncertainty analysis of a GPSDO doesn't involve measuring the time delays introduced by the receiver, antenna, and cables. When measuring frequency, it isn't necessary to measure or compensate for equipment delays because they have no effect on the measurement uncertainty. This differs from a time measurement. For example, if we add extra cable to a GPSDC's antenna without compensating for the additional delay, it causes the time accuracy to permanently change. However, if we add an extra cable to a GPSDO's antenna, it has no effect on the frequency accuracy.

A locked GPSDO is always an accurate source of UTC frequency. As was summarized in Table 12.3, the frequency accuracy of a GPSDO is within parts in 10^{17} when measured over a multi-year period. This means, for all practical purposes, that the frequency difference between a GPSDO and UTC can be regarded as 0. However, as was emphasized in Chapter 6, the frequency accuracy of an oscillator over a given period can never be better than its stability over that same period. Therefore, an uncertainty analysis for a GPSDO always involves estimating its frequency stability with the Allan deviation or a similar statistic. The measurements made to estimate stability should be traceable to the SI, and referenced to either a UTC(k) time scale or to an atomic oscillator of sufficient stability.

By utilizing ADEV, we can now work through an example where we evaluate the expanded frequency uncertainty of the GPSDO whose stability was graphed in Figure 12.18. In this example, we'll first estimate the combined standard uncertainty, u_c, and then multiply by the coverage factor k to obtain the expanded uncertainty, U, for a measurement with a period of 1 day. Only two factors contribute to u_c, and both can be obtained through a Type A evaluation. The first factor is the frequency stability of the GPSDO under test at $\tau = 1$ day with respect to the reference, which in this case was UTC(NIST). The second factor is the frequency stability of UTC(NIST) at $\tau = 1$ day with respect to UTC.

It's important to note that estimating the GPSDO stability at $\tau = 1$ day with a reasonable confidence interval (Section 7.1) usually requires measurements to be made for at least one week, and a measurement period of 10 days or longer is commonly chosen. In our example, the GPSDO under test was measured for more than 10 days, and as previously noted, its stability was estimated as 1.4×10^{-13} at $\tau = 1$ day.

Figure 12.19 shows the stability of UTC(NIST) for the 15-year period from 2009 to 2023, which is estimated by applying ADEV to time difference measurements obtained from the *Circular T*. In this example, τ_0 = 5 days, and UTC(NIST) stability is estimated as 1.34 × 10^{-15} at τ = 5 days. If we make the reasonable and conservative assumption that white phase noise is the dominant noise type in the *Circular T* measurements at averaging periods shorter than five days (see Figure 7.10 for a model of oscillator noise types), we can extrapolate back to estimate ADEV at τ = 1 day by applying a slope of τ^{-1}. This simply means that the stability estimate at one day will be five times larger than the stability estimate at five days, so we can use a value of 6.7 × 10^{-15} at τ = 1 day in our uncertainty calculations.

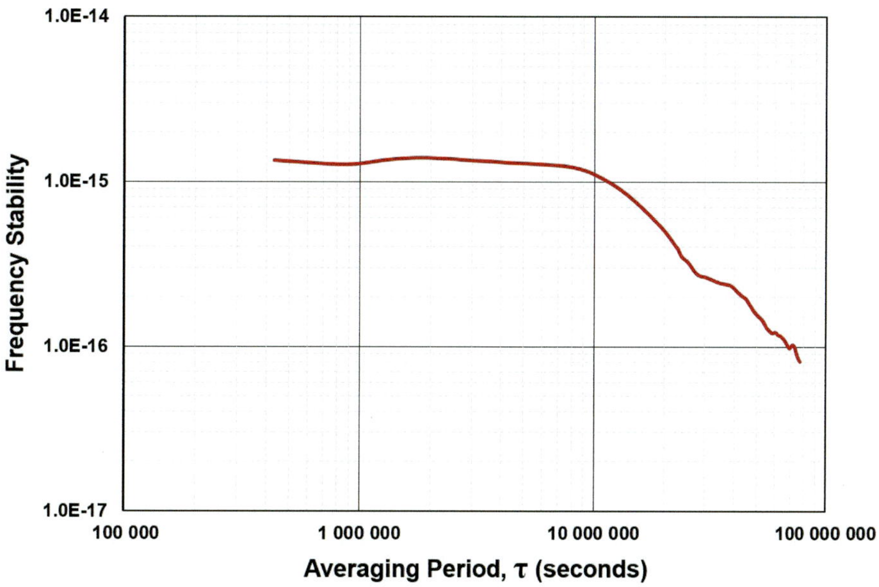

Figure 12.19 The frequency stability of UTC(NIST) with respect to UTC during the 15-year period from 2009 to 2023.

Based on these measurements, the expanded uncertainty, *U*, of the GPSDO under test is estimated with the root sum squares (RSS) method as

$$U = k\sqrt{u_{\text{REF}}^2 + u_{\text{GPSDO}}^2} = 2\sqrt{(1.4\times10^{-13})^2 + (6.7\times10^{-15})^2} = 2.8\times10^{-13}, \qquad (12.3)$$

where k is a coverage factor of 2, u_{REF} is the ADEV of UTC(NIST) with respect to UTC, and u_{GPSDO} is the ADEV of the GPSDO with respect to UTC(NIST).

If no GPSDO measurement data were available, the only available option would be to use the stability specification for τ = 1 day that's provided by the manufacturer. Using the specified stability isn't optimal, because manufacturers seldom disclose how they measured stability, and they sometimes publish the same specifications for GPSDOs of varying quality. The data sheet for the GPSDO in our example conservatively lists its frequency stability as 1 × 10^{-12} at τ = 1 day, which is about 7× less stable than the measured stability. If we use the specified stability as the value for u_{GPSDO}, the expanded uncertainty increases by about one order of magnitude,

$$U = k\sqrt{u_{\text{REF}}^2 + u_{\text{GPSDO}}^2} = 2\sqrt{(1\times10^{-12})^2 + (6.7\times10^{-15})^2} = 2\times10^{-12}. \qquad (12.4)$$

The stability of UTC(NIST) is insignificant in both Eq. (12.3) and Eq. (12.4) and has no influence on the measurement uncertainty. However, with a less stable reference, the u_{REF} term could be significant and might even become the limiting factor.

It's important to distinguish between reasonable and unreasonable frequency uncertainty claims for a GPSDO. For example, a result of $U = 1 \times 10^{-14}$ for a 24-hour measurement is likely incorrect and should be closely scrutinized. However, $U = 1 \times 10^{-13}$ is possible with some GPSDOs for a 24-hour measurement, and almost all GPSDOs are capable of expanded uncertainties near 1×10^{-12}.

12.7 Evaluating the Time Uncertainty of a GPSDC

As we discussed in Section 12.4, the uncertainty of the GPS signals as transmitted in space is near 1 ns with respect to UTC(USNO). However, many sources contribute additional uncertainty to the GPS signals that are received on Earth. In the following sections, we'll describe the eight most common uncertainty sources, and describe the amount of uncertainty that each source can contribute during a 24-hour period. The first of the eight uncertainty sources is evaluated with the Type A method, but the remaining seven are evaluated with the Type B method (Chapter 8). At the end of this section, Table 12.6 lists and summarizes the eight uncertainty sources, and also provides estimates of the expanded uncertainty of a GPSDC for a best, typical, and worst case.

12.7.1 Uncertainty Contributed by Time Stability, u_{AS}

Time stability refers to the variation in the 1 pps output of a GPSDC over a given measurement period. It can be evaluated with the Type A method by applying the Time deviation (TDEV) statistic described in Section 7.2. This method works well, because TDEV is well suited for estimating stability and time transfer noise when the dominant noise type is white phase or flicker phase, which is usually the case with a GPSDC.

Figure 12.20 shows a TDEV graph of a typical GPSDC when compared to UTC(NIST). As indicated by the arrow, TDEV is about 1.4 ns at τ = 1 day, and the stability of most GPSDCs should be similar, likely falling within a range that extends from slightly less than 1 ns to about 2 ns. If a GPSDC is locked and working properly, it would be unlikely for TDEV to ever exceed 5 ns at τ = 1 day. We can therefore consider 5 ns to be the upper limit for u_{AS}.

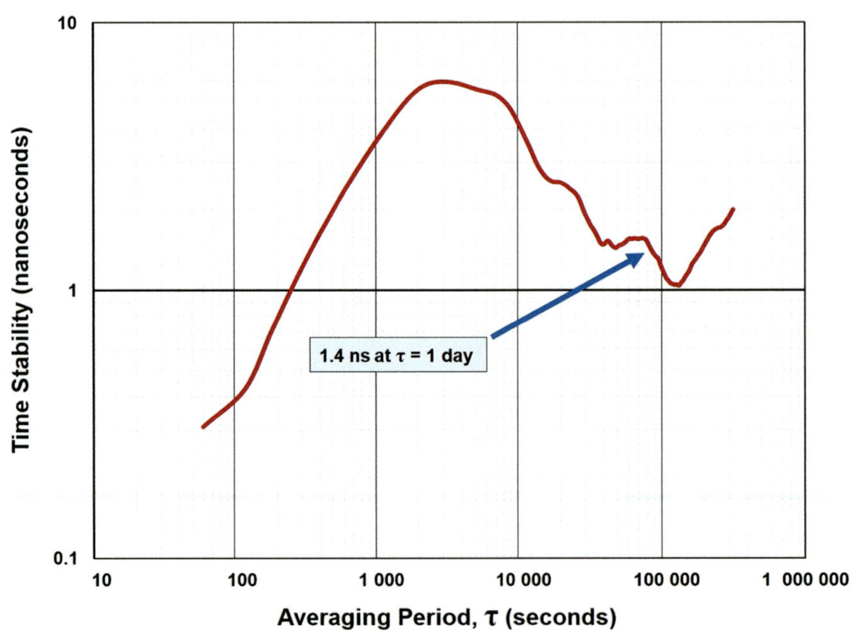

Figure 12.20 The time stability of a typical GPSDC.

12.7.2 Uncertainty Contributed by Hardware Delays, u_{BH}

The fixed delays in the GPSDC hardware—specifically the delays in the receiver, antenna, and antenna cable—can make large contributions to the measurement uncertainty. The dominant hardware delay is usually introduced by the antenna the cable. However, if the cable delay is known, the uncertainty it introduces can be easily compensated, because nearly all GPSDCs allow users to enter a cable delay value that automatically adjusts the 1 pps output.

As was discussed in Section 4.1.3, the cable delay can be estimated if the cable's length and velocity factor are known. Table 12.5 lists several types of antenna cables commonly used by GPSDCs, and the time delay added by each foot or meter of cable. The type of antenna cable used can have a significant impact on the delay. For example, an RG-58 cable of 20 m in length would have a delay of about 101 ns, as opposed to about 78 ns for an LMR-400 cable of the same length.

Table 12.5 Propagation delay through various types of GPS antenna cables

Cable type	Velocity factor (%)	Delay per foot (ns)	Delay per meter (ns)
RG-58 (solid polyethylene)	66	1.54	5.05
RG-58 (foam polyethylene)	73	1.39	4.57
RG-8 (solid polyethylene)	66	1.54	5.05
RG-8 (foam polyethylene)	78	1.30	4.28
LMR-400	85	1.20	3.92

It's always preferable, of course, to measure a cable's delay instead of just making an estimate. If the cable delay is measured before the cable is installed, it not only eliminates errors in the cable length estimate, but it also includes the small delays, typically sub-nanosecond, that are added by cable connectors. Several methods of accurately measuring cable delays are described in Section 4.1.3.

Receiver and antenna delays are more difficult to measure than cable delays. An absolute calibration of the receiver delay can be performed using a GPS simulator, which is an instrument that connects to the antenna input of the GPSDC under test and mimics the signals broadcast by the satellites. The simulator has its own reference clock, so the time signals input to the receiver can be compared to the receiver's output to determine the delay contributed by the hardware. The drawbacks to this method are the high cost of GPS simulators, and the fact that both the antenna and antenna cable delays still need to be calibrated separately, with the antenna delay measurements typically requiring a vector network analyzer and an anechoic chamber [30].

To reduce complexity and cost, a common practice is to calibrate a GPSDC as a system that includes the receiver, antenna, and antenna cable. This is usually done by mounting the antenna in a position with known coordinates, allowing the receiver to lock to the satellites, and then comparing the receiver's 1 pps output to a reference. The reference can be either a reference GPSDC, with the reference shipped to the DUT site or vice versa, or a UTC(k) time scale [22]. Measurements are generally averaged for at least several days to obtain a single delay constant that is keyed into the GPSDC to compensate for the system delays. When a GPSDC is calibrated as a system, the calibration is always relative to the reference. Therefore, any difference between the reference and UTC at the time of the calibration must be included in the delay constant.

The uncertainty contributed by hardware delays, u_{BH}, is the uncertainty that remains after a delay constant has been applied. If the GPS receiver, antenna, and antenna cable are all carefully calibrated, it's possible to reduce u_{BH} to about 2 ns, but it will likely be larger for most GPSDCs. The combined receiver and antenna delays will likely be less than 100 ns, so simply compensating for the antenna cable delay should reduce u_{BH} to less than 100 ns. Therefore, calibrating the antenna cable is always recommended, even when measuring the other delays isn't practical.

The worst-case scenario would be a situation where every hardware delay was ignored and left uncalibrated, and where a very long antenna cable was used. For example, if a GPSDC's antenna has enough gain, it might be able to work with more than 100 m of LMR-400 cable, and even longer antenna cables could be made to work by using in-line amplifiers. In these situations, u_{BH} could possibly reach or exceed 500 ns [29].

12.7.3 Uncertainty Contributed by Antenna Coordinate Errors, u_{BA}

Most GPSDCs have the ability to survey their antenna position by averaging position fixes for a period that ranges from a few hours to one day. Some models allow the user to specify how long the position fixes are averaged, and other models average for a fixed period that can't be changed. If the antenna is stationary, the survey only needs to be done once, and the results are stored in non-volatile memory. From that point forward, the GPSDC operates in a mode where it always holds the same position, and where no further position fixes are calculated.

A GPSDC self-survey can often determine the horizontal position of the antenna—its latitude and longitude—with an uncertainty near ±20 cm. However, because most GPSDCs are single-frequency devices, they often do a very poor job of surveying the vertical position, or altitude, of their antenna. The altitude error is often the largest single contributor to the time uncertainty of a GPSDC.

In the worst case, the 1 pps time offset will equal the altitude error multiplied by the speed of light constant, or about 3.3 ns of time error for every meter of position error. However, this situation can only occur when a GPSDC is receiving signals from a single satellite that's located directly overhead at an elevation angle of 90°. In most cases, the time error per meter of altitude error will be smaller, because GPSDCs obtain the time by averaging signals received from multiple satellites, and the average elevation angle is less than 90°. The error can be estimated by multiplying the speed of light constant by the sine of the average elevation angle, which will equal 1 at a 90° elevation angle and be less than 1 at lower elevations. If we make the reasonable assumption that the average elevation angle is 45°, or halfway between the horizon and the highest point in the sky, the average time offset contributed by altitude error can be roughly estimated as 3.3 × sin(45°), or about 2.3 ns per meter.

Antenna surveys performed with a dual-frequency geodetic receiver and PPP techniques can measure the altitude of an antenna with an uncertainty of about 15 cm, reducing the uncertainty contributed by antenna coordinates to less than 1 ns. An antenna survey performed with a single-frequency receiver will be much worse, and contributes an altitude error that's usually less than 10 m, but that in rare cases can be more than 20 m.

Figure 12.21 shows results obtained from a self-survey performed by a single-frequency GPSDC, which then had its antenna resurveyed using PPP techniques. The resurvey indicated that the initial altitude measurement was in error by 23.4 m. The graph shows the daily time offsets of the GPSDC for 10 days using the self-surveyed coordinates, and for 10 more days after the PPP survey was completed. The difference in the average of the two 10-day segments indicates that the GPSDC time shifted by 49.4 ns, or by about 2.1 ns per meter, revealing that this particular device was less accurate than it initially appeared. The vertical position error in this example represents an extreme case, so it seems reasonable to assume that u_{BA} is not likely to exceed 50 ns.

When the best possible time accuracy is required, the antenna position is either independently surveyed with a dual-frequency geodetic GPS receiver, or a dual-frequency GPSDC is used that can perform an accurate self-survey. Both solutions can reduce the uncertainty of the horizontal and vertical antenna coordinates to less than 30 cm, allowing u_{BA} to be estimated as 1 ns.

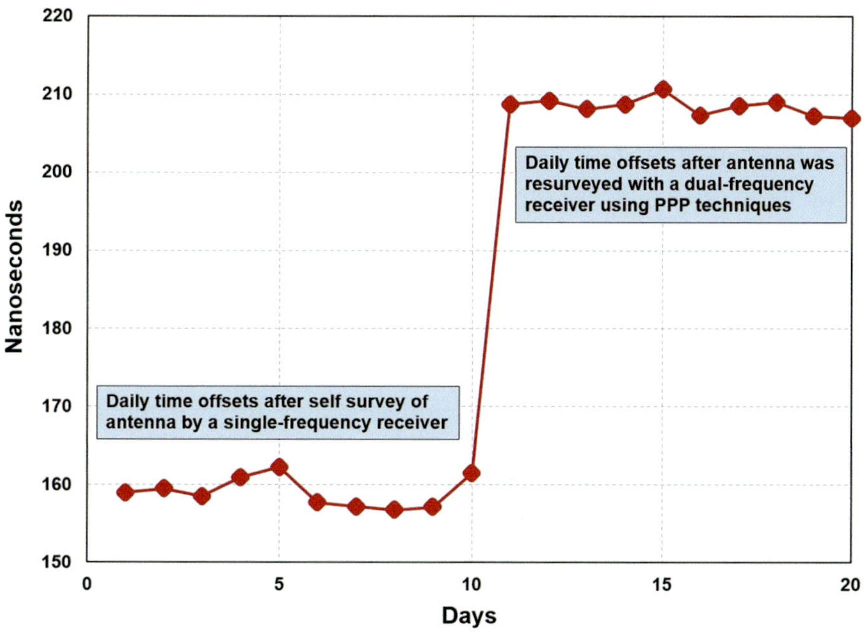

Figure 12.21 The time offset of a GPSDC with respect to a UTC reference, before and after its antenna position was accurately surveyed.

12.7.4 Uncertainty Contributed by Environmental Effects, u_{BE}

The hardware delays of a GPSDC change as a function of temperature and other environmental factors. Somewhat surprisingly, antennas and antenna cables are not likely to experience sudden time shifts, even though they are located outdoors. The temperature coefficient of high-quality antenna cable is usually much less than 1 ps (0.001 ns) per °C per meter [31]. Antennas typically have a temperature coefficient near 10 ps/°C, so their delay variations due to temperature are also small. Therefore, the annual delay variation in the antenna system due to temperature should be less than 1 ns, even at locations where the annual range of outdoor temperatures is large, and even when a large section of the antenna cable is exposed to the outdoor elements.

Snow, ice, and rain can be a larger environmental problem for GPSDC antennas than temperature changes. Thin layers of snow and ice are usually not a concern, but if an antenna is covered by snow and ice, the received signal strength will be reduced due to attenuation and reflections [32]. The reduction in signal strength can increase the delay variations in the pseudorange measurements, which in turn can increase the time uncertainty of a GPSDC. If water gets inside the sealed enclosure of an active antenna, the amplifier will stop working, which can cause signal reception to completely fail. This can sometimes happen during a heavy rainstorm, or if there is a crack in the antenna enclosure. If it does happen, the antenna will probably need to be replaced because its amplifier circuit has likely been damaged.

A GPSDC's receiver and internal oscillator can be more sensitive to temperature changes than its antenna system, even though they are located indoors. For example, if sudden or gradual changes in the indoor temperature occur, the 1 pps output of a GPSDC might experience a time shift or step of several nanoseconds, especially if the LO is a quartz device without temperature compensation. In most cases, the time shift will reverse when the temperature returns to normal.

The total contribution of u_{BE} to the measurement uncertainty—including the delay variations in the receiver, antenna, and antenna cable—typically ranges from 2 ns to 5 ns. This uncertainty is a function of the sensitivity of the GPSDC hardware to temperature changes, and it can fluctuate at various times of the year [29].

12.7.5 Uncertainty Contributed by Ionospheric Delays, u_{BI}

The ionosphere, which was previously discussed in Section 10.2, is the region of the atmosphere that extends from about 50 km to about 650 km above the Earth's surface. The ionosphere is ionized by solar radiation, and it has a large effect on the propagation of terrestrial radio signals. It has a much smaller effect on GPS signals, which originate from above the ionosphere, but it still affects their propagation. GPS signals are refracted or bent slightly as they pass through the ionospheric layers, causing a physical phenomenon known as dispersion. The dispersion contributes to a GPSDC's uncertainty, because it slows down the signal and increases the path delay by a varying amount.

As we continue to note, most GPSDCs are based on single-frequency receivers, and a single-frequency receiver has no way of measuring the ionospheric delay. Therefore, the delay is estimated with a mathematical model that's implemented in the receiver's software, and the software applies a delay correction to the received signals. The Klobuchar model, formulated by the American electrical engineer John Klobuchar (b. 1937), is the best-known and most widely applied ionospheric delay model. Klobuchar was working as a senior physical scientist at the Air Force Cambridge Research Laboratory in Massachusetts when he was asked by the JPO to develop an ionospheric delay model for single-frequency GPS receivers. The first version of his model was published in 1975, which was more than two years before the launch of the first Navstar satellite [33].

Klobuchar's goal was to develop an ionospheric delay model that could be calculated from just a small number of parameters, but that could still remove 50% of the root mean square (RMS) error that the ionosphere contributes to the pseudorange measurement [34]. The inputs to the Klobuchar model are the local time at the receiving site, the latitude and longitude of the receiver, and the elevation angle and azimuth of the satellite being measured. In addition, the model requires eight parameters received from page 18, subframe 4 of the GPS navigation message. Each satellite broadcasts the same ionospheric delay parameters, which were selected to represent the current ionospheric conditions, and which are typically updated once per day.

Almost all single-frequency GPS receivers apply the Klobuchar model by default, and the algorithms for the model are included in the GPS interface specifications [18]. To briefly summarize, the model works by assuming an ionospheric height of 350 km, and it bases its delays estimates on a constant value of 5 ns during the nighttime, and on a half-cosine function during the daytime. Four of the eight parameters contained in the navigation message are called alpha coefficients, α_n, and four are called beta coefficients, β_n. The alpha coefficients are used to calculate the amplitude of the daytime ionospheric delay by using a cosine function that has a fixed peak at 1400 local time. The beta coefficients are used to calculate the period of the cosine function (Figure 12.22). The ionospheric delay from a given satellite is then estimated as a function of its elevation angle with respect to the receiver [10, 34].

At any given time of day or night, the signals from satellites at low elevation angles will require a larger delay correction than the signals from satellites at higher elevation angles, with the correction increasing by about a factor of three from the horizon to the zenith [11]. However, due to the increase in solar radiation, the delay correction for satellites at all elevation angles will always be much larger during the day than it is during the night. This is illustrated by Figure 12.23, which graphs the modeled ionospheric corrections for every GPS satellite received in Boulder, Colorado, during a 24-hour period. The shaded area is the period between sunset and sunrise when the corrections are the smallest, with the minimum nighttime correction equal to 5 ns. The blue markers represent satellites with an elevation angle between 10° and 20°, and the red markers represent satellites with elevation angles above 20°. The largest ionospheric corrections were applied to satellites below 20° during the daytime hours, and some exceeded 40 ns. In accordance with the Klobuchar model, the peak of the large corrections occurred at 2100 UTC, which was 1400 local time in Boulder.

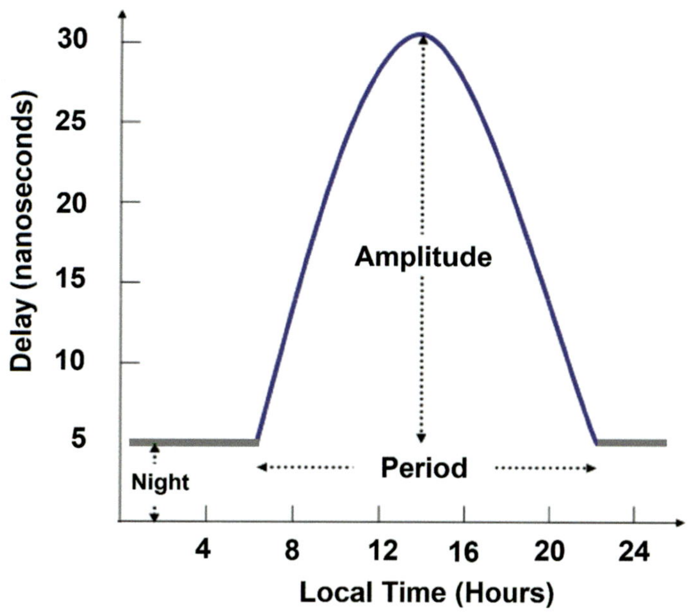

Figure 12.22 The Klobuchar ionospheric delay model.

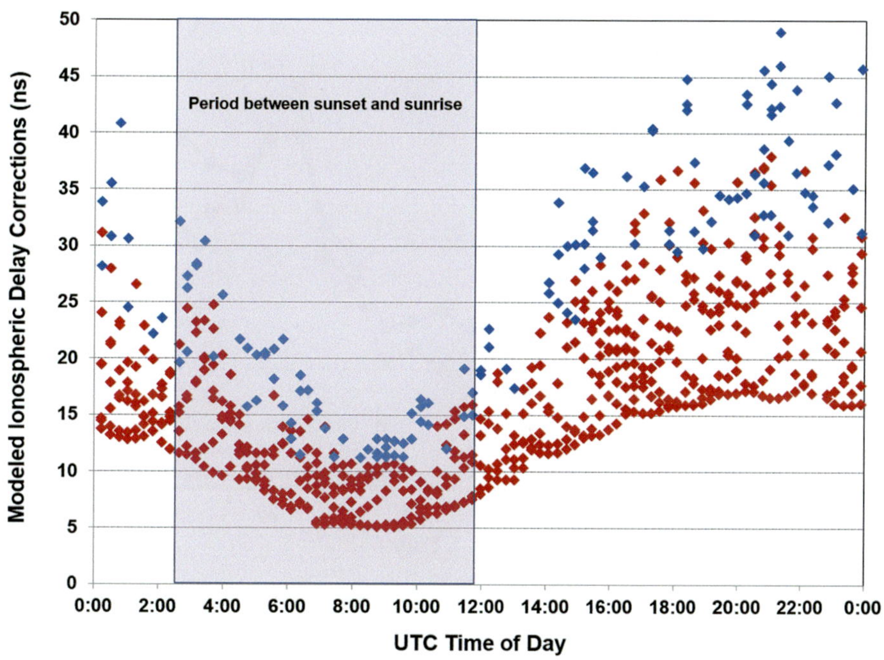

Figure 12.23 The modeled ionospheric delay corrections for all GPS satellites received in Boulder, Colorado during a 24-hour period.

The Klobuchar model involves some necessary compromises. Perhaps the most obvious compromise is the fact that the peak ionospheric delay is assumed to always occur at 1400 local time, which is the midpoint of a six-hour period when the peak is most likely to occur. The actual ionospheric peak can vary from about 1100 to 1700 local time, depending upon the season, the latitude, and the current amount of solar activity [10]. The model almost always achieves its goal of removing 50% of the ionospheric delay, often removing an even higher percentage, but it can remove less than 50% of the delay during periods of high solar activity [35]. The measurements of a single-frequency GPSDC often show that time shifts occur when the ionosphere's electron density increases at sunrise and when it decreases at sunset. In some cases, the time shifts will exceed 10 ns even after the correction is applied. When measurements are averaged for 24 hours the sunrise and sunset effects mostly cancel, and the uncertainty due to ionospheric delay is typically reduced to about 2 ns.

Figure 12.24 illustrates this by showing the average daily time offset of a GPSDC that was compared to UTC(NIST) for 30 days. The GPSDC ran without applying any ionospheric delay corrections, but the resulting measurements were then separately processed with modeled and measured ionospheric delay corrections. The red line shows the results after applying corrections from the Klobuchar model, and the blue line shows the results after applying corrections from an ionospheric delay measurement. The dashed line shows the time difference between the model and the measurement, which varied from day to day, but had an average value of -2.1 ns.. The variation is a function of the receiver's geographic location and the current level of solar activity. When a single-frequency GPSDC is measured for 24 hours with the Klobuchar model applied, some reasonable estimates for u_{BI} are 2 ns in the best case, 5 ns in a typical case, and 10 ns in the worst case.

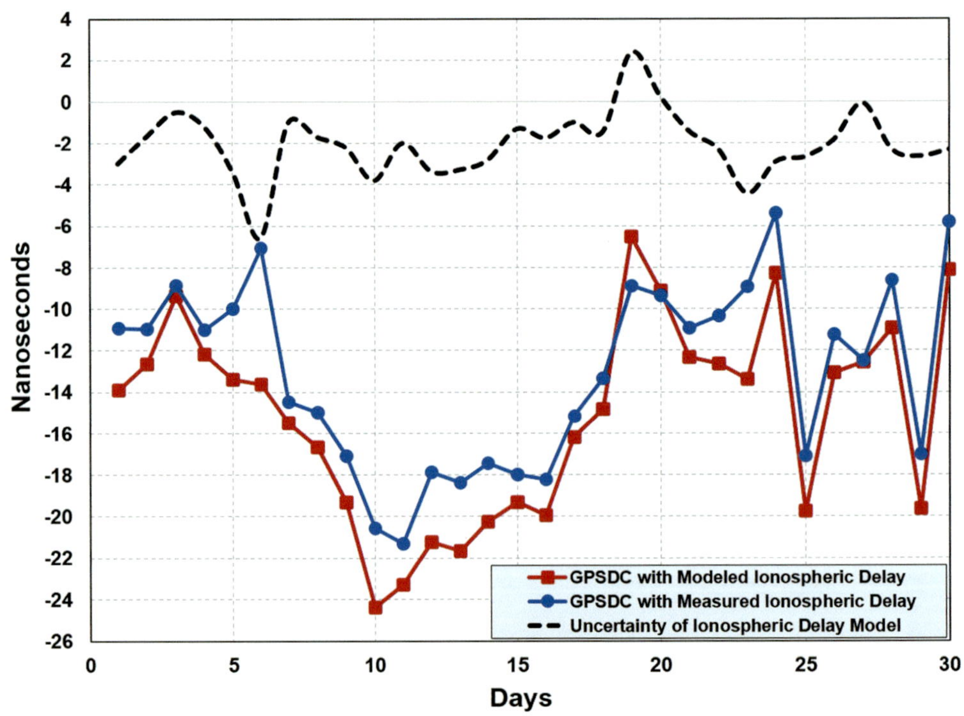

Figure 12.24 Measurements of the same GPSDC when corrected with modeled and measured ionospheric delays during a 30-day period. The dashed line shows the time difference between the model and the measurement.

As we described in Section 9.4.4, the common-view GPS measurements that link UTC(*k*) laboratories to UTC compensate for the ionospheric delay by correcting the measurements after they are recorded. The corrections are obtained from the IGS network. If the corrections from the Klobuchar model were already applied, they're removed before applying the IGS corrections. These post-processing techniques provide the lowest possible measurement uncertainties, and are invaluable for UTC calculations, precise surveying, and other applications. However, it isn't especially helpful to correct measurements after a GPSDC has already provided reference signals to a real-time system.

Fortunately, real-time ionospheric delay compensation can be performed by dual-frequency GPSDCs that compare pseudorange measurements obtained from signals on two different frequencies. The two frequencies are typically L1 and L2, but any two of the three GPS carrier frequencies (L1, L2, and L5) can be used for the comparison. The differential delay between two pseudorange measurements, when both signals originate from the same satellite transmitter, provides enough information to determine what the pseudorange would be if the ionospheric delay was removed. The differential delay method works because the refractive index of the ionosphere varies as a function of the signal frequency, which causes signals transmitted at different frequencies to be delayed by different amounts. The excess phase delay that's incurred by a radio signal as it propagates through the ionosphere, $\Delta\tau_p$, can be calculated in units of seconds as

$$\Delta\tau_p = -\frac{40.3 \times TEC}{cf^2}, \tag{12.5}$$

where TEC is the total electron content of the ionosphere based on the number of free electrons per square meter, c is the speed of light constant, and f is the frequency of the radio signal.

The phase delay is expressed as a negative number because the phase of the carrier frequency is being advanced, while the modulated codes are being delayed by an equal amount. This phenomenon, known as code–carrier divergence, means that the ionospheric delay terms in GPS pseudorange and carrier phase measurements are equal in magnitude but opposite in sign [10, 36]. The f^2 term in the denominator of Eq. (12.5) tells us that an increase in frequency results in a decrease in the ionospheric delay; therefore, the L1 pseudorange measurements have smaller ionospheric delays than the L2 and L5 measurements.

A pseudorange measurement that's mostly free of ionospheric delays can be calculated with an equation that's essentially the same as Eq. (9.2), which was provided while describing the P3 time transfer method that links UTC(k) laboratories to UTC [37]. The ionosphere-free pseudorange, ρ_{IF}, is calculated as

$$\rho_{IF} = \frac{f_{L1}^2}{f_{L1}^2 - f_{L2}^2}\rho_{L1} - \frac{f_{L2}^2}{f_{L1}^2 - f_{L2}^2}\rho_{L2} = 2.546\rho_{L1} - 1.546\rho_{L2}, \tag{12.6}$$

where f_{L1} and f_{L2} are the respective carrier frequencies of L1 and L2, and ρ_{L1} and ρ_{L2} are the L1 and L2 pseudorange measurements [10].

The differential delay method can be applied in other forms that use calculations similar to Eq. (12.6). For example, instead of generating ρ_{IF} as a third independent pseudorange, some methods calculate an ionospheric correction that's applied to the L1 pseudorange. Regardless of the method implemented, removing the ionospheric delay by receiving two frequencies can often reduce u_{BI} to 1 ns or less, without requiring a 24-hour average.

12.7.6 Uncertainty Contributed by Tropospheric Delays, u_{BT}

The troposphere is the lowest layer of the Earth's atmosphere, extending from the Earth's surface to a height ranging from about 6 km to 20 km, depending upon the geographic location and time of year. The weather that we experience on Earth occurs in the tropospheric region. Another atmospheric region, the stratosphere, is above the troposphere and below the ionosphere. For the purposes of this discussion, the delays encountered by the GPS signals as they pass through the stratosphere—and the other layers below the ionosphere—are considered to be part of the tropospheric delay.

Unlike ionospheric delays, tropospheric delays aren't frequency dependent, and they can't be measured by either a single or a dual-frequency GPS receiver. There are also no parameters broadcast by GPS to support a tropospheric delay model, which differs from the support provided to the ionospheric delay model. Fortunately, most tropospheric delay models require only a small amount of input data [38]. The necessary inputs are usually just the elevation angle of the satellite and the altitude of the antenna.

Roughly 90% of the tropospheric delay, known as the dry component, is caused by the signal refraction that results from atmospheric pressure. The dry component can be modeled much more easily than the wet component, which depends on the amount of water vapor in the atmosphere. The wet component is difficult to model, but it accounts for less than 10% of the total tropospheric delay.

A tropospheric delay model that's typically implemented by GPSDCs is the NATO model, which gets its name because it was first published as a North Atlantic Treaty Organization (NATO) standard [38, 39]. This model, in a fashion similar to other tropospheric models, focuses entirely on the dry component—it completely ignores the wet component because no input data are available. It estimates the delay through the troposphere in meters, d_{trop}, as

$$d_{trop} = d_{trop}^{z} \frac{1}{\sin(\varepsilon) + \frac{0.00143}{\tan(\varepsilon) + 0.0455}}, \tag{12.7}$$

where ε is the elevation angle of the satellite in radians. The equation for the d_{trop}^{z} term (total zenith delay) is selected based on the orthometric height (altitude) of the antenna. For a height of less than 1 km,

$$d_{trop}^{z} = \left\{1430 + 732 + \left\{\mathrm{SR} \times (1-H) + 0.5\Delta N(1-H^2)\right\}\right\} \times 10^{-3}, \tag{12.8}$$

and for a height greater than or equal to 1 km and not exceeding 9 km,

$$d_{trop}^{z} = \left\{732 + \left\{\frac{N_1}{c} \times \exp(-c(H-1)) - \exp(-8c)\right\}\right\} \times 10^{-3}, \tag{12.9}$$

where

SR is the global mean sea level refractivity, with a constant of 324.8 recommended by the NATO model;
H is the orthometric height in kilometers;

$\Delta N = 7.32 \times \exp(0.005577 \times SR)$;

$N_1 = SR + \Delta N$; and

$c = \frac{1}{8} \times \log\left(\frac{N_1}{105}\right)$.

The delay correction, d_{trop}, is then converted from meters to nanoseconds using the speed of light constant. Figure 12.25 shows the tropospheric delay corrections generated as a function of elevation angles between 10° and 90° for an antenna height of 1645 m (the altitude of the NIST laboratories in Boulder, Colorado).

Unless it's disclosed by the manufacturer, it is usually not possible to determine which tropospheric model a GPSDC is using. However, even simple models can compensate for about 90% of the delay introduced by the dry component. As Figure 12.25 indicates, the correction is less than 10 ns at an average elevation angle of 45°, which suggests that the NATO model can reduce the uncertainty in the dry component to less than 1 ns. The wet component is ignored, but because it accounts for less than 10% of the total tropospheric delay, its uncertainty contribution should also be less than or near 1 ns for a satellite at an average elevation angle. Therefore, a typical value for u_{BT} is about 2 ns, which could increase or decrease slightly due to the current weather conditions [29].

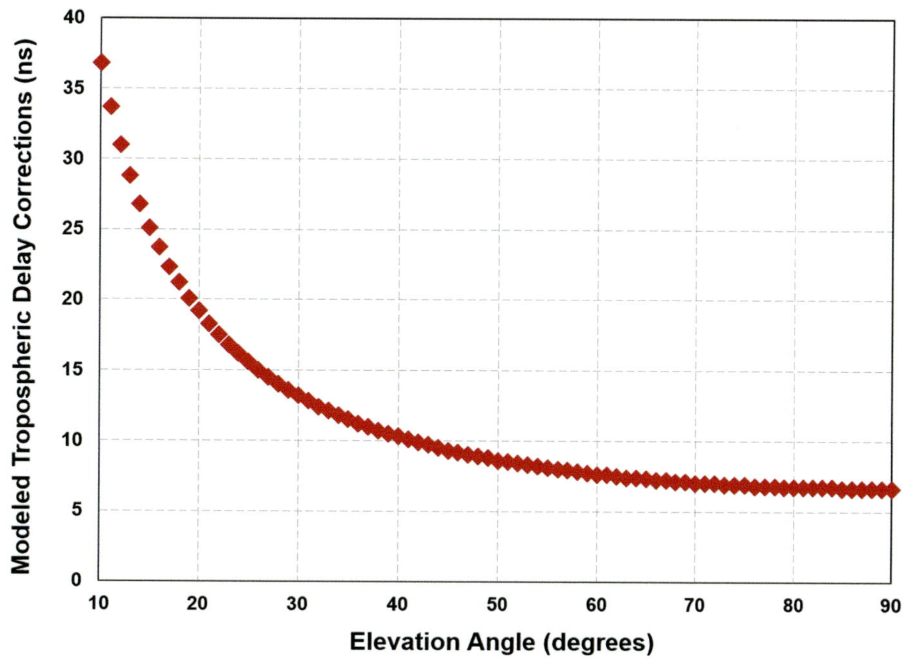

Figure 12.25 The tropospheric delay corrections as a function of elevation angle, generated with the NATO model for Boulder, Colorado.

12.7.7 Uncertainty Contributed by Multipath Signal Reflections, u_{BM}

The GPS signals don't always have an unobstructed path as they travel from the satellite to the antenna. Instead, some signals are reflected by the surfaces and objects that surround the antenna. The reflected signals are called multipath signals, and they arrive at the antenna at different times. They can interfere with, or be mistaken for, the signals that travel across an unobstructed path, and the resulting delay variations contribute to the time uncertainty.

Multipath is one of the most discussed GPS uncertainties, because of the adverse effect it can have on GPS positioning when the antenna is moving [40]. However, the uncertainty contributed by multipath to a GPSDC with a stationary antenna is often insignificant—especially when the antenna has an unobstructed view of the sky. It can be made even less significant by choosing an antenna specifically designed to reduce multipath—including the quadrifilar helix, choke ring, and pinwheel antennas described in Section 12.5. Each of those antennas is less susceptible to multipath than the patch antennas used by most GPSDCs [26].

Multipath reflections can be identified by looking at the individual satellite tracks. If a multipath reflection causes a delay change in the signal from a given satellite, a similar delay change should be noticeable again on the following day. However, the delay change will occur about four minutes earlier each day, because the GPS orbits are based on the sidereal day, which is about four minutes shorter than the UTC or mean solar day. Figure 12.26 shows segments of GPS satellite tracks, based on measurements averaged for one minute, that were recorded from the same satellite at the same time for five consecutive days using a quadrifilar helix antenna. During each flyover, the signals were briefly reflected by a metal structure located near the antenna. The reflected signals caused the peaks shown in the graph, which always occurred four minutes earlier than they did on the previous day. When averaged for one minute, the magnitude of these time shifts is about 50 ns. However, their contribution to the time uncertainty is much smaller than that, because the time shifts are brief, and because only the signals transmitted by satellites located at specific positions in the sky are affected. The uncertainty can be reduced by either moving the antenna away from the metal structure, or by replacing the antenna with one that does a better job of rejecting multipath signals.

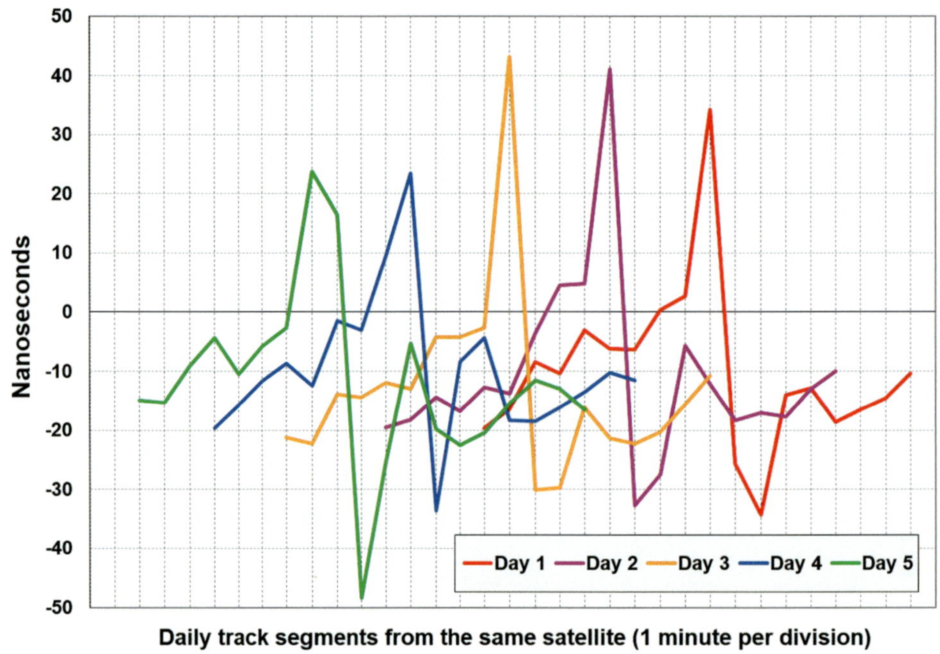

Figure 12.26 Multipath signal reflections from a GPS satellite recorded for five consecutive days.

If the antenna has an unobstructed sky view, the total contribution of u_{BM} to the time uncertainty should be near 1 ns if a multipath mitigating antenna is used, and about 2 ns when using a typical patch antenna. If reasonable care is taken when mounting the antenna—for example, if the antenna is placed on the highest part of the roof, instead of on the side of a building—then u_{BM} should be less than 5 ns.

12.7.8 Uncertainty Contributed by the UTC – USNO(USNO) Time Difference, u_{BU}

The uncertainty of a GPSDC should be estimated with respect to UTC and the SI. As described in Section 12.4, GPSDCs distribute UTC(USNO), so the current time difference between UTC and UTC(USNO) should be included in the uncertainty analysis. This time difference is usually less than 2 ns, and its current value can be obtained from the *Circular T* (Section 9.4.4). The average time difference was 0.4 ns during the 15-year period from 2009 to 2023 (Table 12.3), and as shown in Figure 12.8, the time difference hasn't exceeded 5 ns in recent years. Based on several decades of historical data, 10 ns can be considered as the worst-case scenario.

To summarize, the largest sources of GPSDC uncertainty are usually uncompensated hardware delays and antenna coordinate errors. The other uncertainty sources are relatively small, and typically only attract attention after the hardware delays have been compensated and the antenna has been accurately surveyed.

Table 12.6 shows the best-case, typical, and worst-case expanded uncertainties of a single-frequency GPSDC with respect to UTC. The expanded uncertainty, U, is estimated by accounting for all eight of the uncertainty sources just described, combining them into a single value with the RSS method, and assigning a value of 2 to the coverage factor k. The equation for this example, with all uncertainties expressed in nanoseconds, is

$$U = k\sqrt{u_{AS}^2 + u_{BH}^2 + u_{BA}^2 + u_{BE}^2 + u_{BI}^2 + u_{BT}^2 + u_{BM}^2 + u_{BU}^2}\,. \tag{12.10}$$

Table 12.6 The best-case, typical, and worst-case expanded uncertainties for a single-frequency GPSDC during a 1-day period

Uncertainty source	Best case (geodetic antenna survey, full delay compensated)	Typical (self-surveyed antenna, partial delay compensation)	Worst case (no antenna survey, no delay compensation)
u_{AS}, Time stability	1	2	5
u_{BH}, Hardware delays	2	10	500
u_{BA}, Antenna coordinates	1	20	50
u_{BE}, Environmental effects	2	3	5
u_{BI}, Ionospheric delay	2	5	10
u_{BT}, Tropospheric delay	1	2	3
u_{BM}, Multipath reflections	1	2	5
u_{BU}, UTC – UTC(USNO)	1	5	10
U, $k = 2$	< 10	~50	~1000

As indicated in the table, the best-case expanded uncertainty for a single-frequency GPSDC, when a geodetic antenna survey is performed and when all delays are carefully calibrated, is slightly less than 10 ns. A more typical GPSDC uncertainty is about 50 ns, which is usually achievable when the antenna is self-surveyed and when the hardware delays are compensated to within about 10 ns. The worst-case uncertainty could reach 1 μs (1,000 ns), but that situation usually only occurs when an extremely long antenna cable is used and its delay is uncompensated. As a general rule, a properly working GPSDC will always keep time to within ±1 μs of UTC [29].

12.8 Other Global Navigation Satellite Systems (GNSS)

GPS is an indispensable utility. Its signals are freely available worldwide, and we often use GPS without even knowing it, because its PNT capabilities have been built into so many systems. As was noted in Section 12.1, GPS was first made available for worldwide civilian use in 1983, and since then it has never been deactivated. However, GPS is a United States military system, and its use could potentially be restricted for reasons of national security. This fact, coupled with the tremendous importance of PNT to modern lifestyles, has prompted other nations to develop other global navigation systems that they can control.

There are now four major global navigation satellite systems that provide a high-accuracy, worldwide distribution system for UTC. The four GNSS—GPS, GLONASS, Galileo, and BeiDou—are based on similar technology, and they broadcast signals on similar frequencies. This has made it possible for companies to develop products that can receive signals from all four systems. Multi-constellation receivers are now commonly found in smartphones, and multi-constellation disciplined clocks and oscillators are now utilized for both one-way and common-view time transfer [41]. However, standalone GLONASS, Galileo, or BeiDou disciplined clock and oscillator products aren't especially common. Most products that receive signals from those systems also receive GPS, which continues to dominate the PNT marketplace.

The four GNSS all maintain their own reference time scale, but they each broadcast messages that allow receivers to correct the GNSS time so it agrees with a UTC(k) reference. In addition to the broadcast UTC corrections, the interface control documents for all four GNSS now include format specifications for messages that provide the time differences between the various systems. These messages potentially allow an approximation of the time transmitted by all GNSS to be obtained by receiving the signals from just one GNSS. This should eventually provide a convenient level of interoperability between all of the systems. However, as of 2025, the methods for determining the time differences are still being developed, and not all of the messages have been implemented. The best known of the messages that has been implemented is the Galileo/GPS Time Offset (GGTO) message, which is broadcast by Galileo [42].

The following three sections provide brief descriptions of GLONASS, Galileo, and BeiDou, with the focus placed on their UTC distribution capabilities. These three sections are followed by Table 12.9, which summarizes the characteristics of the four global navigation satellite systems.

12.8.1 GLONASS

The Russian Global Navigation Satellite System (*Global'naya Navigatsionnaya Sputnikovaya Sistema*, GLONASS) was the first GNSS developed outside of the United States. Like GPS, GLONASS was originally designed for military applications, and GLONASS is operated by the Russian Space Force. The first GLONASS satellite was launched in October 1982—less than five years after the launch of the first Navstar GPS satellite—and the system was declared operational in 1993.

The full GLONASS constellation consists of 24 MEO satellites distributed in three orbital planes. The satellites have an orbit inclination of 64.8° and a nominal orbital height of 19,100 km. Their orbital period of 11 hours and 16 minutes is equal to 17 Earth orbits every eight sidereal days, which is one more trip around the Earth than the GPS satellites make during the same period [11, 43].

As of 2025, most of the operational satellites are members of the GLONASS-M series, with the first M-series satellite launched in 2003. The GLONASS-M satellites each carry three cesium clocks. The first member of an improved generation of satellites, called GLONASS-K, was launched in 2011, but the GLONASS-M launches continued until 2022. The lighter GLONASS-K satellites carry two cesium and two rubidium clocks, and have a life expectancy of 10 years, as opposed to seven years for GLONASS-M. All of the satellite clocks have a nominal frequency of 5 MHz, and apply a frequency correction of about -4.36×10^{-10} to compensate for relativistic effects.

The K-series satellites are the first GLONASS satellites to support the broadcasting of CDMA signals. This was done to make the system more commercially viable by improving its interoperability with GPS, Galileo, and BeiDou, which were all designed as CDMA systems. However, the K-series satellites maintain backward compatibility with the frequency-division multiple access (FDMA) signals that GLONASS was originally designed to use.

The use of FDMA instead of CDMA makes GLONASS work differently from GPS. Instead of transmitting a pair of unique PRN codes on the same frequency—like each GPS satellite does when it transmits the C/A and P codes on the L1 carrier—each GLONASS satellite transmits a pair of unique PRN codes on a pair of unique carriers. The unique carrier frequencies are assigned as

$$L1 = 1602 \text{ MHz} + n \times 0.5625 \text{ MHz, and} \tag{12.11}$$

$$L2 = 1246 \text{ MHz} + n \times 0.4375 \text{ MHz,} \tag{12.12}$$

where n is an integer ranging from -7 to +6 that's assigned to an individual satellite. This results in 14 unique frequency channels on both bands. The L1 channels have frequencies ranging from 1598.0625 MHz to 1605.375 MHz, and the L2 channels have frequencies ranging from 1242.9375 MHz to 1248.624 MHz [43]. The use of a 14-channel FDMA system is sufficient for a 24-satellite constellation, because satellites on opposite sides of the Earth can safely broadcast on identical frequencies without causing interference.

The use of FDMA makes GLONASS receivers more complex to design than GPS receivers, because tracking the individual satellites requires synthesizing and processing multiple frequencies. However, FDMA can do a better job of rejecting interference. For example, a narrowband interference source can block all of the satellite signals in a CDMA system, but it might only block the signal from a single GLONASS satellite. In addition, FDMA eliminates the potential CDMA problem of cross-correlation interference between multiple PRN codes on the same frequency [10, 11].

In a similar fashion to GPS, GLONASS uses the BPSK method to modulate a 50 bps navigation message and two PRN codes on to the carrier frequencies. The two PRN codes are similar to the legacy GPS C/A and P codes, but they are transmitted at about half the chipping rate. The standard accuracy code, available to all users, repeats every millisecond. The entire 511 chip code is transmitted every millisecond, so the code frequency is 511 kHz. The much longer high-accuracy code, which is undocumented and only available to authorized users, is truncated so that it repeats every second. It has a code frequency of 5.11 MHz [10, 11, 43].

The time reference for GLONASS is UTC(SU), the national time scale of the Russian Federation, which is operated by the Russian Metrological Institute of Technical Physics and Radio Engineering (VNIIFTRI), located near Moscow. The UTC(SU) time scale was established before the dissolution of the Soviet Union in 1991, which explains the SU acronym. The time scale consists of an ensemble of hydrogen masers that provides frequency stability, and a cesium fountain primary standard that provides accuracy. A TWSTFT system links UTC(SU) to UTC.

Figure 12.27 shows the UTC – UTC(SU) time differences, obtained from the *Circular T*, for the 15-year period from 2009 to 2023. The average time difference during the 15-year period was 0.6 ns. The time difference has progressively decreased and was kept well within ±5 ns during the most recent years shown on the graph.

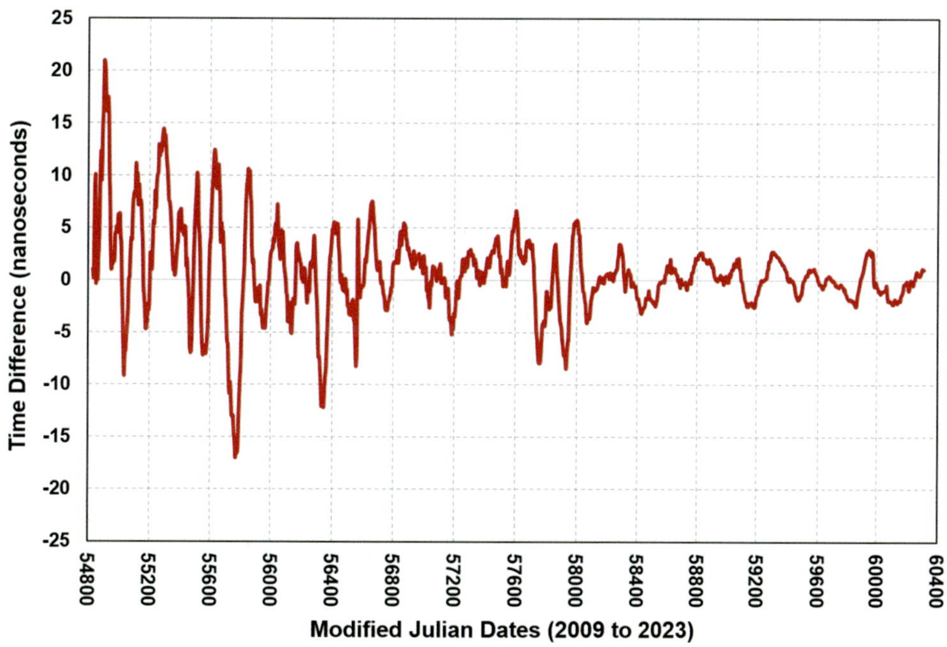

Figure 12.27 The time difference between UTC and UTC(SU) during the 15-year period from 2009 to 2023.

GLONASS time is not calculated from an epoch, but instead is calculated as a fixed offset from UTC(SU). This offset includes a constant three-hour time difference because GLONASS time is set to agree with Moscow local time. This makes GLONASS time equal to UTC(SU) + 3 hours, so receivers need to apply a constant three-hour correction to convert the GLONASS time and date to UTC. The fixed offset from UTC(SU) doesn't include a leap second correction, because GLONASS time is simultaneously corrected for leap seconds along with UTC and UTC(SU). If a leap second is scheduled, the navigation message indicates when it will occur.

Apart from the easily correctable three-hour difference, the system specification states that the difference between GLONASS time and UTC(SU) should not exceed 1 μs [43]. GLONASS time formerly had an offset of about 400 ns with respect to UTC(SU), but it was removed by a correction applied in August 2014 [44]. Since then, the accuracy of the GLONASS time signals transmitted in space has usually been within ±10 ns of UTC(SU), with the time difference seldom exceeding 20 ns [44, 45].

12.8.2 Galileo

Galileo, the GNSS of the European Union, was named after Galileo Galilei, the renowned Italian scientist who was the first to study the periodicity of a pendulum (Section 1.5). The system was developed by the European Space Agency (ESA) and is managed by the European Union Agency for the Space Program (EUSPA), which is headquartered in Prague, Czechoslovakia. The operation of the satellite constellation is handled by two control centers that are located in Oberpfaffenhofen, Germany, and Fucino, Italy.

The first experimental Galileo satellite, named GIOVE-A, was launched in December 2005 (Figure 12.28), and the system was declared operational in December 2016. The full Galileo constellation consists of 30 MEO satellites, with 10 satellites placed in each of three orbital planes. The satellites have an orbit inclination of 56° and a nominal orbital height of 23,222 km. Their orbital period of 14 hours and 5 minutes is equal to 17 Earth orbits every 10 sidereal days, which is three fewer trips around the Earth than the GPS satellites make during the same period [46].

Figure 12.28 The launch of GIOVE-A, the first Galileo satellite, in December 2005 (courtesy of the European Space Agency).

Each Galileo satellite carries four atomic clocks—two rubidium clocks and two passive hydrogen masers (Section 3.10). The Galileo satellite clocks have the same 10.23 MHz nominal frequency as the GPS clocks, and that frequency is multiplied to generate the RF carrier signals. A frequency correction of about -4.72×10^{-10} is applied to the clocks to compensate for relativistic effects. This correction is larger than the GPS relativistic correction, due to the higher orbits of the Galileo satellites.

Like the other GNSS, Galileo transmits on the upper and lower satellite L-bands, but the bands are called E-bands to distinguish them from GPS and GLONASS, and to denote their connection to the European Union. The PNT signals that Galileo makes freely available—called the open service signals—are broadcast on E1 and E5, and a third carrier, E6 (1278.75 MHz), is mostly used for higher accuracy and restricted use services. The E1 signals have the same 1575.42 MHz carrier frequency as the GPS L1 signals, and a similar bandwidth of about 24.5 MHz. The unencrypted signals on E1 include a data signal called E1-B that carries a navigation message, and a data-free signal, E1-C, that is only used for ranging measurements. The PRN code transmitted by both E1-B and E1-C is 4,092 chips long and sent at a rate of 1.023 megachips/s. The code repeats every 4 ms on E1-B, but a 25-chip secondary code on E1-C extends the code period to 100 ms.

The E5 signals have a carrier frequency of 1191.795 MHz and a bandwidth of 51.15 MHz, which is more than twice the bandwidth of E1. The increase in bandwidth is the result of multiplexing the E5a (1176.45 MHz) and E5b (1207.14 MHz) signals together using a modified form of binary offset carrier (BOC) modulation. This technique simultaneously transmits the E5a and E5b signals as sidebands of the carrier. When compared to BPSK modulation, BOC modulation sharpens the transition between the signal's peak and its sidelobes. This makes it easier for a receiver to distinguish between a direct signal and a reflected signal, which can reduce the effects of multipath.

The E5a and E5b signals each have an in-phase (I) and quadrature (Q) component that transmits a 10,230 chip PRN code every millisecond. However, the E5a-Q and E5b-Q codes have a 100-chip secondary code that extends the code period to 100 ms, and secondary codes on E5a-I and E5b-I also extend the code periods, to 20 ms and 4 ms, respectively. Table 12.7 summarizes the characteristics of the Galileo E1 and E5 open service signals [47].

Table 12.7 Characteristics of the Galileo E1 and E5 open service signals

Band	Center frequency (MHz)	Bandwidth (MHz)	Code name	Primary code length (chips)	Secondary code length (chips)	Code period (ms)	Minimum received RF power (dBW)	Data rate (bps)
E1	1575.42	24.552	E1-B	4,092	—	4	−157.0	125
			E1-C	4,092	25	100	−157.0	—
E5	1191.795	51.150	E5a-I	10,230	20	20	−155.0	25
			E5a-Q	10,230	100	100	−155.0	—
			E5b-I	10,230	4	4	−155.0	125
			E5b-Q	10,230	100	100	−155.0	—

The Galileo satellites broadcast two navigation messages that are unencrypted and available to the general public. The F/NAV message, with the "F" denoting that it's freely available, is broadcast on E5a-I at a rate of 25 bps with a frame duration of 600 s. The F/NAV frame is divided into 12 subframes, with each frame divided into five pages. The I/NAV message, known as the integrity message because it includes authentication data, is broadcast on E1-B and E5b-I at a rate of 125 bps with a frame duration of 720 s. The I/NAV frame is divided into 24 subframes, with each subframe divided into 15 pages. The F/NAV and I/NAV messages each broadcast identical time information, which is encoded using formats that are very similar to those used by GPS.

The time broadcast by Galileo is called GST, an acronym for Galileo System Time. GST is continuously adjusted to agree with UTC by using UTC(k) time scales located in France, Germany, Italy, Spain, and Sweden as intermediate references [42]. The GST epoch is defined as 23:59:47 UTC on August 21, 1999, which was 13 s before UTC midnight [47]. This epoch ironically transpired before the launch of the first Galileo satellite, but it was chosen to align GST with GPST. The date coincides with the date of the first GPS week number rollover (Section 12.4), and the 13 s equals the difference between GPST and UTC when the week number rollover occurred.

Like GPST, GST is broadcast as the number of weeks that have elapsed since the epoch, and the number of seconds that have elapsed during the current week. Galileo broadcasts the week number as a 12-bit code, which is longer than the 10-bit code originally used by GPS. However, 12 bits still limits its maximum value to 4095 (2^{12}–1) weeks, meaning that a week number rollover will eventually occur in 2078.

The GGTO correction broadcast by Galileo indicates that the difference between GPST and GST usually falls within ±10 ns [42], and Galileo uses a method similar to GPS (Section 12.4) to correct GST so it agrees with its UTC time scale references. Using information found in the navigation messages, a Galileo receiver calculates and applies two UTC corrections, one large and one small. The large correction, expressed in integer seconds, compensates for the number of leap seconds since have been added to UTC since the GST epoch. The small correction, expressed in nanoseconds, compensates for the most recent measured time difference between GST and UTC.

A noteworthy difference between GPS and Galileo involves the broadcast ionospheric models that they provide for single-frequency receivers. GPS broadcasts parameters for the Klobuchar model (Section 12.7.5) and Galileo broadcasts parameters for an ionospheric model called *NeQuick G*, which is designed to remove at least 70% of the RMS error that the ionosphere contributes to the pseudorange measurements. Unlike the Klobuchar model, which assumes a single-layer ionospheric height of 350 km, the *NeQuick G* model uses the peaks of three ionospheric layers—E, F1, and F2—as its anchor points [48]. This only marginally improves the horizonal position estimate, but it can reduce the altitude error of a single-frequency antenna survey by more than 1 m on average when compared to the Klobuchar model [49]. The improvements introduced by the *NeQuick G* model can reduce the combined uncertainty of a single-frequency disciplined clock by potentially reducing the uncertainty of both the antenna coordinates (Section 12.7.3), and the ionospheric delay (Section 12.7.5).

12.8.3 BeiDou

BeiDou, the GNSS of China, has a name well suited for a navigation system. Its name was inherited from the Chinese word for the Big Dipper, a seven-star group in the Ursa Major constellation. The two stars at one end of the Big Dipper's "bowl" point towards Polaris, the name given to the bright North Star that has guided celestial navigators since the days of antiquity (Section 15.1).

BeiDou is operated by the Chinese National Space Administration (CNSA), which is headquartered in Beijing, China. The original BeiDou system was an RNSS consisting of three GEO satellites, with the first satellite launched in October 2000. The effort to make BeiDou a GNSS began in April 2007, when its first MEO satellite was launched. The current third-generation BeiDou system, known as BDS-3, was declared fully operational in July 2020, becoming a true GNSS while maintaining its RNSS capabilities. It consists of 24 MEO satellites that provide global coverage, three GEO satellites that provide coverage to China, and three inclined geosynchronous orbit (IGSO) satellites that provide coverage to the Asia-Pacific region.

The 24 MEO satellites are arranged in three orbital planes, with eight satellites in each plane. The satellites have an orbit inclination of 55° and a nominal orbital height of 21,528 km. Their orbital period of 12 hours and 53 minutes is equal to 13 Earth orbits every seven sidereal days, which is one less trip around the Earth than the GPS satellites make during the same period [50].

Earlier generations of BeiDou MEO satellites carried only rubidium clocks, but the BDS-3 satellites use passive hydrogen masers (Section 3.10) as the primary clock, with rubidium clocks used as backups [51]. The BeiDou clocks have the same 10.23 MHz nominal frequency as the GPS and Galileo clocks, which is multiplied to generate the RF carrier signals. A frequency correction of about -4.59×10^{-10} is applied to the clocks to compensate for relativistic effects. The nominal orbital height of the BeiDou satellites is higher than GPS, but lower than Galileo. Therefore, the relativistic corrections for BeiDou clocks are larger than those applied to GPS, and smaller than those applied to Galileo.

Like the other GNSS, BeiDou transmits on the upper and lower satellite L-bands, but the bands are called B-bands to denote their connection to BeiDou. The PNT signals that BeiDou makes freely available—called the open service signals—are broadcast on B1, B2, and B3. The BeiDou B1C signals have the same 1575.42 MHz carrier frequency as the GPS L1 and Galileo E1 signals, but a wider bandwidth of 32.736 MHz due to the use of BOC data modulation.

The other open service signals—B1I (1561.098 MHz), B2a (1176.45 MHz), B2b (1207.14 MHz), and B3I (1268.52 MHz)—are modulated with BPSK. The B1I and B3I signals were part of the second-generation BDS-2 system, but the B1C, B2a, and B2b signals weren't made available until BDS-3. They were added to improve BeiDou's interoperability with GPS and Galileo, making it easier to include BeiDou in multi-constellation GNSS receiver designs [52].

With the exception of the B1I signals, which have a narrower bandwidth and a PRN code length of 2,046 chips, the open service BeiDou PRN codes are 10,230 chips in length. The codes are sent at a rate of 1.023 megachips/s on B1C, which matches the rate of the GPS L1 code. On B1I, the codes are sent at 2.046 megachips/s, and a 10.23 megachips/s rate, matching the nominal clock frequency, is used on B2a, B2b, and B3I.

All five of the open service signals include a navigation message. The formats and data rates vary on the different frequency bands, but all of the navigation messages contain the information needed to accurately distribute UTC. The navigation messages also provide the parameters that single-frequency receivers can use to compensate for ionospheric delay [53]. The B1I and B3I signals include the eight parameters needed to support the Klobuchar ionospheric delay model (Section 12.7.5) that was first implemented by GPS. The newer B1C, B2a, and B2b signals provide nine correction parameters used by the BeiDou Global Ionospheric Model (BDGIM) developed by the Chinese Academy of Sciences. A study conducted from 2010 to 2017 found that the BDGIM reduced the residual ionospheric delay error of the Klobuchar model by 10% to 20% [54]. Table 12.8 summarizes the characteristics of the BeiDou open service signals broadcast by MEO satellites.

Table 12.8 Characteristics of the BeiDou open service signals broadcast by MEO satellites

Band	Center frequency (MHz)	Bandwidth (MHz)	Code name	Code length (chips)	Code period (ms)	Code frequency (MHz)	Minimum received RF power (dBW)	Data rate (bps)
B1C	1575.42	32.736	Data	10,230	10	1.023	−159.0	~50
			Pilot	10,230	10	1.023	−159.0	—
B1I	1561.098	4.092	Ranging	2,046	1	2.046	−163.0	50
B2a	1176.45	20.46	Data	10,230	1	10.230	−156.0	~100
			Pilot	10,230	1	10.230	−156.0	—
B2b	1207.14	20.46	Ranging	10,230	1	10.230	−160.0	~500
B3I	1268.52	20.46	Ranging	10,230	1	10.230	−163.0	50

The time broadcast by BeiDou is called BDT, an acronym for BeiDou Time. The BDT epoch is defined as 00:00:00 UTC on January 1, 2006 [50]. BDT ignores leap seconds, and differs from UTC by the integer number of leap seconds that have occurred since its epoch. The reference for BDT is UTC(NTSC), the national time scale of China, operated by the National Time Service Center of the Chinese Academy of Sciences, and located in Xi'an, the capital city of the Shaanxi Province [55]. As of 2025, the UTC(NTSC) time scale includes a large ensemble of more than 20 cesium clocks and more than 10 hydrogen masers, and a continuously running rubidium fountain clock.

BDT is broadcast as the number of weeks that have elapsed since the epoch, and the number of seconds that have elapsed during the current week, the same method established by GPST. The week number is broadcast as a 13-bit code, with a maximum value of 8191 (2^{13}–1) weeks. This prevents the occurrence of a BeiDou week number rollover until the year 2163.

BeiDou receivers apply corrections to BDT so that it agrees with UTC(NTSC), using a method very similar to the GPS method described in Section 12.4. Using information found in the navigation messages, a BeiDou receiver calculates and applies two UTC(NTSC) corrections, one large and one small. The large correction, expressed in integer seconds, compensates for the number of leap seconds that have been added to UTC since the BDT epoch. The small correction, expressed in nanoseconds, compensates for the most recent measured time difference between BDT and UTC(NTSC) [53].

The open service performance standard for BeiDou states that BDT is maintained within ±50 ns (modulo 1 s) of UTC [50]. This specification is easily met, due to the close agreement between UTC and UTC(NTSC). To illustrate this, Figure 12.29 shows the UTC – UTC(NTSC) time differences, obtained from the *Circular T*, for the 15-year period from 2009 to 2023. The average time difference during the 15-year period was 2.1 ns. The time difference has progressively decreased and was kept well within ±5 ns during the most recent years shown on the graph.

The characteristics of the four GNSS—BeiDou, Galileo, GLONASS, and GPS—are summarized in Table 12.9.

Table 12.9 Characteristics of the four global navigation satellite systems

Parameter	Global satellite navigation systems			
	BeiDou	**Galileo**	**GLONASS**	**GPS**
Host Nation(s)	China	European Union	Russia	United States
Orbit Height (km)	21,528	23,222	19,100	20,200
Orbit Inclination	55°	56°	64.8°	55°
Orbit Period (hh:mm)	12:53	14:05	11:16	11:58
Orbit/Sidereal Day Ratio	13:7	17:10	17:8	2:1
MEO Satellites	24	30	24	31
Orbital Planes	3	3	3	6
Carrier frequencies (MHz) and band designations	1575.42 (B1C) 1561.098 (B1I) 1176.45 (B2a) 1207.14 (B2b) 1268.52 (B3I)	1575.42 (E1) 1191.795 (E5) 1278.75 (E6)	1602 (L1)* 1246 (L2)*	1575.42 (L1) 1227.6 (L2) 1176.45 (L5)
Reference Time Scale	UTC(NTSC)	Multiple UTC(*k*) realizations	UTC(SU)	UTC(USNO)
Time Scale Epoch (UTC)	01/01/2006, 00:00:00	08/21/1999, 23:59:47	None	01/06/1980, 00:00:00

*The GLONASS satellites broadcast on multiple FDMA channels centered around these frequencies.

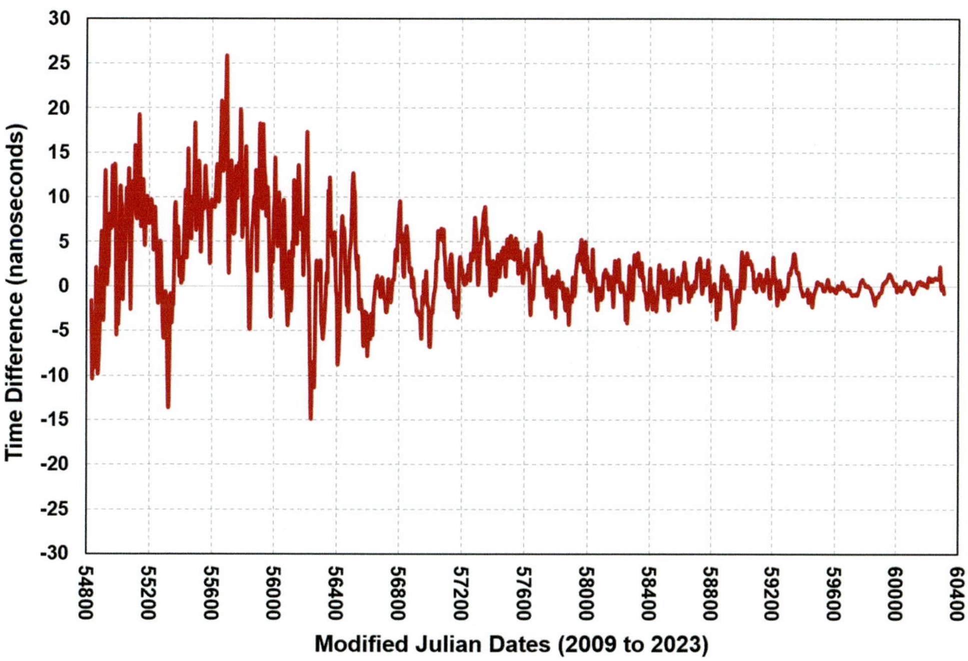

Figure 12.29 The time difference between UTC and UTC(NTSC) during the 15-year period from 2009 to 2023.

12.9 Regional Navigation Satellite Systems (RNSS)

An RNSS is a constellation of navigation satellites that provides PNT services to a specific geographic region, but that doesn't provide global coverage. As just described in Section 12.8.3, BeiDou functions as both an RNSS and a GNSS, with its GEO and IGSO satellites providing the RNSS capability and its MEO satellites providing the GNSS capability.

Some RNSS function as autonomous navigation systems, which means that they can provide navigation services by themselves, without relying on satellites from other constellations. However, most serve as satellite-based augmentation systems (SBAS) that reduce the uncertainty of GNSS by providing additional correction messages. The SBAS work by providing a network of ground stations whose coordinates are a reference for position measurements. The ground stations receive GNSS signals and measure the small variations in their position solutions by comparing them to the reference coordinates. This information is then converted to correction messages that are uplinked to the GEO satellites and then broadcast back to users on Earth.

Both autonomous and non-autonomous RNSS can distribute UTC if their time signals are referenced to a UTC(k) time scale. The following four sections briefly describe four RNSS that serve as UTC distribution systems. Sections 12.9.1 and 12.9.2 describe two non-autonomous RNSS—WAAS and EGNOS—that respectively distribute UTC to the United States and the European Union. Sections 12.9.3 and 12.9.4 describe two autonomous RNSS—QZSS and NavIC—that respectively distribute UTC to Japan and India.

12.9.1 WAAS (United States)

The Wide Area Augmentation System (WAAS) is a navigation aid developed by the U. S. Federal Aviation Administration (FAA) to augment GPS. Planning for WAAS began in 1994, and the system was first activated for general aviation use in 2003. It includes three GEO satellites located in fixed positions above the equator that provide coverage to most of North America (Figure 12.30).

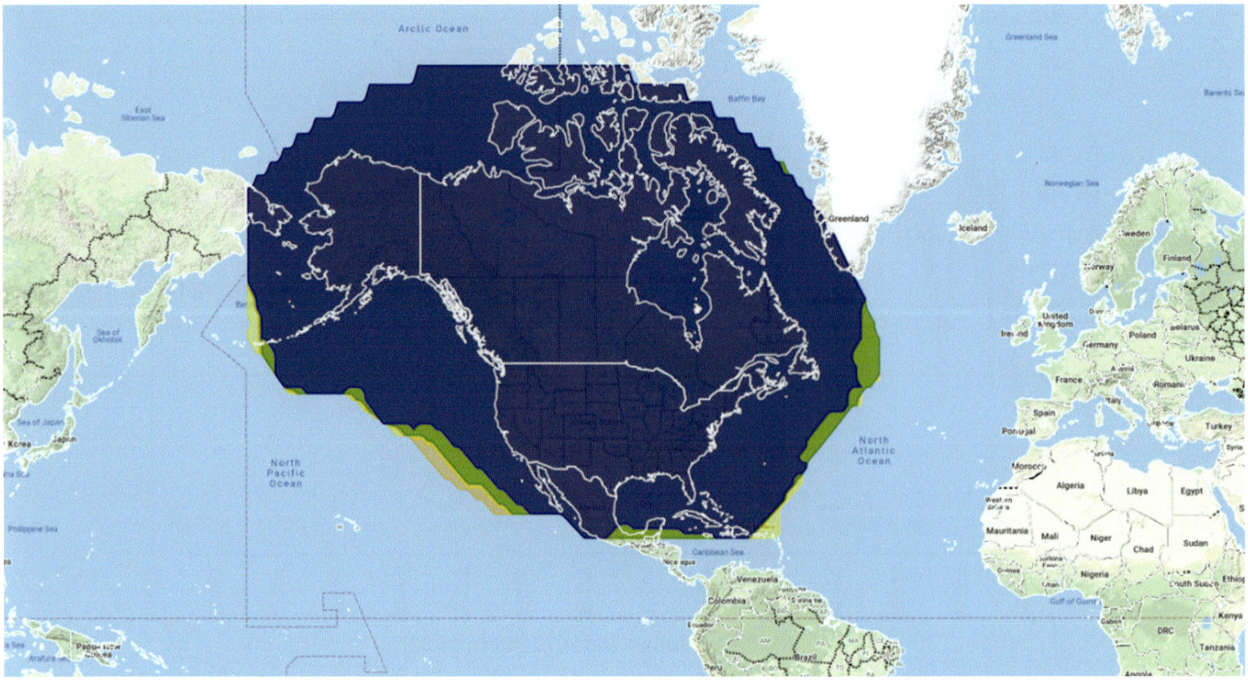

Figure 12.30 The WAAS coverage area (courtesy of the Federal Aviation Administration).

The WAAS satellites broadcast their differential correction messages and time signals as an overlay on the GPS L1 (1575.42 MHz) and L5 (1176.45 MHz) frequencies. The satellites don't carry atomic clocks, and the WAAS Network Time (WNT) that the satellites broadcast is provided by the ground control stations. WNT is typically within ±10 ns of GPST and UTC(USNO) [56].

Like the GPS satellites, the individual WAAS satellites identify themselves by broadcasting unique PRN codes. The WAAS PRN code numbers are typically in the range from 131 to 138. The WAAS PRN codes repeat every millisecond, and have the same chip rate as the GPS codes on the same frequency—1.023 megachips/s on L1 and 10.23 megachips/s on L5. This allows WAAS capability to be easily added to GPS receivers. WAAS is seldom used as a standalone time synchronization source, but it has some potential advantages, because the satellites are located in a fixed location in the sky. This allows North American users to point high-gain directional antennas directly at the satellites, which makes WAAS less susceptible to RF interference than GPS [57].

12.9.2 EGNOS (European Union)

The European Geostationary Navigation Overlay Service (EGNOS) is a navigation aid that was developed by the ESA and first activated in 2005. It was originally designed to only augment GPS, but new capabilities were added to make it also augment Galileo. It includes three GEO satellites located in fixed positions above the equator that provide coverage to the European Union. The satellites broadcast differential correction messages and time signals on the GPS L1 frequency of 1575.42 MHz, and identify themselves with unique PRN codes that are typically numbered in the range from 120 to 136.

The EGNOS satellites broadcast EGNOS Network Time (ENT). The satellites don't carry atomic clocks and ENT is derived from a group of atomic clocks located at the ground control stations. The ground stations clocks are continuously steered towards GPST, and as of 2025 the time difference between ENT and GPST has been less than 20 ns for well over a decade.

EGNOS receivers apply corrections to EDT so that it agrees with UTC(OP), the national time scale of France that's operated by the Paris Observatory. The correction method is very similar to the GPS method described in Section 12.4. Using information found in the navigation messages, an EGNOS receiver calculates and applies two UTC(OP) corrections, one large and one small. The large correction compensates for the integer second difference between ENT and UTC that was caused by leap seconds. The small correction, expressed in nanoseconds, compensates for the most recent measured time difference between ENT and UTC(OP). The expanded uncertainty ($k = 3$) of the UTC(OP) correction is specified as 20 ns [58].

12.9.3 QZSS (Japan)

The Quasi-Zenith Satellite System (QZSS), operated by the Japanese government, provides a UTC distribution service while functioning as both an RNSS and an SBAS. The system was designed to improve navigation in Japan's numerous urban canyons by placing navigation satellites high the sky, because tall buildings block the reception of signals from satellites at low elevation angles. The name "Quasi-Zenith" was given because the satellites appear to be almost directly overhead when viewed from Japan. The first QZSS satellite was launched in 2010, and a constellation consisting of three IGSO satellites and one GEO satellite was declared operational in November 2018. As of 2025, plans are in place to expand the constellation from four to seven satellites.

The goal of positioning the QZSS satellites high in the sky was accomplished by launching them into geosynchronous orbits with a radius of 42,164 km and a period of one sidereal day. However, the IGSO orbits have an inclination angle of 43° with respect to the equator, and this tilted geosynchronous orbit prevents the satellites from staying in a fixed position. Instead, the satellites repeatedly trace an analemma pattern like a true geostationary satellite (Section 10.6), but their pattern covers a much larger area, which in this case is the Asia-Oceanic region, with the focus placed on Japan. The figure-eight-shaped orbits of the QZSS satellites have north-south asymmetry, and the satellites spend about 13 hours in the northern hemisphere and about 11 hours in the southern hemisphere (Figure 12.31). While in the northern hemisphere, the satellites spend most of their time positioned high in the skies above Japan. For example, when viewed from the city of Tokyo, a QZSS satellite is positioned at an elevation angle of 50° or higher for about 12 hours, and at 70° or higher for about eight hours.

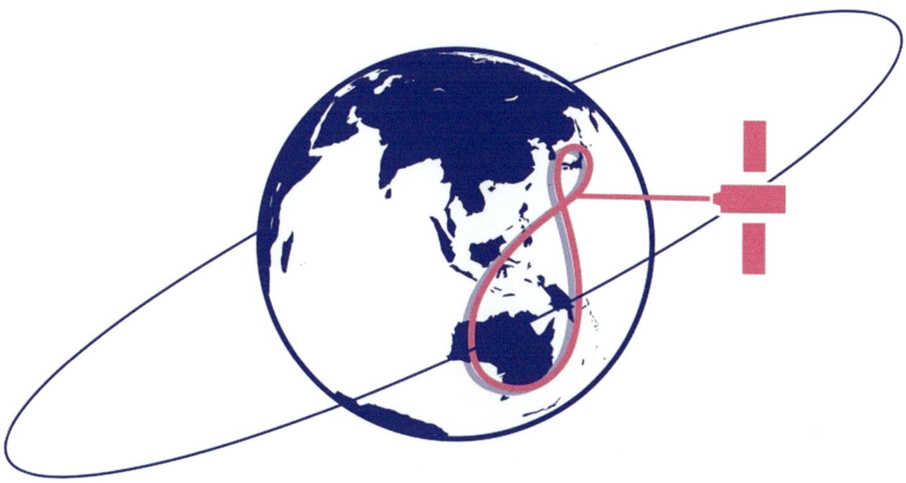

Figure 12.31 The coverage area of a QZSS satellite orbit (courtesy of the National Space Policy Secretariat of Japan).

The QZSS PNT service broadcasts signals that allow the QZSS satellites to act as additional GPS satellites. The QZSS carrier frequencies, as well as the PRN code lengths and periods, are identical to the GPS L1 C/A, L1C, L2CL, L2CM, and L5 I5/Q5 signals that are summarized in Table 12.2. The unique PRN codes broadcast by the QZSS satellites are typically numbered in the range from 193 to 202.

Generating signals compatible with GPS is made possible by the rubidium clocks carried by the QZSS satellites, which conveniently run at the same 10.23 MHz nominal frequency as the GPS satellite clocks. A frequency correction of about -5.4×10^{-10} is applied to the rubidium clocks to compensate for relativistic effects. This correction is necessarily larger than the correction applied to GPS and other GNSS satellite clocks, due to the higher altitude of the geosynchronous QZSS satellites [59].

The QZSS system time, called QZSST, has the same epoch as GPST (January 6, 1980, at 00:00 UTC). QZSST is controlled by atomic clocks referenced to UTC(NICT), the national time scale of Japan that's operated by the National Institute of Information and Communications Technology (NICT) in Tokyo [60]. However, QZSST has historically been synchronized to GPST to ensure interoperability with GPS. Methods for directly synchronizing QZSST to UTC(NICT) have been proposed, because the uncertainty of UTC(NICT) is known to be small enough to meet the interoperability requirements [61].

A QZSS disciplined clock synchronizes to UTC(NICT) in the same way that a GPSDC synchronizes to UTC(USNO). Using information found in the navigation messages, a QZSS receiver calculates and applies two UTC corrections—a large integer second correction that compensates for the number of leap seconds that have occurred since the GPST/QZSST epoch, and a small nanosecond-level correction that compensates for the most recent measured time difference between QZSST and UTC(NICT). The navigation messages also include the eight parameters needed to compute the Klobuchar model (Section 12.7.5), which allows single-frequency QZSS receivers to compensate for most of the ionospheric delay. Unlike GPS, which broadcasts the same ionospheric parameters to the entire world, QZSS broadcasts two sets of parameters—one for receivers in located in Southeast Asia and Oceania, and one for receivers located in Japan [59].

12.9.4 NavIC (India)

The Indian Regional Navigation Satellite System (IRNSS) provides a UTC distribution service while functioning as an autonomous RNSS. The system is commonly called NavIC, an acronym for Navigation with Indian Constellation. The system is operated by the Indian Space Research Organization (ISRO). The first satellite was launched in 2013, and NavIC became an operational service in 2018. An artist's conception of the full NavIC constellation, consisting of four IGSO and three GEO satellites, is shown in Figure 12.32.

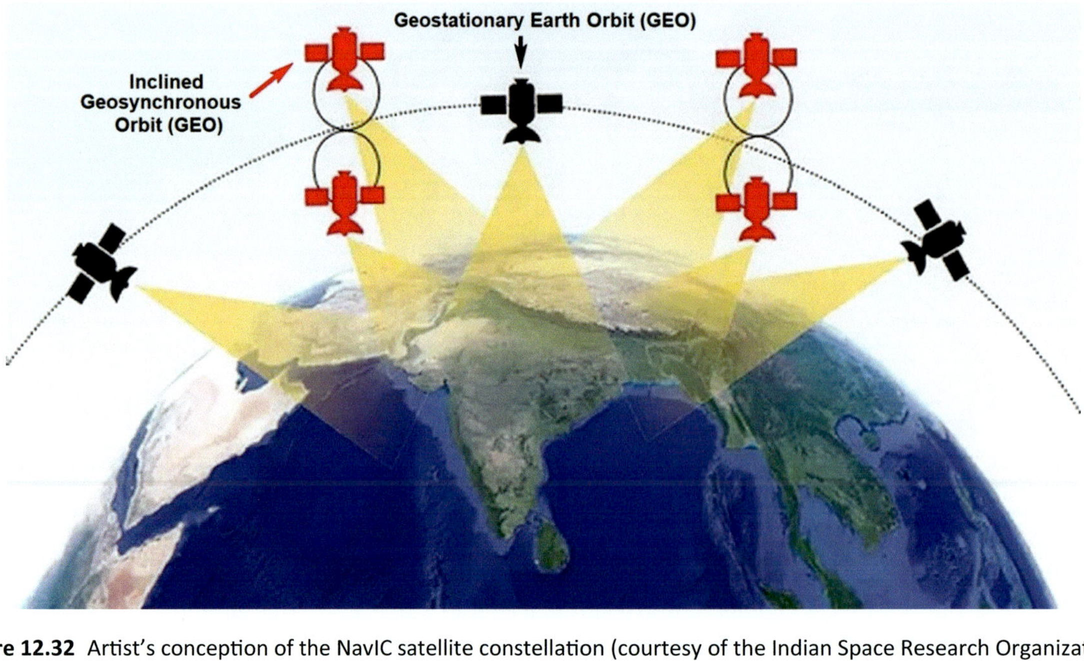

Figure 12.32 Artist's conception of the NavIC satellite constellation (courtesy of the Indian Space Research Organization).

The NavIC signals cover the nation of India, as well as surrounding regions up to 1,500 km from India's borders. Both the IGSO and GEO satellites have an orbital radius of 42,164 km and a period of one sidereal day. However, the IGSO satellites have an inclination angle of 29° with respect to the equator, and this tilted geosynchronous orbit prevents the satellites from staying in a fixed position. The GEO satellites have a slight inclination angle of 5°.

The NavIC satellites broadcast on the L1 and L5 carrier frequencies used by GNSS—1575.42 and 1176.45 MHz, respectively—and on an S-band carrier centered at 2492.028 MHz. The satellites transmit PRN codes of 10,230 chips in length, sent at a rate of 1.023 megachips/s, and repeating every 10 ms. The PRN code numbers were assigned beginning with 1, and have a possible range of 1 to 64. The satellites also transmit a navigation message that's sent at a rate of about 50 bps.

The NavIC satellites carry rubidium clocks that run at a nominal frequency of 10 MHz, with a PLL circuit providing a 10.23 MHz frequency output. The necessary relativistic correction is computed and applied by the receiver, using parameters obtained from the navigation message. The navigation messages also include the eight parameters needed to compute the Klobuchar model (Section 12.7.5), which allows single-frequency NavIC receivers to compensate for most of the ionospheric delay [62].

The NavIC system time has the same epoch as Galileo—23:59:47 UTC on August 21, 1999. The system time is referenced to UTC(NPLI), the national time scale of India that's operated by the National Physical Laboratory (NPL) in New Delhi. This system time is controlled by steering atomic clock ensembles located at ground control stations to UTC(NPLI) via TWSTFT. The ground control stations keep the satellite clocks synchronized to UTC(NPLI) by uploading correction messages. The uncertainty of NavIC system time with respect to UTC(NPLI) is typically 20 ns ($k = 2$) [63].

A NavIC disciplined clock synchronizes to UTC(NPLI) in the same way that a GPSDC synchronizes to UTC(USNO). Using information found in the navigation messages, a NavIC receiver calculates and applies two UTC corrections—a large integer second correction that compensates for the number of leap seconds that have occurred since the Galileo/NavIC epoch, and a small nanosecond-level correction that compensates for the most recent measured time difference between NavIC system time and UTC(NPLI) [62, 63]. Measurements of a NavIC disciplined rubidium clock have demonstrated a time accuracy of ±10 ns and a frequency stability of 6×10^{-14} at $\tau = 1$ day, when compared to the atomic clock ensemble at a NavIC ground control station [64].

12.10 Low Earth Orbit (LEO) Satellite Navigation Systems

A low Earth orbit is usually defined as an orbit with a height of less than 2,000 km above the Earth's surface, which is slightly less than one-third of the Earth's radius. All LEO satellites orbit the Earth more than 11 times per day. The maximum period for a low Earth orbit is about 128 minutes, and typical orbital periods range from about 90 to 120 minutes.

The first satellites used for navigation—the Transit and TIMATION satellites discussed in Section 12.1—were LEO satellites, but the MEO satellites deployed in GNSS constellations have since dominated PNT applications. However, the vast majority of the satellites in space are LEO satellites, due to their extensive use in communication systems. For example, the American company SpaceX launched more than 9,000 LEO satellites for its Starlink constellation during the period from 2019 to 2025, with thousands of additional launches planned. The Starlink satellites already account for about half of the total number of satellites launched since the Space Age began in 1957.

The Starlink satellites haven't offered a PNT service as of 2025, but their potential use for PNT applications has been extensively explored by researchers [65, 66]. Other LEO satellite providers do offer PNT services that are supported by licensing fees. These services differ from government-provided services, where access to the signals is free, and where the specifications needed to design and build new products—such as the GNSS and

RNSS interface control documents—are freely published and distributed [18, 43, 47, 50, 58, 59, 62]. Because signal specifications aren't usually available, the time reference often isn't disclosed to the public, although it's highly likely that most LEO systems are referenced to clocks that are disciplined by GNSS.

The OneWeb system is a notable exception to the usual practice of LEO systems being completely commercialized. The first-generation OneWeb constellation includes 648 LEO satellites that fly at 1,200 km above the Earth's surface, and circle the Earth every 109 minutes while transmitting signals on K_u band frequencies of 10.7 GHz to 12.7 GHz. The current system doesn't provide PNT services, but researchers have performed positioning and time transfer experiments with encouraging results [67, 68]. OneWeb was saved from bankruptcy in 2020 by a transaction that included an injection of funds from the British government, which allowed it to merge in 2022 with the French company Eutelsat. A second-generation OneWeb constellation being developed by Eutelsat is expected to have 300 satellites and to begin providing PNT services in 2028. The British government's minority ownership stake in OneWeb has raised the possibility of a freely available time service that will likely distribute UTC(NPL), the national time scale of the United Kingdom [69].

If a LEO system doesn't provide users with sufficient details about its signal format and time reference, the lack of transparency might make it difficult for users to establish a traceability chain to the SI (Chapter 9). The potential for high-accuracy time distribution is, of course, still there, because any system that provides navigation and positioning services has to be referenced to an accurate time source, and LEO disciplined clocks have routinely demonstrated sub-microsecond accuracy when calibrated for equipment delays.

There are some time transfer disadvantages, however, when LEO satellites are compared to MEO satellites. One disadvantage is that LEO satellites carry smaller payloads than MEO satellites, which usually precludes them from carrying atomic clocks. In most cases, LEO satellites carry only quartz clocks, and when they do carry atomic clocks, they are likely based on CSAC technology (Section 3.11). The lesser stability of the LEO clocks is mostly compensated for by their shorter orbital periods, which allows clock corrections from ground stations to be uplinked more frequently [65].

The time transfer uncertainty of the LEO satellites is also affected by their velocity, which produces larger Doppler frequency shifts than those observed from MEO satellites. A LEO satellite travels at a speed of roughly 8 km/s, which is about twice the speed of a MEO satellite. This means that a satellite is only observable from Earth for a short period—typically for 10 minutes or less—and the Doppler frequency shift of a LEO satellite can vary significantly during its overhead pass. The short period of observation also forces a LEO receiver to switch between satellites more often than an MEO receiver, and the more frequent "handoffs" between satellites can introduce small phase shifts in the data that's used to discipline a clock. All of these uncertainties can be compensated to well below the microsecond level using mathematical models, but the uncertainties of LEO time signals will likely always be slightly larger than those of MEO time signals [65, 68].

Disciplined clocks that receive signals from the Iridium constellation of LEO satellites have been available commercially, with associated licensing fees, since 2017 [70]. The Iridium constellation was developed by the Iridium Satellite Services Company—now named Iridium Communications—which is headquartered in McLean, Virginia. The first satellites were launched in 1997, with Motorola providing the financing and much of the technology. Motorola, along with an international consortium of telecommunication companies, used the satellites to develop a telephone network that provided voice and messaging communications between any two places on Earth.

Iridium gained its name because the early design proposals for the constellation included 77 satellites, and were thought to resemble drawings of the Iridium atom, which has 77 electrons orbiting its nucleus. However, the design that was chosen includes 66 satellites, with 11 satellites in each of six orbital planes (Figure 12.33). The satellites fly at a nominal height of 780 km above the Earth's surface, and travel in a polar orbit with an inclination angle of 86.4° and a period of about 100 minutes. The satellites transmit and receive signals on L-band frequencies ranging from 1616 MHz to 1626.5 MHz, using FMDA and time-division multiple access (TDMA) signal formats [71].

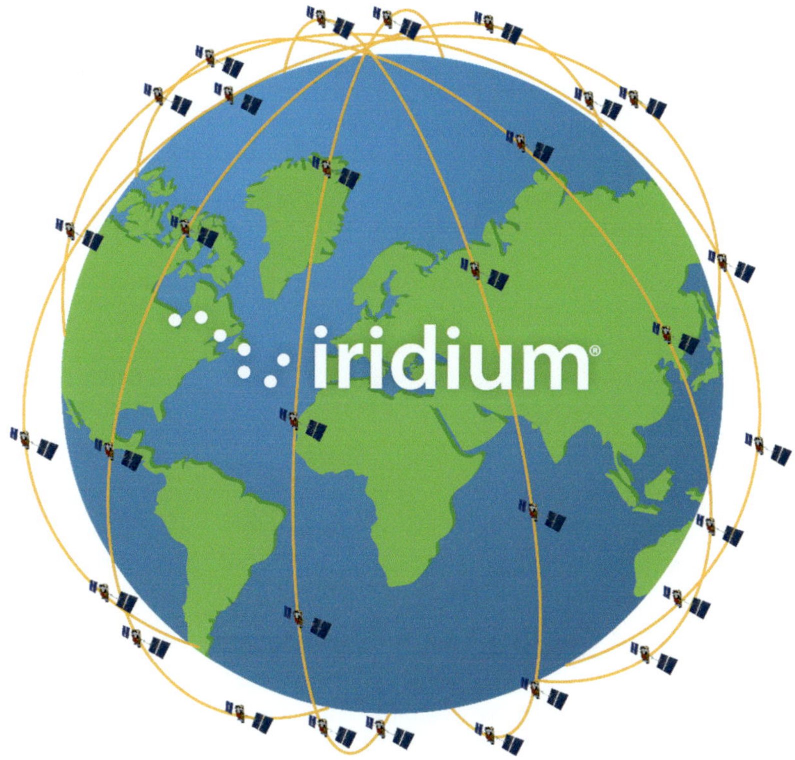

Figure 12.33 The Iridium satellite constellation (courtesy of Iridium Communications).

The Iridium satellites provide a Satellite Time and Location (STL) service that was started by the Satelles Corporation in 2016. Iridium Communications acquired Satelles in 2024, and the service is now called Iridium STL. In comparisons made with UTC(NIST), the STL service has easily demonstrated sub-microsecond accuracy, and the measured performance of STL disciplined clocks has been only slightly worse than a typical GPSDC. For example, a 100-day comparison made in 2023 showed that the peak-to-peak time difference of a delay-calibrated STL disciplined rubidium clock stayed within ±50 ns of UTC(NIST). During this period, the average time difference for data points averaged for 10 minutes, was less than 2 ns. The time stability, as estimated with TDEV, was about 6 ns at τ = 1 day [72].

A proposed PNT service has also been demonstrated using LEO satellites operated by the Globalstar Corporation of Covington, Louisiana [73]. Globalstar began launching its first generation of satellites in 1998 and has since provided global telephone and messaging services similar to those offered by Iridium. The Globalstar satellites fly at a nominal height of 1,410 km above the Earth's surface, with an orbit inclination angle of 52°, and an orbital period of about 114 minutes. The satellites transmit signals on S-band frequencies ranging from 2483.5 MHz to 2500 MHz [74]. The second-generation Globalstar system includes a constellation of at least 24 satellites, and the proposed PNT services are the result of a collaboration with the Parsons Corporation, an American defense technology firm headquartered in Chantilly, Virginia.

The Pulsar constellation of LEO satellites, developed by Xona Space Systems of Burlingame, California, was specifically designed to provide PNT services. The first Pulsar satellite was launched in 2022, coinciding with research and development efforts aimed at adding Pulsar capability to GNSS receivers [75]. The Pulsar satellites fly at a nominal height of 525 km above the Earth's surface, with an orbit inclination angle of 97° and an orbital period of about 95 minutes. The satellites transmit signals on L-band frequencies near 1260 MHz. The full constellation is expected to include 260 satellites [76].

Table 12.10 summarizes the characteristics of the LEO satellite systems discussed in this section. As of 2025, these systems are either currently providing a UTC distribution service, or planning to provide a UTC service in the near future.

Table 12.10 LEO satellite navigation systems that can distribute UTC

Parameter	LEO satellite constellation			
	Globalstar	Iridium	OneWeb (Gen 1)	Pulsar
Operator	Globalstar, Inc.	Iridium Communications	Eutelsat	Xona Space Systems
Orbit Height (km)	1,410	780	1,200	525
Orbit Inclination	52°	86.4°	86.4°	97°
Orbit Period (hh:mm)	1:54	1:40	1:49	1:35
Satellites (full constellation)	24	66	648	260
Downlink signal format	CDMA	FDMA/TDMA	FDMA/TDMA	CDMA
Downlink frequencies and band designation	2483.5 to 2500 MHz (S-band)	1616 to 1626.5 MHz (L-band)	10.7 to 12.7 GHz (K_u band)	~1260 MHz (L-band)

12.11 Vulnerabilities of Satellite PNT Signals

Before GPS was invented, there were no freely available time signals that could provide sub-microsecond accuracy. Once its signals were made publicly available, GPS receivers were quickly incorporated into numerous products, and those products enabled many new technologies. The clocks and oscillators controlled by satellite navigation systems are now an essential part of our critical infrastructure systems, including the telecommunication, electric power, and financial systems that we'll discuss in Chapters 16, 17, and 18. Our dependency on GPS and the other PNT systems has become a subject of international concern. No one knows for sure how much damage would be done if PNT signals disappeared, but nearly everyone agrees that the consequences would be catastrophic.

Three types of interference can cause clocks controlled by satellite PNT signals to fail, even when the satellites and receivers are functioning properly. The first and most common type is unintentional interference, which can originate from any number of RF sources. The second type of interference—called jamming—is intentional and malicious. It involves the transmission of RF signals that block PNT reception by overloading the receiver. The third type of interference, called spoofing, is even more malicious. Spoofing is an advanced form of jamming that blocks PNT reception by broadcasting compatible signals that contain false information [77–79].

Most reception problems are due to unintentional RF interference, which is often intermittent and which usually only blocks reception in a small area. The GNSS signals are especially vulnerable to RF interference due to extremely low power levels that typically range from –155 dBW to –163 dBW when received on Earth, which is less than one femtowatt (10^{-15} W). The loss of all PNT services in a given area can be caused by interfering signals that just a few orders of magnitude more powerful, which means that almost any RF signal source can potentially cause problems. The known sources of unintentional interference include out-of-band harmonics from nearby radio transmitters, in-band interference from radar systems, and high levels of solar radiation. The received signal levels are so small that even the active amplifiers in GNSS antennas can sometimes cause interference, by radiating signals that are large enough to interfere with neighboring receivers [26]. Therefore, maintaining a distance of at least one meter between GNSS antennas is often recommended.

Jamming devices are illegal, but they are still relatively easy to build or buy. They can also be difficult to find and shut down. Because navigation satellites typically broadcast on the same frequency bands, a single jamming device can often block signals from multiple satellite constellations. The type of signal modulation used by a jamming device isn't especially important, because their goal is simply to generate signals at frequencies within the receiver's bandwidth that are large enough to override or distort the incoming satellite signals. Small, inexpensive jammers are sometimes sold as personal privacy devices, and they can typically block GNSS reception over a radius of a few tens of meters, which can prevent the whereabouts of a person or vehicle from being tracked [78]. Because these devices are usually mobile, they usually only cause intermittent problems. For example, a car with a jamming device can block GNSS reception when it passes by an antenna, but typically for only one or two seconds, and the effect that it has on a disciplined clock might not even be noticeable. Far more significant problems are caused by jamming devices that radiate 10 W or more of power and have a range of kilometers instead of meters.

Spoofing devices are far more sophisticated than jamming devices, because the spoofer has to replicate not just the RF carrier, but also the PRN codes and navigation messages of every GNSS system that it tries to spoof. Spoofing can also cause far more damage than jamming, because a receiver that can't distinguish between spoofed and authentic information will produce a false position fix, a false time signal, or both [79].

When reference signals can't be received, or when they are ignored because they are known to be false, a satellite-disciplined clock has to rely on its holdover capability (Section 5.3.2) to maintain synchronization. For many clocks, holdover capability is only provided by the local oscillator, which free runs in the absence of an incoming satellite signal. However, some clocks continue to adjust their oscillator during signal outages, using algorithms that predict which adjustments are needed based on past measurements. A variety of oscillators are found inside of satellite-disciplined clocks—ranging from tiny quartz crystal oscillators that cost just pennies, to rubidium oscillators that can cost several thousand dollars or more—and holdover capability can vary accordingly. In all cases, a time accuracy of ±1 μs can only be maintained for a finite period once the satellite signal is lost. The duration of this period will be less than an hour for most disciplined clocks with low-cost quartz oscillators, and in some cases will be just a few seconds. If the LO is a high-quality OCXO or a rubidium, a disciplined clock can survive much longer outages, but maintaining ±1 μs accuracy for more than a week without satellite reception is unlikely for even the best devices.

Many methods have been demonstrated that make satellite-disciplined clocks more immune to interference. For example, the probability that all satellite signals will be blocked is reduced by receiving signals on multiple frequencies from multiple constellations. Using receivers that track LEO satellites is also beneficial, because the received power of LEO signals is typically about 30 dB higher (1,000×) than MEO GNSS signals, which makes them less likely to be blocked. Receivers within the coverage area of a GEO RNSS signal can guard against interference by using directional antennas, which can be stationary antennas that point to a single satellite, or moving antennas that rotate between satellites. In all cases, a satellite clock's immunity to interference will be improved by keeping multipath signal reflections to a minimum (Section 12.7.7), and by using stable local oscillators with strong holdover capability.

Many methods have also been demonstrated to detect spoofing, but these methods are often proprietary for security reasons. Clocks with firmware that employs receiver autonomous integrity monitoring (RAIM) have some rudimentary spoofing protection built in, because they compare sets of pseudorange measurements and ignore the measurements that lack a believable consistency. However, a sophisticated spoofer can easily defeat a simple RAIM defense by first receiving the GNSS signals. By knowing its own position relative to the position of the receiver being attacked, a spoofer can then broadcast signals that are aligned with the true signals, and drag the attacked receiver away from its true position or time solution. This strategy can avoid detection because the spoofed receiver maintains lock throughout the entire attack. Therefore, spoofing detection methods usually involve checking for differences between the incoming signals and their published definitions, and then looking for signs of interaction between the true signal and the spoofed signal. Some form of interaction should always be detectable, unless the spoofing signal has an unrealistically high power level, which can be another indication that an attack is taking place. In addition to checking the signal's content and power level, some methods can detect more subtle signs of spoofing, including distortion of the correlation function, signals with an unusual power spectral density, and signals that arrive from an unrealistic direction and angle [77, 78, 79].

The vulnerability of satellite PNT signals to interference is an important topic that we've just briefly touched upon. We'll close this chapter by noting that the problems presented by jamming and spoofing will likely drive PNT research and development efforts for years to come, and that these efforts will undoubtedly make future satellite-disciplined clocks less vulnerable to interference.

References

1. R. J. Danchik, "An Overview of Transit Development," *Johns Hopkins APL Technical Digest*, vol. 19, no. 1, pp. 18–26, January-March 1998.
2. T. A. Stansell, "The TRANSIT Navigation Satellite System," *Magnavox R-5933A*, 83 p., 1978.
3. R. E. Beehler, D. D. Davis, J. V. Cateora, A. J. Clements, J. A. Barnes, and E. Méndez-Quiñones, "Time Recovery Measurements Using Operational GOES and TRANSIT Satellites," *Proceedings of the 11th Annual Precise Time and Time Interval Meeting (PTTI)*, pp. 283–309, November 1979.
4. R. L. Easton, "The Role of Time/Frequency in Navy Navigation Satellites," *Proceedings of the IEEE*, vol. 60, no. 5, pp. 557–563, May 1972.
5. R. L. Beard, J. Murray, and J. D. White, "GPS Clock Technology and the Navy PTTI Programs at the U. S. Naval Research Laboratory," *Proceedings of the 18th Annual Precise Time and Time Interval Meeting (PTTI)*, pp. 37–53, December 1986.
6. I. A. Getting, "The Global Positioning System," *IEEE Spectrum*, vol. 30, no. 12, pp. 36–48, 43–47, December 1993.
7. B. W. Parkinson and S. W. Gilbert, "NAVSTAR: Global Positioning System – Ten Years Later," *Proceedings of the IEEE*, vol. 71, no. 10, pp. 1177–1186, October 1983.
8. B. W. Parkinson, Y. T. J. Morton, F. van Diggelen, and J. J. Spilker, Jr., "Introduction, Early History, and Assuring PNT (PTA)," in *Position, Navigation, and Timing Technologies in the 21st Century: Integrated Satellite Navigation, Sensor Systems, and Civil Applications, Volume 1*, pp. 3–41, John Wiley & Sons, 2021.
9. J. Buisson, T. McCaskill, H. Smith, P. Morgan, and J. Woodger, "Precise Worldwide Station Synchronization via the Navstar GPS, Navigation Technology Satellite (NTS-1)," *Proceedings of the 8th Annual Precise Time and Time Interval Meeting (PTTI)*, pp. 55–84, November 1976.
10. P. Misra and P. Enge, *Global Positioning System: Signals, Measurement, and Performance*, 2nd edition, Ganga-Jamuna Press, 2011.
11. E. D. Kaplan and C. J. Hegerty, eds., *Understanding GPS: Principles and Applications*, 2nd edition, Artech House, 2006.
12. A. J. Van Dierendonck, Q. D. Hua, J. R. McLean, and A. R. Denz, "Time Transfer Using NAVSTAR GPS," *Proceedings of the 13*th *Annual Precise Time and Time Interval Meeting (PTTI)*, pp. 389–418, December 1981.
13. K. Putkovich, "Initial Test Results of USNO GPS Time Transfer Unit," *Proceedings of the 34*th *Annual Symposium on Frequency Control,* pp. 326–333, May 1980.
14. R. B. Langley, "Smaller and Smaller: The Evolution of the GPS Receiver," *GPS World*, vol. 11, no. 4, pp. 54–58, April 2000.
15. J. B.-Y. Tsui, *Fundamentals of Global Positioning Receivers: A Software Approach*, 2nd edition, John Wiley & Sons, 2005.
16. F. Vannicola, R. Beard, J. White, K. Senior, M. Largay, and J. Buisson, "GPS Block IIF Atomic Frequency Standard Performance," *Proceedings of the 42nd Annual Precise Time and Time Interval Meeting (PTTI)*, pp. 181–196, November 2010.
17. N. Ashby, "Relativity and the Global Positioning System," *Physics Today*, vol. 55, no. 5, pp. 41–47, May 2002.

18. "NAVSTAR GPS Space Segment/Navigation User Segment Interfaces," *Interface Specification Document IS-GPS-200N*, 248 p., August 2022.
19. R. Gold, "Optimal Binary Sequences for Spread Spectrum Multiplexing," *IEEE Transactions on Information Theory*, vol. 13, no. 4, pp. 619–621, October 1967.
20. M. Miranian, "UTC Dissemination to the Real-Time User: The Role of USNO," *Proceedings of the 27th Annual Precise Time and Time Interval Meeting (PTTI)*, pp. 75–85, November 1995.
21. D. Matsakis, J. Levine, and M. A. Lombardi, "Metrological and Legal Traceability of Time Signals," *Proceedings of the 49*th *Annual Precise Time and Time Interval Meeting (PTTI)*, pp. 59–71, January 2018.
22. M. A. Lombardi, B. R. Patla, A. N. Novick, D. Matsakis, and S. J. Mitchell, "A Simple Method for Calibrating GPS-Disciplined Clocks via Direct Comparison to a UTC(*k*) Time Scale," *URSI Radio Science Bulletin*, no. 378, pp. 51–58, September 2021.
23. P. Defraigne, E. Pinat, G. Petit, and F. Meynadier, "Monitoring of the Offset between UTC and Its Prediction Broadcast by the GNSS," *Metrologia*, vol. 60, no. 6, pp. 1–12, November 2023.
24. V. Zhang, T. E. Parker, R. Bumgarner, J. Hirschauer, A. McKinley, S. Mitchell, E. Powers, J. Skinner, and D. Matsakis, "Recent Calibrations of UTC(NIST) – UTC(USNO)," *Proceedings of the 44th Annual Precise Time and Time Interval Meeting (PTTI)*, pp. 35–42, November 2012.
25. C. C. Kilgus, "Multielement, Fractional Turn Helices," *IEEE Transactions on Antennas and Propagation*, vol. 16, no. 4, pp. 499–500, July 1968.
26. M. A. Lombardi and A. N. Novick, "Effects of the Rooftop Environment on GPS Time Transfer," *Proceedings of the 38th Annual Precise Time and Time Interval Meeting (PTTI*, pp. 449–465, December 2006.
27. L. E. Young, T. K. Meehan, D. J. Spitzmesser, and J. T. Tranquilla, "GPS Antenna Selection: Preliminary Range and Field Test Results," *Proceedings of the 1988 Symposium on Antenna Technology and Applied Electromagnetics*, 8 p., August 1988.
28. W. Kunysz, "High Performance GPS Pinwheel Antenna," *Proceedings of the 13th International Technical Meeting of the Satellite Division of The Institute of Navigation*, pp. 2506–2511, September 2000.
29. M. A. Lombardi, "Evaluating the Frequency and Time Uncertainty of GPS Disciplined Oscillators and Clocks," *NCSLI Measure: The Journal of Measurement Science*, vol. 11, no. 3–4, pp. 30–44, December 2016.
30. J. Plumb, K. M. Larson, J. White, and E. Powers, "Absolute Calibration of a Geodetic Time Transfer System," *IEEE Transactions on Ultrasonics, Ferroelectrics, and Frequency Control*, vol. 52, no. 11, pp. 1904–1911, November 2005.
31. E. Powers, P. Wheeler, D. Judge, and D. Matsakis, "Hardware Delay Measurements and Sensitivities in Carrier Phase Time Transfer," *Proceedings of the 30th Annual Precise Time and Time Interval Meeting (PTTI)*, pp. 293–306, December 1998.
32. K. O'Keefe, J. Stephen, G. Lachapelle, and R. Gonzales, "Effect of Ice Loading of a GPS Antenna," *Geomatics*, vol. 54, no. 1, pp. 63–74, January 2000.
33. J. A. Klobuchar, "A First-Order, Worldwide, Ionospheric, Time-Delay Algorithm," *Air Force Cambridge Research Laboratory Technical Report 75-0502*, 24 p., September 1975.
34. J. A. Klobuchar, "Ionospheric Time-Delay Algorithm for Single-Frequency GPS Users," *IEEE Transactions on Aerospace and Electronic Systems*, vol. AES-23, no. 3, pp. 325–331, May 1987.
35. J. A. Klobuchar, "Present Status and Future Prospects for Ionospheric Propagation Corrections for Precise Time Transfer Using GPS," *Proceedings of the 23rd Annual Precise Time and Time Interval Meeting (PTTI)*, pp. 417–429, December 1991.
36. J.-H. Won and J.-S. Lee, "A Note on the Group Delay and Phase Advance Phenomenon Associated with GPS Signal Propagation through the Ionosphere," *NAVIGATION: Journal of the Institute of Navigation*, vol 52, no. 2, pp. 95–97, Summer 2005.
37. V. Zhang and Z. Li, "Measured Ionospheric Delay Corrections for Code-Based GPS Time Transfer," *NCSLI Measure: The Journal of Measurement Science*, vol. 10, no. 3, pp. 66–71, September 2015.
38. V. B. Mendes, "Modeling the Neutral-Atmosphere Propagation Delay in Radiometric Space Techniques," Ph.D. Dissertation, *Department of Geodesy and Geomatics Engineering Technical Report No. 199*, University of New Brunswick, Canada, 353 p., 1999.

39. North Atlantic Treaty Organization (NATO), "Navstar Global Positioning System (GPS) System Characteristics," *NATO Standardization Agreement (STANAG) 4294*, 1993.

40. T. Kos, I. Markezic, and J. Pokrajcic, "Effects of Multipath Reception on GPS Positioning Performance," *Proceedings of the 52nd International Symposium on Electronics in Marine (ELMAR)*, pp. 399–402, September 2010.

41. K. Verhasselt and P. Defraigne, "Multi-GNSS Time Transfer Based on the CGGTTS," *Metrologia*, vol. 56, no. 6, pp. 065003/1–12, October 2019.

42. I. Sesia, G. Signorile, T. T. Thai, P. Defraigne, and P. Tavella, "GNSS-to-GNSS Time Offsets: Study on the Broadcast of a Common Reference Time," *GPS Solutions*, vol. 25, no. 61, 15 p., February 2021.

43. Russian Institute of Space Device Engineering, *GLONASS Interface Control Document*, edition 5.1, 65 p., 2008.

44. P. Bogdanov, A. Druzhin, T. Primakina, and A. Feoktistov, "Approaches to GLONASS Time Accuracy Improvement," *Proceedings of the 10th Annual Baška GNSS Conference*, pp. 111–116, May 2016.

45. A. Druzhin and T. Primakina, "GNSS System Time Scales and Time Transfer Accuracy," *Proceedings of the 3rd URSI Atlantic and Asia Pacific Radio Science Meeting*, 2 p., June 2022.

46. European Union Agency for the Space Programme, *Galileo Open Service: Service Definition Document (OS SDD)*, no. 1.3, 72 p., November 2023.

47. European Union Agency for the Space Programme, *Galileo Open Service: Signal-in-Space Interface Control Document (OS SIS ICD)*, no. 2.1, 128 p., November 2023.

48. European Union Agency for the Space Programme, *Ionospheric Correction Algorithm for Galileo Single Frequency Users (OS SIS ICD)*, no. 1.2, 96 p., September 2016.

49. U. Ngayap, C. Paparini, M. Porretta, P. Buist, K. S. Jacobsen, M. Dähnn, N. Hanna, D. Halilovic, A. Świątek, and P. Gajdowska, "Comparison of NeQuick G and Klobuchar Model Performances at Single-Frequency User Level," *Engineering Proceedings*, vol. 54, pp. 7/1–10, October 2023.

50. China Satellite Navigation Office, *BeiDou Navigation Satellite System Open Service Performance Standard*, ver. 3.0, 43 p., May 2021.

51. W. Wang, Y. Wang, C. Yu, F. Xu, and X. Dou, "Spaceborne Atomic Clock Performance Review of BDS-3 MEO Satellites," *Measurement*, vol. 175, pp.109075/1–9, January 2021.

52. W. Guang, J. Zhang, H. Yuan, W. Wu, and S. Dong, "Analysis on the Time Transfer Performance of BDS-3 Signals," *Metrologia*, vol. 57, no. 6, pp. 065023/1–17, October 2020.

53. China Satellite Navigation Office, *BeiDou Navigation Satellite System Signal In Space Interface Control Document*, including "Open Service Signal B1C Version 1.0" (December 2017), "Open Service Signal B1I Version 3.0" (February 2019), "Open Service Signal B2a Version 1.0" (December 2017), "Open Service Signal B2b Version 1.0" (July 2020), and "Open Service Signal B3I Version 1.0" (February 2018).

54. N. Wang, Z. Li, Y. Yuan, and X. Huo, "BeiDou Global Ionospheric Delay Correction Model (BDGIM): Performance Analysis During Different Levels of Solar Conditions," *GPS Solutions*, vol. 25, pp. 97/1–13, April 2021.

55. C. Han, Y. Yang, and Z. Cai, "BeiDou Navigation Satellite System and Its Time Scales," *Metrologia*, vol. 48, no. 4, pp. S213–S218, July 2011.

56. W. J. Klepczynski, P. Fenton, and E. Powers, "Time Distribution Capabilities of the Wide Area Augmentation System (WAAS)," *Proceedings of the 33rd Annual Precise Time and Time Interval Meeting (PTTI)*, pp. 111–119, November 2001.

57. D. S. De Lorenzo, S. C. Lo, J. Seo, Y.-H. Chen, and P. K. Enge, "The WAAS/L5 Signal for Robust Time Transfer: Adaptive Beamsteering Antennas for Satellite Time Synchronization," *Proceedings of the 23rd International Technical Meeting of the Satellite Division of The Institute of Navigation*, pp. 2106–2116, September 2010.

58. European Union Agency for the Space Programme, *EGNOS Open Service (OS): Service Definition Document (SDD)*, no. 3.0, 48 p., November 2024.

59. Cabinet Office Government of Japan, "Quasi-Zenith Satellite System Interface Specification: Satellite Positioning, Navigation and Timing Service," *IS-QZSS-PNT-006*, 164 p., July 2024.

60. T. Hobiger, Y. Takahashi, M. Nakamura, T. Gotoh, S. Hama, T. Maruyama, T. Nagatsuma, H. Noda, M. Kishimoto, K. Nakayama, and Y. Ohki "Dissemination of UTC(NICT) by Means of QZSS," *IEEE Transactions on Instrumentation and Measurement*, vol. 62, no. 6, pp. 1537–1544, June 2013.

61. S. Murakami, K. Someya, M. Miyoshi, E. Myojin, and A. Matsumoto, "A Realization of QZSS Time Synchronized with UTC(NICT)," *Proceedings of the 2023 Joint Conference of the European Frequency and Time Forum and IEEE International Frequency Control Symposium (EFTF/IFCS)*, 2 p., May 2023.
62. Indian Space Research Organization, "NavIC Signal in Space ICD for Standard Positioning Service in L1 Frequency," *ISRO-NAVIC-ICD-SPS-L1-1.0*, 111 p., August 2023.
63. A. A. Bhardwajan, S. Dakkumalla, A. Arora, T. S. Ganesh, and A. Sen Gupta, "Navigation with Indian Constellation and its Applications in Metrology," *MAPAN: Journal of Metrology Society of India*, vol. 36, pp. 227–236, June 2021.
64. S. Dakkumalla, A. A. Bhardwajan, P. R. Dubey, R. Sadisivan, T. S. Ganesh, and M. R. Raghavendra, "Timing Performance and Stability Characterization of an in-House Realized NavIC Timing Receiver," *Proceedings of the 2022 URSI Regional Conference on Radio Science (USRI-RCRS)*, pp. 1–3, December 2022.
65. F. Prol, R. Morales Ferre, Z. Saleem, P. Valisuo, C. Pinell, E. S. Lohan, M. Elsanhoury, M. Elmusrati, S. Islam, K. Celikbelik, K. Selvan, J. Yliaho, K. Rutledge, A. Ojala, L. Ferranti, J. Praks, M. Z. H. Bhuiyan, S. Kaasalainen, and H. Kuusniemi, "Position, Navigation, and Timing (PNT) Through Low Earth Orbit (LEO) Satellites: A Survey on Current Status, Challenges, and Opportunities," *IEEE Access*, vol. 10, pp. 83971–84002, July 2022.
66. P. A. Iannucci and T. H. Humphreys, "Fused Low-Earth-Orbit GNSS," *IEEE Transactions on Aerospace and Electronic Systems*, vol. 60, no. 4, pp. 3730–3749, August 2024.
67. S. Kozhaya and Z. M. Kassas, "A First Look at the OneWeb LEO Constellation: Beacons, Beams, and Positioning," *IEEE Transactions on Aerospace and Electronic Systems*, vol. 60, no. 5, pp. 7528–7534, October 2024.
68. T. P. Arora, I. Petrunin, J. Hill-Valler, and E. Anyaegbu, "Improving Time Transfer Performance for Low Earth Orbit Satellites," *Proceedings of the 2024 IEEE International Instrumentation and Measurement Technology Conference*, 6 p., May 2024.
69. R. Jewett, "OneWeb to Collaborate on PNT Capabilities With the UK National Physical Laboratory," *Via Satellite*, September 26, 2023.
70. G. Gutt, D. Lawrence, S. Cobb, and M. O'Connor, "Recent PNT Improvements and Test Results Based on Low Earth Orbit Satellites," *Proceedings of the 49*th *Annual Precise Time and Time Interval Meeting (PTTI)*, pp. 72–79, January 2018.
71. C. E. Fossa, R. A. Raines, G. H. Gunsch, and M. A. Temple, "An Overview of the Iridium Low Earth Orbit (LEO) Satellite System," *Proceedings of the IEEE 1998 National Aerospace and Electronics Conference*, pp. 152–159, July 1998.
72. P. B. Johnson and A. N. Novick, "The Long-Term Timing Performance of Satellite Time and Location Receivers Utilizing Signals from Low Earth Orbit Satellites," *Proceedings of the 55*th *Annual Precise Time and Time Interval Meeting (PTTI)*, pp. 51–64, January 2024.
73. A. Hansen, S. Mackey, H. Wassef, V. Shah, E. Wallisheck, C. Scarpone, M. Barzech, and E. Baskerville, "Complementary PNT and GPS Backup Technologies Demonstration Report," *Department of Transportation Volpe Center Technical Report DOT-VNTSC-20-07*, 457 p., January 2021.
74. F. J. Dietrich, P. Metzen, and P. Monte, "The Globalstar Cellular Satellite System," *IEEE Transactions on Antennas and Propagation*, vol. 46, no. 6, pp. 935–942, June 1998.
75. N. S. Miller, J. T. Koza, S. C. Morgan, S. M. Martin, A. Neish, R. Grayson, and T. Reid, "SNAP: A Xona Space Systems and GPS Software-Defined Receiver," *Proceedings of the 2023 IEEE/ION Position, Location and Navigation Symposium (PLANS)*, pp. 897–904, April 2023.
76. B. Eissfeller, T. Pany, D. Dötterböck, and R. Förstner, "A Comparative Study of LEO-PNT Systems and Concepts," *Proceedings of the Institute of Navigation (ION) 2024 Pacific PNT Meeting*, pp. 758–782, April 2024.
77. J. Zidan, E. I. Adegoke, E. Kampert, S. A. Birrell, C. R. Ford, and M. D. Higgins, "GNSS Vulnerabilities and Existing Solutions: A Review of the Literature," *IEEE Access*, vol. 9, pp. 153960–153976, November 2021.
78. R. T. Ioannides, T. Pany, and G. Gibbons, "Known Vulnerabilities of Global Navigation Satellite Systems, Status, and Potential Mitigation Techniques," *Proceedings of the IEEE*, vol. 104, no. 6, pp. 1174–1194, June 2016.
79. M. L. Psiaki and T. E. Humphreys, "GNSS Spoofing and Detection," *Proceedings of the IEEE*, vol. 104, no. 6, pp. 1258–1270, June 2016.

PART IV

Research Areas

Chapter 13

Optical Clocks

The situation no doubt will have changed again by the time you read this book. Things change faster in physics than in the book publishing business.

— Richard Feynman, "QED: The Strange Theory of Light and Matter"

Optical clocks work the same way as all other clocks—they mark the passage of time by counting the cycles of a frequency produced by an oscillator. The cycles are generated by electromagnetic waves locked to an atomic energy transition, just as they are in a conventional atomic clock. However, conventional atomic clocks resonate at microwave frequencies, and optical clocks resonate at the much higher frequencies of light waves.

The local oscillator in an optical clock is the beam of light produced by a laser. The resonance frequency of the most commonly used atomic transitions ranges from about 411 THz (0.41×10^{15} Hz) for a calcium ion clock, to about 1.267 PHz (1.27×10^{15} Hz) for an indium ion clock [1]. This range of frequencies is about 45,000× to 138,000× higher than the ~9.2 GHz (0.92×10^{10} Hz) microwave frequency that has defined the SI second since 1967.

The huge increase in resonance frequency allows optical clocks to slice time into much smaller increments than microwave clocks. When the laser frequency is stabilized, the increase in resonance frequency provides a correspondingly high increase in an optical clock's Q factor. Higher Q factors lead to better accuracy, and numerous experiments have proven that optical clocks can exceed the accuracy of the best microwave clocks by more than two orders of magnitude. These experiments have established optical clocks as the future of high-accuracy timekeeping and the basis for a new definition of the second. As was noted in Section 9.3, the CGPM will likely approve a new definition of the second based on the optical transitions of atoms in either 2030, 2034, or 2038 [2].

The pending redefinition of the second, along with the promise of unprecedented accuracy, makes optical clock research the most important research area in time and frequency metrology. Optical clocks have already made a huge impact on fundamental physics by putting new boundaries on theories which suggest that fundamental constants could be changing over time, and by testing Einstein's theory of relativity at the millimeter scale. When their designs mature, optical clocks will likely make a similarly huge impact on navigation and communications systems [3–5].

As Time Goes By: A Comprehensive Guide to Time and Frequency Measurements
Michael A. Lombardi

ISBN 978-981-5352-28-3 (Hardcover), 978-1-042-01532-0 (eBook)
www.jennystanford.com

DOI: 10.1201/9781042015320-18

This chapter discusses some of the research and development efforts that have made optical clocks a reality. The first two sections of the chapter provide a concise history of the first measurements of the wavelengths of light, the development of the laser, and the first measurements of laser frequencies. Section 13.3 discusses optical frequency combs, the devices that enabled the development of optical clocks by providing a coherent and bidirectional link between microwave and optical frequencies. The two main types of optical clocks are discussed in Section 13.4, along with a summary of the secondary representations of the SI second that are based on optical transitions. Section 13.5 concludes the chapter by discussing the current and future use of optical clocks as laboratory time and frequency standards.

13.1 Measuring the Wavelengths of Light

The nature of light has always fascinated scientists, and the first measurements of light's wavelengths can be traced back to the discoveries that helped build the foundation of classical and modern physics. In 1671, the English scientist Isaac Newton (1643–1727) demonstrated that white light could be separated into a spectrum of colors with a prism, but Newton believed that light consisted of tiny particles called corpuscles [6]. His reasoning was countered by Christiaan Huygens (1629–1695), the same scientist who had previously invented the pendulum clock (Section 1.5). Huygens published his *Traité de la Lumière* (Treatise on Light) in 1690, which theorized that light consisted of waves that propagated in space. This work provided the mathematical context for deriving the laws of reflection, diffraction, interference, and refraction, which were all principles that Newton's corpuscular theory had struggled to explain [7].

Newton had a tremendous reputation, so Huygens' wave theory was slow to gain acceptance. However, it was finally validated in the early 1800s by the double-slit experiments performed by the English scientist Thomas Young (1773–1829). For his original experiment, Young made a small hole in a window shutter, covered the hole with a thick piece of paper, and perforated the paper with a fine needle. He then brought "a slip of card, about one-thirtieth of an inch in breadth" into the path of a sunbeam, and observed the shadows cast on the wall [8].

Young's experiment is usually reproduced by shining a beam of light through a barrier with two parallel slits, so that it projects images on to a screen (Figure 13.1). The light waves emerging from the two slits combine to form an extended pattern of bright and dark spots. The pattern depends on the way that the light waves interfere with each other. The spots are brightest at the points where the light waves are exactly in phase, because the constructive interference adds them together. However, the light waves cancel each other when they are exactly out of phase, and the destructive interference (Section 5.2.1) causes the light to vanish [9, 10].

The validation of wave theory made it possible to explain the relationship between light and color. Just as the human ear hears sound waves in the audio frequency spectrum (Section 5.1), the human eye sees electromagnetic waves in the visible light spectrum. The light waves are seen in the form of colors, and the different colors correspond to the different wavelengths and frequencies of light. By using data that Newton had collected, Young was able to calculate the wavelengths of light for the seven major colors in the visible spectrum, and his approximations were close to our modern accepted values [11].

When scientists finally accepted Huygens' wave theory, it had been known for more than a century that light traveled at a finite speed. That idea had also been slow to gain acceptance, because it was believed for centuries that light didn't require any time at all to move from one place to another, and that its movement was instantaneous. The first measurements of the speed of light were conducted by the Danish astronomer Ole Christensen Rømer (1644–1710) in 1676, and they disproved the notion that light was instantaneous. Rømer measured the time required for light to travel between the Sun and the Earth by observing the eclipses of what Rømer called the "first satellite of Jupiter" [12]. This satellite was called Io, and it was the innermost of the four large moons of Jupiter that Galileo had discovered earlier that century.

Rømer carefully recorded his timed observations of Io, which eclipsed Jupiter about once every 42 hours, but he didn't use his data to calculate the speed of light. The calculations were left to Huygens, who estimated the speed of light using different units of length than we use today [7]. When converted to modern SI units, Huygens' estimate was roughly 214 km/s. This was much slower than light's actual speed, but it still proved that the speed of light was finite.

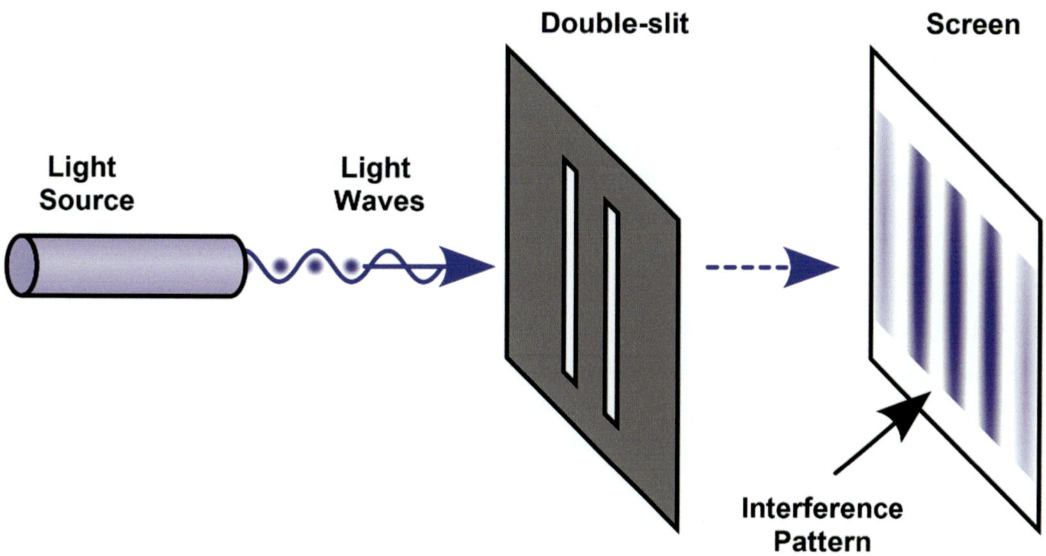

Figure 13.1 The double-slit experiment.

By the mid-19th century, independent measurements conducted by several scientists—including the French physicists Hippolyte Fizeau (1819–1896) and Léon Foucault (1819–1868)—found that the speed of light was somewhere in the neighborhood of 300 km/s. Their results were close to the value of 299,792,458 m/s that we now accept as a fundamental constant.

When it was first known that light consisted of waves traveling at a finite speed, the inherent nature of light waves was still unknown. The relationship between light and magnetism was first demonstrated in 1831 by the English scientist Michael Faraday (1791–1867). Faraday found that when a magnet was passed through a coil of wire, a current flowed in the wire while the magnet was moving. This discovery, called electromagnetic induction, is the basic principle of operation for electric motors and generators. Faraday explained electromagnetic induction with the concept of electric and magnetic "lines of force," which we now call electric and magnetic fields. He applied his force concepts to polarized beams of light in 1845, demonstrating that the direction of light's polarity could be rotated by an angle that was proportional to the magnetic force, a phenomenon now called the Faraday effect. The following year, he suggested that the "lateral vibrations" caused by the lines of electric and magnetic force could explain how light was radiated [13]. Faraday was a skilled experimentalist who could convey his ideas in clear and simple language, but his theories about light waves weren't widely accepted because he didn't support them with formal mathematics.

The necessary formal theory was developed by James Clerk Maxwell (1831–1879)—the same scientist who first suggested an atomic clock (Section 3.1). The close relationship between electricity and magnetism was already well known prior to the time of Maxwell, due to the work of Faraday and other scientists. However, there was no unifying theory that explained how electric and magnetic fields behaved and interacted. This changed in 1865 when Maxwell published a landmark paper entitled *A Dynamical Theory of the Electromagnetic Field* [14]. The paper contained a set of equations that Maxwell had worked on for almost a decade. The equations were more fully developed in *A Treatise on Electricity and Magnetism*, the two-volume textbook that Maxwell published in 1873 [15].

Maxwell's equations not only described the properties of electric and magnetic fields, but they explained how the fields could combine to form electromagnetic waves that propagated through space, which led to the invention of radio a few decades later. His equations also predicted that the speed of electromagnetic waves would be about 300,000 km/s, which coincided with measurements of the speed of light, and which made it easy to conclude that light waves were electromagnetic waves. In Maxwell's own words, the speed of electromagnetic waves is

> *"so nearly that of light, that it seems we have strong reason to conclude that light itself (including radiant heat, and other radiations if any) is an electromagnetic disturbance in the form of waves propagated through the electromagnetic field according to electromagnetic laws."* [14]

Maxwell's theoretical conclusions about the existence and speed of electromagnetic waves were confirmed by Heinrich Hertz (1857–1894), whose name was eventually given to the unit of frequency (Chapter 5). In a series of experiments conducted from 1886 to 1889, Hertz used oscillating electric currents to produce and detect electromagnetic waves at different frequencies. His measurements found that the waves all traveled at the speed of light, just as Maxwell had predicted. By proving that the electromagnetic spectrum encompassed all frequencies from "DC to daylight," Maxwell unified the fields of electricity, magnetism, and optics, and his work ranks as one of the greatest achievements in the history of science.

Optical wavelength metrology was greatly advanced in the 1880s by the American physicist Albert Michelson (1852–1931), who later became the first American scientist to receive a Nobel Prize in physics. Michelson was fascinated with the study of light throughout his life, and he accurately measured the speed of light in the late 1870s while teaching at the United States Naval Academy. In 1881, he invented the instrument that still bears his name, the Michelson interferometer. The interferometer was based on the same basic principles as Young's double-slit experiment. It produced images, called fringe patterns, that were formed by two interfering light waves. The information extracted from an interferometer's fringe patterns can be used to measure the wavelength of a light source. It can also be used to measure the length of other objects if a light source with a known wavelength is used as a reference.

At the time of Michelson's invention, scientists had accepted that light was a wave. However, the general belief was that waves couldn't travel through a vacuum, and that light waves—like the waves in water or the sound waves in air—had to be traveling through some type of medium that filled all space. Theorists called this hypothetical medium the luminiferous ether, or the ether for short. Therefore, Michelson invented the interferometer with a specific goal in mind, which was to find out if the ether existed by making measurements of the speed of light. He reasoned that if the ether was stationary, and if the Earth was moving through it, then the time required for light to move from one point to another on the Earth's surface would depend upon the direction of the light. This hypothesis could be tested by comparing the speed of light in the direction of the Earth's motion to the speed of light at right angles to the Earth's motion.

Figure 13.2 is a drawing of Michelson's original interferometer, which was built while he was working in Germany. The instrument was extremely sensitive to vibrations, so it was originally mounted on a stone pier. That helped, but it still wasn't sufficient, so Michelson had to relocate his interferometer several times before usable measurements could be made. He finally determined that the instrument worked when it was moved into a cellar with circular walls. However, his measurements failed to provide any evidence that the Earth was moving through a stationary ether [16].

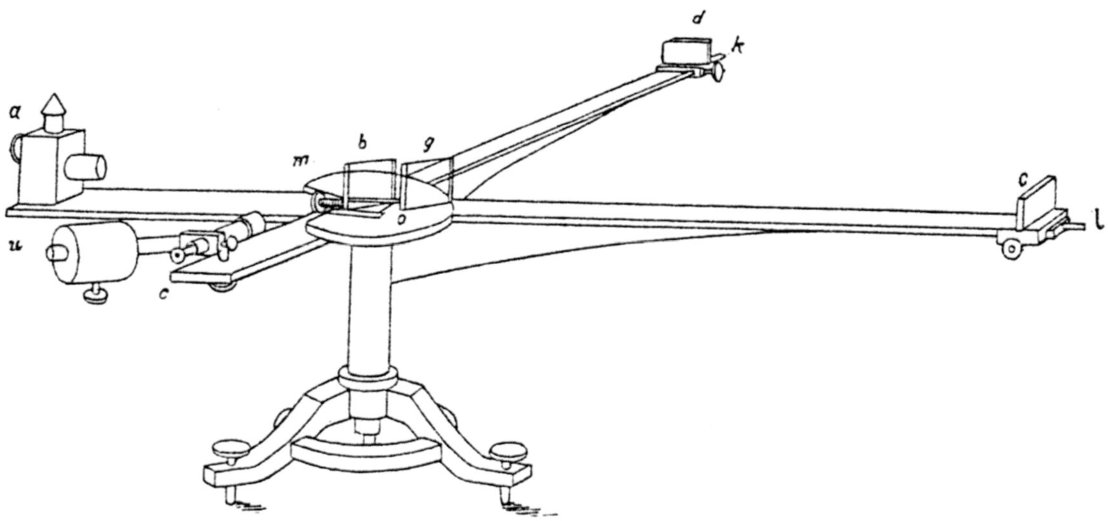

Figure 13.2 A drawing of the original Michelson interferometer, 1881 [16].

Michelson was disappointed with the results and felt that the existence of the ether could still be proven if he could make better measurements. He took a position at the Case School of Applied Science in Cleveland, Ohio (now known as Case Western Reserve University), where he began collaborating with the American scientist Edward Morley (1838–1923) in 1885. The two built a larger and more sensitive interferometer in the basement of a college dormitory. They mounted their instrument on a massive stone block and floated the stone in a pool of mercury so it could be smoothly rotated. For a period of several months in 1887, Michelson and Morley made repeated measurements of the speed of light, rotating the stone block and performing measurements at different times of day to change the alignment of the interferometer with respect to the Earth's orbit. However, they still found no significant differences in their speed of light measurements, and therefore found no evidence that the ether existed [17]. Scientists were puzzled by the null result of the Michelson–Morley experiment, but what seemed to be a paradox was resolved when Einstein published his theory of special relativity in 1905. Einstein abandoned the notion of the ether and asserted that the speed of light was invariant. It was the same for all observers, regardless of the motion of the light source or the observer [18].

After publishing their null result, Michelson and Morley focused on using interferometers to measure the meter, the base unit of length [19]. Today, interferometers are used in nearly all fields of science, and some can measure length with a resolution of less than a nanometer. Figure 13.3 is a simplified diagram that shows how a Michelson interferometer works. A light source sends a light beam to a splitter located near the center of the instrument. The beam splitter, which consists of a mirror that's half reflective and half transparent, splits the incident light into two beams. It reflects half of the incident light into one arm of the interferometer, and it transmits the other half by letting it pass straight through to the other arm. The mirrors located at the end of both arms reflect both beams back to the beam splitter. The beam splitter then acts as a beam combiner by reflecting half of the original transmitted light, and transmitting half of the original reflected light. This recombined beam is directed towards a viewing screen where it produces a circular fringe pattern of alternating bright and dark rings, a pattern caused by the constructive and destructive interference of the light waves.

The two interfering beams of light were split from the same beam, so they were initially in phase. However, their relative phase when they meet at any point on the viewing screen depends on the difference in the lengths of their optical paths. As shown in Figure 13.3, one of the reflecting mirrors is fixed (M1), but the other mirror is movable (M2), which allows the path length of one of the beams to be varied.

The variable beam travels between M2 and the beam splitter twice, so moving M2 one-quarter-wavelength closer to the beam splitter shortens its path by half a wavelength. This causes destructive interference because the two beams are now out of phase. However, if M2 is moved another quarter-wavelength closer to the beam splitter, the variable path will now have been shortened by a full wavelength. This causes constructive interference because the two beams are now back in phase.

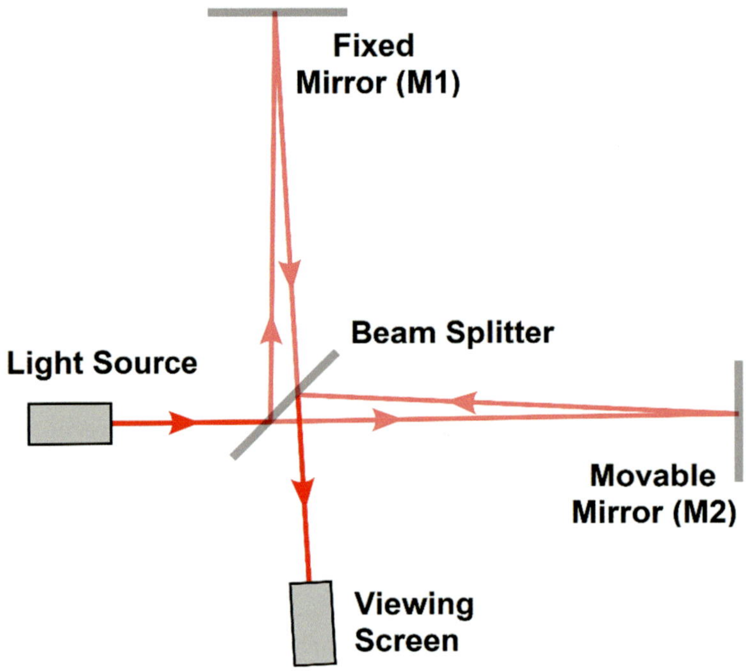

Figure 13.3 A simplified diagram of how an interferometer works.

The fringe patterns change whenever M2 is moved, but they repeat each time M2 is moved by half a wavelength. Therefore, the wavelength of the light source can be calculated by slowly moving M2, measuring the distance that it was moved, and counting the number of times that the pattern repeats. The basic equation is

$$\lambda = \frac{2d}{m}, \tag{13.1}$$

where λ is the wavelength of the incident light, d is the distance that M2 was moved, and m is the number of times that the fringe pattern repeated while M2 was moving. If the incident light has a known wavelength, the distance can be calculated as

$$d = \frac{m\lambda}{2}. \tag{13.2}$$

The electromagnetic spectrum above the RF region was categorized by making wavelength measurements of both visible and non-visible light. Table 13.1 lists the types and sub-types of radiation above the RF region, along with their associated frequencies and wavelengths. This table essentially starts where the RF bands listed in Table 5.4 end. However, the far-infrared region in Table 13.1 is partially overlapped by the THF region (300 GHz to 3 THz) in Table 5.4.

Table 13.1 The electromagnetic spectrum above the RF region

Type of radiation	Sub-type	Frequency range	Unit	Wavelength range	Unit
Infrared	Far-infrared	0.3–20	THz	1,000–15	μm
	Long-wavelength	20–37	THz	15–8	μm
	Mid-wavelength	37–100	THz	8–3	μm
	Short-wavelength	100–214	THz	3–1.4	μm
	Near-infrared	214–400	THz	1.4–0.75	μm
Visible light	Red	400–480	THz	750–625	nm
	Orange	480–510	THz	625–590	nm
	Yellow	510–530	THz	590–565	nm
	Green	530–600	THz	565–500	nm
	Cyan	600–620	THz	500–485	nm
	Blue	620–670	THz	485–450	nm
	Violet	670–750	THz	450–400	nm
Ultraviolet	UVA	750–950	THz	400–315	nm
	UVB	950–1070	THz	315–280	nm
	UVC	1.07–3	PHz	280–100	nm
	Extreme	3–30	PHz	100–10	nm
X-rays	—	30–0.03	PHz	10–0.01	nm
Gamma rays	—	> 30	EHz	< 10	pm

Frequency can be inferred from wavelength measurements, but direct measurements of the frequency of light—measurements that are referenced to frequency standards and traceable to the SI—weren't possible until the 1970s. Direct measurements of the frequency of light required the units of both time and length to be defined based on atomic transitions, and for the speed of light to be established as a fixed and defined number [20]. They also required the invention of the laser, as described in the following section.

13.2 Measuring the Frequencies of the Light Emitted by Lasers

The word laser is an acronym for "light amplification by stimulated emission of radiation." Lasers are based on the same principles utilized in hydrogen masers, the microwave atomic clocks that were discussed in Section 3.10. As was noted in that section, the only difference between the acronyms laser and maser is that laser substitutes "light" for "microwave."

The design principles of lasers resulted from a collaboration between Charles Townes (1915–2015), the inventor of the maser, and the American physicist Arthur Schawlow (1921–1999), who had worked for Townes at Columbia University and who was then working at Bell Labs. When the two published their work in 1958, they called the proposed new device an "optical maser" [21]. Both scientists later became Nobel laureates—Townes in 1964 for his fundamental work in quantum electronics that led to both masers and lasers, and Schawlow in 1981 for his contributions to the development of laser spectroscopy. The science of laser spectroscopy involves using tunable lasers to excite atomic and molecular systems, which allows the transition frequencies between specific energy levels to be measured.

Using the theory proposed by Townes and Schawlow, the laser was invented in 1960 by Theodore Maiman (1927–2007), an American electrical engineer and physicist who was then working at the Hughes Aircraft Company in Malibu, California. The first laser is pictured in Figure 13.4, with its internal components removed from the main housing. It used a photographic flashlamp (bottom left) to optically pump a cylinder-shaped crystal of synthetic ruby (bottom right). Unlike the lasers that were later developed, the first laser didn't produce a continuous wave of light, but instead produced pulses of light each time the flashlamp was fired. The wavelength of the emitted red light was 694 nm [9, 22, 23].

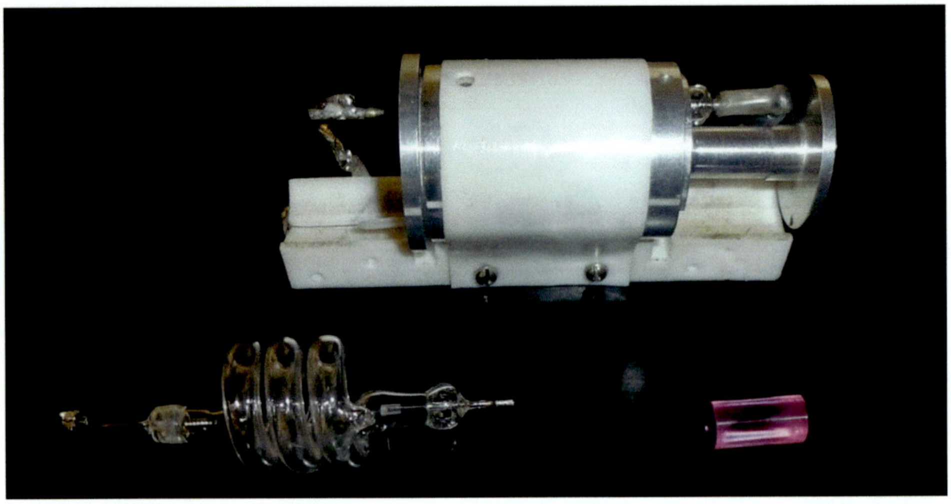

Figure 13.4 The housing and internal components of the ruby laser invented by Maiman in 1960.

Lasers are devices that emit a beam of visible or non-visible light. The light is generated by stimulating atoms or molecules with an optical signal or an electric current, which causes them to release energy by emitting photons. The stimulation causes other atoms or molecules to also emit photons, which amplifies the amount of light. Because light is an oscillatory wave, a laser is both an amplifier and an oscillator, but the acronym only refers to amplification. The acronym was coined in a conference paper presented in 1959, which was the year before the first laser was built. The conference session was chaired by Schawlow, who was known for his sense of humor. He remarked that the new device would have more importance as an oscillator than as an amplifier, but that would unfortunately change the acronym, and the word "loser" barely got out of his mouth before he broke into laughter [23]. Putting humor aside, Schawlow was right. The main function of a laser, not only in frequency and time metrology but in many other areas, is to serve as an optical oscillator.

Figure 13.5 shows the basic elements of a laser. The energy source, or pump, provides energy to the laser system. The pump can be either optical or electrical, sending either light or electric current through the gain medium. The gain medium can be a liquid, a gas, a solid, or a semiconductor. Two parallel mirrors, one at each end of the gain medium, form an optical cavity that provides a feedback system for the light. The mirror at one end of the cavity is called a total reflector, because it reflects all of the light it receives back into the cavity. The mirror at the other end is both a reflector and an output coupler. It reflects the light that hasn't been sufficiently amplified back into the cavity, and each reflection increases the gain. When the gain exceeds a certain threshold, light is output from the cavity in the form of a laser beam.

A laser cavity can resonate at a particular frequency if the wavelength of the light "fits" inside the cavity. The wavelength fits when an integral number of wavelengths is equal to twice the length of the cavity. Laser cavities are, of course, much longer than the wavelengths of light, so a cavity can potentially have hundreds of thousands, or even millions, of resonant oscillation modes. These modes are called longitudinal modes. Each longitudinal mode has a different wavelength, but a laser only emits the modes that exceed the gain threshold.

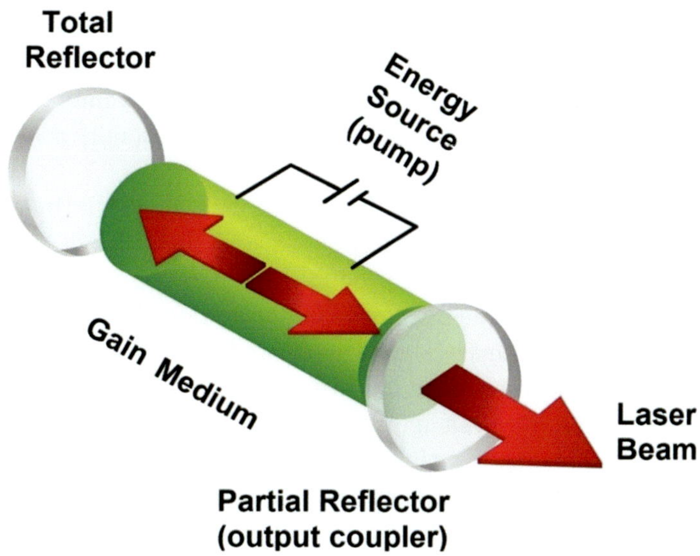

Figure 13.5 The basic elements of a laser.

Lasers emit coherent light, which differs from the incoherent light emitted by the Sun and by ordinary light bulbs. Incoherent light waves span a wide range of frequencies and phases, but coherent light waves have the same frequency and phase. When that frequency is in the visible spectrum, coherent light is monochromatic, which means that it is perceived by the human eye as having a single color. A laser beam has both temporal and spatial coherency. Temporal coherency is a measure of how long the light waves can remain in phase, and spatial coherency is a measure of the size of the area where the light is coherent. Spatial coherency allows laser beams to be focused to a narrow radius. It also allows laser beams to stay narrow, even when they propagate over a long distance [9, 24].

The first measurements of laser frequency that were referenced to frequency standards, instead of being based on interferometry and previous knowledge of the speed of light, were made in the late 1960s and early 1970s [20, 25, 26]. These measurements were motivated by the goal of unifying the standards of time and length. The meter—the base unit of length—was defined by an artifact until 1960, when it was redefined as a specific number of wavelengths of the light released by a specific energy transition of a krypton-86 atom (Section 9.3). This definition was related to light, but only to a specific wavelength, and it was difficult to accurately realize because the wavelength wasn't traceable to a frequency standard. Scientists knew that these problems could be eliminated by redefining the meter as the distance traveled by light in a fixed period of time, but that first required measuring the speed of light with respect to a frequency standard. Because light was a wave, its speed could be measured as

$$c = \lambda f, \tag{13.3}$$

where c is the speed of light in meters per second, λ is the wavelength in meters, and f is the frequency in hertz. The relationship was straightforward, but a measurement of c that was traceable to the SI required building a frequency synthesis chain that linked the microwave frequency of a cesium standard to the optical frequency of a laser. Once this was done, several great things could happen—the speed of light could be measured with less uncertainty than ever before, the meter could be redefined, and the standards of time and length could be unified.

The necessary frequency synthesis chain was built at NBS in Boulder, Colorado, by a team of researchers led by the American physicist Kenneth Evenson (1932–2002). In 1972, Evenson's team used this chain to compare the frequency of a helium-neon (He–Ne) laser with the NBS cesium beam frequency standard. The laser had a 3.39 μm wavelength, and oscillated at the methane (CH_4) absorption frequency. The measured frequency was approximately 88.4 THz, which was then the highest frequency that had ever been measured by direct comparison to a frequency standard. The uncertainty of the initial measurements was large, about 5×10^{-7}, but if was immediately known that the uncertainty could be made much smaller if better methods were used to stabilize the laser [27].

To refine the measurements, Evenson's team collaborated with John Hall (b. 1934) and Richard Barger (1931–1998), two American physicists working at JILA—a joint institute of the University of Colorado and NBS (and later of NIST). The collaboration produced outstanding results, and the uncertainty was reduced by almost three orders of magnitude by the end of the year, to about 6×10^{-10}. The newfound ability to reference a laser's frequency to a primary standard allowed c to be measured with uncertainties that were 100× smaller than the best measurements of the past [28].

Measuring light with even shorter wavelengths could further reduce the uncertainty of c, so the next challenge was to measure the frequency of visible light. This was accomplished in 1979, when Evenson's team collaborated with a team led by Kenneth Baird (1923–2022), a Canadian physicist at NRC, to measure the frequency of the visible yellow-green light produced by a prominent absorption frequency in iodine. This frequency was approximately 520 THz (576 nm wavelength), which was about 6× higher than the ~88 THz frequency first measured in 1972. The uncertainties of the first visible light measurements were correspondingly about 6× smaller, near 1×10^{-10} [29].

Measuring the frequency of visible light required the construction of optical frequency synthesis chains. These chains were built at NBS and at several other national metrology laboratories, but they were large, complex, and difficult to operate, and required constant tweaking by scientists [26]. One such synthesis chain, constructed at NBS in the early 1980s after being refined for more than a decade, connected an iodine-stabilized 520 THz laser to the primary frequency standard. This required multiplying the cesium frequency by a factor of more than 56,500, so the chain included six lasers, four klystrons, and numerous frequency multipliers and mixers. The klystrons were vacuum tubes that acted as high-frequency microwave oscillators or amplifiers, generating or amplifying microwave signals that were mixed with the laser light. The mixing was accomplished with both semiconductor and metal-insulator-metal diodes [30]. The photographs in Figure 13.6 show Evenson adjusting a diode mixer (a), which was just one small part of the large optical frequency synthesis chain that's shown in (b).

The measurements of the frequency of visible light, when linked to primary frequency standards, confirmed that the speed of light in a vacuum was exactly 299,792,458 m/s, the value accepted in the SI as a fundamental constant. This also meant, of course, that the meter was the distance traveled by light in a vacuum in $\frac{1}{299{,}792{,}458}$ s, and the SI meter was redefined based on this relationship in 1983. After the redefinition, the meter was no longer tied to the wavelength of a particular light source, and it can now be realized from the wavelength of any coherent light source whose frequency is known to be traceable to the SI second [31].

Before the meter was redefined, some scientists were already thinking about the possibility of keeping time with optical standards. The first attempt to build an optical clock was made in 1981 by a group led by Veniamin Chebotayev (1938–1992), a physicist at the Siberian Branch of the Russian Academy of Sciences in Novosibirsk, Russia. Instead of controlling a frequency chain from the bottom with a cesium standard, Chebotayev's group followed the conventional clockmaking process—they divided the optical frequency down to a lower frequency that could be used for timekeeping. They accomplished this by controlling their frequency chain from the top with a methane-stabilized He–Ne laser operating at about 88 THz (λ = 3.39 μs). By using a series of seven mixing stages, their frequency chain delivered a microwave output frequency of about 4.1 GHz (λ = 7.36 cm), and their measurements suggested that the output frequency was stable enough to phase lock a conventional RF oscillator that could be used as a time standard [32, 33].

Figure 13.6 Kenneth Evenson of NBS making an adjustment to a diode mixer (a), which was part of the optical frequency synthesis chain (b) used to measure the frequency of visible light.

In a review article about optical frequency measurements published in 1986, Evenson and his colleagues made reference to the work of the Russian scientists. They were sufficiently encouraged to write that

> *"No replacement of the cesium frequency standard is imminent, but a methane-stabilized laser 'clock' has been demonstrated; it provides a less accurate, but interesting alternative to a cesium clock at a much higher frequency."* [25]

They were right, of course, about a replacement of the cesium clock not being imminent. They knew first-hand how difficult it was to make accurate optical frequency measurements, because they had extensive experience building the complex chains of lasers, mixers, and multipliers that linked optical frequencies to a cesium standard. However, the complexity of the synthesis chains would eventually go away, along with the concerns about accuracy. The invention of the laser had enabled optical frequency measurements, but the invention of the optical frequency comb would make the measurements far more accurate and much easier to perform.

13.3 Optical Frequency Combs

An optical frequency comb (OFC) is an instrument that converts optical frequencies, including those in the visible light spectrum, into microwave frequencies that electronic systems can measure. Frequency counters (Section 5.2.2) can directly measure RF frequencies of up to about 40 GHz, but that frequency is still 10,000× lower than 400 THz, which is the lowest frequency in the visible light spectrum. Therefore, optical frequency measurements initially involved frequency synthesis chains, like those just described in Section 13.2, that had numerous conversion stages consisting of frequency mixers, multipliers, and dividers. The conversion stages not only made frequency chains complex and expensive to build, but they also contributed large measurement uncertainties. The OFC solved all of these problems by bridging the gap between microwave and optical frequencies with a single coherent process.

Optical frequency combs were first demonstrated in 1998 at the Max Planck Institute of Quantum Optics (MPQ) in Garching, Germany, by a team of physicists led by Theodor Hänsch (b. 1941) [34–37]. Frequency comb demonstrations began at JILA in Boulder, Colorado, the following year [38, 39], and from there the two laboratories continued to refine their OFC measurements, both collaboratively and independently. Rapid progress was made, and Hänsch and John Hall of JILA each received a portion of the 2005 Nobel Prize for their "contributions to the development of laser-based precision spectroscopy, including the optical frequency comb technique." Numerous other physicists have also made essential contributions to OFC technology, including Thomas Udem (b. 1962) at MPQ, and Scott Diddams (b. 1969) and Tara Fortier (b. 1974) at JILA/NIST.

Frequency combs provide a bidirectional connection between optical and radio frequencies by coherently converting microwaves to optical frequencies by multiplication, or optical frequencies to microwave frequencies by division. They can even convert an optical frequency to a different optical frequency. A frequency comb works by implementing a simple but elegant method that's analogous to a gearbox found in a mechanical device. As depicted in Figure 13.7, which is of course not drawn to scale, a frequency comb can be thought of as a large gear whose teeth smoothly mesh with other gears, without any slippage. At its center, the large gear meshes with a medium-sized gear that's associated with a microwave frequency. At its edge, the large gear meshes with multiple fast-spinning small gears, each associated with an optical frequency. The figure only shows seven of these fast-spinning gears, but an actual frequency comb would include about 100,000 gears. This gives the frequency comb the ability to multiply an input microwave frequency by about 10^5, so it can be output as an array of equally spaced light waves. A frequency comb can also divide the frequency of an incoming light wave by about 10^5, converting it to a microwave frequency [40].

The central component of an OFC is a mode-locked laser (MLL). An MLL is typically designed to have more than 100,000 longitudinal modes, which is made possible by having a sufficiently long cavity and a gain medium with a wide bandwidth. When the longitudinal modes are all in phase, mode locking occurs within the laser cavity, and the constructive interference of the light waves causes the laser to emit an ultrashort pulse of light. The pulse duration is typically less than a few hundred femtoseconds, which is why OFCs are often called "femtosecond frequency combs." The pulses repeat at a period, T, which is the reciprocal of the frequency comb's repetition frequency, f_r, therefore $T = \frac{1}{f_r}$ (Figure 13.8). Locking more modes together shortens the pulses, and shortening the pulses increases the OFC's bandwidth [9, 24].

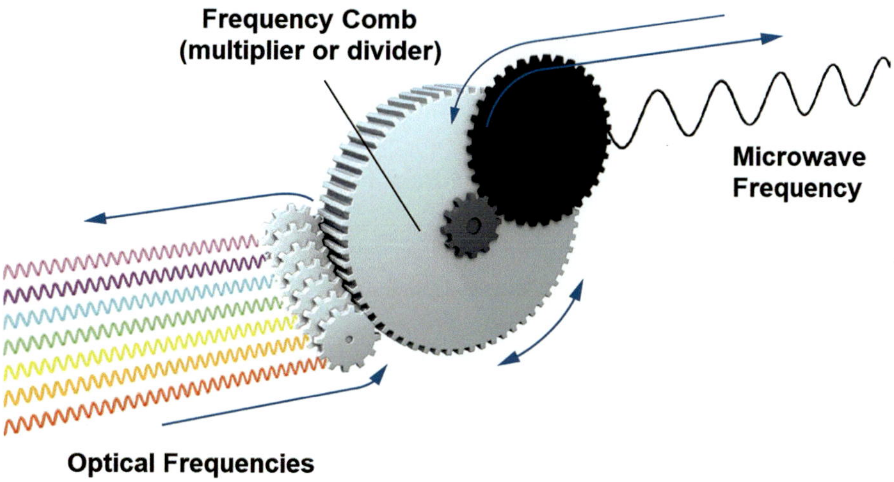

Figure 13.7 A depiction of a frequency comb as a mechanical gearbox, with the "gears" providing a bidirectional connection between optical and microwave frequencies.

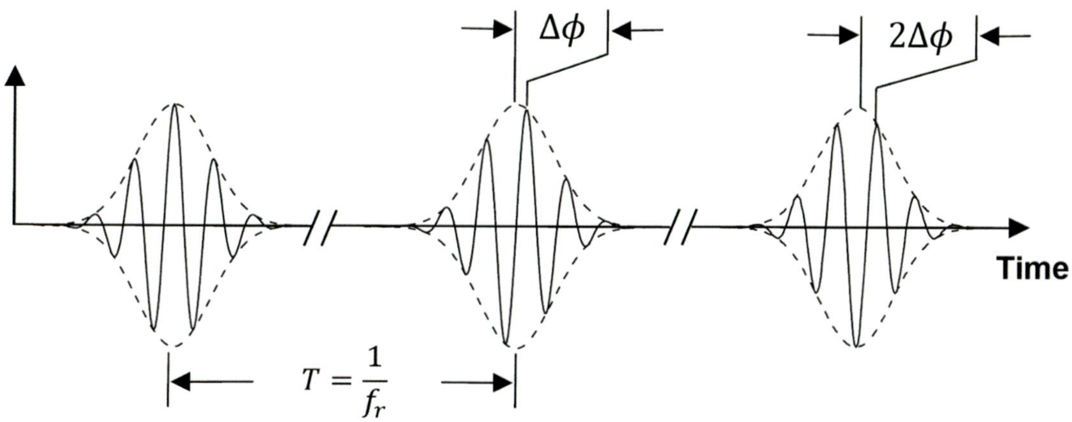

Figure 13.8 A pulse train from a mode-locked laser.

The ultrashort pulses produced in the time domain are transformed through Fourier analysis (Section 5.2) into a comb-like structure of equally spaced mode frequencies (Figure 13.9). The mode spacing is equal to f_r and is typically in the range from 100 MHz to 10 GHz. The modes are shifted by the carrier-envelope offset (CEO) frequency, f_{CEO}, which is the frequency difference between the optical carrier wave and the optical pulse envelope [37, 38, 40, 41].

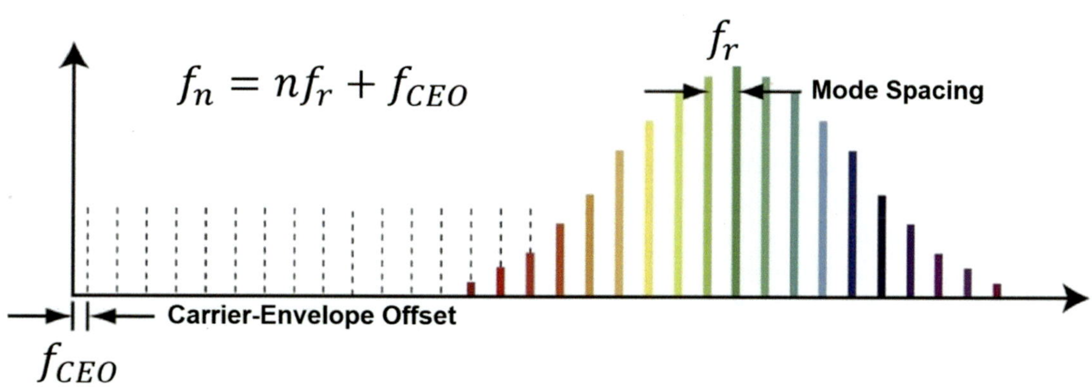

Figure 13.9 The mode spacing and carrier-envelope offset of an optical frequency comb.

The frequency of an individual mode, f_n, is equal to

$$f_n = nf_r + f_{CEO}, \tag{13.4}$$

where n is an integer mode number, f_r is the repetition frequency of the MLL, and f_{CEO} is the carrier-envelope offset frequency.

There are no locking mechanisms between the carrier and the envelope inside of a laser, so the magnitude of f_{CEO} is arbitrary and subject to change. However, frequency combs measure and stabilize f_{CEO} by using a technique called self-referencing. The simplest and most common self-referencing method requires the OFC spectrum to span at least one octave of bandwidth [42]. It involves doubling a frequency at the low end of the OFC spectrum, which is typically the repetition frequency, by using a second-harmonic generator (SHG). The SHG's output frequency, $2v_n$, is then compared to a fundamental frequency with the same wavelength, v_{2n}, that's directly obtained from the high end of the comb. The comparison is performed by heterodyning the two light waves, and the resulting beat frequency reveals the CEO frequency,

$$f_{CEO} = 2v_n - v_{2n} = 2(nf_r + f_{CEO}) - (2nf_r + f_{CEO}). \tag{13.5}$$

To stabilize the comb, the CEO is continuously measured and adjusted using a feedback loop so that it maintains a constant value. This self-referencing process locks the CEO to the MLL's repetition frequency, which makes all of the mode frequencies phase coherent. An OFC can have many thousands of mode frequencies, so self-referencing makes it possible to generate phase-coherent frequencies that span hundreds of terahertz of optical spectrum.

An OFC measures frequency in much the same way that a conventional ruler measures distance, because the known, equally spaced modes of the comb can provide a reference for other frequency measurements. For example, an unknown optical frequency can be determined by measuring its frequency difference, Δf, with respect to the nearest comb mode.

The vast measurement capabilities of frequency combs quickly changed nearly every facet of optical frequency metrology. Within a year of their invention, OFCs were being included in the designs of optical clocks.

13.4 Optical Clock Technology

Shortly after the OFC's invention, scientists began building experimental optical clocks that could outperform the cesium fountain clocks (Section 3.8) that provide the best physical realization of the SI definition of the second. The uncertainty of a cesium fountain clock is typically limited to about one or two parts in 10^{16}, and the absolute accuracy of an optical clock can't officially be smaller than that until the SI second is redefined. However, the relative accuracy of an optical clock can be estimated by directly comparing it to other optical clocks, and by using uncertainty analysis methods similar to those used for cesium fountains (Section 8.4.2), where the systematic uncertainties are evaluated with the Type B method. Using those techniques, scientists have repeatedly demonstrated that optical clocks have measurement uncertainties that are at least one or two orders of magnitude smaller than cesium fountain clocks, reaching parts in 10^{17}, 10^{18}, and even parts in 10^{19} for the most recent measurements.

As was noted at the beginning of this chapter, this dramatic improvement in performance is due to the extremely high Q factor of optical clocks. Consider, for example, that a cesium fountain clock has a Q factor of about 10^{10} if the linewidth is reduced to about 1 Hz. To achieve 1×10^{-16} accuracy, a cesium fountain has to divide the linewidth into one million (10^6) smaller parts. An optical clock with a resonance frequency near 1 PHz (10^{15} Hz) has a tremendous advantage, because the same 1 Hz linewidth results in a Q factor of 10^{15} without any division. If the linewidth is divided into just one thousand (10^3) smaller parts, then 1×10^{-18} accuracy is possible.

Figure 13.10 is a simplified diagram of an optical clock. The clock controls the frequency of a probing laser by locking it to a high-energy atomic transition in the optical spectrum. The probing laser is the clock's local oscillator. Its frequency is first locked to the fringe nearest to the atomic resonance by using a resonant optical cavity that is designed to be as stable as possible. The optical cavity is typically made of glass with an extremely low thermal expansion coefficient, which prevents its dimensions from changing when the temperature varies. The mirrors at each end of the cavity have extremely high reflectivity, which means that they reflect a large portion of the incoming light.

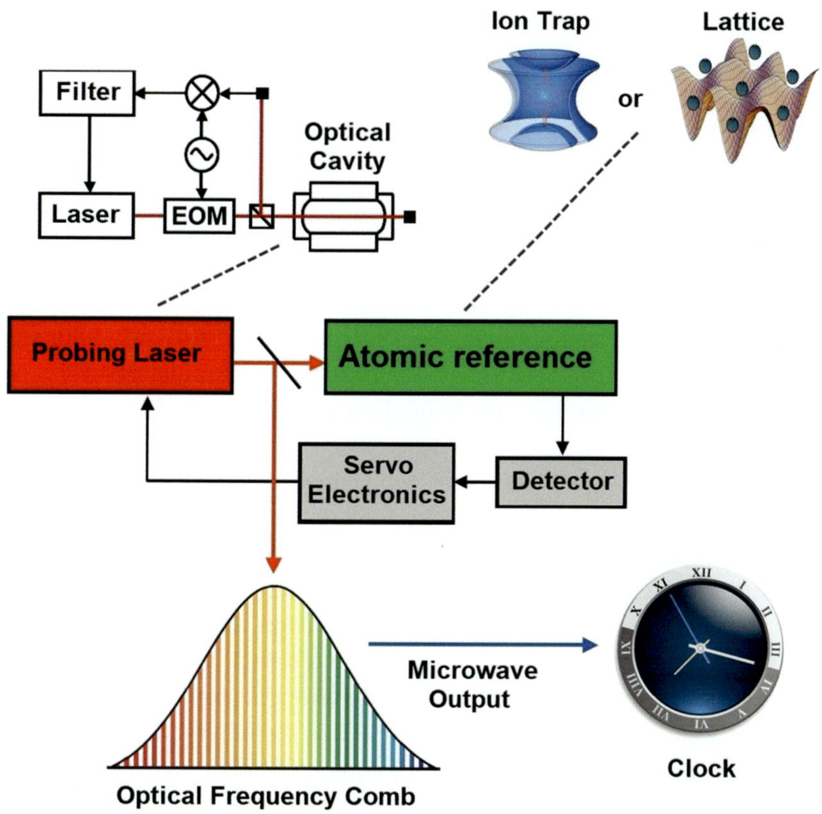

Figure 13.10 A simplified diagram of an optical clock.

The probing laser interrogates the atoms, and the remaining frequency difference between the probing laser and atomic resonance is measured and corrected by a feedback control system. The control system uses AOMs (acousto-optic modulators), EOMs (electro-optic modulators), or a combination of both to modulate and shift the laser frequency until it is locked to the center of the atomic resonance curve [3, 4, 43]. An OFC divides the locked laser frequency down to an ultrastable microwave frequency, which sometimes has an instability of less than 1×10^{-15} at $\tau = 1$ s. The microwave frequency can be measured with conventional electronic systems and used to drive a conventional clock [44].

The development of the OFC was a major leap forward, but optical clock designers have had to overcome numerous other challenges. Finding and accurately measuring the high-energy atomic transitions that correspond to optical frequencies has required many years of experiments, and more years have been spent deciding which optical transition frequencies are best suited for use in a clock. Experimental optical clocks have been designed based on the transition frequencies in both positively charged ions and neutral atoms, using chemical elements that have included aluminum (Al), barium (Ba), beryllium (Be), calcium (Ca), cesium (Cs), hydrogen (H), indium (In), magnesium (Mg), mercury (Hg), and rubidium (Rb); but most of the recent attention has focused on strontium (Sr) and ytterbium (Yb). Reducing the uncertainties of optical clocks and making them work more reliably also requires continued advances in laser technology. Much of the research is motivated by the desire to make optical cavities more stable, by refining their systems of lenses and mirrors.

There are two main types of optical clocks, single-ion and optical lattice, and they use different methods to isolate the atoms and to measure the atomic transition frequency. Single-ion clocks confine a single ion, which is an atom with a net electrical charge, in an ion trap. Optical lattice clocks confine large numbers of neutral atoms—atoms without a net electrical charge—in a lattice configuration. In both types of clocks, the space where the atoms are allowed to move is smaller than an optical wavelength, which effectively eliminates the Doppler frequency shift. The two clock types are further described in the next two sections.

13.4.1 Single-Ion Clocks

The Nobel Prize in physics was all about atomic clocks in 1989. One half of the prize was awarded that year to Norman Ramsey (1915–1911) for his invention of the separated oscillatory fields method that greatly improved the performance of microwave atomic clocks (Section 3.3). The other half was jointly awarded to Wolfgang Paul (1913–1993) at the University of Bonn in Germany, and Hans Dehmelt (1922–2017) at the University of Washington in the United States. Paul and Dehmelt were awarded for introducing and developing the ion trap technique. The trapping technique greatly advanced the study of the properties and spectrums of atoms, because it made it possible to isolate a single ion under constant conditions for a long period.

While working independently in the 1950s, Paul and Dehmelt designed two different types of ion traps. The traps confined an ion in the bottom of an energy well, in a similar fashion to a moving marble that's brought to rest at the bottom of a bowl. Both types of traps utilized a static electric field that was applied to the trap's electrodes, but a static electric field cannot confine ions in all three dimensions by itself. For this reason, Paul's trap design confined ions in one dimension with a static field, and in the other two dimensions with an oscillating RF field, so a Paul trap is also called an RF trap. Dehmelt's design—called a Penning trap—used a combination of static electric and static magnetic fields to confine ions in all three dimensions [45]. Dehmelt named his trap in honor of Frans Penning (1894–1953), the Dutch physicist who invented a vacuum gauge in the 1930s by using magnetic fields to accelerate electrons and measure pressure.

Ion trapping methods were applied to microwave clocks long before being applied to optical clocks. The first demonstration of a trap-based oscillator was in 1973, when Dehmelt and two postdoctoral researchers, David Wineland (b. 1944) and Philip Ekstrom (b. 1941), successfully isolated a single electron in a Penning trap. They called it a "monoelectron oscillator" and noted that it was free of any Doppler frequency shifts [46].

Wineland joined the Time and Frequency division of NBS in 1975. His credentials were exceptional—before his postdoctoral appointment with Dehmelt at the University of Washington, Wineland's doctorate at Harvard was supervised by Ramsey. Shortly after arriving at NBS, he started a new group called "Ion Storage," which had the primary goal of building frequency standards and clocks with the best possible accuracy [45]. As was noted in Section 3.8, Wineland's group successfully demonstrated laser cooling in 1978. The original laser cooling experiments were performed with magnesium ions (Mg^{2+}) confined in a Penning trap [47]. Laser cooling became a crucial part of future trapped-ion clocks, because it reduced the thermal motion of the trapped ions and allowed more accurate frequency measurements. Like his mentors Ramsey and Dehmelt, Wineland was awarded a Nobel Prize in physics in 2012 for his methods of measuring and manipulating individual quantum systems based on trapped ions. These methods not only provided the basis for future-generation atomic clocks, but also for quantum computing systems.

The researchers at NBS/NIST originally investigated the use of both Penning and Paul traps in atomic clocks, and the first demonstration of a laser-cooled frequency standard was based on beryllium ions ($^{9}Be^{+}$) confined in a Penning trap [48]. However, it was found that Penning traps weren't well suited for use in atomic clocks due to their strong magnetic field, which can significantly change the transition frequency. Paul traps were a better choice because they don't rely on magnetic fields, so the attention eventually turned to designing mercury ion ($^{199}Hg^{+}$) clocks based on Paul traps. The electrodes of an RF Paul trap, identical to those used in the NBS mercury ion clock experiments, are shown resting on top of a coin in Figure 13.11. The tiny electrodes had an inside radius of slightly less than 1 mm [45].

Figure 13.11 The electrodes of an RF Paul trap for a mercury ion clock, shown resting on top of a coin [45].

Using Paul traps, the ground state hyperfine transition frequency of a trapped mercury ion was measured to be about 40.5 GHz, which was much higher than the resonance frequency of any previous microwave clock. The concept was considered promising enough for Hewlett-Packard to develop and briefly market a mercury ion clock based on a Paul trap in the 1980s. Their design didn't include laser cooling, which wasn't yet ready for commercialization [49]. By 1998, Wineland's group at NIST had demonstrated a laser-cooled mercury ion clock with a combined standard uncertainty of about 3×10^{-15}, which was smaller than any cesium beam clock, and only about twice as large as the uncertainties that were then being reported for cesium fountain clocks [50]. However, 1998 was the same year that the first OFC was demonstrated, and most of the ion trap research from that point forward was directed towards optical clocks.

Dehmelt predicted in 1982 that the combination of an RF Paul trap and laser cooling could cause all Doppler frequency shifts to vanish, and that a "mono-ion oscillator" could potentially be the "ultimate laser frequency standard," with a frequency resolution of 1×10^{-18} [51]. Due to the rapid progress made after the invention of the OFC, Dehmelt's bold prediction has mostly come true, and single-ion optical clocks are perhaps the purest realization of any atomic clock. The basic concept is simply to tune a laser until it matches the resonance frequency of a solitary ion.

The first single-ion optical clock (Figure 13.12) was demonstrated in 2001 by a team of NIST researchers led by the American physicist James Bergquist (b. 1947). The clock was based on the $^{2}S_{1/2}$ ($F = 0$, $m_F = 0$) ↔ $^{2}D_{5/2}$ ($F = 2$, $m_F = 0$) transition frequency in a single, laser-cooled mercury ion ($^{199}Hg^{+}$). This frequency was approximately 1.064 PHz with a linewidth of about 2 Hz. The local oscillator was a stabilized dye laser with a frequency of 532 THz ($\lambda = 563$ nm), and the clock worked by doubling the laser frequency and locking it to the center of the atomic resonance [52]. A series of improvements reduced the combined standard uncertainty of the mercury ion clock to 7.2×10^{-17} by 2006, which was more than 2× smaller than the uncertainty of a cesium fountain [53].

Figure 13.12 James Bergquist of NIST with the trapped mercury ion optical clock in 2006 (courtesy of NIST, credit: G. Wheeler).

Aluminum ions ($^{27}Al^{+}$) were found to be less sensitive to temperature variations than mercury ions, and this attracted the attention of NIST researchers. In 2009, NIST demonstrated a clock based on trapped aluminum ions with a combined standard uncertainty of 8.8×10^{-18}, almost an order of magnitude smaller than the uncertainty of the mercury ion clock. The $^{1}S_0 \leftrightarrow {}^{3}P_0$ energy transition of an aluminum ion is an attractive candidate for optical clocks because of its high resonance frequency of about 1.121 PHz ($\lambda = 267$ nm), a narrow natural linewidth, and a low sensitivity to electromagnetic disturbances. However, the optical transitions of aluminum ions can't be detected using conventional methods, nor can the ions be directly laser cooled. These problems were solved using quantum logic—the Paul trap confined not only a "clock ion," but also a separate "logic ion" that can be directly laser cooled. When both ions are trapped, the laser-cooled logic ion sympathetically cools the clock ion, and the clock ion transfers its quantum state to the logic ion, where it can be measured.

NIST had previously built quantum logic clocks using beryllium ($^{9}Be^{+}$) as the logic ion, but the excellent results obtained in 2009 were obtained using a magnesium ($^{25}Mg^{+}$) logic ion, whose mass closely matches that of aluminum [54]. After another decade of research and development, the combined standard uncertainty of the aluminum ion quantum logic clock was reduced by almost another order of magnitude, to 9.4×10^{-19}, which fulfilled Dehmelt's vision by reaching the 1×10^{-18} threshold [55].

Single-ion clocks have also been built with excellent results using ytterbium, which is a metal that's about 2.5× denser than aluminum. After several years of refinements, researchers at PTB in Germany demonstrated a single-ion ytterbium ($^{171}Yb^{+}$) clock with a combined standard uncertainty of 3.2×10^{-18} in 2015. The clock confined a single laser-cooled ytterbium ion in a Paul trap and measured the frequency of the $^{2}S_{1/2}$ $(F = 0) \leftrightarrow {}^{2}F_{7/2}$ $(F = 3)$ E3 transition, which is about 642.121 THz [56].

We mentioned earlier in this section that Dehmelt's 1982 prediction has mostly come true. As just discussed, single-ion clocks have reached a frequency resolution of 1×10^{-18} just as Dehmelt had predicted, but the caveat "mostly" is still needed because they might not be the "ultimate laser frequency standard." Optical lattice clocks have demonstrated even lower combined standard uncertainties and much better stability during short averaging periods.

13.4.2 Optical Lattice Clocks

The optical lattice clock was invented in 2001 at the University of Tokyo by the Japanese physicist Hidetoshi Katori (b. 1964). Katori proposed the theory and concept of the clock in 2001 [57], the same year that the first single-ion optical clock was demonstrated at NIST. In 2003, Katori and the research scientist Masao Takamoto (b. 1978), who was then a graduate student, demonstrated an experimental optical lattice clock based on the $^{1}S_{0}$ $(F = \frac{9}{2}) \leftrightarrow {}^{3}P_{0}$ $(F = \frac{9}{2})$ transition in strontium (^{87}Sr) atoms [58, 59]. The invention changed the course of optical clock research and development, and experimental optical lattice clocks have since been built at several other research institutions.

Optical lattice clocks are more complex than single-ion clocks, and their design concept is radically different. Instead of measuring the frequency of a single charged atom confined in a Paul trap, lattice clocks simultaneously measure the frequencies of thousands of neutral atoms. The confinement of the atoms is usually a two-stage process. The first stage rapidly accumulates a large number of atoms, typically 10^{6} or more, from a laser-slowed atomic beam into a magneto-optical trap (MOT), where they are confined with a combination of laser cooling and a varying magnetic field. This process cools the atoms down to a temperature of a few millikelvins. The atoms are then transported to a second-stage MOT where more laser cooling brings their temperature down to the microkelvin region. A fraction of the ultracold atoms, perhaps as many as 10^{5}, are then confined by the lasers into energy wells located inside a vacuum chamber. The wells are arranged in a lattice configuration that's similar to the compartments in an egg carton (Figure 13.13). The depth of the wells, when expressed in units of temperature, is perhaps 10 μK.

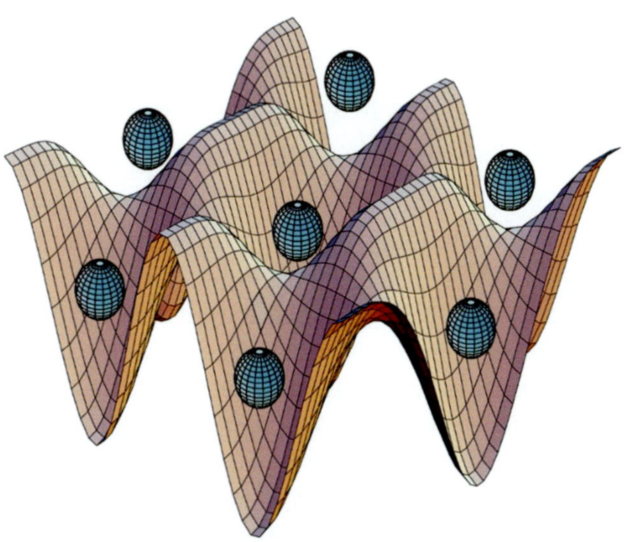

Figure 13.13 An artist's depiction of atoms confined in an optical lattice (courtesy of NIST).

The atoms typically stay in the optical lattice for a few seconds, until they are ejected by collisions with the few room-temperature particles that are still in the vacuum. The clock transition of the atoms is measured during their period of confinement by using the same basic methods employed by all optical clocks (Figure 13.10). A probing laser is first tuned by its optical cavity to be near the atomic transition frequency, and then fine-tuned until its frequency is locked to atomic resonance. This measurement cycle is continuously repeated to maintain a frequency lock [3, 4, 60].

As can be seen from this description, optical lattice clocks use lasers to both trap and probe the atoms. All optical clocks use a laser to probe the atoms, but it was once thought that it wouldn't be possible to build an accurate clock that also used lasers to trap the atoms. This reasoning made sense, because the powerful electromagnetic fields of the trapping lasers can cause a large uncertainty—called a Stark shift or light shift—by altering the energy levels of the atoms and changing their frequency. While working with strontium atoms, Katori solved the light shift problem by making the discovery that made the optical lattice clock possible. He found that if the trapping lasers were tuned to a certain wavelength, the ground ($^1S_0$) and excited ($^3P_0$) states of the atomic transition would be shifted by an equal amount, and the atomic resonance frequency wouldn't change at all. Katori called this wavelength the magic wavelength, because it made the light shift in the clock transition vanish. His calculations and measurements determined that the magic wavelength for strontium atoms was about 813 nm [58, 59].

Most optical lattice clocks are based on isotopes of either strontium (Sr) or ytterbium (Yb), two chemical elements that are both metals. Strontium is an alkaline earth metal—part of the same group of metals that includes other chemical elements used in optical clock research, including beryllium, calcium, and magnesium. Alkaline earth metals differ from alkali metals like cesium and rubidium (Section 3.3), because they have two valence electrons as opposed to just one. Ytterbium is in the lanthanide series of chemical elements, which are commonly referred to as rare-earth elements. The ^{87}Sr, ^{88}Sr, and ^{171}Yb clock transitions are the three that have been most often measured by lattice clock researchers. Both strontium isotopes have a transition frequency of about 429.228 THz (λ = 698 nm). The ^{171}Yb transition frequency is about 518.296 THz (λ = 578 nm), and the magic wavelength for the trapping lasers is about 759 nm.

As described in the previous section, mercury ion and aluminum ion quantum logic clocks both have a resonance frequency greater than 1 PHz, or more than twice as high as strontium or ytterbium lattice clocks. Therefore, the main advantage that lattice clocks have over single-ion clocks is not a higher resonance frequency, but is instead their ability to simultaneously measure the frequency of a large number of atoms. Averaging measurements from a large group of atoms rapidly reduces the measurement noise, which effectively narrows the linewidth and improves the short-term stability [3, 4, 60].

In the short term, a lattice clock can be at least a factor of 10 more stable than a single-ion clock, which allows its performance to be evaluated much faster. For example, the NIST aluminum ion quantum logic clock described in the previous section has a frequency stability of $\frac{1.2 \times 10^{-15}}{\sqrt{\tau}}$, where τ is the averaging period in seconds [55]. In contrast, a ytterbium optical lattice clock that was concurrently developed at NIST has a frequency stability of $\frac{6 \times 10^{-17}}{\sqrt{\tau}}$, a factor of 20 improvement [61]. Figure 13.4 shows a portion of the apparatus of a NIST ytterbium lattice clock in (a), and the composite photograph in (b) shows the clock's illuminated lasers.

A combined standard uncertainty of 8.1 × 10^{-19} was reported in 2024 for measurements of a strontium (^{87}Sr) optical lattice clock built at JILA in Boulder, Colorado. The clock was developed by a team of researchers led by the Chinese-American physicist June Ye (b. 1967), and its optimized design is the result of more than two decades of continued refinements. The reported uncertainty is about 2× smaller than that of any previous strontium lattice clock design, and is the smallest uncertainty reported for any clock at this writing. Much of the improvement over previous designs was due to the use of a shallower lattice, which allows better control of the laser light that traps the atoms, and which also reduces the number of collisions between atoms by packing them less tightly [62].

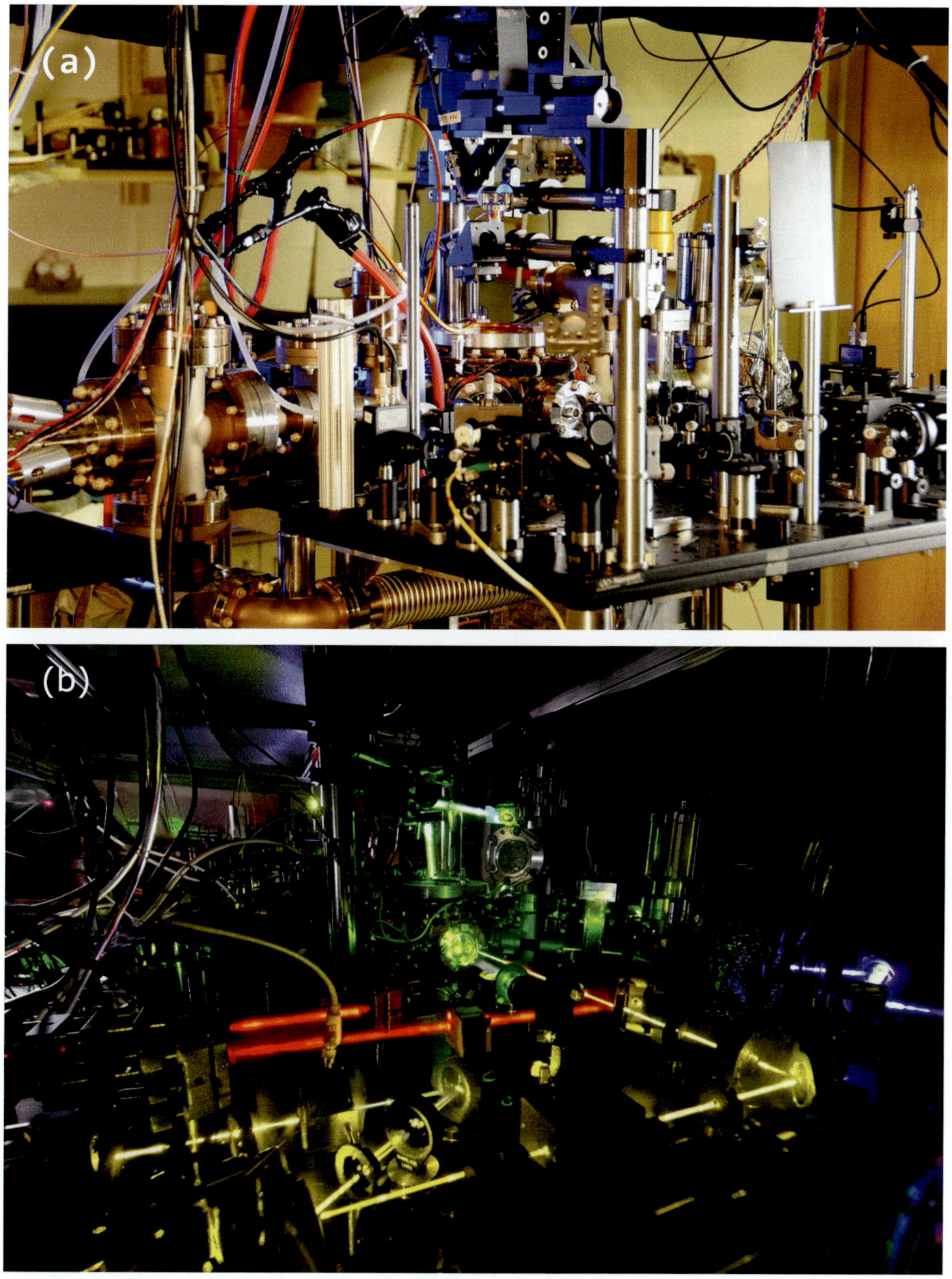

Figure 13.14 A portion of the apparatus (a) of an experimental ytterbium optical lattice clock, and a composite photograph (b) that shows the clock's illuminated lasers (courtesy of NIST).

13.4.3 Secondary Representations of the SI Second Defined by Optical Frequencies

Eleven different optical clock frequencies have been designated by the BIPM as secondary representations of the SI second as of 2021 [1]. These frequencies, ordered from highest to lowest, are listed in Table 13.2, along with their wavelengths rounded to the nearest nanometer, and their relative standard uncertainties. The transition frequencies of mercury, strontium, and ytterbium have been measured for both ions and atoms, and two different atomic transitions, E2 and E3, have been measured for the ytterbium ion. Most of the secondary representations have been compared to the SI second with a relative standard uncertainty of about 2 × 10^{-16}, which is limited by the uncertainty of the best cesium fountain frequency standards. As shown in Figure 13.15, the optical clock transition frequencies fall within the spectrum of either visible or ultraviolet light.

Table 13.2 Recommended values for secondary representations of the SI second based on optical frequencies as of 2021

Atomic species		Frequency (Hz)	Wavelength (nm)	Relative standard uncertainty
Name	Notation			
Indium ion	$^{115}In^+$	1,267,402,452,901,041.30	236	4.3 × 10^{-15}
Mercury atom	^{199}Hg	1,128,575,290,808,154.32	266	2.4 × 10^{-16}
Aluminum ion	$^{27}Al^+$	1,121,015,393,207,859.16	267	1.9 × 10^{-16}
Mercury ion	$^{199}Hg^+$	1,064,721,609,899,146.96	282	2.2 × 10^{-16}
Ytterbium ion	$^{171}Yb^+$ (E2)	688,358,979,309,308.24	436	2.0 × 10^{-16}
Ytterbium ion	$^{171}Yb^+$ (E3)	642,121,496,772,645.12	467	1.9 × 10^{-16}
Ytterbium atom	^{171}Yb	518,295,836,590,863.63	578	1.9 × 10^{-16}
Strontium ion	$^{88}Sr^+$	444,779,044,095,486.30	674	1.3 × 10^{-15}
Strontium 88 atom	^{88}Sr	429,228,066,418,007.01	698	2.0 × 10^{-16}
Strontium 87 atom	^{87}Sr	429,228,004,229,872.99	698	1.9 × 10^{-16}
Calcium ion	$^{40}Ca^+$	411,042,129,776,400.40	729	1.8 × 10^{-15}

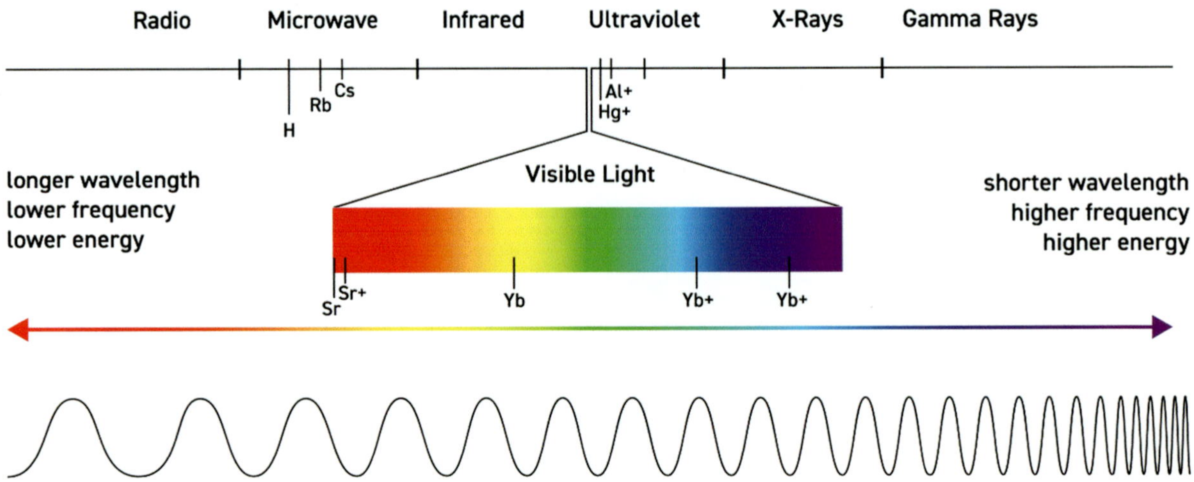

Figure 13.15 A simplified electromagnetic spectrum diagram showing the relative position of the resonance frequencies used by microwave and optical atomic clocks (courtesy of NIST).

As noted in the previous section, optical clocks have demonstrated relative accuracies smaller than 1 × 10^{-18}. This is about eight orders of magnitude smaller than NBS-1, the first microwave primary standard at NBS (Section 3.3), so the rate of improvement in atomic clocks has been more than one order of magnitude per decade. Figure 13.13 graphs the continuous improvement in atomic clock accuracy, comparing optical clocks to some of the microwave primary standards that were previously developed at NBS/NIST.

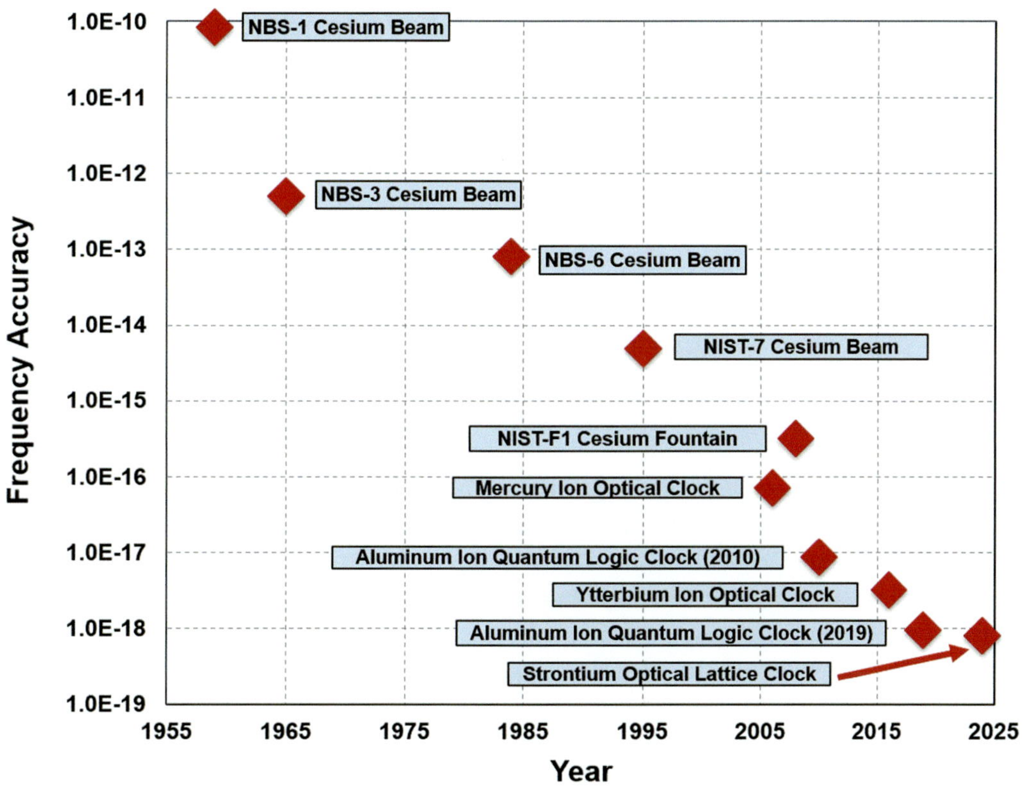

Figure 13.16 The continuous improvement in atomic clock accuracy—about eight orders of magnitude in less than 70 years.

Table 13.3 The relationship between clock accuracy and the years required to gain or lose one second

Frequency accuracy (unitless)	The number of years required to gain or lose one second
1×10^{-8}	3
1×10^{-9}	32
1×10^{-10}	317
1×10^{-11}	3,169
1×10^{-12}	31,689
1×10^{-13}	316,888
1×10^{-14}	3,168,880
1×10^{-15}	31,688,800
1×10^{-16}	316,888,000
1×10^{-17}	3,168,880,000
Age of Universe	~13,700,000,000
1×10^{-18}	31,688,800,000

To put this extraordinary accuracy improvement into better perspective, Table 13.3 lists the number of years that would pass before a clock would gain or lose one second, based on frequency accuracies ranging from 1×10^{-8} to 1×10^{-18}. The estimated age of the universe, which is listed in the table for comparison purposes, is about 13.7 billion years. This means that a clock with a frequency accuracy better than about 2×10^{-18} would have neither gained nor lost a second since the beginning of the universe.

13.5 The Use of Optical Clocks as Laboratory Standards

The large numbers in Table 13.3 are astounding, but it is meaningless in a practical sense to calculate how long it would take for an atomic clock to gain or lose a second, because all human-made clocks fail after a relatively small number of years. Even so, clocks have to be able to keep accurate time for some reasonable period before they can be used as laboratory standards. For example, a commercial cesium beam clock can typically maintain its specified accuracy for about a decade before it finally fails. In contrast, the most accurate optical clocks have been unable to continuously run for periods of more than a day as of 2025. Keeping today's short run times in mind, we'll conclude this chapter by taking a brief look at the prospects of using optical clocks as laboratory frequency and time standards. We'll consider two applications of optical clocks—their use as primary standards that periodically contribute to UTC, and their use as continuously running laboratory standards that could eventually replace cesium beam clocks.

The first application is already happening, because a few laboratories are already reporting their optical clock measurements to the BIPM, and these measurements are contributing to the calculation of UTC. The optical clocks that contribute to UTC are now performing the same function as the cesium fountain clocks that serve as primary frequency standards (Section 9.4.3), and they can do so while running for much shorter periods. This is most evident in the case of optical lattice clocks, which typically have instabilities of just a few parts in 10^{17} at $\tau = 100$ s. These instabilities are so small that the performance of a hydrogen maser-based UTC(k) time scale—even one steered to a continuously running cesium fountain—can be improved by running an optical lattice clock for less than an hour per day, or for just a few hours per week [63, 64].

The second application, making an optical clock continuously run for a long period, is harder to accomplish. The resonating element in an optical clock is a laser, and lasers are much easier to disturb and knock out of alignment than the microwave frequencies that pass through waveguides in conventional atomic clocks. In addition, experimental optical clocks are large and complex systems that can only be operated by specialists. Turning them into a typical laboratory instrument requires an extensive amount of engineering, which is needed not just to make them more reliable, but also to make them smaller and easier to use [65].

Researchers have taken some important first steps to make optical lattice clocks transportable. One example occurred in 2019, when Katori, Takamoto, and their colleagues at the University of Tokyo built two transportable strontium optical lattice clocks to test the gravitational redshift (Section 8.4.2). The clocks, each consisting of several small racks full of equipment, were transported to the Tokyo Skytree, the world's tallest broadcasting tower. One clock was operated on the ground and the other inside the tower at a height of 450 m, with their frequency compared via a fiber optics link [66]. Another example occurred in 2023 when a ytterbium optical lattice clock was shipped by truck from Boulder, Colorado, to the USNO in Washington, DC, a distance of nearly 3,000 km. The clock was built by a team of NIST researchers led by the American physicist Andrew Ludlow (b. 1977) and was enclosed in three 120 cm tall equipment racks. After arriving at the USNO, it successfully served as the reference for a stability measurement of a rubidium fountain microwave clock [67].

The broader benefits of optical clocks won't be realized, of course, until they are widely deployed. As we might expect, the commercialization process is starting to evolve in a way that's reminiscent of what happened with microwave atomic clocks. For the most part, microwave atomic clocks have historically been grouped into three categories—clocks of the highest accuracy which have never been sold commercially (cesium fountains), clocks that directly realize the SI definition of the second but that are sold in limited numbers due to their high cost (cesium beams), and clocks that have many thousands of units deployed each year (conventional rubidium clocks and CSACs). Optical clocks will likely be grouped into similar categories. If this is the case, the types of optical clocks now being built at national metrology institutions will never be sold commercially, but they will replace cesium fountains as primary standards after the SI second is redefined. Smaller, more reliable, and less accurate optical clocks, which are now being built by several companies, will immediately compete with existing cesium beam and rubidium clocks for market share, and could eventually replace them [65].

An impressive demonstration of optical clock reliability was exhibited in 2022 by Vector Atomic, Inc., an American optical clock manufacturer located in Pleasanton, California. The prototype clocks involved in the demonstration had a built-in OFC, and they output several frequency and time signals, including 100 MHz, 10 MHz, and 1 pps. Everything was designed to fit in a rackmount case that was just slightly larger in volume than a 5071A cesium beam clock (Section 3.6). The clocks kept time by measuring the transition frequency of electrons in iodine molecules (I_2). This frequency has a wavelength of about 532 nm, and can be conveniently probed by halving the wavelength (doubling the frequency) of a common 1064 nm industrial laser.

Two of the Vector Atomic optical clocks were brought to NIST in Boulder, Colorado, to have their stability measured with respect to the UTC(NIST) time scale. They were then brought to Pearl Harbor, Hawaii, where they were installed—along with a third nearly identical clock—on a ship that embarked on a 20-day voyage in the Pacific Ocean. The three clocks remarkably ran continuously during their 20 days at sea, in conditions where the temperature, humidity, and vibration control were much worse than they are in a typical laboratory.

The clocks were cross-compared during the entire voyage. Their frequency stability was estimated by using three frequency counters to continuously measure three distinct beat frequencies. When averaged for periods of less than a few hundred seconds, the measurements showed that the stability of the two clocks measured at NIST didn't change at all when they were measured at sea. However, when averaged for longer periods, the stability of the optical clocks was understandably worsened by the ship's environmental conditions. For an averaging period of 100,000 s, the combined stability of the two optical clocks at sea was measured as 8×10^{-15}, which was several times larger than their respective 2×10^{-15} and 4×10^{-15} stabilities that were measured at NIST. Their stability at sea was still 3× smaller than the specified stability of a high-performance commercial cesium beam clock (Table 7.4). This indicates that a commercial optical clock can potentially keep time accurate to within ±400 ps per day in harsh environmental conditions [68].

These measurement results, and others like them, are extremely encouraging. No one knows exactly what the future will hold, but it seems safe to assume that commercially-available optical clocks will continue to get smaller, more reliable, and more accurate. They should also become far more prevalent after the SI second is redefined.

References

1. H. S. Margolis, G. Panfilo, G. Petit, C. Oates, T. Ido, and S. Bize, "The CIPM list 'Recommended Values of Standard Frequencies': 2021 Update," *Metrologia*, vol. 61, no. 3, pp. 035005/1–22, April 2024.
2. N. Dimarcq, M. Gertsvolf, G. Mileti, S. Bize, C. W. Oates, E. Peik, D. Calonico, T. Ido, P. Tavella, F. Meynadier, G. Petit, G. Panfilo, J. Bartholomew, P. Defraigne, E. A. Donley, P. O. Hedekvist, I. Sesia, M. Wouters, P. Dubé, F. Fang, F. Levi, J. Lodewyck, H. S. Margolis, D. Newell, S. Slyusarev, S. Weyers, J.-P. Uzan, M. Yasuda, D.-H. Yu, C. Rieck, H. Schnatz, Y. Hanado, M. Fujieda, P.-E. Pottie, J. Hanssen, A. Malimon, and N. Ashby, "Roadmap towards the Redefinition of the Second," *Metrologia*, vol. 61, no. 1, pp. 1–19, January 2024.
3. N. Poli, C. W. Oates, P. Gill, and G. M. Tino, "Optical Atomic Clocks," *Rivista del Nuovo Cimento*, vol. 36, no. 12, pp. 555–624, December 2013.
4. A. D. Ludlow, M. M. Boyd, J. Ye, E. Peik, and P. O. Schmidt, "Optical Atomic Clocks," *Reviews of Modern Physics*, vol. 87, no. 2, pp. 637–701, April-June 2015.

5. A. Derevianko, K. Gibble, L. Hollberg, N. R. Newbury, C. Oates, M. S. Safronova, L. C. Sinclair, and N. Yu, "Fundamental Physics with a State-of-the-Art Optical Clock in Space," *Quantum Science and Technology*, vol. 7, pp. 044002/1–20, July 2022.
6. I. Newton, "A Letter of Mr. Isaac Newton ... Containing His New Theory about Light and Colors," *Philosophical Transactions of the Royal Society*, vol. 6, no. 80, pp. 3075–3087, February 1671/2.
7. C. Huygens, *Treatise on Light*, 1690, Translated into English by S. P. Thompson, London: McMillan and Co., 1912.
8. T. Young, "Experiments and Calculations Relative to Physical Optics," *Philosophical Transactions of the Royal Society*, vol. 94, pp. 1–16, January 1804.
9. J. Hecht, *Understanding Lasers: An Entry Level Guide*, 4th edition, IEEE Press, 2019.
10. C. Orzel, *A Brief History of Timekeeping*, BenBella Books, 2022.
11. A. Robinson, *The Last Man Who Knew Everything: Thomas Young*, Cambridge: Open Book Publishers, 2023.
12. M. Rømer, "A Demonstration Concerning the Motion of Light, Communicated from Paris, in the *Journal des Scavans*, and here Made English," *Philosophical Transactions of the Royal Society*, vol. 12, no. 136, pp. 893–894, June 1677.
13. M. Faraday, "Thoughts on Ray-Vibrations," *The London, Edinburgh, and Dublin Philosophical Magazine and Journal of Science*, vol. 28, no. 188, pp. 345–350, May 1846.
14. J. C. Maxwell, "A Dynamical Theory of the Electromagnetic Field," *Philosophical Transactions of the Royal Society*, vol. 155, no. 155, pp. 459–512, December 1865.
15. J. C. Maxwell, *A Treatise on Electricity and Magnetism*, volumes I and II, London: McMillan and Co., 1873.
16. A. A. Michelson, "The Relative Motion of the Earth and the Luminiferous Ether," *American Journal of Science*, vol. s3-22, no. 128, pp. 120–129, August 1881.
17. A. A. Michelson and E. W. Morley, "On the Relative Motion of the Earth and the Luminiferous Ether," *American Journal of Science*, vol. s3-34, no. 203, pp. 333–345, November 1887.
18. A. Einstein, "Elektrodynamik bewegter Korper (On the Electrodynamics of Moving Bodies)," *Annalen der Physik*, vol. 17, pp. 891–921, 1905.
19. A. A. Michelson and E. W. Morley, "On the Feasibility of Establishing a Light-Wave as the Ultimate Standard of Length," *American Journal of Science*, vol. s3-38, no. 225, pp. 181–186, September 1889.
20. L. Hollberg, C. W. Oates, G. Wilpers, C. W. Hoyt, Z. W. Barber, S. A. Diddams, W. H. Oskay, and J. C. Bergquist, "Optical Frequency/Wavelength References," *Journal of Physics B: Atomic, Molecular and Optical Physics*, vol. 38, no. 9, pp. S469–S495, April 2005.
21. A. L. Schawlow and C. H. Townes, "Infrared and Optical Masers," *Physical Review*, vol. 112, no. 6, pp. 1940–1949, December 1958.
22. T. Maiman, "Stimulated Optical Radiation in Ruby," *Nature*, vol. 187, pp. 493–494, August 1960.
23. D. F. Nelson, R. J. Collins, and W. Kaiser, "Bell Labs and the Ruby Laser," *Physics Today*, vol. 63, no. 1, pp. 40–45, January 2010.
24. J.-C. Diels and L. Arissian, *Lasers: The Power and Precision of Light*, Wiley-VCH, 2011.
25. D. A. Jennings, K. M. Evenson, and D. J. E. Knight, "Optical Frequency Measurements," *Proceedings of the IEEE*, vol. 74, no. 1, pp. 168–179, January 1986.
26. L. Hollberg, S. Diddams, A. Bartels, T. Fortier, and K. Kim, "The Measurement of Optical Frequencies," *Metrologia*, vol. 42, no. 3, pp. S105–S124, June 2005.
27. K. M. Evenson, G. W. Day, J. S. Wells, and L. O. Mullen, "Extension of Absolute Frequency Measurements to the cw He-Ne Laser at 88 THz (3.39 μ)," *Applied Physics Letters*, vol. 20, no. 3, pp. 133–134, February 1972.
28. K. M. Evenson, J. S. Wells, F. R. Petersen, B. L. Danielson, and G. W. Day, "Speed of Light from Direct Frequency and Wavelength Measurements of the Methane-Stabilized Laser," *Physical Review Letters*, vol. 29, no. 19, pp. 1346–1349, November 1972.
29. K. M. Baird, K. M. Evenson, G. R. Hanes, D. A. Jennings, and F. R. Petersen, "Extension of Absolute-Frequency Measurements to the Visible: Frequencies of Ten Hyperfine Components of Iodine," *Optics Letters*, vol. 4, no. 9, pp. 263–264, September 1979.

30. K. M. Evenson, D. A. Jennings, and F. R. Peterson, "The Frequency Measurement of Visible Light," *Journal de Physique Colloques*, vol. 42, no. C8, pp. C8/473–483, December 1981.

31. D. A. Jennings, R. E. Drullinger, K. M. Evenson, C. R. Pollock, and J. S. Wells, "The Continuity of the Meter: The Redefinition of the Meter and the Speed of Light," *Journal of Research of the National Bureau of Standards*, vol. 92, no. 1, pp. 11–16, January-February 1987.

32. V. P. Chebotayev, "Optical Time Scale," *Journal de Physique Colloques*, vol. 42, no. C8, pp. C8/505–512, December 1981.

33. V. P. Chebotayev, V. G. Goldort, V. M. Klementyev, M. V. Nikitin, B. A. Timchenko, and V. F. Zakharyash, "Development of an Optical Time Scale," *Applied Physics B*, vol. 29, no. 1, pp. 63–65, September 1982.

34. J. Reichert, R. Holzwarth, Th. Udem, and T. W. Hänsch, "Measuring the Frequency of Light with Mode-Locked Lasers," *Optics Communications*, vol. 172, no. 1–6, pp. 59–68, December 1999.

35. Th. Udem, J. Reichert, R. Holzwarth, and T. W. Hänsch, "Absolute Optical Frequency Measurement of the Cesium *D*1 Line with a Mode-Locked Laser," *Physical Review Letters*, vol. 82, no. 18, pp. 3568–3571, May 1999.

36. J. Reichert, M. Niering, R. Holzwarth, M. Weitz, Th. Udem, and T. W. Hänsch, "Phase Coherent Vacuum-Ultraviolet to Radio Frequency Comparison with a Mode-Locked Laser," *Physical Review Letters*, vol. 84, no. 15, pp. 3232–3235, April 2000.

37. Th. Udem, R. Holzwarth, and T. W. Hänsch, "Optical Frequency Metrology," *Nature*, vol. 416, no. 6877, pp. 233–237, March 2002.

38. D. J. Jones, S. A. Diddams, J. K. Ranka, A. Stentz, R. S. Windeler, J. L. Hall, and S. T. Cundiff, "Carrier-Envelope Phase Control of Femtosecond Mode-Locked Lasers and Direct Optical Frequency Synthesis," *Science*, vol. 288, no. 5466, pp. 635–639, April 2000.

39. S. A. Diddams, D. J. Jones, J. Ye, S. T. Cundiff, J. L. Hall, J. K. Ranka, R. S. Windeler, R. Holzwarth, T. Udem, and T. W. Hänsch, "Direct Link between Microwave and Optical Frequencies with a 300 THz Femtosecond Laser Comb," *Physical Review Letters*, vol. 84, no. 22, pp. 5102–5105, May 2000.

40. S. A. Diddams, K. Vahala, and T. Udem, "Optical Frequency Combs: Coherently Uniting the Electromagnetic Spectrum," *Science*, vol. 269, no. 6501, 12 p., July 2020.

41. T. Fortier and E. Baumann, "20 Years of Developments in Optical Frequency Comb Technology and Applications," *Communications Physics*, vol. 2, pp. 153/1–16, December 2019.

42. H. R. Telle, G. Steinmeyer, A. E. Dunlop, J. Stenger, D. H. Sutter, and U. Keller, "Carrier-Envelope Offset Phase Control: A Novel Concept for Absolute Optical Frequency Measurement and Ultrashort Pulse Generation," *Applied Physics B*, vol. 69, no. 4, pp. 327–332, September 1999.

43. L.-S. Ma, "Optical Atomic Clocks: From Dream to Reality," *Optics & Photonics News*, vol. 18, no. 8, pp. 42–47, September 2007.

44. T. M. Fortier, M. S. Kirchner, F. Quinlan, J. Taylor, J. C. Bergquist, T. Rosenband, N. Lemke, A. Ludlow, Y. Jiang, C. W. Oates, and S. A. Diddams, "Generation of Ultrastable Microwaves via Optical Frequency Division," *Nature Photonics*, vol. 5, no. 7, pp. 425–429, July 2011.

45. W. M. Itano, J. C. Bergquist, and D. J. Wineland, "Laser Spectroscopy of Trapped Atomic Ions," *Science*, vol. 237, no. 4815, pp. 612–617, August 1987.

46. D. Wineland, P. Ekstrom, and H. Dehmelt, "Monoelectron Oscillator," *Physical Review Letters*, vol. 31, no. 21, pp. 1279–1282, November 1973.

47. D. J. Wineland, R. E. Drullinger, and F. L. Walls, "Radiation Pressure Cooling of Bound Resonant Absorbers," *Physical Review Letters*, vol. 40, pp. 1639–1642, June 1978.

48. J. J. Bollinger, J. D. Prestage, W. M. Itano, and D. J. Wineland, "Laser-Cooled-Atomic Frequency Standard," *Physical Review Letters*, vol. 54, no. 10, pp. 1000–1003, March 1985.

49. L. S. Cutler, R. P. Giffard, and M. D. McGuire, "A Trapped Mercury 199 Ion Frequency Standard," *Proceedings of the 13*th *Annual Precise Time and Time Interval Meeting (PTTI)*, pp. 563–578, December 1981.

50. D. J. Berkeland, J. D. Miller, J. C. Bergquist, W. M. Itano, and D. J. Wineland, "Laser-Cooled Mercury Ion Frequency Standard," *Physical Review Letters*, vol. 80, no. 10, pp. 2089–2092, March 1998.

51. H. G. Dehmelt, "Mono-Ion Oscillator as Potential Ultimate Laser Frequency Standard," *IEEE Transactions on Instrumentation and Measurement*, vol. IM-31, no. 2, pp. 83–87, June 1982.

52. S. A. Diddams, Th. Udem, J. C. Bergquist, E. A. Curtis, R. E. Drullinger, L. Hollberg, W. M. Itano, W. D. Lee, C. W. Oates, K. R. Vogel, and D. J. Wineland, "An Optical Clock Based on a Single Trapped $^{199}Hg^+$ Ion," *Science*, vol. 293, no. 5531, pp. 825–828, August 2001.

53. W. H. Oskay, S. A. Diddams, E. A. Donley, T. M. Fortier, T. P. Heavner, L. Hollberg, W. M. Itano, S. R. Jefferts, M. J. Delaney, K. Kim, F. Levi, T. E. Parker, and J. C. Bergquist, "Single-Atom Optical Clock with High Accuracy," *Physical Review Letters*, vol. 97, no. 2, pp. 020801/1–4, July 2006.

54. C. W. Chou, D. B. Hume, J. C. J. Koelemeij, D. J. Wineland, and T. Rosenband, "Frequency Comparison of Two High-Accuracy Al^+ Optical Clocks," *Physical Review Letters*, vol. 104, no. 7, pp. 070802/1–4, February 2010.

55. S. M. Brewer, J.-S. Chen, A. M. Hankin, E. R. Clements, C. W. Chou, D. J. Wineland, D. B. Hume, and D. R. Leibrandt, "$^{27}Al^+$ Quantum-Logic Clock with a Systematic Uncertainty below 10^{-18}," *Physical Review Letters*, vol. 123, no. 3, pp. 033201/1–6, July 2019.

56. N. Huntemann, C. Sanner, B. Lipphardt, Chr. Tamm, and E. Peik, "Single-Ion Atomic Clock with 3×10^{-18} Systematic Uncertainty," *Physical Review Letters*, vol. 116, no. 6, pp. 063001/1–5, February 2016.

57. H. Katori, "Spectroscopy of Strontium Atoms in the Lamb-Dicke Confinement," *Proceedings of the 6th Symposium on Frequency Standards and Metrology*, pp. 323–330, September 2001.

58. M. Takamoto and H. Katori, "Spectroscopy of the $^1S_0$–$^3P_0$ Clock Transition of ^{87}Sr in an Optical Lattice," *Physical Review Letters*, vol. 91, no. 22, pp. 223001/1–4, November 2003.

59. M. Takamoto, F.-L. Hong, R. Higashi, and H. Katori, "An Optical Lattice Clock," *Nature*, vol. 435, no. 7040, pp. 321–324, May 2005.

60. C. W. Oates and A. D. Ludlow, "Optical Lattice Clocks," *Optics & Photonics News*, vol. 26, no. 1, pp. 36–43, January 2015.

61. M. Schioppo, R. C. Brown, W. F. McGrew, N. Hinkley, R. J. Fasano, K. Beloy, T. H. Yoon, G. Milani, D. Nicolodi, J. A. Sherman, N. B. Phillips, C. W. Oates, and A. D. Ludlow, "Ultrastable Optical Clock with Two Cold-Atom Ensembles," *Nature Photonics*, vol. 11, no. 1, pp. 48–53, January 2017.

62. A. Aeppli, K. Kim, W. Warfield, M. S. Safronova, and J. Ye, "Clock with 8×10^{-19} Systematic Uncertainty," *Physical Review Letters*, vol. 133, pp. 023401/1–7, July 2024.

63. J. Yao, T. E. Parker, N. Ashby, and J. Levine, "Incorporating an Optical Clock Into a Time Scale," *IEEE Transactions on Ultrasonics, Ferroelectrics, and Frequency Control*, vol. 65, no. 1, pp. 127–134, January 2018.

64. H. Hachisu, N. Nemitz, M. Tønnes, H. Ito, N. Ohtsubo, Y. Miyauchi, M. Morikawa, K. Matsubara, T. Gotoh, and T. Ido, "Contributions of the Optical Lattice Clock NICT-Sr1 to TAI Calibration and UTC(NICT) Steering," *Journal of Physics: Conference Series*, vol. 2889, pp. 012018/1–9, 2024.

65. D. Genkina, "Squeezing an Optical Atomic Clock Into a Briefcase," *IEEE Spectrum*, vol. 61, no. 11, pp. 44–49 & 70, November 2024.

66. M. Takamoto, I. Ushijima, N. Ohmae, T. Yahagi, K. Kokado, H. Shinkai, and H. Katori, "Test of General Relativity by a Pair of Transportable Optical Lattice Clocks," *Nature Photonics*, vol. 14, no. 7, pp. 411–415, July 2020.

67. T. Bothwell, W. Brand, R. Fasano, T. Akin, J. Whalen, T. Grogan, Y.-J. Chen, M. Pomponio, T. Nakamura, B. Rauf, I. Baldoni, M. Giunta, R. Holzwarth, C. Nelson, A. Hati, F. Quinlan, R. Fox, S. Peil, and A. Ludlow, "Deployment of a Transportable Yb Optical Lattice Clock," *Optics Letters*, vol. 50, no. 2, pp. 646–649, January 2025.

68. J. D. Roslund, A. Cingöz, W. D. Lunden, G. B. Partridge, A. S. Kowligy, F. Roller, D. B. Sheredy, G. E. Skulason, J. P. Song, J. R. Abo-Shaeer, and M. M. Boyd, "Optical Clocks at Sea," *Nature*, vol. 628, pp. 736–747, April 2024.

Chapter 14

Optical Time Transfer

In perfect buildings the different members must be in exact symmetrical relations to the whole general scheme.

— Vitruvius, "De architectura libri decem (The Ten Books on Architecture)"

We defined time transfer in Section 4.3 as the process of transferring time signals from one location to another, for the purpose of synchronizing or comparing clocks. In an optical time transfer system, the time signals are transferred by utilizing light waves, but the definition still applies. Optical systems implement the same basic methods as conventional time transfer systems, and they serve the same purpose.

Most of this chapter is devoted to the most common type of optical time transfer, which involves transmitting and receiving time signals sent via optical fibers. An optical fiber is a waveguide made out of flexible glass. As was discussed in Section 4.1.3, optical fibers have several advantages over coaxial cables when used as the transmission medium for a local area time transfer system—they are lighter and more flexible than coax, more immune to electromagnetic interference, and less sensitive to temperature changes. However, their greatest advantage is their low attenuation, which allows them to transfer time signals across distances of thousands of kilometers.

When time signals are transferred via shared optical fibers that carry other traffic, the available bandwidth is limited and large asymmetries can be introduced by the network equipment. The lowest time transfer uncertainties are obtained using dark fiber, which are the unused strands of fiber within a fiber optic cable that are not being used for any other purpose. When dark fiber is leased from a telecommunication network provider, laboratories can compare their clocks by using their own timing equipment to "light" the fiber.

A dedicated fiber optic connection has better symmetry and less noise than any of the RF transmission mediums that are used to transfer time. The inherent symmetry of a fiber connection allows the time delays and phase fluctuations in a time transfer link to be stabilized and almost perfectly compensated, which makes fiber optic links the only medium that's currently suitable for long-distance comparisons of optical clocks. The uncertainties of fiber optic time transfer measurements are usually expressed in picoseconds, and they're usually at least an order of magnitude smaller than the uncertainties of the best satellite-based time transfer measurements [1, 2].

As Time Goes By: A Comprehensive Guide to Time and Frequency Measurements
Michael A. Lombardi

ISBN 978-981-5352-28-3 (Hardcover), 978-1-042-01532-0 (eBook)
www.jennystanford.com

DOI: 10.1201/9781042015320-19

Section 14.1 opens this chapter by providing a concise introduction to fiber optic communications. The discussion of fiber optic time transfer begins in Section 14.2, and it includes descriptions of how clocks are compared over both short and long distances by using either dedicated dark fiber, or shared fiber in public networks. Section 14.3 discusses experimental optical time transfer systems that transmit signals through free space, without using fibers as waveguides. Section 14.4 concludes the chapter by describing a few of the optical clock comparisons that have been made possible by optical time transfer.

14.1 Fiber Optic Communications

The use of optical fibers as a communication medium was first proposed in the mid-1960s, roughly seven years after Schawlow and Townes first proposed the development of lasers (Section 13.2). The proposal was made by the Chinese engineer and physicist Charles Kao (1933–2018), who was then working at Stanford Telecommunication Laboratories in Harlow, England. Kao left China to attend the University of London, where he obtained a doctorate in electrical engineering. After being hired by Stanford, he began working on methods of transmitting digital data using lasers and waveguides made of dielectric materials. The work was originally motivated by the needs of the British Post Office, which needed better technology to send signals between switching stations that were separated by a few kilometers. Kao felt that their needs could be met by using glass as the dielectric material, so he conducted a series of experiments where light from helium-neon and gallium arsenide lasers was transmitted through cladded waveguides made of glass. He concluded in 1965 that the fundamental limit for glass light attenuation was below 20 dB per kilometer, and that glass fibers could be used as a "new form of communication medium" [3].

Reaching the 20 dB/km threshold would mean that 1% of the light that entered the glass waveguide would still remain after traveling a distance of a kilometer, which would make fiber optic communications practical, at least over short distances. Kao was confident that the 20 dB/km threshold was achievable, but his vision of using glass fibers for communications was originally met with disbelief and a lack of interest by engineers and scientists in the telecommunications industry. The telecommunications industry was accustomed to using copper wires as a transmission medium, and their skepticism towards optical fibers wasn't surprising, because the best fibers that were then available reduced the light intensity by 20 dB over a distance of just 20 m.

It was then commonly believed that signal loss due to light scattering was the main reason why optical fibers couldn't be used for communications. Light scattering occurs when the atoms in the glass cause the light to travel in multiple directions, instead of just traveling straight through the fiber. It does cause signal loss, but Kao and his colleagues found that the attenuation was near 1 dB/km at a 1 μm wavelength, or well below the usable threshold for communications. They soon realized that most of the attenuation was due to the absorption of light by impurities in the glass. Solving the absorption problem wasn't possible using ordinary optical glass, because it was made of raw materials that contained numerous impurities. The only solution would be to make glass fibers that were completely transparent and as free from impurities as possible.

By 1968, Kao and his colleagues had focused their research on fused silica, a synthetic glass that is essentially pure silicon dioxide (SiO_2). Their attenuation measurements demonstrated that silica glass would be an ideal material for making optical fibers, easily capable of reaching the 20 dB/km threshold [4]. This research led to the invention of the first low-loss optical fiber for telecommunications in 1970.

The invention of low-loss optical fiber, and the development of the highly pure glass that made it work, was a joint effort of Robert Maurer (1924–2025), Peter Schultz (b. 1942), and Donald Keck (b. 1941), three American scientists at Corning Glass Works in New York. The first fibers that they produced had a cladding made of pure silica, and a core made of silica doped with titanium. Their original measurements showed that the attenuation was just 17 dB/km [4].

When Corning started the mass production of low-loss optical fiber in the late 1970s, the telecommunications industry became far more interested in fiber optics communications. This led to a flurry of research and development that reduced the attenuation of fibers to well below 1 dB/km at certain optical wavelengths. In 1988, the American physicist Linn Mollenauer (1937–2021) of Bell Laboratories conducted an experiment where laser light, at a 1,600 nm wavelength, was recirculated numerous times through a 42 km loop of single-mode fiber. The experiment made use of Raman scattering, an effect discovered by the Indian physicist and Nobel laureate Chandrasekhara Venkata Raman (1888–1970), which causes incident light that's scattered by deflections to release energy at a higher or lower frequency. Mollenauer periodically compensated for the small attenuation in the fiber, which was about 0.2 dB/km, using Raman amplification, a method that transfers energy from a strong laser beam at one wavelength to a weaker beam at a second wavelength. His experiment proved that optical fiber could be used to transmit light pulses over a distance of more than 4,000 km, without having to regenerate the light [5].

This experiment, and many others like it, firmly established optical fiber as a long-distance communications medium. From there, its use expanded exponentially, and it eventually enabled the worldwide Internet access and media streaming services that we enjoy today. According to estimates made in 2023, more than six billion kilometers of optical fiber have been deployed, which is enough fiber to make about 20 round trips between the Earth and the Sun [6]. Kao played a formative role in making all of this happen, and his achievements were honored during his lifetime. He received a Nobel Prize in physics in 2009 for his "groundbreaking achievements concerning the transmission of light in fibers for optical communication," and he was knighted in 2010 by Queen Elizabeth II.

Figure 14.1 is a simplified diagram showing how light is injected into an optical fiber. The light is refracted, or bent, at the interface between the core and the cladding. The maximum acceptance angle for the light is at the limit between refraction and reflection. If the acceptance angle is exceeded, the light is refracted through the cladding and not captured by the core. However, if the angle of the light is less than the acceptance angle, the light is reflected into the core. The light then propagates through the fiber, because the reflections inside the core prevent it from escaping.

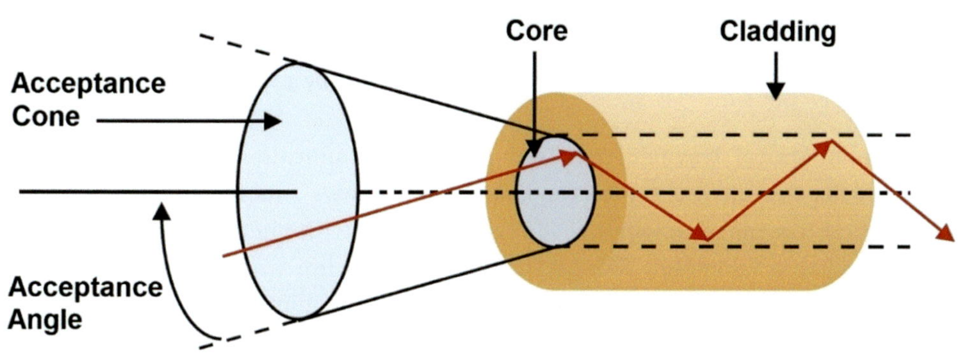

Figure 14.1 The injection of light into an optical fiber (adapted from [7]).

The speed of light in an optical fiber is typically about 68% of the speed of light in a vacuum. As previously seen in Eq. (4.10), the velocity factor for an optical fiber is equal to $\frac{1}{n_{\text{eff}}}$, where n_{eff} is the effective group index of refraction, which is a weighted average of all of the refractive indices that light encounters when it travels within a fiber. The effective group index of refraction is typically near 1.47 for the wavelengths used in telecommunication systems [7].

As was previously described in Section 4.1.3, optical fiber can be either single-mode or multi-mode. However, multi-mode fiber isn't generally used for high-accuracy time transfer, due to the uncertainty introduced by spreading the light into multiple paths of different lengths. This effect, called modal dispersion, is completely eliminated by single-mode fibers. A single-mode fiber can simultaneously carry multiple signals at different wavelengths by implementing a method called wavelength division multiplexing (WDM) [7]. The use of WDM enables two-way time transfer, because two optical signals, each with a different wavelength, can simultaneously travel in opposite directions through the same fiber.

For the purposes of illustration, Figure 14.2 shows a spool of single-mode fiber that's being illuminated by green light from a laser that operates at the visible wavelength of 561 nm. However, nearly all fiber optic communication and time transfer systems are designed to use longer wavelength infrared light (Table 13.1), which is invisible to the human eye. Infrared light is a better choice than visible light for fiber optic communications, because increasing the wavelength decreases the attenuation. This allows infrared light to travel further than visible light before it needs to be amplified or regenerated.

Figure 14.2 A spool of single-mode optical fiber, illuminated by green laser light (561 nm).

The most common optical wavelengths used in telecommunications are 850 nm, 1,310 nm, and 1,550 nm, which are located in the three transmission "windows" shown in Figure 14.3. The 850 nm wavelength is in the first transmission window. It's typically used with multi-mode fibers, for communication over distances of less than a kilometer. The 1,310 nm and 1,550 nm wavelengths in the second and third transmission windows are typically used for longer distance communications via single-mode fibers. Optical fibers at these wavelengths have very little attenuation due to absorption. According to ITU standards, the maximum allowable attenuation is 0.4 dB/km for wavelengths of 1,310 nm to 1,625 nm. The lowest attenuation is in the third transmission window, sometimes called the "zero loss window," which is the range of wavelengths from 1,530 nm to 1,565 nm. The maximum allowable attenuation at these wavelengths is 0.3 dB/km [8].

The "water peaks" shown in Figure 14.3 are caused by hydrogen and hydroxide ions in the fiber optic material, which increases the fiber attenuation at certain wavelengths. These ions are introduced by the presence of water, which can enter the cable due to chemical reactions during the manufacturing process, or due to environmental humidity. The most prominent water peak is at the 1,383 nm wavelength. The water peaks are significantly reduced, but not completely eliminated, in higher-grade fiber optic cables that cost more than conventional fibers [7].

Even though the attenuation of optical fibers has been reduced to very low levels, optical signals still need to be periodically amplified. The most common type of amplifier used in long-distance fiber optic communications is the erbium-doped fiber amplifier (EDFA), which primarily amplifies light in the third transmission window. An EFDA uses a special type of fiber that's been "doped" by intentionally introducing erbium ions (Er^{3+}) into the core. These ions absorb energy from a pump laser, and then amplify the optical signals that pass through the fiber by releasing their stored energy as photons.

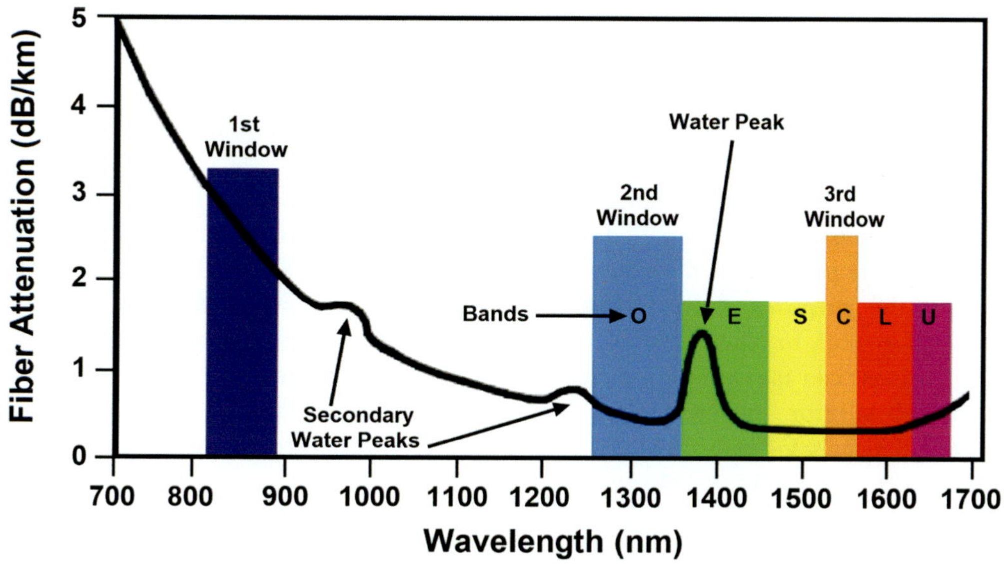

Figure 14.3 Optical fiber attenuation at different wavelengths.

Table 14.1 lists the ITU optical wavelength bands, which are categorized by the letter designations O, E, S, C, L, and U. The O and C bands are the most widely used, and they respectively occupy the second and third transmission windows shown in Figure 14.3. The O band got its name because it was the original band used for long-distance communications after fiber was invented in the 1970s. Its lower limit of 1,260 nm is the cut-off wavelength for single-mode fibers—the use of shorter wavelengths requires multi-mode fibers. The O band's upper limit of 1,360 nm was chosen to avoid the rising edge of the 1,383 nm water peak.

The C band is known as the conventional band. It's also called the erbium window, because its range, from 1,530 nm to 1,565 nm, was chosen to match the effective amplification range of the first EFDAs. The E, S, and L bands are also occupied by fiber optic transmission channels. However, the use of the "ultralong wavelength" U band is usually reserved for fiber measurements and diagnostic tests.

Table 14.1 The optical wavelength bands for fiber communications

Band	Description	Wavelength range (nm)
O	Original	1,260 to 1,360
E	Extended	1,360 to 1,460
S	Short wavelengths	1,460 to 1,530
C	Conventional (erbium window)	1,530 to 1,565
L	Long wavelengths	1,565 to 1,625
U	Ultralong wavelengths	1,625 to 1,675

14.2 Fiber Optic Time Transfer

The use of fiber optics as a distribution medium for high-accuracy time signals was first demonstrated in 1980 by George Lutes (1937–2020), an American electrical engineer who worked at JPL in Pasadena, California. Lutes conducted two-way frequency transfer measurements over a round-trip distance of 3 km using two multi-mode fibers that were welded together. The optical transmitters and receivers operated at a wavelength of about 830 nm. His results showed that the fiber optic link was far more stable than a hydrogen maser at averaging periods longer than 100 s, reaching a stability near 1×10^{-15} at τ = 100 s and near 1×10^{-16} at τ = ~1 hour. Lutes was encouraged by these results, and he pointed out the many advantages that optical fibers had over previous transmission mediums. He also correctly predicted that single-mode fibers, which weren't quite ready for use in 1980, would be "better in nearly every respect" than multi-mode fibers when used for time transfer [9].

Technically speaking, the systems that synchronize computer clocks using NTP (Section 11.2.4) and other time code protocols could be classified as fiber optic time transfer systems, because there is a very high probability that the path traveled by the time code packets includes optical fibers. Generally speaking, however, and for the purposes of this chapter, fiber optic time transfer refers to systems that have uncertainties small enough to allow accurate comparisons of free-running atomic clocks or UTC(*k*) time scales. The comparisons can measure either frequency or time differences, and the clocks being compared typically provide 10 MHz or 100 MHz frequency signals and 1 pps time signals. The clock signals are converted to optical signals for transmission via fiber, and then converted back to electrical signals at the other end of the time transfer link.

The clock signals are typically conveyed on to an optical carrier by modulating the intensity of the light generated by a telecommunication-grade laser. The optical carrier is transmitted through the fiber, and a receiver at the other end detects the intensity of the light by using a photodiode, which allows it to demodulate the optical carrier and regenerate the original electrical signal [1, 10, 11]. The back-and-forth conversion between electrical and optical signals can be accomplished by using unidirectional devices that operate as transmitters or receivers, or by using bidirectional devices that operate as transceivers. Network devices that support the PTP or WR protocols (Sections 11.2.5 and 11.2.6) are commonly used as transceivers, and they can usually accept 10 MHz and 1 pps input signals [12, 13].

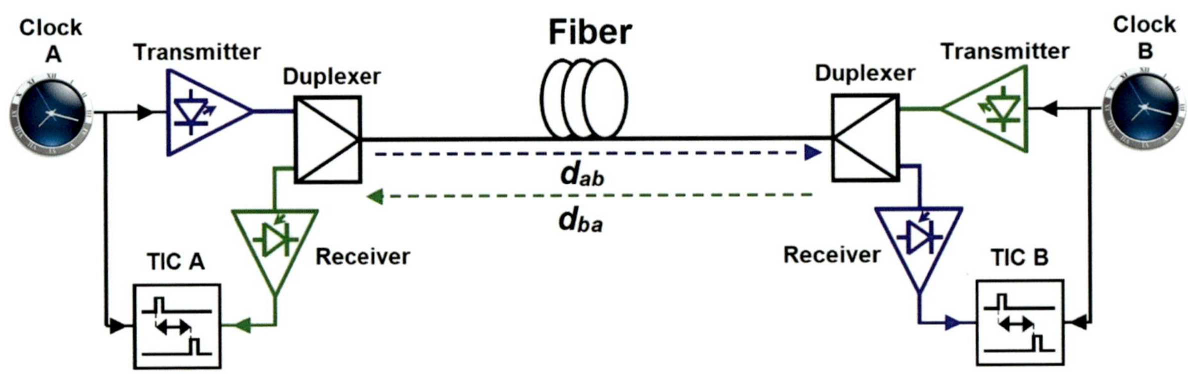

Figure 14.4 Fiber optic two-way time transfer (adapted from [1]).

The two-way method, which was previously described for time transfer via geostationary satellites (Section 4.3.4), is also used for fiber optic time transfer. It allows two clocks to be compared by transmitting two signals at different wavelengths through a single fiber. As shown in the diagram in Figure 14.4, time signals from clocks A and B are split into two paths. One path leads to an optical transmitter that converts the clock signal to an optical signal and sends it through the fiber, and the other path leads to the start or stop channel of a time interval counter (TIC). The fiber has a duplexer at both ends, which is often a device called an optical circulator, that separates the signals traveling in opposite directions. This allows signals from both clocks to be simultaneously transmitted on the same fiber. When the optical signals reach the end of the fiber, receivers convert them back to electrical signals and direct them to a TIC, which measures the time difference between the local clock and the remote clock [1, 14–16].

The two TICs record the following time differences,

$$TIC_A = Clock\,A - (Clock\,B + d_{ba}),\ \text{and} \tag{14.1}$$

$$TIC_B = Clock\,B - (Clock\,A + d_{ab}), \tag{14.2}$$

where d_{ba} is the path delay from B to A, and d_{ab} is the path delay from A to B. The time difference between clocks A and B is calculated as

$$Clock\,A - Clock\,B = \frac{TIC_A - TIC_B}{2} - \frac{d_{ab} - d_{ba}}{2}. \tag{14.3}$$

If all of the differential delays were fully compensated and the time transfer link was perfectly symmetric, only the left side of the equation, $\frac{TIC_A - TIC_B}{2}$, would be necessary. The right side of the equation, $\frac{d_{ab} - d_{ba}}{2}$, is needed to account for any delay differences that introduce asymmetry. The path delays in both directions include the delays introduced by the transmitters and receivers, which can differ at the two clock sites. However, in most cases, a larger differential path delay is introduced by the difference in velocity between two optical signals that are transmitted at different wavelengths.

The difference in velocity is due to chromatic dispersion, a phenomenon that causes different wavelengths of light to travel at different velocities through a fiber, and to spread out, or disperse, over time. As previously noted, single-mode fibers eliminate the effects of modal dispersion, but they are still affected by chromatic dispersion. The specifications for a single-mode fiber include the zero dispersion wavelength, λ_0, where the chromatic dispersion is near 0. The dispersion values for other wavelengths are expressed in units of picoseconds per nanometer, normalized to a distance of one kilometer. The effective group index of refraction discussed earlier in this chapter accounts for chromatic dispersion, because it's based on a weighted average of refractive indices, and these indices vary with wavelength. Therefore, the average delay through a fiber can be estimated as

$$d = L\left(\frac{n_{\text{eff}}}{c}\right), \tag{14.4}$$

where d is the delay through the fiber in one direction, L is the length of the fiber in meters, n_{eff} is the effective group index of refraction for a given wavelength, and c is the fundamental constant for the speed of light. For OS2 9/125 single-mode fiber (Section 4.1.3), the refractive index at standard telecommunication wavelengths of 1,310 nm and 1,550 nm might be 1.4676 and 1.4682, respectively. This difference appears to be slight, but it causes two signals that propagate through the same fiber to have average delays that differ by about 2 ns per kilometer. This delay difference has to be compensated to reach the low measurement uncertainties that are needed for atomic clock comparisons.

The two-way method is based on the premise that the propagation delays in both directions are equal, or can be made equal through compensation. The duration of the delays is unimportant because it doesn't affect the clock comparison; the delays can be long or short as long as they are equal and cancel out. The situation is different when the time from a reference clock, such as a UTC(k) time scale, is distributed over fiber for the purpose of synchronizing other clocks. In that case, the delay from the reference clock not only has to be known, but it also has to be stabilized so that it varies as little as possible. Systems that distribute time over fiber can be classified as one-way time transfer systems (Section 4.3.1), because the time from the reference clock is sent in one direction to a remote clock. However, they also usually involve some type of bidirectional communication that's similar to loop-back time transfer (Section 4.3.2).

An example of a system that distributes time from a reference clock over fiber is shown in the diagram in Figure 14.5. The time from the reference clock *A* is transmitted over fiber for the purpose of synchronizing the remote clock *B*. The goal is to constantly stabilize the delay of the time transmitted by the reference clock, which is accomplished by using a delay-locked loop (DLL). The reference clock signal is split into two paths. One path leads to an optical transmitter that converts the clock signal to an optical signal and sends it through the fiber, and the other path leads to a TIC. When the optical time signal reaches the remote clock site, a receiver converts it back to an electrical signal that is again split into two paths. The signal on one path is used to synchronize the remote clock. The other path leads to an optical transmitter that sends the signal back to the reference clock site, where the TIC measures the round-trip path delay. The optical time signal is then compensated to account for the one-way delay from the reference clock, and actively stabilized with a variable optical delay system that's based on electronically controlled delay lines [1, 17]. The one-way delay is assumed to be one half of the round-trip delay, so the uncertainty of this method is still limited by any asymmetry in the bidirectional fiber link.

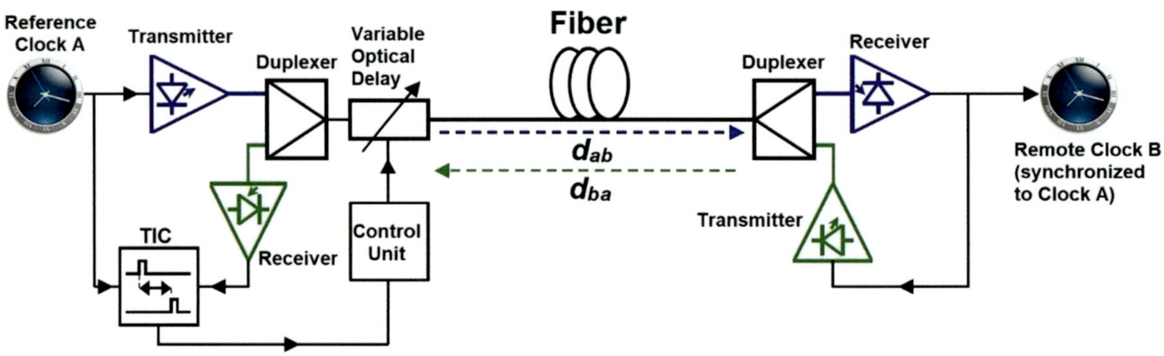

Figure 14.5 Fiber optic one-way time transfer, compensated for path delays (adapted from [1]).

A one-way time transfer method that doesn't require a bidirectional fiber link has been demonstrated by researchers at the SP Technical Research Institute of Sweden. This method sends signals from a reference clock on two different wavelengths over the same fiber. The difference in arrival times is measured at the remote clock site, and the difference in velocity between the two known wavelengths is used to estimate the path delay in one of the wavelength channels. This method is especially useful for transferring time in asymmetric networks, where one-way delay estimates based on round-trip measurements can have large uncertainties [18, 19].

14.2.1 Limiting Factors of Measurement Uncertainty

Any factor that causes the optical signal delays to change will contribute uncertainty to a fiber optic time transfer measurement. Two-way time transfer via a single fiber can effectively cancel most uncertainties, because the delay variations are nearly the same in both directions. However, some asymmetry will always be present, and generally speaking, the longer the fiber link, the larger the measurement uncertainty. There are several factors that usually limit the combined standard uncertainty to about 10 ps for bidirectional links of less than 1,000 km in length, with the uncertainty typically increasing linearly for longer links [20]. We'll briefly examine three of those factors in this section: temperature, light scattering, and polarization mode dispersion (PMD).

Temperature is usually the largest contributor to fiber optic time transfer uncertainty. The temperature variations in a fiber change the group index of refraction, which in turn changes the path delay. Fibers are typically buried in the ground at a depth of one or two meters, but the soil temperature can still vary by about ±10 °C per year at some locations [14]. The delay in a single-mode fiber can vary by about 40 picoseconds per kilometer per degree Celsius (40 ps/km/°C), so path delays of a long link can easily vary by multiple nanoseconds due to temperature variations. However, when the time transfer is bidirectional over the same fiber, temperature has an almost equal effect on both paths. This reduces the uncertainty by about four orders of magnitude, to an insignificant level of about 2 fs/km/°C [1].

The delay variations that temperature causes in optical transceivers and measurement instruments are often more of a concern than the delay variations in the fiber. The same type of hardware is typically used at both ends of a two-way time transfer link, but the environmental temperature isn't exactly the same at both sites, so the hardware delay variations won't completely cancel. The asymmetry introduced by hardware delay variations can contribute tens of picoseconds of measurement uncertainty [21, 22]. To reduce these uncertainties as much as possible, researchers have designed optical transceivers with temperature coefficients of less than 1 ps/°C [16].

As mentioned in Section 14.1, it was first thought that light scattering, and not light absorption, was the main cause of attenuation in glass fibers. After the impurities that caused absorption were mostly removed from fibers, light scattering was the primary cause of attenuation, especially Rayleigh scattering, which is the dominant form of scattering when light interacts with particles smaller than its wavelength. Rayleigh scattering was named to honor the British physicist and Nobel laureate John William Strutt (1842–1919), who was later known as Lord Rayleigh. The phenomenon of Rayleigh scattering is why the daytime sky appears to be blue—the scattering of sunlight has a wavelength dependency, and the shorter wavelengths of blue light are scattered more strongly than the longer wavelengths of other colors (Figure 14.6).

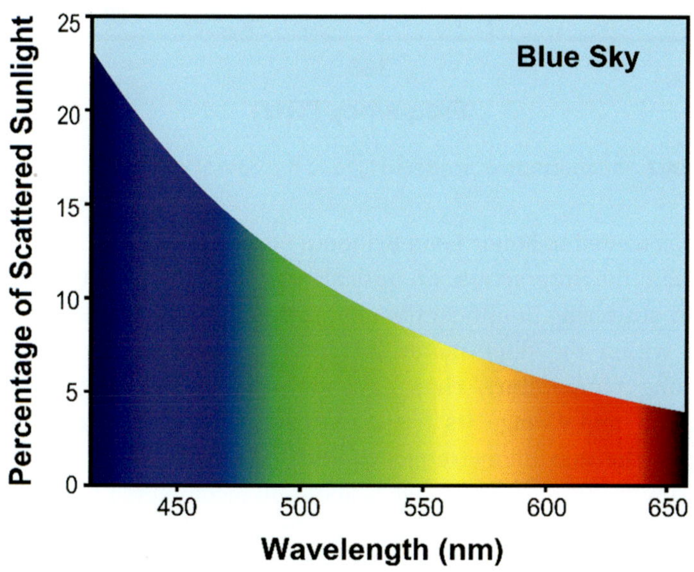

Figure 14.6 The Rayleigh scattering of sunlight makes the daytime sky appear to be blue.

Rayleigh scattering is caused by density fluctuations that are frozen into fused silica glass during the fiber manufacturing process. Because the density of the glass varies from point to point, the refractive indices also vary. This causes the light in a fiber to scatter in all directions, including back towards its source. Rayleigh backscattering causes most of the attenuation in optical fibers, and it sets a fundamental limit on how small the signal losses can be. However, backscattering can also be useful, because measuring the time delay of backscattered light is the basic principle behind the optical time domain reflectometer (OTDR), an instrument that's commonly used to characterize fibers. In addition to measuring propagation delay, an OTDR can locate faults in a fiber and can measure a fiber's length and attenuation [23].

As shown in Figure 14.7, Rayleigh light scattering has a higher intensity than Raman scattering (Section 14.1) and Brillouin scattering. Rayleigh scattering is elastic (linear), which means that the scattered light has the same frequency as the incident light, but Raman and Brillouin scattering are inelastic (non-linear), which means that the scattered light has a higher or lower frequency than the incident light. The light waves shifted to a lower frequency and a longer wavelength are called Stokes waves, and the light waves shifted to a higher frequency and a shorter wavelength are called anti-Stokes waves. The name honors the Irish physicist George Stokes (1819–1903), who studied changes in the wavelength of light as it passed through various mediums.

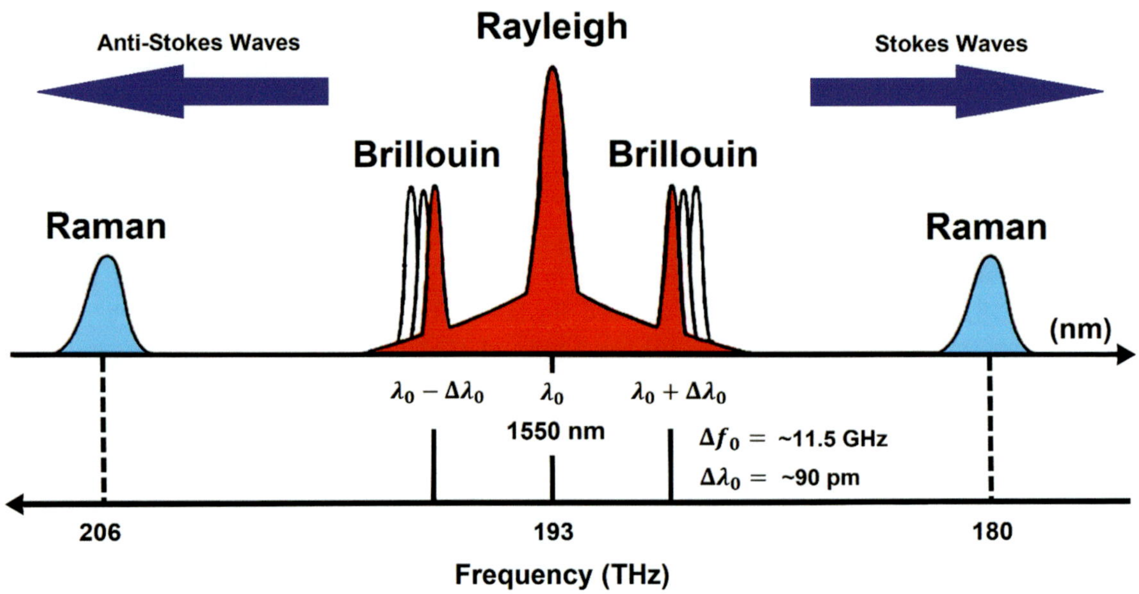

Figure 14.7 The spectrum of scattered light at the optical wavelength of 1,550 nm.

Brillouin scattering was named to honor Léon Brillouin (1889–1969), the French physicist who predicted and theoretically described the interaction of optical waves with acoustic waves in a medium [24]. The spontaneous Brillouin scattering in optical fibers is caused by the interaction of light with the acoustic phonons (sound waves) within the fiber material. The sound waves can be generated by changes in the environment—including temperature changes and mechanical vibrations—that interact with the light and cause it to scatter and change its frequency. For light traveling through a single-mode fiber at a 1,550 nm wavelength, the maximum frequency shift due to Brillouin scattering is about 11.5 GHz, which means that the maximum wavelength change is about 90 picometers [25].

Raman scattering results from the interaction of light with the vibrational or rotational modes of molecules in the fiber. The peak Raman gain is about 13 THz from the incident light, so it can shift a 1,550 nm wavelength by about 100 nm [25]. However, Raman scattered light is weaker and easier to suppress than Rayleigh or Brillouin scattered light, and it usually has a smaller effect on the time transfer uncertainty.

The jitter introduced by light scattering always adds some uncertainty to the TOA measurements of optical signals, and much of the problem is related to the backscattering that occurs when light is amplified. The EDFAs and other amplifiers that compensate for attenuation are usually unidirectional, and use optical isolators that act like a one-way valve to protect against the backscattering of the amplified light.

However, if unidirectional amplifiers are used for two-way time transfer, two fibers are needed, and the use of single fibers is preferred to reduce asymmetry. Therefore, single-path bidirectional amplifiers (SPBAs) have been developed that allow the use of a single fiber, while still mostly avoiding the effects of backscattering. These devices operate in the low-loss optical C band at wavelengths near 1,550 nm. They've been designed using various methods, but they typically implement either WDM [1, 10, 26, 27] or time division multiplexing (TDM) [20].

An example of a bidirectional optical amplifier is shown in the simplified diagram in Figure 14.8. Two time transfer signals at slightly different wavelengths, λ_1 and λ_2, are both amplified by the same EDFA, but isolated from each other. This is accomplished by dividing λ_1 and λ_2 with WDM filters and having them pass through different optical isolators. This configuration suppresses the backscattered light from both amplified wavelengths [26]. An SPBA is inherently symmetric and insensitive to variations in temperature, so its contribution to the measurement uncertainty is usually insignificant. However, if the gain of an SPBA is too high, it can propagate unwanted signals caused by backscattering or reflections from optical connectors. Therefore, the measurement uncertainty of a long-distance link might be lowered if the number of SPBAs is increased and their gain is decreased [10, 27].

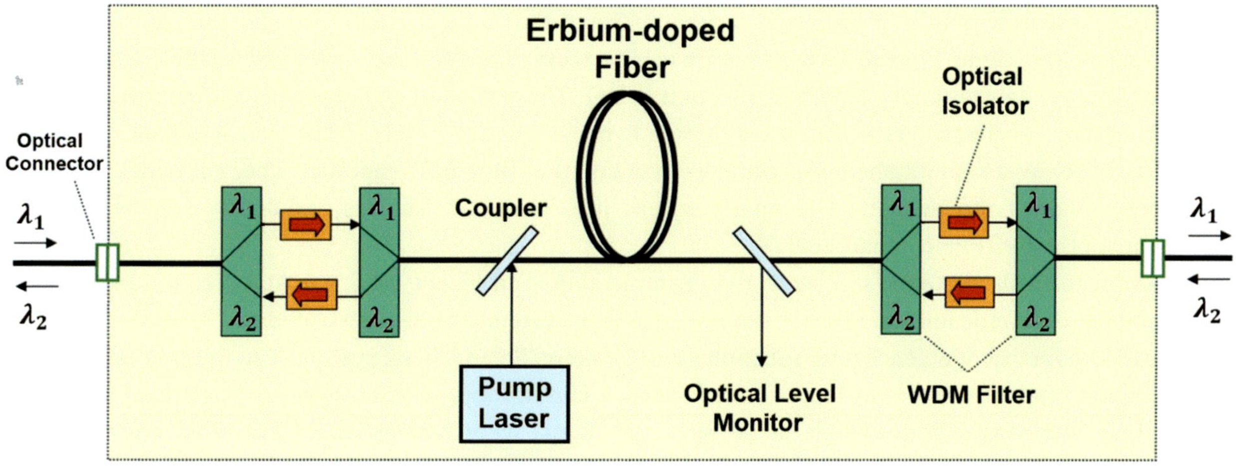

Figure 14.8 A simplified diagram of a single-path bidirectional amplifier (adapted from [26]).

Polarization mode dispersion is a basic property of an optical fiber that splits the electromagnetic energy into two perpendicular polarization states. If the fiber was perfectly cylindrical and perfectly straight, the two polarization states would travel at the same speed. This isn't the case, because flaws in the fiber can make it non-cylindrical, and the fiber can be strained, bent, or twisted at various points along its path. Therefore, the two polarization states propagate at slightly different velocities, and their separation in time causes the light pulses to broaden [7].

The differential delay between the two polarization modes is always fluctuating, but optical fiber manufacturers specify a PMD coefficient, in units of picoseconds, that estimates the mean value of the delay. Using this value, the asymmetry contributed to a time transfer link by PMD can be estimated as

$$u_{\mathrm{PMD}} = \frac{\mathrm{PMD}_Q(\sqrt{L})}{2}, \tag{14.5}$$

where u_{PMD} is the uncertainty contributed to a time transfer link, PMD_Q is the PMD coefficient specified by the fiber manufacturer, and L is the length of the fiber in kilometers. The PMD coefficient is typically near 0.05 ps for a single-mode fiber, so u_{PMD} is typically about 0.6 ps for a 500 km link, and about 1.4 ps for a 3,000 km link. These uncertainties are small, but they usually can't be compensated due to their randomness, so they impose some limits on the time transfer uncertainty. Because PMD is related to the strain placed on fibers, it usually won't affect the stability of links where the fibers are buried underground and motionless, but it can make a link less stable if the fibers are disturbed [20, 28].

As was noted at the beginning of this chapter, these uncertainties are all much smaller than the uncertainties of the other transmission mediums that are used for long-distance atomic clock comparisons. The next two sections provide examples of atomic clock comparisons performed via dark fiber that was used exclusively for time transfer, and via shared fiber in public networks.

14.2.2 Time Transfer via Dark Fiber

Atomic clocks are routinely compared over short distances using dedicated optical fibers and transceivers that implement the PTP or WR protocols. White Rabbit has been used, for example, to compare atomic clocks in the UTC(NIST) time scale ensemble that are located on different parts of the campus. The clocks are separated by distances of about 100 m, and their signals are transmitted using standard optical wavelengths of 1,310 nm and 1,550 nm. The TDEV of the time transfer links is below 20 ps at all averaging periods [12].

In 2012, researchers at PTB in Germany demonstrated that time could be transferred over a dedicated fiber by using the same type of two-way spread spectrum modems that are used to link UTC(k) time scales to UTC via geostationary satellites (Section 9.4.4). These modems, along with the necessary optical transmitters and receivers, were used to compare a hydrogen maser to UTC(PTB) over a distance of 73 km. The two clocks were simultaneously compared using the GPS PPP method. The combined standard uncertainty of the two-way optical comparison was just 74 ps, which was about 10× smaller than the uncertainty of the GPS PPP comparison [29].

Delay-locked loop systems, similar to the example shown in Figure 14.5, can stabilize the propagation delays of long fiber optic links. A notable example of a DLL system was demonstrated in 2012 by researchers at the AGH University of Science and Technology in Kraków, Poland. This system was initially tested using a single-mode fiber of 480 km in length that formed a closed loop, with both ends of the fiber accessible in the same laboratory. Clock signals of 10 MHz and 1 pps were converted to optical wavelengths near 1,550 nm and transmitted through the fiber, and custom-built SFBAs compensated for the fiber attenuation. Measurements of the link's stability at τ = 1 day produced an ADEV of about 4×10^{-17} and a TDEV of less than 1 ps. When the tests were completed, the system was installed as a continuous time transfer link between the two UTC(k) time scales located in Poland, UTC(AOS) in Borowiec and UTC(PL) in Warsaw. The fiber optic link had less measurement noise than an existing GPS PPP link between the two time scales, and the time differences measured via fiber were within the calibration uncertainty of the GPS measurements [27].

Fiber optic time transfer based on the DLL method was demonstrated over longer distances by researchers at NTSC in China in 2022. Their initial laboratory tests used a fiber of 1,100 km in length, where the delays were compensated using a temperature-controlled system that was based on a VCO and a PID controller. After the laboratory testing was completed, a link of 988 km in length was installed between two cities in the Shanxi province, as part of a development effort to distribute UTC(NTSC) throughout China via a network spanning 20,000 km of optical fiber. Measurements showed that the laboratory link was roughly twice as stable as the "real" link between the two cities, because the TDEV increased from ~2.5 ps to ~5 ps at τ = 10,000 s, but the results were still excellent in both cases [30].

The first international comparisons of primary frequency standards via optical fiber were made in 2015, when a dedicated link with a length of 1,415 km was used to compare four fountain clocks for a period of 21 days. The clocks were located at PTB and at the Time-Space Reference Systems laboratory (*Sytèmes de Référence Temps Espace*, SYRTE) at the Paris Observatory, and were separated by a geographic distance of about 700 km. The clocks included three cesium fountains, two at PTB and one at SYRTE, as well as a dual fountain at SYRTE that operated simultaneously with both cesium and rubidium atoms.

The comparison was performed by using the fountain clocks at both laboratories to measure the frequency of hydrogen masers, and then using optical frequency combs to link the maser frequencies to lasers that operate at wavelengths of 1,542 nm. The laser signals from both sites were injected into stabilized fiber links that met at a connection point in Strasbourg, France, where their frequency difference was measured with a counter. The measured frequency differences between the cesium fountains were all within a range of $\pm 3 \times 10^{-16}$, which agreed with their estimated combined standard uncertainties. The measured frequency difference of the SYRTE rubidium fountain with respect to the cesium fountains resulted in the best estimate of the ^{87}Rb resonance frequency (Section 3.9) that had yet been made [31].

14.2.3 Time Transfer via Shared Fiber in Public Networks

The PTP and WR protocols can be used to compare atomic clocks over long distances via public telecommunication networks, if the fiber optic link is periodically calibrated using an independent method, such as common-view GNSS. The fiber link is usually more stable than the GNSS link, but calibrations are needed to compensate for link asymmetries. One example of this is a measurement performed in 2014, when the UTC(MIKE) time scale located at the Center for Metrology in Espoo, Finland, was compared to a cesium beam clock located in Kajaani, Finland. The GPS PPP method was used to initially calibrate the fiber link, which was 950 km in length, so the asymmetry could be compensated. After this was done, the two clocks were compared for a period of three months using both GPS PPP and WR, and the two time transfer methods agreed to within ±2 ns [13].

Shared fibers in public networks that carry other traffic are inherently noisier than dedicated dark fibers, but they can still be more stable than satellite-based time transfer links. This was demonstrated by measurements performed in France in 2012, when optical frequency and time signals were simultaneously transferred via a public telecommunication network that also carried Internet traffic. In this experiment, both ends of a fiber loop with a length of 540 km were accessible at the Laser Physics Laboratory at the University of Paris. An ultrastable laser, phase locked to a local clock and operating at a wavelength near 1,550 nm, transmitted optical signals through one end of the fiber. When received at the far end of the fiber, the light signal was regenerated and returned by a narrow-line-width laser that was phase locked to the incoming light, but offset in frequency by 76 MHz. This intentional offset allowed the frequency stability of the link to be characterized by measuring the beat frequency between the outgoing and incoming optical signals. The ADEV for the link was 5×10^{-15} at $\tau = 1$ s, dominated by white phase noise that averaged down to about 3×10^{-19} at $\tau = 30{,}000$ s [32]. Time was transferred via the same link by using a pair of TWSTFT spread spectrum modems to generate two modulated RF carriers, each kept in phase with 1 pps and 5 MHz signals from a common clock, and each converted to optical time signals that were sent in opposite directions through the fiber. The TDEV of the link was less than 20 ps for all averaging periods longer than 100 s [33].

Another example of a time transfer over a public fiber optic communication network was demonstrated in 2018 by researchers in Sweden, who developed a delay-compensated system for the distribution of UTC(SP). The stability of the communication network was measured using a fiber loop with a length of 817 km that began and ended in Borås, Sweden, the city where UTC(SP) is located. The loop was first tested by transmitting optical time signals that weren't compensated for path delay, and then measuring the delay variations with a TIC. This measurement showed that the path delay varied by about 550 ns peak-to-peak over a 6-day period. When delay compensation was added to the loop, the peak-to-peak variation was reduced to 0.75 ns for a 24-day period. This improved the link stability by more than three orders of magnitude, reducing TDEV at $\tau = 1$ day from about 30 ns to less than 30 ps [34].

14.3 Free-Space Optical Time Transfer

In order for a time transfer link to be useful, it has to have a lower noise floor—after a reasonably short measurement period—than the clocks being compared. Fiber optic links have lower noise floors than links based on RF transmissions from satellites, and as we just discussed in Section 14.2.2, their noise floors are low enough to easily measure the frequency differences between cesium fountain primary standards [31]. The noise floors of fiber optic links are also low enough, after a reasonable period of averaging, to measure the frequency differences between optical clocks.

Fiber optic links have, however, two obvious limitations. Unlike GNSS and the other RF links, they are completely inaccessible from most places on Earth. They also can't be used to measure clocks that are airborne or on moving platforms. These drawbacks have led researchers to investigate what could be a better solution—the transmission of optical time signals through free space. Free-space optical time transfer potentially offers the best of both worlds, because it combines the accessibility of an RF system with the low noise floor of a fiber optic system.

The free-space optical time transfer methods that have been investigated so far can be divided into three categories: satellite laser ranging (SLR) methods, continuous wave (CW) optical carrier methods, and optical frequency comb (OFC) methods [35]. We'll briefly discuss the first two methods before focusing on frequency comb-based time transfer, which is the free-space method that has the lowest potential measurement uncertainty over long distances.

Satellite laser ranging works like the other ranging methods that measure the distance between ground stations and orbiting satellites (Section 12.1), except that SLR involves the transmission of ultra-short laser pulses instead of RF signals. The laser pulses are sent from ground stations to satellites equipped with retroreflectors, which are devices that reflect the light directly back towards its source with a minimum amount of scattering. The TOA of the reflected light is used to calculate the distance to the satellite, and that information is used to determine the satellite's orbit.

The SLR method was applied to free-space optical time transfer from June 2008 until the end of 2017, which was the period of the Time Transfer by Laser Link (T2L2) experiment conducted by the French National Center for Space Studies (*Centre national d'études spatiales*, CNES). The T2L2 experiment used a space instrument built by CNES that flew onboard a LEO satellite named Jason-2, at an altitude of 1,336 km. The satellite also carried an ultrastable quartz oscillator, and the laser ranging was performed by a network of more than 40 ground stations that were each referenced to an atomic clock that provided a reference frequency and a 1 pps time signal. A simplified diagram of T2L2 ranging measurements made from a single ground station is shown in Figure 14.9.

The ground stations typically used lasers with a 532 nm wavelength, and used a telescope to transmit and receive a train of laser pulses. The pulse width was about 100 ps and the pulse repetition rate was between 10 Hz and 2 kHz. Time stamps were recorded when the pulses left the ranging station, when they were retroreflected back to Earth by the Jason-2 satellite, and when they arrived back at the ranging station. The three time stamps made it possible to calculate the time difference between the ground station clock and the satellite clock. From 2008 to 2017, the T2L2 instrument carried by Jason-2 recorded about 800 million time stamps. An analysis of a portion of this data showed that when two ground clocks were compared using T2L2 common-view measurements, the expanded measurement uncertainty (k = 2) was about 140 ps. This was more than one order of magnitude smaller than the lowest uncertainties reported for the UTC(k) links (Section 9.4.4) based on GNSS common-view methods [36, 37]. These low uncertainties were still too large to accurately compare optical clocks, but they will likely be reduced by future SLR-based systems.

Smaller uncertainties have been demonstrated by using CW lasers to transmit an optical carrier frequency through free space. This method is similar to a phase-stabilized fiber optic link, because the phase fluctuations in the link are measured and cancelled using a feedback control loop. In this case, however, the phase fluctuations aren't caused by the acoustic and thermal fluctuations seen in optical fibers, but are instead caused by the turbulence in the air. The turbulence results from changes in the weather, which causes pockets of the air to have different temperature and humidity levels. This changes the refractive index, and it causes the intensity of the light passing through the air to randomly fluctuate, due to the effects of refraction and scattering. This same general effect causes the light from the stars to appear to flicker or twinkle.

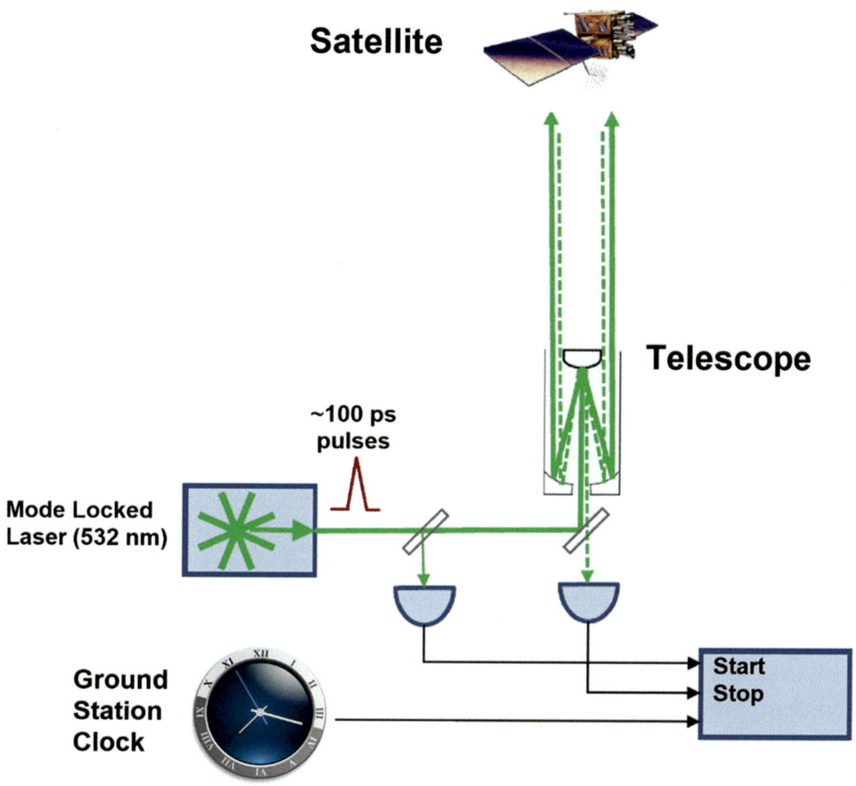

Figure 14.9 A simplified diagram of the T2L2 satellite ranging method.

In 2009, researchers at SYRTE demonstrated that a CW optical signal could be used for free-space frequency transfer. Their multi-day experiment used a CW laser with a 1,064 nm wavelength, and a telescope that transmitted 275 mW of optical power across a 2.5 km distance. The optical signal was then reflected back to its source, traveling a round-trip distance of 5 km, where it was received with a photodiode at a reduced power level of about 20 mW. The received signal was used to phase lock a VCO. The link wasn't compensated for delay fluctuations, but the stability at τ = 100 s still reached 2 × 10^{-15}. This was encouraging, because it was similar to the stability of the uncompensated 100 km optical fiber links that were then being measured [38].

The CW technique was later refined to use phase-stabilized optical links that reduced the uncertainty. The effectiveness of phase stabilization was demonstrated in 2020, when frequency was transferred across a short free-space link of 265 m between two buildings on the CNES campus in Toulouse, France. The link reached a stability of 3 × 10^{-19} at τ = 40 s, as estimated with Mod ADEV, and its stability was verified by using an underground optical fiber that ran between the buildings. This stability is low enough to enable short-term comparisons of optical clocks, because it's at least one order of magnitude lower than optical clock stability over the same averaging period [39]. Across short distances, a free-space CW optical link can transfer frequency with uncertainties comparable to those obtained via fiber optic links, but they are less reliable due to signal fading caused by atmospheric turbulence. Signal fading makes free-space links highly sensitive to interruptions, which usually limits the distance between the clocks being compared to just a few kilometers. It also limits the duration of a measurement, which makes it difficult to measure a link's long-term stability [35].

The OFC method is more complex and requires more equipment than the CW method, but it has many advantages. It works over longer distances, it's more reliable, and it can transfer both time and frequency. The CW method isn't well suited for time transfer, because the optical carrier phase would have to be tracked unambiguously. This is very difficult to do because the CW method can lose phase information due to cycle slips during interruptions—caused by the turbulence of air or any other reason—that are longer than the period of the laser frequency. This makes it extremely sensitive to link interruptions, because of the period of the laser frequency is about 5 fs for a laser operating at a wavelength near 1,550 nm. The OFC method is far less sensitive to signal interruptions, because it can resume a measurement without losing phase information if the period of the interruption is less than the period of the OFC's repetition frequency. This means that for a typical repetition frequency of 200 MHz, the OFC method won't be affected by link interruptions of less than 5 ns.

The use of OFCs (Section 13.3) in free-space time transfer systems was first demonstrated in 2012 by a team of NIST researchers led by the American physicist Nathan Newbury (b. 1964). In the original experiments, optical transceivers were located at two sites on the NIST campus in Boulder, Colorado. Each site had its own OFC and laser, but both were referenced to the same optical clock via a fiber connection. The OFCs generated pulse trains at repetition rates near 100 MHz, differing by about 3 kHz at the two sites, and the two phase-locked lasers transmitted optical signals through free space at a power level of ~4 mW, and at respective frequencies of 192.1 THz and 195.3 THz. The pulses were exchanged between the two sites by reflecting optical signals off a mirror that was located on a mesa that was higher in elevation than the NIST laboratories. The round-trip distance to the mirror was about 2 km. Measurements made over a 24-hour period showed that the link was exceptionally stable—after averaging for 1,000 s, ADEV was less than 1×10^{-18} and TDEV was less than 1 fs [40].

This experiment used a common clock to demonstrate the concept and to measure the link stability, but like all time transfer methods, the free-space OFC method can be used to either compare or synchronize a remote clock to a reference clock. The method works by locking a self-referenced OFC at each site to the probing laser of the optical clock that's being measured. This stabilizes the repetition frequency of the comb's pulses, and it also stabilizes the underlying carrier frequency of the comb's output (Section 13.3). The lasers locked to the comb's output are typically chosen to be compatible with standard telecommunication components, which means that they usually operate in the C band, with frequencies near 193.5 THz and wavelengths near 1,550 nm [35].

The time transfer is performed using the two-way method, and it's conceptually similar to the method used with satellites (Section 4.3.4). The comb lasers at both sites send optical pulses bidirectionally through the air, and their arrival times are measured with respect to the other site's clock. The two-way propagation cancels the delays that are common to both signal paths, so the result of the measurement is the time difference between the two clocks. As is always the case, the frequency difference between the clocks can be obtained from the rate of change of the time difference.

Even though the concepts are similar to two-way satellite time transfer, the details are different. The two main differences are that the transmitted signals have optical frequencies instead of radio frequencies, and that the signals are relayed to the other clock site using mirrors instead of satellite transponders. Some other differences are illustrated in Figure 14.10, which is a simplified diagram of a free-space system developed at NIST that can be used to synchronize a clock at site *B* to a reference clock at site *A*. This system combines the light from the clock comb and the modulated light from a distributed feedback (DFB) laser in the same single-mode fiber, and uses an optical transceiver to transmit the signal into free space towards a reflecting mirror. The DFB is phase modulated with a pseudorandom binary sequence that contains time stamps and other information that's needed to complete the synchronization.

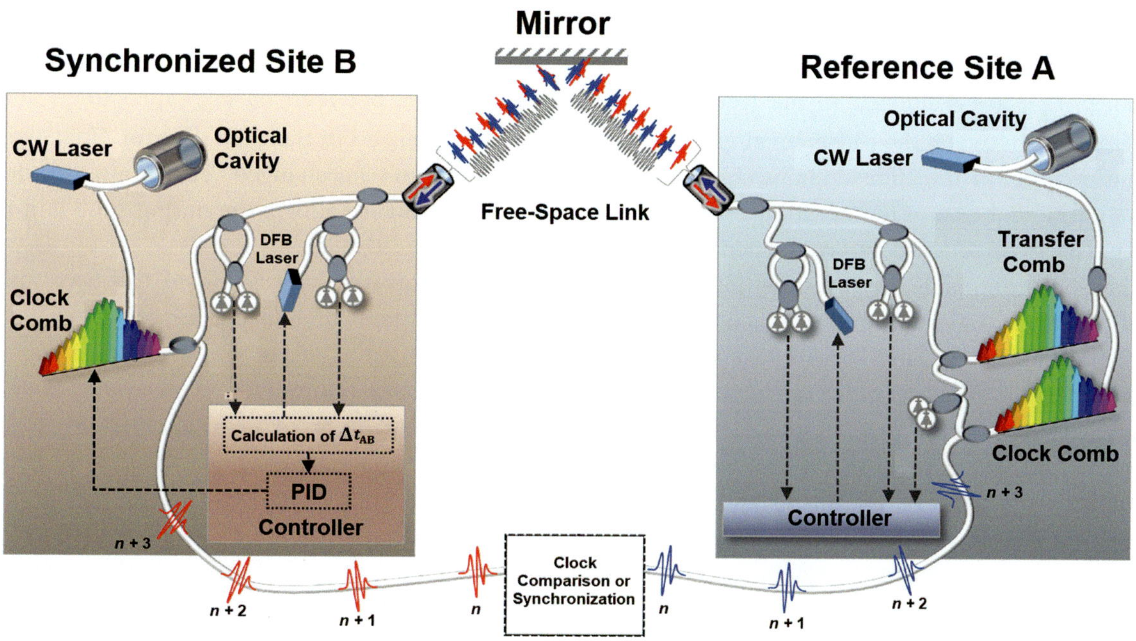

Figure 14.10 An optical two-way time transfer system that synchronizes two clocks via a free-space link (adapted from [41]).

In the simplest form of two-way time transfer, timing pulses can be directly exchanged, but that doesn't work well with optical signals, because the photodetection of the incoming pulses introduces uncertainties of picoseconds, which are far larger than the femtosecond-level jitter on the optical clock pulses being compared. This problem is solved by including a third OFC, a "transfer" comb that's located at site *A*. The time differences between the transfer comb and the clock combs at both sites are measured using the heterodyne method, and these values are applied as corrections to the standard two-way measurement to obtain Δt_{AB}, which the time difference between the two clocks. This value is then sent to a PID controller that synchronizes the site *B* clock to the site *A* reference and compensates for all future time differences. Measurements conducted at NIST in 2015 proved that this system was exceptionally stable. When time was transferred across a 4 km path, the TDEV was less than 1 fs for all averaging periods up to 6,500 s [41].

The light attenuation caused by the turbulence of air limited the distance of the early free-space experiments using the OFC method to about 20 km. Most of the experiments conducted during the first decade of research, including the system shown in Figure 14.10, transmitted optical signals at power levels of 5 mW or less. It was known, of course, that increasing the power would also increase the distance, but power increases weren't easy to accomplish. The optical amplifiers had to be carefully designed and placed to prevent excessive widening and distortion of the comb pulses, and the heat generated by the higher power levels had to be dissipated to avoid damaging the equipment. A collaborative research effort, led by scientists at the University of Science and Technology of China, overcame some of these difficulties by designing a system that allowed the OFCs to output a full watt of optical power. This system produced chirped pulses with widths of 60 ps to 90 ps to reduce the power density, and used water-cooled EFDAs to dissipate the excess heat. The high-power OFCs extended the range of their free-space time transfer experiments to 113 km, which was the distance between two laboratories in the Xinjiang Province of China. A series of two-way measurements were made via this link in 2021 and 2022, using lasers with respective wavelengths of 1,545 nm and 1,563 nm. The results consistently showed that the link stability, as estimated with Mod ADEV, averaged down to $\sim 4 \times 10^{-19}$ at $\tau = 10{,}000$ s [42].

The time-programmable frequency comb (TPFC), which was first described by NIST researchers in 2022, has since reduced the amount of optical power needed for free-space time transfer by a factor of about 10,000 [43]. Conventional frequency combs are inherently accurate, but a TPFC is both agile and accurate, because it allows the output time of a comb pulse to be adjusted by issuing digital commands. As was described in Section 13.3, a comb's repetition frequency, f_r, and its carrier-envelope offset frequency, f_{CEO}, are stabilized in the frequency domain. The period between pulses, equal to $\frac{1}{f_r}$, and the relative phase between the pulse envelope and the carrier, are stabilized in the time domain. However, in a conventional comb the time domain components are rigid and can't be changed. This limitation doesn't apply to a TPFC, because the addition of digital electronics and firmware allows the output time and phase of the pulses to be selected and controlled with an accuracy of ±2 attoseconds.

Time-programmable frequency combs reduce the necessary optical power due to the way that the time differences between combs are measured. The heterodyne method is the most accurate way to measure comb time differences, and it works by mixing the light from two combs, each with a different repetition frequency and a different pulse spacing. When the mixed light hits a photodiode, it generates a peak in the interference signal when the pulses overlap in time (left side of Figure 14.11). This method is photon inefficient and has a poor signal-to-noise ratio, because there is a large amount of dead time between the signal peaks. A TPFC eliminates the dead time, because it can shift its pulses so that they track the pulses from the other comb by using the signal amplitude as feedback (right side of Figure 14.11). This makes much more efficient use of the received light. The tracking style detection method operates near the quantum limit—also called the photon-shot-noise limit—where the measurement's signal-to-noise ratio is determined by the number of received photons. The quantum limit represents the best measurement possible for a given amount of optical power [43, 44].

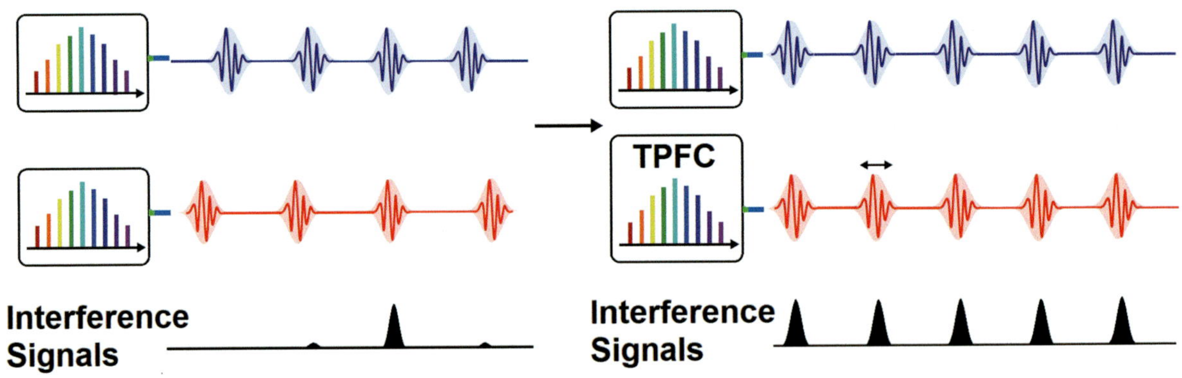

Figure 14.11 The signal-to-noise difference in heterodyne measurements between conventional frequency combs (left) and time-programmable frequency combs (right), adapted from [44].

The use of TPFCs allowed NIST researchers to conduct optical time transfer experiments over a distance of 300 km in 2022. The free-space path was between two mountaintops in the state of Hawaii, each with an elevation of over 3,000 m, that are separated by a distance of about 150 km. The measurements compared optical clocks at two sites (*A* and *B*) that were both located in the Mauna Loa Observatory on Hawaii's largest island. The clock signals transmitted from each site were returned to Mauna Loa by a retroflecting mirror on the Haleakala Summit on the island of Maui, and then received at the other clock site (Figure 14.12).

Each clock site had two TPFCs; a clock TPFC that generated the outgoing signal, and a tracking TPFC that acquired, tracked, and measured the incoming signal. When this system was operated with 4 mW of transmitted optical power, the median received power was 14 picowatts, and the link stability, as estimated with Mod ADEV, reached a noise floor of ~3 × 10^{-19} after about 2,000 s of averaging. Due to the advantages of the TPFC's tracking detection at near the quantum limit, it was found that the link was still usable and stable when the transmitted power was reduced to 40 µW, and the median received power was just 150 femtowatts. A signal loss of 100 dB reduces the power of a 4 mW signal to 400 femtowatts, so the experiments proved that the link could tolerate more than 100 dB of attenuation when just 4 mW of optical power was transmitted [45].

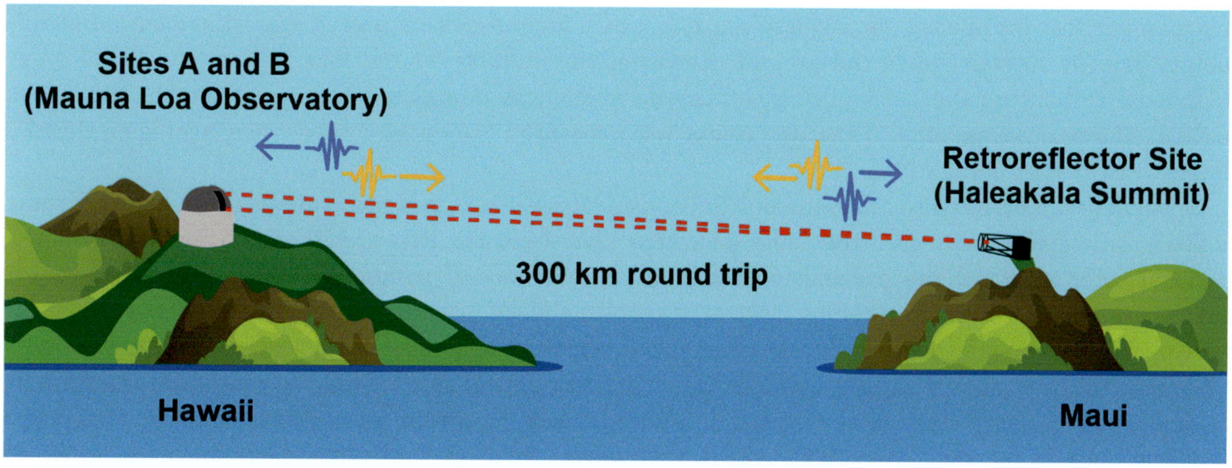

Figure 14.12 Two-way free-space optical time transfer via a 300 km terrestrial link in Hawaii (courtesy of NIST).

The ability to tolerate more than 100 dB of attenuation suggests that it's entirely feasible to apply the OFC free-space method to both one-way and common-view time transfer from satellites, including satellites in geostationary orbits. The distance from Earth to a GEO satellite is 35,786 km (Section 10.6), which is more than 100× further than the 300 km link used in Hawaii. However, the turbulence integrated along the satellite path is roughly 100× smaller than that of the Hawaii link, and the estimated ground-to-GEO attenuation is just 91 dB on average. Based on these estimates, a GEO satellite transmitting 4 mW of optical power could easily reach receivers on Earth, with a margin of safety of about 10 dB [45, 46].

14.4 Comparing Optical Clocks with Optical Time Transfer Systems

Fiber optic time transfer has so far been the primary method used for optical clock comparisons. Because optical clocks typically run for only short periods, a dedicated dark fiber link is generally needed to reduce the measurement noise to an acceptable level after only a short period of averaging. Therefore, the distance between the clocks being compared is usually limited by the availability and cost of dedicated fiber links.

One of the first examples of a long-distance optical clock comparison was performed in 2009, when two optical clocks at PTB—one based on calcium atoms and the other on ytterbium ions—were compared using a phase-stabilized dark fiber link. Both clocks were located on the PTB campus, but an optical signal from the calcium clock was transmitted to the University of Hanover and back to PTB via optical fiber, a round-trip distance of 146 km. An OFC was referenced to the ~456 THz output of the calcium atom clock, and a CW laser with a 1,542 nm wavelength was phase locked to the comb. The optical carrier sent through the fiber was compared at the other end to an OFC referenced to the ytterbium ion clock. The link was stable enough to easily measure the frequency difference between the two clocks. The link stability, as estimated with ADEV, was 3×10^{-15} at $\tau = 1$ s, and it averaged down as white phase noise to less than 1×10^{-19} at $\tau = 30{,}000$ s [47].

An optical clock comparison using a similar method was performed in 2011 by researchers at NICT in Japan. However, in this case, the comparison was between two strontium (^{87}Sr) optical lattice clocks that didn't reside in the same laboratory, but instead were geographically separated by 24 km, and linked by a phase-stabilized fiber of 60 km in length. One clock was located at NICT, and it transmitted its frequency through the fiber link to the University of Tokyo, where the other clock was located, and where the frequency difference between the clocks was measured. The fiber was routed through an urban environment that introduced some phase noise, because for most of its path it was either buried along a subway line or located above the ground. However, the stability of the link was still well below the short-term stability of the clocks being compared, which was dominated by noise from the clock laser at NICT. The relative stability of the two clocks was measured as 5×10^{-15} at $\tau = 1$ s, averaging down to about 4×10^{-16} at $\tau = 1{,}000$ s [48].

A comparison between two strontium (^{87}Sr) optical lattice clocks was made over a much longer distance in 2015. This comparison used the same dedicated fiber between PTB and SYRTE, 1,415 km in length, that was later used for the previously described comparisons of cesium fountains [31]. At both PTB and SYRTE, the light from the probing laser of a strontium clock was sent to an OFC that phase coherently bridged the frequency gap between the ~429 THz clock resonance frequency, and the ~194.4 THz laser frequency that was transmitted through the fiber. The two optical clock signals met at a connection point in Strasbourg, France, where their frequency difference was measured for multiple days. The combined standard uncertainty was 5×10^{-17}, which was lower than any previous uncertainty recorded for a long-distance clock comparison. The insignificant contribution of the fiber link to the combined uncertainty was estimated as 2.5×10^{-19} [49].

As part of the collaboration between NIST and JILA, three optical clocks located in Boulder, Colorado—aluminum ion ($^{27}Al^+$) and ytterbium lattice ($^{27}Al^+$) clocks at NIST and a strontium lattice (^{87}Sr) clock at JILA—have been compared via the Boulder Atomic Clock Optical Network (BACON). This network was designed to use both free-space and fiber optic time transfer. Because the optical clocks are located in the same city, the respective lengths of the free-space and fiber optic links are just 1.5 km and 3.6 km (Figure 14.13). The frequency ratio of the NIST ytterbium clock to the JILA strontium clock was simultaneously measured via both links for a period of about 30 hours, spread over six days, in 2018. The free-space and fiber optic measurements agreed to within an uncertainty of 6×10^{-19} [50].

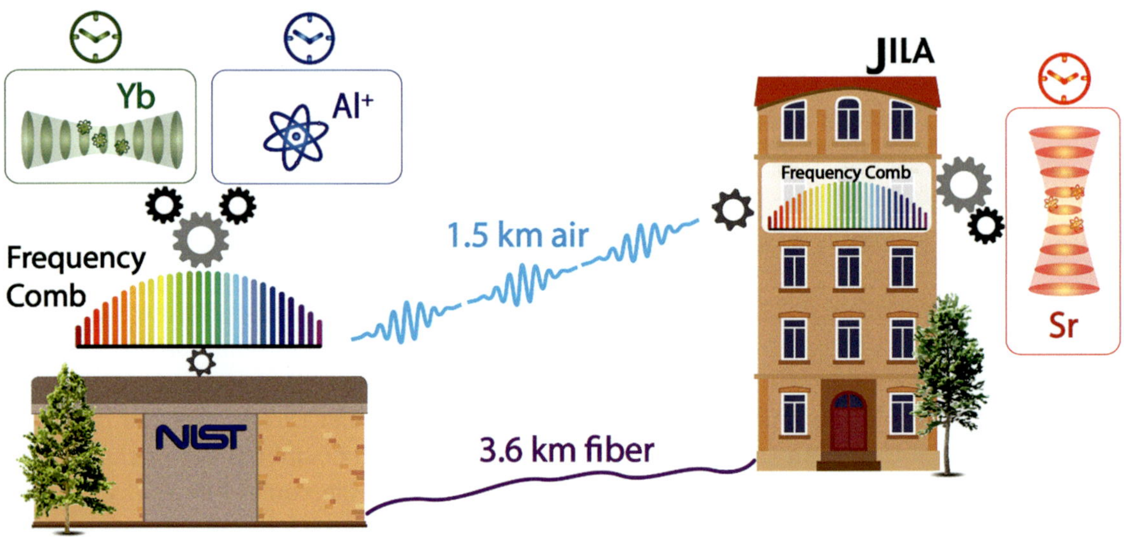

Figure 14.13 Optical clock comparisons between NIST and JILA in Boulder, Colorado via fiber optic and free-space optical time transfer (courtesy of Menlo Systems, adapted from [50]).

Optical time transfer systems should eventually have a huge impact on international timekeeping, because reducing the uncertainty of clock comparisons improves both the accuracy and stability of UTC. As of 2025, the existing global networks of microwave atomic clocks are mostly linked together by GNSS signals that can be received anywhere on Earth. It seems likely that the future global networks of optical clocks will be mostly linked together by some form of free-space optical time transfer that can provide the same accessibility, perhaps by a future GNSS that's based on optical signals.

References

1. L. Śliwczyński, P. Krehlik, and M. Lipiński, "Optical Fibers in Time and Frequency Transfer," *Measurement Science and Technology*, vol. 21, no. 7, pp. 075302/1–11, May 2010.
2. Z. Jiang, A. Czubla, J. Nawrocki, W. Lewandowski, and E. F. Arias, "Comparing a GPS Time Link Calibration with an Optical Fibre Self-Calibration with 200 ps Accuracy," *Metrologia*, vol. 52, no. 2, pp. 384–391, March 2015.
3. K. C. Kao and G. A. Hockham, "Dielectric-Fibre Surface Waveguides for Optical Frequencies," *Proceedings of the IEEE*, vol. 113, no. 7, pp. 1151–1158, July 1966.
4. J. Hecht, *City of Light: The Story of Fiber Optics*, Oxford University Press, 1999.
5. L. F. Mollenauer and K. Smith, "Demonstration of Soliton Transmission Over More than 4000 km in Fiber with Loss Periodically Compensated by Raman Gain," *Optics Letters*, vol. 13, no. 8, pp. 675–677, August 1988.
6. A. Sullivan, P. Tandon, R. McCool, A. Diaz, and C. Herrmann, "A Sustainable Future with Optical Fiber," *Corning White Paper 1000*, 7 p., March 2023.
7. J. Laferriere, G. Lietaert, R. Taws, and S. Wolszczak, *Reference Guide to Fiber Optic Testing – Volume 1*, 2nd ed., JDS Uniphase Corporation, 2011.
8. International Telecommunications Union, "Characteristics of a Single-Mode Optical Fibre and Cable," *Recommendation ITU-T G.652*, August 2024.
9. G. F. Lutes, "Optical Fibers for the Distribution of Frequency and Timing References," *Proceedings of the 12th Annual Precise Time and Time Interval Meeting (PTTI)*, pp. 663–680, December 1980.
10. L. Sliwczynski and J. Kolodziej, "Bidirectional Optical Amplification in Long-Distance Two-Way Fiber-Optic Time and Frequency Transfer Systems," *IEEE Transactions on Instrumentation and Measurement*, vol. 62, no. 1, pp. 253–262, January 2013.
11. P. Banerjee and D. Matsakis, "Optical Time and Frequency Transfer," Chapter 12 in *An Introduction to Modern Timekeeping and Time Transfer*, Springer, 2023.
12. J. Savory, J. Sherman, and S. Romisch, "White Rabbit-Based Time Distribution at NIST," *Proceedings of the 2018 IEEE International Frequency Control Symposium*, 5 p., May 2018.
13. E. F. Dierikx, A. E. Wallin, T. Fordell, J. Myyry, P. Koponen, M. Merimaa, T. J. Pinkert, J. C. J. Koelemeij, H. Z. Peek, and R. Smets, "White Rabbit Precision Time Protocol on Long-Distance Fiber Links," *IEEE Transactions on Ultrasonics, Ferroelectrics, and Frequency Control*, vol. 63, no. 7, pp. 945–952, July 2016.
14. P. O. Hedekvist and S.-C. Ebenhag, "Time and Frequency Transfer in Optical Fibers," Chapter 17 in *Recent Progress in Optical Fiber Research*, pp. 371–386, M. Yasin, S. W. Harun, and H. Arof, eds., InTech, 2011.
15. W.-H. Tseng and S.-Y. Lin, "A Survey of Time Transfer via a Bidirectional Fiber Link for Precise Calibration Services," *NCSLI Measure: The Journal of Measurement Science*, vol. 8, no. 2, pp. 70–77, June 2013.
16. J. Kodet, P. Pánek, and I. Prochazka, "Two-Way Time Transfer via Optical Fiber Providing Subpicosecond Precision and High Temperature Stability," *Metrologia*, vol. 53, no. 1, pp. 18–26, December 2015.
17. L. Sliwczynski, P. Krehlik, L. Buczek, and M. Lipinski, "Active Propagation Delay Stabilization for Fiber-Optic Frequency Distribution Using Controlled Electronic Delay Lines," *IEEE Transactions on Instrumentation and Measurement*, vol. 60, no. 4, pp. 1480–1488, April 2011.
18. S.-C. Ebenhag, P. O. Hedekvist, and K. Jaldehag, "Single-Way Fiber-Based Time Transfer with Active Detection of Time Transfer Variations," *Proceedings of the 42nd Annual Precise Time and Time Interval Meeting (PTTI)*, pp. 413–425, November 2010.
19. S.-C. Ebenhag, P. O. Hedekvist, and K. Jaldehag, "Two-Color One-Way Frequency Transfer in a Metropolitan Optical Fiber Data Network," *NCSLI Measure: The Journal of Measurement Science*, vol. 8, no. 2, pp. 52–61, June 2013.

20. H. Zhang, G. Wu, X. Li, and J. Chen, "Uncertainty Analysis of BTDM-SFSW Based Fiber-Optic Time Transfer," *Metrologia*, vol. 54, no. 1, pp. 94–101, January 2017.

21. S.-C. Ebenhag, P. O. Hedekvist, P. Jarlemark, R. Emardson, K. Jaldehag, C. Rieck, and P. Löthberg, "Measurements and Error Sources in Time Transfer Using Asynchronous Fiber Network," *IEEE Transactions on Instrumentation and Measurement*, vol. 59, no. 7., pp. 1918–1924, July 2010.

22. L. Sliwczynski, P. Krehlik, J. Kolodziej, H. Schnatz, D. Piester, A. Bauch, H. Imlau, and H. Ender, "Calibrated Optical Time Transfer of UTC(*k*) for Supervision of Telecom Networks," *Metrologia*, vol. 56, no. 1, pp. 015006/1–13, November 2018.

23. X. Bao and Y. Wang, "Recent Advancements in Rayleigh Scattering-Based Distributed Fiber Sensors," *Advanced Devices and Instrumentation*, vol. 2021, article no. 8696571, 17 p., March 2021.

24. M. Merklein, I. V. Kabakova, A. Zarifi, and B. J. Eggleton, "100 Years of Brillouin Scattering: Historical and Future Perspectives," *Applied Physics Reviews*, vol. 9, no. 4, pp. 041306/1–27, December 2022.

25. Y. Muanenda, C. J. Oton, and F. Di Pasquale, "Application of Raman and Brillouin Scattering Phenomena in Distributed Optical Fiber Sensing," *Frontiers in Physics*, vol. 7, article 155, 14 p., October 2019.

26. M. Amemiya, M. Imae, Y. Fujii, T. Suzuyama, F.-L. Hong, and M. Takamoto, "Precise Frequency Comparison System Using Bidirectional Optical Amplifiers," *IEEE Transactions on Instrumentation and Measurement*, vol. 59, no. 3., pp. 631–640, March 2010.

27. L. Sliwczynski, P. Krehlik, A. Czubla, and L. Buczek, "Dissemination of Time and RF Frequency via a Stabilized Fiber Optic Link Over a Distance of 420 km," *Metrologia*, vol. 50, no. 2, pp. 133–145, February 2013.

28. T. Fordell, "Open-Loop Polarization Mode Dispersion Mitigation for Fibre-Optic Time and Frequency Transfer," *Optics Express*, vol. 30, no. 4, pp. 6311–6319, February 2022.

29. M. Rost, D. Piester, W. Yang, T. Feldmann, T. Wübbena, and A. Bauch, "Time Transfer through Optical Fibres Over a Distance of 73 km with an Uncertainty below 100 ps," *Metrologia*, vol. 49, no. 6, pp. 772–778, October 2012.

30. B. Liu, X. Guo, W. Kong, T. Liu, R. Dong, and S. Zhang, "Stabilized Time Transfer via a 1000 km Optical Fiber Link Using High-Precision Delay Compensation System," *Photonics*, vol. 9, no. 8, pp. 522/1–15, July 2022.

31. J. Guéna, S. Weyers, M. Abgrall, C. Grebing, V. Gerginov, P. Rosenbusch, S. Bize, B. Lipphardt, H. Denker, N. Quintin, S. M. F. Raupach, D. Nicolodi, F. Stefani, N. Chiodo, S. Koke, A. Kuhl, F. Wiotte, F. Meynadier, E. Camisard, C. Chardonnet, Y. Le Coq, M. Lours, G. Santarelli, A. Amy-Klein, R. Le Targat, O. Lopez, P. E. Pottie, and G. Grosche, "First International Comparison of Fountain Primary Frequency Standards via a Long Distance Optical Fiber Link," *Metrologia*, vol. 54, no. 3, pp. 348–354, May 2017.

32. O. Lopez, A. Haboucha, B. Chanteau, C. Chardonnet, A. Amy-Klein, and G. Santarelli, "Ultra-Stable Long Distance Optical Frequency Distribution Using the Internet Fiber Network," *Optics Express*, vol. 20, no. 21, pp. 23518–23526, October 2012.

33. O. Lopez, A. Kanj, P. E. Pottie, D. Rovera., J. Achkar, C. Chardonnet, A. Amy-Klein, and G. Santarelli, "Simultaneous Remote Transfer of Accurate Timing and Optical Frequency Over a Public Fiber Network," *Applied Physics B*, vol. 110, pp. 3–6, October 2012.

34. S.-C. Ebenhag, P. O. Hedekvist, C. Rieck, M. Bergroth, P. Krehlik, L. Sliwczynski, "Evaluation of Fiber Optic Time and Frequency Distribution System in a Coherent Communication Network," *Proceedings of the 2019 Joint Conference of the IEEE International Frequency Control Symposium and the European Frequency and Time Forum (IFCS/EFTF)*, 5 p., April 2019.

35. E. D. Caldwell, T. M. Triano, and L. C. Sinclair, "High-Precision Optical Time and Frequency Transfer," *Advances in Optics and Photonics*, vol. 17, no. 2, pp. 375–440, June 2025.

36. E. Samain, P. Exertier, C. Courde, P. Fridelance, P. Guillemot, M. Laas-Bourez, and J.-M. Torre, "Time Transfer by Laser Link: A Complete Analysis of the Uncertainty Budget," *Metrologia*, vol. 52, no. 2, pp. 423–432, April 2015.

37. P. Exertier, E. Samain, C. Courde, M. Aimar, J. M. Torre, G. D. Rovera, M. Abgrall, P. Uhrich, R. Sherwood, G. Herold, U. Schreiber, and P. Guillemot, "Sub-ns Time Transfer Consistency: A Direct Comparison between GPS CV and T2L2," *Metrologia*, vol. 53, no. 6, pp. 1395–1401, November 2016.

38. K. Djerroud, O. Acef, A. Clairon, P. Lemonde, C. N. Man, E. Samain, and P. Wolf, "Coherent Optical Link through the Turbulent Atmosphere," *Optics Letters*, vol. 35, no. 9, pp. 1479–1481, 2010.

39. B. P. Dix-Matthews, S. W. Schediwy, D. R. Gozzard, E. Savalle, F.-X. Esnault, T. Lévèque, C. Gravestock, D. D'Mello, S. Karpathakis, M. Tobar, and P. Wolf, "Point-to-Point Stabilized Optical Frequency Transfer with Active Optics," *Nature Communications*, vol. 12, no. 1, article no. 515, pp. 1–8, January 2021.

40. F. R. Giorgetta, W. C. Swann, L. C. Sinclair, E. Baumann, I. Coddington, and N. R. Newbury, "Optical Two-Way Time and Frequency Transfer Over Free Space," *Nature Photonics*, vol. 7, no. 6, pp. 434–438, June 2013.

41. J.-D. Deschênes, L. C. Sinclair, F. R. Giorgetta, W. C. Swann, E. Baumann, H. Bergeron, M. Cermak, I. Coddington, and N. R. Newbury, "Synchronization of Distant Optical Clocks at the Femtosecond Level," *Physical Review X*, vol. 6, no. 2, pp. 021016/1–15, May 2016.

42. Q. Shen, J.-Y. Guan, J.-G. Ren, T. Zeng, L. Hou, M. Li, Y. Cao, J.-J. Han, M.-Z. Lian, Y.-W. Chen, X.-X. Peng, S.-M. Wang, D.-Y. Zhu, X.-P. Shi, Z.-G. Wang, Y. Li, W.-Y. Liu, G.-S. Pan, Y. Wang, Z.-H. Li, J.-C. Wu, Y.-Y. Zhang, F.-X. Chen, C.-Y. Lu, S.-K. Liao, J. Yin, J.-J. Jia, C.-Z. Peng, H.-F. Jiang, Q. Zhang, and J.-W. Pan, "Free-Space Dissemination of Time and Frequency with 10^{-19} Instability Over 113 km," *Nature*, vol. 610, no. 7933, pp. 661–666, October 2022.

43. E. D. Caldwell, L. C. Sinclair, N. R. Newbury, and J.-D. Deschenes, "The Time-Programmable Frequency Comb and Its Use in Quantum-Limited Ranging," *Nature*, vol. 610, pp. 667–673, no. 7933, October 2022.

44. E. Caldwell and L. Sinclair, "Comparing Clocks by Using Pulses of Light," *Physics Today*, vol. 77, no. 6, pp. 54–55, June 2024.

45. E. D. Caldwell, J. D. Deschenes, J. Ellis, W. C. Swann, B. K. Stuhl, H. Bergeron, N. R. Newbury, and L. C. Sinclair, "Quantum-Limited Optical Time Transfer for Future Geosynchronous Links," *Nature*, vol. 618, no. 7966, pp. 721–726, June 2023.

46. E. D. Caldwell, L. C. Sinclair, J. D. Deschenes, F. Giorgetta, and N. R. Newbury, "Application of Quantum-Limited Optical Time Transfer to Space-Based Optical Clock Comparisons and Coherent Networks," *APL Photonics*, vol. 9, pp. 016112/1–19, January 2024.

47. O. Terra, G. Grosche, K. Predehl, R. Holzwarth, T. Legero, U. Sterr, B. Lipphardt, and H. Schnatz, "Phase-Coherent Comparison of Two Optical Frequency Standards Over 146 km Using a Telecommunication Fiber Link," *Applied Physics B*, vol. 97, pp. 541–551, July 2009.

48. M. Fujieda, M. Kumagai, S. Nagano, A. Yamaguchi, H. Hachisu, and T. Ido, "All-Optical Link for Direct Comparison of Distant Optical Clocks," *Optics Express*, vol. 19, no. 17, pp. 16498–16507, August 2011.

49. C. Lisdat, G. Grosche, N. Quintin, C. Shi, S. M. F. Raupach, C. Grebing, D. Nicolodi, F. Stefani, A. Al-Masoudi, S. Dorscher, S. Hafner, J.-L. Robyr, N. Chiodo, S. Bilicki, E. Bookjans, A. Koczwara, S. Koke, A. Kuhl, F. Wiotte, F. Meynadier, E. Camisard, M. Abgrall, M. Lours, T. Legero, H. Schnatz, U. Sterr, H. Denker, C. Chardonnet, Y. Le Coq, G. Santarelli, A. Amy-Klein, R. Le Targat, J. Lodewyck, O. Lopez, and P.-E. Pottie, "A Clock Network for Geodesy and Fundamental Science," *Nature Communications*, vol. 7, no. 1, article no. 12443, pp. 1–7, August 2016.

50. Boulder Atomic Clock Optical Network (BACON) Collaboration, "Frequency Ratio Measurements at 18-Digit Accuracy Using an Optical Clock Network," *Nature*, vol. 591, no. 7851, pp. 564–569, March 2021.

PART V

Applications

Chapter 15

Time and Navigation

Well, I made it.

— Charles Lindbergh, after completing the first solo trans-Atlantic flight in 1927.

Time and navigation have always been intertwined. Our knowledge of "where," the things that we want to know about our current location and next destination, has always depended on our knowledge of "when," which is based on the things that we already know about time.

We navigate whenever we move from one place to another. Most of the routes that we travel are repetitive and well established, so we don't always think about how we navigate. However, when we do think about navigation, our thoughts always include time, because of the triangular relationship between speed, time, and distance. As shown in Figure 15.1, speed is measured by dividing distance by time, time is measured by dividing distance by speed, and distance is measured by multiplying speed by time.

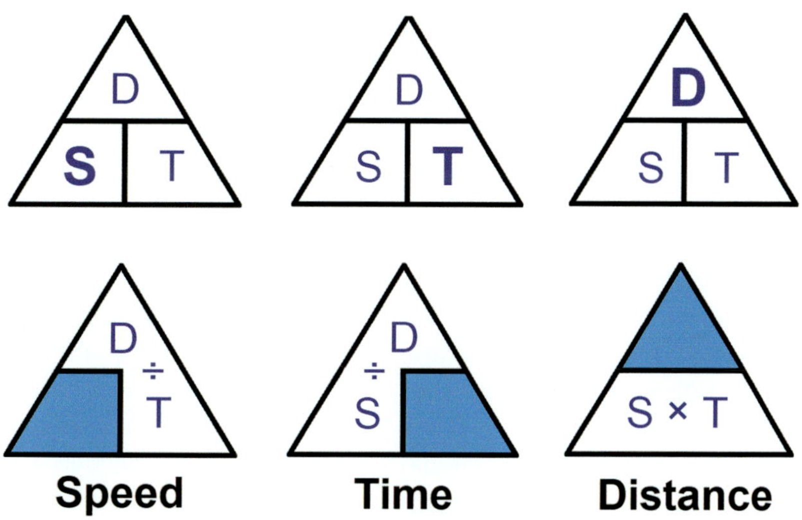

Figure 15.1 The triangular relationship between speed, time, and distance.

As Time Goes By: A Comprehensive Guide to Time and Frequency Measurements
Michael A. Lombardi

ISBN 978-981-5352-28-3 (Hardcover), 978-1-042-01532-0 (eBook)
www.jennystanford.com

DOI: 10.1201/9781042015320-21

The relationships between speed, time, and distance are fundamental to navigation. The relationships are so familiar that we calculate them subconsciously, which makes it easy to estimate how long it will take to reach our destination. Consider a simple example where we drive a car from Point A to Point B. If we know that Point B is 80 km away, and also know that we have to be there in an hour, we immediately know that we'll have to drive at an average speed of 80 km/hour to get there on time, because speed is distance divided by time. If we know that our average speed during the drive was 80 km/hour but didn't bother to measure our travel time, we can estimate that we drove for an hour, because time is distance divided by speed. Finally, if we don't know the distance to Point B, but do know that we made it there after driving for an hour at a speed of 80 km/hour, we can estimate the distance as 80 km, because distance is speed multiplied by time.

The tight connection between location and time—between "where" and "when"—means that accurate navigation requires accurate clocks [1]. It also means that the uncertainty of a navigation system always depends on the uncertainty of its time measurements. In the sections that follow, we'll describe the role that time measurements play in celestial, acoustic, terrestrial radio, and satellite radio navigation systems.

15.1 Celestial Navigation

Celestial navigation is the practice of finding a location on Earth by observing the positions of celestial bodies such as the Sun, the Moon, or the stars, and then using their positions as reference points. For at least several thousand years, mariners practiced celestial navigation by visually observing the positions of the Sun and the stars, without any extensive use of charts or instruments. They also relied on the knowledge gained from past experience, including the seasonal patterns that affected the wind and the waves, the behavior of birds and fishes, and the routes that had been successfully traveled in the past.

Around 1040 AD, Chinese mariners began using rudimentary compasses for direction finding at sea. At first, compasses consisted of a magnetized needle floating in a bowl of water, with the poles of the needle pointing towards the magnetic poles of the Earth. By the beginning of the period that historians call the Age of Exploration, which extended from about the mid-15th century to the early 17th century, compasses became more sophisticated and more suitable for shipboard use. Their needles were balanced on a fixed point so they could rotate freely, and they indicated direction by pointing to a fixed printed scale. The needle and its scale were enclosed in a case that protected them from the weather. The entire compass was often mounted on a pivoted support that adjusted its level as the ship rolled.

The Age of Exploration was also the period when a mapmaking revolution took place in Europe. Maps became more accurate, and the fanciful maps decorated with painted illustrations and literary references were replaced by maps based on actual measurements of the Earth [1]. Producing accurate maps required agreeing upon coordinate systems, a process that slowly evolved over a period of many centuries. During the third and second centuries BC, the Greek astronomers Eratosthenes and Hipparchus devised both north-south (latitude) and east-west (longitude) coordinate systems (Section 1.3). These coordinate systems were refined and put to use by Ptolemy, who completed a major work called the *Geographia* after he completed the *Almagest* in the mid-2nd century AD. The *Geographia* listed the latitude and longitude of almost 10,000 sites in Africa, Asia, and Europe, the three continents that were then known [2].

Despite the extraordinary work of the ancient astronomers, maps with lines of latitude and longitude weren't widely circulated until the *Geographia* was finally translated into Latin, an event that didn't occur until the early 15th century. In keeping with Ptolemy's terminology, the map lines that divide the Earth into north-south degrees of latitude are called parallels, because they run parallel to the equator. The map lines that divide the Earth into east-west degrees of longitude are called meridians, a word derived from the Latin *meridies*, which means midday or noon.

Compasses and accurate maps enabled a form of navigation called dead reckoning, which worked fairly well at sea, at least over short distances. The compass indicated the direction that a ship was heading, and by using a map to measure distance and a sandglass (Section 1.4) to measure time, a navigator could measure the ship's speed. The ship was steered by estimating the time spent headed in a given direction at a given speed [1]. Dead reckoning was the primary means of navigation for most mariners during the Age of Exploration, including the Italian explorer and navigator Christopher Columbus (1451–1506). However, it was usually supplemented by increasingly sophisticated forms of celestial navigation that used angle-measuring instruments to determine latitude and longitude.

Celestial navigators express the results of angle measurements in units of degrees, minutes, and seconds. Like the hours measured by a clock, the degrees are divided by 60 to obtain a resolution of minutes, and divided by 60 again to obtain a resolution of seconds (Section 1.3). A nautical mile was historically defined as the length of a meridian arc (Section 9.2) that corresponds to one minute of latitude, or one-sixtieth of a degree, which is the approximation used by celestial navigators. However, because the Earth isn't a perfect sphere, a minute of latitude has a length of about 1,843 m at the poles and about 1,862 m at the equator. To avoid confusion, the nautical mile was eventually defined as being exactly 1,852 m.

The quadrant was the most accurate angle-measuring instrument of the 15th century. It was based on a circular-faced astronomical instrument called the astrolabe that dated back to ancient times, but the quadrant condensed the astrolabe's functions into an area one-fourth the size. A quadrant consisted of a metal plate in the shape of a quarter-circle. A weighted string hung from the center of the quarter-circle, and the string crossed a curved scale. By using a quadrant, a navigator could measure a celestial body's elevation angle above the horizon with an accuracy of one or two degrees. The measurement was made by sighting a celestial body along one edge of the quadrant, and then marking the point where the string crossed the scale. A horary quadrant was marked with a scale that indicated the time of day when the elevation angle of the Sun was measured, essentially performing the same function as a sundial (Section 1.2). A drawing of a horary quadrant from an 18th-century textbook is shown in Figure 15.2, with the scale showing the hours of the day, along with elevation angles of up to 90° [3].

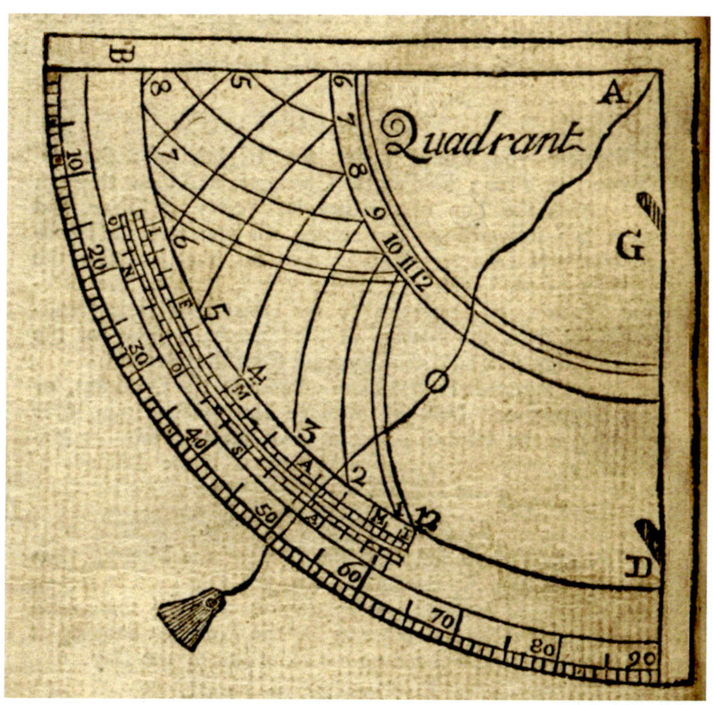

Figure 15.2 An 18th century drawing of a horary quadrant.

The quadrant was replaced by the octant and sextant in the 18th century [1]. As their names imply, an octant measures angles of up to 45°, or one-eighth of a circle, and a sextant measures angles of up to 60°, or one-sixth of a circle. Both instruments use two mirrors to reflect the path of light to the observer and double the angle, which respectively increases the largest angles they can measure to 90° and 120°. The invention of the reflecting quadrant, which evolved into the octant and sextant, is usually credited to the English mathematician John Hadley (1682–1744), who published a description of the instrument in 1731 [4].

Octants were produced in large numbers from the 1730s until about 1766, which was the year when the RGO in England published the first edition of *The Nautical Almanac*. The almanac included the information needed to implement the lunar distance method, which allowed celestial navigators to obtain the current time by measuring the angular distance between the Moon and another celestial body [5]. Determining the lunar distance sometimes required measuring angles of up to 120°. That was too large of an angle for an octant to measure, so octants were mostly replaced by sextants by the end of the 18th century.

Figure 15.3 provides a diagram (a) and a photograph (b) of a sextant. An arc scale is printed on the curved part of the sextant's frame. An index arm, a sighting telescope, a horizon mirror, and sun shades that protect the navigator's eyes are attached to the frame. An index mirror is attached to the index arm, along with a micrometer drum gauge that expands the arc scale and provides higher resolution measurements.

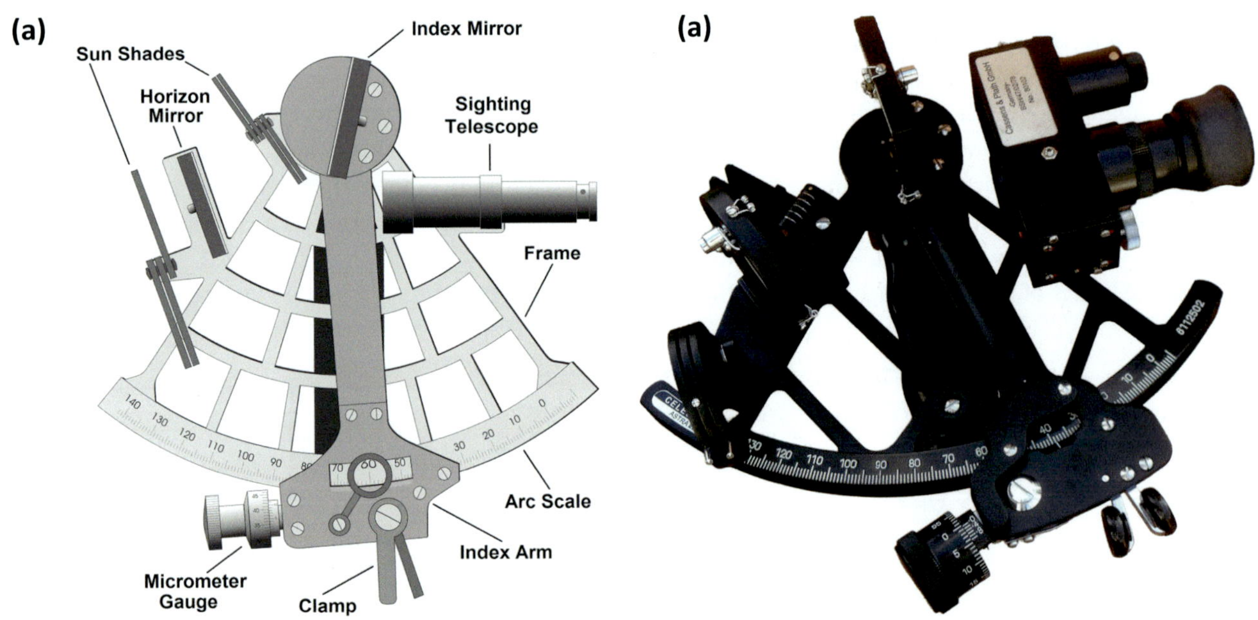

Figure 15.3 A diagram (a) and photograph (b) of a sextant.

Angle measurements are made by holding the sextant vertically and looking through the telescope towards the horizon mirror. One side of the horizon mirror is clear glass, and the other side is reflective. This provides a split-view, where the horizon is directly viewed through the clear glass section, and the celestial body is viewed by double reflection from the horizon and index mirrors. The angle of the index mirror is adjusted by rotating the index arm until the images of the horizon and the celestial body appear side by side. The lower end of the index arm rides over the arc scale, so when the images do appear side by side, the scale indicates the celestial body's elevation angle with respect to the horizon.

A modern sextant can have resolution as small as 0.1 minutes of arc angle, which translates in distance to 0.1 nautical miles (~185 m). This is more than sufficient, because an accuracy of 0.5 nautical miles is usually considered excellent, and anything less than five nautical miles is usually considered acceptable. The sextant's use is generally limited to safely navigating through open waters, because its measurements simply aren't accurate enough to ensure a safe landfall. Other navigation techniques, including steering towards visual landmarks, are typically practiced once the ship is near the shore.

Celestial navigators in the northern hemisphere can easily convert a sextant's angle measurements to latitude. This is due to the presence of Polaris, a bright star aligned over the Earth's north pole that's also called the North Star. Polaris has a nearly fixed position in the sky that directly corresponds to an observer's latitude. For example, an observer viewing the nighttime sky from Boulder, Colorado, will find Polaris at 40° above the horizon, because the 40th parallel passes directly through Boulder.

Polaris is one of a group of seven stars in the Ursa Minor constellation that are collectively known as the Little Dipper, and it can be located by its position relative to the Big Dipper in the Ursa Major constellation. As shown in Figure 15.4, the Little and Big Dippers each have the shape of a ladle, and the two stars at one end of the Big Dipper's "bowl" point towards Polaris, which is at the end of the Little Dipper's "handle." The positions of the stars shown in Figure 15.4 were arbitrarily chosen, because the other stars in the Ursa Major and Minor constellations appear to rotate around Polaris as the Earth spins. Meanwhile, Polaris appears to be almost stationary, due to its relative alignment with the Earth's rotational axis.

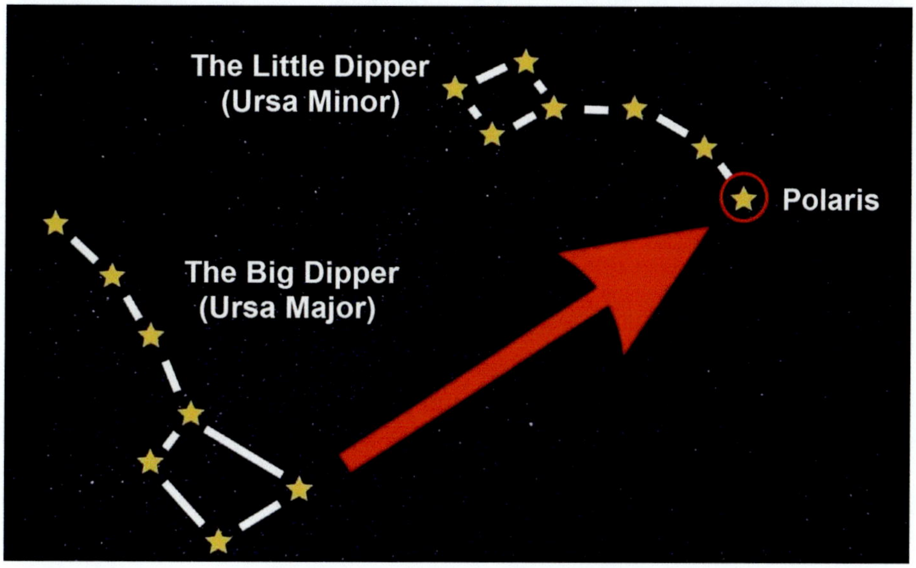

Figure 15.4 Locating Polaris, the North Star.

Celestial navigators can measure latitude using any celestial body, not just Polaris, if they apply mathematical corrections to their angle measurements. These corrections are obtained from tables in nautical almanacs that include the angular coordinates of the Sun, the Moon, the planets, and 57 navigational stars that were chosen for their brightness and ease of identification. The positions of the Sun, the Moon, and the planets change rapidly when observed from Earth, so almanac tables provide their angular coordinates for every hour of every day. In contrast, the angular coordinates of the navigational stars change very slowly, so the almanac tables only provide updates every three days.

The almanac tables also provide the declination angle, which is the angular distance north or south of the Earth's equator, for every listed celestial body. For the Sun, the Moon, and the planets, the almanac tables provide the Greenwich Hour Angle (GHA), which is the angular distance from the prime meridian. For navigational stars, the almanac tables instead list the Sidereal Hour Angle (SHA), which is the star's angular distance from the vernal equinox. However, the tables also provide hourly updates of the GHA of Aries, which provide a reference point for the vernal equinox, and which can be added to the SHA to obtain the GHA of a star. A portion of a nautical almanac table is shown in Figure 15.5 as an example. The portion includes hourly values for the GHA of Aries, and for the GHA and declination angle of the Sun. It also includes the SHA and declination angle for the first 20 navigational stars (listed alphabetically).

UTC		ARIES	SUN		STARS		
		GHA	GHA	Dec	Name	SHA	Dec
4	00	103° 04'.5	178° 52'.5	22° 47'.7 S	Acamar	315° 14'.7	40° 13'.8 S
	01	118° 06'.9	193° 52'.2	47'.4	Achernar	335° 23'.3	57° 08'.6 S
	02	133° 09'.4	208° 51'.9	47'.2	Acrux	173° 04'.2	63° 12'.2 S
	03	148° 11'.9	223° 51'.6	46'.9	Adhara	255° 08'.6	29° 00'.1 S
S	04	163° 14'.3	238° 51'.4	46'.7	Aldebaran	290° 43'.9	16° 32'.8 N
a	05	178° 16'.8	253° 51'.1	46'.4			
t	06	193° 19'.3	268° 50'.8	46'.2	Alioth	166° 16'.7	55° 50'.9 N
u	07	208° 21'.7	283° 50'.5	45'.9	Alkaid	152° 55'.4	49° 12'.7 N
	08	223° 24'.2	298° 50'.2	45'.7	Alnair	27° 38'.4	46° 52'.1 S
r	09	238° 26'.6	313° 49'.9	45'.4	Alnilam	275° 41'.5	1° 11'.5 S
d	10	253° 29'.1	328° 49'.7	45'.2	Alphard	217° 51'.4	8° 44'.7 S
a	11	268° 31'.6	343° 49'.4	44'.9			
y	12	283° 34'.0	358° 49'.1	22° 44'.6 S	Alphecca	126° 07'.4	26° 38'.8 N
	13	298° 36'.5	13° 48'.8	44'.4	Alpheratz	357° 38'.9	29° 12'.1 N
	14	313° 39'.0	28° 48'.5	44'.1	Altair	62° 04'.2	8° 55'.3 N
	15	328° 41'.4	43° 48'.2	43'.9	Ankaa	353° 11'.3	42° 12'.2 S
	16	343° 43'.9	58° 47'.9	43'.6	Antares	112° 21'.0	26° 28'.4 S
	17	358° 46'.4	73° 47'.7	43'.3			
	18	13° 48'.8	88° 47'.4	43'.1	Arcturus	145° 51'.8	19° 04'.7 N
	19	28° 51'.3	103° 47'.1	42'.8	Atria	107° 19'.3	69° 03'.5 S
	20	43° 53'.8	118° 46'.8	42'.6	Avior	234° 15'.6	59° 34'.3 S
	21	58° 56'.2	133° 46'.5	42'.3	Bellatrix	278° 26'.9	6° 21'.9 N
	22	73° 58'.7	148° 46'.2	42'.0	Betelgeuse	270° 56'.1	7° 24'.5 N
	23	89° 01'.1	163° 46'.0	41'.8			

Figure 15.5 A portion of a nautical almanac table.

Latitude is determined by first measuring the elevation angle of a celestial body with a sextant, and then applying corrections to convert the celestial body's sextant height, H_S, to its observed height, H_0. Some of these corrections are related to the way the sextant is operated. For example, a dip correction might be needed to compensate for the height of the observer's eye above sea level, which can make the horizon appear to be lower than it actually is. Other corrections are related to the celestial body itself. In the case of the Sun, corrections are applied for the refraction of its light, which can make it appear higher in the sky than it actually is. Corrections are also applied for the Sun's radius, because the observer often measures the angle of the Sun's edge rather than its center. If the latitude is greater than the celestial body's declination angle, then

$$Latitude = (90° - H_0) + Declination. \tag{15.1}$$

If the celestial body's declination angle is greater than the latitude, then

$$Latitude = Declination - (90° - H_0). \tag{15.2}$$

Celestial measurements of longitude are more demanding than latitude measurements. Longitude and time have a very close connection due to the Earth's rotation on its axis, so measuring longitude requires an accurate clock in addition to a sextant and almanac. The clock should be synchronized to astronomical time (UT1) at the prime meridian, which serves as the reference longitude. For most practical purposes, this means that the clock can be synchronized to UTC, but a DUT1 correction can be applied if needed to obtain more accuracy (Section 9.4.1). An accurate clock is needed to record the difference between the local time when the angle measurement was made, and the time at the prime meridian.

Celestial navigators commonly use the meridian crossing method to determine longitude. This method involves determining the time, as closely as possible, when the observed Sun reaches its highest point in the sky. When the Sun reaches its highest point, it crosses the observer's meridian of longitude, and the time of the crossing is called local apparent noon (LAN). The time of LAN is determined by making angle measurements with a sextant, and recording the time of each measurement with a clock.

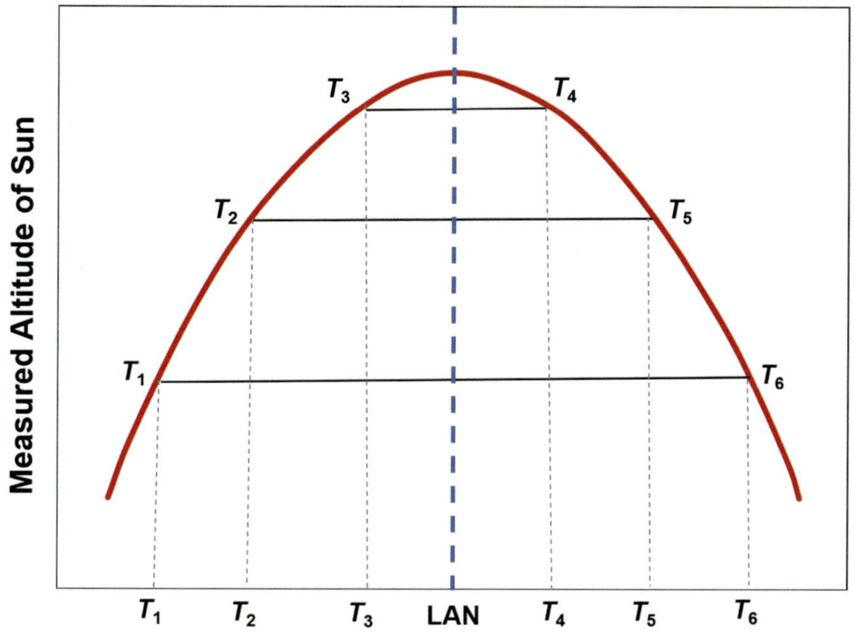

Figure 15.6 Determining local apparent noon from a series of timed angle measurements.

A single sextant measurement isn't sufficient for determining LAN, because the Sun appears to stay at its maximum altitude for a relatively long period. Therefore, multiple measurements are made to get a more accurate estimate, typically starting about 30 minutes before the estimated time of LAN, and continuing for about 30 minutes afterwards. The number of measurements needed depends on the experience of the navigator, but if three timed angle measurements are made before (T_1, T_2, T_3) and after (T_4, T_5, T_6) the Sun reaches its highest altitude, LAN can be estimated from a graph similar to the one shown in Figure 15.6. In this example,

$$LAN = \frac{T_A + T_B + T_C}{3}, \tag{15.3}$$

where

$$T_A = \frac{T_1 + T_6}{2}, \text{ and}$$

$$T_B = \frac{T_2 + T_5}{2}, \text{ and}$$

$$T_C = \frac{T_3 + T_4}{2}.$$

In the western hemisphere, longitude is equal to the Sun's GHA at local apparent noon. In the eastern hemisphere, the Sun's GHA has to first be subtracted from 360° to determine longitude. The Sun's GHA is obtained from the nautical almanac tables. The daily almanac tables, such as the one shown in Figure 15.5, lack sufficient resolution because they only provide the GHA for whole hours. Other almanac tables, which are typically labelled as "Increments and Corrections," are used to interpolate the Sun's hourly GHA down to a resolution of the nearest second [6].

The uncertainty of a celestial navigator's clock will, of course, contribute to the uncertainty of their longitude estimate. The Earth makes a full rotation of 360° every 24 hours, so the Sun appears to move across the sky at a rate of 15° per hour (360/24), which is the longitudinal width of a time zone (Section 9.6.1). This means that the Sun's position changes by 1° of arc angle every 240 s, and by one minute of arc angle every 4 s. A nautical mile, as noted earlier, is the approximate length of one minute of arc angle. Therefore, a clock that differs from UTC by 4 s can cause a position error of one nautical mile.

As we discussed in Chapter 1, building a clock that was accurate enough to determine longitude at sea was one of the greatest scientific problems of the 18th century, a problem that was eventually solved by the British clockmaker John Harrison (1693–1776). Like all free-running clocks, Harrison's chronometers would accumulate a time error, so he established the practice of determining their rate, or frequency accuracy, by comparing them to a reference clock on land before going to sea. The chronometer was synchronized to the reference clock prior to a voyage, and its rate was periodically adjusted during the voyage based on the knowledge of its past performance. The rate adjustments reduced the time error accumulated during the voyage by as much as possible, but they could introduce other errors. Celestial navigators relied upon these same methods until the early 1900s, when radio time signals first allowed them to accurately resynchronize their chronometers while they were still at sea (Section 10.1).

Before the work of Harrison, and for decades afterwards, celestial navigators didn't have accurate clocks, so they obtained the time using the lunar distance method that was briefly mentioned earlier. The lunar distance method was ingenious for its era because it worked without a clock and instead relied on the motion of the Moon. The Moon moves about 30 arc minutes, or half a degree, in one hour, which allows it to serve as a moving clock hand that can tell time against the background sky. However, telling time by lunar distance is a slow and complex process. It requires making multiple measurements of the elevation angles of the Moon and Sun (or another star), having an accurate knowledge of the motion of the Moon and stars obtained from almanac tables, and working through some tedious equations [5]. Despite all of the work involved, the lunar distance method still wasn't very accurate. Uncertainties in the angle measurements and almanac tables made it difficult for navigators to measure lunar distance to better than about 0.5 arc minutes, which was equivalent to a time uncertainty of about ±1 minute. Due to these factors, the lunar distance method became obsolete when accurate clocks were available.

Marine navigators are still trained to use sextants as a backup to their electronic systems, and new nautical almanacs are still published annually. As previously noted, the first edition of *The Nautical Almanac* was published in England in 1766, containing data for 1767. Beginning in 1855, the United States Navy began the annual publication of its own nautical almanac. After slightly more than a century of being separately maintained, the two almanacs were merged, and *The Nautical Almanac* has been a joint product of the United States and England since 1958.

Nautical almanacs contain some explanatory material, but they mostly consist of a large collection of tables. An encyclopedic guide to celestial navigation—containing extensive instructions, tables, and examples—was first published by the American mathematician Nathaniel Bowditch (1773–1838) in 1802 [7]. Bowditch's monumental work, called the *American Practical Navigator*, became a nautical institution. It's still being updated and distributed by the National Geospatial-Intelligence Agency (NGA) as of 2025, and it of course now covers both electronic and celestial navigation.

Celestial navigation and time measurements have always been related, but Harrison's yeoman efforts forged a permanent link between navigation and human-made clocks. Some 25 years after Harrison's death, Bowditch included a short section in the first edition of the *American Practical Navigator* entitled "To find the longitude by a perfect time-keeper." He opened the section with the following passage, which nicely summarized the relationship between clocks and longitude:

> *"If a clock or watch was so contrived, as to go uniformly in all seasons; and in all places, the longitude might easily be deduced therefrom, by comparing the time shewn by the watch, regulated to some given meridian, to the mean time at the place of observation; for the difference would be the difference of longitude between that meridian and the place of observation.*
>
> *When you mean to use a time-keeper, you must examine its rate of going before you leave the land, and adjust it to the meridian of the place from which you reckon your longitude."* [7]

15.2 Acoustic Navigation

Acoustic navigation is the practice of determining location by using sound waves. It can work in either air or water, and it can be either active or passive. In an active system, the navigator emits sounds that produce echoes when they are reflected by objects. A sensor detects when the echoes return, and the time difference between their emission and detection is used to determine the distance between a moving vehicle and the reflecting objects. A series of distance/range measurements can determine a vehicle's position. In a passive system, the object being located emits sounds, and the relative difference in the sound's arrival determines whether the navigator is moving towards or away from the object.

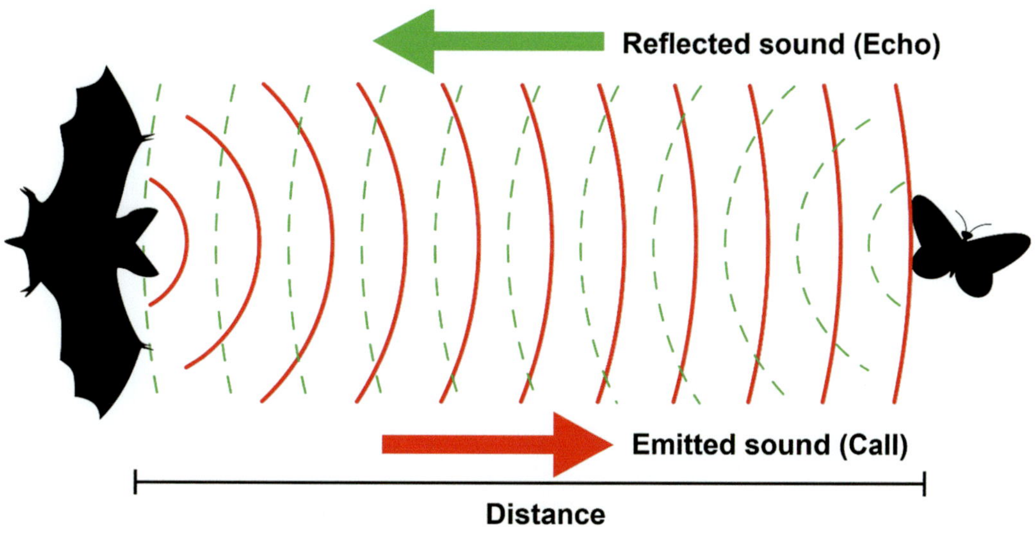

Figure 15.7 Determining distance via echolocation.

Echolocation is a natural form of acoustic navigation that some animals rely upon to sense their surroundings. Echolocation allows these animals to essentially "see" with sound, so they can move about freely when visibility is poor. For example, bats use echolocation to navigate at night, to avoid colliding with obstacles, and to locate their prey (Figure 15.7). They do so by emitting calls and then listening for their echoes. The frequency of a bat's call is usually ultrasonic—typically between 20 kHz and 60 kHz. The echoed sound is much quieter than the call, often orders of magnitude quieter, but bats adjust the duration and frequency of their calls as necessary, which allows their highly sophisticated ears to hear the echo. Sound through air travels from the reflecting object and back again at a speed of about 340 m/s, so a delay of 2 ms equals a distance of 34 cm. Remarkably, some bats may be able to distinguish between time differences of less than a microsecond, which translates to distances of less than a millimeter [8].

Even though echolocation is inherent in bats and other animals, it's ironically sometimes called biosonar, which is a reference to the human invention of underwater sound navigation and ranging that didn't occur until the latter part of World War I (Section 2.1). The acronym sonar, which stands for sound navigation and ranging, was eventually coined in 1942 by Frederick Vinton Hunt (1905–1972), an American physicist and acoustic engineer at Harvard University. Hunt's acronym paid homage to radar, an acronym for radio detection and ranging that had been coined just a few years earlier by the United States Navy. Both acronyms have become common nouns, and sonar is often called "underwater radar." However, sonar isn't always limited to underwater use. It's sometimes used in ground-based systems with well-controlled acoustic environments, perhaps most notably in the field of robotics [9].

An active sonar system measures the distance between a reference point where sound waves are both transmitted and received, and a reflection point that echoes a portion of the transmitted sound waves back to the receiver. Therefore, an active sonar system makes active (transmit and receive) ranging measurements, which differ, for example, from the passive (receive only) ranging measurements made by a GPS receiver (Section 12.1). Transducers serve as both transmitters and receivers. The transducers transmit sound waves by converting electrical signals into acoustic pulses or "pings." When the echoed sound waves return, the transducers convert them back to electrical signals (Figure 15.8).

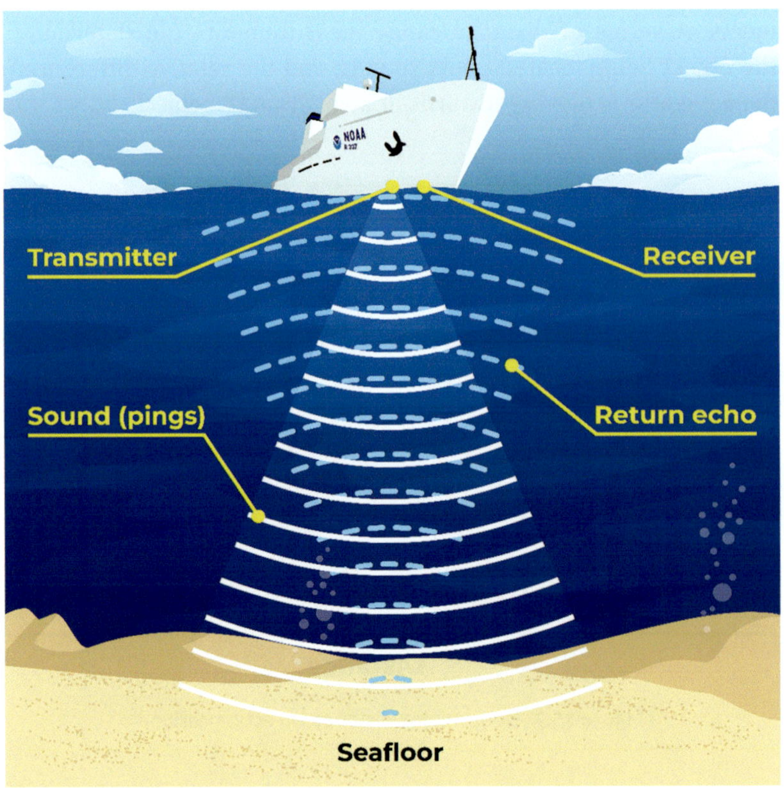

Figure 15.8 Measuring the depth of the seafloor with an active sonar system (courtesy of NOAA).

The time interval between the transmission and reception of the sound—called the time of flight (TOF)—is measured and divided by two. The time interval measurement results are converted to the distance between the reference point and the reflecting object, as

$$Distance = 0.5 \times v_{suw} \times TOF, \tag{15.4}$$

where v_{suw} is the velocity of sound underwater. The velocity varies as a function of the water's temperature and salinity, but is typically about 1,500 m/s on average, or more than 4× faster than the velocity of sound through air (Section 5.1). The depth finders found on nearly every sailing vessel are active sonar systems that determine depth by using the water's surface as the reference point and its floor as the reflecting object. More elaborate multibeam sonar systems use an array of transducers that are mounted on a ship's hull and pointed at different angles. The transducers generate a wide fan-shaped swath of acoustic pulses, and record hundreds of measurements for every echoed ping. This information is used to produce high-resolution and three-dimensional bathymetric maps that show the depths and shape of underwater terrain, in the same way that a topographic map shows the terrain of a land area [10].

Doppler sonar is commonly used for maritime navigation. It's one of several navigational tools used by surface vehicles, and it's an essential tool for underwater vehicles because GNSS signals can't penetrate water. A Doppler sonar navigation system utilizes instruments called Doppler Velocity Logs (DVLs) that are rigidly attached to the vehicle. The DVLs transmit sound waves at a fixed ultrasonic frequency, which is typically in the range from 150 kHz to 600 kHz. They then receive the sound that's echoed by the seafloor. Only a small percentage of the transmitted sound is echoed back to the DVLs, and the echoed sound has had its frequency shifted by the Doppler effect. The Doppler frequency shift, f_d, is calculated as

$$f_d = 2 \times f_s \times \left(\frac{C}{V}\right), \tag{15.5}$$

where f_s is the frequency of the transmitted sound, C is the speed of sound in meters per second, and V is the relative velocity in meters per second between the transmitter and the reflecting object. Multiplying by two is necessary because the same moving transducer is both transmitting sound and receiving the echo. This doubles the Doppler frequency shift because the relative motion between the transducer and the reflecting object is considered twice—once when the sound wave travels to the reflection point, and again when the echoed sound wave travels back to the transducer [11]. The same Doppler doubling effect occurs when law enforcement officers use radar devices to measure vehicle speed (Section 5.1.2). However, the situation is reversed in that case, because the reflecting object is moving and the transmitter and receiver are stationary.

The DVL transducers are usually arranged in a Janus configuration, which gets its name from a god in Roman mythology who was usually depicted as having two faces looking in opposite directions. The name fits, because the transducers in a Janus configuration essentially "face" all directions, which allows Doppler measurements of the forward, backward, and sideways motion of the vehicle with respect to the seafloor. Four ultrasonic beams are transmitted from transducers that are separated by 90°. Each beam points downward at a fixed angle from vertical, which is typically about 30° (Figure 15.9).

A DVL measures the Doppler frequency shift in the direction of each of the four beams, and then combines the information to determine the vehicle's velocity. The accuracy of the velocity estimates depends on factors that include the seafloor's terrain and the water's depth, but their 1σ uncertainty is typically less than 1%. The velocity estimates are converted to a coordinate frame that's referenced to the vehicle's starting position. Because distance equals speed multiplied by time, the vehicle's position is continuously updated by integrating the velocity estimates over the DVL measurement period. When coupled with a heading estimate, which can be obtained from a compass or an inertial navigation system, the velocity information obtained by Doppler sonar measurements allows an underwater vehicle to navigate by dead reckoning, without the use of any celestial or radio navigation aids [11, 12].

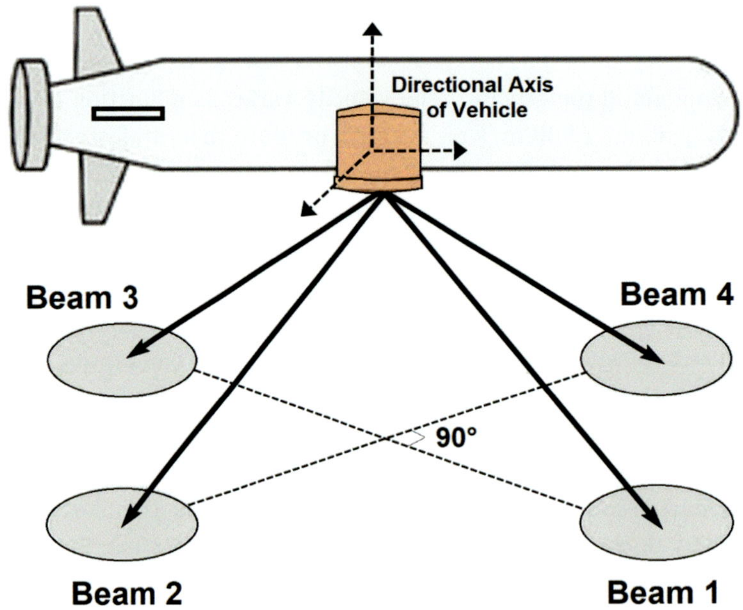

Figure 15.9 An underwater vehicle navigating with Doppler sonar.

15.3 Terrestrial Radio Navigation

Radio transmitters have been used for navigation since the early 1900s. Their first use was serving as direction-finding beacons that fulfilled the same role as a lighthouse or a passive underwater acoustic system. A direction-finding beacon transmits signals from a known location, which directs navigators who receive the signals towards that location. Hundreds of thousands of direction-finding beacons—typically installed near airports, on offshore platforms, or in remote areas—were once deployed worldwide strictly for navigational purposes [13]. The non-directional beacon (NDB) was the most common type, and their signals sometimes carried no information other than a continuously repeated station identification that was transmitted in Morse code. The NDB frequencies usually ranged from 190 kHz to 535 kHz in North America, which is directly below the AM broadcast band. The number of NDBs in use likely peaked shortly after 1995, when GPS was declared to be fully operational (Section 12.1). Their numbers have since dropped dramatically, but some are still operational as of 2025.

A more sophisticated radio navigation method—called hyperbolic navigation—was first practiced in the early 1940s. Its development was motivated by the events of World War II, and it was enabled by the rapid advances then being made in quartz clocks. Hyperbolic systems use time difference measurements to plot a line of position (LOP) in the shape of a hyperbola, which is a smooth, curved line that becomes less curved as it moves further away from its center.

Plotting a LOP requires a pair of synchronized radio transmitters (A and B) with known locations (Figure 15.10). The two transmitters broadcast signals at the same time, but the navigator receives them at different times, and the receiver measures their time difference of arrival (TDOA). A LOP is plotted by drawing a line through all of the locations that match the measured TDOA. Having just one LOP isn't very useful, however, because the navigator's position could be anywhere on the set of points that forms the hyperbola. A hyperbolic position fix is calculated by measuring the TDOA from two pairs of stations, plotting two LOPs, and then finding the point where the two hyperbolas intersect. This requires three synchronized transmitters, as shown in Figure 15.11.

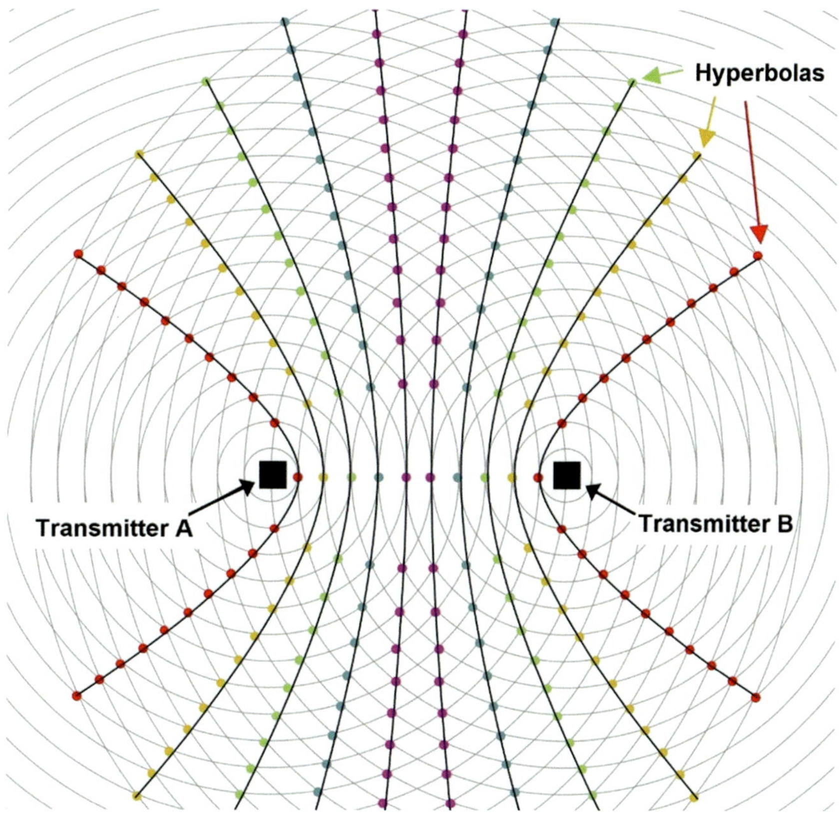

Figure 15.10 Hyperbolic lines of position.

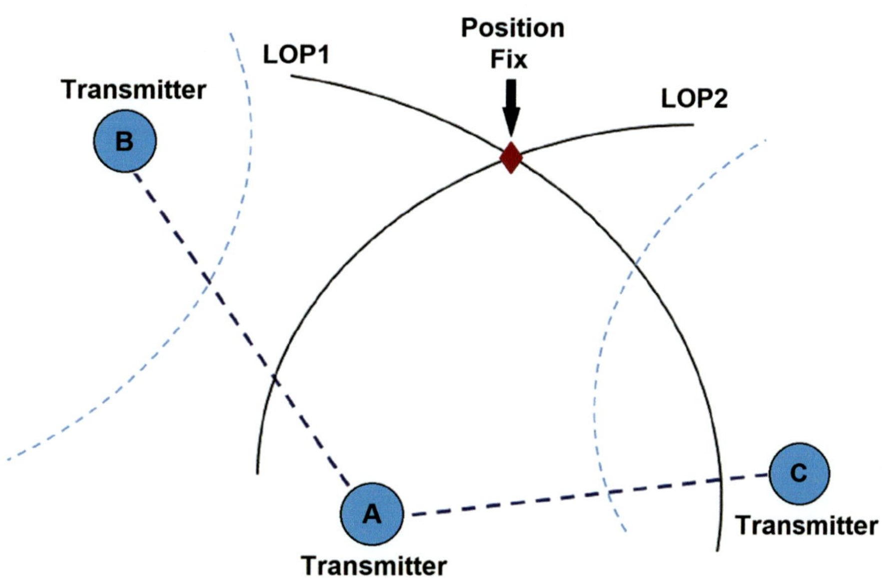

Figure 15.11 Determining a hyperbolic position fix from three synchronized radio transmitters.

The best-known and most successful hyperbolic navigation system is Loran, an acronym for long-range navigation. Loran was invented by Alfred Lee Loomis (1887–1975), an American attorney, investment banker, scientist, and inventor. In the late 1920s, Loomis used the considerable wealth he'd accumulated from his various careers to build a personal laboratory in Tuxedo Park, New York, where he could pursue his scientific interests. Two of his main interests were accurate timekeeping and the art and science of navigation. Loomis purchased several Shortt pendulum clocks (Section 1.8) for his laboratory in 1928, and for the next decade he conducted numerous timekeeping experiments with both pendulum and quartz clocks. He then began to pursue the problem of synchronizing clocks at radio transmitter sites that were separated by hundreds of kilometers, so that a radio navigation system might be built [14]. His solution to the problem resulted in Loran, the hyperbolic, long-range navigation system that he would eventually patent [15].

By the end of the 1930s, Loomis was also heavily involved in radar detection studies, a new area of research that had become a critical part of military defense efforts. This work led to his appointment as chair of the Microwave Committee, a group of representatives from MIT, government, and industry who were concerned with military radio operations during World War II. Loomis proposed the development of his hyperbolic navigation system to the committee in October 1940, and his proposal was quickly accepted as a method that could meet the navigation requirements of the Army Air Corps. The first pair of experimental Loran stations were built and demonstrated in New York and Delaware in 1942. The following year, the leadership of the Loran research effort was assumed by John Alvin Pierce (1907–1996), the physicist and radio researcher who later invented the disciplined oscillator (Section 5.3.2). Progress was rapid from that point forward, and about 60 Loran stations were operational when the war ended in 1945 [16].

The United States Coast Guard (USCG) was involved in the development and maintenance of Loran from the beginning, and it assumed responsibility for all station operations in 1958. By then, Loran had evolved into several forms, and the Coast Guard began using letter designations to distinguish between them. The original system that was started in 1942 was named Loran-A, and an experimental, classified system that was started in the late 1940s was named Loran-B. The name Loran-C was given to a new system that began operations in 1957. The Loran-C stations operated at 100 kHz, in the LF portion of the radio spectrum. This differed from Loran-A, which operated in the MF spectrum at frequencies slightly below 2 MHz [17]. The transition from Loran-A to Loran-C wasn't completed until the 1970s, but Loran-C eventually became the standard version of Loran. All of the Loran-A stations in North America were turned off by 1980, but Loran-C continued to operate until February 2010.

Loran-C consisted of a TDMA network of radio stations that continuously transmitted on a carrier frequency of 100 kHz. The signals were referenced to cesium clocks that were synchronized to UTC(USNO). To make their signals cover a large area, the stations usually radiated more than 250 kW and sometimes more than one megawatt of power. The circular patterns in Figure 15.12 show the approximate groundwave coverage of the 29 Loran-C stations that operated in North America before the 2010 shutdown.

The Loran-C stations were organized into groups called chains. Each chain provided navigation coverage for a specific region, and some stations were included in two different chains. All chains had a master station (*M*), and most had three secondary stations (*X*, *Y*, and *Z*), with a few chains having either two or four secondaries. The chains were identified by a unique Group Repetition Interval (GRI) that indicated how often the stations transmitted pulses. By convention, the GRIs were expressed in 10 μs increments. This meant that the stations in the 7980 chain transmitted pulses every 79,800 μs, which was a pulse frequency of about 12.5 Hz. Depending on the chain, the GRI could range from 4990 to 9990, so the pulse frequencies ranged from about 10 Hz to about 20 Hz.

As shown in Figure 15.13, the master station transmitted the first of its nine pulses at the start of each GRI. The first eight pulses were spaced at 1 ms intervals, and the ninth pulse that identified the master was sent 2 ms after the eighth pulse. The secondary stations all marked the arrival time of the first pulse they received from the master station. They then waited for a fixed period—called the emission delay—before sending their own group of eight pulses. Both the GRI and the emission delays were carefully chosen based on the chain's number of stations and the distances between them. The GRI had to be long enough for each station to transmit its pulses, and the emission delays had to be long enough to prevent the pulse groups from interfering with each other.

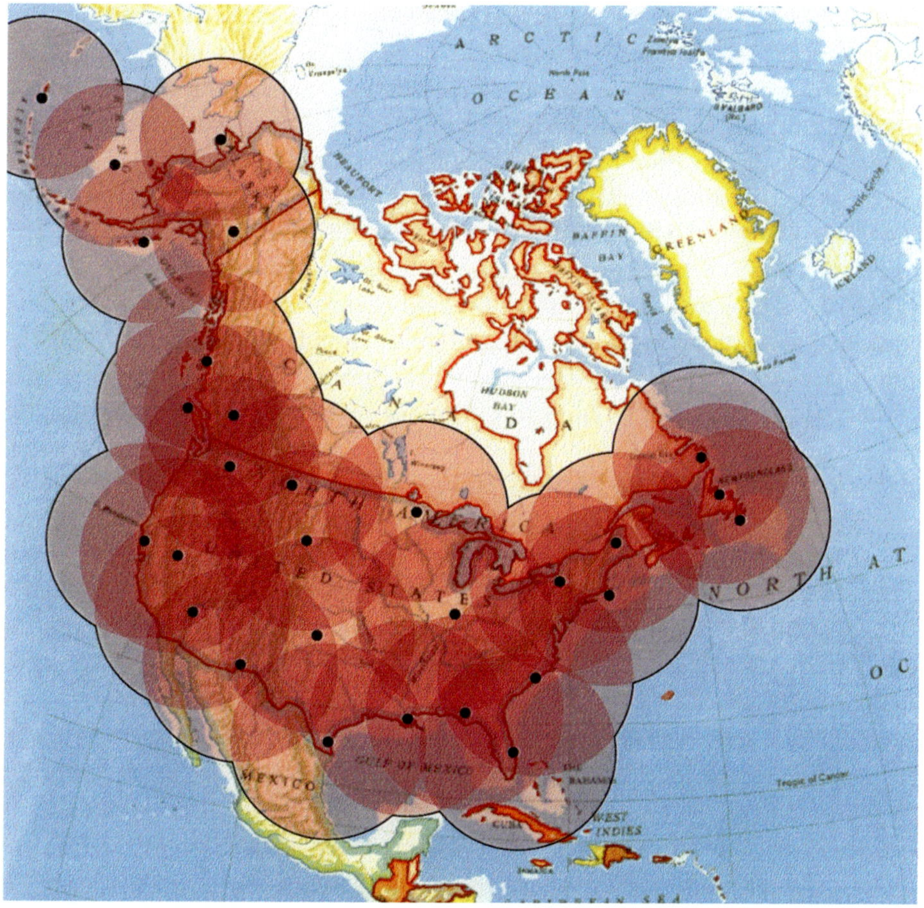

Figure 15.12 North American Loran-C coverage prior to 2010.

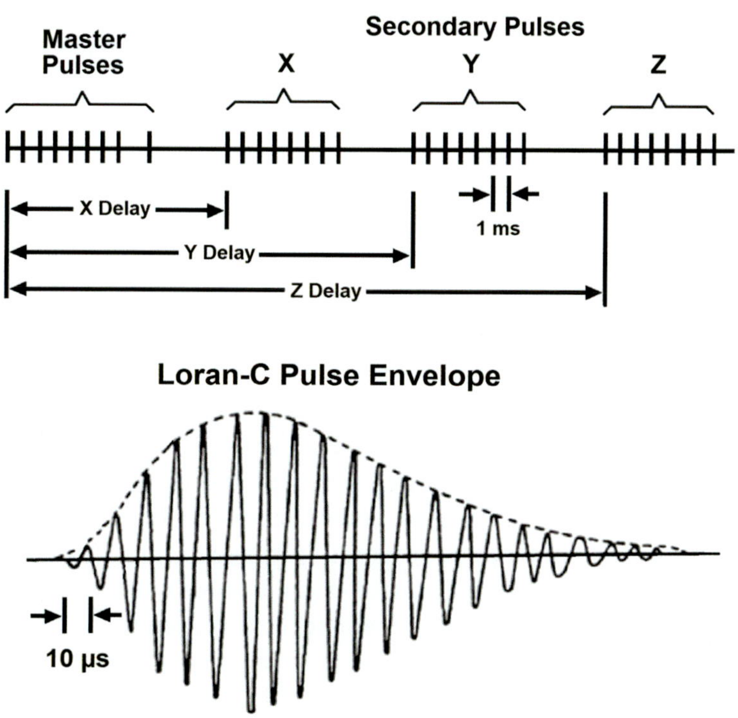

Figure 15.13 The Loran-C pulse sequence and pulse envelope.

The pulse envelope format of Loran-C is also shown in Figure 15.13. Each pulse was sent by increasing and decreasing the carrier amplitude over a period of about 20 cycles (200 μs). To help distinguish between groundwave and skywave, the tracking point for a Loran-C receiver was defined as the third positive zero crossing of the pulse envelope cycle. The third cycle arrived early in the pulse, so it was likely to be groundwave. It was also easier to track than the first and second cycles because it had more amplitude. Cycle slips occurred if the receiver was unable to track the third cycle and had to move deeper into the pulse. Each cycle slip delayed the arrival of the signal by 10 μs [13, 18, 19].

The reception of signals from three stations in the same chain allowed a Loran-C receiver to make two TDOA measurements, $T_M - T_X$ and $T_M - T_Y$. The two time differences resulted in two hyperbolic LOPs, and a position was calculated from their intersection. Additional secondary stations resulted in more lines of position and a more accurate position fix. The absolute accuracy of a Loran-C position fix was typically limited to a few hundred meters, but the repeatable accuracy was much better, typically ranging from 18 m to 90 m [13, 18].

Loran-C didn't broadcast a time code, but its pulse transmissions were aligned with UTC. The Loran time scale had the same epoch as atomic time, which was January 1, 1958, at 0 hours, 0 minutes, and 0 s UT2 (Section 9.4). Therefore, the start of a GRI—which was marked by the first pulse radiated from the master station—would periodically coincide with the start of the UTC second. This event was called the time of coincidence (TOC), and its periodicity was easily determined by looking at the first three digits of a chain's GRI. For odd-numbered GRIs the first three digits equaled the TOC period in seconds, and dividing the first three digits by two produced the TOC period for even-numbered GRIs. This meant, for example, that a TOC occurred every 897 s for a GRI of 8970, and every 497 s for a GRI of 9940. Some Loran-C receivers were specifically designed to be time standards, and they relied on the TOC method to synchronize a 1 pps output. The groundwave path of a Loran-C signal was stable, so if the path and hardware delays between the transmitter and receiver were compensated, the 1 pps uncertainty wouldn't exceed one half of the period of the 100 kHz carrier (±5 μs).

The development and testing of an enhanced version of Loran-C—called eLoran—began in the early 2000s. The eLoran system was designed to be backward compatible with existing Loran-C equipment, but added a data channel that transmitted information about the station (including its name, coordinates, and status), a UTC time code, and correction messages designed to reduce the navigation and time uncertainty [20]. However, due to the 2010 shutdown, eLoran never became an operational system in North America.

As of 2025, Loran stations are still operational in a few countries—including China, Saudi Arabia, and South Korea—and eLoran experiments are still being conducted from several sites in the United States. The experiments are part of a concerted effort to bring Loran back as an alternative or backup PNT system, because its high-power terrestrial signals are much harder to jam than GNSS signals (Section 12.11). If an operational Loran system does eventually return to the United States, it might not support navigation, because that would require rebuilding nearly 30 stations. However, a relatively small number of stations could potentially benefit United States infrastructure by providing an accurate UTC reference signal that could still be received when GNSS reception isn't possible.

The Omega navigation system also deserves mention as the first and only terrestrial radio navigation system that had global coverage. The concept and proposal for Omega were among the many achievements of John Pierce, and the system was developed and tested by the United States Navy during the 1960s. The name Omega was perhaps chosen as a reference to the last letter in the Greek alphabet, which symbolizes the end or the ultimate limit, because the system was designed to use the longest wavelengths in the radio spectrum [21]. Omega was established several decades after Loran, and its operational life span, from 1971 to 1997, was relatively short. It was operated as a cooperative effort between seven nations with strong maritime navigation interests—Argentina, Australia, France, Japan, Liberia, Norway, and the United States [22]. Omega became obsolete when GPS was declared fully operational in 1995, and it lasted for only two years afterwards.

The Omega system consisted of eight VLF radio stations that were each synchronized to within about ±1 μs of UTC. The stations transmitted CW signals in a time/frequency multiplex format that repeated every 10 s. The 10 s period was divided into eight time slots, and all stations used four of the slots to transmit the same sequence of frequencies—10.2 kHz, 13.6 kHz, 11.33 kHz, and 11.05 kHz. The remaining four slots were used to transmit a fifth frequency that was unique to the station, with the unique frequencies ranging from 11.8 kHz to 13.1 kHz. The time slots were arranged in a way that permitted a given frequency to be transmitted by only one station at any given time.

As was the case with Loran, navigating with Omega required the reception of two stations to obtain a hyperbolic LOP, and three stations to obtain a position fix from two LOPs. However, Omega receivers obtained an LOP by measuring the phase difference between two 10.2 kHz signals, and Loran measured the time difference between two pulses. The accuracy of an Omega position fix was typically limited to about two nautical miles (3,704 m), which was much worse than Loran [13, 22]. Its primary virtue was the long wavelengths of its VLF signals, which were only attenuated by a small amount over long distances. This allowed just eight stations to cover nearly 100% of the Earth's surface. The long wavelengths could also penetrate water and reached ocean depths of about 30 m, which allowed submarines to use Omega while staying submerged.

Since the demise of Omega and Loran, most terrestrial radio navigation systems are based on large numbers of multi-purpose transmitters, with each transmitter covering only a small area. The main application of these systems is obtaining an indoor position fix, which is something that usually can't be done with GNSS, because satellite signals are heavily attenuated by roofs and walls. The transmitters are often the cellular base stations in mobile phone networks (Section 16.3), which have accurate clocks synchronized to GNSS [23]. However, many types of signals are used for indoor navigation, including Wi-Fi, Bluetooth, RF identification systems, and ultra-wideband communication systems [24]. The localized indoor systems are based on the same basic principles used by wide-area navigation systems. The receiver—which is typically a smartphone—estimates the distance or range to the fixed-position transmitters by making either a time of arrival (TOA) or a TDOA measurement. The receiver's location is determined by finding the places where the ranging estimates intersect.

15.4 Satellite Radio Navigation

Satellite navigation systems pay homage to celestial navigation systems because their reference signals come from the sky. They also pay homage to acoustic and terrestrial radio navigation systems because they measure time to determine distance. Their development is a classic example of the way measurement science typically advances—a combination of old methods leads to a new method that has unprecedented accuracy.

The roots of satellite navigation can be traced back to the 1957 launch of *Sputnik 1*, the first artificial satellite. Immediately after the launch, researchers at APL in the United States began tracking *Sputnik 1*'s radio signals. They soon found that the satellite's orbit parameters and position could be calculated from Doppler measurements made from a single ground station. This led to the realization that the process could be inverted—if measurements made from known locations on Earth could determine the position of a satellite, then measurements of satellites with known locations in space could determine a position on Earth. This breakthrough idea is credited to the Canadian-American chemist Frank McClure (1916–1973) and it soon led to the satellite systems that now dominate world navigation. McClure started the Transit satellite navigation program (Section 12.1) at APL in 1958, paving the way for GPS and the other systems that followed [25, 26].

A satellite navigation receiver finds its location by using a method called trilateration. The term "lateration" is used in engineering to refer to methods where the position of an object is located by measuring distances and calculating geometric intersections. The prefix "tri" indicates that the distances are measured with respect to three or more fixed locations.

Trilateration is similar to hyperbolic positioning, but it differs in several ways. One key difference is that it obtains a position fix from the intersecting points of circles, and not from hyperbolas. Figure 15.14 shows why at least three transmitters are needed to locate a receiver. In example (a), a range measurement is made with just one transmitter. This tells us that the receiver has to be located somewhere on a circle centered around the transmitter, but the receiver's actual location is unknown. In example (b), range measurements are made with two transmitters. This results in two circles with two points of intersection, providing both a true and a false receiver location. Example (c) shows how trilateration eliminates the ambiguity. When range measurements are made with three transmitters, the three circles have a single point of intersection, which is the point where the receiver is located.

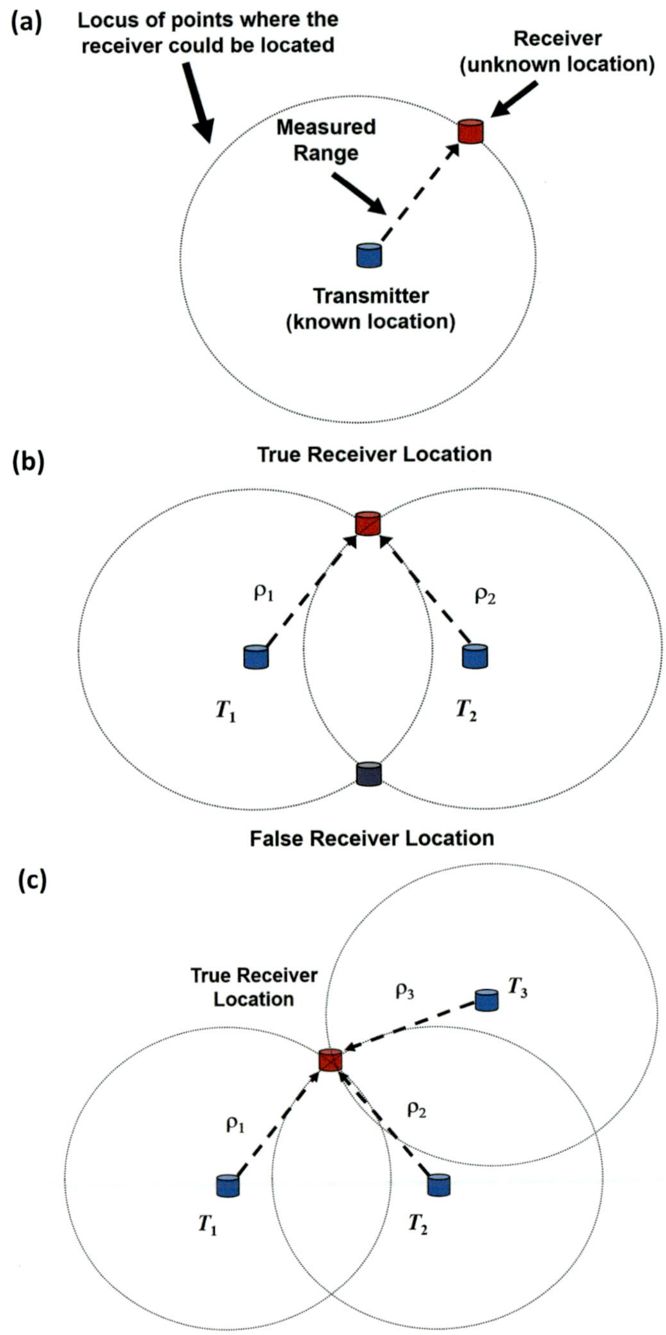

Figure 15.14 Range measurements with (a) one, (b) two, and (c) three transmitters (trilateration).

The range measurements depicted in Figure 15.14 are TOA measurements, which leads us to the next key difference between trilateration and hyperbolic positioning. Trilateration uses TOA measurements to estimate the absolute distance between the receiver and each transmitter, while the hyperbolic method uses TDOA measurements to estimate the relative distance between the receiver and a pair of transmitters. A receiver that measures TOA has to synchronize its clock to the transmitter clock, but that's not a requirement for a receiver that measures TDOA.

A final key difference is that trilateration is capable of 3D positioning if at least one of the transmitters has a large elevation angle with respect to the receiver, which is always the case with satellite navigation. Because the height of their transmitters was always limited, hyperbolic navigation systems like Loran were limited to 2D positioning, which meant that they could only locate a receiver's horizontal position [27]. Loran produced (x, y) Cartesian coordinates that were converted to longitude and latitude. A 3D positioning system like GPS locates a receiver's horizontal and vertical position, and produces (x, y, z) Cartesian coordinates that are converted to longitude, latitude, and altitude.

As shown in Figure 15.15, GPS and other satellite navigation systems are based on an Earth-centered Earth-fixed (ECEF) coordinate system. The (x, y, z) coordinates are expressed in meters, and are obtained from distance measurements along three axes—an x-axis that points towards the intersection of the equator and prime meridian, a y-axis that's perpendicular to the x-axis, and a z-axis that points towards the north pole. The origin for the three axes is the Earth's center of mass.

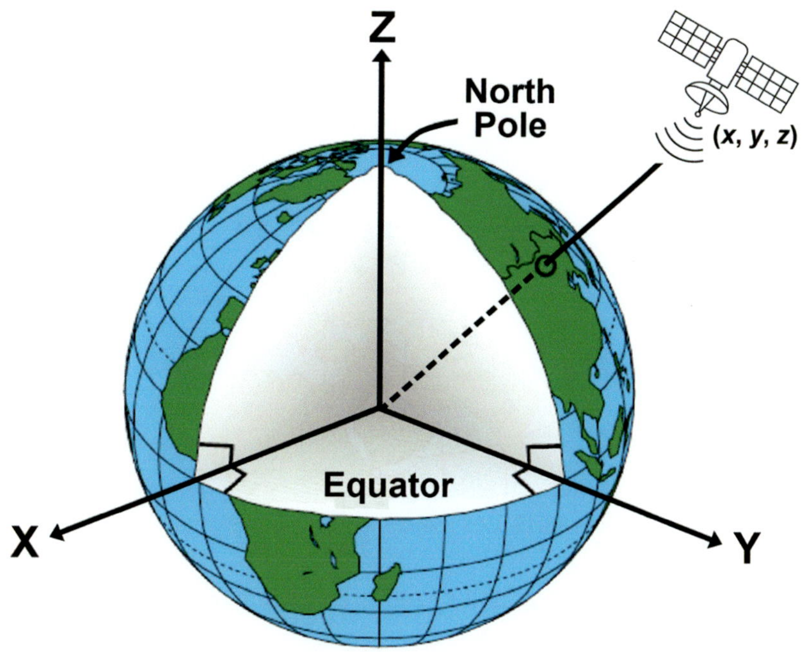

Figure 15.15 A diagram of an Earth-centered, Earth-fixed coordinate system used for satellite navigation.

The basic principles and signal formats of satellite navigation systems, along with the many factors that contribute to their uncertainty, were covered while discussing GPS disciplined clocks and oscillators in Chapter 12. Therefore, we'll limit the discussion here to a simplified description of how a GPS receiver calculates a position fix from range measurements. In satellite navigation parlance, range measurements are called pseudorange measurements, because a variety of factors contribute to measurement uncertainty (Sections 12.5 and 12.7) and prevent the measured range from being exactly equal to the true geometric range. Without considering uncertainty, the receiver-to-satellite geometric range, r_k, is

$$r_k = \sqrt{(x_k - x_r)^2 + (y_k - y_r)^2 + (z_k - z_r)^2}, \tag{15.6}$$

where x_k, y_k, and z_k are the ECEF coordinates for a given satellite k, and x_r, y_r, and z_r are the ECEF coordinates for the receiver.

The GPS navigation messages (Section 12.3) contain everything a receiver needs to calculate both the ECEF coordinates for a given satellite, k, and the time difference between the satellite's clock and GPS time (t_k). However, when a receiver is first turned on, its own position is unknown and its own clock is unsynchronized. To find its location, a receiver has to solve for four unknowns, which are its coordinates (x_r, y_r, z_r) and its clock offset from GPS time (t_r). As shown in Figure 15.16, this requires measuring the pseudorange (p) between the receiver and four different satellites (Figure 15.16). Only three satellites are needed for a trilateration position fix, but a fourth is needed to correct the receiver's clock offset. Correcting the clock offset isn't absolutely necessary for navigation purposes, but without a synchronized clock, a receiver couldn't easily schedule its internal measurements, nor could it serve as a time standard [27].

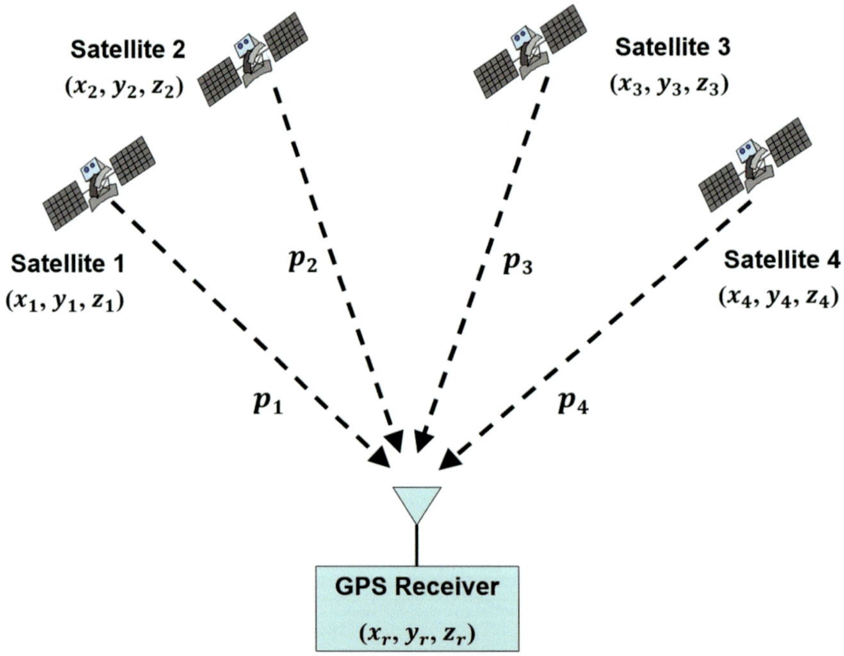

Figure 15.16 GPS pseudorange measurements.

Solving for the four unknowns (x_r, y_r, z_r, and t_r) requires solving four non-linear equations, one for each pseudorange:

$$p_1 = \sqrt{(x_1 - x_r)^2 + (y_1 - y_r)^2 + (z_1 - z_r)^2} + ct_r, \tag{15.7}$$

$$p_2 = \sqrt{(x_2 - x_r)^2 + (y_2 - y_r)^2 + (z_2 - z_r)^2} + ct_r, \tag{15.8}$$

$$p_3 = \sqrt{(x_3 - x_r)^2 + (y_3 - y_r)^2 + (z_3 - z_r)^2} + ct_r, \tag{15.9}$$

$$p_4 = \sqrt{(x_4 - x_r)^2 + (y_4 - y_r)^2 + (z_4 - z_r)^2} + ct_r, \tag{15.10}$$

where c is the speed of light constant. These equations can be solved with iterative techniques based on linearization, by closed-form solutions, or by Kalman filtering [27–29]. They are repeatedly solved after a short observation period, and each solution provides the receiver with a new position fix.

Some GPS receivers can update their position at rates that exceed 20 Hz. Fast position updates allow fast-moving vehicles to easily navigate. For example, a 20 Hz update rate provides a new position fix every meter for a vehicle moving at a speed of 72 km/hour. The accuracy of each position fix depends on the uncertainty of the pseudorange measurements (Section 12.7.5). The single-frequency receivers found in most consumer products are usually accurate to within ±5 m, with most of the uncertainty caused by the variation in ionospheric delays. Dual-frequency receivers compensate for almost all of the ionospheric delay variations, and can be accurate to within a few centimeters.

Satellite radio navigation systems have billions of users and don't require any special equipment, because their receivers, antennas, and software mapping systems are integrated into every smartphone. It's easy to forget, but important to remember, the incredible impact that atomic clocks and satellites have had on navigation systems. Before atomic clocks were flown in space, only highly trained professionals had access to navigation systems, and the best systems were only accurate to within hundreds of meters. Less than a half century later, nearly everyone has personal access to navigation systems that are least 100× more accurate and completely intuitive to use.

References

1. A. K. Johnston, R. D. Connor, C. E. Stephens, and P. E. Ceruzzi, *Time and Navigation: The Untold Story of Getting from Here to There*, Smithsonian Books, 2015.
2. E. Livieratos, A. Tsorlini, C. Boutoura, and M. Manoledakis, "Ptolemy's *Geographia* in digits," *e-Perimetron*, vol. 3, no. 1, pp. 22–39, March 2008.
3. G. Fisher, *The Instructor: or, Young Man's Best Companion*, 7th ed., London: C. Hitch, 1744.
4. J. Hadley, "The Description of a New Instrument for Taking Angles," *Philosophical Transactions of the Royal Society*, vol. 37, no. 420, pp. 147–157, September 1731.
5. D. G. Jablonski, "The Circle Game: The Use of the Lunar Distance and Related Measurements for Celestial and Satellite-Based Navigation and Timekeeping," *NCSLI Measure: The Journal of Measurement Science*, vol. 8, no. 2, pp. 18–31, June 2013.
6. National Geospatial-Intelligence Agency (NGA), "The Almanacs," Chapter 18 in the *American Practical Navigator: The Epitome of Navigation*, NGA Publication 9, vol. I, pp. 331–340, 2024.
7. N. Bowditch, *The New American Practical Navigator*, Newbury Port, Massachusetts: Edwin M. Blunt, 1802.
8. G. Jones, "Echolocation," *Current Biology*, vol. 15, no. 13, pp. R484–R488, July 2005.
9. L. Kleeman and R. Kuc, "Sonar Sensing," Chapter 21 in Springer *Handbook of Robotics*, B. Siciliano and O. Khatib, eds., Springer, 2008.
10. I. Russel and R. G. Wright, "Navigation Sonar: More Than Underwater Radar," *International Hydrographic Review*, no. 17, pp. 41–60, May 2017.
11. National Geospatial-Intelligence Agency (NGA), "Doppler Sonar Navigation," Chapter 25 in the *American Practical Navigator: The Epitome of Navigation*, NGA Publication 9, vol. I, pp. 415–423, 2024.
12. L. Paull, S. Saeedi, M. Seto, and H. Li, "AUV Navigation and Localization: A Review," *IEEE Journal of Oceanic Engineering*, vol. 39, no. 1, pp. 131–149, January 2014.
13. B. J. Uttam, D. H. Amos, J. M. Covino, and P. Morris, "Terrestrial Radio-Navigation Systems," Chapter 4 in *Avionics Navigation Systems*, M. Kayton and W. R. Fried, eds., pp. 99–177, John Wiley & Sons, 1997.
14. L. W. Alvarez, "Alfred Lee Loomis – Last Great Amateur of Science," *Physics Today*, vol. 36, no. 1, pp. 25–34, January 1983.

15. A. L. Loomis, "Long Range Navigation System," *United States Patent 2,884,628,* filed July 3, 1945, granted April 28, 1959.
16. J. A. Pierce, A. A. McKenzie, and R. H. Woodward, eds., *Loran: Long Range Navigation*, McGraw-Hill, 1948.
17. G. Hefley, "The Development of Loran-C Navigation and Timing," *National Bureau of Standards Monograph 129*, 144 p., October 1972.
18. National Geospatial-Intelligence Agency (NGA), "Loran Navigation," Chapter 25 in the *American Practical Navigator: The Epitome of Navigation*, NGA Publication 9, vol. I, pp. 441–455, 2024.
19. S. C. Lo, B. B. Peterson, and P. K. Enge, "Loran Data Modulation: A Primer," *IEEE Aerospace and Electronic Systems Magazine*, vol. 22, no. 9, pp. 31–51, September 2007.
20. K. Dykstra and B. Peterson, "The Loran Data Channel: Progress to Date and Future Plans," *Proceedings of the 35th Annual Technical Symposium of the International Loran Association (ILA)*, 8 p., October 2006.
21. J. A. Pierce, "Omega," *IEEE Transactions on Aerospace and Electronic Systems*, vol. AES-1, no. 3, pp. 206–215, December 1965.
22. J. F. Kasper, Jr., and C. E. Hutchinson, "The Omega Navigation System: An overview," *IEEE Communications Society Magazine*, vol. 16, no. 3, pp. 23–35, May 1978.
23. Z. M. Kassas, "Navigation with Cellular Signals of Opportunity," Chapter 38 in *Position, Navigation, and Timing Technologies in the 21st Century, Volume 2*, J. Y. Morton, F. van Diggelen, J. J. Spilker Jr., and B. W. Parkinson, eds., pp. 1171–1223, IEEE Press, 2021.
24. Y. Yang, M. Chen, Y. Blankenship, J. Lee, Z. Ghassemlooy, J. Cheng, and S. Mao, "Positioning Using Wireless Networks: Applications, Recent Progress, and Future Challenges," *IEEE Journal on Selected Areas in Communications*, vol. 42, no. 9, pp. 2149–2178, September 2024.
25. T. A. Stansell, "The TRANSIT Navigation Satellite System," *Magnavox R-5933A*, 83 p., 1978.
26. R. E. Gibson, "Frank Trelford McClure, 1916–1973," *APL Technical Digest*, vol. 13, no. 4, pp. 2–7, October-December 1974.
27. P. Misra and P. Enge, *Global Positioning System: Signals, Measurement, and Performance*, 2nd edition, Ganga-Jamuna Press, 2011.
28. E. D. Kaplan and C. J. Hegerty, eds., *Understanding GPS: Principles and Applications*, 2nd edition, Artech House, 2006.
29. J. B.-Y. Tsui, *Fundamentals of Global Positioning Receivers: A Software Approach*, 2nd edition, John Wiley & Sons, 2005.

Chapter 16

Time and Telecommunications

I can't talk to you in terms of time—your time and my time are different.

— Graham Greene, "Under the Garden"

The word telecommunication is derived from the Greek prefix, *tele*, which means "far off" or "from a distance," and the Latin word, *communicare*, which means to "make something common" or to "share." The field of telecommunications is exceptionally broad, because it includes all electronic systems that exchange or share information across a distance by using transmitters and receivers. The information is exchanged or shared via a pathway called the transmission medium. In a wired system, the transmission medium is usually a cable or optical fiber. In a wireless system, the transmission medium is usually air.

We've emphasized throughout this book that a group of free-running clocks can't keep exactly the same time because their oscillators can't run at exactly the same frequencies. The clocks and oscillators in a telecommunication network don't have to keep exactly the same time, but they do have to stay within a specified frequency or time tolerance so that information can be reliably exchanged. The required tolerances are often small, sometimes within $\pm 1 \times 10^{-12}$ for frequency and ±100 ns for time. If the clocks are out of tolerance, telecommunication systems will lose some or all of the information that's being shared. Clock failures can reduce a network's capacity by causing some of its nodes to shut down. In extreme cases, clock failures can even cause a shutdown of the entire network.

We use our electronic telecommunication systems so often that it's hard to imagine being without them, and our safety and well-being depends on them during emergencies. When we think about how important these systems are, it's easy to understand why telecommunications has been the motivating force behind many advancements and innovations in both engineering and physics. In fact, the development of most of the oscillator and frequency control technology described in previous chapters was motivated by the need to make telecommunication systems more capable and more reliable.

This chapter explains why accurate clocks and oscillators are needed for reliable telecommunications. Section 16.1 introduces some basic concepts in radio communications and summarizes how the oscillator circuits in receivers have evolved since the early days of radio. Sections 16.2 and 16.3 respectively discuss the importance of accurate frequency and time to wired and wireless telephone networks, and summarizes the accuracy requirements. Section 16.4 concludes the chapter by describing how telephone networks meet their accuracy requirements.

As Time Goes By: A Comprehensive Guide to Time and Frequency Measurements
Michael A. Lombardi

ISBN 978-981-5352-28-3 (Hardcover), 978-1-042-01532-0 (eBook)
www.jennystanford.com

DOI: 10.1201/9781042015320-22

16.1 Radio Communications

We don't always think of them as such, but broadcast radio and television stations are telecommunication systems. They send information in one direction from a transmitter to an unlimited number of receivers within their coverage area. The frequency accuracy requirements for radio stations were difficult to meet in the early days of the radio industry, but they later became almost trivial due to advances in quartz oscillator technology (Section 2.2). The FCC requirements for carrier frequency accuracy are modest; ±20 Hz for AM, ±2,000 Hz for FM, and ±1,000 Hz for OTA television, which translates in all cases to a unitless frequency accuracy of parts in 10^5 [1].

The modest accuracy requirements are acceptable for AM, FM, and TV, because the signal bandwidths are wide enough to not be affected if the transmitter or receiver frequency is off by a few hertz. Better frequency accuracy is needed, however, for radio communication that involves narrow bandwidth signals, especially when the signals are transmitted at high frequencies. For example, if we need to establish radio communications at 1 GHz using signals with a 10 Hz bandwidth, even a 1 Hz offset would cause communication problems. Therefore, the frequency accuracy of both the transmitter and receiver would need to be less than 1×10^{-9}.

In the previous example, the accuracy requirement is influenced by both the high carrier frequency and the narrow bandwidth of the signal. As a general rule, inaccurate frequency becomes more of a problem when the carrier frequency goes up. To understand why, let's suppose that we have a quartz oscillator with a frequency accuracy of $\pm 1 \times 10^{-6}$. If we bring that oscillator to an AM radio station and use it as the reference for a 1 MHz carrier, everything should work fine, because the ±1 Hz frequency offset will be too small for any of the station's listener to notice. However, if we bring that same oscillator to a satellite ground station, and use it as the reference for the 14 GHz uplink signal in a two-way satellite time transfer system (Section 9.4.4), the resulting ±14 kHz frequency offset would prevent the system from working at all, because the satellite transponder wouldn't be able to detect the signal.

Most radio receivers have a variable frequency oscillator (VFO) circuit that allows tuning to different stations by adjusting the oscillator frequency over a specified range. In early receiver designs, the VFO circuit included an LC oscillator, which is a tuned circuit that uses a combination of inductors (L) and capacitors (C) to oscillate at a specific frequency.

The Colpitts oscillator is the most common type of LC oscillator found in VFO circuits. The oscillator was named after Edwin Colpitts (1872–1949), the Canadian-American electrical engineer who invented it 1918 while working at Western Electric [2]. Colpitts' work with oscillators was motivated by his desire to duplicate the function of a telephone without using wires—what was then called radio telephony—so that people could talk to each other over the air instead of being limited to the use of Morse code. In 1919, Colpitts and his engineering colleague Edward Craft (1881–1929) wrote that

> *"The possibility of communication by speech between any two individuals in the civilized world is one of the most desirable ends for which engineering can strive. For this reason it is particularly desirable to form some opinion of the part which radio telephony may play in securing this universal service. It is clear that the elimination of the Morse operator, which is accomplished by the use of radio telephony rather than radio telegraphy, is necessary for universal and direct communication."* [3]

An LC oscillator normally outputs a fixed frequency, but a VFO circuit allows the frequency to be changed by including a variable capacitor [4]. Air-gapped variable capacitors, such as the one pictured in Figure 16.1, are commonly found in old radios. They consist of a group of semicircular metal plates that rotate when a shaft is turned. The shaft is connected—either directly or by a geared mechanism—to the receiver's tuning knob. The capacitance is highest when the plates overlap the most, and lowest when they overlap the least, and the change in frequency is inversely proportional to the change in capacitance.

Figure 16.1 An air-gapped variable capacitor from the VFO circuit of a 1940's radio.

Receivers that use LC oscillators were, and still are, stable enough to reliably receive wide bandwidth AM and FM signals. However, when narrow-band signals are used, reliable radio communication requires receivers whose frequency is stabilized by a quartz oscillator. To better illustrate this, we'll briefly describe a long-established method for reducing signal bandwidth, called single side band (SSB) modulation. As we touched upon in Section 2.6, the development of the TCXO was likely motivated by the need to make SSB communication more reliable.

Figure 16.2 shows how SSB modulation differs from AM. In both cases, the information being modulated is called the baseband signal. An AM signal contains an unmodulated carrier, and two sidebands that are centered around the carrier. The sidebands carry identical copies of the baseband. In SSB modulation, the carrier is suppressed and one of the sidebands is eliminated, which leaves only a lower side band (LSB) or an upper side band (USB). Suppressing the carrier reduces the transmitted power by about 67%, and using just one sideband reduces the bandwidth by more than half [5].

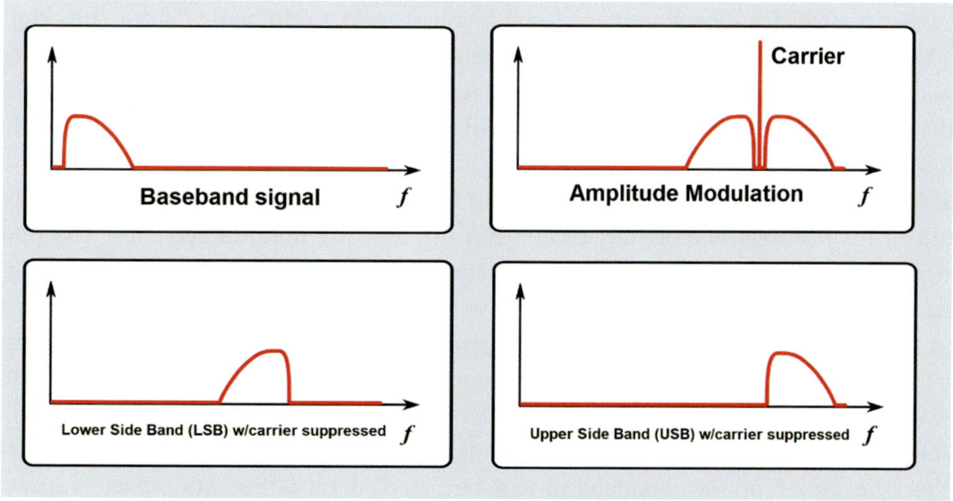

Figure 16.2 A comparison of AM and lower and upper side band modulation.

When demodulating an AM radio signal, a receiver can use the carrier as a reference point for detecting the audio. However, an incoming SSB signal can't be demodulated in the same way, because the carrier has been suppressed. Analog superheterodyne receivers (Section 5.2.4) typically demodulate SSB using the filtering method, which requires an additional oscillator. This oscillator has historically been called a beat frequency oscillator (BFO), a name that originated when Morse code was the dominant form of radio communications. However, for SSB voice communications, a better and more descriptive name is carrier insertion oscillator (CIO), because the oscillator essentially restores the carrier that was suppressed at the transmitter.

Figure 16.3 is a simplified diagram of an SSB receiver that employs the filtering method to demodulate an audio signal. The incoming radio signal is heterodyned with the signal from the VFO to produce an IF signal. The IF then passes through a fixed-bandwidth filter that removes everything from the signal except the desired sideband. The filter's output is then heterodyned with the signal from the CIO to recover the baseband audio.

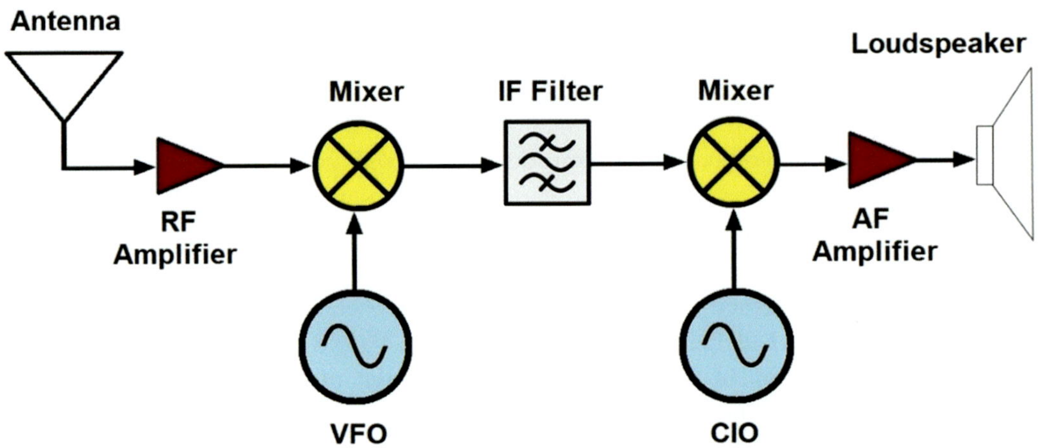

Figure 16.3 A simplified diagram of an SSB receiver that implements the filtering method.

The fixed bandwidth of the filter is typically within the range of 2.4 kHz to 2.8 kHz, which is just slightly narrower than the typical 3 kHz audio bandwidth of a telephone. This means that the SSB audio quality can be comparable to that of a telephone call, but only if the CIO frequency is accurate. Any uncertainty in the CIO will cause the transmitted baseband audio to be shifted in frequency when it's demodulated. Even a small frequency shift can make voices sound odd, and a large shift can make voices unintelligible. As a general rule, the CIO frequency needs to be within ±30 Hz of the suppressed carrier to ensure reasonably good voice quality. That translates to a frequency tolerance of $\pm 1 \times 10^{-6}$ at the top of the HF band (30 MHz), with the unitless tolerance becoming proportionally smaller at higher frequencies [5–7].

Two other long-established methods—the phasing method and the Weaver method—were invented to eliminate the need for the fixed-bandwidth filters that the filtering method requires. The phasing method was described in 1948 by the American engineer Donald Norgaard (1914-1982), who was then working for the General Electric Company in Schenectady, New York [8]. The phasing method applies the incoming SSB signal to a wide-band phase-shift network that produces two copies of the signal, each with the same amplitude, but differing in phase by 90°. The in-phase and quadrature phase signals are each mixed with the receiver's LO to produce two IF signals. When the two IF signals are either added or subtracted, only one sideband remains because the other sideband is cancelled out by the phase difference. Therefore, either the upper or lower sideband can be selected by either adding or subtracting the two IF signals.

The Weaver method was invented in 1956 by the American electrical engineer Donald Weaver (1924–1998), who was then working at the Stanford Research Institute in Menlo Park, California [9]. It not only eliminated the need for fixed-bandwidth filters, but also eliminated the need for a wide-band phase-shift network, because it only required four mixers and two LPFs. The first mixer stage translates the incoming RF signal down to its baseband, resulting in two AF signals in phase quadrature. The AF signals are each filtered to select the desired sideband, and then mixed with a frequency generated at half of the bandwidth of the sideband, which is typically about 1.4 kHz. The resulting signals are then combined to produce the receiver's audio output (Figure 16.4).

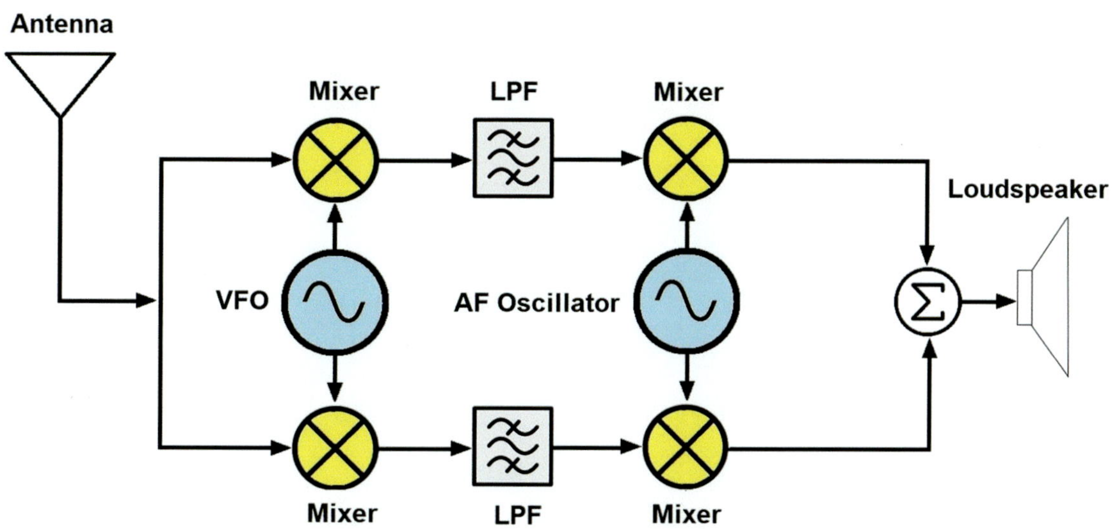

Figure 16.4 A simplified diagram of an SSB receiver that implements the Weaver method.

The Weaver method was well ahead of its time. Until the development of software-defined radios and DSP techniques, most receiver designs continued to use the traditional filtering method to demodulate SSB signals. However, due to the proliferation of SDRs—which employ accurate digital mixers and filters and the resulting I and Q data streams (Section 5.2.5)—implementations of the Weaver method are now common.

Radio communications became far more reliable after quartz oscillators were added to receiver circuits. Radio operators, who were long accustomed to using receivers that constantly drifted off frequency, called the early quartz-based receivers "rock bound." This catch-all reference acknowledged that a "rock" was now the resonator, that the frequency was "stable as a rock," and that the tuning range was bound to the band segments that had crystals available. The band segments were typically no wider than 500 kHz, and were each stabilized by their own crystal. Tuning was a two-step process—the desired band segment was first chosen with a preselector knob, then the VFO knob was adjusted to tune across the segment and find the desired signal. This type of design made it costly to build receivers that could cover the entire HF band, because their design required so many crystals. For example, Figure 16.5 is a photograph of a bank of quartz crystals, each housed in a metal casing, that was found inside of an early-1970s communications receiver. Despite the large number of crystals, this receiver still had significant gaps in its HF coverage.

The Wadley loop was a significant breakthrough in VFO design because it allowed receivers with just a single quartz crystal to tune across the entire HF band. The circuit was invented in the mid-1940s by the South African electrical engineer Trevor Wadley (1920–1981). Wadley developed the circuit while working on the design of an ionosonde, which is a type of radar system that measures the ionosphere by transmitting HF signals across a range of frequencies, and then receiving and analyzing the reflected signals. His work led to a crystal-controlled frequency generator that covered the range from 0 to 20 MHz with a single 100 kHz crystal. By the mid-1950s, the range of the circuit was expanded to 30 MHz with a single 1 MHz crystal [10].

Figure 16.5 A bank of metal-encased quartz crystals in a 1970's HF communications receiver.

Wadley's circuit utilized the comb of harmonics generated by a reference quartz oscillator. In the traditional fashion, a Wadley loop mixes the VFO signal with the incoming RF signal to generate an IF, but it also simultaneously mixes the VFO signal with a selected harmonic from the reference oscillator. Mixing with the harmonic generates a frequency correction signal. A feedback loop applies the correction signal to the IF, which compensates for the VFO's inaccuracy. The method resulted in reasonably good frequency stability across the entire HF band, and Wadley loop circuits were included in some communication receiver designs from 1957 until the early 1980s.

All aspects of radio communications changed and improved after the development of PLL-based frequency synthesizers (Section 5.3.1). The first communication receivers that could tune across the entire HF band using PLL-based synthesizers appeared in the late 1970s, and they gradually replaced nearly all of the receivers based on older VFO technology. The synthesizer's LO is typically a TCXO or OCXO, with some units having the ability to phase lock the LO to an external reference. It's now common for the local oscillators in HF receivers and transceivers to have the stability needed to support tuning in 1 Hz increments.

Receivers with digital frequency displays preceded digitally synthesized receivers by several years. However, the first receivers with digital displays still had analog VFOs, but they included frequency counter circuits that measured and displayed the VFO's output. Figure 16.6 shows the tuning dial on one such receiver, where an analog scale is beneath the digital display. The frequency counter was no longer needed after digital frequency synthesizers became common in the 1980s. Receivers could then simply display the frequency specified by a digital command. The command was issued by either turning a knob that served as a rotary encoder, or by directly entering the frequency from a keypad.

Figure 16.6 A VFO with both a digital readout and an analog scale, found on a 1970s communication receiver.

The FT8 communication protocol used by amateur radio operators is an example of a digital SSB system that requires both accurate time and frequency, albeit with modest accuracy requirements. The protocol was invented in 2017 by two American scientists who were also radio amateurs—the electrical engineer Steven Franke (b. 1957) and the astrophysicist and Nobel laureate Joseph Taylor (b. 1941). The initials "FT" refer to the last names of the two inventors, and the number "8" refers to the eight-tone continuous-phase frequency shift keying (CPFSK) method that the protocol uses to transmit messages [11].

When amateur radio stations communicate via FT8, they alternate between transmitting and receiving signals every 15 s. The 15 s segments are referenced to UTC, with a new segment beginning at the top of each minute, and at 15 s, 30 s, and 45 s afterwards. Therefore, every station's clock has to be synchronized to within about ±1 s of UTC. This synchronization is usually accomplished with either NTP (Section 11.2.4) or GPS.

The bandwidth of an individual FT8 signal is just 50 Hz. Multiple stations transmit and receive signals on the upper side band of a selected frequency, which typically has a total bandwidth of slightly less than 3 kHz. That's enough bandwidth to potentially allow almost 60 stations to simultaneously transmit without having their signals overlap, but there are often more stations than that, so each station operator uses a spectrum display to find an open frequency where they can transmit. Figure 16.7 shows a typical scrolling spectrum display, where the spectral lines are FT8 signals transmitted at a frequency of about 24.9 MHz. The scale at the top of the display lists the audio frequencies in hertz, and the horizontal lines mark the 30 s periods where each station alternates between transmitting and receiving. The display is divided into 10 vertical segments, so it covers a period of five minutes.

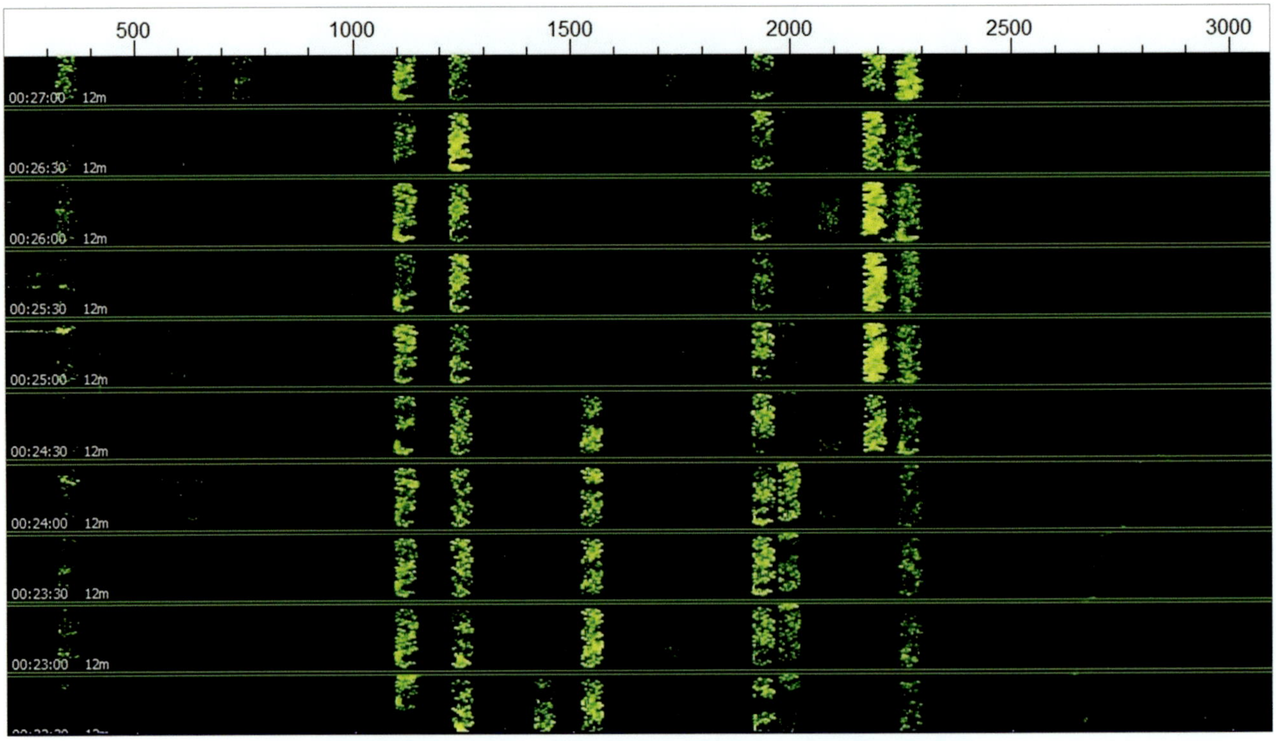

Figure 16.7 A spectrum display of digital FT8 signals.

In order to decode digital signals from other stations, and to allow its own signals to be decoded, each FT8 station has to control its frequency more tightly than a station engaged in voice communications. The frequency has to stay within about ±1 Hz for a period of a few minutes, an allowable uncertainty of $\pm 4 \times 10^{-8}$ at a nominal frequency of 25 MHz. The straight vertical lines in the spectrum display indicate that the transmitting stations (and the receiving station) are well within this tolerance, and some of the small variations can likely be attributed to the instability of HF radio propagation. This example demonstrates that even low-cost radio equipment can now have very good frequency stability, which can be credited to more than a century of progress in VFO design.

16.2 Wired Telephone Networks

Telephone service has been available since the early 20th century (Chapter 11), but the telephone networks in the United States didn't need synchronization standards until January 1, 1984, which was the date of the AT&T divestiture. The divestiture resulted from an antitrust settlement with the U. S. government that broke up AT&T's monopoly of local telephone service. Before the divestiture, all U. S. telephone calls—about 600 million per day—were routed through Bell System telephone companies that AT&T owned [12]. The frequency reference for the entire Bell System consisted of a group of cesium clocks located underground in Hillsboro, Missouri. The clocks were measured by comparing them to Loran-C signals referenced to UTC(USNO), and they maintained a frequency accuracy of $\pm 1 \times 10^{-11}$ [12, 13].

The divestiture divided the Bell System into seven regional operating companies. Each of the seven companies now had to maintain their own primary reference clock (PRC) that had accuracy equivalent to the Bell System reference. The PRCs maintained by the regional companies were located in different parts of the country and ran independently of each other, but they had to stay close enough in frequency be able to exchange traffic with another carrier at the boundary between two networks (Figure 16.8).

This situation resulted in network connections that were plesiochronous, which meant that the networks were almost—but not quite—synchronized with each other. To keep the plesiochronous connections working with a minimal amount of data loss, every PRC had to be kept within a specified frequency accuracy tolerance.

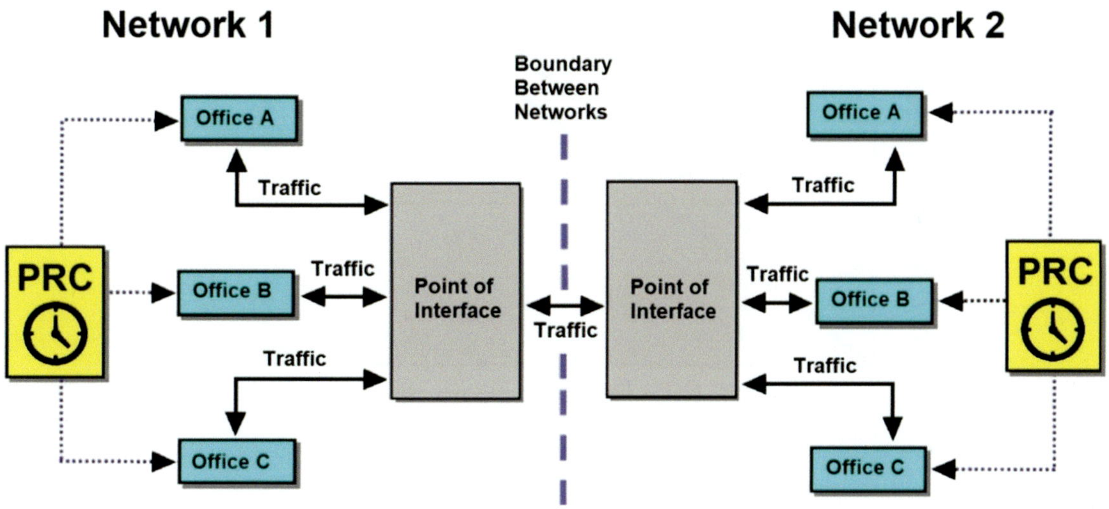

Figure 16.8 A plesiochronous connection between telephone networks that each maintains its own primary reference clock.

To understand the importance of frequency accuracy in a plesiochronous network connection, consider an example where the traffic exchanged between networks consists of Digital Signal 1 (DS1) signals transmitted on a T1 line. The T1 carrier system was first put into service in 1962 [14], and is still the primary digital telephone standard in the United States, Canada, and Japan. It consists of a digital data stream clocked at a frequency of 1.544 MHz. The data stream is divided into 24 voice channels, each with 64 kHz of bandwidth. A total of 8,000 DS1 frames are sent per second, so the frame period is 125 μs. Each frame contains 193 bits—one framing bit and eight bits for each of the 24 channels. The bit period is about 648 ns, which is the period of the 1.544 MHz clock frequency.

A frame slip occurs if the accumulated time difference between two PRCs exceeds the period of the sampling rate. When this happens, the frames are misaligned and a complete frame can be lost. Frame slips cause garbled audio, and they can sometimes even cause calls to be dropped. The frame slip period, FS_{per}, is calculated as

$$FS_{\text{per}} = \frac{F_{\text{per}}}{f_{\text{diff}}}, \tag{16.1}$$

where F_{per} is the frame period, and f_{diff} is the frequency difference between the two PRCs. If each PRC has a frequency accuracy of 1×10^{-11}, with one high and one low in frequency with respect to UTC, the period between frame slips is calculated as

$$FS_{\text{per}} = \frac{125\times10^{-6}\,\text{s}}{2\times10^{-11}} = 6{,}250{,}000\ \text{s} = 72.5\ \text{days}. \tag{16.2}$$

Figure 16.9 depicts a frame slip where the PRC in Network 1 has no discernible frequency inaccuracy, but the PRC in Network 2 is only accurate to 2×10^{-11}, causing a frame slip to occur every 72.5 days. This example shows why it's essential for all PRCs to maintain frequency accuracy within a specified tolerance.

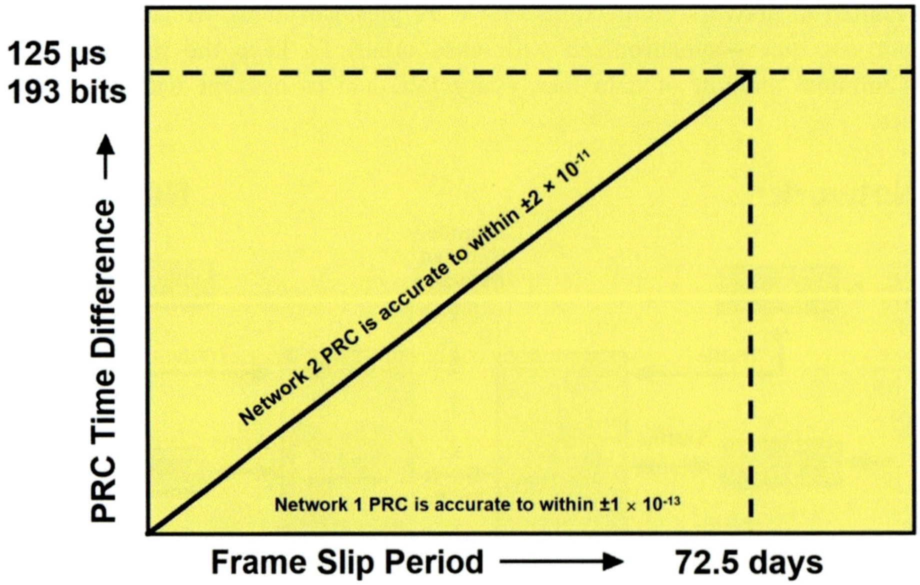

Figure 16.9 The frame slip period between two networks that differ in frequency by 2×10^{-11}.

The phone companies originally felt that frequency differences of up to 3.5×10^{-9} between PRCs were acceptable, because that reduced the frame slip period to about 10 hours [15]. However, a frequency accuracy of $\pm 1 \times 10^{-11}$ became the international PRC standard in 1997. The ITU specifies that this accuracy has to be verified with a UTC reference, and that it has to be maintained for observation periods greater than one week, over all applicable operational conditions [16]. Another ITU standard defines an enhanced primary reference clock (EPRC) as being accurate to within $\pm 1 \times 10^{-12}$ of UTC for observation periods greater than one week [17]. The 10× improvement in the frequency accuracy of an EPRC would increase the period between frame slips from 72.5 days to about two years.

16.3 Wireless Telephone Networks

The first public demonstration of a wireless telephone was made by Motorola in April 1973, when the American electrical engineer Martin Cooper (b. 1928) made a successful telephone call from a New York City sidewalk. Perhaps not surprisingly, Cooper's first call was to Joel Engel (b. 1936), who was then the head of research at Bell Labs [18]. The use of wireless mobile phones was commonplace by the 1990s, and they have since become the dominant type of telephone system.

The number of mobile phone subscriptions in the United States increased from about 100 million in the year 2000 to almost 400 million in 2023, as shown in Figure 16.10. The figure also shows the sharp decline in the number of fixed phone subscriptions that utilize wired networks. In the year 2000, the fixed-to-mobile phone ratio was almost 2:1. However, mobile phone subscriptions surpassed fixed phone subscriptions in 2004, and by the end of 2023 the mobile-to-fixed subscription ratio was more than 4:1 [19]. By 2023, an estimated 90% of U. S. adults not only had a mobile phone subscription, but they also had a smartphone [20]. As we discussed in Section 10.7, smartphones do a lot more than just make telephone calls. Their many telecommunication features include providing wireless Internet access and functioning as radio-controlled clocks.

Mobile phones, also called cell phones, are radio transceivers that connect to the telephone network by communicating with a base transceiver station (BTS). The area where service is provided is divided into small geographical areas called cells, with each cell serviced by a separate BTS that includes a multi-channel radio transceiver and antenna. The place where the BTS resides is called a cellular site, and it's typically located on a radio tower or on the roof of a building. The RF signals received and transmitted by mobile phones have historically been in the UHF band, typically within the range of 400 MHz to 3 GHz. However, the most recent generations of mobile phones can operate at frequencies higher than 50 GHz.

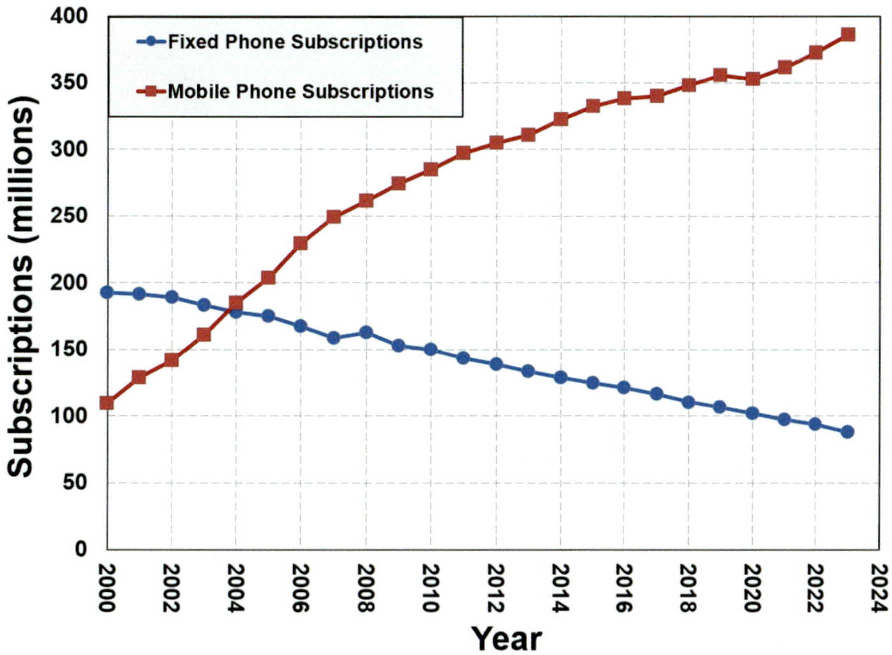

Figure 16.10 The number of mobile and fixed phone subscriptions in the United States from 2000 to 2023.

The first mobile phone networks were analog, and are now known as first generation, or 1G. The 1G networks were designed for voice telephone calls and didn't support any data services. These networks were rapidly replaced by a second generation of digital networks, now called 2G, that were designed to transfer data in addition to making phone calls. The digital networks offered much better privacy because the telephone calls were now encrypted, unlike the wireless analog calls that could be heard by any listener with an appropriate UHF receiver. In addition, the ability of digital networks to transfer data enabled the services now offered by all mobile phones, such as text messaging and Internet access. The 2G networks utilized two standard methods to communicate with cellular sites—GSM, which is an acronym for Global System for Mobile Communications, and CDMA. The use of GSM was far more common, and more than 90% of 2G phones have GSM capability.

The first GSM network was launched in Finland in 1991. The GSM standard is based on a time-division multiple access (TDMA) system that allows users to share the same frequency channel at different times. The frequency channel has a bandwidth of 200 kHz, and it's divided into frames with a duration of about 4.6 ms. The frames are divided into eight time slots, and one time slot is allocated to each user's phone. During the duration of their time slot, each user has access to the entire frequency channel, but the access only lasts for about 577 µs (Figure 16.11). However, each user seems to have exclusive access to the frequency channel, because the process of switching between phones is so fast. The process is analogous to the way a video system works; it displays images at a fast enough rate to produce the illusion of continuous motion. The GSM standard requires the BTS to use a single frequency source with an accuracy better than 0.05 ppm (5×10^{-8}) for both the RF carrier and for clocking the time base [21]. The standard doesn't require time synchronization, but the BTS clocks are usually synchronized because GPSDCs are installed at nearly every cellular site.

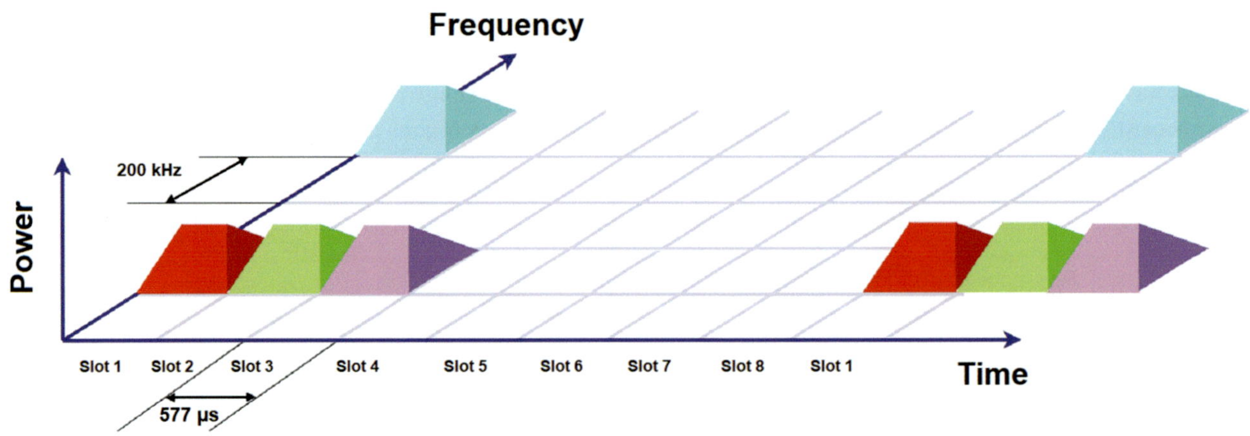

Figure 16.11 Time-division multiple access as implemented by the GSM standard, where mobile phone users share the same frequency at different times.

In a code-division multiple access (CDMA) system, all users share the same frequency channel and there are no time slots to allocate. This method became well known after it was implemented by GPS to uniquely identify the signals broadcast by each satellite (Section 12.3). When CDMA is implemented in a mobile phone network, the analog voice information is digitized and spread over a wider bandwidth at a much lower power level, and a code generator assigns a unique pseudorandom code to each user of the channel (Figure 16.12). The receiver includes a correlator circuit that uniquely identifies a specific code, and it only recovers the information that was intended for its user. The remaining information is treated as random noise and ignored.

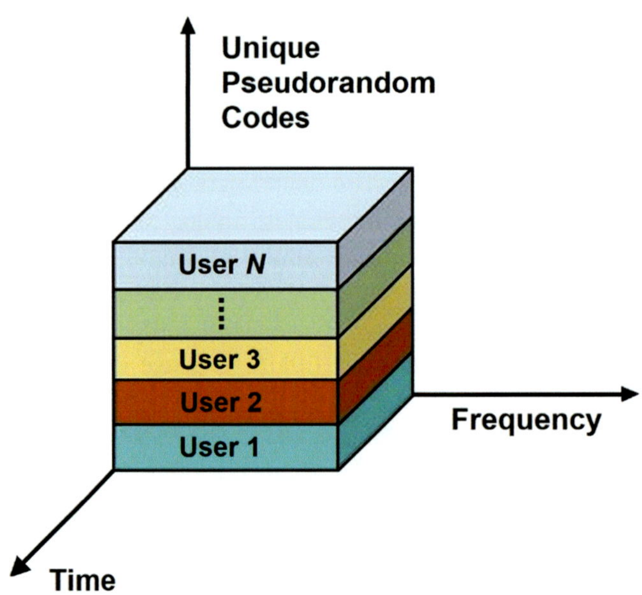

Figure 16.12 Code-division multiple access, where mobile phone users share the same frequency channel at the same time.

The CDMA method requires all BTS clocks to be synchronized to the same system time, which is nearly always accomplished by installing GPSDCs at every cellular site. The synchronization should always be within ±3 µs, and when the base stations support multiple simultaneous CDMA channels their clocks should be within ±1 µs of each other. The holdover requirement is ±10 µs for a period of not less than eight hours, even when the external source of CDMA system time is unavailable [22].

As of 2025, very few mobile phone networks still support 2G, because it's been superseded by the third, fourth, and fifth generation standards that are respectively called 3G, 4G, and 5G. The newer standards transfer data at higher speeds and are much better suited for use as mobile Internet platforms. They aren't backward compatible with each other, but many networks still support all three standards to keep older phones operational.

The first 3G networks were launched in Japan in 2001. A typical download speed for a 3G network is about 3 Mbps, about 30× faster than 2G, with the speed increase achieved by using significantly more bandwidth. The 3G networks typically implement a wide-band code-division multiple access (WCDMA) communications method that can operate in either a frequency-division duplex (FDD) or a time-division duplex (TDD) mode. For WCDMA-TDD communications, the 3G standard states that the relative time difference between base station clocks can't exceed 2.5 μs [23]. To meet this requirement, every base station clock has to be accurate to within ±1.25 μs.

To obtain much faster data transmission speeds, the 4G and 5G network designers abandoned the idea of using spread spectrum techniques to share the same carrier frequency. Instead, they transmitted data across multiple subcarrier frequencies using a method called orthogonal frequency-division multiple access (OFDMA). This method was invented at Bell Labs in 1966 by the Taiwanese-American electrical engineer Robert Chang (1933–1992), who was working on the problem of simultaneously sending multiple messages through a bandwidth-limited medium without causing interference [24]. Crosstalk and interference are prevented by placing the subcarriers orthogonal to each other, which is accomplished by aligning the peak amplitude of each subcarrier with the zero crossings of the adjacent subcarriers (Figure 16.13). The OFDMA method also conserves bandwidth and makes better use of the available spectrum than conventional FDMA methods, because the subcarriers don't have to be separated by unused guard bands (Figure 16.14).

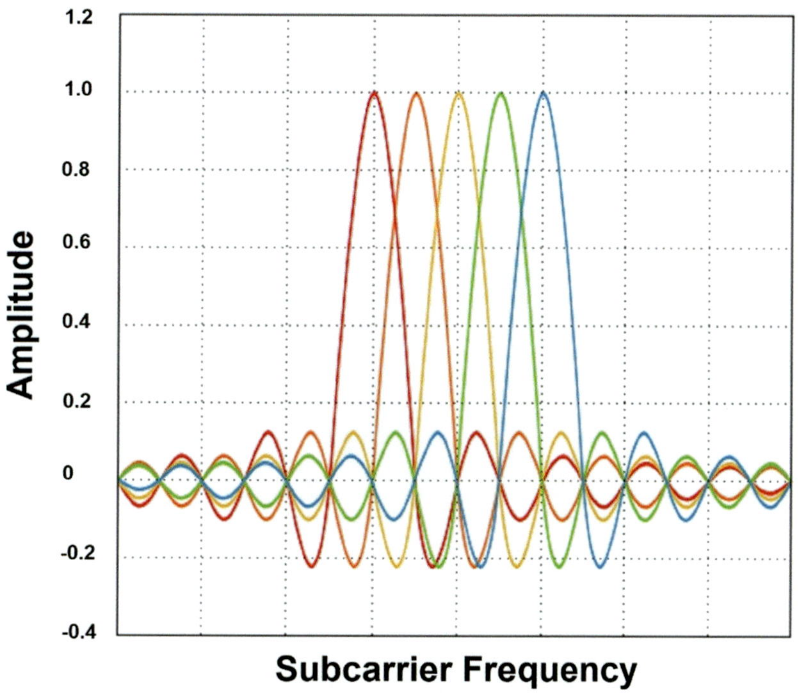

Figure 16.13 Orthogonal frequency-division multiplexing.

The first 4G networks were deployed in Sweden in 2009, and the first 5G deployment occurred in South Korea in 2019. The 4G standards are based on various Long Term Evolution (LTE) protocols that were designed to download data at peak speeds of about 100 Mbps, more than 30× faster than a typical 3G network. Under ideal conditions, a 5G network is potentially about 100× faster than 4G, with download speeds exceeding 10 Gbps [25]. In both 4G and 5G networks, the downlink from the base station to the mobile device utilizes OFDMA. In a 4G network, the uplink from the mobile phone to the base station utilizes single-carrier FDMA (SC-FDMA), but depending upon how the network is configured, a 5G uplink can use either SC-FDMA or OFDMA.

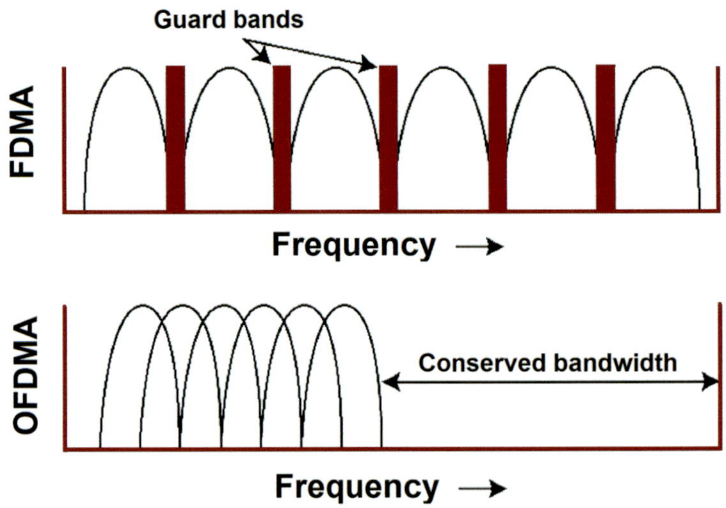

Figure 16.14 Bandwidth conservation of OFDMA compared to conventional FDMA.

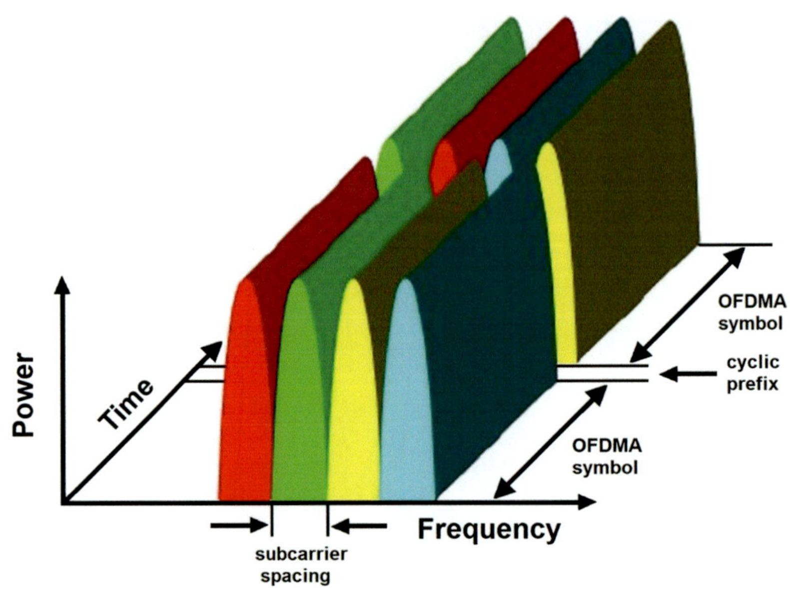

Figure 16.15 A simplified diagram of OFDMA symbol transmission.

An OFDMA symbol represents a single unit of data, and Figure 16.15 is a simplified diagram that shows how a symbol is sent. For 4G, the bandwidth of a single subcarrier is standardized as 15 kHz, which means that each subcarrier is spaced 15 kHz from its adjacent subcarriers [26]. Because a 5G network is designed to send data at higher bandwidths, the bandwidth and subcarrier spacing can be 15 kHz, but it can also be 15 kHz multiplied by 2^n. This means that it can be 30 kHz, 60 kHz, 120 kHz, 240 kHz, or even higher [27]. To keep the OFDMA symbols from interfering with each other, they are separated by a given number of subcarrier cycles called the cyclic prefix (CP). The CP is needed because the symbols are transmitted over multiple paths that each have different delays.

The duration of the CP is carefully chosen to both mitigate symbol interference and to minimize network overhead. It's typically set to be less than 10% of the period of the subcarrier bandwidth. At 15 kHz bandwidth, the CP duration is about 4.7 µs, or roughly 7% of the 66.67 µs period, and it gets correspondingly shorter at higher bandwidths. The CP duration is also related to the time accuracy requirements of a 4G or 5G network. For 4G-LTE networks, the maximum allowable time alignment error (TAE) for base stations covering a radius of less than 3 km is 3 µs [26]. This means that the 4G base station clocks have to be accurate to within ±1.5 µs.

To widen the transmission channel and increase data speeds, 5G networks assign multiple frequency blocks, called component carriers, to the same user. This technique is called carrier aggregation (CA), and the 5G time accuracy requirement depends on the type of CA. As shown in Figure 16.16, the component carriers can be in the same frequency band (intra-band) or in different frequency bands (inter-band). Intra-band CA can be either non-contiguous, which means that the component carriers are separated by a frequency gap, or contiguous, which means that there isn't a frequency gap.

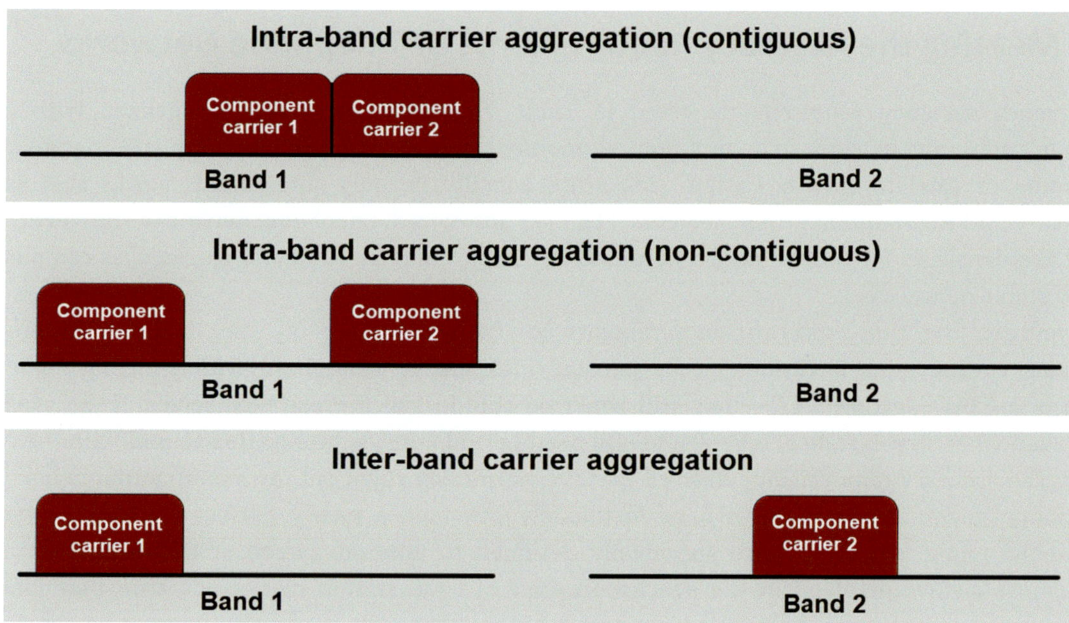

Figure 16.16 Intra-band and inter-band carrier aggregation.

For inter-band CA, or for non-contiguous intra-band CA, the maximum allowable TAE is 3 µs. Therefore, the same ±1.5 µs base station accuracy requirement that applies to 4G also applies to 5G. However, for intra-band contiguous CA, preventing OFDMA symbol interference requires reducing the maximum allowable TAE to 260 ns, which in turn reduces the accuracy requirement for base station clocks to ±130 ns [27, 28].

Table 16.1 summarizes the frequency and time accuracy requirements of wired and wireless telephone networks. It's worthwhile to note that the wired telephone networks described in Section 16.2 only have requirements for frequency accuracy. However, with the exception of GSM, the 2G to 5G wireless telephone networks have accuracy requirements for both frequency and time.

Table 16.1 A summary of frequency and time accuracy requirements for wired and wireless telephone networks

Application	Accuracy requirements	
	Frequency	Time
Primary reference clock, PRC	$\pm 1 \times 10^{-11}$	—
Enhanced primary reference clock, EPRC	$\pm 1 \times 10^{-12}$	—
GSM	$\pm 5 \times 10^{-8}$, RF carrier	—
2G (CDMA2000)	$\pm 5 \times 10^{-8}$, RF carrier	±3 μs, general requirement ±1 μs, multiple channels ±10 μs, 8-hour holdover
3G (WCDMA-TDD)	$\pm 5 \times 10^{-8}$, RF carrier	±1.25 μs
4G (LTE)	$\pm 5 \times 10^{-8}$, RF carrier	±1.5 μs
5G	$\pm 5 \times 10^{-8}$, RF carrier	±1.5 μs, non-contiguous CA ±130 ns, contiguous CA

16.4 Meeting the Accuracy Requirements of Telephone Networks

The frequency accuracy requirements listed in Table 16.1 can potentially be realized with either a free-running autonomous clock or with a non-autonomous clock that's disciplined by UTC signals received from a radio or satellite system. Cesium clocks are usually the only autonomous clocks that can meet either PRC or EPRC frequency requirements [16, 17] for wired telephone networks. However, the RF accuracy requirements for mobile phone networks are easily met by rubidium clocks, and can sometimes be met by stable quartz clocks.

In contrast, the time accuracy requirements in Table 16.1 are all very difficult to meet with autonomous clocks. Even if cesium clocks become inexpensive enough to widely deploy—perhaps in the form of a CSAC (Section 3.11)—they still won't be able to synchronize themselves. If the clocks were synchronized when first installed, they would still need periodic resynchronization to maintain the required accuracy. The accuracy requirements near ±1 μs have essentially ruled out the use of autonomous clocks in mobile phone networks, but GPSDCs have fortunately provided a nearly perfect solution. In fact, most digital mobile phone networks were specifically designed to take advantage of GPS time accuracy, and GPS is almost always involved in the synchronization of base station clocks. It's true that many base station clocks are synchronized via PTP (Section 11.2.5), but the PTP clocks also have to be synchronized, and their synchronization source is usually GPS or another GNSS time signal.

The downside, of course, is the inescapable fact that mobile phone networks are heavily dependent on GPS, perhaps more dependent than any other industrial sector. Because of this dependency, the vulnerabilities of GPS and other GNSS signals are a huge concern for network providers. If satellite time signals can't be received in a given area, the cellular sites in that same area are likely to fail. The failures will start to occur after a period that depends on the holdover capability of the clock. That period will likely be less than one day, and some clocks might fail after just a few seconds (Section 12.11).

It's impossible to know exactly how many GPSDCs are installed in mobile phone networks, but we can make some rough estimates for the United States. According to industry estimates, about 432,000 cell sites were operational in the United States at the end of 2023 [29]. These sites are typically shared by multiple network providers, but their equipment usually isn't shared, which means that each provider has to have their own reference clock. If the average number of GPSDCs installed at each cell site is between one and three, which seems to be a plausible assumption, we can estimate that the total number of units is somewhere between 432,000 and 1.3 million. Whatever the actual numbers are, they are almost certain to go up for the foreseeable future, due to the ever-increasing demand for wireless communication devices.

References

1. United States Code of Federal Regulations, "Carrier frequency departure tolerances," *47 § 73.1545*, 2004.
2. E. H. Colpitts, "Oscillation Generator," *United States Patent 1,624,537*, filed February 1, 1918, granted April 12, 1927.
3. E. B. Craft and E. H. Colpitts, "Radio Telephony," *Transactions of the American Institute of Electrical Engineers*, vol. 38, no. 1, pp. 305–343, January 1919.
4. H. W. Silver, ed., "Oscillators and Synthesizers," Chapter 9 in *The ARRL Handbook for Radio Communications*, 100th edition, The American Radio Relay League, 2022–2023.
5. H. W. Silver, ed., "Modulation" and "Receiving," Chapters 11 and 12 in *The ARRL Handbook for Radio Communications*, 100th edition, The American Radio Relay League, 2022–2023.
6. Philco Corporation, "Single Sideband Communications," *Philco Training Manual – NAVSHIPS 93224*, 164 p., 1957.
7. U. L. Rohde, J. Whitaker, and T. T. N. Bucher, *Communications Receivers*, 2nd edition, McGraw-Hill, 1997.
8. D. E. Norgaard, "A New Approach to Single Sideband," *QST*, vol. 32, no. 6, pp. 36–42, June 1948.
9. D. K. Weaver, Jr., "A Third Method of Generation and Detection of Single-Sideband Signals," *Proceedings of the Institute of Radio Engineers (IRE)*, vol. 44, no. 12, pp. 1703–1705, December 1956.
10. T. L. Wadley, "Variable-Frequency Crystal-Controlled Receivers and Generators," *Transactions of the South African Institute of Electrical Engineers*, vol. 45, no. 2, pp. 77–90, February 1954.
11. S. Franke, B. Somerville, and J. Taylor, "The FT4 and FT8 Communication Protocols," *QEX*, no. 321, pp. 7–17, July/August 2020.
12. D. V. Glen, "Networks, Signaling, and Switching for Post-Divestiture and the ISDN," *NTIA Technical Report 86-191*, National Telecommunications and Telecommunication Administration (NTIA), 120 p., February 1986.
13. R. Powers, "Reference Transmission Over Bell System Radio and Coaxial Facilities," *Proceedings of the 28th Annual Symposium on Frequency Control*, pp. 373–378, May 1974.
14. K. E. Fultz and D. B. Penick, "The T1 Carrier System," *The Bell System Technical Journal*, vol. 44, no. 7, pp. 1405–1451, September 1965.
15. J. W. Pan, "Present and Future of Synchronization in the US Telephone Network," *IEEE Transactions on Ultrasonics, Ferroelectrics, and Frequency Control*, vol. UFFC-34, no. 6, pp. 629–638, November 1987.
16. International Telecommunication Union (ITU), "Timing Characteristics of Primary Reference Clocks: Amendment 1," *ITU-T G.811 Amendment 1*, April 2016.
17. International Telecommunication Union (ITU), "Timing Characteristics of Enhanced Primary Reference Clocks," *ITU-T G.811.1*, August 2017.
18. D. Steinbock, *The Mobile Revolution: The Making of Mobile Services Worldwide*, Kogan Page, 2005.
19. International Telecommunication Union, "ITU DataHub," (https://datahub.itu.int), last accessed in January 2025.
20. R. Gelles-Watnick, "Americans' Use of Mobile Technology and Home Broadband," *Pew Research Center*, 28 p., January 2024.
21. 3rd Generation Partnership Project (3GPP), "Digital Cellular Telecommunications System (Phase 2+) (GSM); GSM/EDGE Radio Subsystem Synchronization," *3GPP TS 45.010 version 18.0.0*, 38 p., May 2024.
22. 3rd Generation Partnership Project 2 (3GPP2), "Recommended Minimum Performance Standards for cdma2000 Spread Spectrum Base Stations: Release C," *C.S0010-C, Version 2.0*, 153 p., February 2006.
23. 3rd Generation Partnership Project (3GPP), "Universal Mobile Telecommunications Systems (UMTS); Synchronisation in UTRAN Stage 2," *3GPP TS 25.402 V18.0.0*, 53 p., April 2024.
24. R. W. Chang, "Synthesis of Band-Limited Orthogonal Signals for Multichannel Data Transmission," *The Bell System Technical Journal*, vol. 45, no. 10, pp. 1775–1796, December 1966.

25. H. Hodara and E. Skaljo, "From 1G to 5G," *Fiber and Integrated Optics*, vol. 40, no. 2–3, pp. 85–183, December 2021.
26. 3rd Generation Partnership Project (3GPP), "LTE; Evolved Universal Terrestrial Radio Access (E-UTRA); Requirements for Support of Radio Resource Management," *3GPP TS 36.133 V18.7.0,* 3694 p., October 2024.
27. 3rd Generation Partnership Project (3GPP), "5G; NR; Base Station (BS) Radio Transmission and Reception," *3GPP TS 38.104 V18.8.0,* 420 p., January 2025.
28. S. Ruffini, M. Johansson, B. Pohlman, and M. Sandgren, "5G Synchronization Requirements and Solutions," *Ericsson Technology Review*, vol. 103, no. 1, pp. 52–63, January 2021.
29. CTIA, "2024 Annual Survey Highlights," 10 p., 2024.

Chapter 17

Time and Electricity

Electricity is really just organized lightning.

— George Carlin

Reliable electric power distribution is a top priority for every nation. This is appropriate, because the citizens of all nations depend on electric power for their safety, comfort, and quality of life. A long power outage can be more than just an inconvenience—it can cause lives to be lost. It can also have devastating effects on both national security and the economy.

Electric power is distributed through large systems called power grids, and keeping power grids reliable requires keeping accurate time. Power grids are spread across large geographic areas, and they need reference time signals that reach all parts of the grid. Keeping all of a power grid's resources synchronized to UTC allows the electricity allocations to be adjusted in real-time, so that enough electric power is delivered to all areas. Synchronization to UTC also makes it possible to quickly find and repair grid failures, which reduces the likelihood of a long and widespread power outage.

This chapter explains why the electric power industry needs accurate clocks and oscillators. Electric power is distributed as alternating current (AC), which is the topic of Section 17.1. The need for accurate time in electric power grids is discussed in Section 17.2. Synchrophasor measurements, which capture information about the health of a power grid, are the topic of Section 17.3. Section 17.4 concludes the chapter by summarizing the time accuracy requirements of electric power grids, and by describing the types of reference clocks that the grids rely upon.

17.1 Alternating Current

Alternating current is the standard form of electricity delivered to homes and businesses, and it's the source of power for all electrical devices that are plugged into wall sockets. The direct current (DC) that batteries produce doesn't involve frequency because it flows continually in the same direction, but AC does involve frequency. The direction of the current is reversed, or alternated, a specific number of times per second.

As Time Goes By: A Comprehensive Guide to Time and Frequency Measurements
Michael A. Lombardi

ISBN 978-981-5352-28-3 (Hardcover), 978-1-042-01532-0 (eBook)
www.jennystanford.com

DOI: 10.1201/9781042015320-23

It is interesting to note that DC was briefly the standard form of power when electricity was first distributed in the 1880s. The chief advocate for DC power distribution was the American inventor and businessman Thomas Edison (1847–1931). In 1879, Edison invented the incandescent light bulb, and he immediately began to patent and promote DC distribution systems to power the new electric lights. In 1882, Edison's company built a power station that distributed 110 volts of direct current to consumers in New York City. Edison soon had competition from the American inventor and industrialist George Westinghouse (1846–1914), who entered the electric lighting business in 1884. Westinghouse originally planned to develop his own DC power distribution system while trying not to infringe on Edison's patents. He contracted the American physicist William Stanley (1858–1916) to work on his system, but Stanley had read about the alternating current systems that were then being developed in Europe, so their attention soon shifted from DC to AC.

The distribution of DC power had inherent limitations, mostly because DC wasn't easy to convert to higher or lower voltages. This prevented it from being used for long-distance power transmissions. Alternating current solved that problem, because it could be easily converted to different voltages by using a transformer. Stanley built the first practical AC transformer in 1885, and the following year he demonstrated an AC current distribution system in Great Barrington, Massachusetts. Stanley's power station was slightly more than a kilometer away from Great Barrington's business district, so he raised the generator voltage from 500 volts to 3,000 volts for the trip downtown, and then lowered it back to 500 volts so it could be distributed through the streets, where it illuminated lights that were hung from the trees. Stanley's method was based on the same concept still used in present-day power systems, a generator "step-up" transformer is used to raise the system voltage to a very high level for long-distance transmission, and then "step-down" transformers at power substations lower the voltage to some intermediate level for local distribution. By the end of 1886, the newly formed Westinghouse Electric Company had built its first commercial AC power plant in Buffalo, New York [1].

Needless to say, Edison eventually lost the battle that some historians call "The War of the Currents." Westinghouse licensed or bought numerous patents—including those of the Serbian-American inventor Nikola Tesla (1856–1943)—and his company continuously worked on ways of making AC power distribution safer and more reliable. With Westinghouse's financial backing, AC was firmly entrenched as the standard form of electric power distribution by the mid-1890s.

The AC frequency is either 50 Hz or 60 Hz, depending on the region of the world. At least 75% of the world's population uses 50 Hz, but 60 Hz is used in North America, in most of Central and South America, and in a few parts of Asia. The RMS amplitude of the AC sine wave is usually either 220, 230, or 240 volts in the regions that use 50 Hz, and either 100, 110, or 120 volts in the regions that use 60 Hz. However, some regions can supply dual voltages—for example, 120 and 240 volts—with the higher voltage provided for devices that need more power. Figure 17.1 depicts 50 Hz (blue line) and 60 Hz (red line) sine waves, and their respective RMS amplitudes of 230 and 120 volts are indicated by the dotted lines. Their respective peak voltages of 325 and 170 volts are obtained by multiplying the RMS voltage by $\sqrt{2}$.

The AC frequency can regulate the time of an electric clock, as Tesla first demonstrated at the Chicago World's Fair in 1893. Since then, countless numbers of electric clocks that are always plugged into an outlet—including alarm clocks and clocks on kitchen appliances—have kept time by using the AC frequency as their oscillator.

Clocks controlled by the AC frequency became much less common after inexpensive quartz clocks became available, but the method can still provide accurate time if the power grid supports a feature called time error correction (TEC). The TEC feature is necessary for timekeeping, because the AC frequency can't always be distributed at its nominal value, due to the constant variations in the amount of electricity that's generated and consumed. As we discussed in Section 6.2.2, any frequency inaccuracy in a clock's oscillator will cause the clock to accumulate a time difference. For example, a clock that's designed to be controlled by 60 Hz signals will gain 28.8 s every 24 hours (86,400 s), if it receives signals that have an average frequency of 60.02 Hz:

$$\frac{60.02-60}{60} \times 86{,}400\ \text{s} = 28.8\ \text{s}. \tag{17.1}$$

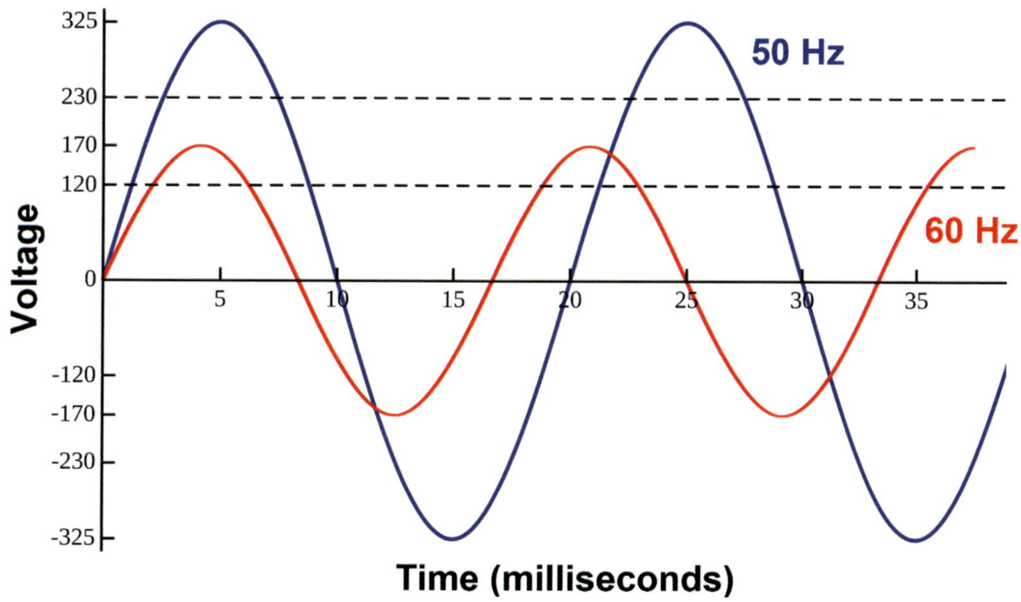

Figure 17.1 Examples of 50 Hz and 60 Hz AC sine waves.

When TEC is implemented, the long-term average frequency is kept very close to its nominal value by increasing or decreasing the frequency when the accumulated time difference becomes too large. In the above example, where the frequency is 0.02 Hz higher than nominal, corrections can be applied by generating a frequency that is 0.02 Hz lower than nominal, or 59.98 Hz. The frequency corrections usually begin on the hour or the half-hour, and they continue for periods that are long enough to keep the accumulated time difference below a specified threshold, which is usually ±10 s [2].

Figure 17.2 shows the time difference (a) and frequency difference (b) between the AC frequency received in Boulder, Colorado and UTC(NIST) during a period of 180 days. A clock controlled by this frequency would typically have been accurate to within a few seconds, but would occasionally have been fast or slow by as much as 8 s. Even though the average frequency difference for the 180-day period was just 42 μHz, the range of frequency differences was slightly larger than ±0.4 Hz.

Time error corrections are applied by electric power utilities in North America, Europe, Japan, Australia, and elsewhere, and they allow clocks controlled by the AC frequency to have long-term accuracy that's usually sufficient for everyday timekeeping purposes [3]. The AC frequency is an unmodulated sine wave without a time code, so all clocks controlled by AC need to be initially synchronized to another reference. Once an AC-controlled clock is synchronized, multi-year measurements conducted on TEC-enabled grids have shown that time accurate to within ±10 s can be kept indefinitely [3, 4].

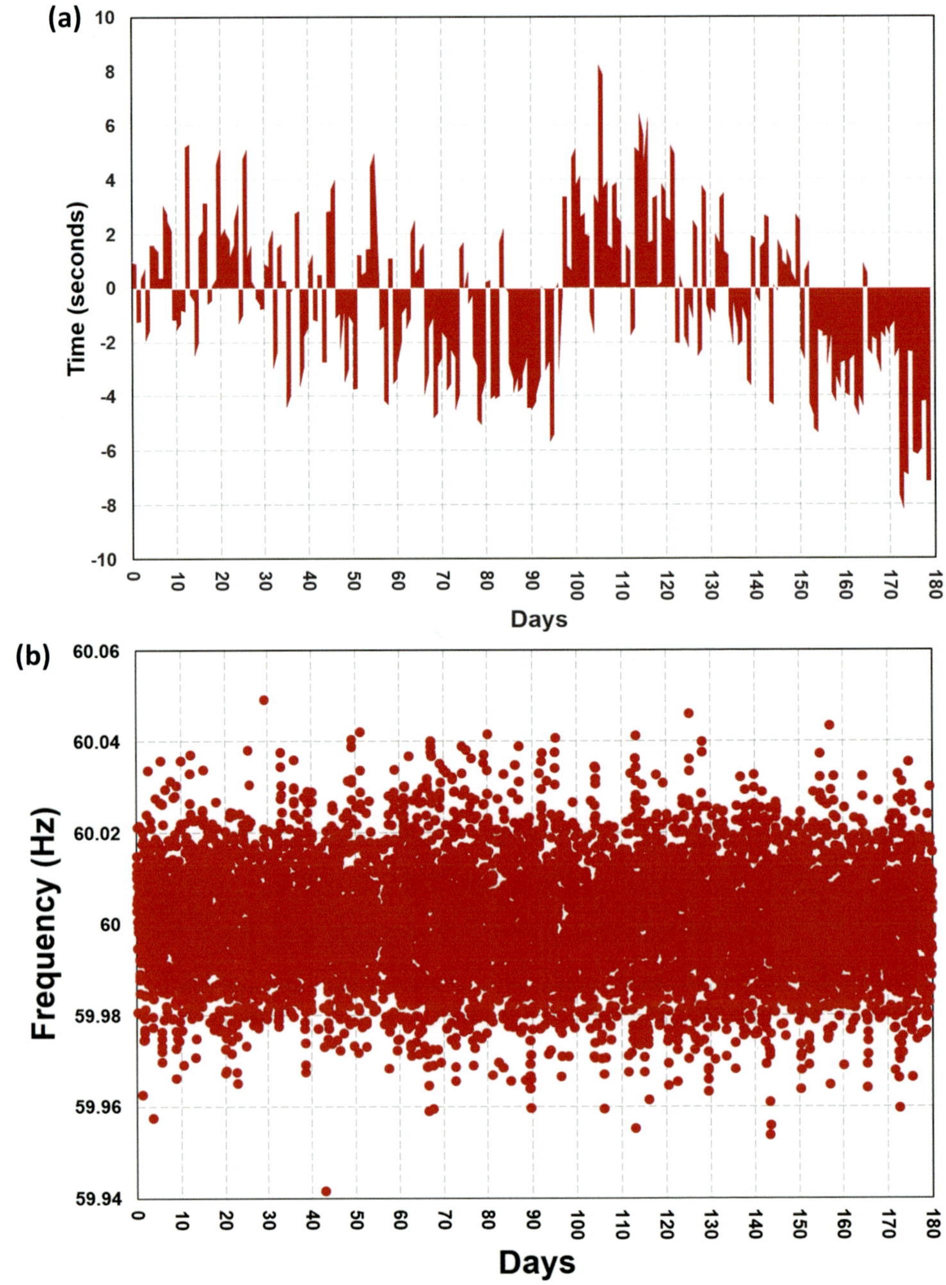

Figure 17.2 Time (a) and frequency (b) differences of the 60 Hz AC frequency during a 180-day comparison to UTC(NIST).

17.2 The Need for Accurate Time in Electric Power Grids

The power grids that distribute electricity are both large and complex, and their size and complexity vary in different parts of the world. For the purposes of example, we'll briefly describe the North American power grid, which distributes 120 and 240 volts of electric power at a nominal frequency of 60 Hz. A non-profit corporation called the North American Electric Reliability Corporation (NERC) is responsible for the reliability and security of this grid, which consists of the four distinct interconnections shown in Figure 17.3. The four interconnections provide electric power to nearly 400 million citizens as of 2025, with the bulk of the power provided by the Eastern and Western interconnections. The Eastern interconnection serves more than 200 million people in all or part of 36 states, and the Western interconnection serves more than 90 million people in all or part of 14 states and two Canadian provinces. The state of Texas is covered by a third interconnection, and a fourth interconnection covers the Canadian province of Quebec. The grid systems in Alaska and Hawaii are not part of the interconnections shown in Figure 17.3, but the coordinating body in Alaska is an affiliate of NERC [5]. In the United States alone, the power distribution system consists of more than 360,000 miles (~580,000 km) of transmission lines, including about 180,000 miles (~290,000 km) of high-voltage lines, with the lines connected to about 7,000 power plants [6].

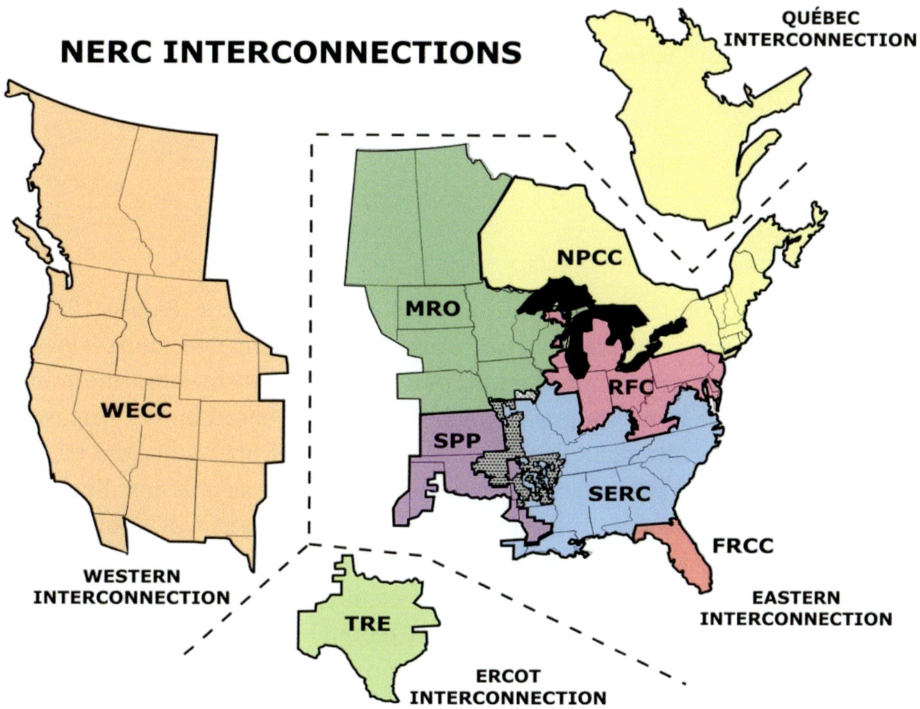

Figure 17.3 The four interconnections of the North American power grid (courtesy of NERC).

Power grid operators need accurate clocks so they can quickly identify and respond to grid incidents and failures. If the response time is fast enough, it can prevent an outage from happening at all, and fast responses can also shorten outages and prevent them from spreading when they inevitably do occur. Protective relays are the workhorse devices that protect power grid equipment. When triggered by a power grid incident or failure, relays have to respond quickly to prevent significant damage, and they do so by tripping a circuit breaker, by opening a line, or by isolating a power plant. The typical response time of a protective relay ranges from less than 1 ms to about 2 ms, and the relays are required to have clocks that are accurate to within ±2 ms [7]. Protective relays respond to events whenever necessary, and they don't rely on accurate time from their internal clock when taking action such as tripping a circuit breaker. However, their clock records the time of their response, which allows an accurate failure analysis to be performed later [8].

The equipment that monitors and records the time of power grid disturbances—including event and digital fault recorders—also has time accuracy requirements. Event recorders monitor multiple input signals, and they time stamp everything they record. Their reports provide power grid operators with a continuous log of the sequence of events. Digital fault recorders are diagnostic tools that monitor and time stamp measurements of the power grid's current and voltage. The time-stamped data is then aligned and correlated with data from other recorders to analyze when and where a fault occurred. This analysis allows grid operators to find, for example, the time and place of a lightning strike.

The NERC time accuracy requirement for disturbance monitoring equipment is ±2 ms, which again allows reasonably accurate postmortem failure analysis [7]. However, power companies often operate equipment that conforms to the more stringent requirements of the International Electrotechnical Commission (IEC) standards. As listed in Table 17.1, the IEC categorizes the time stamping of events into five performance classes—T1 through T5—with accuracy requirements ranging from ±1 ms to ±1 μs [9]. The T3 class of ±25 μs accuracy is usually sufficient for event and digital fault recorders [10].

Table 17.1 IEC time accuracy requirements [9]

Performance class	Time accuracy requirement
T1	±1 ms
T2	±100 μs
T3	±25 μs
T4	±4 μs
T5	±1 μs

The quick location of faults on transmission lines supported by towers (Figure 17.4) usually requires T5 class accuracy. There are many events that can cause transmission line faults—including equipment failures, lightning strikes, bad weather, animal interference, and vandalism—and there are many ways to locate faults. A common way, called traveling wave fault detection, is based on the premise that a traveling wave will always be generated at the point of a fault. During normal operation, the traveling voltage and current waves are related by the characteristic impedance of the transmission line. However, when a fault occurs, the voltage drops and moves towards zero, while the current wave doubles. This generates a high-frequency pulse that travels down the transmission line in both directions at nearly the speed of light. The arrival time of this pulse is measured at both ends of the line, and is time stamped by clocks that have sub-microsecond resolution and accuracy [11]. This method makes it possible to quickly locate the tower nearest to the fault. For example, if the transmission line is supported by towers spaced 300 m apart, ±1 μs accuracy can find the nearest tower, because 1 μs is the period of a signal with a 300 m wavelength [12].

Figure 17.4 High-voltage transmission lines supported by towers.

17.3 Synchrophasor Measurements

Most of the attention in power grid synchronization is now focused on synchrophasor measurements. These measurements capture voltage, current, and phase angle readings at high data rates, and they provide a more comprehensive picture of the health of a power grid than all previous monitoring methods. Synchrophasor measurements inform grid operators of problems like voltage instability and power swings as soon as they start to occur. This greatly improves the grid's reliability, because it allows grid operators to be proactive and to respond to problems before an outage occurs.

Synchrophasor measurements are performed by instruments called phasor measurement units (PMUs). The PMU was invented by the electrical engineer Arun Phadke (b. 1938) and his colleagues at Virginia Tech University in Blacksburg, Virginia. Phadke and his team built the first prototype PMU—based on a GPSDC—in 1988 [13]. Commercial PMUs were made available shortly afterwards, and they are now deployed internationally in nearly all electric power grids [14]. A simplified block diagram of a PMU is shown in Figure 17.5.

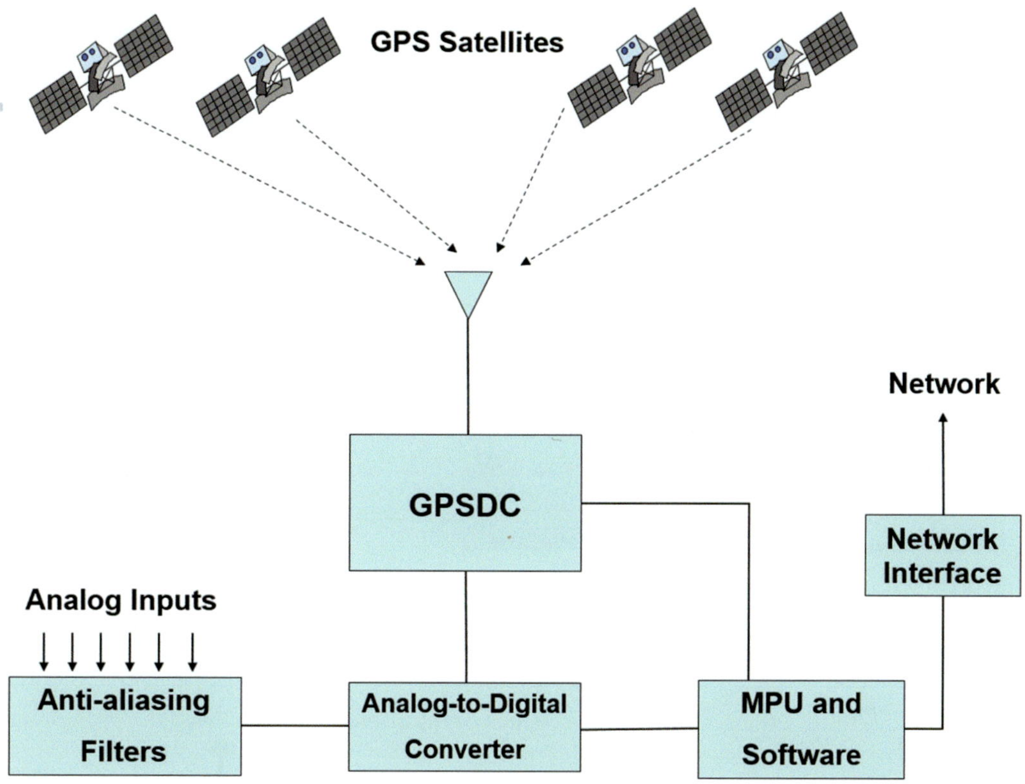

Figure 17.5 A simplified block diagram of a phasor measurement unit.

To better understand synchrophasor measurements, we should first realize that a phasor is just a vector that represents a sine wave, and that a complete sine wave can be constructed by a single vector that rotates counter clockwise in a circle. This relationship works, because a sine wave is really just a projection of circle's rotating radius onto a vertical axis. The rotational rate of a phasor is

$$\omega = 2\pi f, \tag{17.2}$$

where ω is the angular velocity expressed in radians per second, and f is the frequency, which will be either 50 Hz or 60 Hz.

In the electric power industry, the term phasor is normally used to refer to a stationary phasor rather than a rotating phasor. If a stationary phasor is recorded at a given point in time, it will indicate the amplitude and direction of the sine wave. Figure 17.6 is a simplified diagram where the red vectors inside the circle are stationary phasors, or "snapshots" of the sine wave that were taken at a given time. The horizontal blue lines connect the stationary phasor to the corresponding point on the sine wave.

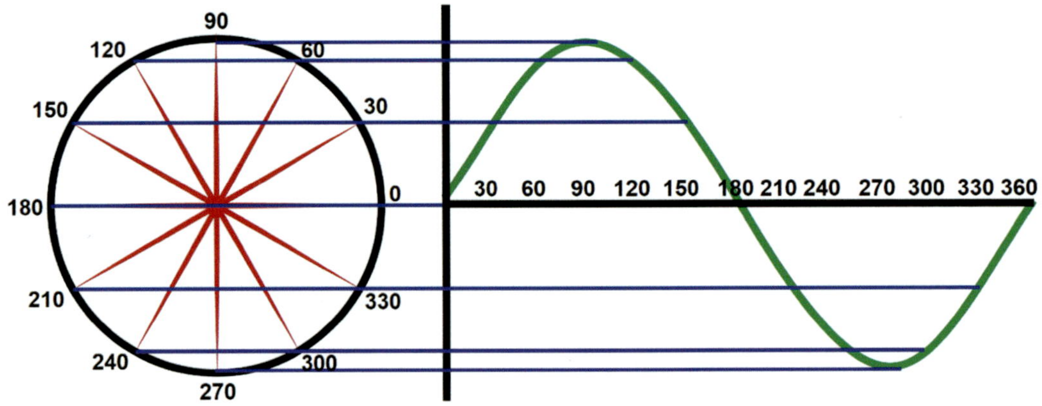

Figure 17.6 The relationship between a phasor and a sine wave.

Synchrophasors got their name because they are time stamped by clocks that are synchronized to UTC, such as the GPSDC in Figure 17.5. A synchrophasor measurement is illustrated in Figure 17.7, which shows the phase of a sine wave that was sampled at two time-stamped points, δ_1 and δ_2, and also shows how the two samples are represented by two stationary phasors inside of a circle. The phase angle difference between the two stationary phasors is $\Delta\delta$, and Δt is the inverse of the sampling rate. The frequency, f, is derived as

$$f = f_{\text{nom}} + \Delta f, \tag{17.3}$$

where f_{nom} is the nominal frequency of the sine wave, and

$$\Delta f = \frac{\Delta\delta}{360° \times \Delta t}. \tag{17.4}$$

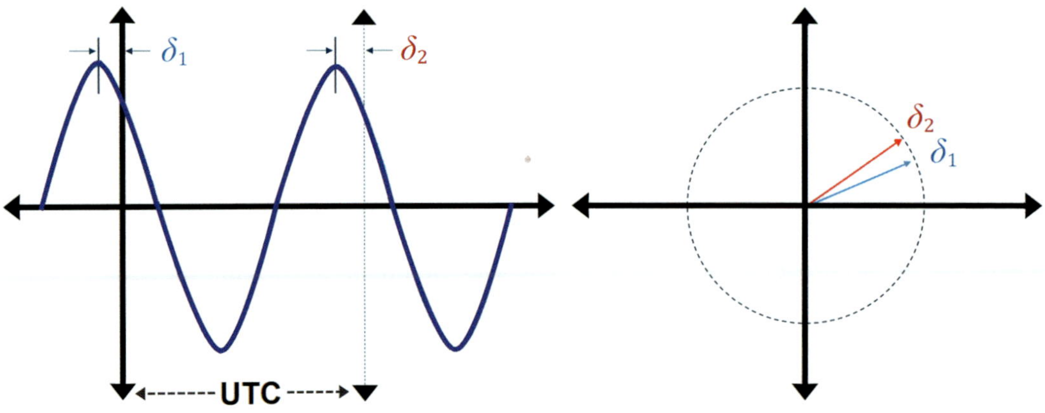

Figure 17.7 A synchrophasor measurement.

The primary function of a PMU is to measure the total vector error (TVE), which is the difference between the phasor under test and a phasor referenced to UTC. The word "total" is used because the TVE includes the error in the phasor angle and amplitude, in addition to the time synchronization error. The maximum acceptable TVE is 1%.

Figure 17.8 represents a 1% TVE as a circle drawn around the end of a phasor, with the size of the circle exaggerated for clarity. If the angle error exceeds 0.573°, the TVE will exceed 1% even if there is no amplitude error. A PMU can be calibrated by inputting a test signal of known accuracy, and then ensuring that the measurement results that it outputs fall inside the circle.

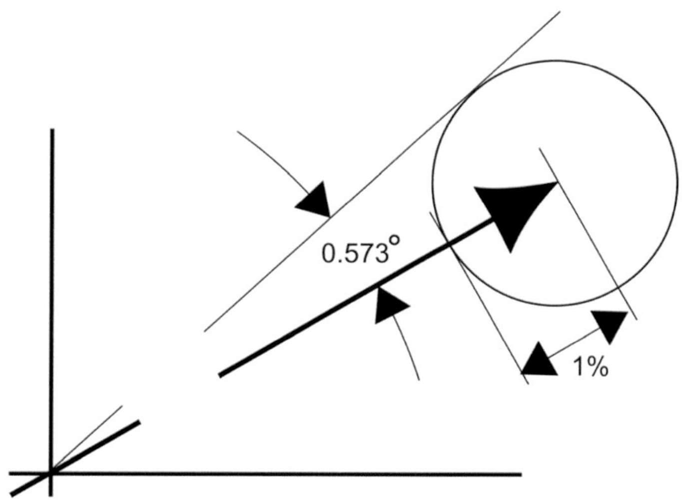

Figure 17.8 A total vector error (TVE) of 1%.

The time accuracy requirement for a PMU, t_{ar}, is calculated as

$$t_{ar} = \frac{AC_{per}}{360^\circ / 0.573^\circ}, \tag{17.5}$$

where AC_{per} is the period of the AC frequency, 360 is the number of degrees in a circle, and 0.573° is the maximum acceptable phase angle error. Therefore, the time accuracy requirement is 31.8 μs for 50 Hz AC (20 ms period) and 26.5 μs for 60 Hz AC (16.67 ms period). As discussed in Chapter 6, verifying an accuracy requirement typically requires a reference with a TAR of 10:1. This means that the reference clock needs to be accurate to within ±3.2 μs for a 50 Hz system and ±2.6 μs for a 60 Hz system [15]. These values are near 1 μs, so it's customary to state that ±1 μs is the required time accuracy.

A PMU typically time stamps the voltage and current vector values at a rate of 30 to 120 samples per second. Measurements from multiple networked PMUs are continuously sent to a data center, where the readings are aligned and compared to provide a near real-time indication of the health of the power grid. This data not only provides advance notice of impending problems, but it also influences decisions made about the power allocations within the grid. These decisions can be made automatically in a configuration known as a smart grid, or they can be made based on the analysis and experience of power grid operators. In either case, the information provided by PMUs makes it easier to keep the grid running smoothly.

Figure 17.9 provides a map showing the locations of networked PMUs in the North American power grid as of 2017 [7]. Tens of thousands of PMUs are operational worldwide as of 2025, with more than 2,500 operational in the United States alone.

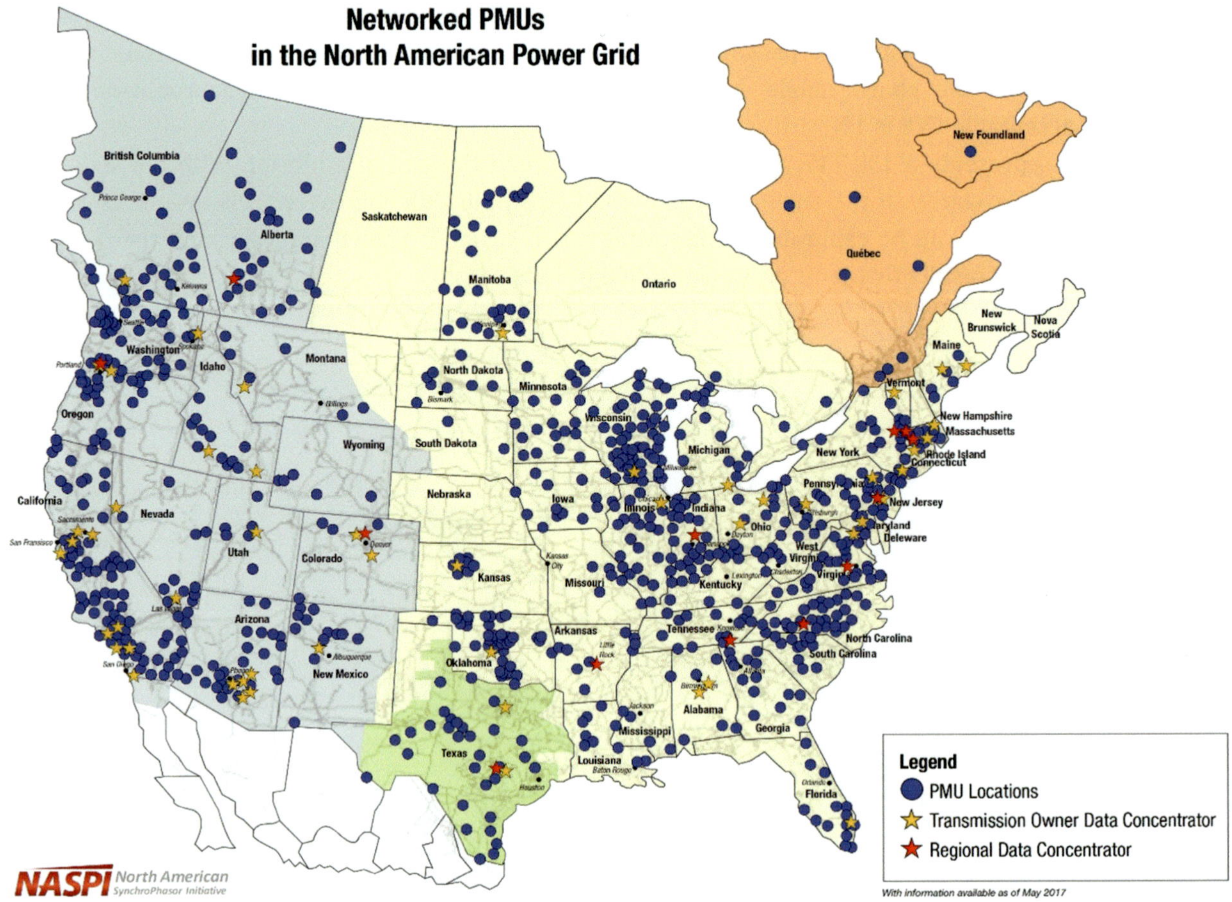

Figure 17.9 A map of PMUs in the North American power grid as of 2017 (courtesy of the North American Synchrophasor Initiative).

17.4 Power Grid Accuracy Requirements and Reference Clocks

Power grids have evolved for many years, and they are comprised of numerous systems that have different time accuracy requirements. The older systems have less stringent requirements, because they were designed using clocks and time transfer methods that predated GPS. The newer systems have more stringent requirements—especially the PMUs used for synchrophasor measurements—because they were designed from the beginning to take advantage of GPSDC accuracy. Table 17.2 summarizes the timing requirements of the systems described in the previous two sections.

Table 17.2 Summary of timing requirements for electric power distribution systems

Application	Timing requirements
Protective relays	Time accuracy of ±2 ms Response time of 1 ms to 2 ms
Event and digital fault recorders	Time accuracy of ±25 μs to ±2 ms
Synchrophasor measurements	Time accuracy of ±2.6 μs (60 Hz system) Time accuracy of ±3.2 μs (50 Hz system)
Traveling wave fault location	Time accuracy of 0.1 μs to 1 μs Time resolution of 0.1 μs

All of the accuracy requirements in Table 17.2 can be easily met by using GPSDCs or other types of GNSS disciplined clocks. However, without GNSS, some of the requirements are very difficult to meet, especially those for synchrophasor measurements and traveling wave fault location. For this reason, the reference time signals in a power grid nearly always originate either solely from GPS, or from some combination of GPS, GLONASS, Galileo, and BeiDou (Section 12.8). When the reference time signals aren't directly distributed by a GNSS-disciplined clock, they're usually distributed by a system that's been synchronized to GNSS. That system can be an IRIG time code system (Section 4.2.1), but it's usually a NTP or PTP time server (Section 11.2) that's connected to a LAN [16, 17].

Power grid operators know that GNSS disciplined clocks can easily provide the accuracy that they need, so they worry more about reliability than accuracy, due to the vulnerabilities of satellite signals (Section 12.11). Their concerns are well founded because the control of a power grid is based on a constant series of decisions that are either made automatically by software systems, or made manually by operators. Bad decisions can be made if the reference clocks are unavailable or wrong. For example, if the time signal received by a PMU is being jammed or spoofed, the resulting time error will cause false calculations of phase angle and a misalignment of measured grid conditions relative to other PMUs. Any decisions that are made based on false information could potentially cause a grid failure.

To safeguard against grid failures, the electric power industry has committed to developing resilient time distribution systems that have multiple UTC references. The references can include disciplined clocks that receive signals from multiple GNSS constellations, network time signals that originate from GNSS clocks located outside of the current area of vulnerability, and network time signals from UTC(k) time scales that are received via optical fiber. Ideally, all of the available UTC references will be continuously intercompared, so that any reference found to be inaccurate can be ignored [18]. The IEEE guidelines for designing power station time-distribution systems stress the importance of having diverse UTC references, because it's unlikely that all of the references will simultaneously fail. The guidelines also emphasize the importance of using clocks with good holdover capability [19].

In conclusion, it's important to note that most extended power grid failures are caused by severe weather events that do extensive damage to a grid's resources—such as floods, hurricanes, and ice storms. Time synchronization problems occasionally do cause outages, but power grids usually continue to operate at reduced capacity and efficiency when synchronization problems occur. Even so, the vulnerabilities of GNSS time signals are one of the main factors that have kept power grids from becoming totally automated. Everyone knows that long power outages can have potentially devastating effects, so the risks associated with total grid automation will likely continue to outweigh the potential benefits.

References

1. T. P. Hughes, *Networks of Power: Electrification in Western Society, 1880–1930*, Johns Hopkins University Press, 1983.
2. NERC, "Balancing and Frequency Control," 53 p., January 2011.
3. Y. Zhang, W. Yao, S. You, W. Yu, L. Wu, Y. Cui, and Y. Liu, "Impacts of Power Grid Frequency Deviation on Time Error of Synchronous Electric Clock and Worldwide Power System Practices on Time Error Correction," *Energies*, vol. 10, pp. 1283/1–15, August 2017.
4. J. E. Hardis, B. Fonville, and D. Matsakis, "Time and Frequency from Electrical Power Lines," *Proceedings of the 48*th *Annual Precise Time and Time Interval Meeting (PTTI)*, pp. 372–386, January 2017.
5. NERC, "2023 Long-Term Reliability Assessment," 135 p., December 2023.
6. U. S. Department of Energy, "United States Electric Industry Primer," Office of Electricity Delivery and Energy Reliability, *DOE/OE-0017*, 94 p., July 2015.
7. NERC, "Disturbance Monitoring and Reporting Requirements," *NERC Standard PRC-002-2*, 38 p., September 2015.
8. North American Synchrophasor Initiative (NASPI), "Time Synchronization in the Electric Power System," *NASPI Technical Report PNNL-26331*, 54 p., March 2017.

9. IEC, "Communication Networks and Systems for Power Utility Automation – Part 5: Communication Requirements for Functions and Device Models," *IEC 61850-5:2013*, 306 p., January 2013.
10. A. Apostolov, "Time in IEC 61850 Based Substation Protection and Control Systems," *Protection, Automation, and Control World Magazine*, no. 52, pp. 18–25, June 2020.
11. IEEE, "IEEE Guide for Determining Fault Location on AC Transmission and Distribution Lines," *IEEE Standard C37.114-2014*, 76 p., December 2014.
12. H. Lee and A. M. Mousa, "GPS Travelling Wave Fault Locator Systems: Investigation into the Anomalous Measurements Related to Lightning Strikes," *IEEE Transactions on Power Delivery*, vol. 11, no. 3, pp. 1214–1223, July 1996.
13. A. G. Phadke, "Synchronized Phasor Measurements ~ A Historical Overview," *Proceedings of the 2002 IEEE/PES Transmission and Distribution Conference and Exhibition*, pp. 476–479, October 2002.
14. A. G. Phadke and T. Bi, "Phasor Measurement Units, WAMS, and Their Applications in Protection and Control of Power Systems," *Journal of Modern Power Systems and Clean Energy*, vol. 6, pp. 619–629, July 2018.
15. IEEE, "IEEE Standard for Synchrophasor Measurements for Power Systems," *IEEE Standard C37.118.1-2011*, 61 p., December 2011.
16. H. Guo and P. Crossley, "Design of a Time Synchronization System Based on GPS and IEEE 1588 for Transmission Substations," *IEEE Transactions on Power Delivery*, vol. 32, no. 4, pp. 2091–2100, August 2017.
17. D. Macii and S. Rinaldi, "Time Synchronization for Smart Grids Applications: Requirements and Uncertainty Issues," *IEEE Instrumentation & Measurement Magazine*, vol. 25, no. 6, pp. 1–18, September 2022.
18. P. Robertson, K. Fodero, C. Huntley, M. Elshafi, and D. Williams, "Precise Time, All the Time: A Resilient Architecture for the Electric Power Industry and Beyond," *IEEE Power & Energy Magazine*," vol. 21, no. 5, pp. 27–37, September/October 2023.
19. IEEE, "IEEE Guide for Designing a Time Synchronization System for Power Substations," *IEEE Standard 2030.101-2018*, 118 p., March 2018.

Chapter 18

Time and Money

Remember that time is money.

— Benjamin Franklin, "Advice to a Young Tradesman"

The connection between time and money is well understood by all citizens in a capitalistic society. As everyone knows, we get paid for time when we work for an employer, and we pay for time when we rent something. The amount of money we receive for work is usually determined by an hourly rate, where our pay is based on the number of hours worked, or by a fixed salary, where our pay is based on working an agreed-upon number of hours over the course of a year. The list of things we can rent for a specified amount of time includes homes, hotel rooms, cars, parking spaces, washers and dryers in laundromats, and subscriptions to all types of services.

There are rules and regulations that apply when we get paid for time or when we pay for time, and they're usually specified in the fine print of some type of contract. Most of these regulations don't involve high-accuracy timekeeping, so we can comply with them by just using common sense. For example, if our job requires us to check in and out with a punch clock, checking in 15 minutes early or late might annoy our employer, but checking in 15 s early or late probably won't matter at all. The same common-sense logic applies when we rent a hotel room. If we stay for a few extra hours after the check-out time we'll probably be charged for an extra day, but it usually doesn't matter if we stay for a few extra minutes.

To protect consumers, the regulations for some time interval measurements are enforced by local or national governments, and the guidelines are specified in published standards for weights and measures. The measurements usually have very coarse requirements that are defined using terms such as "underregistration" and "overregistration." Underregistration refers to situations where the measured time interval is shorter than its nominal value, for example, when a parking meter indicates that an hour has elapsed after just 55 minutes. This causes consumers to pay too much for time. Overregistration refers to the opposite situation, where the measured time interval is longer than its nominal value, which causes consumers to pay too little. In the United States, there are no regulations placed on the overregistration of parking meters—any operator can give their customers a more than fair deal if they want to—but underregistration of more than seven minutes per hour (11.7%) is illegal. The punch clocks that record the hours worked by employees have an allowable tolerance of 3 s per hour for both

As Time Goes By: A Comprehensive Guide to Time and Frequency Measurements
Michael A. Lombardi

ISBN 978-981-5352-28-3 (Hardcover), 978-1-042-01532-0 (eBook)
www.jennystanford.com

DOI: 10.1201/9781042015320-24

underregistration and overregistration (0.08%), which is stricter than the requirement for parking meters, but still coarse enough to be easily checked by using a stopwatch as a reference [1].

The high-accuracy time measurements that relate to money mostly apply to financial markets, and the regulations for these measurements are relatively new. In the rest of this chapter, we'll discuss why financial markets need accurate time. We'll also discuss the time accuracy requirements of stock exchanges, and the methods that the exchanges use to synchronize their clocks and to verify their accuracy.

18.1 The Need for Accurate Time in Financial Markets

Financial markets are places where investors buy and sell financial instruments. There are three main categories of financial instruments—equities, derivatives, and debt. Equity investors buy and sell stocks, which are shares in the ownership of a company. Derivative investors buy and sell contracts that derive their value from an underlying asset, which typically is stock. For example, a common type of derivative is a stock option, which is a contract that gives its owner the right to either buy or sell shares of stock at a specific price. Debt investors typically buy and sell bonds, where the money they invest is loaned to a company or government, in exchange for interest payments and the eventual return of their original loan amount.

When a financial instrument is bought or sold, the transaction is called a trade, and the various forms of equities, derivatives, and debt each have their own trading facilities. Regulatory agencies require these facilities to record the time of each trade with a specified level of accuracy. For the purposes of our discussion, we'll focus solely on the time accuracy requirements of stock exchanges, the facilities where stocks are traded. Due to the large volume of trades, stock exchanges have a greater need for accurate time than any other type of financial market.

The world's two largest stock exchanges—both headquartered in New York City—are the New York Stock Exchange (NYSE) and the Nasdaq stock exchange. The name "Nasdaq" is a registered trademark, but it originated as an unwieldy acronym for the National Association of Securities Dealers Automated Quotations. At least 60 other stock exchanges operate internationally, with some listing their own stocks, and others handling a percentage of the trades for stocks listed elsewhere.

Stock exchanges are now mostly automated due to electronic trading. The decisions about which shares to buy and sell are often made by software algorithms, and the trades often execute within a small number of microseconds. In this high-speed environment, inaccurate time can make illegal trades and market manipulation very difficult to detect. We'll begin this section with a brief description of how stocks are traded. We'll then describe some of the changes brought about by electronic trading, and explain how accurate time can help prevent illegal trading activity.

18.1.1 The Fundamentals of a Stock Trade

Stock trades involve buyers, sellers, and market makers, which are the organizations that accept the order and facilitate the trade. Trades can only happen when a price is agreed upon, and there are always two prices listed, which are called the bid and the ask. The bid is the share price that a seller receives from the market maker when they sell stock, and the ask is the share price that a buyer pays to the market maker when they buy stock. The market maker is always required to make the bid and ask prices publicly available. This required transparency is sometimes called an open outcry; a name that originated when stocks were still traded face to face, by buyers, sellers, and market makers who shouted out prices to each other on the stock exchange floor [2]. Regardless of whether the prices are being shouted out in person or displayed on a computer screen, the bid and the ask are constantly changing, because they reflect the current level of demand between buyers and sellers.

Experienced investors know the basic rule for all stock trades, which is simply that you buy at the ask and you sell at the bid. To better understand this, consider a situation where ACME Corporation stock has an ask of $10.05 and a bid of $10.00. This means that a buyer of ACME stock has to pay $10.05 per share, and that a seller will only receive $10.00 per share. However, when an investor places an order to either buy or sell, they can place either a limit order or a market order. A buyer who submits a market order will immediately buy the stock at $10.05, and a seller who submits a market order will immediately sell their stock at $10.00. If the buyer submits a limit order by offering to pay a maximum of $10 per share, the trade won't immediately execute. It might go through later if the market maker lowers their ask to meet the offer, or if the buyer raises their offer to meet the ask, but the trade won't execute if neither situation occurs. As we can see from this example, market makers do exactly the opposite of what investors do; they buy at the bid and sell at the ask.

The difference between the bid and the ask is called the spread (Figure 18.1). Investors can often determine the liquidity of the stock, or how easy it is to buy or sell, by looking at the size of the spread. A small spread, for example $0.01 on a $10 stock, usually means that a stock is very liquid, because the demand from both buyers and sellers is high. A large spread, especially one that exceeds 1% of the share price, usually means that the stock is thinly traded, and that the few buyers and sellers who are currently interested can't agree with the market maker on a price.

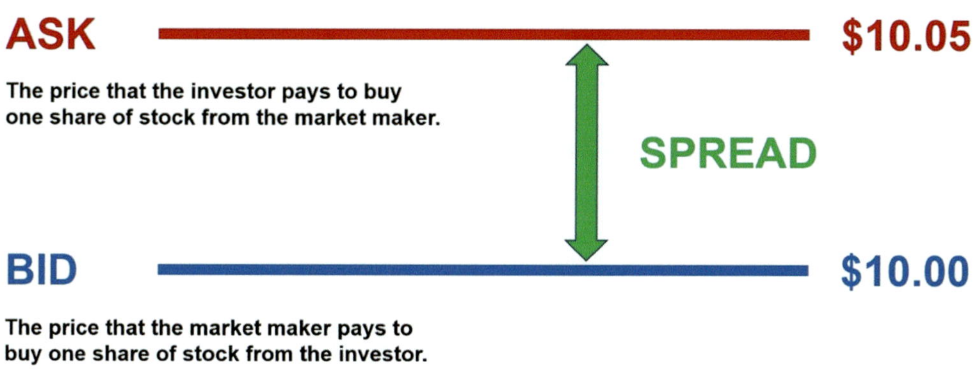

Figure 18.1 The basics of a stock trade—the bid, the ask, and the spread.

It follows, of course, that the spread is the revenue that the market maker receives for facilitating the trade. If the market maker buys ACME Corporation stock from one investor for $10 per share and sells it to another investor for $10.05, they receive $0.05 per share. This means that market makers are always motivated to adjust the spread in ways that maximize their profit, by either increasing the spread and making fewer trades, or by decreasing the spread and making more trades [2–4].

18.1.2 High-Frequency Trading

The accuracy of stock exchange clocks became an area of concern after the advent of electronic trading. Electronic trading was made possible by computers and computer networks, and began in the 1970s. However, individual investors didn't have access to electronic trading platforms until the 1990s, when online brokerage services began to appear on the Internet. From that point forward, electronic trading platforms permanently changed the way stocks were traded, because the market makers were no longer individuals who executed trades by telephone or by face-to-face negotiations. Instead, the trades were automatically executed by computers that bought and sold stocks to each other, based on software algorithms. This automation caused the number of stock exchanges to increase, with each exchange offering their own bid/ask prices to attract business. It also caused the period required to complete a trade, called the execution time, to decrease. In short, automation made the stock market more complex while simultaneously causing it to move faster.

Automation also led to smaller spreads. By 2001, this resulted in the decimalization of stock prices on both the NYSE and Nasdaq. Before decimalization, stock prices were quoted using fractional parts of a dollars—for example, a price of $25.25 was quoted as $25¼. This kept the spreads large, because stocks commonly traded in price increments of 1/16 of a dollar, which meant that the minimum price increment was 6.25 cents. After decimalization, the minimum price increment was reduced to just one cent [5]. The smaller spreads brought about by decimalization were favorable to investors because they made the market more liquid. However, they also reduced the profits of market makers, which gave them the incentive to execute more trades [6].

Each of these factors contributed to the now common practice of high-frequency trading (HFT), which increased the volume of trades and reduced execution times to microseconds [7]. To illustrate the impact of HFT, Figure 18.2 shows the daily average number of shares traded by U. S. stock exchanges from 1996 to 2018. Less than one billion shares per day were traded before electronic trading platforms became prevalent in the late 1990s, but the volume increased to almost 10 billion shares per day by 2008. The contribution of HFT to the total trading volume was first noticeable in 2004, and it has accounted for at least half of the volume since 2008. It's interesting to note that the trading volume decreased and leveled off after the peak years of 2008 to 2010, which coincided with the period of a U. S. financial crisis, and remained below eight billion shares per day until 2020 (not shown on the graph). Even so, the post-2010 volume is still more than 10× larger than the pre-electronic trading volume [8].

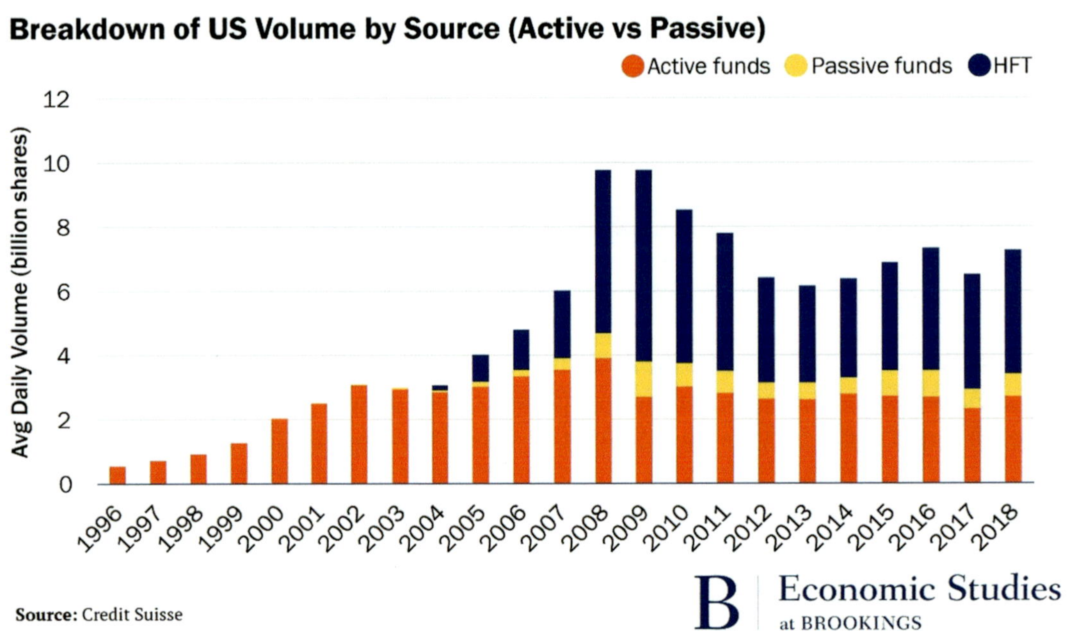

Figure 18.2 The large contribution of HFT to the average daily volume of U. S. stock exchanges (courtesy of Brookings Institution and Credit Suisse).

18.1.3 Preventing Illegal Trading Activity with Accurate Time

High-frequency trading made it essential for all stock exchanges to synchronize their clocks to UTC, and to be able to verify that the time stamps attached to their trades were accurate, so they could prove that illegal trading activity wasn't taking place. To illustrate the importance of accurate time to a fair stock market—which is a market that never favors one investor over another—consider a situation where an individual investor is about to place an order for 100 shares of ACME Corporation stock. Before submitting their order, the investor checks their computer display and notes that the current bid is $9.23, and that the current ask is $9.25. The display also indicates that no one else is in line to buy the stock. Therefore, they understandably assume that the ask of $9.25 per share is the price they will pay, so they click a button and submit their order. A fraction of a second later, a large investment bank submits an order to buy 1,000,000 shares of ACME. The market maker receives the small order before they receive the large order, but for reasons unknown to the individual investor, they decide to execute the large trade first.

After the large trade is executed, the market maker immediately raises both the bid and ask prices by 20 cents, to $9.43 and $9.45, respectively. This action causes the individual investor to instantly lose money, regardless of whether they entered a market or limit order. If they entered a market order, they'll pay $9.45 for the stock they were expecting to buy at $9.25. If they entered a limit order, indicating that $9.25 was the most they would pay, they completely missed out on the trade because their order was never filled. Meanwhile, the large investor who was allowed to cut ahead in line has the option of selling some of their shares back to the market maker at the new bid price. This allows the large investor to take an instant profit. It also produces more revenue for the market maker, because they can immediately resell their newly acquired shares to other unsuspecting investors whose pending market orders were conveniently skipped over [9].

When computers are automatically executing the trades, these illegal practices are essentially impossible to prove or to regulate without accurate time stamps, because everything happens so fast. To ensure that trades are executed in the order that they're received, every stock exchange clock has to be synchronized to a UTC reference clock. In addition, the resolution of every time stamp has to be smaller than the trade's execution time. For example, a time stamp resolution of 1 s is unacceptable for HFT, because many thousands of trades can execute during the same second, and all of them would have the same time stamp. Allowing inaccurate clocks or low-resolution time stamps makes it possible for market makers to execute trades in any order that suits their purpose, because there won't be a way for auditors to prove that the trades were executed in the wrong order [9, 10].

Trading firms haven't always been interested in providing accurate time stamps, but they've always been interested in measuring time intervals. This interest stems from their desire to reduce the delay, or latency, of the trade requests that they send to stock exchanges. Every firm wants their own trade requests to arrive at the stock exchange before the requests of their competitors, and in pursuit of this goal they've historically explored every communications method that might give them a competitive edge. The pursuit perhaps started with the invention of telegraph, which was immediately embraced by trading firms. At first, telegraph lines didn't cover all trading routes, so carrier pigeons were used to bridge the gaps between telegraph stations on the route between Berlin and Paris in the late 1840s. The telephone was also immediately embraced by trading firms after its invention, as was nearly every type of computer network, and nearly every form of radio communication.

The lowest-latency trading networks are based on line-of-sight radio signals that travel at the speed of light, and a number of custom microwave networks cover the trading routes between New York and Chicago, as well as other routes. Due to land use restrictions, the towers usually aren't aligned in a way that provides the shortest possible path. For example, the theoretical latency between the New York and Chicago financial markets is about 3.9 ms, but the latency of the various microwave networks is larger, ranging from about 4.2 ms to 5.2 ms [11]. The information sent via radio will still get there faster than information sent via optical fibers, because the fibers also aren't routed over the shortest possible path, and because the velocity of propagation through a fiber is about 68% of the speed of light (Section 4.1.3). However, optical fibers are the most common transmission medium used by trading networks due to several advantages—they are easy to obtain and lease, they are fairly easy to reroute, and undersea fibers can send information across an ocean.

The continuous efforts made by trading firms to reduce latency is a much-discussed topic that perhaps attracts more attention than it should, because seeking a competitive advantage by employing more advanced technology isn't illegal. Therefore, it isn't a concern for the agencies that regulate stock exchanges. As we'll describe in the next section, the time-related regulations for stock exchanges have nothing to do with latency, and everything to do with time stamp accuracy. These regulations arose from the genuine concern that market makers didn't have the best interests of investors in mind.

18.2 Time Accuracy Requirements for Stock Exchanges

Stocks were traded without any regulatory time accuracy requirements for nearly four centuries after the first stock exchange was founded in early 17th-century Europe, and for more than two centuries after the NYSE was founded in 1792. During the many years before electronic training, there were situations where only the date of a trade was recorded, and the time was completely ignored. When the time was recorded, it was usually physically stamped in ink onto a ticker tape or some other form of paper document. The time stamp was often obtained from a clock of unknown accuracy that hadn't been recently synchronized. The resolution was sometimes limited to 1 minute, but the clocks often used a decimal time format. When that was the case, the time stamp resolution was either 1/100th of an hour (36 s) or 1/10th of a minute (6 s).

The catalyst for the regulation of stock exchange clocks was a report published in 1996 by the United States Securities and Exchange Commission (SEC). The report resulted from an investigation into the practices of Nasdaq market makers. Its findings included evidence of unfair discrimination between market participants, which went against the principles of a free and open market. One of the unfair practices cited was the issuance of trade reports with inaccurate execution times [12]. After reaching a financial settlement, the SEC required the securities dealers who operated the Nasdaq to develop an Order Audit Trail System (OATS) with a set of clearly defined rules. This eventually led to the formation of the Financial Industry Regulatory Authority (FINRA), which regulates its member firms by requiring them to record and report the event order of their market transactions, as shown in Figure 18.3.

Figure 18.3 A simplified diagram of the FINRA order audit trail system.

The OATS rules first went into effect in 1998. One of them, OATS Rule 6953, was entitled "Synchronization of Member Business Clocks." This rule required the clocks at all member firms—including their computer clocks and their mechanical time stamp clocks—to always be synchronized before the stock exchange opened. The chosen time reference was "the NIST atomic clock," and the required accuracy with respect to NIST time was ±3 s. This meant, for example, that a clock synchronized to within 1 s of UTC(NIST) before the market opened would be allowed up to 2 s of drift during the trading day [13].

The ±3 s accuracy requirement originally applied only to the Nasdaq, but the SEC enforced the same requirement on the NYSE in 2003. It was a very modest requirement, especially when compared to the accuracy requirements of other industries, but it was significant for several reasons. Most importantly, it finally forced stock exchanges to synchronize their clocks. It also forced them to get rid of the old clocks that couldn't record integer seconds, such as the decimal time clocks discussed earlier. Finally, it established UTC(NIST) as the time reference for U. S. stock exchanges.

It's interesting to note that while the OATS rules were being written, some firms were using clocks that synchronized to the GOES satellite time code service (Section 10.6), which was then still being operated by NIST. These clocks were thought of by some in the financial sector as the most accurate clocks of all, so the original drafts of Rule 6953 listed the "GOES clock" as the time reference. However, the authors changed the wording to the "NIST atomic clock" after learning that UTC(NIST) provided the reference not only for GOES, but for numerous other time code services [9].

The ±3 s regulation remained in effect until 2011. It was then modified to require every clock that recorded the time of any stock market event to be within ±1 s of UTC(NIST) whenever the markets were open [14]. The ±1 s accuracy requirement is still in effect for manual orders as of 2025, but it was soon considered too coarse to prevent the manipulation of automated orders. Therefore, a stricter accuracy requirement of ±50 ms became effective in 2017 that applied to all stock exchange clocks, except for those used to time stamp manual orders [15]. The SEC approved the ±50 ms accuracy requirement, and strengthened the regulations by requiring the time stamp resolution of all clocks used for automated orders to be 1 ms or less. To ensure a proper trade sequence, the SEC also required the audit trail to include a series of time stamps that record when an order was placed, routed, received, modified or cancelled, and executed [16].

The time accuracy regulations for European stock exchanges are published by the European Securities and Markets Authority (ESMA). Manual orders in the European Union have the same ±1 s accuracy requirement as the United States, but other time accuracy requirements went into effect in 2018 that are much stricter. Automated orders require ±1 ms accuracy, and HFT orders require an accuracy of ±100 μs with ±1 μs resolution. The reference can be any UTC(*k*) time scale [17]. The ESMA HFT requirements raised the bar substantially, because they are 500× more stringent for accuracy and 1,000× more stringent for resolution than the SEC requirements for automated orders.

The SEC might eventually enforce regulations that are equivalent to, or even more stringent, than the ESMA regulations. This possibility exists because concerns about illegal trading activity are always present, and because some trading firms and stock exchanges already maintain microsecond-level synchronization in their internal operations [16]. However, there are practical limits on how strict the accuracy regulations can be, because they apply to every clock involved in a trade, which includes all of the clocks in the networked client computers that time stamp trades.

The accuracy of the computer time stamps depends on several factors. One factor is asymmetry, which is present in all computer networks (Section 11.2.1). The delay introduced by asymmetry is a small percentage—typically less than 5%—of the round-trip delay between a server and client computer. If a client computer inside a stock exchange data center synchronizes its clock to a time server via the public Internet, the asymmetry alone might make it impossible to comply with a ±100 μs accuracy regulation. Reducing the asymmetry to a level with an acceptable margin of safety typically requires small round-trip network delays, which means that the client computers and the time server often have to reside on the same LAN.

Even when the time server and its clients are on the same LAN, some asymmetry will still be present, and it will likely result in the time accuracy being worse than 1 μs, even at the moment of synchronization. The type of client computer also matters. If the client is an ordinary personal or business computer, it's clock is typically based on a quartz oscillator with poor stability that rapidly accumulates a time error (Section 11.2). These clocks require frequent synchronization, and they're typically synchronized with a software-only implementation of either NTP or PTP, so their time stamp accuracy is limited not only by asymmetry and oscillator instability, but also by fluctuations in the load placed on the computer's

OS. The load fluctuations can cause changes in several types of OS delays—including interrupt handing delays, process scheduling delays, and the processing delays between the network interface card and the device driver. Client computers that synchronize via NTP nearly always depend on software-only solutions, so software delays always limit the accuracy of their time stamps. Software delays can also contribute uncertainties of tens of microseconds, or even larger, to software-only implementations of PTP [18, 19].

Client computers equipped with dedicated clock hardware designed to support PTP (Section 11.2.5) or White Rabbit (Section 11.2.6) eliminate most of the uncertainty due to software delays, and can provide time stamps with sub-microsecond accuracy. Even then, delays of multiple microseconds can be introduced when time stamps from the hardware clock are used to synchronize the OS system time or other software applications [20, 21]. As a very general rule, the client computers that connect to time servers via a LAN should be able to comply with a ±100 μs accuracy requirement if they have dedicated clock hardware, and they might be able to comply if they have standard clock hardware. However, any client computer with standard clock hardware that connects to a time server via the Internet probably won't be able to comply.

As we can see from this discussion, none of the client computer synchronization issues are insurmountable, and complying with a ±100 μs accuracy requirement could become routine if the proper network configuration and clock hardware are in place. However, if the accuracy regulations for HFT were reduced to ±1 μs, or even to ±10 μs, they would be difficult for most firms to comply with, and probably even more difficult for auditors to verify. It's also true that from a practical perspective, neither an accuracy or a resolution of 1 μs is needed to execute trades in their proper order, because a single stock exchange is unlikely to receive one million orders in the same second. For these and other reasons, including the inherent resistance to regulation that most trading firms have, a ±100 μs accuracy requirement for HFT seems stringent enough for the foreseeable future.

Table 18.1 summarizes the time accuracy and resolution requirements of stock exchange clocks in the United States, the European Union, Australia [22], and Canada [23] as of 2025. Both Australia and Canada have accuracy requirements similar to those in the United States.

Table 18.1 A summary of time accuracy and resolution requirements for stock exchange clocks

		Required time accuracy and resolution		
Region	**Time reference**	**Trade type**	**Accuracy**	**Resolution**
United states	UTC(NIST)	Automated	±50 ms	1 ms
		Manual	±1 s	1 s
European union	Any UTC(k) time scale	Manual	±1 s	1 s
		High frequency	±100 μs	1 μs
		All other	±1 ms	1 ms
Australia	UTC(AUS)	All	±20 ms	Not specified
Canada	UTC(NRC), or any UTC(k) time scale	Automated	±50 ms	Not specified
		Manual	±1 s	Not specified

Now that we've summarized the time accuracy requirements for stock exchanges, we can turn our attention to their actual timekeeping practices. The following sections describe some of the methods that stock exchanges use to synchronize their clocks.

18.3 Synchronization Methods for Stock Exchanges

As Table 18.1 indicates, UTC(NIST) is the legal time reference for all stock exchange clocks in the United States, but any UTC(*k*) time scale can legally provide the time reference in Europe or Canada. This section describes how UTC(*k*) is delivered to a stock exchange, and how the accuracy of stock exchange clocks is validated.

18.3.1 Delivering UTC(*k*) to a Stock Exchange

The accuracy requirements for both manual trades and automated trades, with the exception of the ±100 μs European requirement for HFT, can be complied with by using the free time code services provided by UTC(*k*) laboratories, which include their broadcasts of LF radio time codes (Section 10.2) and their publicly accessible NTP servers (Section 11.2.4). The radio broadcasts are often not practical to use, because their reception inside stock exchange data centers can be poor or intermittent, especially during the daytime when the markets are open. In contrast, the public NTP servers operated by UTC(*k*) providers are a convenient synchronization option, because NTP client software is available for essentially every type of computer. Standalone clocks that synchronize to NTP after being plugged into to an Ethernet connection are also available (Figure 18.4), and they can meet audit trail requirements for manual orders by physically time-stamping the documents that record a trade.

Figure 18.4 A mechanical time stamp clock that synchronizes to UTC(NIST) via NTP (courtesy of Amano).

The free UTC(*k*) time code services are valuable to trading firms, but they lack the accuracy needed to synchronize the reference clocks that the stock exchanges depend upon. The reference clocks are usually GNSS disciplined clocks—predominantly GPSDCs—and they synchronize the NTP and PTP time servers located inside the data centers, which in turn synchronize the client computers. The financial markets are highly dependent on GPSDCs, which is evidenced by the large number of GPS antennas that are found on the roofs of most stock exchange data centers.

All GNSS time signals are referenced to UTC(*k*) time scales (Table 12.9), so their use is considered acceptable when complying with time accuracy regulations in the European Union, provided that any offset from UTC—in particular the leap second offset—is removed from the GNSS time stamp [24]. The situation is a bit more complicated in the United States because the regulations specifically call for stock exchange clocks to be synchronized to NIST time, and GPST is referenced to UTC(USNO) and not to UTC(NIST). Even so, from a technical perspective, a GPSDC can satisfy the NIST synchronization requirement. As was discussed in Chapter 12, the time difference between UTC(NIST) and UTC(USNO) has not exceeded ±5 ns in recent years (Figure 12.9), and the uncertainty of a GPSDC is unlikely to ever exceed ±1 μs, even when its hardware delays are uncompensated (Table 12.6). These time differences are too small to matter, because they are orders of magnitude smaller than the accuracy required by the stock exchanges.

If a trading firm or stock exchange elects to rely exclusively on GNSS time, the potential concerns usually have nothing to do with time accuracy, but instead are related to GNSS vulnerability (Section 12.11) and to the lack of documentation associated with GNSS measurements. The time signals broadcast by GNSS are freely available to anyone with a receiver, and there are no guarantees that the signals will be used properly. For this reason, stock exchange auditors typically request documentation that explains how the clocks are synchronized and how the time stamps are verified. If a legal dispute arises, more detailed documentation describing the time measurements made by specific systems during a specific period might be requested by a court of law.

To obtain measurement documentation that they can show to auditors—and to courts of law if necessary—trading firms and stock exchanges often subscribe to paid measurement services that link them directly to a UTC(*k*) time scale. The direct connection is usually provided via a dedicated optical fiber, or via a common-view disciplined clock (CVDC). As of 2025, several UTC(*k*) laboratories offer one or both of these subscription-based services.

Services that link financial markets to UTC(NPL) via optical fiber were introduced in the United Kingdom in 2015 [25]. These services are completely independent of GNSS time signals, and distribute UTC(NPL) via PTP over dedicated fiber links that originate from the NPL laboratories in Teddington. For purposes of verification and redundancy, independent hydrogen masers are located at hubs outside of Teddington, and these masers can serve as the UTC reference if the other fiber connections break, or if the primary UTC(NPL) time scale fails. The two main service nodes are located in data centers in Reading and the London Docklands, and are connected to UTC(NPL) via dark fibers with respective lengths of 80 km and 74 km (Figure 18.5). Each data center is located near numerous financial firms that can receive time signals with a certified accuracy of ±1 μs by subscribing to the NPL service [26, 27].

Services that link financial markets to UTC via the CVDC method were introduced by NIST in 2011. A CVDC is an instrument that's delivered to a financial market customer by NIST or another UTC(*k*) provider, and it's calibrated for equipment delays before being delivered. Once installed at the customer's site, a CVDC is synchronized to UTC(*k*) using the common-view time transfer method (Section 4.3.3), with the common-view signals provided by GPS or other GNSS satellites. A CVDC can employ one or more of the several variations of common-view measurements that link UTC(*k*) time scales to UTC (Section 9.4.4).

Figure 18.6 is a simplified diagram that shows how a CVDC system works. As illustrated in the figure, a UTC reference clock, which can be any UTC(*k*) time scale, is compared to a common-view signal (CVS) by measuring the *REF* – *CVS* time difference every second. A CVDC located at a stock exchange site simultaneously measures the *CVDC* – *CVS* time difference. The measurements are periodically sent to a cloud server, where software performs a simple calculation, (*REF* – *CVS*) – (*CVDC* – *CVS*), that produces the time difference *REF* – *CVDC*. The CVDC then retrieves *REF* – *CVDC* from the cloud, and converts the time difference to a frequency correction. The frequency correction is then applied to the CVDC's local oscillator by use of an adaptive PID controller or a similar control loop. The control loop runs continuously, and keeps the CVDC locked to the UTC reference.

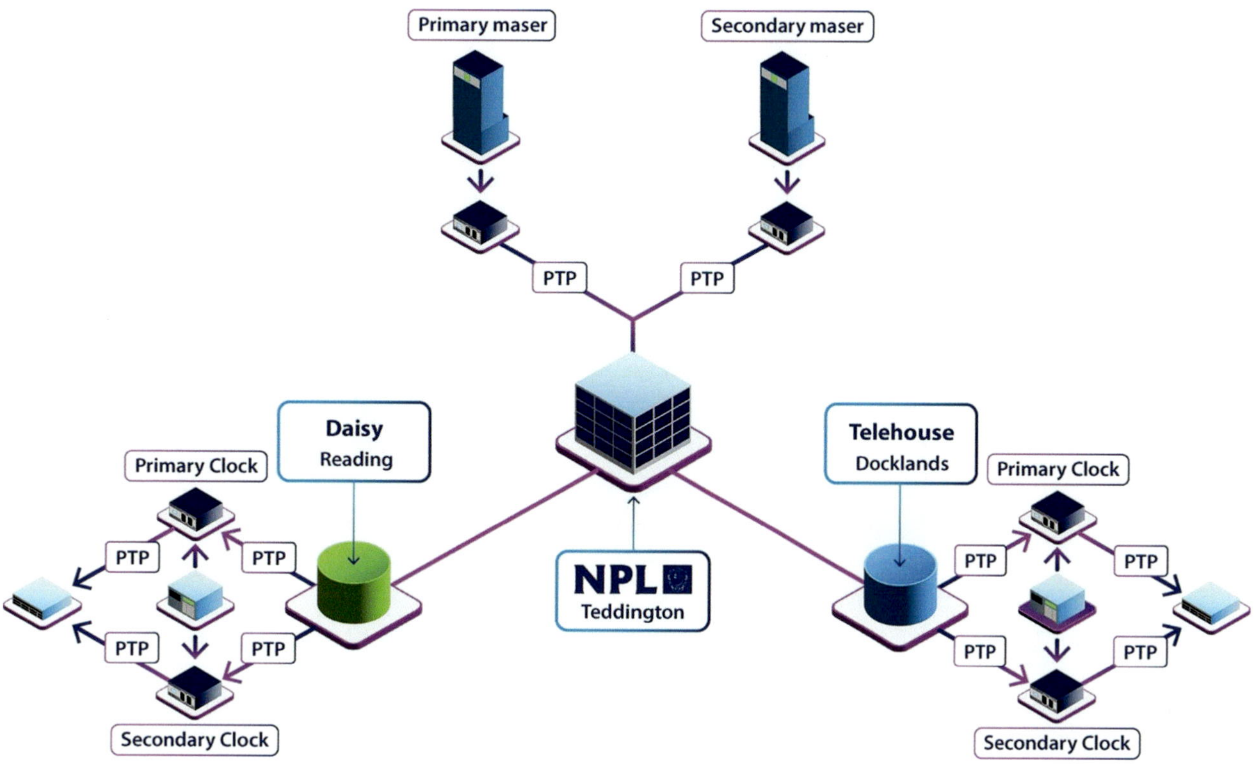

Figure 18.5 A simplified diagram of UTC(NPL) distribution to financial markets via optical fiber (courtesy of NPL).

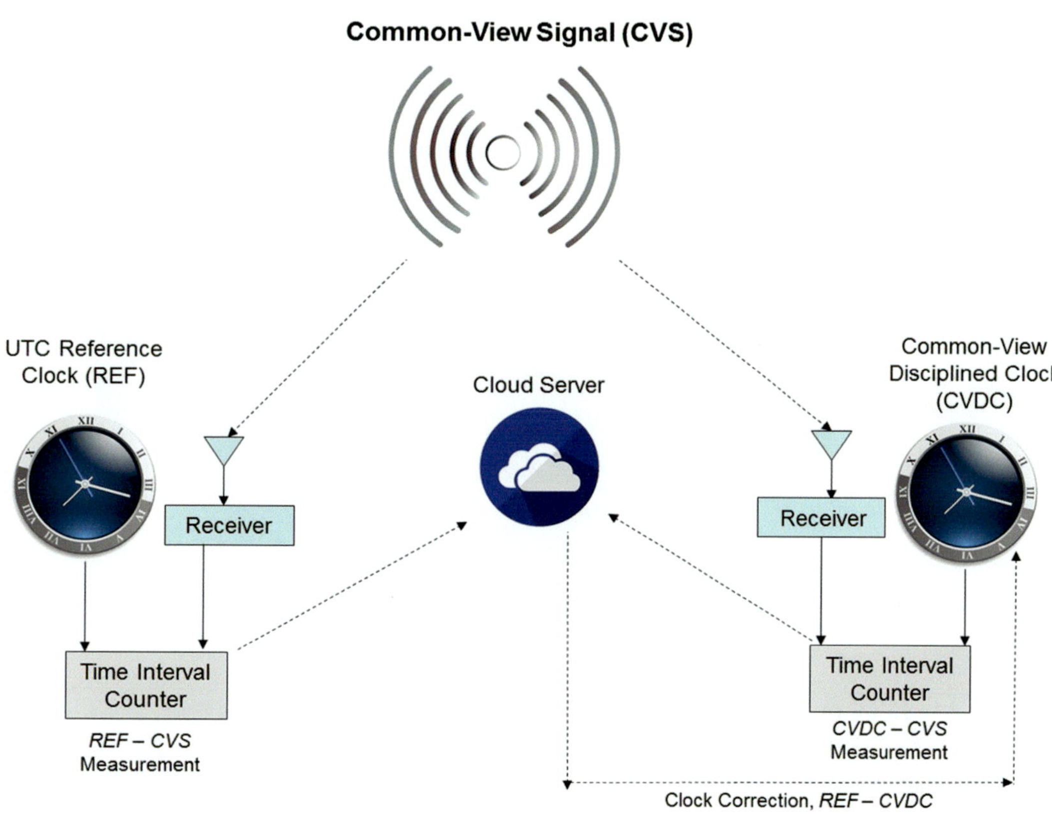

Figure 18.6 A simplified diagram of the common-view disciplined clock method.

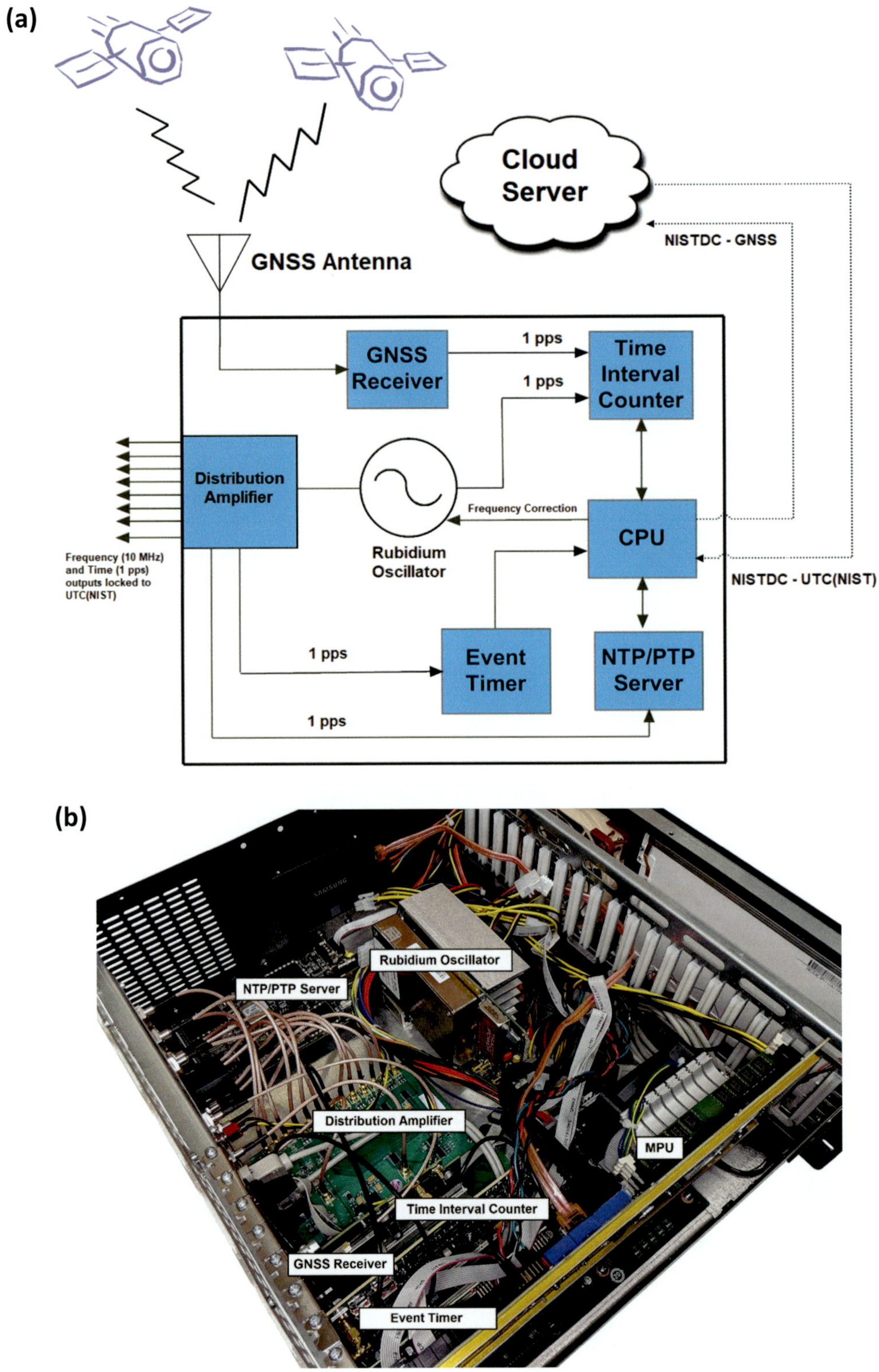

Figure 18.7 Block diagram (a) and interior view (b) of a NIST disciplined clock.

The CVDC that NIST delivers to stock exchange customers is called a NIST disciplined clock (NISTDC). Its basic function is to replicate UTC(NIST) near the site of a stock exchange, so it's usually located in the same data center that houses the stock exchange computers. A NISTDC requires electric power, an Internet connection, and a connection via a delay-calibrated cable to a GNSS antenna that's installed on the roof of the data center. If these basic requirements are met, it can work anywhere on Earth, because the cloud software can compute common-view differences using either the classic common-view method or the all-in-view method (Section 9.4.4). This feature has allowed NISTDCs to provide synchronization not only to the NYSE, Nasdaq, and other U. S. stock exchanges, but also to stock exchanges in London, Frankfurt, and Tokyo [9, 28].

A fully configured NISTDC includes an MPU with a network interface, a rubidium local oscillator, a GNSS receiver, a time interval counter, a time server that supports both NTP and PTP, an event timer that measures the accuracy of time stamps sent by other time servers, and a distribution amplifier that provides multiple 10 MHz and 1 pps outputs that are locked and synchronized to UTC(NIST). Figure 18.7 includes both a block diagram (a) and an interior view (b) of a NISTDC.

A CVDC can potentially lock to any UTC(k) time scale that makes their common-view measurement data in real time. This allows NIST and other UTC(k) providers to distribute their time signals by using navigation satellites as a relay. A multi-source common-view disciplined clock (MSCVDC) is an enhanced CVDC that adds multiple layers of redundancy to the basic design (Figure 18.8). This means that if a UTC(k) source is unavailable, an MSCVDC can switch to another UTC(k) source by simply requesting different data from the cloud server. If a CVS is unavailable, an MSCVDC with the proper receiving equipment can switch to another CVS by switching between GNSS constellations, or can even switch to a geostationary satellite or a terrestrial radio station if common-view data for those signals are available. If the network connection is unavailable, an MSCVDC can switch to another network if one is available. Or, it can lock its clock directly to the CVS signal, becoming, for instance, a GPSDC instead of a CVDC [29].

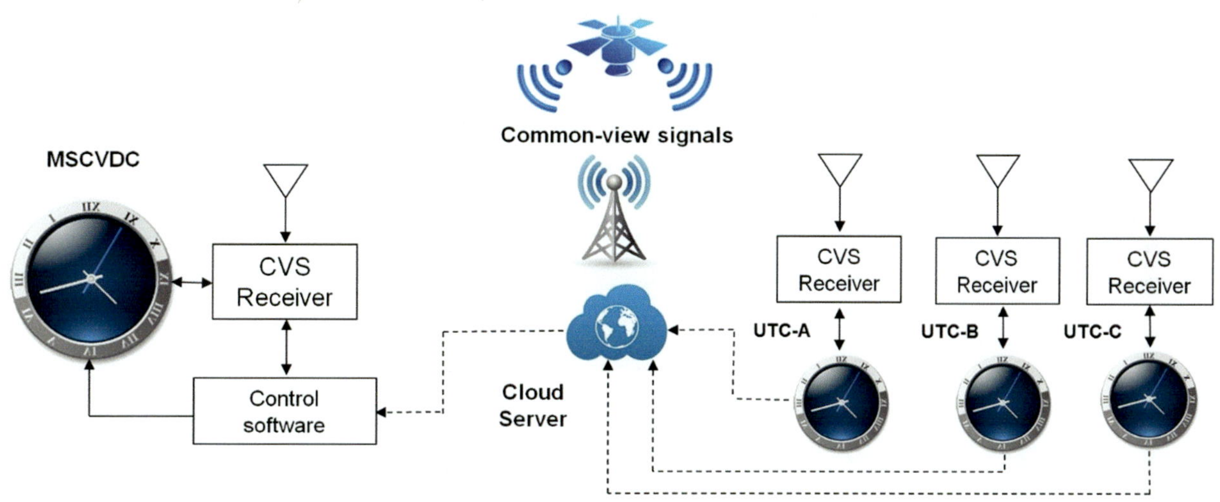

Figure 18.8 A simplified diagram of the multi-source common-view disciplined clock method.

A NISTDC implements some MSCVDC features, including the ability to switch between the UTC(NIST) primary and secondary time scales, and the ability to become a GPSDC if the network connection is lost. In an extreme situation where all synchronization sources are lost, the rubidium oscillator provides enough holdover capability to maintain ±1 μs accuracy for at least 48 hours.

18.3.2 Verifying the Accuracy of Stock Exchange Clocks

To comply with regulations, stock exchanges require every clock involved in a stock trade to be within its accuracy tolerance whenever the markets are open. Therefore, compliance involves the verification of clock accuracy from endpoint to endpoint [9, 26]. The verification process is basically the same process followed when establishing a measurement traceability chain (Section 9.7), but it's more demanding because it has to be done in real-time. Consider, for example, a situation where a glitch in the accuracy of a single stock exchange clock causes trades to be executed out of sequence. Once the problem is noticed, it can be fixed going forward, but there is no way to go back and undo the improper trades. To avoid this situation, the accuracy of every clock has to be continuously measured, and providing endpoint-to-endpoint verification in real-time is an essential part of the services offered to stock exchanges by UTC(k) laboratories.

For the purpose of example, we'll briefly describe the endpoint-to-endpoint verification that NIST provides for its NISTDC service. The true endpoints are the definition of the SI second—realized by UTC—at one end, and the computer clock that time stamps a trade on the other end. However, because the link between the UTC(k) time scales and UTC was described in detail in Section 9.4.4, we'll skip that part here and describe how NIST verifies the accuracy of the NISTDC and the stock exchange time servers.

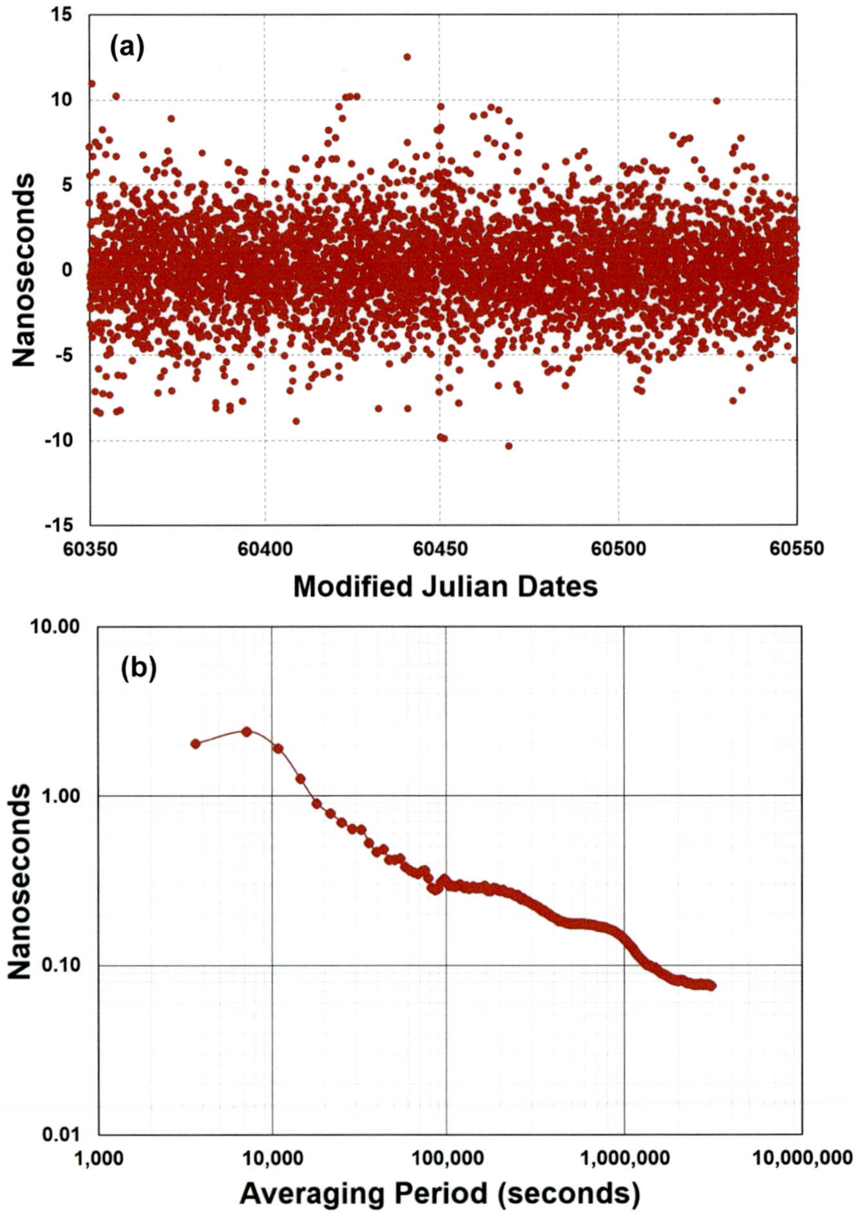

Figure 18.9 The accuracy (a) and stability (b) of a NISTDC in a stock exchange data center.

The NISTDC's accuracy is verified by common-view measurements that compare its 1 pps output to UTC(NIST). These measurements are automated, and they never stop unless the equipment fails, continuing even when the stock exchange is closed. To illustrate this, Figure 18.9(a) shows the results of a continuous 200-day measurement made in 2024. This measurement compared UTC(NIST) in Boulder, Colorado to a NISTDC that's located in a stock exchange data center in New York City. The reported average time difference for the 200-day period was less than 0.2 ns, but the actual average time difference was probably slightly larger, because the expanded uncertainty (k = 2) of the common-view measurements was about ±10 ns.

Figure 18.9(b) shows the stability of the NISTDC during the same 200-day period, as estimated with TDEV. The stability of the NISTDC was about 2 ns for an averaging period of one hour, and it improved to about 0.3 ns for an averaging period of one day. The measurement results that report the accuracy and stability of a NISTDC are updated every 10 minutes, and are made available to customers via an Internet portal. The customers also receive a signed monthly report that lists the average daily time differences and their associated measurement uncertainties [9].

A NISTDC can also synchronize the other time servers that reside on the same LAN by providing them with a 1 pps output, along with packets transmitted by its own internal NTP/PTP time server. When this is done, all of the time servers share the same UTC(NIST) common clock. The NISTDC then verifies the accuracy of the time servers by acting as a client computer that sends each server a request every 10 s. When the servers respond, the time stamps in the received packets are compared to the time kept by the NISTDC. These measurements reveal the uncertainty contributed to the server's time stamps by network asymmetry and other factors.

Figure 18.10 shows the measured time differences between a NTP server and a NISTDC during a 100-day period in 2024. Each data point was averaged for one hour and rounded to the nearest microsecond. In this example, the average round-trip delay between the NISTDC and the server was 300 μs, and the average time difference was −12 μs. Like the NISTDC measurements, the results of all time server measurements are updated every 10 minutes and made available to customers via an Internet portal.

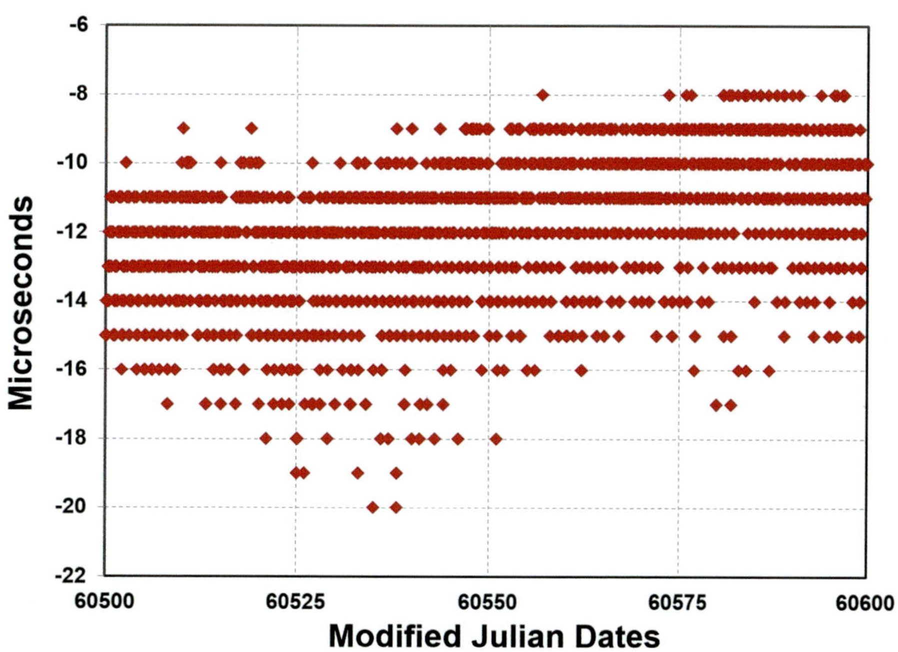

Figure 18.10 The time stamp accuracy of an NTP server located at a stock exchange, as measured by a NISTDC.

18.4 The Effects of the Loss of Stock Exchange Synchronization

Stock exchange activities have a huge impact on the global economy because a tremendous amount of wealth changes hands during each trading day. During each of the years from 2021 to 2023, the average daily trading volume for U. S. stock exchanges exceeded 11 billion shares, and the value of the shares was more than $500 billion. The NYSE and Nasdaq handle about 80% of the trading volume, and make the shares of nearly 6,000 companies available for public trading. The combined market capitalization of the NYSE and Nasdaq—calculated by multiplying the share price by the total number of shares—was nearly $49 trillion in 2023 [30].

The stakes are clearly high, and as we've described throughout this chapter, stock exchanges need accurate time to keep the markets fair and equitable. The stock exchange clocks should always be synchronized to protect investors, but the stock market's need for accurate time differs from the needs of the industries described in the previous three chapters. No one likes to lose money, but the loss of money pales in comparison to the loss of human safety and life that can result from failures in navigation, communication, and electric power distribution systems. In addition, when investors do lose money, it's easy to allege, but difficult to prove, that their losses were due to illegal trading. No trading firm is likely to ever confirm the allegations, because they can always say that trading stocks is an inherently risky activity.

Another difference is that the stock exchanges won't provide warning signs when synchronization is lost. The loss of synchronization in other industries causes immediately noticeable problems—accidents occur when vehicles veer off course, phone calls are disconnected or don't go through at all, and electrical equipment shuts down during power outages. In contrast, an unsynchronized stock exchange will cause an order audit trail system to lose its integrity, but it probably won't cause trading to stop or markets to crash. The absence of warning signs is one of the main reasons why regulations are needed to protect investors, and why the continuous verification of clock accuracy is needed to ensure compliance with the regulations. Without verifiably accurate clocks, stock exchanges can be easily manipulated in ways that are invisible to investors.

References

1. J. Konijnenburg, G. D. Lee, L. B. Minnich, J. S. Williams, L. Warfield, and K. A. Lippa, "Specifications, Tolerances, and Other Technical Requirements for Weighing and Measuring Devices," *National Institute of Standards and Technology Handbook 44*, 537 p., November 2023.
2. L. Engel and B. Boyd, *How to Buy Stocks*, 7th ed., Little Brown & Company, 1982.
3. L. Harris, *Trading and Exchanges: Market Microstructure for Practitioners*, Oxford University Press, 2003.
4. M. Sincere, *Understanding Stocks*, McGraw-Hill, 2004.
5. H. Bessembinder, "Trade Execution Costs and Market Quality after Decimalization," *The Journal of Financial and Quantitative Analysis*, vol. 38, no. 4, pp. 747–777, December 2003.
6. M. Goldstein, P. Kumar, and F. Graves, "Computerized and High-Frequency Trading," *Financial Review*, vol. 49, no. 2, pp. 177–202, May 2014.
7. R. P. Barlett III and J. McCrary, "How Rigged Are Stock Markets? Evidence from Microsecond Timestamps," *Journal of Financial Markets*, vol. 45, pp. 37–60, September 2019.
8. A. Avramovic, "We're All High Frequency Traders Now," *Credit Suisse Trading Strategy Report*, 8 p., March 2017.
9. M. A. Lombardi, A. N. Novick, G. Neville-Neil, and B. Cooke, "Accurate, Traceable, and Verifiable Time Synchronization for World Financial Markets," *Journal of Research of the National Institute of Standards and Technology*, vol. 121, pp. 436–463, 2016.
10. J. J. Angel, "When Finance Meets Physics: The Impact of the Speed of Light on Financial Markets and Their Regulation," *The Financial Review*, vol. 49, no. 2, pp. 271–281, May 2014.

11. G. Laughlin, A. Aguirre, and J. Grundfest, "Information Transmission between Financial Markets in Chicago and New York," *The Financial Review*, vol. 49, no. 2, pp. 283–312 May 2014.
12. United States Securities and Exchange Commission (SEC), *Report Pursuant to Section 21(a) of the Securities Exchange Act of 1934 Regarding the NASD and the NASDAQ Market*," 140 p., August 1996.
13. National Association of Securities Dealers (NASD), *OATS Reporting Technical Specifications*, March 1998.
14. Financial Industry Regulatory Authority (FINRA), *OATS Reporting Technical Specifications*, 290 p., May 3, 2011.
15. Financial Industry Regulatory Authority (FINRA), "Rule 6820. Clock Synchronization," effective March 15, 2017.
16. United States Securities and Exchange Commission (SEC), "Joint Industry Plan; Order Approving the National Market System Plan Governing the Consolidated Audit," *SEC Release No. 34-79318, File No. 4-698*, 979 p., November 15, 2016.
17. European Union, "Commission Delegated Regulation (EU) 2017/574 of 7 June 2016 supplementing Directive 2014/65/EU of the European Parliament and of the Council with Regard to Regulatory Technical Standards for the Level of Accuracy of Business Clocks," *Official Journal of the European Union*, legislation series 87, vol. 60, pp. 148–151, March 31, 2017.
18. T. Kovácsházy and B. Ferencz, "Performance Evaluation of PTPd, a IEEE 1588 implementation, on the x86 Linux platform for Typical Application Scenarios," *Proceedings of the IEEE Instrumentation and Measurement Technology Conference*, pp. 2548–2552, May 2012.
19. P. Ferrari, A. Flammini, S. Rinaldi, A. Bondavalli, and F. Brancati, "Experimental Characterization of Uncertainty Sources in a Software-Only Synchronization System," *IEEE Transactions on Instrumentation and Measurement*, vol. 61, no. 5, pp. 1512–1521, May 2012.
20. R. Cochran, C. Marinescu, and C. Riesch, "Synchronizing the Linux System Time to a PTP Hardware Clock," *Proceedings of the International IEEE Symposium on Precision Clock Synchronization for Measurement Control and Communication*, pp. 87–92, September 2011.
21. E. Kyriakakis, J. Sparsø, and M. Schoeberl, "Hardware Assisted Clock Synchronization with the IEEE 1588–2008 Precision Time Protocol," *Proceedings of the 26th International Conference on Real-Time Networks and Systems*, pp. 51–60, October 2018.
22. Australian Securities & Investment Commission (ASIC), "Financial Markets: Domestic and Overseas Operators," *ASIC Regulatory Guide 172*, 157 p., August 2022.
23. Investment Industry Regulatory Organization of Canada (IIROC), "Guidance on Time Synchronization," *IIROC Notice 16-0022*, February 2016.
24. European Securities and Markets Authority (ESMA), "Guidelines: Transaction Reporting, Order Record Keeping and Clock Synchronisation Under MiFID II," *ESMA/2016/1452*, 291 p., July 2017.
25. D. Hicks, P. Whibberley, E. Laier-English, L. Lobo, T. Lee, and A. Austin, "NPL*Time*®—UTC Traceable Time for the Financial Sector," *Proceedings of the 47*th *Annual Precise Time and Time Interval Meeting (PTTI)*, pp. 167–173, January 2016.
26. E. Laier-English, P. Whibberley, C. Langham, D. Hicks, and L. Lobo, "Update on the NPL*Time*® and Future Developments with White Rabbit," *Proceedings of the 48th Annual Precise Time and Time Interval (PTTI) Meeting*, pp. 80–87, January 2017.
27. J. Cartwright, "Time Traders," *Physics World*, vol. 31, no. 7, pp. 24–27, July 2018.
28. M. A. Lombardi, "Synchronizing Stock Market Clocks to UTC(NIST)," *Proceedings of the XXXIII General Assembly and Scientific Symposium of the International Union of Radio Science (URSI)*, pp. 1–4, August 2020.
29. M. A. Lombardi, "Multi-Source Common-View Disciplined Clock: A Fail-Safe Clock for Critical Infrastructure Systems," *Journal of Research of the National Institute of Standards and Technology*, vol. 126, article no. 126041, pp. 1–29, January 2022.
30. K. Kolchin, J. Romulus, and M. Paluzzi, *2024 Capital Markets Fact Book*, SIFMA Research, July 2024.

Epilogue

For everything there is a season, and a time for every matter under heaven.

— Ecclesiastes 3:1

Now that we've reached the end of our narrative, we can examine time measurements from a broad perspective. As we've discussed in the preceding chapters, we began measuring time by observing the natural order of our world and universe. Watching the light turn to darkness and back again provided a natural order of days, and the repeating cycles of days provided a natural order of seasons and years. It was soon found that the passage of time during the day could be measured by looking at the length of the shadows cast by the Sun. This led to the realization that the nature of time could be better understood by looking at the sky, because we could tell time by observing the positions of the Sun in the daytime and the Moon and the stars at night.

The first human-made clocks that could measure time when the Sun wasn't visible were based on steady, continuous processes such as the gradual melting of a candle, or the flow of sand or water. By the 14th century, it was known that more accurate time could be kept by counting the repetitive cycles of devices that oscillated at a constant frequency, just as the Earth rotated at a constant frequency.

For more than 600 years, the best oscillators were mechanisms—such as crown wheels and pendulums—that oscillated at frequencies near 1 Hz. After the advent of electricity, things dramatically changed. For a period of about two decades in the 20th century, the most accurate clocks counted the cycles of quartz crystals that oscillated at frequencies measured in units of kilohertz or megahertz. In the mid-20th century, we began keeping time by counting the energy transitions of atoms. In a sense, this brought everything full circle, because atoms, like the Earth, are a natural form of oscillator.

The atomic clocks that we rely on today measure the radio frequency transitions of atoms in units of gigahertz, but the atomic clocks of the future will measure the optical frequency transitions of atoms in units of terahertz or petahertz. Faster oscillations lead to more accurate clocks, and timekeeping accuracy, once relegated to hours, minutes, and seconds, has now reached minuscule fractions of a second—nanoseconds, picoseconds, and femtoseconds—and is continuing to advance.

We could ask ourselves, was all of this even necessary? After all, civilizations advanced for hundreds of thousands of years while using nothing other than the Sun as their timekeeper. The answer, however, has to be a resounding "yes" due to all of the benefits that accurate clocks have brought to society. These benefits go far beyond everyday timekeeping, and once we realize that nearly everything is connected to time and frequency, we can easily find them everywhere we look. The alternating current that flows out of our wall outlets and provides our electricity, as well as the electronic systems that we rely upon to communicate and navigate, were all made possible by accurate time and frequency measurements.

Apart from the societal benefits, it's also inherently satisfying to accurately measure time and frequency. The satisfaction comes from the pursuit of truth, because measurements bring us closer to fundamental truth than any guess, opinion, or theory. The value of measurements was nicely summarized by the computer scientist Grace Hopper, who said in 1947 that "one accurate measurement is worth a thousand expert opinions." Pursuing truth also invokes a sense of discovery, mystery, and adventure—a feeling

DOI: 10.1201/9781042015320-25

articulated by Allen Astin, the director of the National Bureau of Standards, who spoke of the "romance in precision measurement" in 1953.

Keeping both fundamental truth and romance in mind, it seems fitting to close by quoting a lyric from the song that lent its title to this book. The song was composed in 1931 by Herman Hupfeld, and it was later immortalized by the movie *Casablanca*. Hupfeld was speaking of love and not metrology, but his words ring true in either case—

"The fundamental things apply as time goes by."

Name Index

Subject Index

For Product Safety Concerns and Information please contact our EU representative GPSR@taylorandfrancis.com Taylor & Francis Verlag GmbH, Kaufingerstraße 24, 80331 München, Germany

Batch number: 10706711

Printed by Printforce, the Netherlands